P9-AFA-130

THIRD EDITION

APPLIED ECONOMETRIC TIME SERIES

Walter Enders
University of Alabama

www.wiley.com/college/enders

VP & Publisher *George Hoffman*
Associate Publisher *Judith Joseph*
Associate Editor *Jennifer Manias*
Editorial Assistant *Emily McGee*
Associate Director of Marketing *Amy Scholz*
Assistant Marketing Manager *Diane Mars*
Marketing Assistant *Laura Finley*
Media Editor *Greg Chaput*
Senior Production Editor *Trish McFadden*
Design Director *Harry Nolan*
Senior Designer *James O'Shea*
Cover Designer *Arthur Medina*

This book was set in Times 10/12 by Aptara®, Inc. and printed and bound by Courier Kendallville.
The cover was printed by Courier Kendallville.

This book is printed on acid free paper. ∞

ISBN-13 978-0470-50539-7
ISBN-10 0-470-50539-7

Printed in the United States of America

10 9 8 7 6 5

To the family: Linda, Lola, Justin, Jennifer, Daniell, Kya and Baby B.

PREFACE

When I began writing the first edition, my intent was to write a text in time-series macroeconometrics. Fortunately, a number of my colleagues convinced me to broaden the focus. Applied microeconomists have embraced time-series methods, and the political science journals have become more quantitative. As in the first edition, examples are drawn from macroeconomics, agricultural economics, international finance, and my work with Todd Sandler on the study of transnational terrorism. You should find that the examples in the text provide a reasonable balance between macroeconomic and microeconomic applications.

The text is intended for those with some background in multiple regression analysis. I presume the reader understands the assumptions underlying the use of ordinary least squares. All of my students are familiar with the concepts of correlation and covariation; they also know how to use t-tests and F-tests in a regression framework. I use terms such as *mean square error, significance level,* and *unbiased estimate* without explaining their meaning. Two chapters of the text examine multiple time-series techniques. To work through these chapters, it is necessary to know how to solve a system of equations using matrix algebra. Chapter 1, "Difference Equations," is the cornerstone of the text. In my experience, this material and a knowledge of regression analysis are sufficient to bring students to the point where they are able to read the professional journals and to embark on a serious applied study. Nevertheless, one unfortunate reader wrote, "I did everything you said in your book, and my article still got rejected."

I have tried to be careful about the trade-off between being complete and being concise. Textbook bloat has often ruined revisions of formerly fine manuscripts. No one wants to read an encyclopedic treatment of a topic or technique that has fallen out of style. Nevertheless, there have been a number of important developments in time-series methods that needed to be included in the text. In deciding which topics to include, I relied heavily on the e-mail messages I received from instructors and from students. Many were concerned about issues pertaining to coefficient instability and structural change. The new material in Chapter 2 discusses some of the formal tests for structural change, examples of recursive estimation, and the CUSUM test. Some of the issues involved with testing for endogenous breaks (i.e., potential breaks occurring at an unknown date) are deferred until Chapter 7. It seems that many readers want to utilize the ability of all professional software packages to estimate multivariate GARCH models. As such, Chapter 3 contains a number of new developments in ARCH modeling with special emphasis on multivariate GARCH models. Some of the technical details are left to the appendix of the chapter. I revamped much of the discussion in Chapter 4 concerning the appropriate choice of the deterministic regressors in unit root tests. Instead, much of the focus is on the new detrending methods used in LM and DF-GLS unit root tests. These tests have far better power than standard unit root tests, and the issue of the appropriate deterministic regressors is less critical. In Chapter 6, I tried to improve on the discussion of the general-to-specific methodology in testing for cointegration. The text now focuses on error-correction and ADL tests for cointegration.

I take the term *applied* that appears in the title very seriously. Towards this end, I believe in teaching by induction. The method is to take a simple example and build toward more general and more complicated models. Detailed examples of each procedure are provided. Each concludes with a step-by-step summary of the stages typically employed in using that procedure. The approach is one of learning by doing. A large number of solved problems are included in the body of each chapter. The Questions and Exercises section at the end of each chapter is especially important. You are encouraged to work through as many of the examples and exercises as possible. To help you through the material, there are three supplements to the text.

An *Instructors' Manual* is available to those adopting the text for their class. The manual contains the answers to all of the mathematical questions. It also contains programs that can be used to reproduce most of the results reported in the text and all of the models indicated in the "Questions and Exercises" sections. Versions of the manual are available for EVIEWS, RATS, SAS, and STATA users.

Since it was necessary to exclude some topics from the text, I also prepared a *Supplementary Manual*. This manual contains material that I deemed important (or interesting), but not sufficiently important for all readers, to include in the text. Often the text refers you to this *Supplementary Manual* to obtain additional information on a topic. Nevertheless, it is my intent to continually add new topics to the manual as I receive inquiries from interested readers. As such, you might want to check that you have the most current version of this manual.

Some of the techniques illustrated in the text need to be explicitly programmed. Structural VARs need to be estimated using a package that has the capacity to manipulate matrices. Monte Carlo methods are very computer-intensive. Nonlinear models need to be estimated using a package that can perform nonlinear least squares and maximum likelihood estimation. Completely menu-driven software packages are not able to estimate every form of time-series model. As I tell my students, by the time a procedure appears on the menu of an econometric software package, it's not new. To assist you in your programming, I have written a *RATS Programming Manual* to accompany this text. Of course, it is impossible for me to have versions of the guide for every possible platform. Most programmers should be able to transcribe a program written in RATS into the language used by their personal software package.

Wiley makes all three manuals available to faculty who use the text for their class. The *Supplementary Manual* and *Programming Manual* can be downloaded (at no charge) from the Wiley Web site at www.wiley.com/college/enders or from my personal Web page: www.cba.ua.edu/~wenders. The *Programming Manual* can also be downloaded from the ESTIMA Web site: www.estima.com.

In spite of all my efforts, some errors have undoubtedly crept into the text. If the first two editions are any guide, the number is embarrassingly large. I will keep an updated list of typos and corrections in the *Supplementary Manual*.

Many people made valuable suggestions for improving the organization, style, and clarity of the manuscript. Reviewers of the Third Edition provided many helpful comments and suggestions. They include Kyle Beardsley of Emory University; Randall C. Campbell of Mississippi State University; Andre J. D. Crawford of Virginia Tech; Can Erbil of Brandeis University; Raphael Franck of George Mason University; Tarron Khemraj of New College of Florida; Ali Kuton of Southern Illinois University, Edwardsville; Keil Manfred of Claremont McKenna College and Claremont Graduate University; Andre V. Mollick of University of Texas, Pan American; and, Jan Mutl of University at Albany and Frankfurt University. I am grateful to my students, who kept me challenged and were quick to point out errors. Especially helpful were my current and former students Karl Boulware, Pin Chung, Selahattin Dibooglu, HyeJin Lee, Jing Li, Eric Olson, Ling Shao, and Jingan Yuan. Pierre Siklos and Mark Wohar made a number of important suggestions for earlier editions of the text. I learned so much about time series from Barry Falk and Junsoo Lee that they deserve a special mention. I would like to thank my loving wife, Linda, for putting up with me while I was working on the manuscript.

Just before writing this preface, I learned that Clive Granger had died. A few months before I was to take a sabbatical at the University of Minnesota, I had the opportunity to present a seminar at UCSD. At the time, I was working with overlapping generations models and had no thoughts about being an applied econometrician. However, when I first met Clive, he stated, "It will be 100 degrees warmer here than in Minnesota next winter. Why not do the sabbatical here?" I changed my plans, thinking that I would work with the math-econ types at University of California at San Diego. Fortunately, I happened to sit through one of his classes (team-taught with Robert Engle) and fell in love with time-series econometrics. I know that it tickled Clive to tell people the story of how his class clearly changed my career.

ABOUT THE AUTHOR

Walter Enders holds the Bidgood Chair of Economics and Finance at the University of Alabama. He received his doctorate in economics in 1975 from Columbia University in New York. Dr. Enders has published numerous research articles in such journals as the *Review of Economics and Statistics, Quarterly Journal of Economics,* and the *Journal of International Economics*. He has also published articles in the *American Economic Review* (a journal of the American Economic Association), the *Journal of Business and Economic Statistics* (a journal of the American Statistical Association), and the *American Political Science Review* (a journal of the American Political Science Association).

Dr. Enders (along with Todd Sandler) received the National Academy of Sciences' Estes Award for Behavioral Research Relevant to the Prevention of Nuclear War. The award recognizes "basic research in any field of cognitive or behavioral science that has employed rigorous formal or empirical methods, optimally a combination of these, to advance our understanding of problems or issues relating to the risk of nuclear war." The National Academy presented the award for their "joint work on transnational terrorism using game theory and time-series analysis to document the cyclic and shifting nature of terrorist attacks in response to defensive counteractions."

BRIEF CONTENTS

CONTENTS

CHAPTER 1

DIFFERENCE EQUATIONS

INTRODUCTION

The theory of difference equations underlies all of the time-series methods employed in later chapters of this text. It is fair to say that time-series econometrics is concerned with the estimation of difference equations containing stochastic components. The traditional use of time-series analysis was to forecast the time path of a variable. Uncovering the dynamic path of a series improves forecasts since the predictable components of the series can be extrapolated into the future. The growing interest in economic dynamics has given a new emphasis to time-series econometrics. Stochastic difference equations arise, quite naturally, from dynamic economic models. Appropriately estimated equations can be used for the interpretation of economic data and for hypothesis testing.

This introductory chapter has three aims:

1. Explain how stochastic difference equations can be used for forecasting, and illustrate how such equations can arise from familiar economic models. The chapter is not meant to be a treatise on the theory of difference equations; only those techniques essential to the appropriate estimation of *linear* time-series models are presented. This chapter focuses on single equation models; multivariate models are considered in Chapters 5 and 6.
2. Explain what it means to *solve* a difference equation. The solution will determine whether a variable has a stable or an explosive time path. Knowledge of the stability conditions is essential to understanding recent innovations in time-series econometrics. Contemporary time-series literature pays special attention to the issue of stationary versus nonstationary variables. The stability conditions underlie the conditions for stationarity.
3. Demonstrate how to find the solution to a stochastic difference equation. There are several different techniques that can be used; each has its own relative merits. A number of examples are presented to help you understand the different methods. Try to work through each example carefully. For extra practice, you should do the exercises at the end of the chapter.

1. TIME-SERIES MODELS

The task facing the modern time-series econometrician is to develop reasonably simple models capable of forecasting, interpreting, and testing hypotheses concerning economic data. The challenge has grown over time; the original use of time-series analysis was primarily as an aid to forecasting. As such, a methodology was developed to decompose a series into a trend, a seasonal, a cyclical, and an irregular component. The trend component represented the long-term behavior of the series and the cyclical component represented the regular periodic movements. The irregular component was stochastic and the goal of the econometrician was to estimate and forecast this component.

Suppose you observe the 50 data points shown in Figure 1.1 and are interested in forecasting the subsequent values. Using the time-series methods discussed in the next several chapters, it is possible to decompose this series into the trend, seasonal, and irregular components shown in the lower panel of the figure. As you can see, the trend changes the mean of the series, and the seasonal component imparts a regular cyclical pattern with peaks occurring every 12 units of time. In practice, the trend and seasonal components will not be the simplistic deterministic functions shown in this figure. The modern view maintains that a series contains stochastic elements in the trend, seasonal, and irregular components. For the time being, it is wise to sidestep these complications so that the projection of the trend and seasonal components into periods 51 and beyond is straightforward.

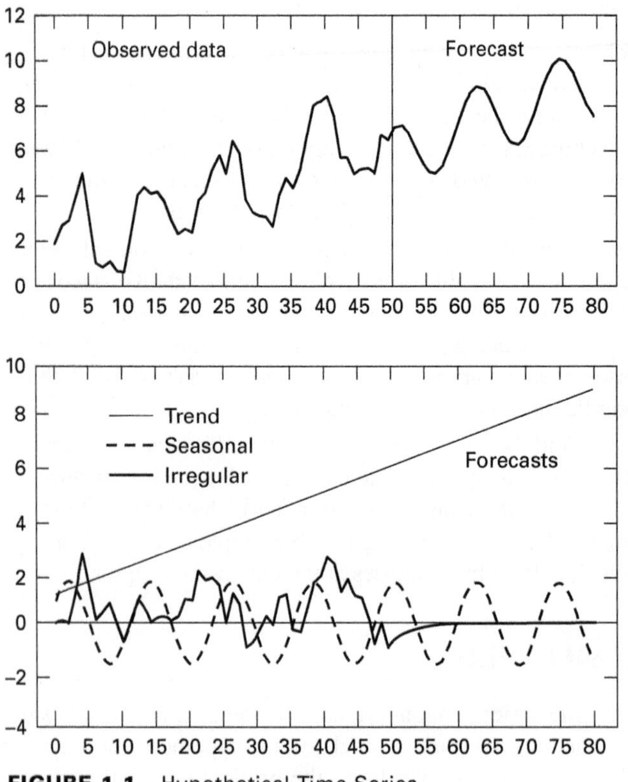

FIGURE 1.1 Hypothetical Time Series

Notice that the irregular component, while lacking a well-defined pattern, is somewhat predictable. If you examine the figure closely, you will see that the positive and negative values occur in runs; the occurrence of a large value in any period tends to be followed by another large value. Short-run forecasts will make use of this positive correlation in the irregular component. Over the entire span, however, the irregular component exhibits a tendency to revert to zero. As shown in the lower part, the projection of the irregular component past period 50 rapidly decays toward zero. The overall forecast, shown in the top part of the figure, is the sum of each forecasted component.

The general methodology used to make such forecasts entails finding the *equation of motion* driving a stochastic process and using that equation to predict subsequent outcomes. Let y_t denote the value of a data point at period t; if we use this notation, the example in Figure 1.1 assumes we observed y_1 through y_{50}. For $t = 1$ to 50, the equations of motion used to construct components of the y_t series are

$$\text{Trend: } T_t = 1 + 0.1t$$

$$\text{Seasonal: } S_t = 1.6\sin(t\pi/6)$$

$$\text{Irregular: } I_t = 0.7\,I_{t-1} + \varepsilon_t$$

where: T_t = value of the trend component in period t
S_t = value of the seasonal component in t
I_t = the value of the irregular component in t
ε_t = a pure random disturbance in t

Thus, the irregular disturbance in t is 70 percent of the previous period's irregular disturbance plus a random disturbance term.

Each of these three equations is a type of **difference equation.** In its most general form, a difference equation expresses the value of a variable as a function of its own lagged values, time, and other variables. The trend and seasonal terms are both functions of time and the irregular term is a function of its own lagged value and of the stochastic variable ε_t. The reason for introducing this set of equations is to make the point that *time-series econometrics is concerned with the estimation of difference equations containing stochastic components.* The time-series econometrician may estimate the properties of a single series or a vector containing many interdependent series. Both univariate and multivariate forecasting methods are presented in the text. Chapter 2 shows how to estimate the irregular part of a series. Chapter 3 considers estimating the variance when the data exhibit periods of volatility and tranquility. Estimation of the trend is considered in Chapter 4, which focuses on the issue of whether the trend is deterministic or stochastic. Chapter 5 discusses the properties of a vector of stochastic difference equations, and Chapter 6 is concerned with the estimation of trends in a multivariate model. Chapter 7 introduces the new and growing area of research involving nonlinear time-series models.

Although forecasting has always been the mainstay of time-series analysis, the growing importance of economic dynamics has generated new uses for time-series analysis. Many economic theories have natural representations as stochastic difference equations. Moreover, many of these models have testable implications concerning the time path of a key economic variable. Consider the following four examples:

1. **The Random Walk Hypothesis:** In its simplest form, the random walk model suggests that day-to-day changes in the price of a stock should have a mean value of zero. After all, if it is known that a capital gain can be made by buying a share on day t and selling it for an expected profit the very next day, efficient speculation will drive up the current price. Similarly, no one will want to hold a stock if it is expected to depreciate. Formally, the model asserts that the price of a stock should evolve according to the stochastic difference equation

$$y_{t+1} = y_t + \varepsilon_{t+1}$$

or

$$\Delta y_{t+1} = \varepsilon_{t+1}$$

where y_t = the price of a share of stock on day t, and ε_{t+1} = a random disturbance term that has an expected value of zero.

Now consider the more general stochastic difference equation

$$\Delta y_{t+1} = \alpha_0 + \alpha_1 y_t + \varepsilon_{t+1}$$

The random walk hypothesis requires the testable restriction: $\alpha_0 = \alpha_1 = 0$. Rejecting this restriction is equivalent to rejecting the theory. Given the information available in period t, the theory also requires that the mean of ε_{t+1} be equal to zero; evidence that ε_{t+1} is predictable invalidates the random walk hypothesis. Again, the appropriate estimation of a single equation model is considered in Chapters 2 through 4.

2. **Reduced-Form and Structural Equations:** It is often useful to collapse a system of difference equations into separate single-equation models. To illustrate the key issues involved, consider a stochastic version of Samuelson's (1939) classic model:

$$y_t = c_t + i_t \qquad (1.1)$$

$$c_t = \alpha y_{t-1} + \varepsilon_{ct} \qquad 0 < \alpha < 1 \qquad (1.2)$$

$$i_t = \beta(c_t - c_{t-1}) + \varepsilon_{it} \qquad \beta > 0 \qquad (1.3)$$

where y_t, c_t, and i_t denote real GDP, consumption, and investment in time period t, respectively. In this Keynesian model, y_t, c_t, and i_t are endogenous variables. The previous period's GDP and consumption, y_{t-1} and c_{t-1}, are called predetermined or lagged endogenous variables. The terms ε_{ct} and ε_{it} are zero mean random disturbances for consumption and investment, and the coefficients α and β are parameters to be estimated.

The first equation equates aggregate output (GDP) with the sum of consumption and investment spending. The second equation asserts that consumption spending is proportional to the previous period's GDP plus a random disturbance term. The third equation illustrates the accelerator principle. Investment spending is proportional to the change in consumption; the idea is that growth in consumption necessitates new investment spending. The error terms ε_{ct} and ε_{it} represent the portions of consumption and investment not explained by the behavioral equations of the model.

Equation (1.3) is a **structural equation** since it expresses the endogenous variable i_t as being dependent on the current realization of another endogenous variable, c_t. A **reduced-form equation** is one expressing the value of a variable in terms of its own lags, lags of other endogenous variables, current and past values of exogenous variables, and disturbance terms. As formulated, the consumption function is already in reduced form; current consumption depends only on lagged income and the current value of the stochastic disturbance term ε_{ct}. Investment is not in reduced form because it depends on current period consumption.

To derive a reduced-form equation for investment, substitute (1.2) into the investment equation to obtain

$$\begin{aligned} i_t &= \beta[\alpha y_{t-1} + \varepsilon_{ct} - c_{t-1}] + \varepsilon_{it} \\ &= \alpha\beta y_{t-1} - \beta c_{t-1} + \beta\varepsilon_{ct} + \varepsilon_{it} \end{aligned}$$

Notice that the reduced-form equation for investment is not unique. You can lag (1.2) one period to obtain: $c_{t-1} = \alpha y_{t-2} + \varepsilon_{ct-1}$. Using this expression, the reduced-form investment equation can also be written as

$$\begin{aligned} i_t &= \alpha\beta y_{t-1} - \beta(\alpha y_{t-2} + \varepsilon_{ct-1}) + \beta\varepsilon_{ct} + \varepsilon_{it} \\ &= \alpha\beta(y_{t-1} - y_{t-2}) + \beta(\varepsilon_{ct} - \varepsilon_{ct-1}) + \varepsilon_{it} \end{aligned} \tag{1.4}$$

Similarly, a reduced-form equation for GDP can be obtained by substituting (1.2) and (1.4) into (1.1):

$$\begin{aligned} y_t &= \alpha y_{t-1} + \varepsilon_{ct} + \alpha\beta(y_{t-1} - y_{t-2}) + \beta(\varepsilon_{ct} - \varepsilon_{ct-1}) + \varepsilon_{it} \\ &= \alpha(1 + \beta)y_{t-1} - \alpha\beta y_{t-2} + (1 + \beta)\varepsilon_{ct} + \varepsilon_{it} - \beta\varepsilon_{ct-1} \end{aligned} \tag{1.5}$$

Equation (1.5) is a **univariate** reduced-form equation; y_t is expressed solely as a function of its own lags and disturbance terms. A univariate model is particularly useful for forecasting since it enables you to predict a series based solely on its own current and past realizations. It is possible to estimate (1.5) using the univariate time-series techniques explained in Chapters 2 through 4. Once you have obtained estimates of α and β, it is straightforward to use the observed values of y_1 through y_t to predict all future values in the series (i.e., $y_{t+1}, y_{t+2}, \ldots$).

Chapter 5 considers the estimation of multivariate models when all variables are treated as jointly endogenous. The chapter also discusses the restrictions needed to recover (i.e., identify) the structural model from the estimated reduced-form model.

3. **Error-Correction: Forward and Spot Prices:** Certain commodities and financial instruments can be bought and sold on the spot market (for immediate delivery) or for delivery at some specified future date. For example,

suppose that the price of a particular foreign currency on the spot market is s_t dollars and that the price of the currency for delivery one period into the future is f_t dollars. Now, consider a speculator who purchased forward currency at the price f_t dollars per unit. At the beginning of period $t + 1$, the speculator receives the currency and pays f_t dollars per unit received. Since spot foreign exchange can be sold at s_{t+1}, the speculator can earn a profit (or loss) of $s_{t+1} - f_t$ per unit transacted.

The Unbiased Forward Rate (UFR) hypothesis asserts that expected profits from such speculative behavior should be zero. Formally, the hypothesis posits the following relationship between forward and spot exchange rates:

$$s_{t+1} = f_t + \varepsilon_{t+1} \tag{1.6}$$

where ε_{t+1} has a mean value of zero from the perspective of time period t.

In (1.6), the forward rate in t is an unbiased estimate of the spot rate in $t + 1$. Thus, suppose you collected data on the two rates and estimated the regression

$$s_{t+1} = \alpha_0 + \alpha_1 f_t + \varepsilon_{t+1}$$

If you were able to conclude that $\alpha_0 = 0$, $\alpha_1 = 1$, and that the regression residuals ε_{t+1} have a mean value of zero from the perspective of time period t, the UFR hypothesis could be maintained.

The spot and forward markets are said to be in *long-run equilibrium* when $\varepsilon_{t+1} = 0$. Whenever s_{t+1} turns out to differ from f_t, some sort of adjustment must occur to restore the equilibrium in the subsequent period. Consider the adjustment process

$$s_{t+2} = s_{t+1} - \alpha[s_{t+1} - f_t] + \varepsilon_{st+2} \qquad \alpha > 0 \tag{1.7}$$

$$f_{t+1} = f_t + \beta[s_{t+1} - f_t] + \varepsilon_{ft+1} \qquad \beta > 0 \tag{1.8}$$

where ε_{st+2} and ε_{ft+1} both have a mean value of zero.

Equations (1.7) and (1.8) illustrate the type of simultaneous adjustment mechanism considered in Chapter 6. This dynamic model is called an **error-correction** model because the movement of the variables in any period is related to the previous period's gap from long-run equilibrium. If the spot rate s_{t+1} turns out to equal the forward rate f_t, (1.7) and (1.8) state that the spot rate and forward rates are expected to remain unchanged. If there is a positive gap between the spot and forward rates so that $s_{t+1} - f_t > 0$, (1.7) and (1.8) lead to the prediction that the spot rate will fall and the forward rate will rise.

4. **Nonlinear Dynamics:** All of the equations considered thus far are linear (in the sense that each variable is raised to the first power) with constant coefficients. Chapter 7 considers the estimation of models that allow for more complicated dynamic structures. Recall that equation (1.3) assumes investment is always a constant proportion of the change in consumption. It might be more realistic to assume investment responds more to positive

than to negative changes in consumption. After all, firms might want to take advantage of positive consumption growth but simply let the capital stock decay in response to declines in consumption. Such behavior can be captured by modifying (1.3) such that the coefficient on $(c_t - c_{t-1})$ is not constant. Consider the specification

$$i_t = \beta_1(c_t - c_{t-1}) - \lambda_t\beta_2(c_t - c_{t-1}) + e_{it}$$

where $\beta_1 > \beta_2 > 0$ and is an indicator function such that $\lambda_t = 1$ if $(c_t - c_{t-1}) < 0$, otherwise $\lambda_t = 0$. Hence, if $(c_t - c_{t-1}) \geq 0$, $\lambda_t = 0$ so that $i_t = \beta_1(c_t - c_{t-1}) + e_{it}$ and if $(c_t - c_{t-1}) < 0$, $\lambda_t = 1$ so that $i_t = (\beta_1 - \beta_2)(c_t - c_{t-1}) + e_{it}$. If $\beta_1 - \beta_2 > 0$, investment is more responsive to positive than negative changes in consumption.

2. DIFFERENCE EQUATIONS AND THEIR SOLUTIONS

Although many of the ideas in the previous section were probably familiar to you, it is necessary to formalize some of the concepts used. In this section, we will examine the type of difference equation used in econometric analysis and make explicit what it means to "solve" such equations. To begin our examination of difference equations, consider the function $y = f(t)$. If we evaluate the function when the independent variable t takes on the specific value t^*, we get a specific value for the dependent variable called y_{t^*}. Formally, $y_{t^*} = f(t^*)$. Using this same notation, y_{t^*+h} represents the value of y when t takes on the specific value $t^* + h$. The first difference of y is defined as the value of the function when evaluated at $t = t^* + h$ minus the value of the function evaluated at t^*:

$$\begin{aligned} \Delta y_{t^*+h} &\equiv f(t^* + h) - f(t^*) \\ &\equiv y_{t^*+h} - y_{t^*} \end{aligned} \tag{1.9}$$

Differential calculus allows the change in the independent variable (i.e., the term h) to approach zero. Since most economic data is collected over discrete periods, however, it is more useful to allow the length of the time period to be greater than zero. Using difference equations, we normalize units so that h represents a unit change in t (i.e., $h = 1$) and consider the sequence of equally spaced values of the independent variable. Without any loss of generality, we can always drop the asterisk on t^*. We can then form the **first differences:**

$$\begin{aligned} \Delta y_t &= f(t) - f(t - 1) \equiv y_t - y_{t-1} \\ \Delta y_{t+1} &= f(t + 1) - f(t) \equiv y_{t+1} - y_t \\ \Delta y_{t+2} &= f(t + 2) - f(t + 1) \equiv y_{t+2} - y_{t+1} \end{aligned}$$

Often it will be convenient to express the entire sequence of values $\{\ldots y_{t-2}, y_{t-1}, y_t, y_{t+1}, y_{t+2}, \ldots\}$ as $\{y_t\}$. We can then refer to any particular value in the sequence as y_t. Unless specified, the index t runs from $-\infty$ to $+\infty$. In time-series econometric models, we use t to represent "time" and h to represent the length of a time period. Thus, y_t and y_{t+1} might represent the realizations of the $\{y_t\}$ sequence in the first and second quarters of 2009, respectively.

In the same way, we can form the **second difference** as the change in the first difference. Consider

$$\Delta^2 y_t \equiv \Delta(\Delta y_t) = \Delta(y_t - y_{t-1}) = (y_t - y_{t-1}) - (y_{t-1} - y_{t-2}) = y_t - 2y_{t-1} + y_{t-2}$$
$$\Delta^2 y_{t+1} \equiv \Delta(\Delta y_{t+1}) = \Delta(y_{t+1} - y_t) = (y_{t+1} - y_t) - (y_t - y_{t-1}) = y_{t+1} - 2y_t + y_{t-1}$$

The *n*th difference (Δ^n) is defined analogously. At this point, we risk taking the theory of difference equations too far. As you will see, the need to use second differences rarely arises in time-series analysis. It is safe to say that third- and higher-order differences are never used in applied work.

Since most of this text considers linear time-series methods, it is possible to examine only the special case of an *n*th-order linear difference equation with constant coefficients. The form for this special type of difference equation is given by

$$y_t = a_0 + \sum_{i=1}^{n} a_i y_{t-i} + x_t \tag{1.10}$$

The order of the difference equation is given by the value of n. The equation is linear because all values of the dependent variable are raised to the first power. Economic theory may dictate instances in which the various a_i are functions of variables within the economy. However, as long as they do not depend on any of the values of y_t or x_t, we can regard them as parameters. The term x_t is called the **forcing process.** The form of the forcing process can be very general; x_t can be any function of time, current and lagged values of other variables, and/or stochastic disturbances. From an appropriate choice of the forcing process, we can obtain a wide variety of important macroeconomic models. Reexamine equation (1.5), the reduced-form equation for real GDP. This equation is a second-order difference equation since y_t depends on y_{t-2}. The forcing process is the expression $(1 + \beta)\varepsilon_{ct} + \varepsilon_{it} - \beta\varepsilon_{ct-1}$. You will note that (1.5) has no intercept term corresponding to the expression a_0 in (1.10).

An important special case for the $\{x_t\}$ sequence is

$$x_t = \sum_{i=0}^{\infty} \beta_i \varepsilon_{t-i}$$

where the β_i are constants (some of which can equal zero) and the individual elements of the sequence $\{\varepsilon_t\}$ are not functions of the y_t. At this point, it is useful to allow the $\{\varepsilon_t\}$ sequence to be nothing more than a sequence of unspecified exogenous variables. For example, let $\{\varepsilon_t\}$ be a random error term and set $\beta_0 = 1$ and $\beta_1 = \beta_2 = \ldots = 0$; in this case, (1.10) becomes the autoregression equation

$$y_t = a_0 + a_1 y_{t-1} + a_2 y_{t-2} + \ldots + a_n y_{t-n} + \varepsilon_t$$

Let $n = 1$, $a_0 = 0$, and $a_1 = 1$ to obtain the random walk model. Notice that equation (1.10) can be written in terms of the **difference operator** (Δ). Subtracting y_{t-1} from (1.10), we obtain

$$y_t - y_{t-1} = a_0 + (a_1 - 1)y_{t-1} + \sum_{i=2}^{n} a_i y_{t-i} + x_t$$

or defining $\gamma = (a_1 - 1)$, we get

$$\Delta y_t = a_0 + \gamma y_{t-1} + \sum_{i=2}^{n} a_i y_{t-i} + x_t \tag{1.11}$$

Clearly, equation (1.11) is simply a modified version of (1.10).

A **solution** to a difference equation expresses the value of y_t as a function of the elements of the $\{x_t\}$ sequence and t (and possibly some given values of the $\{y_t\}$ sequence called **initial conditions**). Examining (1.11) makes it clear that there is a strong analogy to integral calculus, where the problem is to find a primitive function from a given derivative. We seek to find the primitive function $f(t)$, given an equation expressed in the form of (1.10) or (1.11). Notice that a solution is a function rather than a number. The key property of a solution is that it satisfies the difference equation for all permissible values of t and $\{x_t\}$. Thus, the substitution of a solution into the difference equation must result in an identity. For example, consider the simple difference equation $\Delta y_t = 2$ (or $y_t = y_{t-1} + 2$). You can easily verify that a solution to this difference equation is $y_t = 2t + c$, where c is any arbitrary constant. By definition, if $2t + c$ is a solution, it must hold for all permissible values of t. Thus, for period $t - 1$, $y_{t-1} = 2(t - 1) + c$. Now substitute the solution into the difference equation to form

$$2t + c \equiv 2(t - 1) + c + 2 \tag{1.12}$$

It is straightforward to carry out the algebra and verify that (1.12) is an identity. This simple example also illustrates that the solution to a difference equation need not be unique; there is a solution for any arbitrary value of c.

Another useful example is provided by the irregular term shown in Figure 1.1; recall that the equation for this expression is $I_t = 0.7I_{t-1} + \varepsilon_t$. You can verify that the solution to this first-order equation is

$$I_t = \sum_{i=0}^{\infty} (0.7)^i \varepsilon_{t-i} \tag{1.13}$$

Since (1.13) holds for all time periods, the value of the irregular component in $t - 1$ is given by

$$I_{t-1} = \sum_{i=0}^{\infty} (0.7)^i \varepsilon_{t-1-i} \tag{1.14}$$

Now substitute (1.13) and (1.14) into $I_t = 0.7I_{t-1} + \varepsilon_t$ to obtain

$$\begin{aligned} &\varepsilon_t + 0.7\varepsilon_{t-1} + (0.7)^2\varepsilon_{t-2} + (0.7)^3\varepsilon_{t-3} + \dots \\ &= 0.7[\varepsilon_{t-1} + 0.7\varepsilon_{t-2} + (0.7)^2\varepsilon_{t-3} + (0.7)^3\varepsilon_{t-4} + \dots] + \varepsilon_t \end{aligned} \tag{1.15}$$

The two sides of (1.15) are identical; this proves that (1.13) is a solution to the first-order stochastic difference equation $I_t = 0.7I_{t-1} + \varepsilon_t$. Be aware of the distinction between reduced-form equations and solutions. Since $I_t = 0.7I_{t-1} + \varepsilon_t$ holds for all values of t, it follows that $I_{t-1} = 0.7I_{t-2} + \varepsilon_{t-1}$. Combining these two equations yields

$$\begin{aligned} I_t &= 0.7[0.7I_{t-2} + \varepsilon_{t-1}] + \varepsilon_t \\ &= 0.49I_{t-2} + 0.7\varepsilon_{t-1} + \varepsilon_t \end{aligned} \tag{1.16}$$

Equation (1.16) is a reduced-form equation since it expresses I_t in terms of its own lags and disturbance terms. However, (1.16) does not qualify as a solution because it contains the "unknown" value of I_{t-2}. To qualify as a solution, (1.16) must express I_t in terms of the elements x_t, t, and any given initial conditions.

3. SOLUTION BY ITERATION

The solution given by (1.13) was simply postulated. The remaining portions of this chapter develop the methods you can use to obtain such solutions. Each method has its own merits; knowing the most appropriate to use in a particular circumstance is a skill that comes only with practice. This section develops the method of iteration. Although iteration is the most cumbersome and time-intensive method, most people find it to be very intuitive.

If the value of y in some specific period is known, a direct method of solution is to iterate forward from that period to obtain the subsequent time path of the entire y sequence. Refer to this known value of y as the initial condition or the value of y in time period 0 (denoted by y_0). It is easiest to illustrate the iterative technique using the first-order difference equation

$$y_t = a_0 + a_1 y_{t-1} + \varepsilon_t \tag{1.17}$$

Given the value of y_0, it follows that y_1 will be given by

$$y_1 = a_0 + a_1 y_0 + \varepsilon_1$$

In the same way, y_2 must be

$$\begin{aligned} y_2 &= a_0 + a_1 y_1 + \varepsilon_2 \\ &= a_0 + a_1[a_0 + a_1 y_0 + \varepsilon_1] + \varepsilon_2 \\ &= a_0 + a_0 a_1 + (a_1)^2 y_0 + a_1\varepsilon_1 + \varepsilon_2 \end{aligned}$$

Continuing the process in order to find y_3, we obtain

$$\begin{aligned} y_3 &= a_0 + a_1 y_2 + \varepsilon_3 \\ &= a_0[1 + a_1 + (a_1)^2] + (a_1)^3 y_0 + a_1^2\varepsilon_1 + a_1\varepsilon_2 + \varepsilon_3 \end{aligned}$$

You can easily verify that for all $t > 0$, repeated iteration yields

$$y_t = a_0 \sum_{i=0}^{t-1} a_1^i + a_1^t y_0 + \sum_{i=0}^{t-1} a_1^i \varepsilon_{t-i} \tag{1.18}$$

Equation (1.18) is a solution to (1.17) since it expresses y_t as a function of t, the forcing process $x_t = \Sigma(a_1)^i\varepsilon_{t-i}$, and the known value of y_0. As an exercise, it is useful to show that iteration from y_t back to y_0 yields exactly the formula given by (1.18). Since $y_t = a_0 + a_1 y_{t-1} + \varepsilon_t$, it follows that

$$\begin{aligned} y_t &= a_0 + a_1[a_0 + a_1 y_{t-2} + \varepsilon_{t-1}] + \varepsilon_t \\ &= a_0(1 + a_1) + a_1\varepsilon_{t-1} + \varepsilon_t + a_1^2[a_0 + a_1 y_{t-3} + \varepsilon_{t-2}] \end{aligned}$$

Continuing the iteration back to period 0 yields equation (1.18).

Iteration without an Initial Condition

Suppose you were not given the initial condition for y_0. The solution given by (1.18) would no longer be appropriate because the value of y_0 is an unknown. You could not select this initial value of y and iterate forward, nor could you iterate backward from y_t and simply choose to stop at $t = t_0$. Thus, suppose we continued to iterate backward by substituting $a_0 + a_1y_{-1} + \varepsilon_0$ for y_0 in (1.18):

$$\begin{aligned} y_t &= a_0\sum_{i=0}^{t-1} a_1^i + a_1^t(a_0 + a_1y_{-1} + \varepsilon_0) + \sum_{i=0}^{t-1} a_1^i\varepsilon_{t-i} \\ &= a_0\sum_{i=0}^{t} a_1^i + \sum_{i=0}^{t} a_1^i\varepsilon_{t-i} + a_1^{t+1}y_{-1} \end{aligned} \tag{1.19}$$

Continuing to iterate backward another m periods, we obtain

$$y_t = a_0\sum_{i=0}^{t+m} a_1^i + \sum_{i=0}^{t+m} a_1^i\varepsilon_{t-i} + a_1^{t+m+1}y_{-m-1} \tag{1.20}$$

Now examine the pattern emerging from (1.19) and (1.20). If $|a_1| < 1$, the term a_1^{t+m+1} approaches zero as m approaches infinity. Also, the infinite sum $[1 + a_1 + (a_1)^2 + \ldots]$ converges to $1/(1 - a_1)$. Thus, if we temporarily assume that $|a_1| < 1$, after continual substitution, (1.20) can be written as

$$y_t = a_0/(1 - a_1) + \sum_{i=0}^{\infty} a_1^i\varepsilon_{t-i} \tag{1.21}$$

You should take a few minutes to convince yourself that (1.21) is a solution to the original difference equation (1.17); substitution of (1.21) into (1.17) yields an identity. However, (1.21) is not a unique solution. For any arbitrary value of A, a solution to (1.17) is given by

$$y_t = Aa_1^t + a_0/(1 - a_1) + \sum_{i=0}^{\infty} a_1^i\varepsilon_{t-i} \tag{1.22}$$

To verify that (1.22) is a solution *for any arbitrary value of* A, substitute (1.22) into (1.17) to obtain

$$\begin{aligned} y_t &= Aa_1^t + a_0/(1 - a_1) + \sum_{i=0}^{\infty} a_1^i\varepsilon_{t-i} \\ &= a_0 + a_1\left[Aa_1^{t-1} + a_0/(1 - a_1) + \sum_{i=0}^{\infty} a_1^i\varepsilon_{t-1-i}\right] + \varepsilon_t \end{aligned}$$

Since the two sides are identical, (1.22) is necessarily a solution to (1.17).

Reconciling the Two Iterative Methods

Given the iterative solution (1.22), suppose that you are now given an initial condition concerning the value of y in the arbitrary period t_0. It is straightforward to show that we can impose the initial condition on (1.22) to yield the same solution as

(1.18). Since (1.22) must be valid for all periods (including t_0), when $t = 0$, it must be true that

$$y_0 = A + a_0/(1 - a_1) + \sum_{i=0}^{\infty} a_1^i \varepsilon_{-i} \text{ so that}$$

$$A = y_0 - a_0/(1 - a_1) - \sum_{i=0}^{\infty} a_1^i \varepsilon_{-i} \tag{1.23}$$

Since y_0 is given, we can view (1.23) as the value of A that renders (1.22) a solution to (1.17) given the initial condition. Hence, the presence of the initial condition eliminates the arbitrariness of A. Substituting this value of A into (1.22) yields

$$y_t = \left[y_0 - a_0/(1 - a_1) - \sum_{i=0}^{\infty} a_1^i \varepsilon_{-i} \right] a_1^t + a_0/(1 - a_1) + \sum_{i=0}^{\infty} a_1^i \varepsilon_{t-i} \tag{1.24}$$

Simplification of (1.24) results in

$$y_t = [(y_0 - a_0)/(1 - a_1)]a_1^t + a_0/(1 - a_1) + \sum_{i=0}^{t-1} a_1^i \varepsilon_{t-i} \tag{1.25}$$

It is a worthwhile exercise to verify that (1.25) is identical to (1.18).

Nonconvergent Sequences

Given that $|a_1| < 1$, (1.21) is the limiting value of (1.20) as m grows infinitely large. What happens to the solution in other circumstances? If $|a_1| > 1$, it is not possible to move from (1.20) to (1.21) because the expression $|a_1|^{t+m}$ grows infinitely large as $t + m$ approaches infinity.[1] However, if there is an initial condition, there is no need to obtain the infinite summation. Simply select the initial condition y_0 and iterate forward; the result will be (1.18):

$$y_t = a_0 \sum_{i=0}^{t-1} a_1^i + a_1^t y_0 + \sum_{i=0}^{t-1} a_1^i \varepsilon_{t-i}$$

Although the successive values of the $\{y_t\}$ sequence will become progressively larger in absolute value, all values in the series will be finite.

A very interesting case arises if $a_1 = 1$. Rewrite (1.17) as

$$y_t = a_0 + y_{t-1} + \varepsilon_t$$

or

$$\Delta y_t = a_0 + \varepsilon_t$$

As you should verify by iterating from y_t back to y_0, a solution to this equation is[2]

$$y_t = a_0 t + \sum_{i=1}^{t} \varepsilon_i + y_0 \tag{1.26}$$

After a moment's reflection, the form of the solution is quite intuitive. In every period t, the value of y_t changes by $a_0 + \varepsilon_t$ units. After t periods, there are t such changes; hence, the total change is ta_0 plus the t values of the $\{\varepsilon_t\}$ sequence. Notice that the solution contains summation of all disturbances from ε_1 through ε_t. Thus, when $a_1 = 1$, each disturbance has a permanent nondecaying effect on the value of y_t. You should compare this result to the solution found in (1.21). For the case in which $|a_1| < 1$, $|a_1|^t$ is a decreasing function of t so that the effects of past disturbances become successively smaller over time.

The importance of the magnitude of a_1 is illustrated in Figure 1.2. Thirty random numbers with a theoretical mean equal to zero were computer-generated and denoted by ε_1 through ε_{30}. Then the value of y_0 was set equal to unity and the next 25 values of the $\{y_t\}$ sequence were constructed using the formula $y_t = 0.9y_{t-1} + \varepsilon_t$. The result is shown by the thin line in Panel (a) of Figure 1.2. If you substitute $a_0 = 0$ and $a_1 = 0.9$

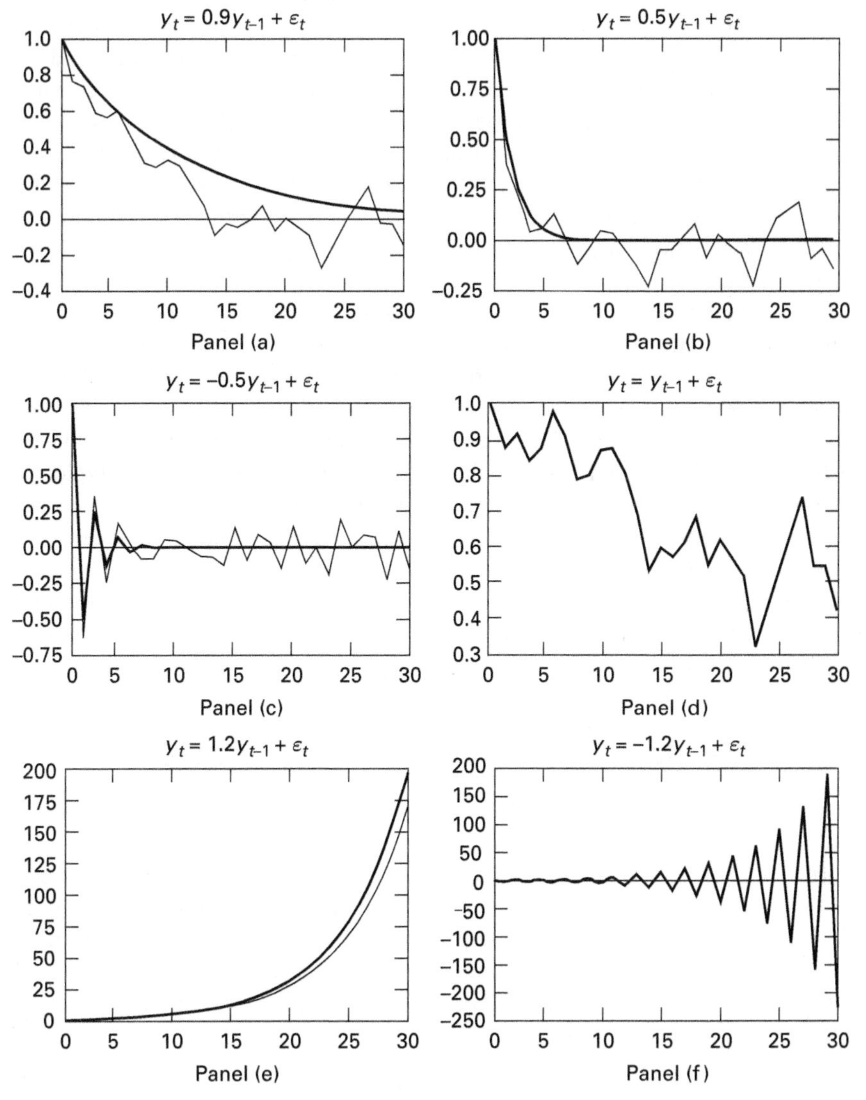

FIGURE 1.2 Convergent and Nonconvergent Sequences

into (1.18), you will see that the time path of $\{y_t\}$ consists of two parts. The first part, 0.9^t, is shown by the slowly decaying thick line in the panel. This term dominates the solution for relatively small values of t. The influence of the random part is shown by the difference between the thin and the thick line; you can see that the first several values of $\{\varepsilon_t\}$ are negative. Notice that as t increases, the influence of the initial value $y_0 = 1$ becomes less pronounced.

Using the previously drawn random numbers, we again set y_0 equal to unity and a second sequence was constructed using the formula $y_t = 0.5y_{t-1} + \varepsilon_t$. This second sequence is shown by the thin line in Panel (b) of Figure 1.2. The influence of the expression 0.5^t is shown by the rapidly decaying thick line. Again, as t increases, the random portion of the solution becomes more dominant in the time path of $\{y_t\}$. When we compare the first two panels, it is clear that reducing the magnitude of $|a_1|$ increases the rate of convergence. Moreover, the discrepancies between the simulated values of y_t and the thick line are less pronounced in the second panel. As you can see in (1.18), each value of ε_{t-i} enters the solution for y_t with a coefficient of $(a_1)^i$. The smaller value of a_1 means that the past realizations of ε_{t-i} have a smaller influence on the current value of y_t.

Simulating a third sequence with $a_1 = -0.5$ yields the thin line shown in Panel (c). The oscillations are due to the negative value of a_1. The expression $(-0.5)^t$, shown by the thick line, is positive when t is even and negative when t is odd. Since $|a_1| < 1$, the oscillations are dampened.

The next three panels in Figure 1.2 all show nonconvergent sequences. Each uses the initial condition $y_0 = 1$ and the same thirty values of $\{\varepsilon_t\}$ used in the other simulations. The line in Panel (d) shows the time path of $y_t = y_{t-1} + \varepsilon_t$. Since each value of ε_t has an expected value of zero, Panel (d) illustrates a random walk process. Here $\Delta y_t = \varepsilon_t$ so that the change in y_t is purely random. The nonconvergence is shown by the tendency of $\{y_t\}$ to meander. In Panel (e), the thick line representing the explosive expression $(1.2)^t$ dominates the random portion of the $\{y_t\}$ sequence. Also notice that the discrepancy between the simulated $\{y_t\}$ sequence and the thick line widens as t increases. The reason is that past values of ε_{t-i} enter the solution for y_t with the coefficient $(1.2)^i$. As i increases, the importance of these previous discrepancies becomes increasingly important. Similarly, setting $a_1 = -1.2$ results in the exploding oscillations shown in the lower-right panel of the figure. The value $(-1.2)^t$ is positive for even values of t and negative for odd values of t.

4. AN ALTERNATIVE SOLUTION METHODOLOGY

Solution by the iterative method breaks down in higher-order equations. The algebraic complexity quickly overwhelms any reasonable attempt to find a solution. Fortunately, there are several alternative solution techniques that can be helpful in solving the nth-order equation given by (1.10). If we use the principle that you should learn to walk before you learn to run, it is best to step through the first-order equation given by (1.17). Although you will be covering some familiar ground, the first-order case illustrates the general methodology extremely well. To split the procedure into its component parts, consider only the homogeneous portion of (1.17):[3]

$$y_t = a_1 y_{t-1} \tag{1.27}$$

The solution to this homogeneous equation is called the **homogeneous solution;** at times it will be useful to denote the homogeneous solution by the expression y_t^h. Obviously, the trivial solution $y_t = y_{t-1} = \ldots = 0$ satisfies (1.27). However, this solution is not unique. By setting a_0 and all values of $\{\varepsilon_t\}$ equal to zero, (1.18) becomes $y_t = a_1^t y_0$. Hence, $y_t = a_1^t y_0$ must be a solution to (1.27). Yet, even this solution does not constitute the full set of solutions. It is easy to verify that the expression a_1^t multiplied by any arbitrary constant A satisfies (1.27). Simply substitute $y_t = Aa_1^t$ and $y_{t-1} = Aa_1^{t-1}$ into (1.27) to obtain

$$Aa_1^t = a_1 A a_1^{t-1}$$

Since $a_1^t = a_1 a_1^{t-1}$, it follows that $y_t = Aa_1^t$ also solves (1.27). With the aid of the thick lines in Figure 1.2, we can classify the properties of the homogeneous solution as follows:

1. If $|a_1| < 1$, the expression a_1^t converges to zero as t approaches infinity. Convergence is direct if $0 < a_1 < 1$ and oscillatory if $-1 < a_1 < 0$.
2. If $|a_1| > 1$, the homogeneous solution is not stable. If $a_1 > 1$, the homogeneous solution approaches infinity as t increases. If $a_1 < -1$, the homogeneous solution oscillates explosively.
3. If $a_1 = 1$, any arbitrary constant A satisfies the homogeneous equation $y_t = y_{t-1}$. If $a_1 = -1$, the system is *meta-stable*: $a_1^t = 1$ for even values of t and -1 for odd values of t.

Now consider (1.17) in its entirety. In the last section, you confirmed that (1.21) is a valid solution to (1.17). Equation (1.21) is called a **particular solution** to the difference equation; all such particular solutions will be denoted by the term y_t^p. The term "particular" stems from the fact that a solution to a difference equation may not be unique; hence, (1.21) is just one particular solution out of the many possibilities. The particular solution satisfies the original difference equation without imposing the restriction that it be homogeneous.

In moving to (1.22) you verified that the particular solution was not unique. The homogeneous solution Aa_1^t plus the particular solution given by (1.21) constituted the complete solution to (1.17). The **general solution** to a difference equation is defined to be a particular solution plus all homogeneous solutions. Once the general solution is obtained, the arbitrary constant A can be eliminated by imposing an initial condition for y_0.

The Solution Methodology

The results of the first-order case are directly applicable to the nth-order equation given by (1.10). In this general case, it will be more difficult to find the particular solution and there will be n distinct homogeneous solutions. Nevertheless, the solution methodology will always entail the following four steps:

STEP 1: Form the homogeneous equation and find all n homogeneous solutions;

STEP 2: Find a particular solution;

STEP 3: Obtain the general solution as the sum of the particular solution and a linear combination of all homogeneous solutions;

STEP 4: Eliminate the arbitrary constant(s) by imposing the initial condition(s) on the general solution.

Before we address the various techniques that can be used to obtain homogeneous and particular solutions, it is worthwhile to illustrate the methodology using the equation

$$y_t = 0.9y_{t-1} - 0.2y_{t-2} + 3 \tag{1.28}$$

Clearly, this second-order equation is in the form of (1.10) with $a_0 = 3$, $a_1 = 0.9$, $a_2 = -0.2$, and $x_t = 0$. Beginning with the first of the four steps, form the homogeneous equation

$$y_t - 0.9y_{t-1} + 0.2y_{t-2} = 0 \tag{1.29}$$

In the first-order case of (1.17), the homogeneous solution was Aa_1^t. Section 6 will show you how to find the complete set of homogeneous solutions. For now, it is sufficient to assert that the two homogeneous solutions are $y_{1t}^h = (0.5)^t$ and $y_{2t}^h = (0.4)^t$. To verify the first solution, note that $y_{1t-1}^h = (0.5)^{t-1}$ and $y_{1t-2}^h = (0.5)^{t-2}$. Thus, y_{1t}^h is a solution if it satisfies

$$(0.5)^t - 0.9(0.5)^{t-1} + 0.2(0.5)^{t-2} = 0$$

If we divide by $(0.5)^{t-2}$, the issue is whether

$$(0.5)^2 - 0.9(0.5) + 0.2 = 0$$

Carrying out the algebra, $0.25 - 0.45 + 0.2$ does equal zero so that $(0.5)^t$ is a solution to (1.29). In the same way, it is easy to verify that $y_{2t}^h = (0.4)^t$ is a solution since

$$(0.4)^t - 0.9(0.4)^{t-1} + 0.2(0.4)^{t-2} = 0$$

Divide by $(0.4)^{t-2}$ to obtain $(0.4)^2 - 0.9(0.4) + 0.2 = 0.16 - 0.36 + 0.2 = 0$.

The second step is to obtain a particular solution; you can easily confirm that the particular solution $y_t^p = 10$ solves (1.28) as: $10 = 0.9(10) - 0.2(10) + 3$.

The third step is to combine the particular solution and a linear combination of both homogeneous solutions to obtain

$$y_t = A_1(0.5)^t + A_2(0.4)^t + 10$$

where A_1 and A_2 are arbitrary constants.

For the fourth step, assume you have two initial conditions for the $\{y_t\}$ sequence. So that we can keep our numbers reasonably round, suppose that $y_0 = 13$ and $y_1 = 11.3$. Thus, for periods zero and one, our solution must satisfy

$$13 = A_1 + A_2 + 10$$
$$11.3 = A_1(0.5) + A_2(0.4) + 10$$

Solving simultaneously for A_1 and A_2, you should find $A_1 = 1$ and $A_2 = 2$. Hence, the solution is

$$y_t = (0.5)^t + 2(0.4)^t + 10$$

You can substitute $y_t = (0.5)^t + 2(0.4)^t + 10$ into (1.28) to verify that the solution is correct.

Generalizing the Method

To show that this method is applicable to higher-order equations, consider the homogeneous part of (1.10):

$$y_t = \sum_{i=1}^{n} a_i y_{t-i} \tag{1.30}$$

As shown in Section 6, there are n homogeneous solutions that satisfy (1.30). For now, it is sufficient to demonstrate the following proposition: *If y_t^h is a homogeneous solution to (1.30), Ay_t^h is also a solution for any arbitrary constant A*. By assumption, y_t^h solves the homogeneous equation so that

$$y_t^h = \sum_{i=1}^{n} a_i y_{t-i}^h \tag{1.31}$$

The expression Ay_t^h is also a solution if

$$Ay_t^h = \sum_{i=1}^{n} a_i A y_{t-i}^h \tag{1.32}$$

We know (1.32) is satisfied because dividing each term by A yields (1.31). Now suppose that there are two separate solutions to the homogeneous equation denoted by y_{1t}^h and y_{2t}^h. It is straightforward to show that for any two constants A_1 and A_2, the linear combination $A_1 y_{1t}^h + A_2 y_{2t}^h$ is also a solution to the homogeneous equation. If $A_1 y_{1t}^h + A_2 y_{2t}^h$ is a solution to (1.30), it must satisfy

$$\begin{aligned} A_1 y_{1t}^h + A_2 y_{2t}^h &= a_1(A_1 y_{1t-1}^h + A_2 y_{2t-1}^h) + a_2(A_1 y_{1t-2}^h + A_2 y_{2t-2}^h) \\ &\quad + \ldots + a_n(A_1 y_{1t-n}^h + A_2 y_{2t-n}^h) \end{aligned}$$

Regrouping terms, we want to know if

$$\left[A_1 y_{1t}^h - \sum_{i=1}^{n} A_1 a_i y_{1t-i}^h\right] + \left[A_2 y_{2t}^h - \sum_{i=1}^{n} A_2 a_i y_{2t-i}^h\right] = 0$$

Since $A_1 y_{1t}^h$ and $A_2 y_{2t}^h$ are separate solutions to (1.30), each of the expressions in brackets is zero. Hence, the linear combination is necessarily a solution to the homogeneous equation. This result easily generalizes to all n homogeneous solutions to an nth-order equation.

Finally, the use of Step 3 is appropriate since *the sum of any particular solution and any linear combination of all homogeneous solutions is also a solution*. To prove this proposition, substitute the sum of the particular and homogeneous solutions into (1.10) to obtain

$$y_t^p + y_t^h = a_0 + \sum_{i=1}^{n} a_i(y_{t-i}^p + y_{t-i}^h) + x_t \tag{1.33}$$

Recombining the terms in (1.33), we want to know if

$$\left[y_t^p - a_0 - \sum_{i=1}^{n} a_i y_{t-i}^p - x_t\right] + \left[y_t^h - \sum_{i=1}^{n} a_i y_{t-i}^h\right] = 0 \tag{1.34}$$

Since y_t^p solves (1.10), the expression in the first bracket of (1.34) is zero. Since y_t^h solves the homogeneous equation, the expression in the second bracket is zero. Thus, (1.34) is an identity; the sum of the homogeneous and particular solutions solves (1.10).

5. THE COBWEB MODEL

An interesting way to illustrate the methodology outlined in the previous section is to consider a stochastic version of the traditional **cobweb model.** Since the model was originally developed to explain the volatility in agricultural prices, let the market for a product—say, wheat—be represented by

$$d_t = a - \gamma p_t \qquad \gamma > 0 \tag{1.35}$$

$$s_t = b + \beta p_t^* + \varepsilon_t \qquad \beta > 0 \tag{1.36}$$

$$s_t = d_t \tag{1.37}$$

where: d_t = demand for wheat in period t
s_t = supply of wheat in t
p_t = market price of wheat in t
p_t^* = price that farmers expect to prevail at t
ε_t = a zero mean stochastic supply shock

and parameters a, b, γ, and β are all positive such that $a > b$.[4]

The nature of the model is such that consumers buy as much wheat as is desired at the market clearing price p_t. At planting time, farmers do not know the price prevailing at harvest time; they base their supply decision on the expected price (p^*). The actual quantity produced depends on the planned quantity $b + \beta p_t^*$ plus a random supply shock ε_t. Once the product is harvested, market equilibrium requires that the quantity supplied equal the quantity demanded. Unlike the actual market for wheat, the model does not allow for the possibility of storage. The essence of the cobweb model is that farmers form their expectations in a naive fashion; let farmers use last year's price as the expected market price

$$p_t^* = p_{t-1} \tag{1.38}$$

Point E in Figure 1.3 represents the long-run equilibrium price and quantity combination. Note that the equilibrium concept in this stochastic model differs from that of the traditional cobweb model. If the system is stable, successive prices will *tend* to converge to point E. However, the nature of the stochastic equilibrium is such that the ever-present supply shocks prevent the system from remaining at E. Nevertheless, it is useful to solve for the long-run price. If we set all values of the $\{\varepsilon_t\}$ sequence equal to zero, set $p_t = p_{t-1} = \ldots = p$, and equate supply and demand, the long-run equilibrium price is given by $p = (a - b)/(\gamma + \beta)$. Similarly, the equilibrium quantity (s) is given by $s = (a\beta + \gamma b)/(\gamma + \beta)$.

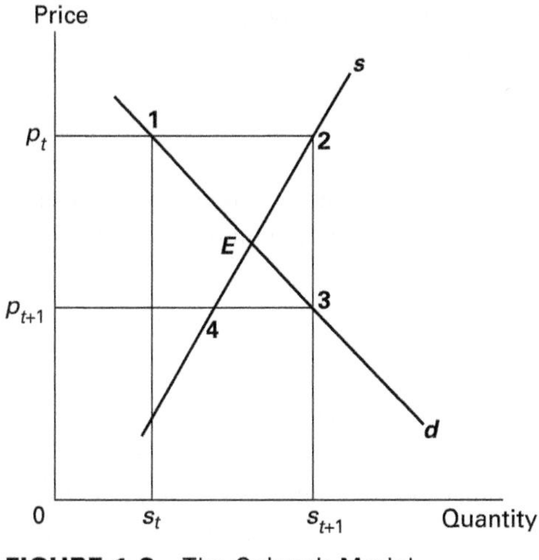

FIGURE 1.3 The Cobweb Model

To understand the dynamics of the system, suppose that farmers in t plan to produce the equilibrium quantity s. However, let there be a negative supply shock such that the actual quantity produced turns out to be s_t. As shown by point 1 in Figure 1.3, consumers are willing to pay p_t for the quantity s_t; hence, market equilibrium in t occurs at point 1. Updating one period allows us to see the main result of the cobweb model. For simplicity, assume that all subsequent values of the supply shock are zero (i.e., $\varepsilon_{t+1} = \varepsilon_{t+2} = \ldots = 0$). At the beginning of period $t + 1$, farmers expect the price at harvest time to be the price of the previous period; thus, $p^*_{t+1} = p_t$. Accordingly, they produce quantity s_{t+1} (see point 2 in the figure); consumers, however, are willing to buy quantity s_{t+1} only if the price falls to that indicated by p_{t+1} (see point 3 in the figure). The next period begins with farmers expecting to be at point 4. The process continually repeats until the equilibrium point E is attained.

As drawn, Figure 1.3 suggests that the market will always converge to the long-run equilibrium point. This result does not hold for all demand and supply curves. To formally derive the stability condition, combine (1.35) through (1.38) to obtain

$$b + \beta p_{t-1} + \varepsilon_t = a - \gamma p_t$$

or

$$p_t = (-\beta/\gamma)p_{t-1} + (a - b)/\gamma - \varepsilon_t/\gamma \tag{1.39}$$

Clearly, (1.39) is a stochastic first-order linear difference equation with constant coefficients. To obtain the general solution, proceed using the four steps listed at the end of the last section:

1. Form the homogeneous equation $p_t = (-\beta/\gamma)p_{t-1}$. In the next section you will learn how to find the solution(s) to a homogeneous equation. For now, it is sufficient to verify that the homogeneous solution is

$$p_t^h = A(-\beta/\gamma)^t$$

where A is an arbitrary constant.

2. If the ratio β/γ is less than unity, you can iterate (1.39) backward from p_t to verify that the particular solution for the price is

$$p_t^p = \frac{a-b}{\gamma+\beta} - \frac{1}{\gamma}\sum_{i=0}^{\infty}(-\beta/\gamma)^i\varepsilon_{t-i} \tag{1.40}$$

If $\beta/\gamma \geq 1$, the infinite summation in (1.40) is not convergent. As discussed in the last section, it is necessary to impose an initial condition on (1.40) if $\beta/\gamma \geq 1$.

3. The general solution is the sum of the homogeneous and particular solutions; combining these two solutions, the general solution is

$$p_t = \frac{a-b}{\gamma+\beta} - \frac{1}{\gamma}\sum_{i=0}^{\infty}(-\beta/\gamma)^i\varepsilon_{t-i} + A(-\beta/\gamma)^t \tag{1.41}$$

4. In (1.41), A is an arbitrary constant that can be eliminated if we know the price in some initial period. For convenience, let this initial period have a time subscript of zero. Since the solution must hold for every period, *including period zero*, it must be that case that

$$p_0 = \frac{a-b}{\gamma+\beta} - \frac{1}{\gamma}\sum_{i=0}^{\infty}(-\beta/\gamma)^i\varepsilon_{-i} + A(-\beta/\gamma)^0$$

Since $(-\beta/\gamma)^0 = 1$, the value of A is given by

$$A = p_0 - \frac{a-b}{\gamma+\beta} + \frac{1}{\gamma}\sum_{i=0}^{\infty}(-\beta/\gamma)^i\varepsilon_{-i}$$

Substituting this solution for A back into (1.41) yields

$$p_t = \frac{a-b}{\gamma+\beta} - \frac{1}{\gamma}\sum_{i=0}^{\infty}(-\beta/\gamma)^i\varepsilon_{t-i} + \left(-\frac{\beta}{\gamma}\right)^t\left[p_0 - \frac{a-b}{\gamma+\beta} + \frac{1}{\gamma}\sum_{i=0}^{\infty}(-\beta/\gamma)^i\varepsilon_{-i}\right]$$

and, after simplification of the two summations,

$$p_t = \frac{a-b}{\gamma+\beta} - \frac{1}{\gamma}\sum_{i=0}^{t-1}(-\beta/\gamma)^i\varepsilon_{t-i} + \left(-\frac{\beta}{\gamma}\right)^t\left[p_0 - \frac{a-b}{\gamma+\beta}\right] \tag{1.42}$$

We can interpret (1.42) in terms of Figure 1.3. In order to focus on the stability of the system, temporarily assume that all values of the $\{\varepsilon_t\}$ sequence are zero. Subsequently, we will return to a consideration of the effects of supply shocks. If the system begins in long-run equilibrium, the initial condition is such that $p_0 = (a - b)/(\gamma + \beta)$. In this case, inspection of equation (1.42) indicates that $p_t = (a - b)/(\gamma + \beta)$. Thus, if we begin the process at point E, the system remains in long-run equilibrium. Instead, suppose that the process begins at a price below long-run equilibrium: $p_0 < (a - b)/(\gamma + \beta)$. Equation (1.42) tells us that p_1 is

$$p_1 = (a-b)/(\gamma+\beta) + [p_0 - (a-b)/(\gamma+\beta)]\,(-\beta/\gamma)^1 \tag{1.43}$$

Since $p_0 < (a - b)/(\gamma + \beta)$ and $-\beta/\gamma < 0$, it follows that p_1 will be above the long-run equilibrium price $(a - b)/(\gamma + \beta)$. In period 2,

$$p_2 = (a-b)/(\gamma+\beta) + [p_0 - (a-b)/(\gamma+\beta)](-\beta/\gamma)^2$$

Although $p_0 < (a - b)/(\gamma + \beta)$, $(-\beta/\gamma)^2$ is positive; hence, p_2 is below the long-run equilibrium. For the subsequent periods, note that $(-\beta/\gamma)^t$ will be positive for even values of t and negative for odd values of t. Just as we found graphically, the successive values of the $\{p_t\}$ sequence will oscillate above and below the long-run equilibrium price. Since $(\beta/\gamma)^t$ goes to zero if $\beta < \gamma$ and explodes if $\beta > \gamma$, the magnitude of β/γ determines whether the price actually converges toward the long-run equilibrium. If $\beta/\gamma < 1$, the oscillations will diminish in magnitude, and if $\beta/\gamma > 1$, the oscillations will be explosive.

The economic interpretation of this stability condition is straightforward. The slope of the supply curve (i.e., $\partial p_t/\partial s_t$) is $1/\beta$ and the absolute value of the slope of the demand curve [i.e., $-\partial p_t/\partial d_t$] is $1/\gamma$. If the supply curve is steeper than the demand curve, $1/\beta > 1/\gamma$ or $\beta/\gamma < 1$ so that the system is stable. This is precisely the case illustrated in Figure 1.3. As an exercise, you should draw a diagram with the demand curve steeper than the supply curve and show that the price oscillates and diverges from the long-run equilibrium.

Now consider the effects of the supply shocks. The contemporaneous effect of a supply shock on the price of wheat is the partial derivative of p_t with respect to ε_t; from (1.42):

$$\frac{\partial p_t}{\partial \varepsilon_t} = -\frac{1}{\gamma} \tag{1.44}$$

Equation (1.44) is called the **impact multiplier** since it shows the impact effect of a change in ε_t on the price in t. In terms of Figure 1.3, a negative value of ε_t implies a price above the long-run price p; the price in t rises by $1/\gamma$ units for each unit decline in current period supply. Of course, this terminology is not specific to the cobweb model; in terms of the nth-order model given by (1.10), the impact multiplier is the partial derivative of y_t with respect to the partial change in the forcing process.[5]

The effects of the supply shock in t persist into future periods. Updating (1.42) by one period yields the **one-period multiplier:**

$$\frac{\partial p_{t+1}}{\partial \varepsilon_t} = -\frac{1}{\gamma}(-\beta/\gamma) = \beta/\gamma^2$$

Point 3 in Figure 1.3 illustrates how the price in $t + 1$ is affected by the negative supply shock in t. It is straightforward to derive the result that the effects of the supply shock decay over time. Since $\beta/\gamma < 1$, the absolute value of $\partial p_t/\partial \varepsilon_t$ exceeds $\partial p_{t+1}/\partial \varepsilon_t$. All of the multipliers can be derived analogously; updating (1.42) by two periods:

$$\partial p_{t+2}/\partial \varepsilon_t = -(1/\gamma)(-\beta/\gamma)^2$$

and after n periods:

$$\partial p_{t+n}/\partial \varepsilon_t = -(1/\gamma)(-\beta/\gamma)^n$$

The time path of all such multipliers is called the **impulse response function.** This function has many important applications in time-series analysis because it shows how the entire time path of a variable is affected by a stochastic shock. Here, the impulse function traces the effects of a supply shock in the wheat market. In other economic applications, you may be interested in the time path of a money supply shock or a productivity shock on real GDP.

In actuality, the function can be derived without updating (1.42) because it is always the case that

$$\frac{\partial p_{t+j}}{\partial \varepsilon_t} = \frac{\partial p_t}{\partial \varepsilon_{t-j}}$$

To find the impulse response function, simply find the partial derivative of (1.42) with respect to the various ε_{t-j}. These partial derivatives are nothing more than the coefficients of the $\{\varepsilon_{t-j}\}$ sequence in (1.42).

Each of the three components in (1.42) has a direct economic interpretation. The deterministic portion of the particular solution $(a - b)/(\gamma + \beta)$ is the long-run equilibrium price; if the stability condition is met, the $\{p_t\}$ sequence tends to converge to this long-run value. The stochastic component of the particular solution captures the short-run price adjustments due to the supply shocks. The ultimate decay of the coefficients of the impulse response function guarantees that the effects of changes in the various ε_t are of a short-run duration. The third component is the expression $(-\beta/\gamma)^t A = (-\beta/\gamma)^t[p_0 - (a - b)/(\gamma + \beta)]$. The value of A is the initial period's deviation of the price from its long-run equilibrium level. Given that $\beta/\gamma < 1$, the importance of this initial deviation diminishes over time.

6. SOLVING HOMOGENEOUS DIFFERENCE EQUATIONS

Higher-order difference equations arise quite naturally in economic analysis. Equation (1.5)—the reduced-form GDP equation resulting from Samuelson's (1939) model—is an example of a second-order difference equation. Moreover, in time-series econometrics it is quite typical to estimate second- and higher-order equations. To begin our examination of homogeneous solutions, consider the second-order equation

$$y_t - a_1 y_{t-1} - a_2 y_{t-2} = 0 \tag{1.45}$$

Given the findings in the first-order case, you should suspect that the homogeneous solution has the form $y_t^h = A\alpha^t$. Substitution of this trial solution into (1.45) yields

$$A\alpha^t - a_1 A\alpha^{t-1} - a_2 A\alpha^{t-2} = 0 \tag{1.46}$$

Clearly, any arbitrary value of A is satisfactory. If you divide (1.46) by $A\alpha^{t-2}$, the problem is to find the values of α that satisfy

$$\alpha^2 - a_1\alpha - a_2 = 0 \tag{1.47}$$

Solving this quadratic equation—called the **characteristic equation**—yields two values of α, called the **characteristic roots.** Using the quadratic formula, we find that the two characteristic roots are

$$\begin{aligned}\alpha_1, \alpha_2 &= \frac{a_1 \pm \sqrt{a_1^2 + 4a_2}}{2} \\ &= (a_1 \pm \sqrt{d})/2\end{aligned} \tag{1.48}$$

where d is the discriminant $[(a_1)^2 + 4a_2]$.

Each of these two characteristic roots yields a valid solution for (1.45). Again, these solutions are not unique. In fact, for any two arbitrary constants A_1 and A_2, the linear combination $A_1(\alpha_1)^t + A_2(\alpha_2)^t$ also solves (1.45). As proof, simply substitute $y_t = A_1(\alpha_1)^t + A_2(\alpha_2)^t$ into (1.45) to obtain

$$A_1(\alpha_1)^t + A_2(\alpha_2)^t = a_1[A_1(\alpha_1)^{t-1} + A_2(\alpha_2)^{t-1}] + a_2[A_1(\alpha_1)^{t-2} + A_2(\alpha_2)^{t-2}]$$

Now, regroup terms as follows:

$$A_1[(\alpha_1)^t - a_1(\alpha_1)^{t-1} - a_2(\alpha_1)^{t-2}] + A_2[(\alpha_2)^t - a_1(\alpha_2)^{t-1} - a_2(\alpha_2)^{t-2}] = 0$$

Since α_1 and α_2 each solve (1.45), both terms in brackets must equal zero. As such, the complete homogeneous solution in the second-order case is

$$y_t^h = A_1(\alpha_1)^t + A_2(\alpha_2)^t$$

Without knowing the specific values of a_1 and a_2, we cannot find the two characteristic roots α_1 and α_2. Nevertheless, it is possible to characterize the nature of the solution; three possible cases are dependent on the value of the discriminant d.

CASE 1

If $a_1^2 + 4a_2 > 0$, d is a real number and there will be two distinct real characteristic roots. Hence, there are two separate solutions to the homogeneous equation denoted by $(\alpha_1)^t$ and $(\alpha_2)^t$. We already know that any linear combination of the two is also a solution. Hence,

$$y_t^h = A_1(\alpha_1)^t + A_2(\alpha_2)^t$$

It should be clear that if the absolute value of *either* α_1 or α_2 exceeds unity, the homogeneous solution will explode. Worksheet 1.1 examines two second-order equations showing real and distinct characteristic roots. In the first example, $y_t = 0.2y_{t-1} + 0.35y_{t-2}$, the characteristic roots are shown to be $\alpha_1 = 0.7$ and $\alpha_2 = -0.5$. Hence, the full homogeneous solution is $y_t^h = A_1(0.7)^t + A_2(-0.5)^t$. Since both roots are less than unity in absolute value, the homogeneous solution is convergent. As you can see in the graph on the bottom left-hand side of Worksheet 1.1, convergence is not monotonic because of the influence of the expression $(-0.5)^t$.

In the second example, $y_t = 0.7y_{t-1} + 0.35y_{t-2}$. The worksheet indicates how to obtain the solution for the two characteristic roots. Given that one characteristic root is 1.037, the $\{y_t\}$ sequence explodes. The influence of the negative root ($\alpha_2 = -0.337$) is responsible for the nonmonotonicity of the time path. Since $(-0.337)^t$ quickly approaches zero, the dominant root is the explosive value 1.037.

CASE 2

If $a_1^2 + 4a_2 = 0$, it follows that $d = 0$ and $\alpha_1 = \alpha_2 = a_1/2$. Hence, a homogeneous solution is $a_1/2$. However, when $d = 0$, there is a second homogeneous solution given by $t(a_1/2)^t$. To demonstrate that $y_t^h = t(a_1/2)^t$ is a homogeneous solution, substitute it into (1.45) to determine whether

$$t(a_1/2)^t - a_1[(t-1)(a_1/2)^{t-1}] - a_2[(t-2)(a_1/2)^{t-2}] = 0$$

WORKSHEET 1.1

SECOND-ORDER EQUATIONS

Example 1: $y_t = 0.2y_{t-1} + 0.35y_{t-2}$. Hence: $a_1 = 0.2$ and $a_2 = 0.35$

Form the homogeneous equation: $y_t - 0.2y_{t-1} - 0.35y_{t-2} = 0$

A check of the discriminant reveals: $d = a_1^2 + 4a_2$ so that $d = 1.44$. Given that $d > 0$, the roots will be real and distinct.

Let the trial solution have the form: $y_t = \alpha^t$. Substitute the trial solution into the homogenous equation to obtain: $\alpha^t - 0.2\alpha^{t-1} - 0.35\alpha^{t-2} = 0$

Divide by α^{t-2} to obtain the characteristic equation: $\alpha^2 - 0.2\alpha - 0.35 = 0$

Compute the two characteristic roots:

$$\alpha_1 = 0.5(a_1 + d^{1/2}) \qquad \alpha_2 = 0.5(a_1 - d^{1/2})$$
$$\alpha_1 = 0.7 \qquad \alpha_2 = -0.5$$

The homogeneous solution is: $A_1(0.7)^t + A_2(-0.5)^t$. The first graph shows the time path of this solution for the case in which the arbitrary constants equal unity and t runs from 1 to 20.

Example 2: $y_t = 0.7y_{t-1} + 0.35y_{t-2}$. Hence: $a_1 = 0.7$ and $a_2 = 0.35$

Form the homogeneous equation: $y_t - 0.7y_{t-1} - 0.35y_{t-2} = 0$

A check of the discriminant reveals: $d = a_1^2 + 4a_2$ so that $d = 1.89$. Given that $d > 0$, the roots will be real and distinct.

Form the characteristic equation $\alpha^t - 0.7\alpha^{t-1} - 0.35\alpha^{t-2} = 0$

Divide by α^{t-2} to obtain the characteristic equation: $\alpha^2 - 0.7\alpha - 0.35 = 0$

Compute the two characteristic roots:

$$\alpha_1 = 0.5(a_1 + d^{1/2}) \qquad \alpha_2 = 0.5(a_1 - d^{1/2})$$
$$\alpha_1 = 1.037 \qquad \alpha_2 = -0.337$$

The homogeneous solution is: $A_1(1.037)^t + A_2(-0.337)^t$. The second graph shows the time path of this solution for the case in which the arbitrary constants equal unity and t runs from 1 to 20.

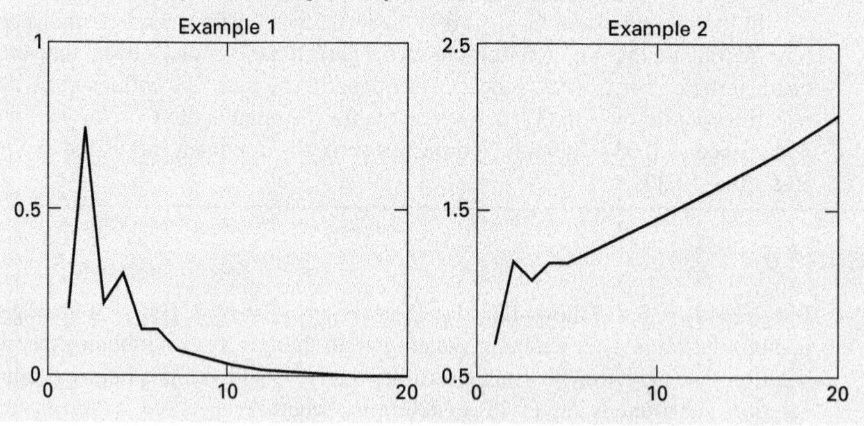

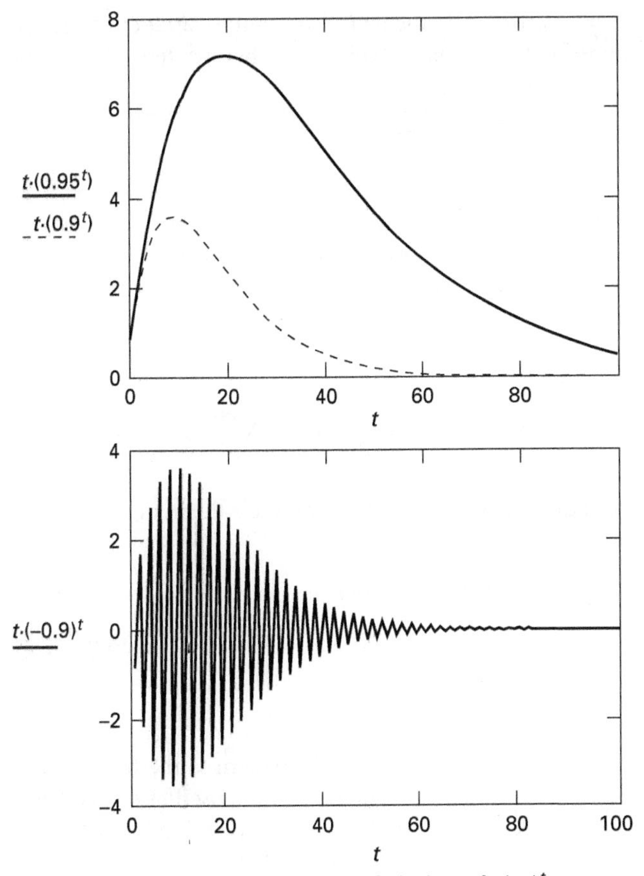

FIGURE 1.4 The Homogeneous Solution of $t(a_1)^t$

Divide by $(a_1/2)^{t-2}$ and form

$$-[(a_1^2/4) + a_2]t + [(a_1^2/2) + 2a_2] = 0$$

Since we are operating in the circumstance where $a_1^2 + 4a_2 = 0$, each bracketed expression is zero; hence, $t(a_1/2)^t$ solves (1.45). Again, for arbitrary constants A_1 and A_2, the complete homogeneous solution is

$$y_t^h = A_1(a_1/2)^t + A_2 t(a_1/2)^t$$

Clearly, the system is explosive if $|a_1| > 2$. If $|a_1| < 2$, the term $A_1(a_1/2)^t$ converges, but you might think that the effect of the term $t(a_1/2)^t$ is ambiguous [since the diminishing $(a_1/2)^t$ is multiplied by t]. This ambiguity is correct in the limited sense that the behavior of the homogeneous solution is not monotonic. As illustrated in Figure 1.4 for $a_1/2 = 0.95$, 0.9, and -0.9, as long as $|a_1| < 2$, $\lim[t(a_1/2)^t]$ is necessarily zero as $t \to \infty$; thus, there is always convergence. For $0 < a_1 < 2$, the homogeneous solution appears to explode before ultimately converging to zero. For $-2 < a_1 < 0$, the behavior is wildly erratic; the homogeneous solution appears to oscillate explosively before the oscillations dampen and finally converge to zero.

CASE 3

If $a_1^2 + 4a_2 < 0$, it follows that d is negative so that the characteristic roots are imaginary. Since $a_1^2 \geq 0$, imaginary roots can occur only if $a_2 < 0$. Although this might be hard to interpret directly, if we switch to polar coordinates it is possible to transform the roots into more easily understood trigonometric functions. The technical details are presented in Appendix 1.1. For now, write the two characteristic roots as

$$\alpha_1 = (a_1 + i\sqrt{-d})/2 \qquad \alpha_2 = (a_1 - i\sqrt{-d})/2$$

where $i = \sqrt{-1}$.

As shown in Appendix 1.1, you can use de Moivre's theorem to write the homogeneous solution as

$$y_t^h = \beta_1 r^t \cos(\theta t + \beta_2) \tag{1.49}$$

where β_1 and β_2 are arbitrary constants, $r = (-a_2)^{1/2}$, and the value of θ is chosen so as to satisfy

$$\cos(\theta) = a_1/[2(-a_2)^{1/2}] \tag{1.50}$$

The trigonometric functions impart a wavelike pattern to the time path of the homogeneous solution; note that the frequency of the oscillations is determined by θ. Since $\cos(\theta t) = \cos(2\pi + \theta t)$, the stability condition is determined solely by the magnitude of $r = (-a_2)^{1/2}$. If $|a_2| = 1$, the oscillations are of unchanging amplitude; the homogeneous solution is periodic. The oscillations will dampen if $|a_2| < 1$ and explode if $|a_2| > 1$.

Example: It is worthwhile to work through an exercise using an equation with imaginary roots. The left-hand side of Worksheet 1.2 examines the behavior of the equation $y_t = 1.6y_{t-1} - 0.9y_{t-2}$. A quick check shows that the discriminant d is negative so that the characteristic roots are imaginary. If we transform to polar coordinates, the value of r is given by $(0.9)^{1/2} = 0.949$. From (1.50), $\cos(\theta) = 1.6/(2 * 0.949) = 0.843$. You can use a trig table or a calculator to show that $\theta = 0.567$ (i.e., if $\cos(\theta) = 0.843$, $\theta = 0.567$). Thus, the homogeneous solution is

$$y_t^h = \beta_1(0.949)^t \cos(0.567t + \beta_2) \tag{1.51}$$

The graph on the left-hand side of Worksheet 1.2 sets $\beta_1 = 1$ and $\beta_2 = 0$ and plots the homogeneous solution for $t = 1, \ldots, 30$. Case 2 uses the same value of a_2 (hence, $r = 0.949$) but sets $a_1 = -0.6$. Again, the value of d is negative; however, for this set of calculations, $\cos(\theta) = -0.316$ so that θ is 1.89. Comparing the two graphs, you can see that increasing the value of θ acts to increase the frequency of the oscillations.

WORKSHEET 1.2

IMAGINARY ROOTS

Example 1	**Example 2**
$y_t - 1.6y_{t-1} + 0.9y_{t-2}$	$y_t + 0.6y_{t-1} + 0.9y_{t-2}$

(a) Check the discriminant $d = (a_1)^2 + 4a_2$

$d = (1.6)^2 + 4(-0.9)$	$d = (-0.6)^2 + 4(-0.9)$
$= -1.04$	$= -3.24$

Hence, the roots are imaginary. The homogeneous solution has the form

$$y_t^h = \beta_1 r^t \cos(\theta t + \beta_2)$$

where β_1 and β_2 are arbitrary constants.

(b) Obtain the value of $r = (-a_2)^{1/2}$

$r = (0.9)^{1/2}$	$r = (0.9)^{1/2}$
$= 0.949$	$= 0.949$

(c) Obtain θ from $\cos(\theta) = a_1/[2(-a_2)^{1/2}]$

$\cos(\theta) = 1.6/[2(0.9)^{1/2}]$	$\cos(\theta) = -0.6/[2(0.9)^{1/2}]$
$= 0.843$	$= -0.316$

Given $\cos(\theta)$, use a trig table or a calculator to find θ:

$\theta = 0.567$	$\theta = 1.89$

(d) Form the homogeneous solution: $y_t^h = \beta_1 r^t \cos(\theta t + \beta_2)$

$y_t^h = \beta_1(0.949)^t \cos(0.567t + \beta_2)$	$y_t^h = \beta_1(0.949)^t \cos(1.89t + \beta_2)$

For $\beta_1 = 1$ and $\beta_2 = 0$, the time paths of the homogeneous solutions are:

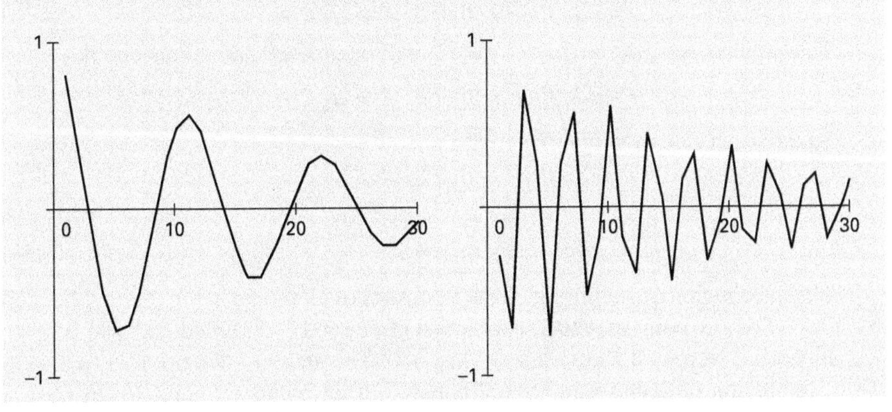

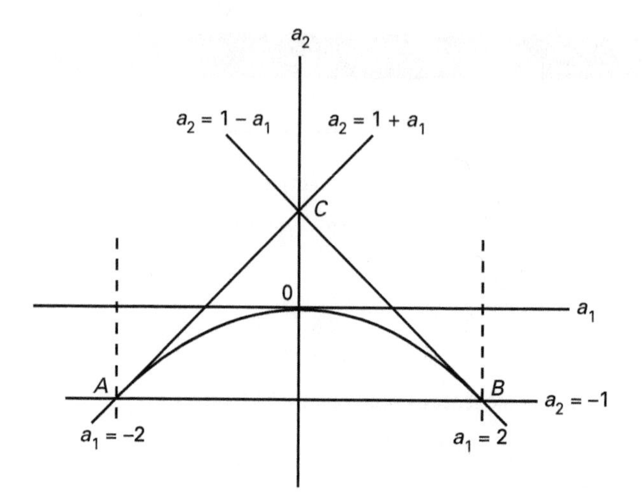

FIGURE 1.5 Characterizing the Stability Conditions

Stability Conditions

The general stability conditions can be summarized using triangle ABC in Figure 1.5. Arc $A0B$ is the boundary between Cases 1 and 3; it is the locus of points where $d = a_1^2 + 4a_2 = 0$. The region above $A0B$ corresponds to Case 1 (since $d > 0$), and the region below $A0B$ corresponds to Case 3 (since $d < 0$).

In Case 1 (in which the roots are real and distinct), stability requires that the largest root be less than unity and the smallest root be greater than -1. The largest characteristic root, $\alpha_1 = (a_1 + \sqrt{d})/2$, will be less than unity if

$$a_1 + (a_1^2 + 4a_2)^{1/2} < 2 \quad \text{or} \quad (a_1^2 + 4a_2)^{1/2} < 2 - a_1$$

Square each side to obtain the condition:

$$a_1^2 + 4a_2 < 4 - 4a_1 + a_1^2$$

or

$$a_1 + a_2 < 1 \tag{1.52}$$

The smallest root, $\alpha_2 = (a_1 - \sqrt{d})/2$, will be greater than minus one if

$$a_1 - (a_1^2 + 4a_2)^{1/2} > -2 \quad \text{or} \quad 2 + a_1 > (a_1^2 + 4a_2)^{1/2}$$

Square each side to obtain the condition:

$$4 + 4a_1 + a_1^2 > a_1^2 + 4a_2$$

or

$$a_2 < 1 + a_1 \tag{1.53}$$

Thus, the region of stability in Case 1 consists of all points in the region bounded by $A0BC$. For any point in $A0BC$, conditions (1.52) and (1.53) hold and $d > 0$.

In Case 2 (repeated roots), $a_1^2 + 4a_2 = 0$. The stability condition is $|a_1| < 2$. Thus, the region of stability in Case 2 consists of all points on arc $A0B$. In Case 3 ($d < 0$), the stability condition is $r = (-a_2)^{1/2} < 1$. Hence,

$$-a_2 < 1 \qquad (\text{where } a_2 < 0) \tag{1.54}$$

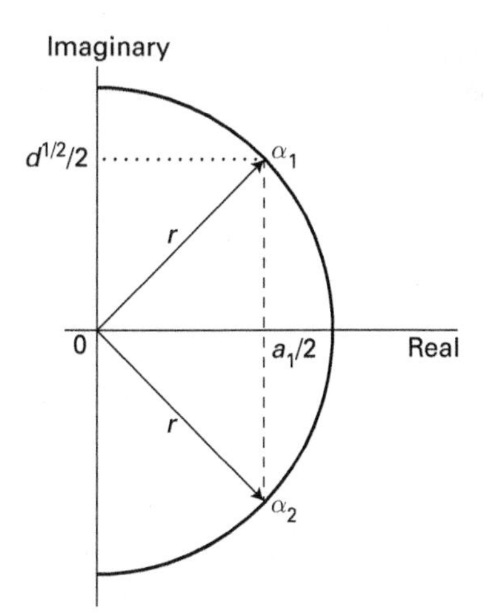

FIGURE 1.6 Characteristic Roots and the Unit Circle

Thus, the region of stability in Case 3 consists of all points in region $A0B$. For any point in $A0B$, (1.54) is satisfied and $d < 0$.

A succinct way to characterize the stability conditions is to state that the characteristic roots must lie within the unit circle. Consider the semicircle drawn in Figure 1.6. Real numbers are measured on the horizontal axis and imaginary numbers are measured on the vertical axis. If the characteristic roots α_1 and α_2 are both real, they can be plotted on the horizontal axis. Stability requires that they lie within a circle of radius one. Complex roots will lie somewhere in the complex plane. If $a_1 > 0$, the roots $\alpha_1 = (a_1 + i\sqrt{d})/2$ and $\alpha_2 = (a_1 - i\sqrt{d})/2$ can be represented by the two points shown in Figure 1.6. For example, α_1 is drawn by moving $a_1/2$ units along the real axis and $\sqrt{d}/2$ units along the imaginary axis. Using the distance formula, the length of the radius r is given by

$$r = \sqrt{(a_1/2)^2 + (d^{1/2}i/2)^2}$$

and, using the fact that $i^2 = -1$, we obtain

$$r = (-a_2)^{1/2}$$

The stability condition requires that $r < 1$. Therefore, when plotted on the complex plane, the two roots α_1 and α_2 must lie within a circle of radius equal to unity. In the time-series literature it is simply stated that *stability requires that all characteristic roots lie within the unit circle.*

Higher-Order Systems

The same method can be used to find the homogeneous solution to higher-order difference equations. The homogeneous equation for (1.10) is

$$y_t - \sum_{i=1}^{n} a_i y_{t-i} = 0 \tag{1.55}$$

Given the results in Section 4, you should suspect each homogeneous solution to have the form $y_t^h = A\alpha^t$ where A is an arbitrary constant. Thus, to find the value(s) of α, we seek the solution for

$$A\alpha^t - \sum_{i=1}^{n} a_i A\alpha^{t-i} = 0 \tag{1.56}$$

or, dividing through by α^{t-n}, we seek the values of α that solve

$$\alpha^n - a_1\,\alpha^{n-1} - a_2\,\alpha^{n-2} \ldots - a_n = 0 \tag{1.57}$$

This nth-order polynomial will yield n solutions for α. Denote these n characteristic roots by $\alpha_1, \alpha_2, \ldots, \alpha_n$. As in Section 4, the linear combination $A_1\alpha_1^t + A_2\alpha_2^t + \ldots + A_n\alpha_n^t$ is also a solution. The arbitrary constants A_1 through A_n can be eliminated by imposing n initial conditions on the general solution. The α_i may be real or complex numbers. Stability requires that all real valued α_i be less than unity in absolute value. Complex roots will necessarily come in pairs. Stability requires that all roots lie within the unit circle shown in Figure 1.6.

In most circumstances there is little need to directly calculate the characteristic roots of higher-order systems. Many of the technical details are included in Appendix 1.2 to this chapter. However, there are some useful rules for checking the stability conditions in higher-order systems.

1. In an nth-order equation, a necessary condition for all characteristic roots to lie inside the unit circle is

$$\sum_{i=1}^{n} a_i < 1$$

2. Since the values of the a_i can be positive or negative, a sufficient condition for all characteristic roots to lie inside the unit circle is

$$\sum_{i=1}^{n} |a_i| < 1$$

3. At least one characteristic root equals unity if

$$\sum_{i=1}^{n} a_i = 1$$

Any sequence that contains one or more characteristic roots that equal unity is called a **unit root** process.

4. For a third-order equation, the stability conditions can be written as

$$1 - a_1 - a_2 - a_3 > 0$$
$$1 + a_1 - a_2 + a_3 > 0$$
$$1 - a_1a_3 + a_2 - a_3^2 > 0$$
$$3 + a_1 + a_2 - 3a_3 > 0 \text{ or } 3 - a_1 + a_2 + 3a_3 > 0$$

Given that the first three inequalities are satisfied, either of the last two can be checked. One of the last conditions is redundant, given that the other three hold.

7. PARTICULAR SOLUTIONS FOR DETERMINISTIC PROCESSES

Finding the particular solution to a difference equation is often a matter of ingenuity and perseverance. The appropriate technique depends heavily on the form of the $\{x_t\}$ process. We begin by considering those processes that contain only deterministic components. Of course, in econometric analysis, the forcing process will contain both deterministic and stochastic components.

CASE 1

$\boldsymbol{x_t = 0}$**.** When all elements of the $\{x_t\}$ process are zero, the difference equation becomes

$$y_t = a_0 + a_1 y_{t-1} + a_2 y_{t-2} + \dots + a_n y_{t-n} \qquad (1.58)$$

Intuition suggests that an unchanging value of y (i.e., $y_t = y_{t-1} = \dots = c$) should solve the equation. Substitute the trial solution $y_t = c$ into (1.58) to obtain

$$c = a_0 + a_1 c + a_2 c + \dots + a_n c$$

so that

$$c = a_0/(1 - a_1 - a_2 - \dots - a_n) \qquad (1.59)$$

As long as $(1 - a_1 - a_2 - \dots - a_n)$ does not equal zero, the value of c given by (1.59) is a solution to (1.58). Hence, the particular solution to (1.58) is given by $y_t^p = a_0/(1 - a_1 - a_2 - \dots - a_n)$.

If $1 - a_1 - a_2 - \dots - a_n = 0$, the value of c in (1.59) is undefined; it is necessary to try some other form for the solution. The key insight is that $\{y_t\}$ is a unit root process if $\Sigma a_i = 1$. Since $\{y_t\}$ is not convergent, it stands to reason that the constant solution does not work. Instead, recall equations (1.12) and (1.26); these solutions suggest that a linear time trend can appear in the solution of a unit root process. As such, try the solution $y_t^p = ct$. For ct to be a solution it must be the case that

$$ct = a_0 + a_1 c(t-1) + a_2 c(t-2) + \dots + a_n c(t-n)$$

or, combining like terms,

$$(1 - a_1 - a_2 - \dots - a_n)ct = a_0 - c(a_1 + 2a_2 + 3a_3 + \dots + na_n)$$

Since $1 - a_1 - a_2 - \dots - a_n = 0$, select the value of c such that

$$c = a_0/(a_1 + 2a_2 + 3a_3 + \dots + na_n)$$

For example, let

$$y_t = 2 + 0.75 y_{t-1} + 0.25 y_{t-2}$$

Here, $a_1 = 0.75$ and $a_2 = 0.25$; $\{y_t\}$ is a unit root process because $a_1 + a_2 = 1$. The particular solution has the form ct, where $c = 2/[0.75 + 2(0.25)] = 1.6$.

In the event that the solution ct fails, sequentially try the solutions $y_t^p = ct^2, ct^3, \ldots, ct^n$. For an nth-order equation, one of these solutions will always be the particular solution.

CASE 2

The Exponential Case. Let x_t have the exponential form $b(d)^{rt}$, where b, d, and r are constants. Since r has the natural interpretation as a growth rate, we would expect to encounter this type of forcing process case in a growth context. We illustrate the solution procedure using the first-order equation

$$y_t = a_0 + a_1 y_{t-1} + bd^{rt} \tag{1.60}$$

To try to gain an intuitive feel for the form of the solution, notice that if $b = 0$, (1.60) is a special case of (1.58). Hence, you should expect a constant to appear in the particular solution. Moreover, the expression d^{rt} grows at the constant rate r. Thus, you might expect the particular solution to have the form $y_t^p = c_0 + c_1 d^{rt}$, where c_0 and c_1 are constants. If this equation is actually a solution, you should be able to substitute it back into (1.60) and obtain an identity. Making the appropriate substitutions, we get

$$c_0 + c_1 d^{rt} = a_0 + a_1[c_0 + c_1 d^{r(t-1)}] + bd^{rt} \tag{1.61}$$

For this solution to work, it is necessary to select c_0 and c_1 such that

$$c_0 = a_0/(1 - a_1) \text{ and } c_1 = [bd^r]/(d^r - a_1)$$

Thus, a particular solution is

$$y_t^p = \frac{a_0}{1 - a_1} + \frac{bd^r}{d^r - a_1} d^{rt}$$

The nature of the solution is that y_t^p equals the constant $a_0/(1 - a_1)$ plus an expression that grows at the rate r. Note that for $|d^r| < 1$, the particular solution converges to $a_0/(1 - a_1)$.

If either $a_1 = 1$ or $a_1 = d^r$, use the trick suggested in Case 1. If $a_1 = 1$, try the solution $c_0 = ct$, and if $a_1 = d^r$, try the solution $c_1 = tb$. Use precisely the same methodology in higher-order systems.

CASE 3

Deterministic Time Trend. In this case, let the $\{x_t\}$ sequence be represented by the relationship $x_t = bt^d$ where b is a constant and d is a positive integer. Hence,

$$y_t = a_0 + \sum_{i=1}^{n} a_i y_{t-i} + bt^d \tag{1.62}$$

Since y_t depends on t^d, it follows that y_{t-1} depends on $(t-1)^d$, y_{t-2} depends on $(t-2)^d$, and so on. As such, the particular solution has the form

$y_t^p = c_0 + c_1t + c_2t^2 + \ldots + c_dt^d$. To find the values of the c_i, substitute the particular solution into (1.62). Then select the value of each c_i that results in an identity. Although various values of d are possible, in economic applications it is common to see models incorporating a linear time trend ($d = 1$). For illustrative purposes, consider the second-order equation $y_t = a_0 + a_1y_{t-1} + a_2y_{t-2} + bt$. Posit the solution $y_t^p = c_0 + c_1t$ where c_0 and c_1 are undetermined coefficients. Substituting this "challenge solution" into the second-order difference equation yields

$$c_0 + c_1t = a_0 + a_1[c_0 + c_1(t-1)] + a_2[c_0 + c_1(t-2)] + bt \qquad (1.63)$$

Now select values of c_0 and c_1 so as to force equation (1.63) to be an identity for all possible values of t. If we combine all constant terms and all terms involving t, the required values of c_0 and c_1 are

$$c_1 = b/(1 - a_1 - a_2)$$
$$c_0 = [a_0 - (2a_2 + a_1)c_1]/(1 - a_1 - a_2)$$

so that

$$c_0 = [a_0/(1 - a_1 - a_2)] - [b/(1 - a_1 - a_2)^2](2a_2 + a_1)$$

Thus, the particular solution will also contain a linear time trend. You should have no difficulty foreseeing the solution technique if $a_1 + a_2 = 1$. In this circumstance—which is applicable to higher-order cases, as well—try multiplying the original challenge solution by t.

8. THE METHOD OF UNDETERMINED COEFFICIENTS

At this point, it is appropriate to introduce the first of two useful methods for finding particular solutions when there are stochastic components in the $\{y_t\}$ process. The key insight of the **method of undetermined coefficients** is that linear equations have linear solutions. Hence, the particular solution to a linear difference equation is necessarily linear. Moreover, the solution can depend only on time, a constant, and the elements of the forcing process $\{x_t\}$. Thus, it is often possible to know the exact form of the solution even though the coefficients of the solution are unknown. The technique involves positing a solution—called a **challenge solution**—that is a linear function of all terms thought to appear in the actual solution. The problem becomes one of finding the set of values for those undetermined coefficients that solve the difference equation.

The actual technique for finding the coefficients is straightforward. Substitute the challenge solution into the original difference equation and solve for the values of the undetermined coefficients that yield an identity for all possible values of the included variables. If it is not possible to obtain an identity, the form of the challenge solution is incorrect. Try a new trial solution and repeat the process. In fact, we used the method of undetermined coefficients when positing the challenge solutions $y_t^p = c_0 + c_1d^{rt}$ and $y_t^p = c_0 + c_1t$ for Cases 2 and 3 in Section 7.

To begin, reconsider the simple first-order equation $y_t = a_0 + a_1 y_{t-1} + \varepsilon_t$. Since you have solved this equation using the iterative method, the equation is useful for illustrating the method of undetermined coefficients. The nature of the $\{y_t\}$ process is such that the particular solution can depend only on a constant term, time, and the individual elements of the $\{\varepsilon_t\}$ sequence. Since t does not explicitly appear in the forcing process, t can be in the particular solution only if the characteristic root is unity. Since the goal is to illustrate the method, posit the challenge solution:

$$y_t = b_0 + b_1 t + \sum_{i=0}^{\infty} \alpha_i \varepsilon_{t-i} \tag{1.64}$$

where b_0, b_1, and all the α_i are the coefficients to be determined.

Substitute (1.64) into the original difference equation to form

$$b_0 + b_1 t + \alpha_0 \varepsilon_t + \alpha_1 \varepsilon_{t-1} + \alpha_2 \varepsilon_{t-2} + \dots = a_0 + a_1 [b_0 + b_1 (t-1) + \alpha_0 \varepsilon_{t-1} + \alpha_1 \varepsilon_{t-2} + \dots] + \varepsilon_t$$

Collecting like terms, we obtain

$$(b_0 - a_0 - a_1 b_0 + a_1 b_1) + b_1(1 - a_1)t + (\alpha_0 - 1)\varepsilon_t + (\alpha_1 - a_1\alpha_0)\varepsilon_{t-1} + (\alpha_2 - a_1\alpha_1)\varepsilon_{t-2} + (\alpha_3 - a_1\alpha_2)\varepsilon_{t-3} + \dots = 0 \tag{1.65}$$

Equation (1.65) must hold for all values of t and all possible values of the $\{\varepsilon_t\}$ sequence. Thus, each of the following conditions must hold:

$$\begin{aligned} \alpha_0 - 1 &= 0 \\ \alpha_1 - a_1\alpha_0 &= 0 \\ \alpha_2 - a_1\alpha_1 &= 0 \\ &\cdot \\ &\cdot \\ b_0 - a_0 - a_1 b_0 + a_1 b_1 &= 0 \\ b_1 - a_1 b_1 &= 0 \end{aligned}$$

Notice that the first set of conditions can be solved for the α_i recursively. The solution of the first condition entails setting $\alpha_0 = 1$. Given this solution for α_0, the next equation requires $\alpha_1 = a_1$. Moving down the list, $\alpha_2 = a_1\alpha_1$ or $\alpha_2 = a_1^2$. Continuing the recursive process, we find $\alpha_i = a_1^i$. Now consider the last two equations. There are two possible cases depending on the value of a_1. If $a_1 \neq 1$, it immediately follows that $b_1 = 0$ and $b_0 = a_0/(1 - a_1)$. For this case, the particular solution is

$$y_t = \frac{a_0}{1 - a_1} + \sum_{i=0}^{\infty} a_1^i \varepsilon_{t-i}$$

Compare this result to (1.21); you will see that it is precisely the same solution found using the iterative method. The general solution is the sum of this particular solution plus the homogeneous solution Aa_1^t. Hence, the general solution is

$$y_t = \frac{a_0}{1 - a_1} + \sum_{i=0}^{\infty} a_1^i \varepsilon_{t-i} + Aa_1^t$$

Now, if there is an initial condition for y_0, it follows that

$$y_0 = \frac{a_0}{1-a_1} + \sum_{i=0}^{\infty} a_1^i \varepsilon_{-i} + A$$

Combining these two equations so as to eliminate the arbitrary constant A, we obtain

$$y_t = \frac{a_0}{1-a_1} + \sum_{i=0}^{\infty} a_1^i \varepsilon_{t-i} + a_1^t \left[y_0 - a_0/(1-a_1) - \sum_{i=0}^{\infty} a_1^i \varepsilon_{-i} \right]$$

so that

$$y_t = \frac{a_0}{1-a_1} + \sum_{i=0}^{t-1} a_1^i \varepsilon_{t-i} + a_1^t [y_0 - a_0/(1-a_1)] \tag{1.66}$$

It can be easily verified that (1.66) is identical to (1.25). Instead, if $a_1 = 1$, b_0 can be any arbitrary constant and $b_1 = a_0$. The improper form of the solution is

$$y_t = b_0 + a_0 t + \sum_{i=0}^{\infty} \varepsilon_{t-i}$$

The form of the solution is “improper” because the sum of the $\{\varepsilon_t\}$ sequence may not be finite. Therefore, it is necessary to impose an initial condition. If the value y_0 is given, it follows that

$$y_0 = b_0 + \sum_{i=0}^{\infty} \varepsilon_{-i}$$

Imposing the initial condition on the improper form of the solution yields (1.26)

$$y_t = y_0 + a_0 t + \sum_{i=1}^{t} \varepsilon_i$$

To take a second example, consider the equation

$$y_t = a_0 + a_1 y_{t-1} + \varepsilon_t + \beta_1 \varepsilon_{t-1} \tag{1.67}$$

Again, the solution can depend only on a constant, the elements of the $\{\varepsilon_t\}$ sequence, and t raised to the first power. As in the previous example, t does not need to be included in the challenge solution if the characteristic root differs from unity. To reinforce this point, use the challenge solution given by (1.64). Substitute this tentative solution into (1.67) to obtain

$$b_0 + b_1 t + \sum_{i=0}^{\infty} \alpha_i \varepsilon_{t-i} = a_0 + a_1 \left[b_0 + b_1(t-1) + \sum_{i=0}^{\infty} \alpha_i \varepsilon_{t-1-i} \right] + \varepsilon_t + \beta_1 \varepsilon_{t-1}$$

Matching coefficients on all terms containing $\varepsilon_t, \varepsilon_{t-1}, \varepsilon_{t-2}, \ldots$, yields

$\alpha_0 = 1$

$\alpha_1 = a_1 \alpha_0 + \beta_1$ [so that $\alpha_1 = a_1 + \beta_1$]

$\alpha_2 = a_1 \alpha_1$ [so that $\alpha_2 = a_1(a_1 + \beta_1)$]

$$\alpha_3 = a_1\alpha_2 \qquad [\text{so that } \alpha_3 = (a_1)^2(a_1 + \beta_1)]$$

$$\ldots$$

$$\alpha_i = a_1\alpha_{i-1} \qquad [\text{so that } \alpha_i = (a_1)^{i-1}(a_1 + \beta_1)]$$

Matching coefficients of intercept terms and coefficients of terms containing t, we get

$$b_0 = a_0 + a_1 b_0 - a_1 b_1$$
$$b_1 = a_1 b_1$$

Again, there are two cases. If $a_1 \neq 1$, then $b_1 = 0$ and $b_0 = a_0/(1 - a_1)$. The particular solution is

$$y_t = \frac{a_0}{1 - a_1} + \varepsilon_t + (a_1 + \beta_1)\sum_{i=1}^{\infty} a_1^{i-1}\varepsilon_{t-i}$$

The general solution augments the particular solution with the term Aa_1^t. You are left with the exercise of imposing the initial condition for y_0 on the general solution. Now consider the case in which $a_1 = 1$. The undetermined coefficients are such that $b_1 = a_0$ and b_0 is an arbitrary constant. The improper form of the solution is

$$y_t = b_0 + a_0 t + \varepsilon_t + (1 + \beta_1)\sum_{i=1}^{\infty}\varepsilon_{t-i}$$

If y_0 is given, it follows that

$$y_0 = b_0 + \varepsilon_0 + (1 + \beta_1)\sum_{i=1}^{\infty}\varepsilon_{-i}$$

Hence, imposing the initial condition, we obtain

$$y_t = y_0 + a_0 t + \varepsilon_t + (1 + \beta_1)\sum_{i=1}^{t-1}\varepsilon_{t-i}$$

Higher-Order Systems

The identical procedure is used for higher-order systems. As an example, let us find the particular solution to the second-order equation

$$y_t = a_0 + a_1 y_{t-1} + a_2 y_{t-2} + \varepsilon_t \tag{1.68}$$

Since we have a second-order equation, we use the challenge solution

$$y_t = b_0 + b_1 t + b_2 t^2 + \alpha_0\varepsilon_t + \alpha_1\varepsilon_{t-1} + \alpha_2\varepsilon_{t-2} + \ldots$$

where b_0, b_1, b_2, and the α_i are the undetermined coefficients.

Substituting the challenge solution into (1.68) yields

$$\begin{aligned}
[b_0 + b_1 t + b_2 t^2] + \alpha_0\varepsilon_t + \alpha_1\varepsilon_{t-1} + \alpha_2\varepsilon_{t-2} + \ldots \\
= a_0 + a_1[b_0 + b_1(t-1) + b_2(t-1)^2 \\
+ \alpha_0\varepsilon_{t-1} + \alpha_1\varepsilon_{t-2} + \alpha_2\varepsilon_{t-3} + \ldots] \\
+ a_2[b_0 + b_1(t-2) + b_2(t-2)^2 \\
+ \alpha_0\varepsilon_{t-2} + \alpha_1\varepsilon_{t-3} + \alpha_2\varepsilon_{t-4} + \ldots] + \varepsilon_t
\end{aligned}$$

There are several necessary and sufficient conditions for the values of the α_i's to render the equation above an identity for all possible realizations of the $\{\varepsilon_t\}$ sequence:

$$\alpha_0 = 1$$

$$\alpha_1 = a_1\alpha_0 \qquad [\text{so that } \alpha_1 = a_1]$$

$$\alpha_2 = a_1\alpha_1 + a_2\alpha_0 \qquad [\text{so that } \alpha_2 = (a_1)^2 + a_2]$$

$$\alpha_3 = a_1\alpha_2 + a_2\alpha_1 \qquad [\text{so that } \alpha_3 = (a_1)^3 + 2a_1a_2]$$

$$\cdot$$

$$\cdot$$

Notice that for any value of $j \geq 2$, the coefficients solve the second-order difference equation $\alpha_j = a_1\alpha_{j-1} + a_2\alpha_{j-2}$. Since we know α_0 and α_1, we can solve for all the α_j iteratively. The properties of the coefficients will be precisely those discussed when considering homogeneous solutions:

1. Convergence necessitates that $|a_2| < 1$, $a_1 + a_2 < 1$, and that $a_2 - a_1 < 1$. Notice that convergence implies that past values of the $\{\varepsilon_t\}$ sequence ultimately have a successively smaller influence on the current value of y_t.
2. If the coefficients converge, convergence will be direct or oscillatory if $(a_1^2 + 4a_2) > 0$, will follow a sine/cosine pattern if $(a_1^2 + 4a_2) < 0$, and will "explode" and then converge if $(a_1^2 + 4a_2) = 0$. Appropriately setting the α_i, we are left with the remaining expression:

$$b_2(1 - a_1 - a_2)t^2 + [b_1(1 - a_1 - a_2) + 2b_2(a_1 + 2a_2)]t +$$
$$[b_0(1 - a_1 - a_2) - a_0 + a_1(b_1 - b_2) + 2a_2(b_1 - 2b_2)] = 0 \qquad (1.69)$$

Equation (1.69) must equal zero for all values of t. First, consider the case in which $a_1 + a_2 \neq 1$. Since $(1 - a_1 - a_2)$ does not vanish, it is necessary to set the value of b_2 equal to zero. Given that $b_2 = 0$ and that the coefficient of t must equal zero, it follows that b_1 must also be set equal to zero. Finally, given that $b_1 = b_2 = 0$, we must set $b_0 = a_0/(1 - a_1 - a_2)$. Instead, if $a_1 + a_2 = 1$, the solutions for the b_i depend on the specific values of a_0, a_1, and a_2. The key point is that *the stability condition for the homogeneous equation is precisely the condition for convergence of the particular solution. If any characteristic root of the homogeneous equation is equal to unity, a polynomial time trend will appear in the particular solution. The order of the polynomial is the number of unitary characteristic roots.* This result generalizes to higher-order equations.

If you are really clever, you can combine the discussion of the last section with the method of undetermined coefficients. Find the deterministic portion of the particular solution using the techniques discussed in the last section. Then use the method of undetermined coefficients to find the stochastic portion of the particular solution. In (1.67), for example, set $\varepsilon_t = \varepsilon_{t-1} = 0$ and obtain the solution $a_0/(1 - a_1)$. Now use the method of undetermined coefficients to find the particular solution of $y_t = a_1y_{t-1} + \varepsilon_t + \beta_1\varepsilon_{t-1}$. Add the deterministic and stochastic components to obtain all components of the particular solution.

A Solved Problem

To illustrate the methodology using a second-order equation, augment (1.28) with the stochastic term ε_t so that

$$y_t = 3 + 0.9y_{t-1} - 0.2y_{t-2} + \varepsilon_t \tag{1.70}$$

You have already verified that the two homogeneous solutions are $A_1(0.5)^t$ and $A_2(0.4)^t$ and that the deterministic portion of the particular solution is $y_t^p = 10$. To find the stochastic portion of the particular solution, form the challenge solution

$$y_t = \sum_{i=0}^{\infty} \alpha_i \varepsilon_{t-i}$$

In contrast to (1.64), the intercept term b_0 is excluded (since we have already found the deterministic portion of the particular solution) and the time trend b_1t is excluded (since both characteristic roots are less than unity). For this challenge to work, it must satisfy

$$\alpha_0\varepsilon_t + \alpha_1\varepsilon_{t-1} + \alpha_2\varepsilon_{t-2} + \alpha_3\varepsilon_{t-3} + \ldots = 0.9[\alpha_0\varepsilon_{t-1} + \alpha_1\varepsilon_{t-2} + \alpha_2\varepsilon_{t-3} + \alpha_3\varepsilon_{t-4} + \ldots]$$
$$- 0.2\,[\alpha_0\varepsilon_{t-2} + \alpha_1\varepsilon_{t-3} + \alpha_2\varepsilon_{t-4} + \alpha_3\varepsilon_{t-5} + \ldots] + \varepsilon_t \tag{1.71}$$

Since (1.71) must hold for all possible realizations of ε_t, ε_{t-1}, ε_{t-2}, ..., each of the following conditions must hold:

$$\alpha_0 = 1$$
$$\alpha_1 = 0.9\alpha_0$$

so that $\alpha_1 = 0.9$, and for all $i \geq 2$,

$$\alpha_i = 0.9\alpha_{i-1} - 0.2\alpha_{i-2} \tag{1.72}$$

Now, it is possible to solve (1.72) iteratively so that $\alpha_2 = 0.9\alpha_1 - 0.2\alpha_0 = 0.61$, $\alpha_3 = 0.9(0.61) - 0.2(0.9) = 0.369$, and so forth. A more elegant solution method is to view (1.72) as a second-order difference equation in the $\{\alpha_i\}$ sequence with initial conditions $\alpha_0 = 1$ and $\alpha_1 = 0.9$. The solution to (1.72) is

$$\alpha_i = 5(0.5)^i - 4(0.4)^i \tag{1.73}$$

To obtain (1.73), note that the solution to (1.72) is: $\alpha_i = A_3(0.5)^i + A_4(0.4)^i$ where A_3 and A_4 are arbitrary constants. Imposing the conditions $\alpha_0 = 1$ and $\alpha_1 = 0.9$ yields (1.73). If we use (1.73), it follows that: $\alpha_0 = 5(0.5)^0 - 4(0.4)^0 = 1$; $\alpha_1 = 5(0.5)^1 - 4(0.4)^1 = 0.9$; $\alpha_2 = 5(0.5)^2 - 4(0.4)^2 = 0.61$; and so on.

The general solution to (1.70) is the sum of the two homogeneous solutions and the deterministic and stochastic portions of the particular solution:

$$y_t = 10 + A_1(0.5)^t + A_2(0.4)^t + \sum_{i=0}^{\infty} \alpha_i \varepsilon_{t-i} \tag{1.74}$$

where the α_i are given by (1.73).

Given initial conditions for y_0 and y_1, it follows that A_1 and A_2 must satisfy

$$y_0 = 10 + A_1 + A_2 y_0 + \sum_{i=0}^{\infty} \alpha_i \varepsilon_{-i} \tag{1.75}$$

$$y_1 = 10 + A_1(0.5) + A_2(0.4) + \sum_{i=0}^{\infty} \alpha_i \varepsilon_{1-i} \tag{1.76}$$

Although the algebra gets messy, (1.75) and (1.76) can be substituted into (1.74) to eliminate the arbitrary constants:

$$y_t = 10 + (0.4)^t[5(y_0 - 10) - 10(y_1 - 10)] + (0.5)^t[10(y_1 - 10) - 4(y_0 - 10)] + \sum_{i=0}^{t-2} \alpha_i \varepsilon_{t-i}$$

9. LAG OPERATORS

If it is not important to know the actual values of the coefficients appearing in the particular solution, it is often more convenient to use lag operators rather than the method of undetermined coefficients. The **lag operator** L is defined to be a *linear* operator such that for any value y_t

$$L^i y_t \equiv y_{t-i} \tag{1.77}$$

Thus, L^i preceding y_t simply means to lag y_t by i periods. It is useful to consider the following properties of lag operators:

1. The lag of a constant is a constant: $Lc = c$.
2. The distributive law holds for lag operators. We can set $(L^i + L^j)y_t = L^i y_t + L^j y_t = y_{t-i} + y_{t-j}$.
3. The associative law of multiplication holds for lag operators. We can set $L^i L^j y_t = L^i(L^j y_t) = L^i y_{t-j} = y_{t-i-j}$. Similarly, we can set $L^i L^j y_t = L^{i+j} y_t = y_{t-i-j}$. Note that $L^0 y_t = y_t$.
4. L raised to a negative power is actually a *lead* operator: $L^{-i} y_t = y_{t+i}$. To explain, define $j = -i$ and form $L^j y_t = y_{t-j} = y_{t+i}$.
5. For $|a| < 1$, the infinite sum $(1 + aL + a^2L^2 + a^3L^3 + \ldots)y_t = y_t/(1 - aL)$. This property of lag operators may not seem intuitive, but it follows directly from properties 2 and 3 above.

   Proof: Multiply each side by $(1 - aL)$ to form $(1 - aL)(1 + aL + a^2L^2 + a^3L^3 + \ldots)y_t = y_t$. Multiply the two expressions to obtain $(1 - aL + aL - a^2L^2 + a^2L^2 - a^3L^3 + \ldots)y_t = y_t$. Given that $|a| < 1$, the expression $a^nL^ny_t$ converges to zero as $n \to \infty$. Thus, the two sides of the equation are equal.
6. For $|a| > 1$, the infinite sum $[1 + (aL)^{-1} + (aL)^{-2} + (aL)^{-3} + \ldots]y_t = -aLy_t/(1 - aL)$. Thus,

$$y_t/(1 - aL) = -(aL)^{-1} \sum_{i=0}^{\infty} (aL)^{-i} y_t$$

Proof: Multiply by $(1 - aL)$ to form $(1 - aL)[1 + (aL)^{-1} + (aL)^{-2} + (aL)^{-3} + \ldots]y_t = -aLy_t$. Perform the indicated multiplication to obtain $[1 - aL + (aL)^{-1} - 1 + (aL)^{-2} - (aL)^{-1} + (aL)^{-3} - (aL)^{-2} \ldots]y_t = -aLy_t$. Given that $|a| > 1$, the expression $a^{-n}L^{-n}y_t$ converges to zero as $n \rightarrow \infty$. Thus, the two sides of the equation are equal.

Lag operators provide a concise notation for writing difference equations. Using lag operators, we can write the pth-order equation $y_t = a_0 + a_1y_{t-1} + \ldots + a_py_{t-p} + \varepsilon_t$ as

$$(1 - a_1L - a_2L^2 - \ldots - a_pL^p)y_t = a_0 + \varepsilon_t$$

or, more compactly, as

$$A(L)y_t = a_0 + \varepsilon_t$$

where $A(L)$ is the polynomial $(1 - a_1L - a_2L^2 - \ldots - a_pL^p)$

Since $A(L)$ can be viewed as a polynomial in the lag operator, the notation $A(1)$ is used to denote the sum of the coefficients

$$A(1) = 1 - a_1 - a_2 \ldots - a_p$$

As a second example, lag operators can be used to express the equation $y_t = a_0 + a_1y_{t-1} + \ldots + a_py_{t-p} + \varepsilon_t + \beta_1\varepsilon_{t-1} + \ldots + \beta_q\varepsilon_{t-q}$ as

$$A(L)y_t = a_0 + B(L)\varepsilon_t$$

where $A(L)$ and $B(L)$ are polynomials of orders p and q, respectively.

It is straightforward to use lag operators to solve linear difference equations. Again consider the first-order equation $y_t = a_0 + a_1y_{t-1} + \varepsilon_t$ where $|a_1| < 1$. Use the definition of L to form

$$y_t = a_0 + a_1Ly_t + \varepsilon_t \tag{1.78}$$

Solving for y_t, we obtain

$$y_t = \frac{a_0 + \varepsilon_t}{1 - a_1L} \tag{1.79}$$

From property 1, we know that $La_0 = a_0$, so that $a_0/(1 - a_1L) = a_0 + a_1a_0 + a_1^2a_0 + \ldots = a_0/(1 - a_1)$. From property 5, we know that $\varepsilon_t/(1 - a_1L) = \varepsilon_t + a_1\varepsilon_{t-1} + a_1^2\varepsilon_{t-2} + \ldots$. Combining these two parts of the solution, we obtain the particular solution given by (1.21).

For practice, we can use lag operators to solve (1.67): $y_t = a_0 + a_1y_{t-1} + \varepsilon_t + \beta_1\varepsilon_{t-1}$, where $|a_1| < 1$. Use property 2 to form $(1 - a_1L)y_t = a_0 + (1 + \beta_1L)\varepsilon_t$. Solving for y_t yields

$$y_t = [a_0 + (1 + \beta_1L)\varepsilon_t]/(1 - a_1L)$$

so that

$$y_t = [a_0/(1 - a_1)] + [\varepsilon_t/(1 - a_1L)] + [\beta_1\varepsilon_{t-1}/(1 - a_1L)] \tag{1.80}$$

Expanding the last two terms of (1.80) yields the same solution found using the method of undetermined coefficients.

Now suppose $y_t = a_0 + a_1y_{t-1} + \varepsilon_t$ but $|a_1| > 1$. The application of property 5 to (1.79) is inappropriate because it implies that y_t is infinite. Instead, expand (1.79) using property 6:

$$y_t = \frac{a_0}{1 - a_1} - (a_1L)^{-1}\sum_{i=0}^{\infty}(a_1L)^{-i}\varepsilon_t$$

$$= \frac{a_0}{1 - a_1} - \frac{1}{a_1}\sum_{i=0}^{\infty}(a_1L)^{-i}\varepsilon_{t+1} \tag{1.81}$$

$$= \frac{a_0}{1 - a_1} - \left(\frac{1}{a_1}\right)\sum_{i=0}^{\infty}a_1^{-i}\varepsilon_{t+1+i} \tag{1.82}$$

Lag Operators in Higher-Order Systems

We can also use lag operators to transform the nth-order equation $y_t = a_0 + a_1y_{t-1} + a_2y_{t-2} + \ldots + a_ny_{t-n} + \varepsilon_t$ into

$$(1 - a_1L - a_2L^2 - \ldots - a_nL^n)y_t = a_0 + \varepsilon_t$$

or

$$y_t = (a_0 + \varepsilon_t)/(1 - a_1L - a_2L^2 - \ldots a_nL^n)$$

From our previous analysis (also see Appendix 1.2), we know that the stability condition is such that the characteristic roots of the equation $\alpha^n - a_1\alpha^{n-1} - \ldots - a_n = 0$ all lie *within* the unit circle. Notice that the values of α solving the characteristic equation are the reciprocals of the values of L that solve the equation $1 - a_1L \ldots - a_nL^n = 0$. In fact, the expression $1 - a_1L \ldots - a_nL^n$ is often called the *inverse characteristic equation*. Thus, in the literature, it is often stated that the stability condition is for the characteristic roots of $(1 - a_1L \ldots - a_nL^n)$ to lie *outside* of the unit circle.

In principle, one could use lag operators to actually obtain the coefficients of the particular solution. To illustrate using the second-order case, consider $y_t = (a_0 + \varepsilon_t)/(1 - a_1L - a_2L^2)$. If we knew the factors of the quadratic equation were such that $(1 - a_1L - a_2L^2) = (1 - b_1L)(1 - b_2L)$, we could write

$$y_t = (a_0 + \varepsilon_t)/[(1 - b_1L)(1 - b_2L)]$$

If both b_1 and b_2 are less than unity in absolute value, we can apply property 5 to obtain

$$y_t = \frac{[a_0/(1 - b_1)] + \sum_{i=0}^{\infty}b_1^i\varepsilon_{t-i}}{1 - b_2L}$$

Reapply the rule to $a_0/(1 - b_1)$ and to each of the elements in the summation $\Sigma b_1{}^i\,\varepsilon_{t-i}$ to obtain the particular solution. If you want to know the actual coefficients of the process, it is preferable to use the method of undetermined coefficients. The beauty of lag operators is that they can be used to denote such particular solutions succinctly. The general model

$$A(L)y_t = a_0 + B(L)\varepsilon_t$$

has the particular solution

$$y_t = a_0/A(L) + B(L)\varepsilon_t/A(L)$$

As suggested by (1.82), there is a **forward-looking** solution to any linear difference equation. This text will not make much use of the forward-looking solution since future realizations of stochastic variables are not directly observable. Some of the details of forward-looking solutions can be found in the *Supplementary Manual* to this text available at www.cba.ua.edu/~wenders or from Wiley.

10. SUMMARY

Time-series econometrics is concerned with the estimation of difference equations containing stochastic components. Originally, time-series models were used for forecasting. Uncovering the dynamic path of a series improves forecasts because the predictable components of the series can be extrapolated into the future. The growing interest in economic dynamics has given a new emphasis to time-series econometrics. Stochastic difference equations arise, quite naturally, from dynamic economic models. Appropriately estimated equations can be used for the interpretation of economic data and for hypothesis testing.

This introductory chapter focused on methods of "solving" stochastic difference equations. Although iteration can be useful, it is impractical in many circumstances. The solution to a linear difference equation can be divided into two parts: a *particular* solution and a *homogeneous* solution. One complicating factor is that the homogeneous solution is not unique. The *general* solution is a linear combination of the particular solution and all homogeneous solutions. Imposing n initial conditions on the general solution of an nth-order equation yields a unique solution.

The homogeneous portion of a difference equation is a measure of the *disequilibrium* in the initial period(s). The homogeneous equation is especially important in that it yields the characteristic roots; an nth-order equation has n such characteristic roots. If all of the characteristic roots lie within the unit circle, the series will be convergent. As you will see in Chapter 2, there is a direct relationship between the stability conditions and the issue of whether an economic variable is stationary or nonstationary.

The method of undetermined coefficients and the use of lag operators are powerful tools for obtaining the particular solution. The particular solution will be a linear function of the current and past values of the forcing process. In addition, this solution may contain an intercept term and a polynomial function of time. Unit roots and characteristic roots outside of the unit circle require the imposition of an initial condition for the particular solution to be meaningful. Some economic models allow for forward-looking solutions; in such circumstances, anticipated future events have consequences for the present period.

The tools developed in this chapter are aimed at paving the way for the study of time-series econometrics. It is a good idea to work all of the exercises presented below. Characteristic roots, the method of undetermined coefficients, and lag operators will be encountered throughout the remainder of the text.

QUESTIONS AND EXERCISES

1. Consider the difference equation $y_t = a_0 + a_1 y_{t-1}$ with the initial condition y_0. Jill solved the difference equation by iterating backward:

$$\begin{aligned} y_t &= a_0 + a_1 y_{t-1} \\ &= a_0 + a_1(a_0 + a_1 y_{t-2}) \\ &= a_0 + a_0 a_1 + a_0 a_1^2 + \ldots + a_0 a_1^{t-1} + a_1^t y_0 \end{aligned}$$

Bill added the homogeneous and particular solutions to obtain $y_t = a_0/(1 - a_1) + a_1^t [y_0 - a_0/(1 - a_1)]$.

a. Show that the two solutions are identical for $|a_1| < 1$.

b. Show that for $a_1 = 1$, Jill's solution is equivalent to $y_t = a_0 t + y_0$. How would you use Bill's method to arrive at this same conclusion in the case that $a_1 = 1$?

2. The cobweb model in Section 5 assumed *static* price expectations. Consider an alternative formulation called *adaptive expectations*. Let the expected price in t (denoted by p_t^*) be a weighted average of the price in $t-1$ and the price expectation of the previous period. Formally,

$$p_t^* = \alpha p_{t-1} + (1 - \alpha)p_{t-1}^* \qquad 0 < \alpha \leq 1$$

Clearly, when $\alpha = 1$, the static and adaptive expectations schemes are equivalent. An interesting feature of this model is that it can be viewed as a difference equation expressing the expected price as a function of its own lagged value and the forcing variable p_{t-1}.

a. Find the homogeneous solution for p_t^*.

b. Use lag operators to find the particular solution. Check your answer by substituting your answer into the original difference equation.

3. Suppose that the money supply process has the form $m_t = m + \rho m_{t-1} + \varepsilon_t$, where m is a constant and $0 < \rho < 1$.

a. Show that it is possible to express m_{t+n} in terms of the known value m_t and the sequence $\{\varepsilon_{t+1}, \varepsilon_{t+2}, \ldots, \varepsilon_{t+n}\}$.

b. Suppose that all values of ε_{t+i} for $i > 0$ have a mean value of zero. Explain how you could use your result in part a to forecast the money supply n periods into the future.

4. Find the particular solutions for each of the following:

a. $y_t = a_1 y_{t-1} + \varepsilon_t + \beta_1 \varepsilon_{t-1}$

b. $y_t = a_1 y_{t-1} + \varepsilon_{1t} + \beta_1 \varepsilon_{2t}$ (*Hint*: The form of the solution is $y_t = \Sigma c_i \varepsilon_{1t-i} + \Sigma d_i \varepsilon_{2t-i}$)

5. The *unit root problem* in time-series econometrics is concerned with characteristic roots that are equal to unity. In order to preview the issue:

a. Find the homogeneous solution to each of the following (*Hint*: Each has at least one unit root):

i. $y_t = 1.5y_{t-1} - 0.5y_{t-2} + \varepsilon_t$

ii. $y_t = y_{t-2} + \varepsilon_t$

iii. $y_t = 2y_{t-1} - y_{t-2} + \varepsilon_t$

iv. $y_t = y_{t-1} + 0.25y_{t-2} - 0.25y_{t-3} + \varepsilon_t$

b. Show that each of the backward-looking solutions is not convergent.

c. Show that Equation *i* can be written entirely in first differences; that is, $\Delta y_t = 0.5\Delta y_{t-1} + \varepsilon_t$. Find the particular solution for Δy_t. (*Hint*: Find the particular solution for the $\{\Delta y_t\}$ sequence in terms of the $\{\varepsilon_t\}$ sequence.)

d. Similarly transform the other equations into their first-difference form. Find the particular solution, if it exists, for the transformed equations.

e. Write equations *i* through *iv* using lag operators.

f. Given an initial condition y_0, find the solution for: $y_t = a_0 - y_{t-1} + \varepsilon_t$.

6. For each of the following, calculate the characteristic roots and the discriminant d in order to describe the adjustment process. Also, for each let $y_1 = y_2 = 10$. Use a spreadsheet program to calculate and plot the next 25 realizations of the series above.

i. $y_t = 0.75y_{t-1} - 0.125y_{t-2}$ **ii.** $y_t = 1.5y_{t-1} - 0.75y_{t-2}$

iii. $y_t = 1.8y_{t-1} - 0.81y_{t-2}$ **iv.** $y_t = 1.5y_{t-1} - 0.5625y_{t-2}$

7. A researcher estimated the following relationship for the inflation rate (π_t):

$$\pi_t = -0.05 + 0.7\pi_{t-1} + 0.6\pi_{t-2} + \varepsilon_t$$

a. Suppose that in periods 0 and 1, the inflation rate was 10 percent and 11 percent, respectively. Find the homogeneous, particular, and general solutions for the inflation rate.

b. Discuss the shape of the impulse response function. Given that the United States is not headed for runaway inflation, why do you believe that the researcher's equation is poorly estimated?

8. Consider the stochastic process $y_t = a_0 + a_2y_{t-2} + \varepsilon_t$.

a. Find the homogeneous solution and determine the stability condition.

b. Find the particular solution using the method of undetermined coefficients.

c. Show that lag operators yield the same solution.

9. For each of the following, verify that the posited solution satisfies the difference equation. The symbols c, c_0, and a_0 denote constants.

Equation	Solution
a. $y_t - y_{t-1} = 0$	$y_t = c$
b. $y_t - y_{t-1} = a_0$	$y_t = c + a_0t$
c. $y_t - y_{t-2} = 0$	$y_t = c + c_0(-1)^t$
d. $y_t - y_{t-2} = \varepsilon_t$	$y_t = c + c_0(-1)^t + \varepsilon_t + \varepsilon_{t-2} + \varepsilon_{t-4} + \ldots$

10. Part 1: For each of the following, determine whether $\{y_t\}$ represents a stable process. Determine whether the characteristic roots are real or imaginary and whether the real parts are positive or negative.

a. $y_t - 1.2y_{t-1} + 0.2y_{t-2} = 0$ **b.** $y_t - 1.2y_{t-1} + 0.4y_{t-2} = 0$

c. $y_t - 1.2y_{t-1} - 1.2y_{t-2} = 0$ **d.** $y_t + 1.2y_{t-1} = 0$

e. $y_t - 0.7y_{t-1} - 0.25y_{t-2} + 0.175y_{t-3} = 0$

[*Hint*: $(x - 0.5)(x + 0.5)(x - 0.7) = x^3 - 0.7x^2 - 0.25x + 0.175$.]

Part 2: Write each of the above equations using lag operators. Determine the characteristic roots of the inverse characteristic equation.

11. Consider the stochastic difference equation

$$y_t = 0.8y_{t-1} + \varepsilon_t - 0.5\varepsilon_{t-1}$$

a. Suppose that the initial conditions are such that $y_0 = 0$ and $\varepsilon_0 = \varepsilon_{-1} = 0$. Now suppose that $\varepsilon_1 = 1$. Determine the values y_1 through y_5 by forward iteration.

b. Find the homogeneous and particular solutions.

c. Impose the initial conditions in order to obtain the general solution.

d. Trace out the time path of an ε_t shock on the entire time path of the $\{y_t\}$ sequence.

12. In equation (1.5), determine the restrictions on α and β necessary to ensure that the $\{y_t\}$ process is stable.

13. Consider the following two stochastic difference equations:

i. $y_t = 3 + 0.75y_{t-1} - 0.125y_{t-2} + \varepsilon_t$ **ii.** $y_t = 3 + 0.25y_{t-1} + 0.375y_{t-2} + \varepsilon_t$

a. Use the method of undetermined coefficients to find the particular solution for each equation.

b. Find the homogeneous solutions for each equation.

c. For each process, suppose that $y_0 = y_1 = 8$ and that all values of ε_t for $t = 1, 0, -1, -2, \ldots = 0$. Use the method illustrated by equations (1.75) and (1.76) to find the values of the constants A_1 and A_2.

ENDNOTES

1. Another possibility is to obtain the forward-looking solution. Since we are dealing with forecasting equations, forward-looking solutions are not important for our purposes. Some of the details concerning forward-looking solutions are included in the *Supplementary Manual* available on my Web site: www.cba.ua.edu/~wenders.
2. Alternatively, you can substitute (1.26) into (1.17). Note that when ε_t is a pure random disturbance, $y_t = a_0 + y_{t-1} + \varepsilon_t$ is called a random walk plus drift model.
3. Any linear equation in the variables z_1 through z_n is homogeneous if it has the form $a_1z_1 + a_2z_2 + \ldots + a_nz_n = 0$. To obtain the homogeneous portion of (1.10), simply set the intercept term a_0 and the forcing process x_t equal to zero. Hence, the homogeneous equation for (1.10) is $y_t = a_1y_{t-1} + a_2y_{t-2} + \ldots + a_ny_{t-n}$.
4. If $b > a$, the demand and supply curves do not intersect in the positive quadrant. The assumption $a > b$ guarantees that the equilibrium price is positive.
5. For example, if the forcing process is $x_t = \varepsilon_t + \beta_1\varepsilon_{t-1} + \beta_2\varepsilon_{t-2} + \ldots$, the impact multiplier is the partial derivative of y_t with respect to ε_t.

APPENDIX 1.1: IMAGINARY ROOTS AND DE MOIVRE'S THEOREM

Consider a second-order difference equation $y_t = a_1y_{t-1} + a_2y_{t-2}$ such that the discriminant d is negative [i.e., $d = a_1^2 + 4a_2 < 0$]. From Section 6, we know that the full homogeneous solution can be written in the form

$$y_t^h = A_1\alpha_1^t + A_2\alpha_2^t \tag{A1.1}$$

where the two imaginary characteristic roots are

$$\alpha_1 = (a_1 + i\sqrt{-d})/2 \text{ and } \alpha_2 = (a_1 - i\sqrt{-d})/2 \tag{A1.2}$$

The purpose of this appendix is to explain how to rewrite and interpret (A1.1) in terms of standard trigonometric functions. You might first want to refresh your memory concerning two useful trig identities. For any two angles θ_1 and θ_2,

$$\sin(\theta_1 + \theta_2) = \sin(\theta_1)\cos(\theta_2) + \cos(\theta_1)\sin(\theta_2)$$
$$\cos(\theta_1 + \theta_2) = \cos(\theta_1)\cos(\theta_2) - \sin(\theta_1)\sin(\theta_2) \qquad \text{(A1.3)}$$

If $\theta_1 = \theta_2$, we can drop subscripts and form

$$\sin(2\theta) = 2\sin(\theta)\cos(\theta)$$
$$\cos(2\theta) = \cos(\theta)\cos(\theta) - \sin(\theta)\sin(\theta) \qquad \text{(A1.4)}$$

The first task is to demonstrate how to express imaginary numbers in the complex plane. Consider Figure A1.1 in which the horizontal axis measures real numbers and the vertical axis measures imaginary numbers. The complex number $a + bi$ can be represented by the point a units from the origin along the horizontal axis and b units from the origin along the vertical axis. It is convenient to represent the distance from the origin by the length of the vector denoted by r. Consider angle θ in triangle $0ab$ and note that $\cos(\theta) = a/r$ and $\sin(\theta) = b/r$. Hence, the lengths a and b can be measured by

$$a = r\cos(\theta) \quad \text{and} \quad b = r\sin(\theta)$$

In terms of (A1.2), we can define $a = a_1/2$ and $b = \sqrt{-d}/2$. Thus, the characteristic roots α_1 and α_2 can be written as:

$$\alpha_1 = a + bi = r[\cos(\theta) + i\sin(\theta)]$$
$$\alpha_2 = a - bi = r[cos(\theta) - i\sin(\theta)] \qquad \text{(A1.5)}$$

The next step is to consider the expressions $\alpha_1{}^t$ and $\alpha_2{}^t$. Begin with the expression $\alpha_1{}^2$ and recall that $i^2 = -1$:

$$\alpha_1{}^2 = \{r[\cos(\theta) + i\sin(\theta)]\}\{r[\cos(\theta) + i\sin(\theta)]\}$$
$$= r^2[\cos(\theta)\cos(\theta) - \sin(\theta)\sin(\theta) + 2i\sin(\theta)\cos(\theta)]$$

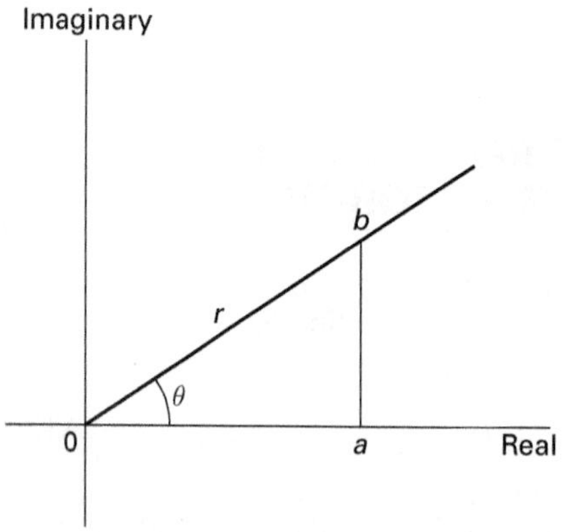

Imaginary

b

r

θ

0 a Real

FIGURE A1.1 A Graphical Representation of Complex Numbers

From (A1.4),

$$\alpha_1^2 = r^2[\cos(2\theta) + i\sin(2\theta)] \tag{A1.6}$$

If we continue in this fashion, it is straightforward to demonstrate that

$$\alpha_1^t = r^t[\cos(t\theta) + i\sin(t\theta)] \quad \text{and} \quad \alpha_2^t = r^t[\cos(t\theta) - i\sin(t\theta)]$$

Since y_t^h is a real number and α_1 and α_2 are complex, it follows that A_1 and A_2 must be complex. Although A_1 and A_2 are arbitrary complex numbers, they must have the form

$$A_1 = B_1[\cos(B_2) + i\sin(B_2)] \quad \text{and} \quad A_2 = B_1[\cos(B_2) - i\sin(B_2)] \tag{A1.7}$$

where B_1 and B_2 are arbitrary real numbers measured in radians.

In order to calculate $A_1(\alpha_1^t)$, use (A1.6) and (A1.7) to form

$$\begin{aligned} A_1\alpha_1^t &= B_1[\cos(B_2) + i\sin(B_2)]r^t[\cos(t\theta) + i\sin(t\theta)] \\ &= B_1r^t[\cos(B_2)\cos(t\theta) - \sin(B_2)\sin(t\theta) + i\cos(t\theta)\sin(B_2) \\ &\quad + i\sin(t\theta)\cos(B_2)] \end{aligned}$$

Using (A1.3), we obtain

$$A_1\alpha_1^t = B_1r^t[\cos(t\theta + B_2) + i\sin(t\theta + B_2)] \tag{A1.8}$$

You should use the same technique to convince yourself that

$$A_2\alpha_2^t = B_1r^t[\cos(t\theta + B_2) - i\sin(t\theta + B_2)] \tag{A1.9}$$

Since the homogeneous solution y_t^h is the sum of (A1.8) and (A1.9),

$$\begin{aligned} y_t^h &= B_1r^t[\cos(t\theta + B_2) + i\sin(t\theta + B_2)] + B_1r^t[\cos(t\theta + B_2) - i\sin(t\theta + B_2)] \\ &= 2B_1r^t\cos(t\theta + B_2) \end{aligned} \tag{A1.10}$$

Since B_1 is arbitrary, the homogeneous solution can be written in terms of the arbitrary constants B_2 and B_3

$$y_t^h = B_3r^t\cos(t\theta + B_2) \tag{A1.11}$$

Now imagine a circle with a radius of unity superimposed on Figure A1.1. The stability condition is for the distance $r = 0b$ to be less than unity. Hence, in the literature it is said that the stability condition is for the characteristic root(s) to lie within this unit circle.

APPENDIX 1.2: CHARACTERISTIC ROOTS IN HIGHER-ORDER EQUATIONS

The characteristic equation to an nth-order difference equation is

$$\alpha^n - a_1\alpha^{n-1} - a_2\alpha^{n-2} \ldots - a_n = 0 \tag{A1.12}$$

As stated in Section 6, the n values of α which solve this characteristic equation are called the **characteristic roots.** Denote the n solutions by $\alpha_1, \alpha_2, \ldots \alpha_n$. Given the results in Section 4, the linear combination $A_1\alpha_1^t + A_2\alpha_2^t + \ldots + A_n\alpha_n^t$ is also a solution to (A1.12).

A priori, the characteristic roots can take on any values. There is no restriction that they be real versus complex nor any restriction concerning their sign or magnitude. Consider the possibilities:

1. **All the α_i are real and distinct.** There are several important subcases. First suppose that each value of α_i is less than unity in absolute value. In this case, the homogeneous solution (A1.12) converges since the limit of each α_i^t equals zero as t approaches infinity. For a negative value of α_i, the expression α_i^t is positive for even values of t and negative for odd values of t. Thus, if any of the α_i are negative (but less than one in absolute value), the solution will tend to exhibit some oscillation. If any of the α_i are greater than unity in absolute value, the solution will diverge.
2. **All of the α_i are real but $m \leq n$ of the roots are repeated.** Let the solution be such that $\alpha_1 = \alpha_2 = \ldots = \alpha_m$. Call the single distinct value of this root $\overline{\alpha}$ and let the other *n-m* roots be denoted by α_{m+1} through α_n. In the case of a second-order equation with a repeated root, you saw that one solution was $A_1\overline{\alpha}^t$ and the other was $A_2t\overline{\alpha}^t$. With m repeated roots, it is easily verified that $t\overline{\alpha}^t, t^2\overline{\alpha}^t, \ldots, t^{m-1}\overline{\alpha}^t$ are also solutions to the homogeneous equation. With m repeated roots, the linear combination of all these solutions is

$$A_1\overline{\alpha}^t + A_2t\overline{\alpha}^t + A_3t^2\overline{\alpha}^t + \ldots + A_mt^{m-1}\overline{\alpha}^t + A_{m+1}\alpha_{m+1}^t + \ldots + A_n\alpha_n^t$$

3. **Some of the roots are complex.** Complex roots (which necessarily come in conjugate pairs) have the form $\alpha_i \pm i\theta$, where α_i and θ are real numbers and i is defined to be $\sqrt{-1}$. For any such pair, a solution to the homogeneous equation is: $A_1(\alpha_1 + i\theta)^t + A_2(\alpha_1 - i\theta)^t$ where A_1 and A_2 are arbitrary constants. Transforming to polar coordinates, the associated two solutions can be written in the form: $\beta_1r^t\cos(\theta t + \beta_2)$ with arbitrary constants β_1 and β_2. Here stability hinges on the magnitude of r^t; if $|r| < 1$, the system converges. However, even if there is convergence, convergence is not direct because the sine and cosine functions impart oscillatory behavior to the time path of y_t. For example, if there are three roots, two of which are complex, the homogeneous solution has the form

$$\beta_1r^t\cos(\theta t + \beta_2) + A_3(\alpha_3)^t$$

Stability of Higher-Order Systems: In practice, it is difficult to find the actual values of the characteristic roots. Unless the characteristic equation is easily factored, it is necessary to use numerical methods to obtain the characteristic roots. Fortunately, software packages such as Mathematica, Maple or Mathcad can easily obtain the characteristic roots of difference equations. At one time it was popular to use the **Schur Theorem** to determine whether all of the roots lie within the unit circle. Rather than calculate all of these determinants, it is often possible to use the simple rules discussed in Section 6. Those of you familiar with matrix algebra may wish to consult the *Supplementary Manual* to this text available at my Web site www.cba.ua.edu/~wenders or Samuelson (1941) for the appropriate conditions.

CHAPTER 2

STATIONARY TIME-SERIES MODELS

The theory of linear difference equations can be extended to allow the forcing process $\{x_t\}$ to be stochastic. This class of linear stochastic difference equations underlies much of the theory of time-series econometrics. Especially important is the Box–Jenkins (1976) methodology for estimating time-series models of the form

$$y_t = a_0 + a_1 y_{t-1} + \ldots + a_p y_{t-p} + \varepsilon_t + \beta_1 \varepsilon_{t-1} + \ldots + \beta_q \varepsilon_{t-q}$$

Such models are called autoregressive integrated moving-average (ARIMA) time-series models. This chapter has three aims:

1. Present the theory of stochastic linear difference equations and consider the time-series properties of stationary ARIMA models. A stationary ARIMA model is called an autoregressive moving-average (ARMA) model. It is shown that the stability conditions described in the previous chapter are necessary conditions for stationarity.
2. Develop the tools used in estimating ARMA models. Especially useful are the autocorrelation and partial autocorrelation functions. It is shown how the Box–Jenkins methodology relies on these tools to estimate an ARMA model from sample data.
3. Consider various test statistics to check for model adequacy. Several examples of estimated ARMA models are analyzed in detail. It is shown how a properly estimated model can be used for forecasting.

1. STOCHASTIC DIFFERENCE EQUATION MODELS

In this chapter we continue to work with **discrete,** rather than continuous, time-series models. Recall from the discussion in Chapter 1 that we can evaluate the function $y = f(t)$ at t_0 and $t_0 + h$ to form

$$\Delta y = f(t_0 + h) - f(t_0)$$

As a practical matter, most economic time-series data are collected for discrete time periods. Thus, we consider only the equidistant intervals $t_0, t_0 + h, t_0 + 2h, t_0 + 3h, \ldots$ and conveniently set $h = 1$. Be careful to recognize, however, that a discrete time series implies that t, but not necessarily y_t, is discrete. For example, although Scotland's annual rainfall is a continuous variable, the sequence of such annual rainfall totals for years 1 through t is a discrete time series. In many economic applications, t refers to "time" so that h represents the change in time. However, t need not refer to the type of time interval that is measured by a clock or calendar. Instead of allowing our measurement units to be minutes, days, quarters, or years, t can refer to an ordered event number. We could let y_t denote the outcome of spin t on a roulette wheel; y_t can then take on any of the 38 values 00, 0, 1, ..., 36.

A discrete variable y is said to be a **random** variable (i.e., stochastic) if for any real number r there exists a probability $p(y \leq r)$ that y will take on a value less than or equal to r. This definition is fairly general; in common usage, it is typically implied that there is at least one value of r for which $0 < p(y = r) < 1$. If there is some r for which $p(y = r) = 1$, y is deterministic rather than random.

It is useful to consider the elements of an observed time series $\{y_0, y_1, y_2, ..., y_t\}$ as being realizations (i.e., outcomes) of a stochastic process. As in Chapter 1, we continue to let the notation y_t to refer to an element of the entire sequence $\{y_t\}$. In our roulette example, y_t denotes the outcome of spin t on a roulette wheel. If we observe spins 1 through T, we can form the sequence $y_1, y_2, ..., y_T$ or, more compactly, $\{y_t\}$. In the same way, the term y_t could be used to denote gross domestic product (GDP) in time period t. Since we cannot forecast GDP perfectly, y_t is a random variable. Once we learn the value of GDP in period t, y_t becomes one of the realized values from a stochastic process. (Of course, measurement error may prevent us from ever knowing the "true" value of GDP.)

For discrete variables, the probability distribution of y_t is given by a formula (or table) that specifies each possible realized value of y_t and the probability associated with that realization. If the realizations are linked across time, there exists the joint probability distribution $p(y_1 = r_1, y_2 = r_2, ..., y_T = r_T)$ where r_i is the realized value of y in period i. Having observed the first t realizations, we can form the expected value of $y_{t+1}, y_{t+2}, ...,$ conditioned on the observed values of y_1 through y_t. This conditional mean, or expected value, of y_{t+i} is denoted by $E_t[y_{t+i}|y_t, y_{t-1}, ..., y_1]$, or $E_t y_{t+i}$.

Of course, if y_t refers to the outcome of spinning a fair roulette wheel, the probability distribution is easily characterized. In contrast, we may never be able to completely describe the probability distribution for GDP. Nevertheless, the task of economic theorists is to develop models that capture the essence of the true data-generating process. Stochastic difference equations are one convenient way of modeling dynamic economic processes. To take a simple example, suppose that the Federal Reserve's money supply target grows 3 percent each year. Hence,

$$m_t^* = 1.03m_{t-1}^* \tag{2.1}$$

so that, given the initial condition m_0^*, the particular solution is

$$m_t^* = (1.03)^t m_0^*$$

where: m_t^* = the logarithm of the money supply target in year t

m_0^* = the initial condition for the target money supply in period zero

Of course, the actual money supply (m_t) and the target need not be equal. Suppose that at the end of period $t - 1$, there exists m_{t-1} outstanding dollars that are carried forward into period t. Hence, at the beginning of t there are m_{t-1} dollars so that the gap between the target and the actual money supply is $m_t^* - m_{t-1}$. Suppose that the Fed cannot perfectly control the money supply but attempts to change the money supply by ρ percent ($\rho < 100\%$) of any gap between the desired and actual money supply. We can model this behavior as

$$\Delta m_t = \rho[m_t^* - m_{t-1}] + \varepsilon_t$$

Or, using (2.1), we obtain

$$m_t = \rho(1.03)^t m_0^* + (1 - \rho)m_{t-1} + \varepsilon_t \tag{2.2}$$

where ε_t is the uncontrollable portion of the money supply.

We assume the mean of ε_t is zero in all time periods.

Although the economic theory is overly simple, the model does illustrate the key points discussed above. Note the following:

1. Although the money supply is a continuous variable, (2.2) is a discrete difference equation. Since the forcing process $\{\varepsilon_t\}$ is stochastic, the money supply is stochastic; we can call (2.2) a linear stochastic difference equation.
2. If we knew the distribution of $\{\varepsilon_t\}$, we could calculate the distribution of each element in the $\{m_t\}$ sequence. Since (2.2) shows how the realizations of the $\{m_t\}$ sequence are linked across time, we would be able to calculate the various joint probabilities. Notice that the distribution of the money supply sequence is completely determined by the parameters of the difference equation (2.2) and the distribution of the $\{\varepsilon_t\}$ sequence.
3. Having observed the first t observations in the $\{m_t\}$ sequence, we can make forecasts of $m_{t+1}, m_{t+2}, \ldots$. For example, updating (2.2) by one period and taking the conditional expectation, the forecast of m_{t+1} is $E_t m_{t+1} = \rho(1.03)^{t+1} m_0^* + (1 - \rho)m_t$.

Before we proceed too far along these lines, let's go back to the basic building block of discrete stochastic time-series models: the **white-noise** process. A sequence $\{\varepsilon_t\}$ is a white-noise process if each value in the sequence has a mean of zero, has a constant variance, and is uncorrelated with all other realizations. Formally, if the notation $E(x)$ denotes the theoretical mean value of x, the sequence $\{\varepsilon_t\}$ is a white-noise process if, for each time period t,

$$E(\varepsilon_t) = E(\varepsilon_{t-1}) = \ldots = 0$$

$$E(\varepsilon_t^2) = E(\varepsilon_{t-1}^2) = \ldots = \sigma^2 \qquad [\text{or } \operatorname{var}(\varepsilon_t) = \operatorname{var}(\varepsilon_{t-1}) = \ldots = \sigma^2]$$

$$E(\varepsilon_t \varepsilon_{t-s}) = E(\varepsilon_{t-j}\varepsilon_{t-j-s})$$

$$= 0 \text{ for all } j \text{ and } s \qquad [\text{or } \operatorname{cov}(\varepsilon_t, \varepsilon_{t-s}) = \operatorname{cov}(\varepsilon_{t-j}, \varepsilon_{t-j-s}) = 0]$$

In the remainder of this text, $\{\varepsilon_t\}$ will always refer to a white-noise process and σ^2 will refer to the variance of that process. When it is necessary to refer to two or more white-noise processes, symbols such as $\{\varepsilon_{1t}\}$ and $\{\varepsilon_{2t}\}$ will be used. Now, use a white-noise process to construct the more interesting time series

$$x_t = \sum_{i=0}^{q} \beta_i \varepsilon_{t-i} \tag{2.3}$$

For each period t, x_t is constructed by taking the values $\varepsilon_t, \varepsilon_{t-1}, \ldots, \varepsilon_{t-q}$ and multiplying each by the associated value of β_i. A sequence formed in this manner is called a **moving average** of order q and is denoted by MA(q). To illustrate a typical moving-average process, suppose you win \$1 if a fair coin shows heads and lose \$1 if it shows tails. Denote the outcome on toss t by ε_t (i.e., for toss t, ε_t is either +\$1 or −\$1). If you want to keep track of your hot streaks, you might want to calculate your average winnings on the last four tosses. For each coin toss t, your average payoff on the last four tosses is $1/4\varepsilon_t + 1/4\varepsilon_{t-1} + 1/4\varepsilon_{t-2} + 1/4\varepsilon_{t-3}$. In terms of (2.3), this sequence is a moving-average process such that $\beta_i = 0.25$ for $i \leq 3$ and 0 otherwise.

Although the $\{\varepsilon_t\}$ sequence is a white-noise process, the constructed $\{x_t\}$ sequence will *not* be a white-noise process if two or more of the β_i differ from zero. To illustrate using an MA(1) process, set $\beta_0 = 1$, $\beta_1 = 0.5$, and all other $\beta_i = 0$. In this circumstance, $E(x_t) = E(\varepsilon_t + 0.5\varepsilon_{t-1}) = 0$ and $\text{var}(x_t) = \text{var}(\varepsilon_t + 0.5\varepsilon_{t-1}) = 1.25\sigma^2$. You can easily convince yourself that $E(x_t) = E(x_{t-s})$ and that $\text{var}(x_t) = \text{var}(x_{t-s})$ for all s. Hence, the first two conditions for $\{x_t\}$ to be a white-noise process are satisfied. However, $E(x_t x_{t-1}) = E[(\varepsilon_t + 0.5\varepsilon_{t-1})(\varepsilon_{t-1} + 0.5\varepsilon_{t-2})] = E(\varepsilon_t\varepsilon_{t-1} + 0.5(\varepsilon_{t-1})^2 + 0.5\varepsilon_t\varepsilon_{t-2} + 0.25\varepsilon_{t-1}\varepsilon_{t-2}) = 0.5\sigma^2$. Given that there exists a value of $s \neq 0$ such that $E(x_t x_{t-s}) \neq 0$, the $\{x_t\}$ sequence is not a white-noise process.

Exercise 1 at the end of this chapter asks you to find the mean, variance, and covariance of your hot streaks in coin tossing. For practice, you should work that exercise before continuing. If you are a bit "rusty" on the algebra of finding means, variances, and covariances, you should also work through Exercises 2 and 3 and consult the *Supplementary Manual* to this text.

2. ARMA MODELS

It is possible to combine a moving-average process with a linear difference equation to obtain an autoregressive moving-average model. Consider the pth order difference equation

$$y_t = a_0 + \sum_{i=1}^{p} a_i y_{t-i} + x_t \tag{2.4}$$

Now let $\{x_t\}$ be the MA(q) process given by (2.3), so that we can write

$$y_t = a_0 + \sum_{i=1}^{p} a_i y_{t-i} + \sum_{i=0}^{q} \beta_i \varepsilon_{t-i} \tag{2.5}$$

We follow the convention of normalizing units so that β_0 is always equal to unity. If the characteristic roots of (2.5) are all in the unit circle, $\{y_t\}$ is called an **autoregressive moving-average** (ARMA) model for y_t. The autoregressive part of the model is the *difference equation* given by the homogeneous portion of (2.4), and the moving-average part is the $\{x_t\}$ sequence. If the homogeneous part of the difference equation contains p lags and the model for x_t contains q lags, the model is called an ARMA (p, q) model. If $q = 0$, the process is called a pure autoregressive process denoted by AR(p), and if $p = 0$, the process is a pure moving-average process denoted by MA(q). In an ARMA model, it is perfectly permissible to allow p and/or q to be infinite. In this chapter we consider only models in which all of the characteristic roots of (2.5) are within the unit circle. However, if one or more characteristic roots of (2.5) is greater than or equal to unity, the $\{y_t\}$ sequence is said to be an **integrated** process and (2.5) is called an autoregressive integrated moving-average (ARIMA) model.

Treating (2.5) as a difference equation suggests that we can "solve" for y_t in terms of the $\{\varepsilon_t\}$ sequence. The solution of an ARMA(p, q) model expressing y_t in terms of the $\{\varepsilon_t\}$ sequence is the **moving-average representation** of y_t. The procedure is no different from that discussed in Chapter 1. For the AR(1) model $y_t = a_0 + a_1 y_{t-1} + \varepsilon_t$, the moving-average representation was shown to be

$$y_t = a_0/(1 - a_1) + \sum_{i=0}^{\infty} a_1^i \varepsilon_{t-i}$$

For the general ARMA(p, q) model, rewrite (2.5) using lag operators so that

$$\left(1 - \sum_{i=1}^{p} a_i L^i\right) y_t = a_0 + \sum_{i=0}^{q} \beta_i \varepsilon_{t-i}$$

So, the *particular* solution for y_t is

$$y_t = \left(a_0 + \sum_{i=0}^{q} \beta_i \varepsilon_{t-i}\right) \Big/ \left(1 - \sum_{i=1}^{p} a_i L^i\right) \tag{2.6}$$

Fortunately, it will not be necessary for us to expand (2.6) to obtain the specific coefficient for each element in $\{\varepsilon_t\}$. The important point to recognize is that the expansion will yield an MA(∞) process. The issue is whether such an expansion is convergent so that the stochastic difference equation given by (2.6) is stable. As you will see in the next section, the stability condition is that the roots of the polynomial $(1 - \Sigma a_i L^i)$ must lie outside the unit circle. It is also shown that *if y_t is a linear stochastic difference equation, the stability condition is a necessary condition for the time-series $\{y_t\}$ to be stationary.*

3. STATIONARITY

Suppose that the quality-control division of a manufacturing firm samples four machines each hour. Every hour, quality control finds the mean of the machines' output levels. The plot of each machine's hourly output is shown in Figure 2.1. If y_{it} represents machine y_i's output at hour t, the means ($\bar{y}_t$) are readily calculated as

$$\bar{y}_t = \sum_{i=1}^{4} y_{it}/4$$

For hours 5, 10, and 15, these mean values are 4.61, 5.14, and 5.03, respectively.

The sample variance for each hour can similarly be constructed. Unfortunately, applied econometricians do not usually have the luxury of being able to obtain an **ensemble** (i.e., multiple time-series data of the same process over the same time period). Typically, we observe only one set of realizations for any particular series. Fortunately, if $\{y_t\}$ is a **stationary** series, the mean, variance, and autocorrelations can usually be

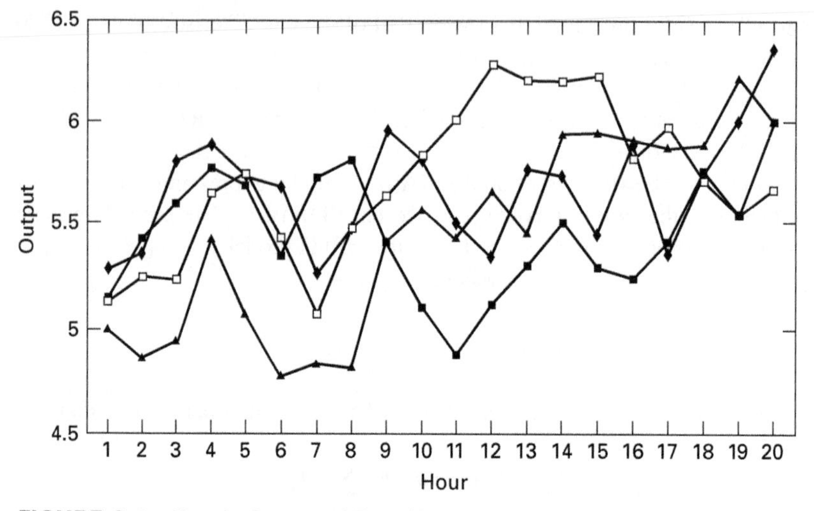

FIGURE 2.1 Hourly Output of Four Machines

well approximated by sufficiently long **time averages** based on the single set of realizations. Suppose you observed only the output of machine 1 for 20 periods. If you knew that the output was stationary, you could approximate the mean level of output by

$$\bar{y}_t \cong \sum_{t=1}^{20} y_{1t}/20$$

In using this approximation, you would be assuming that the mean was the same for each period. Formally, a stochastic process having a finite mean and variance is **covariance-stationary** if for all t and $t - s$,

$$E(y_t) = E(y_{t-s}) = \mu \tag{2.7}$$

$$E[(y_t - \mu)^2] = E[(y_{t-s} - \mu)^2] = \sigma_y^2 \qquad [\text{var}(y_t) = \text{var}(y_{t-s}) = \sigma_y^2] \tag{2.8}$$

$$E[(y_t - \mu)(y_{t-s} - \mu)] = E[(y_{t-j} - \mu)(y_{t-j-s} - \mu)] = \gamma_s$$
$$[\text{cov}(y_t, y_{t-s}) = \text{cov}(y_{t-j}, y_{t-j-s}) = \gamma_s] \tag{2.9}$$

where μ, σ_y^2, and γ_s are all constants.

In (2.9), allowing $s = 0$ means that γ_0 is equivalent to the variance of y_t. Simply put, a time-series is covariance-stationary if its mean and all autocovariances are unaffected by a change of time origin. In the literature, a covariance-stationary process is also referred to as a weakly stationary, second-order-stationary, or wide-sense-stationary process. (Note that a **strongly** stationary process need not have a finite mean and/or variance.) The text considers only covariance-stationary series so that there is no ambiguity in using the terms stationary and covariance-stationary interchangeably. One further word about terminology: In multivariate models, the term autocovariance is reserved for the covariance between y_t and its own lags. Cross-covariance refers to the covariance between one series and another. In univariate time-series models, there is no ambiguity and the terms autocovariance and covariance are used interchangeably.

For a covariance-stationary series, we can define the **autocorrelation** between y_t and y_{t-s} as

$$\rho_s \equiv \gamma_s/\gamma_0$$

where γ_0 and γ_s are defined by (2.9).

Since γ_s and γ_0 are time-independent, the autocorrelation coefficients ρ_s are also time-independent. Although the autocorrelation between y_t and y_{t-1} can differ from the autocorrelation between y_t and y_{t-2}, the autocorrelation between y_t and y_{t-1} must be identical to that between y_{t-s} and y_{t-s-1}. Obviously, $\rho_0 = 1$.

Stationarity Restrictions for an AR(1) Process

For expositional convenience, first consider the necessary and sufficient conditions for an AR(1) process to be stationary. Let

$$y_t = a_0 + a_1 y_{t-1} + \varepsilon_t$$

where ε_t = white noise.

Suppose that the process started in period zero, so that y_0 is a deterministic initial condition. In Section 3 of the last chapter, it was shown that the solution to this equation is (see also Question 4 at the end of this chapter):

$$y_t = a_0 \sum_{i=0}^{t-1} a_1^i + a_1^t y_0 + \sum_{i=0}^{t-1} a_1^i \varepsilon_{t-i} \tag{2.10}$$

Taking the expected value of (2.10), we obtain

$$Ey_t = a_0 \sum_{i=0}^{t-1} a_1^i + a_1^t y_0 \tag{2.11}$$

Updating by s periods yields

$$Ey_{t+s} = a_0 \sum_{i=0}^{t+s-1} a_1^i + a_1^{t+s} y_0 \tag{2.12}$$

Comparing (2.11) and (2.12), it is clear that both means are time-dependent. Since Ey_t is not equal to Ey_{t+s}, the sequence cannot be stationary. However, if t is large, we can consider the limiting value of y_t in (2.10). If $|a_1| < 1$, the expression $(a_1)^t y_0$ converges to zero as t becomes infinitely large and the sum $a_0[1 + a_1 + (a_1)^2 + (a_1)^3 + \ldots]$ converges to $a_0/(1 - a_1)$. Thus, as $t \to \infty$ and if $|a_1| < 1$,

$$\lim y_t = \frac{a_0}{1 - a_1} + \sum_{i=0}^{\infty} a_1^i \varepsilon_{t-i} \tag{2.13}$$

Now take expectations of (2.13) so that for sufficiently large values of t, $Ey_t = a_0/(1 - a_1)$. Thus, the mean value of y_t is finite and time-independent so that $Ey_t = Ey_{t-s} = a_0/(1 - a_1) \equiv \mu$ for all t. Turning to the variance, we find

$$\begin{aligned} E(y_t - \mu)^2 &= E[(\varepsilon_t + a_1\varepsilon_{t-1} + (a_1)^2\varepsilon_{t-2} + \ldots)^2] \\ &= \sigma^2[1 + (a_1)^2 + (a_1)^4 + \ldots] = \sigma^2/(1 - (a_1)^2) \end{aligned}$$

which is also finite and time-independent. Finally, it is easily demonstrated that the limiting values of all autocovariances are finite and time-independent:

$$\begin{aligned} E[(y_t - \mu)(y_{t-s} - \mu)] &= E\{[\varepsilon_t + a_1\varepsilon_{t-1} + (a_1)^2\varepsilon_{t-2} + \dots] \\ &\quad [\varepsilon_{t-s} + a_1\varepsilon_{t-s-1} + (a_1)^2\varepsilon_{t-s-2} + \dots]\} \\ &= \sigma^2(a_1)^s[1 + (a_1)^2 + (a_1)^4 + \dots] \\ &= \sigma^2(a_1)^s/[1 - (a_1)^2] \end{aligned} \quad (2.14)$$

In summary, if we can use the limiting value of (2.10), the $\{y_t\}$ sequence will be stationary. For any given y_0 and $|a_1| < 1$, it follows that t must be sufficiently large. Thus, if a sample is generated by a process that has recently begun, the realizations may not be stationary. It is for this very reason that many econometricians assume that the data-generating process has been occurring for an infinitely long time. In practice, the researcher must be wary of any data generated from a "new" process. For example, $\{y_t\}$ could represent the daily change in the dollar/mark exchange rate beginning immediately after the demise of the Bretton Woods fixed exchange rate system. Such a series may not be stationary due to the fact there were deterministic initial conditions (exchange rate changes were essentially zero in the Bretton Woods era). The careful researcher wishing to use stationary series might consider excluding some of these earlier observations from the period of analysis.

Little would change were we not given the initial condition. Without the initial value y_0, the sum of the homogeneous and particular solutions for y_t is

$$y_t = a_0/(1 - a_1) + \sum_{i=0}^{\infty} a_1^i\varepsilon_{t-i} + A(a_1)^t \quad (2.15)$$

where A = an arbitrary constant.

If you take the expectation of (2.15), it is clear that the $\{y_t\}$ sequence cannot be stationary unless the expression $A(a_1)^t$ is equal to zero. Either the sequence must have started infinitely long ago (so that $a_1^t = 0$) or the arbitrary constant A must be zero. Recall that the arbitrary constant is interpreted as a deviation from long-run equilibrium. The stability conditions can be stated succinctly:

1. The homogeneous solution must be zero. Either the sequence must have started infinitely far in the past or the process must always be in equilibrium (so that the arbitrary constant is zero).
2. The characteristic root a_1 must be less than unity in absolute value.

These two conditions readily generalize to all ARMA(p, q) processes. We know that the homogeneous solution to (2.5) has the form

$$\sum_{i=1}^{p} A_i\alpha_i^t$$

or, if there are m repeated roots,

$$\alpha\sum_{i=1}^{m} A_i t^i + \sum_{i=m+1}^{p} A_i\alpha_i^t$$

where the A_i are all arbitrary constants, α is the repeated root, and the α_i are the distinct characteristic roots.

If any portion of the homogeneous equation is present, the mean, variance, and all covariances will be time-dependent. Hence, for any ARMA(p, q) model, stationarity necessitates that the homogeneous solution be zero. The next section addresses the stationarity restrictions for the particular solution.

4. STATIONARITY RESTRICTIONS FOR AN ARMA(p, q) MODEL

As a prelude to the stationarity conditions for the general ARMA(p, q) model, first consider the restrictions necessary to ensure that an ARMA(2, 1) model is stationary. Since the magnitude of the intercept term does not affect the stability (or stationarity) conditions, set $a_0 = 0$ and write

$$y_t = a_1 y_{t-1} + a_2 y_{t-2} + \varepsilon_t + \beta_1 \varepsilon_{t-1} \tag{2.16}$$

From the previous section, we know that the homogeneous solution must be zero. As such, it is only necessary to find the particular solution. Using the method of undetermined coefficients, we can write the challenge solution as

$$y_t = \sum_{i=0}^{\infty} c_i \varepsilon_{t-i} \tag{2.17}$$

For (2.17) to be a solution of (2.16), the various c_i must satisfy

$$c_0\varepsilon_t + c_1\varepsilon_{t-1} + c_2\varepsilon_{t-2} + c_3\varepsilon_{t-3} + \ldots = a_1(c_0\varepsilon_{t-1} + c_1\varepsilon_{t-2} + c_2\varepsilon_{t-3} + c_3\varepsilon_{t-4} + \ldots) + a_2(c_0\varepsilon_{t-2} + c_1\varepsilon_{t-3} + c_2\varepsilon_{t-4} + c_3\varepsilon_{t-5} + \ldots) + \varepsilon_t + \beta_1\varepsilon_{t-1}$$

To match coefficients on the terms containing ε_t, ε_{t-1}, ε_{t-2}, ..., it is necessary to set

1. $c_0 = 1$
2. $c_1 = a_1 c_0 + \beta_1 \qquad \Rightarrow c_1 = a_1 + \beta_1$
3. $c_i = a_1 c_{i-1} + a_2 c_{i-2} \qquad$ for all $i \geq 2$

The key point is that for $i \geq 2$, the coefficients satisfy the difference equation $c_i = a_1 c_{i-1} + a_2 c_{i-2}$. If the characteristic roots of (2.16) are within the unit circle, the $\{c_i\}$ must constitute a convergent sequence. For example, reconsider the case in which $a_1 = 1.6$ and $a_2 = -0.9$, and let $\beta_1 = 0.5$. Worksheet 2.1 shows that the coefficients satisfying (2.17) are 1, 2.1, 2.46, 2.046, 1.06, −0.146, ... (also see Worksheet 1.2 of the previous chapter).

To verify that the $\{y_t\}$ sequence generated by (2.17) is stationary, take the expectation of (2.17) to form $Ey_t = Ey_{t-i} = 0$ for all t and i. Hence, the mean is finite and time-invariant. Since the $\{\varepsilon_t\}$ sequence is assumed to be a white-noise process, the variance of y_t is constant and time-independent; that is,

$$\begin{aligned}\operatorname{var}(y_t) &= E[(c_0\varepsilon_t + c_1\varepsilon_{t-1} + c_2\varepsilon_{t-2} + c_3\varepsilon_{t-3} + \ldots)^2] \\ &= \sigma^2 \sum_{i=0}^{\infty} c_i^2\end{aligned}$$

WORKSHEET 2.1

COEFFICIENTS OF THE ARMA(2, 1) PROCESS: $y_t = 1.6y_{t-1} - 0.9y_{t-2} + \varepsilon_t + 0.5\varepsilon_{t-1}$

If we use the method of undetermined coefficients, the c_i must satisfy

$$c_0 = 1$$
$$c_1 = 1.16 + 0.5 \qquad \text{hence } c_1 = 2.1$$
$$c_i = 1.6c_{i-1} - 0.9c_{i-2} \qquad \text{for all } i = 2, 3, 4, \ldots$$

Notice that the coefficients follow a second-order difference equation with imaginary roots. If we use de Moivre's Theorem, the coefficients will satisfy

$$c_i = 0.949^i \cdot \beta_1 \cos(0.567i + \beta_2)$$

Imposing the initial conditions for c_0 and c_1 yields

$$1 = \beta_1 \cos(\beta_2) \qquad \text{and} \qquad 2.1 = 0.949\beta_1 \cos(0.567 + \beta_2)$$

Since $\beta_1 = 1/\cos(\beta_2)$, we seek the solution to

$$\cos(\beta_2) - (0.949/2.1)\cos(0.567 + \beta_2) = 0$$

You can use a trig table to verify that the solution for β_2 is –1.197 and the solution for β_1 is 2.739. Hence, the c_i must satisfy

$$(2.739) \cdot 0.949^i \cdot \cos(0.567i - 1.197)$$

Alternatively, we can use the initial values of c_0 and c_1 to find the other c_i by iteration. The sequence of the c_i is shown in the graph below.

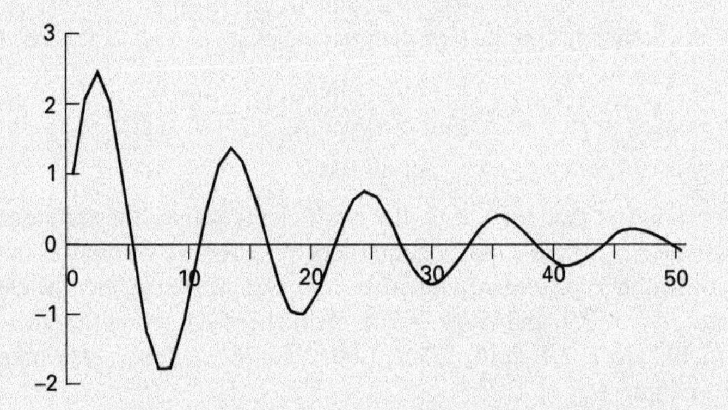

You can use a spreadsheet to verify that the values of c_0 through c_{10} are:

i	0	1	2	3	4	5	6	7	8	9	10
c_i	1.00	2.10	2.46	2.046	1.06	−0.146	1.187	−1.786	−1.761	−1.226	−0.378

Hence, $\text{var}(y_t) = \text{var}(y_{t-s})$ for all t and s. Finally, the covariance between y_t and y_{t-s} is

$$\text{cov}(y_t, y_{t-1}) = E[(\varepsilon_t + c_1\varepsilon_{t-1} + c_2\varepsilon_{t-2} + \ldots)(\varepsilon_{t-1} + c_1\varepsilon_{t-2} + c_2\varepsilon_{t-3} + c_3\varepsilon_{t-4} + \ldots)]$$

$$= \sigma^2(c_1 + c_2c_1 + c_3c_2 + \ldots)$$

$$\text{cov}(y_t, y_{t-2}) = E[(\varepsilon_t + c_1\varepsilon_{t-1} + c_2\varepsilon_{t-2} + \ldots)(\varepsilon_{t-2} + c_1\varepsilon_{t-3} + c_2\varepsilon_{t-4} + c_3\varepsilon_{t-5} + \ldots)]$$

$$= \sigma^2(c_2 + c_3c_1 + c_4c_2 + \ldots)$$

so that

$$\text{cov}(y_t, y_{t-s}) = \sigma^2(c_s + c_{s+1}c_1 + c_{s+2}c_2 + \ldots) \tag{2.18}$$

Thus, $\text{cov}(y_t, y_{t-s})$ is constant and independent of t. Conversely, if the characteristic roots of (2.16) do not lie within the unit circle, the $\{c_i\}$ sequence will *not* be convergent. As such, the $\{y_t\}$ sequence cannot be convergent.

It is not too difficult to generalize these results to the entire class of ARMA(p, q) models. Begin by considering the conditions ensuring the stationarity of a pure MA(∞) process. By appropriately restricting the β_i, all of the finite-order MA(q) processes can be obtained as special cases. Consider

$$x_t = \sum_{i=0}^{\infty} \beta_i \varepsilon_{t-i}$$

where $\{\varepsilon_t\}$ = a white-noise process with variance σ^2.

We have already determined that $\{x_t\}$ is not a white-noise process; now the issue is whether $\{x_t\}$ is covariance-stationary. Given conditions (2.7), (2.8), and (2.9), we ask the following:

1. Is the mean finite and time-independent? Take the expected value of x_t and remember that the expectation of a sum is the sum of the individual expectations. Therefore,

$$E(x_t) = E(\varepsilon_t + \beta_1\varepsilon_{t-1} + \beta_2\varepsilon_{t-2} + \ldots)$$
$$= E\varepsilon_t + \beta_1 E\varepsilon_{t-1} + \beta_2 E\varepsilon_{t-2} + \ldots = 0$$

Repeat the procedure with x_{t-s}:

$$E(x_{t-s}) = E(\varepsilon_{t-s} + \beta_1\varepsilon_{t-s-1} + \beta_2\varepsilon_{t-s-2} + \ldots) = 0$$

Hence, all elements in the $\{x_t\}$ sequence have the same finite mean ($\mu = 0$).

2. Is the variance finite and time-independent? Form $\text{var}(x_t)$ as

$$\text{var}(x_t) = E[(\varepsilon_t + \beta_1\varepsilon_{t-1} + \beta_2\varepsilon_{t-2} + \ldots)^2]$$

Square the term in parentheses and take expectations. Since $\{\varepsilon_t\}$ is a white-noise process, all terms $E\varepsilon_t\varepsilon_{t-s} = 0$ for $s \neq 0$. Hence,

$$\begin{aligned}\text{var}(x_t) &= E(\varepsilon_t)^2 + (\beta_1)^2E(\varepsilon_{t-1})^2 + (\beta_2)^2E(\varepsilon_{t-2})^2 + \ldots \\ &= \sigma^2[1 + (\beta_1)^2 + (\beta_2)^2 + \ldots]\end{aligned}$$

As long as $\sum(\beta_i)^2$ is finite, it follows that var(x_t) is finite. Thus, $\sum(\beta_i)^2$ being finite is a necessary condition for $\{x_t\}$ to be stationary. To determine whether var(x_t) = var(x_{t-s}), form

$$\begin{aligned}\text{var}(x_{t-s}) &= E[(\varepsilon_{t-s} + \beta_1\varepsilon_{t-s-1} + \beta_2\varepsilon_{t-s-2} + \ldots)^2] \\ &= \sigma^2[1 + (\beta_1)^2 + (\beta_2)^2 + \ldots]\end{aligned}$$

Thus, var(x_t) = var(x_{t-s}) for all t and $t - s$.

3. Are all autocovariances finite and time-independent? First form $E(x_tx_{t-s})$ as

$$E[x_tx_{t-s}] = E[(\varepsilon_t + \beta_1\varepsilon_{t-1} + \beta_2\varepsilon_{t-2} + \ldots)(\varepsilon_{t-s} + \beta_1\varepsilon_{t-s-1} + \beta_2\varepsilon_{t-s-2} + \ldots)]$$

Carrying out the multiplication and noting that $E(\varepsilon_t\varepsilon_{t-s}) = 0$ for $s \neq 0$, we get

$$E(x_tx_{t-s}) = \sigma^2(\beta_s + \beta_1\beta_{s+1} + \beta_2\beta_{s+2} + \ldots)$$

Restricting the sum $\beta_s + \beta_1\beta_{s+1} + \beta_2\beta_{s+2} + \ldots$ to be finite means that $E(x_tx_{t-s})$ is finite. Given this second restriction, it is clear that the covariance between x_t and x_{t-s} only depends on the number of periods separating the variables (i.e., the value of s) but *not* on the time subscript t.

In summary, the necessary and sufficient conditions for any MA process to be covariance-stationary are for the sums $\sum(\beta_i)^2$ and $(\beta_s + \beta_1\beta_{s+1} + \beta_2\beta_{s+2} + \ldots)$ to be finite. For an infinite-order process, these conditions must hold for all $s \geq 0$. Some of the details involved with maximum likelihood estimation of MA processes are discussed in Appendix 2.1 of this chapter.

Stationarity Restrictions for the Autoregressive Coefficients

Now consider the pure autoregressive model

$$y_t = a_0 + \sum_{i=1}^{p} a_iy_{t-i} + \varepsilon_t \tag{2.19}$$

If the characteristic roots of the homogeneous equation of (2.19) all lie inside the unit circle, it is possible to write the particular solution as

$$y_t = a_0\Big/\left[1 - \sum_{i=1}^{p} a_i\right] + \sum_{i=0}^{\infty} c_i\varepsilon_{t-i} \tag{2.20}$$

where the c_i = undetermined coefficients.

Although it is possible to find the undetermined coefficients $\{c_i\}$, we know that (2.20) is a convergent sequence as long as the characteristic roots of (2.19) are inside the unit circle. To sketch the proof, we can use the method of undetermined coefficients to write the particular solution in the form of (2.20). We also know that the sequence $\{c_i\}$ will eventually solve the difference equation

$$c_i - a_1 c_{i-1} - a_2 c_{i-2} - \ldots - a_p c_{i-p} = 0 \tag{2.21}$$

If the characteristic roots of (2.21) are all inside the unit circle, the $\{c_i\}$ sequence will be convergent. Although (2.20) is an infinite-order moving-average process, the convergence of the MA coefficients implies that Σc_i^2 is finite. Thus, we can use (2.20) to check the three conditions for stationarity. From (2.20),

$$Ey_t = Ey_{t-s} = a_0/(1 - \Sigma a_i)$$

You should recall from Chapter 1 that a necessary condition for all characteristic roots to lie inside the unit circle is $1 - \Sigma a_i > 0$. Hence, the mean of the sequence is finite and time-invariant.

$$\mathrm{var}(y_t) = E[(\varepsilon_t + c_1\varepsilon_{t-1} + c_2\varepsilon_{t-2} + c_3\varepsilon_{t-3} + \ldots)^2] = \sigma^2 \Sigma c_i^2$$

and

$$\mathrm{var}(y_{t-s}) = E[(\varepsilon_{t-s} + c_1\varepsilon_{t-s-1} + c_2\varepsilon_{t-s-2} + c_3\varepsilon_{t-s-3} + \ldots)^2] = \sigma^2 \Sigma c_i^2$$

Given that Σc_i^2 is finite, the variance is finite and time-independent.

$$\begin{aligned}\mathrm{cov}(y_t, y_{t-s}) &= E[(\varepsilon_t + c_1\varepsilon_{t-1} + c_2\varepsilon_{t-2} + \ldots)(\varepsilon_{t-s} + c_1\varepsilon_{t-s-1} + c_2\varepsilon_{t-s-2} + \ldots)] \\ &= \sigma^2(c_s + c_1 c_{s+1} + c_2 c_{s+2} + \ldots)\end{aligned}$$

Thus, the covariance between y_t and y_{t-s} is constant and time-invariant for all t and $t - s$. Nothing of substance is changed by combining the AR(p) and MA(q) models into the general ARMA(p, q) model:

$$\begin{aligned} y_t &= a_0 + \sum_{i=1}^{p} a_i y_{t-i} + x_t \\ x_t &= \sum_{i=0}^{q} \beta_i \varepsilon_{t-i} \end{aligned} \tag{2.22}$$

If the roots of the inverse characteristic equation lie outside the unit circle [i.e., if the roots of the homogeneous form of (2.22) lie inside the unit circle] and if the $\{x_t\}$ sequence is stationary, the $\{y_t\}$ sequence will be stationary. Consider:

$$y_t = \frac{a_0}{1 - \sum_{i=1}^{p} a_i} + \frac{\varepsilon_t}{1 - \sum_{i=1}^{p} a_i L^i} + \frac{\beta_1 \varepsilon_{t-1}}{1 - \sum_{i=1}^{p} a_i L^i} + \frac{\beta_2 \varepsilon_{t-2}}{1 - \sum_{i=1}^{p} a_i L^i} + \ldots \tag{2.23}$$

With very little effort, you can convince yourself that the $\{y_t\}$ sequence satisfies the three conditions for stationarity. Each of the expressions on the right-hand side of (2.23) is stationary as long as the roots of $1 - \Sigma a_i L^i$ are outside the unit circle. Given that $\{x_t\}$ is stationary, only the roots of the autoregressive portion of (2.22) determine whether the $\{y_t\}$ sequence is stationary.

5. THE AUTOCORRELATION FUNCTION

The autocovariances and autocorrelations of the type found in (2.18) serve as useful tools in the Box–Jenkins (1976) approach to identifying and estimating time-series models. We illustrate by considering four important examples: the AR(1), AR(2), MA(1), and ARMA(1, 1) models. For the AR(1) model, $y_t = a_0 + a_1 y_{t-1} + \varepsilon_t$, (2.14) shows

$$\gamma_0 = \sigma^2/[1 - (a_1)^2]$$
$$\gamma_s = \sigma^2(a_1)^s/[1 - (a_1)^2]$$

Forming the autocorrelations by dividing each γ_s by γ_0, we find that $\rho_0 = 1, \rho_1 = a_1, \rho_2 = (a_1)^2, \ldots, \rho_s = (a_1)^s$. For an AR(1) process, a necessary condition for stationarity is $|a_1| < 1$. Thus, the plot of ρ_s against s—called the autocorrelation function (ACF) or **correlogram**—should converge to zero geometrically if the series is stationary. If a_1 is positive, convergence will be direct, and if a_1 is negative, the autocorrelations will follow a dampened oscillatory path around zero. The first two graphs on the left-hand side of Figure 2.2 show the theoretical autocorrelation functions for $a_1 = 0.7$ and $a_1 = -0.7$, respectively. Here, ρ_0 is not shown since its value is necessarily unity.

The Autocorrelation Function of an AR(2) Process

Now consider the more complicated AR(2) process $y_t = a_1 y_{t-1} + a_2 y_{t-2} + \varepsilon_t$. We omit an intercept term (a_0) since it has no effect on the ACF. For the second-order process to be stationary, we know that it is necessary to restrict the roots of $(1 - a_1 L - a_2 L^2)$ to be outside the unit circle. In Section 4, we derived the autocovariances of an ARMA(2, 1) process by use of the method of undetermined coefficients. Now we want to illustrate an alternative technique using the **Yule–Walker** equations. Multiply the second-order difference equation by y_{t-s} for $s = 0, s = 1, s = 2, \ldots$ and take expectations to form

$$Ey_t y_t = a_1 Ey_{t-1} y_t + a_2 Ey_{t-2} y_t + E\varepsilon_t y_t$$
$$Ey_t y_{t-1} = a_1 Ey_{t-1} y_{t-1} + a_2 Ey_{t-2} y_{t-1} + E\varepsilon_t y_{t-1}$$
$$Ey_t y_{t-2} = a_1 Ey_{t-1} y_{t-2} + a_2 Ey_{t-2} y_{t-2} + E\varepsilon_t y_{t-2}$$
$$\cdot$$
$$\cdot$$
$$Ey_t y_{t-s} = a_1 Ey_{t-1} y_{t-s} + a_2 Ey_{t-2} y_{t-s} + E\varepsilon_t y_{t-s} \tag{2.24}$$

By definition, the autocovariances of a stationary series are such that $Ey_t y_{t-s} = Ey_{t-s} y_t = Ey_{t-k} y_{t-k-s} = \gamma_s$. We also know that $E\varepsilon_t y_t = \sigma^2$ and $E\varepsilon_t y_{t-s} = 0$. Hence, we can use the equations in (2.24) to form

$$\gamma_0 = a_1\gamma_1 + a_2\gamma_2 + \sigma^2 \tag{2.25}$$
$$\gamma_1 = a_1\gamma_0 + a_2\gamma_1 \tag{2.26}$$
$$\gamma_s = a_1\gamma_{s-1} + a_2\gamma_{s-2} \tag{2.27}$$

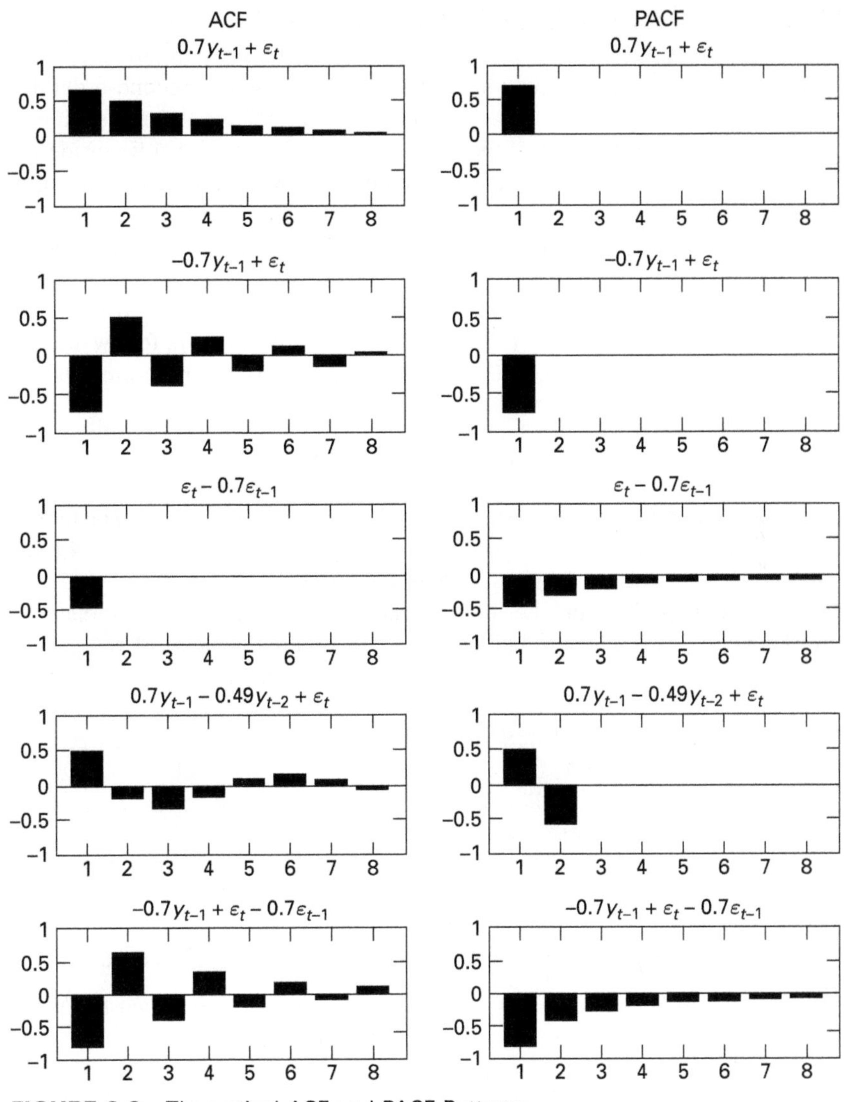

FIGURE 2.2 Theoretical ACF and PACF Patterns

Dividing (2.26) and (2.27) by γ_0 yields

$$\rho_1 = a_1\rho_0 + a_2\rho_1 \tag{2.28}$$

$$\rho_s = a_1\rho_{s-1} + a_2\rho_{s-2} \tag{2.29}$$

We know that $\rho_0 = 1$, so from (2.28), $\rho_1 = a_1/(1 - a_2)$. Hence, we can find all ρ_s for $s \geq 2$ by solving the difference equation (2.29). For example, for $s = 2$ and $s = 3$,

$$\rho_2 = (a_1)^2/(1 - a_2) + a_2$$

$$\rho_3 = a_1[(a_1)^2/(1 - a_2) + a_2] + a_2a_1/(1 - a_2)$$

Although the values of the ρ_s are cumbersome to derive, we can easily characterize their properties. Given the solutions for ρ_0 and ρ_1, the key point to note is that the ρ_s all satisfy the difference equation (2.29). As in the case of a second-order difference equation, the solution may be oscillatory or direct. Note that the stationarity condition for y_t necessitates that the characteristic roots of (2.29) lie inside the unit circle. Hence, the $\{\rho_s\}$ sequence must be convergent. The correlogram for an AR(2) process must be such that $\rho_0 = 1$ and that ρ_1 be determined by (2.28). These two values can be viewed as *initial values* for the second-order difference equation (2.29).

The fourth panel on the left-hand side of Figure 2.2 shows the ACF for the process $y_t = 0.7y_{t-1} - 0.49y_{t-2} + \varepsilon_t$. The properties of the various ρ_s follow directly from the homogeneous equation $y_t - 0.7y_{t-1} + 0.49y_{t-2} = 0$. The roots are obtained from the solution to

$$\alpha = \{0.7 \pm [(-0.7)^2 - 4(0.49)]^{1/2}\}/2$$

Since the discriminant $d = (-0.7)^2 - 4(0.49)$ is negative, the characteristic roots are imaginary, so that the solution oscillates. However, since $a_2 = -0.49$, the solution is convergent and the $\{y_t\}$ sequence is stationary.

Finally, we may wish to find the autocovariances rather than the autocorrelations. Since we know all of the autocorrelations, if we can find the variance of y_t (i.e., γ_0), we can find all of the other γ_s. To find γ_0, use (2.25) and note that $\rho_i = \gamma_i/\gamma_0$ so that

$$\gamma_0(1 - a_1\rho_1 - a_2\rho_2) = \sigma^2$$

Substitution for ρ_1 and ρ_2 yields

$$\gamma_0 = \text{var}(y_t) = [(1 - a_2)/(1 + a_2)]\left(\frac{\sigma^2}{(a_1 + a_2 - 1)(a_2 - a_1 - 1)}\right)$$

The Autocorrelation Function of an MA(1) Process

Next consider the MA(1) process $y_t = \varepsilon_t + \beta\varepsilon_{t-1}$. Again, we can obtain the Yule–Walker equations by multiplying y_t by each y_{t-s} and take expectations

$$\gamma_0 = \text{var}(y_t) = Ey_ty_t = E[(\varepsilon_t + \beta\varepsilon_{t-1})(\varepsilon_t + \beta\varepsilon_{t-1})] = (1 + \beta^2)\sigma^2$$
$$\gamma_1 = Ey_ty_{t-1} = E[(\varepsilon_t + \beta\varepsilon_{t-1})(\varepsilon_{t-1} + \beta\varepsilon_{t-2})] = \beta\sigma^2$$

and

$$\gamma_s = Ey_ty_{t-s} = E[(\varepsilon_t + \beta\varepsilon_{t-1})(\varepsilon_{t-s} + \beta\varepsilon_{t-s-1})] = 0 \qquad \text{for all } s > 1$$

Hence, by dividing each γ_s by γ_0, it can be immediately seen that the ACF is simply $\rho_0 = 1$, $\rho_1 = \beta/(1 + \beta^2)$, and $\rho_s = 0$ for all $s > 1$. The third graph on the left-hand side of Figure 2.2 shows the ACF for the MA(1) process $y_t = \varepsilon_t - 0.7\varepsilon_{t-1}$. As an exercise, you should demonstrate that the ACF for an MA(2) process has two spikes and then cuts to zero.

The Autocorrelation Function of an ARMA(1, 1) Process

Finally, let $y_t = a_1 y_{t-1} + \varepsilon_t + \beta_1 \varepsilon_{t-1}$. Using the now-familiar procedure, we find the Yule–Walker equations

$$Ey_t y_t = a_1 Ey_{t-1} y_t + E\varepsilon_t y_t + \beta_1 E\varepsilon_{t-1} y_t \Rightarrow \gamma_0 = a_1\gamma_1 + \sigma^2 + \beta_1(a_1 + \beta_1)\sigma^2 \quad (2.30)$$

$$Ey_t y_{t-1} = a_1 Ey_{t-1} y_{t-1} + E\varepsilon_t y_{t-1} + \beta_1 E\varepsilon_{t-1} y_{t-1} \Rightarrow \gamma_1 = a_1\gamma_0 + \beta_1\sigma^2 \quad (2.31)$$

$$Ey_t y_{t-2} = a_1 Ey_{t-1} y_{t-2} + E\varepsilon_t y_{t-2} + \beta_1 E\varepsilon_{t-1} y_{t-2} \Rightarrow \gamma_2 = a_1\gamma_1 \quad (2.32)$$

$$Ey_t y_{t-s} = a_1 Ey_{t-1} y_{t-s} + E\varepsilon_t y_{t-s} + \beta_1 E\varepsilon_{t-1} y_{t-s} \Rightarrow \gamma_s = a_1\gamma_{s-1} \quad (2.33)$$

Solving (2.30) and (2.31) simultaneously for γ_0 and γ_1 yields

$$\gamma_0 = \frac{1 + \beta_1^2 + 2a_1\beta_1}{(1 - a_1^2)}\sigma^2$$

$$\gamma_1 = \frac{(1 + a_1\beta_1)(a_1 + \beta_1)}{(1 - a_1^2)}\sigma^2$$

Hence,

$$\rho_1 = \frac{(1 + a_1\beta_1)(a_1 + \beta_1)}{(1 + \beta_1^2 + 2a_1\beta_1)} \quad (2.34)$$

and $\rho_s = a_1\rho_{s-1}$ for all $s \geq 2$.

Thus, the ACF for an ARMA(1, 1) process is such that the magnitude of ρ_1 depends on both a_1 and β_1. Beginning with this value of ρ_1, the ACF of an ARMA(1, 1) process looks like that of the AR(1) process. If $0 < a_1 < 1$, convergence will be direct, and if $-1 < a_1 < 0$, the autocorrelations will oscillate. The ACF for the function $y_t = -0.7y_{t-1} + \varepsilon_t - 0.7\varepsilon_{t-1}$ is shown as the last graph on the left-hand side of Figure 2.2. The top portion of Worksheet 2.2 derives these autocorrelations.

We leave you with the exercise of deriving the correlogram of the ARMA(2, 1) process used in Worksheet 2.1. You should be able to recognize the point at which the correlogram can reveal the pattern of the autoregressive coefficients. For an ARMA (p, q) model beginning after lag q, the values of the ρ_i will satisfy

$$\rho_i = a_1\rho_{i-1} + a_2\rho_{i-2} + \ldots + a_p\rho_{i-p}$$

The previous p values can be treated as initial conditions that satisfy the Yule–Walker equations. For these lags, the shape of the ACF is determined by the characteristic equation.

6. THE PARTIAL AUTOCORRELATION FUNCTION

In an AR(1) process, y_t and y_{t-2} are correlated even though y_{t-2} does not directly appear in the model. The correlation between y_t and y_{t-2} (i.e., ρ_2) is equal to the correlation between y_t and y_{t-1} (i.e., ρ_1) multiplied by the correlation between y_{t-1} and y_{t-2} (i.e., ρ_1 again) so that $\rho_2 = (\rho_1)^2$. It is important to note that all such *indirect* correlations are present in the ACF of any autoregressive process. In contrast, the

WORKSHEET 2.2

CALCULATION OF THE PARTIAL AUTOCORRELATIONS OF $y_t = -0.7y_{t-1} + \varepsilon_t - 0.7\varepsilon_{t-1}$

Step 1: Calculate the autocorrelations. Use (2.34) to calculate ρ_1 as

$$\rho_1 = \frac{(1 + 0.49)(-0.7 - 0.7)}{1 + 0.49 + 2(0.49)} = -0.8445$$

The remaining autocorrelations decay at the rate $\rho_i = -0.7\rho_{i-1}$ so that

$\rho_2 = 0.591, \rho_3 = -0.414, \rho_4 = 0.290, \rho_5 = -0.203, \rho_6 = 0.142, \rho_7 = -0.099, \rho_8 = 0.070$

Step 2: Calculate the first two partial autocorrelations using (2.35) and (2.36). Hence,

$$\phi_{11} = \rho_1 = -0.8445$$
$$\phi_{22} = [0.591 - (-0.8445)^2]/[1 - (-0.8445)^2] = -0.426$$

Step 3: Calculate all remaining ϕ_{ss} iteratively using (2.37). To find ϕ_{33}, note that $\phi_{21} = \phi_{11} - \phi_{22}\phi_{11} = -1.204$ and form

$$\phi_{33} = \left(\rho_3 - \sum_{j=1}^{2} \phi_{2j}\rho_{3-j}\right)\left(1 - \sum_{j=1}^{2} \phi_{2j}\rho_j\right)^{-1}$$
$$= [-0.414 - (-1.204)(0.591) - (-0.426)(-0.8445)]/[1 - (-1.204)(-0.8445) - (-0.426)(0.591)]$$
$$= -0.262$$

Similarly, to find ϕ_{44} use

$$\phi_{44} = \left(\rho_4 - \sum_{j=1}^{3} \phi_{3j}\rho_{4-j}\right)\left(1 - \sum_{j=1}^{3} \phi_{3j}\rho_j\right)^{-1}$$

Since $\phi_{3j} = \phi_{2j} - \phi_{33}\phi_{2,2-j}$, it follows that $\phi_{31} = -1.315$ and $\phi_{32} = -0.74$. Hence,

$$\phi_{44} = -0.173$$

If we continue in this fashion, it is possible to demonstrate that $\phi_{55} = -0.117$, $\phi_{66} = -0.081$, $\phi_{77} = -0.056$, and $\phi_{88} = -0.039$.

partial autocorrelation between y_t and y_{t-s} eliminates the effects of the intervening values y_{t-1} through y_{t-s+1}. As such, in an AR(1) process the partial autocorrelation between y_t and y_{t-2} is equal to zero. The most direct way to find the partial autocorrelation function is to first form the series $\{y_t^*\}$ by subtracting the mean of the series (i.e., μ) from each observation to obtain $y_t^* \equiv y_t - \mu$. Next, form the first-order autoregression

$$y_t^* = \phi_{11}y_{t-1}^* + e_t$$

where e_t is an error term.

Here the symbol $\{e_t\}$ is used since this error process may not be white noise.

Since there are no intervening values, ϕ_{11} is both the autocorrelation and the partial autocorrelation between y_t and y_{t-1}. Now form the second-order autoregression equation

$$y_t^* = \phi_{21}y_{t-1}^* + \phi_{22}y_{t-2}^* + e_t$$

Here ϕ_{22} is the partial autocorrelation coefficient between y_t and y_{t-2}. In other words, ϕ_{22} is the correlation between y_t and y_{t-2} controlling for (i.e., "netting out") the effect of y_{t-1}. Repeating this process for all additional lags s yields the partial autocorrelation function (PACF). In practice, with sample size T, only $T/4$ lags are used in obtaining the sample PACF.

Since most statistical computer packages perform these transformations, there is little need to elaborate on the computational procedure. However, it should be pointed out that a simple computational method relying on the Yule–Walker equations is available. One can form the partial autocorrelations from the autocorrelations as

$$\phi_{11} = \rho_1 \tag{2.35}$$

$$\phi_{22} = (\rho_2 - \rho_1^2)/(1 - \rho_1^2) \tag{2.36}$$

and for additional lags,

$$\phi_{ss} = \frac{\rho_s - \sum_{j=1}^{s-1} \phi_{s-1,j}\rho_{s-j}}{1 - \sum_{j=1}^{s-1} \phi_{s-1,j}\rho_j}, \quad s = 3, 4, 5, \ldots \tag{2.37}$$

where $\phi_{sj} = \phi_{s-1,j} - \phi_{ss}\phi_{s-1,s-j}, j = 1, 2, 3, \ldots, s - 1$.

For an AR(p) process, there is no direct correlation between y_t and y_{t-s} for $s > p$. Hence, for $s > p$, all values of ϕ_{ss} will be zero and the PACF for a pure AR(p) process should cut to zero for all lags greater than p. This is a useful feature of the PACF that can aid in the identification of an AR(p) model. In contrast, consider the PACF for the MA(1) process: $y_t = \varepsilon_t + \beta\varepsilon_{t-1}$. As long as $\beta \neq -1$, we can write $y_t/(1 + \beta L) = \varepsilon_t$, which we know has the infinite-order autoregressive representation

$$y_t - \beta y_{t-1} + \beta^2 y_{t-2} - \beta^3 y_{t-3} + \ldots = \varepsilon_t$$

As such, the PACF will *not* jump to zero since y_t will be correlated with all of its own lags. Instead, the PACF coefficients exhibit a geometrically decaying pattern. If $\beta < 0$, decay is direct, and if $\beta > 0$, the PACF coefficients oscillate.

Worksheet 2.2 illustrates the procedure used in constructing the PACF for the ARMA(1, 1) model shown in the fifth panel on the right-hand side of Figure 2.2:

$$y_t = -0.7y_{t-1} + \varepsilon_t - 0.7\varepsilon_{t-1}$$

First, calculate the autocorrelations. Clearly, $\rho_0 = 1$; use equation (2.34) to calculate as $\rho_1 = -0.8445$. Thereafter, the ACF coefficients decay at the rate $\rho_i = (-0.7)\rho_{i-1}$ for $i \geq 2$. Using (2.35) and (2.36), we obtain $\phi_{11} = -0.8445$ and $\phi_{22} = -0.4250$. All subsequent ϕ_{ss} and ϕ_{sj} can be calculated from (2.37) as in Worksheet 2.2.

More generally, the PACF of a stationary ARMA(p, q) process must ultimately decay toward zero beginning at lag p. The decay pattern depends on the coefficients

Table 2.1 Properties of the ACF and PACF

Process	ACF	PACF
White noise	All $\rho_s = 0$ ($s \neq 0$)	All $\phi_{ss} = 0$
AR(1): $a_1 > 0$	Direct geometric decay: $\rho_s = a_1^s$	$\phi_{11} = \rho_1$; $\phi_{ss} = 0$ for $s \geq 2$
AR(1): $a_1 < 0$	Oscillating decay: $\rho_s = a_1^s$	$\phi_{11} = \rho_1$; $\phi_{ss} = 0$ for $s \geq 2$
AR(p)	Decays toward zero. Coefficients may oscillate.	Spikes through lag p. All $\phi_{ss} = 0$ for $s > p$.
MA(1): $\beta > 0$	Positive spike at lag 1. $\rho_s = 0$ for $s \geq 2$	Oscillating decay: $\phi_{11} > 0$.
MA(1): $\beta < 0$	Negative spike at lag 1. $\rho_s = 0$ for $s \geq 2$	Geometric decay: $\phi_{11} < 0$.
ARMA(1, 1) $a_1 > 0$	Geometric decay beginning after lag 1. Sign ρ_1 = sign($a_1 + \beta$)	Oscillating decay after lag 1. $\phi_{11} = \rho_1$
ARMA(1, 1) $a_1 < 0$	Oscillating decay beginning after lag 1. Sign ρ_1 = sign($a_1 + \beta$)	Geometric decay beginning after lag 1. $\phi_{11} = \rho_1$ and sign(ϕ_{ss}) = sign(ϕ_{11}).
ARMA(p, q)	Decay (either direct or oscillatory) beginning after lag q.	Decay (either direct or oscillatory) beginning after lag p.

of the polynomial $(1 + \beta_1 L + \beta_2 L^2 + \ldots + \beta_q L^q)$. Table 2.1 summarizes some of the properties of the ACF and PACF for various ARMA processes. Also, the graphs on the right-hand side of Figure 2.2 show the partial autocorrelation functions of the five indicated processes.

For stationary processes, the key points to note are the following:

1. The ACF of an ARMA(p, q) process will begin to decay after lag q. After lag q, the coefficients of the ACF (i.e., the ρ_i) will satisfy the difference equation ($\rho_i = a_1\rho_{i-1} + a_2\rho_{i-2} + \ldots + a_p\rho_{i-p}$). Since the characteristic roots are inside the unit circle, the autocorrelations will decay after lag q. Moreover, the pattern of the autocorrelation coefficients will mimic that suggested by the characteristic roots.
2. The PACF of an ARMA(p, q) process will begin to decay after lag p. After lag p, the coefficients of the PACF (i.e., the ϕ_{ss}) will mimic the ACF coefficients from the model $y_t/(1 + \beta_1 L + \beta_2 L^2 + \ldots + \beta_q L^q)$.

We can illustrate the usefulness of the ACF and PACF functions using the model $y_t = a_0 + 0.7y_{t-1} + \varepsilon_t$. If we compare the top two graphs in Figure 2.2, the ACF shows the monotonic decay of the autocorrelations while the PACF exhibits the single spike at lag 1. Suppose that a researcher collected sample data and plotted the ACF and PACF functions. If the actual patterns compared favorably to the theoretical patterns, the researcher might try to estimate data using an AR(1) model. Correspondingly, if the ACF exhibited a single spike and the PACF exhibited monotonic decay (see the third graph for the model $y_t = \varepsilon_t - 0.7\varepsilon_{t-1}$), the researcher might try an MA(1) model.

7. SAMPLE AUTOCORRELATIONS OF STATIONARY SERIES

In practice, the theoretical mean, variance, and autocorrelations of a series are unknown to the researcher. Given that a series is stationary, we can use the sample mean, variance, and autocorrelations to estimate the parameters of the actual data-generating process. Let there be T observations labeled y_1 through y_T. We can let $\bar{y}, \hat{\sigma}^2$ and r_s be estimates of μ, σ^2, and ρ_s, respectively, where[1]

$$\bar{y} = (1/T)\sum_{t=1}^{T} y_t \tag{2.38}$$

$$\hat{\sigma}^2 = (1/T)\sum_{t=1}^{T} (y_t - \bar{y})^2 \tag{2.39}$$

and for each value of $s = 1, 2, \ldots,$

$$r_s = \frac{\sum_{t=s+1}^{T} (y_t - \bar{y})(y_{t-s} - \bar{y})}{\sum_{t=1}^{T} (y_t - \bar{y})^2} \tag{2.40}$$

The sample autocorrelation function [i.e., the ACF derived from (2.40)] and the sample PACF can be compared to various theoretical functions to help identify the actual nature of the data-generating process. Box and Jenkins (1976) discuss the distribution of the sample values of r_s under the null that y_t is stationary with normally distributed errors. Allowing var(r_s) to denote the sampling variance of r_s, they obtain

$$\begin{aligned} \text{var}(r_s) &= T^{-1} && \text{for } s = 1 \\ &= T^{-1}\left(1 + 2\sum_{j=1}^{s-1} r_j^2\right) && \text{for } s > 1 \end{aligned} \tag{2.41}$$

if the true value of $r_s = 0$ [i.e., if the true data-generating process is an MA($s - 1$) process]. Moreover, in large samples (i.e., for large values of T), r_s will be normally distributed with a mean equal to zero. For the PACF coefficients, under the null hypothesis of an AR(p) model (i.e, under the null that all $\phi_{p+i,p+i}$ are zero), the variance of the $\hat{\phi}_{p+i,p+i}$ is approximately $1/T$.

In practice, we can use these sample values to form the sample autocorrelation and partial autocorrelation functions and test for significance using (2.41). For example, if we use a 95 percent confidence interval (i.e., two standard deviations), and the calculated value of r_1 exceeds $2T^{-1/2}$, it is possible to reject the null hypothesis that the first-order autocorrelation is not statistically different from zero. Rejecting this hypothesis means rejecting an MA($s - 1$) = MA(0) process and accepting the alternative $q > 0$. Next, try $s = 2$; var(r_2) is $(1 + 2r_1^2)/T$. If r_1 is 0.5 and T is 100, the variance of r_2 is 0.015 and the standard deviation is about 0.123. Thus, if the calculated value of r_2 exceeds 2(0.123), it is possible to reject the hypothesis $r_2 = 0$. Here, rejecting the null means accepting the alternative that $q > 1$. Repeating for the various values of s is helpful in identifying the order to the process. The maximum number of sample autocorrelations and partial autocorrelations to use is typically set equal to $T/4$.

Within any large group of autocorrelations, some will exceed two standard deviations as a result of pure chance even though the true values in the data-generating process are zero. The Q-statistic can be used to test whether a group of autocorrelations is significantly different from zero. Box and Pierce (1970) used the sample autocorrelations to form the statistic

$$Q = T\sum_{k=1}^{s} r_k^2$$

Under the null hypothesis that all values of $r_k = 0$, Q is asymptotically χ^2 distributed with s degrees of freedom. The intuition behind the use of the statistic is that high sample autocorrelations lead to large values of Q. Certainly, a white-noise process (in which all autocorrelations should be zero) would have a Q value of zero. If the calculated value of Q exceeds the appropriate value in a χ^2 table, we can reject the null of no significant autocorrelations. Note that rejecting the null means accepting an alternative that at least one autocorrelation is not zero.

A problem with the Box–Pierce Q-statistic is that it works poorly even in moderately large samples. Ljung and Box (1978) reported superior small-sample performance for the modified Q-statistic calculated as

$$Q = T(T+2)\sum_{k=1}^{s} r_k^2/(T-k) \tag{2.42}$$

If the sample value of Q calculated from (2.42) exceeds the critical value of χ^2 with s degrees of freedom, then *at least* one value of r_k is statistically different from zero at the specified significance level. The Box–Pierce and Ljung–Box Q-statistics also serve as a check to see if the *residuals* from an estimated ARMA(p, q) model behave as a white-noise process. However, when the s correlations from an estimated ARMA(p, q) model are formed, the degrees of freedom are reduced by the number of estimated coefficients. Hence, using the residuals of an ARMA(p, q) model, Q has a χ^2 with $s - p - q$ degrees of freedom (if a constant is included, the degrees of freedom are $s - p - q - 1$). Nevertheless, a common practice is to simply report results using s degrees of freedom.

Model Selection Criteria

One natural question to ask of any estimated model is: How well does it fit the data? Adding additional lags for p and/or q will necessarily reduce the sum of squares of the estimated residuals. However, adding such lags entails the estimation of additional coefficients and an associated loss of degrees of freedom. Moreover, the inclusion of extraneous coefficients will reduce the forecasting performance of the fitted model.

As discussed in some detail in Appendix 2.2 of this chapter, there exist various model selection criteria that trade off a reduction in the sum of squares of the residuals for a more **parsimonious** model. The two most commonly used model selection criteria are the Akaike Information Criterion (AIC) and the Schwartz Bayesian Criterion (SBC). In the text, we will use the following formulas:

$$\text{AIC} = T\ln(\text{sum of squared residuals}) + 2n$$
$$\text{SBC} = T\ln(\text{sum of squared residuals}) + n\ln(T)$$

where: n = number of parameters estimated ($p + q$ + possible constant term)
T = number of usable observations

When you estimate a model using lagged variables, some observations are lost. To adequately compare the alternative models, T should be kept fixed. Otherwise, you will be comparing the performance of the models over different sample periods. Moreover, decreasing T has the direct effect of reducing the AIC and the SBC; the goal is not to select a model because it has the smallest number of usable observations. For example, with 100 data points, estimate an AR(1) and an AR(2) using only the last 98 observations in each estimation. Compare the two models using $T = 98$.

Ideally, the AIC and SBC will be as small as possible (note that both can be negative). As the fit of the model improves, the AIC and SBC will approach $-\infty$. We can use these criteria to aid in selecting the most appropriate model; model A is said to fit better than model B if the AIC (or SBC) for A is smaller than for model B. In using the criteria to compare alternative models, we must estimate them over the same sample period so that they will be comparable. For each, increasing the number of regressors increases n but should have the effect of reducing the sum of squared residuals. Thus, if a regressor has no explanatory power, adding it to the model will cause both the AIC and the SBC to increase. Since $\ln(T)$ will be greater than 2, the SBC will always select a more parsimonious model than will the AIC; the marginal cost of adding regressors is greater with the SBC than with the AIC.

An especially useful feature of the model selection criteria is for comparing nonnested models. For example, suppose you want to compare an AR(2) model to an MA(3) model. Neither is a restricted form of the other. You would not want to estimate an ARMA(2, 3) model and perform F-tests to determine whether $a_1 = a_2 = 0$ or whether $\beta_1 = \beta_2 = \beta_3 = 0$. As discussed in Appendix 2.1, the estimation of ARMA models necessitates computer-based solution methods. If the AR(2) and MA(3) models are each reasonable, the nonlinear search algorithms required to estimate an ARMA(2, 3) model are not likely to converge to a solution. Moreover, the values of y_{t-1} and y_{t-2} are clearly correlated with the values of ε_{t-1}, ε_{t-2}, and ε_{t-3}. It is quite possible that both hypotheses could be accepted (or rejected). However, it is straightforward to compare the estimated AR(2) and MA(3) models using the AIC or the SBC.

Of the two criteria, the SBC has superior large-sample properties. Let the true order of the data-generating process be (p^*, q^*) and suppose that we use the AIC and SBC to estimate all ARMA models of order (p, q) where $p \geq p^*$ and $q \geq q^*$. Both the AIC and the SBC will select models of orders greater than or equal to (p^*, q^*) as the sample size approaches infinity. However, the SBC is asymptotically consistent while the AIC is biased toward selecting an overparameterized model. However, in small samples, the AIC can work better than the SBC. You can be quite confident in your results if both the AIC and the SBC select the same model. If they select different models, you need to proceed cautiously. Since SBC selects the more parsimonious model, you should check to determine if the residuals appear to be white noise. Since the AIC can select an overparameterized model, the t-statistics of all coefficients should be significant at conventional levels. A number of other diagnostic checks that can be used to compare alternative models are presented in Sections 8 and 9. Nevertheless, it is wise to retain a healthy skepticism of your estimated models. Section 10 illustrates the situation in which it is not possible to find one model that clearly dominates all others. There is nothing wrong with reporting the results and the forecasts using alternative estimations.

Before proceeding, be aware that a number of different ways are used to report the AIC and the SBC. For example, the software packages EViews and SAS report values for the AIC and SBC using

$$\text{AIC}^* = -2\ln(L)/T + 2n/T$$
$$\text{SBC}^* = -2\ln(L)/T + n\ln(T)/T$$

where n and T are as defined above, and L = maximized value of the log of the likelihood function.

For a normal distribution, $-2\ln(L) = T\ln(2\pi) + T\ln(\sigma^2) + (1/\sigma^2)$ (sum of squared residuals). The reason for the plethora of reporting methods is that many software packages (such as OX, RATS, and GAUSS) do not display any model selection criteria, so users must calculate these values by themselves. Programmers quickly find that coding all of the parameters contained in the formulas is unnecessary, and they simply report the shortened versions. In point of fact, it does not matter which method you use. If you work through Question 13 at the end of this chapter, it should be clear that the model with the smallest value for AIC will always have the smallest AIC*. Specifically, Question 13 asks you to write down the formula for $\ln(L)$ and show that the equation for AIC* is a monotonic transformation of that for AIC. Hence, whether you use the formula for AIC or AIC*, you will always be selecting the same model as the one selected in the text. The identical relationship holds between SBC* and SBC; the model yielding the smallest value for SBC will always have the smallest value for SBC*.

Estimation of an AR(1) Model

Let us use a specific example to see how the sample autocorrelation function and partial autocorrelation function can be used as an aid in identifying an ARMA model. A computer program was used to draw 100 normally distributed random numbers with a theoretical variance equal to unity. Call these random variates ε_t, where t runs from 1 to 100. Beginning with $t = 1$, values of y_t were generated using the formula $y_t = 0.7y_{t-1} + \varepsilon_t$ and the initial condition $y_0 = 0$. Note that the problem of nonstationarity is avoided since the initial condition is consistent with long-run equilibrium. Panel (a) of Figure 2.3 shows the sample correlogram and Panel (b) shows the sample PACF. You

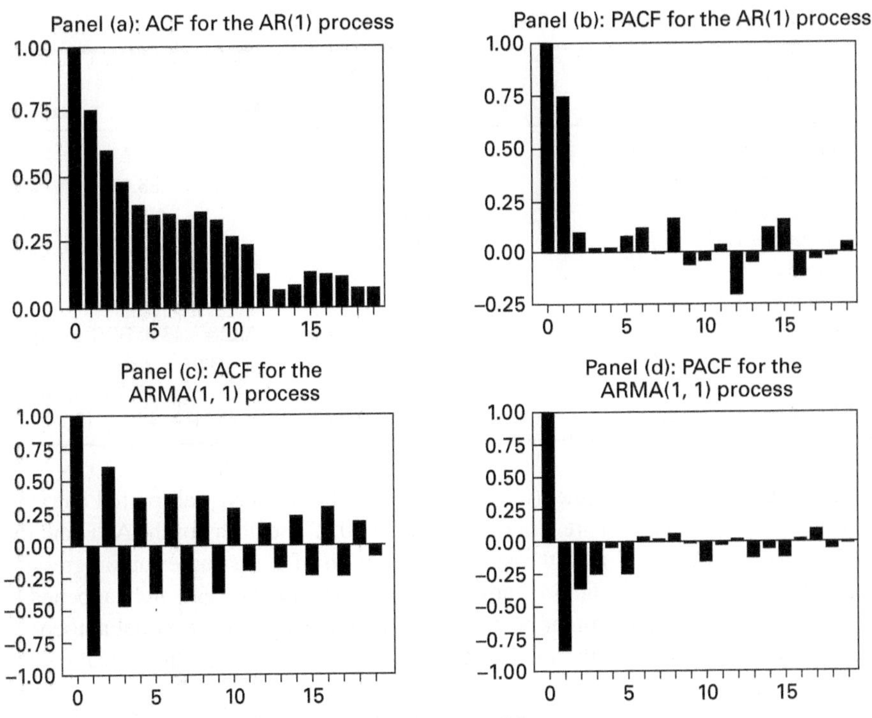

FIGURE 2.3 ACF and PACF for Two Simulated Processes

should take a minute to compare the ACF and PACF to those of the theoretical processes shown in Figure 2.2.

In practice, we never know the true data-generating process. As an exercise, suppose we were presented with these 100 sample values and were asked to uncover the true process using the Box–Jenkins methodology. The first step might be to compare the sample ACF and PACF to those of the various theoretical models. The decaying pattern of the ACF and the single large spike at lag 1 in the sample PACF suggest an AR(1) model. The first three autocorrelations are $r_1 = 0.74$, $r_2 = 0.58$, and $r_3 = 0.47$ [which are somewhat greater than the theoretical values of 0.7, 0.49 ($0.7^2 = 0.49$), and 0.343]. In the PACF, there is a sizable spike of 0.74 at lag 1, and all other partial autocorrelations (except for lag 12) are very small.

Under the null hypothesis of an MA(0) process, the standard deviation of r_1 is $T^{-1/2} = 0.1$. Since the sample value of $r_1 = 0.74$ is more than seven standard deviations from zero, we can reject the null hypothesis that r_1 equals 0. The standard deviation of r_2 is obtained by applying (2.41) to the sampling data, where $s = 2$:

$$\text{var}(r_2) = (1 + 2(0.74)^2)/100 = 0.021$$

Since $(0.021)^{1/2} = 0.1449$, the sample value of r_2 is more than three standard deviations from zero; at conventional significance levels, we can reject the null hypothesis that r_2 equals zero. We can similarly test the significance of the other values of the autocorrelations.

Table 2.2 Estimates of an AR(1) Model

	Model 1 $y_t = a_1 y_{t-1} + \varepsilon_t$	**Model 2** $y_t = a_1 y_{t-1} + \varepsilon_t + \beta_{12}\varepsilon_{t-12}$
Degrees of freedom	98	97
Sum of squared residuals	85.10	85.07
Estimated a_1 (standard error)	0.7904 (0.0624)	0.7938 (0.0643)
Estimated β (standard error)		−0.325 (0.1141)
AIC; SBC	AIC = 441.9; SBC = 444.5	AIC = 443.9; SBC = 449.1
Ljung–Box Q-statistics for the residuals (significance level in parentheses)	$Q(8)$ = 6.43 (0.490) $Q(16)$ = 15.86 (0.391) $Q(24)$ = 21.74 (0.536)	$Q(8)$ = 6.48 (0.485) $Q(16)$ = 15.75 (0.400) $Q(24)$ = 21.56 (0.547)

As you can see in the Panel (b) of the figure, other than ϕ_{11}, all partial autocorrelations (except for lag 12) are less than $2T^{-1/2} = 0.2$. The decay of the ACF and the single spike of the PACF give the strong impression of a first-order autoregressive model. Nevertheless, if we did not know the true underlying process, and happened to be using monthly data, we might be concerned with the significant partial autocorrelation at lag 12. After all, with monthly data we might expect some direct relationship between y_t and y_{t-12}.

Although we know that the data were actually generated from an AR(1) process, it is illuminating to compare the estimates of two different models. Suppose we estimate an AR(1) model and also try to capture the spike at lag 12 with an MA coefficient. Thus, we can consider the two tentative models:

$$\text{Model 1: } y_t = a_1 y_{t-1} + \varepsilon_t$$
$$\text{Model 2: } y_t = a_1 y_{t-1} + \varepsilon_t + \beta_{12}\varepsilon_{t-12}$$

Table 2.2 reports the results of the two estimations. The coefficient of Model 1 satisfies the stability condition $|a_1| < 1$ and has a low standard error (the associated t-statistic for a null of zero is more than 12). As a useful diagnostic check, we plot the correlogram of the **residuals** of the fitted model in Figure 2.4. The Q-statistics for these residuals indicate that each one of the autocorrelations is less than two standard deviations from zero. The Ljung–Box Q-statistics of these residuals indicate that *as a group*, lags 1 through 8, 1 through 16, and 1 through 24 are not significantly different from zero. This is strong evidence that the AR(1) model "fits" the data well. After all, if residual autocorrelations were significant, the AR(1) model would not utilize all available information concerning movements in the $\{y_t\}$ sequence. For example, suppose we wanted to forecast y_{t+1} conditioned on all available information up to and including period t. With Model 1, the value of y_{t+1} is $y_{t+1} = a_1 y_t + \varepsilon_{t+1}$. Hence, the forecast from Model 1 is $a_1 y_t$. If the residual autocorrelations had been significant, this forecast would not capture all of the available information set.

In examining the results for Model 2, note that both models yield similar estimates for the first-order autoregressive coefficient and the associated standard error.

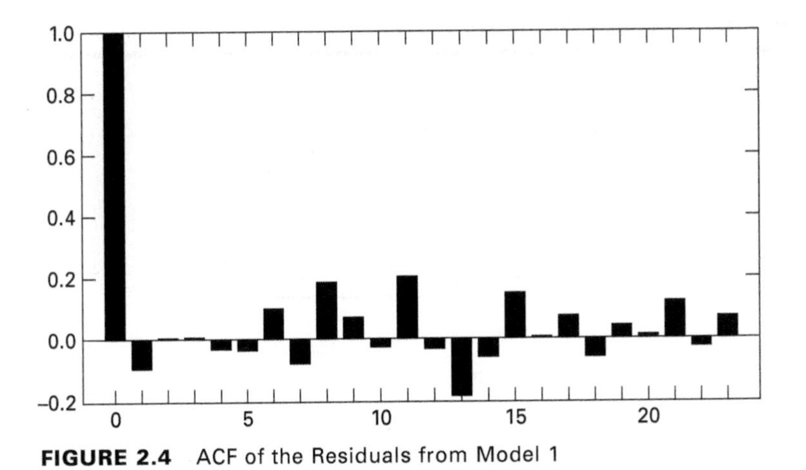

FIGURE 2.4 ACF of the Residuals from Model 1

However, the estimate for β_{12} is of poor quality; the insignificant t value suggests that it should be dropped from the model. Moreover, comparing the AIC and the SBC values of the two models suggests that any benefit of a reduced sum of squared residuals is overwhelmed by the detrimental effects of estimating an additional parameter. All of these indicators point to the choice of Model 1.

Exercise 8 at the end of this chapter entails various estimations using this series. The series is denoted by Y1 in the file SIM2.XLS. In this exercise you are asked to show that the AR(1) model performs better than some alternative specifications. It is important that you complete this exercise.

Estimation of an ARMA(1, 1) Model

A second $\{y_t\}$ sequence in the file SIM2.XLS was constructed to illustrate the estimation of an ARMA(1, 1). Given 100 normally distributed values of $\{\varepsilon_t\}$, 100 values of $\{y_t\}$ were generated using

$$y_t = -0.7y_{t-1} + \varepsilon_t - 0.7\varepsilon_{t-1}$$

where y_0 and ε_0 were both set to equal zero.

Both the sample ACF and the PACF from the simulated data (see the second set of graphs in Figure 2.3) are roughly equivalent to those of the theoretical model shown in Figure 2.2. However, if the true data-generating process were unknown, the researcher might be concerned about certain discrepancies. An AR(2) model could yield a sample ACF and PACF similar to those in the figure. Table 2.3 reports the results of estimating the data using the following three models:

$$\text{Model 1: } y_t = a_1y_{t-1} + \varepsilon_t$$

$$\text{Model 2: } y_t = a_1y_{t-1} + \varepsilon_t + \beta_1\varepsilon_{t-1}$$

$$\text{Model 3: } y_t = a_1y_{t-1} + a_2y_{t-2} + \varepsilon_t$$

Table 2.3 Estimates of an ARMA(1, 1) Model

	Estimates[1]	Q-Statistics[2]	AIC/SBC[3]
Model 1	a_1: −0.835 (.053)	$Q(8)$ = 26.19 (.000)	AIC = 496.5
		$Q(24)$ = 41.10 (.001)	SBC = 499.0
Model 2	a_1: −0.679 (.076)	$Q(8)$ = 3.86 (.695)	AIC = 471.0
	β_1: −0.676 (.081)	$Q(24)$ = 14.23 (.892)	SBC = 476.2
Model 3	a_1: −1.16 (.093)	$Q(8)$ = 11.44 (.057)	AIC = 482.8
	a_2: −0.378 (.092)	$Q(24)$ = 22.59 (.424)	SBC = 487.9

Notes:

[1] Standard errors in parentheses.

[2] Ljung–Box Q-statistics of the residuals from the fitted model. The significance levels are in parentheses.

[3] For comparability, the AIC and SBC values are reported for estimations that used only observations 3 through 100. If the AR(1) is estimated using 99 observations, the AIC and SBC are 502.3 and 504.9, respectively. If the ARMA(1, 1) is estimated using 99 observations, the AIC and SBC are 476.6 and 481.1, respectively.

In examining Table 2.3, notice that all of the estimated values of a_1 are highly significant; each of the estimated values is *at least* eight standard deviations from zero. It is clear that the AR(1) model is inappropriate. The Q-statistics for Model 1 indicate that there is significant autocorrelation in the residuals. The estimated ARMA(1, 1) model does not suffer from this problem. Moreover, both the AIC and the SBC select Model 2 over Model 1.

The same type of reasoning indicates that Model 2 is preferred to Model 3. Note that for each model, the estimated coefficients are highly significant, and the point estimates imply convergence. Although the Q-statistic at 24 lags indicates that these two models do not suffer from correlated residuals, the Q-statistic at 8 lags indicates serial correlation in the residuals of Model 3. Thus, the AR(2) model does not capture short-term dynamics as well as the ARMA(1, 1) model. Also note that the AIC and SBC both select Model 2.

Estimation of an AR(2) Model

A third data series was simulated as

$$y_t = 0.7y_{t-1} - 0.49y_{t-2} + \varepsilon_t$$

The estimated ACF and the PACF of the series are:

Lags	Autocorrelations									
1–10	0.47	−0.16	−0.32	−0.11	−0.05	−0.16	−0.10	0.13	0.18	0.03
11–20	−0.09	−0.11	−0.16	−0.06	0.12	0.25	0.05	−0.17	−0.15	0.01
	Partial Autocorrelations									
1–10	0.47	−0.48	0.02	0.05	−0.25	−0.12	0.10	0.04	−0.08	0.02
11–20	−0.02	−0.14	−0.17	0.21	0.01	0.09	−0.22	0.01	−0.02	−0.03

Note the large autocorrelation at lag 16 and the large partial autocorrelations at lags 14 and 17. Given the way the process was simulated, the presence of these autocorrelations is due to nothing more than chance. However, an econometrician unaware of the actual data-generating process might be concerned about these autocorrelations. The estimated AR(2) model (with t-statistics in parentheses) is

$$y_t = \underset{(7.73)}{0.692}y_{t-1} - \underset{(-5.37)}{0.481}y_{t-2} \qquad \text{AIC} = 219.87, \text{SBC} = 225.04$$

Overall, the model appears to be adequate. However, the two AR(2) coefficients are unable to capture the correlations at very long lags. For example, the partial autocorrelations of the *residuals* for lags 14 and 17 are both greater than 0.2 in absolute value. The calculated Ljung–Box statistic for 16 lags is 24.6248 (which is significant at the 0.038 level). At this point, it might be tempting to try to model the correlation at lag 16 by including the moving average term $\beta_{16}\varepsilon_{t-16}$. Such an estimation results in:[2]

$$y_t = \underset{(7.87)}{0.717}y_{t-1} - \underset{(-5.11)}{0.465}y_{t-2} + \underset{(2.78)}{0.306}\varepsilon_{t-16} \quad \text{AIC} = 213.40, \text{SBC} = 221.16$$

All estimated coefficients are significant and the Ljung–Box Q-statistics for the residuals are all insignificant at conventional levels. In conjunction with the fact that the AIC and SBC both select this second model, the researcher unaware of the true process might be tempted to conclude that the data-generating process includes a moving average term at lag 16.

A useful model check is to split the sample into two parts. If a coefficient is present in the data-generating process, its influence should be seen in both subsamples. If the simulated series is split into two parts, the ACF and PACF using observations 50 through 100 follow:

Lags					**Autocorrelations**					
1–10	0.46	−0.21	−0.28	0.03	0.10	−0.15	−0.13	0.10	0.18	0.03
11–20	−0.01	0.01	−0.06	−0.09	0.04	0.21	0.06	−0.16	−0.18	−0.05
					Partial Autocorrelations					
1–10	0.46	−0.53	0.19	0.06	−0.20	−0.13	0.23	−0.08	0.00	0.06
11–20	0.15	−0.26	0.03	0.15	0.04	0.00	−0.05	−0.01	−0.14	−0.08

As you can see, the size of the partial autocorrelations at lags 14 and 17 is diminished. Now, estimating a pure AR(2) model over this second part of the sample yields

$$y_t = \underset{(5.92)}{0.714}y_{t-1} - \underset{(-4.47)}{0.538}y_{t-2}$$

$$Q(8) = 7.83;\ Q(16) = 15.93;\ Q(24) = 26.06$$

All estimated coefficients are significant, and the Ljung–Box Q-statistics do not indicate any significant autocorrelations in the residuals. The significance levels of $Q(8)$, $Q(16)$, and $Q(24)$ are 0.251, 0.317, and 0.249, respectively. In fact, this model does capture the actual data-generating process quite well. In this example, the large, spurious autocorrelations of the long lags can be eliminated by changing the sample period. Thus, it is hard to maintain that the correlation at lag 16 is meaningful. Most sophisticated practitioners warn against trying to fit any model to the very long lags.

As you can infer from (2.41), the variance of r_s can be sizable when s is large. Moreover, in small samples, a few "unusual" observations can create the appearance of significant autocorrelations at long lags. Since econometric estimation involves unknown data-generating processes, the more general point is that we always need to be wary of our estimated model. Fortunately, Box and Jenkins (1976) established a set of procedures that can be used to check a model's adequacy.

8. BOX–JENKINS MODEL SELECTION

The estimates of the AR(1), ARMA(1, 1), and AR(2) models in the previous section illustrate the Box–Jenkins (1976) strategy for appropriate model selection. Box and Jenkins popularized a three-stage method aimed at selecting an appropriate model for the purpose of estimating and forecasting a univariate time series. In the **identification stage,** the researcher visually examines the time plot of the series, the autocorrelation function, and the partial correlation function. Plotting the time path of the $\{y_t\}$ sequence provides useful information concerning outliers, missing values, and structural breaks in the data. Nonstationary variables may have a pronounced trend or appear to meander without a constant long-run mean or variance. Missing values and outliers can be corrected at this point. At one time, the standard practice was to first difference any series deemed to be nonstationary. Currently, a large body of literature is evolving that develops formal procedures to check for nonstationarity. We defer this discussion until Chapter 4 and assume that we are working with stationary data. A comparison of the sample ACF and PACF to those of various theoretical ARMA processes may suggest several plausible models. In the **estimation stage,** each of the tentative models is fit and the various a_i and β_i coefficients are examined. In this second stage, the goal is to select a stationary and parsimonious model that has a good fit. The third stage involves **diagnostic checking** to ensure that the residuals from the estimated model mimic a white-noise process.

Parsimony

A fundamental idea in the Box–Jenkins approach is the principle of **parsimony.** Parsimony (meaning sparseness or stinginess) should come as second nature to economists. Incorporating additional coefficients will necessarily increase fit (e.g., the value of R^2 will increase) at a cost of reducing degrees of freedom. Box and Jenkins argue that parsimonious models produce better forecasts than overparameterized models. A parsimonious model fits the data well without incorporating any needless coefficients. Certainly, forecasters do not want to project poorly estimated coefficients into the future. The aim is to approximate the true data-generating process but not to pin down the exact process. The goal of parsimony suggests eliminating the MA(12) coefficient in the simulated AR(1) model above.

In selecting an appropriate model, the econometrician needs to be aware that several different models may have similar properties. As an extreme example, note that the AR(1) model $y_t = 0.5y_{t-1} + \varepsilon_t$ has the equivalent infinite-order moving-average representation of $y_t = \varepsilon_t + 0.5\varepsilon_{t-1} + 0.25\varepsilon_{t-2} + 0.125\varepsilon_{t-3} + 0.0625\varepsilon_{t-4} + \ldots$. In most samples, approximating this MA(∞) process with an MA(2) or MA(3) model will give a very good fit. However, the AR(1) model is the more parsimonious model and is preferred. As an exercise, you should show that this AR(1) model has the equivalent representation of $y_t = 0.25y_{t-2} + 0.5\varepsilon_{t-1} + \varepsilon_t$.

Also, be aware of the **common factor problem.** Suppose we wanted to fit the ARMA(2, 3) model

$$(1 - a_1L - a_2L^2)y_t = (1 + \beta_1L + \beta_2L^2 + \beta_3L^3)\varepsilon_t \tag{2.43}$$

Also suppose that $(1 - a_1L - a_2L^2)$ and $(1 + \beta_1L + \beta_2L^2 + \beta_3L^3)$ can each be factored as $(1 + cL)(1 + aL)$ and $(1 + cL)(1 + b_1L + b_2L^2)$, respectively. Since $(1 + cL)$ is a common factor to each, (2.43) has the equivalent, but more parsimonious, form:[3]

$$(1 + aL)y_t = (1 + b_1L + b_2L^2)\varepsilon_t \tag{2.44}$$

If you correctly did the exercise suggested above, you know that $(1 - 0.25L^2)y_t = (1 + 0.5L)\varepsilon_t$ is equivalent to $(1 + 0.5L)(1 - 0.5L)y_t = (1 + 0.5L)\varepsilon_t$, so that $y_t = 0.5y_{t-1} + \varepsilon_t$. In practice, the polynomials will not factor exactly. However, if the factors are similar, you should try a more parsimonious form.

In order to ensure that the model is parsimonious, the various a_i and β_i should all have t-statistics of 2.0 or greater (so that each coefficient is significantly different from zero at the 5 percent level). Moreover, the coefficients should not be strongly correlated with each other. Highly collinear coefficients are unstable; usually one or more can be eliminated from the model without reducing forecast performance.

Stationarity and Invertibility

The distribution theory underlying the use of the sample ACF and PACF as approximations to those of the true data-generating process assumes that the $\{y_t\}$ sequence is stationary. Moreover, t-statistics and Q-statistics also presume that the data are stationary. The estimated autoregressive coefficients should be consistent with this underlying assumption. Hence, we should be suspicious of an AR(1) model if the estimated value of a_1 is close to unity. For an ARMA(2, q) model, the characteristic roots of the estimated polynomial $(1 - a_1L - a_2L^2)$ should lie outside the unit circle.

As discussed in greater detail in Appendix 2.1, the Box–Jenkins approach also necessitates that the model be **invertible.** Formally, $\{y_t\}$ is invertible if it can be represented by a finite-order or convergent autoregressive process. Invertibility is important because the use of the ACF and PACF implicitly assumes that the $\{y_t\}$ sequence can be represented by an autoregressive model. As a demonstration, consider the simple MA(1) model:

$$y_t = \varepsilon_t - \beta_1\varepsilon_{t-1} \tag{2.45}$$

so that if $|\beta_1| < 1$,

$$y_t/(1 - \beta_1L) = \varepsilon_t$$

or

$$y_t + \beta_1y_{t-1} + \beta_1^2y_{t-2} + \beta_1^3y_{t-3} + \ldots = \varepsilon_t \tag{2.46}$$

If $|\beta_1| < 1$, (2.46) can be estimated using the Box–Jenkins method. However, if $|\beta_1| \geq 1$, the $\{y_t\}$ sequence cannot be represented by a finite-order AR process; as such, it is not invertible. More generally, for an ARMA model to have a convergent AR representation, the roots of the polynomial $(1 + \beta_1 L + \beta_2 L^2 + \ldots + \beta_q L^q)$ must lie outside the unit circle. Note that there is nothing improper about a noninvertible model. The $\{y_t\}$ sequence implied by $y_t = \varepsilon_t - \varepsilon_{t-1}$ is stationary in that it has a constant time-invariant mean $[Ey_t = Ey_{t-s} = 0]$, a constant time-invariant variance $[\text{var}(y_t) = \text{var}(y_{t-s}) = \sigma^2(1 + \beta_1^2) + 2\sigma^2]$, and the autocovariances $\gamma_1 = -\beta_1\sigma^2$ and all other $\gamma_s = 0$. The problem is that the technique does not allow for the estimation of such models. If $\beta_1 = 1$, (2.46) becomes

$$y_t + y_{t-1} + y_{t-2} + y_{t-3} + y_{t-4} + \ldots = \varepsilon_t$$

Clearly, the autocorrelations and partial autocorrelations between y_t and y_{t-s} will never decay.

Goodness of Fit

A good model will fit the data well. Obviously, R^2 and the average of the residual sum of squares are common *goodness-of-fit* measures in ordinary least squares. The problem with these measures is that the fit necessarily improves as more parameters are included in the model. Parsimony suggests using the AIC and/or SBC as more appropriate measures of the overall fit of the model. Also be cautious of estimates that fail to converge rapidly. Most software packages estimate the parameters of an ARMA model using a nonlinear search procedure. If the search fails to converge rapidly, it is possible that the estimated parameters are unstable. In such circumstances, adding an additional observation or two can greatly alter the estimates.

Post-Estimation Evaluation

The third stage of the Box–Jenkins methodology involves **diagnostic checking.** The standard practice is to plot the residuals to look for outliers and for evidence of periods in which the model does not fit the data well. One common practice is to create the standardized residuals by dividing each residual, ε_t, by its estimated standard deviation, σ. If the residuals are normally distributed, the plot of the ε_t/σ series should be such that no more than 5 percent lie outside the band from -2 to $+2$. If the standardized residuals seem to be much larger in some periods than in others, it may be evidence of structural change. If all plausible ARMA models show evidence of a poor fit during a reasonably long portion of the sample, it is wise to consider using intervention analysis, transfer function analysis, or any other of the multivariate estimation methods discussed in later chapters. If the variance of the residuals is increasing, a logarithmic transformation may be appropriate. Alternatively, you may wish to actually model any tendency of the variance to change using the ARCH techniques discussed in Chapter 3.

It is particularly important that the residuals from an estimated model be serially uncorrelated. Any evidence of serial correlation implies a systematic movement in the $\{y_t\}$ sequence that is not accounted for by the ARMA coefficients included in the model. Hence, any of the tentative models yielding nonrandom residuals should be eliminated from consideration. To check for correlation in the residuals, construct the ACF and the PACF of the *residuals* of the estimated model. You can then use (2.41) and (2.42) to determine whether any or all of the residual autocorrelations or partial autocorrelations are statistically significant.[4] Although there is no significance level that is deemed "most appropriate," be wary of any model yielding (1) several residual correlations that are marginally significant and (2) a Q-statistic that is barely significant at the 10 percent level. In such circumstances, it is usually possible to formulate a better-performing model.

Similarly, a model can be estimated over only a portion of the data set. The estimated model can then be used to forecast the known values of the series. Comparing the sum of the squared forecast errors is a useful way to evaluate the adequacy of alternative models. Those models with poor *out-of-sample* forecasts should be eliminated. Some of the details of constructing out-of-sample forecasts are discussed in the next section.

9. PROPERTIES OF FORECASTS

Perhaps the most important use of an ARMA model is to forecast future values of the $\{y_t\}$ sequence. To simplify the discussion, it is assumed that the actual data-generating process and the current and past realizations of the $\{\varepsilon_t\}$ and $\{y_t\}$ sequences are known to the researcher. First consider the forecasts from the AR(1) model $y_t = a_0 + a_1 y_{t-1} + \varepsilon_t$. Updating one period, we obtain

$$y_{t+1} = a_0 + a_1 y_t + \varepsilon_{t+1}$$

If you know the coefficients a_0 and a_1, you can forecast y_{t+1} conditioned on the information available at period t as

$$E_t y_{t+1} = a_0 + a_1 y_t \tag{2.47}$$

where $E_t y_{t+j}$ is a shorthand way to write the conditional expectation of y_{t+j} given the information available at t. Formally, $E_t y_{t+j} = E(y_{t+j} \mid y_t, y_{t-1}, y_{t-2}, \ldots, \varepsilon_t, \varepsilon_{t-1}, \ldots)$.

In the same way, since $y_{t+2} = a_0 + a_1 y_{t+1} + \varepsilon_{t+2}$, the forecast of y_{t+2} conditioned on the information available at period t is

$$E_t y_{t+2} = a_0 + a_1 E_t y_{t+1}$$

and using (2.47),

$$E_t y_{t+2} = a_0 + a_1(a_0 + a_1 y_t)$$

Thus, the forecast of y_{t+1} can be used to forecast y_{t+2}. The point is that forecasts can be constructed using forward iteration; the forecast of y_{t+j} can be used to forecast y_{t+j+1}. Since $y_{t+j+1} = a_0 + a_1 y_{t+j} + \varepsilon_{t+j+1}$, it immediately follows that

$$E_t y_{t+j+1} = a_0 + a_1 E_t y_{t+j} \tag{2.48}$$

From (2.47) and (2.48) it should be clear that it is possible to obtain the entire sequence of j-step-ahead forecasts by forward iteration. Consider:

$$E_t y_{t+j} = a_0(1 + a_1 + a_1^2 + \ldots + a_1^{j-1}) + a_1^j y_t$$

This equation, called the **forecast function,** expresses all of the the j-step-ahead forecasts as a function of the information set in period t. Unfortunately, the quality of the forecasts declines as we forecast further out into the future. Think of (2.48) as a first-order difference equation in the $\{E_t y_{t+j}\}$ sequence. Since $|a_1| < 1$, the difference equation is stable, and it is straightforward to find the particular solution to the difference equation. If we take the limit of $E_t y_{t+j}$ as $j \to \infty$, we find that $E_t y_{t+j} \to a_0/(1 - a_1)$. This result is really quite general: *For any stationary ARMA model, the conditional forecast of y_{t+j} converges to the unconditional mean as $j \to \infty$.*

Because the forecasts from an ARMA model will not be perfectly accurate, it is important to consider the properties of the forecast errors. Forecasting from time period t, we can define the j-step-ahead forecast error, called $e_t(j)$, as the difference between the realized value of y_{t+j} and the forecasted value:

$$e_t(j) \equiv y_{t+j} - E_t y_{t+j}$$

Since the one-step-ahead forecast error is equivalent to $e_t(1) = y_{t+1} - E_t y_{t+1} = \varepsilon_{t+1}$, $e_t(1)$ is precisely the "unforecastable" portion of y_{t+1} given the information available in t.

To find the two-step-ahead forecast error, we need to form $e_t(2) = y_{t+2} - E_t y_{t+2}$. Since $y_{t+2} = a_0 + a_1 y_{t+1} + \varepsilon_{t+2}$ and $E_t y_{t+2} = a_0 + a_1 E_t y_{t+1}$, it follows that

$$e_t(2) = a_1(y_{t+1} - E_t y_{t+1}) + \varepsilon_{t+2} = \varepsilon_{t+2} + a_1 \varepsilon_{t+1}$$

You should take a few moments to demonstrate that for the AR(1) model, the j-step-ahead forecast error is given by

$$e_t(j) = \varepsilon_{t+j} + a_1 \varepsilon_{t+j-1} + a_1^2 \varepsilon_{t+j-2} + a_1^3 \varepsilon_{t+j-3} + \ldots + a_1^{j-1} \varepsilon_{t+1} \quad (2.49)$$

Since the mean of (2.49) is zero, the forecasts are unbiased estimates of each value y_{t+j}. The proof is trivial. Since $E_t \varepsilon_{t+j} = E_t \varepsilon_{t+j-1} = \ldots = E_t \varepsilon_{t+1} = 0$, the conditional expectation of (2.49) is $E_t e_t(j) = 0$. Since the expected value of the forecast error is zero, the forecasts are unbiased.

Although unbiased, the forecasts from an ARMA model are necessarily inaccurate. To find the variance of the forecast error, continue to assume that the elements of the $\{\varepsilon_t\}$ sequence are independent, with a variance equal to σ^2. Hence, from (2.49), the variance of the forecast error is

$$\text{var}[e_t(j)] = \sigma^2 [1 + a_1^2 + a_1^4 + a_1^6 + \ldots + a_1^{2(j-1)}] \quad (2.50)$$

Thus, the one-step-ahead forecast error variance is σ^2, the two-step-ahead forecast error variance is $\sigma^2(1 + a_1^2)$, and so forth. The essential point to note is that the variance of the forecast error is an increasing function of j. As such, you can have more confidence in short-term forecasts than in long-term forecasts. In the limit as $j \to \infty$, the forecast error variance converges to $\sigma^2/(1 - a_1^2)$; hence, the forecast error variance converges to the unconditional variance of the $\{y_t\}$ sequence.

Moreover, assuming the $\{\varepsilon_t\}$ sequence is normally distributed, you can place confidence intervals around the forecasts. The one-step-ahead forecast of y_{t+1} is $a_0 + a_1 y_t$ and the forecast error is σ^2. As such, the 95 percent confidence interval for the one-step-ahead forecast can be constructed as

$$a_0 + a_1 y_t \pm 1.96\sigma$$

We can construct a confidence interval for the two-step-ahead forecast error in the same way. From (2.48), the two-step-ahead forecast is $a_0(1 + a_1) + a_1^2 y_t$, and (2.50) indicates that var$[e_t(2)]$ is $\sigma^2(1 + a_1^2)$. Thus, the 95 percent confidence interval for the two-step-ahead forecast is

$$a_0(1 + a_1) + a_1^2 y_t \pm 1.96\sigma(1 + a_1^2)^{1/2}$$

Higher-Order Models

To generalize the discussion, it is possible to use the iterative technique to derive the forecasts for any ARMA(p, q) model. To keep the algebra simple, consider the ARMA(2, 1) model

$$y_t = a_0 + a_1 y_{t-1} + a_2 y_{t-2} + \varepsilon_t + \beta_1 \varepsilon_{t-1} \quad (2.51)$$

Updating one period yields

$$y_{t+1} = a_0 + a_1 y_t + a_2 y_{t-1} + \varepsilon_{t+1} + \beta_1 \varepsilon_t$$

If we continue to assume that (1) all coefficients are known; (2) all variables subscripted $t, t - 1, t - 2, \dots$ are known at period t; and (3) $E_t \varepsilon_{t+j} = 0$ for $j > 0$, the conditional expectation of y_{t+1} is

$$E_t y_{t+1} = a_0 + a_1 y_t + a_2 y_{t-1} + \beta_1 \varepsilon_t \quad (2.52)$$

Equation (2.52) is the one-step-ahead forecast of y_{t+1}. The one-step-ahead forecast error is the difference between y_{t+1} and $E_t y_{t+1}$ so that $e_t(1) = \varepsilon_{t+1}$. To find the two-step-ahead forecast, update (2.51) by two periods:

$$y_{t+2} = a_0 + a_1 y_{t+1} + a_2 y_t + \varepsilon_{t+2} + \beta_1 \varepsilon_{t+1}$$

The conditional expectation of y_{t+2} is

$$E_t y_{t+2} = a_0 + a_1 E_t y_{t+1} + a_2 y_t \quad (2.53)$$

Equation (2.53) expresses the two-step-ahead forecast in terms of the one-step-ahead forecast and the current value of y_t. Combining (2.52) and (2.53) yields

$$\begin{aligned} E_t y_{t+2} &= a_0 + a_1[a_0 + a_1 y_t + a_2 y_{t-1} + \beta_1 \varepsilon_t] + a_2 y_t \\ &= a_0(1 + a_1) + [a_1^2 + a_2] y_t + a_1 a_2 y_{t-1} + a_1 \beta_1 \varepsilon_t \end{aligned}$$

To find the two-step-ahead forecast error, subtract (2.53) from y_{t+2}. Thus,

$$e_t(2) = a_1(y_{t+1} - E_t y_{t+1}) + \varepsilon_{t+2} + \beta_1 \varepsilon_{t+1} \quad (2.54)$$

Since $y_{t+1} - E_t y_{t+1}$ is the one-step-ahead forecast error, we can write the forecast error as

$$e_t(2) = (a_1 + \beta_1)\varepsilon_{t+1} + \varepsilon_{t+2} \quad (2.55)$$

Finally, all j-step-ahead forecasts can be obtained from

$$E_t y_{t+j} = a_0 + a_1 E_t y_{t+j-1} + a_2 E_t y_{t+j-2}, \quad j \geq 2 \quad (2.56)$$

Equation (2.56) demonstrates that the forecasts will satisfy a second-order difference equation. As long as the characteristic roots of (2.56) lie inside the unit circle, the forecasts will converge to the unconditional mean: $a_0/(1 - a_1 - a_2)$. We can use (2.56) to find the j-step-ahead forecast errors. Since $y_{t+j} = a_0 + a_1y_{t+j-1} + a_2y_{t+j-2} + \varepsilon_{t+j} + \beta_1\varepsilon_{t+j-1}$, the j-step-ahead forecast error is

$$\begin{aligned} e_t(j) &= a_1(y_{t+j-1} - E_ty_{t+j-1}) + a_2(y_{t+j-2} - E_ty_{t+j-2}) + \varepsilon_{t+j} + \beta_1\varepsilon_{t+j-1} \\ &= a_1e_t(j-1) + a_2e_t(j-2) + \varepsilon_{t+j} + \beta_1\varepsilon_{t+j-1} \end{aligned}$$

It should be clear that forecasts from any stationary ARMA(p, q) process will eventually satisfy the pth-order difference equation comprising the homogeneous portion of the model. As such, the multi-step-ahead forecasts will converge to the long-run mean of the series.

Forecast Evaluation

Now that you have estimated a series and have forecasted its future values, the obvious question is, "How good are my forecasts?" Typically, there will be several plausible models that you can select to use for your forecasts. Do not be fooled into thinking that the one with the best fit is the one that will forecast the best. To make a simple point, suppose you wanted to forecast the future values of the ARMA(2, 1) process given by (2.51). If you could forecast the value of y_{T+1} using (2.52), you would obtain the one-step-ahead forecast error

$$e_T(1) = y_{T+1} - a_0 - a_1y_T - a_2y_{T-1} - \beta_1\varepsilon_T = \varepsilon_{T+1} \tag{2.57}$$

Since the forecast error is the pure *unforecastable* portion of y_{T+1}, no other ARMA model can provide you with superior forecasting performance. As such, it appears that the "true" model will provide superior forecasts to those from any other possible model. In practice, you will not know the actual order of the ARMA process or the actual values of the coefficients of that process. Instead, to create out-of-sample forecasts, it is necessary to use the estimated coefficients from what you believe to be the most appropriate form of an ARMA model. Let a "hat" or caret (^) over a parameter denote the estimated value of a parameter, and let $\{\hat{\varepsilon}_t\}$ denote the residuals of the estimated model. Hence, if you use the estimated model, the one-step-ahead forecast will be

$$E_Ty_{T+1} = \hat{a}_0 + \hat{a}_1y_T + \hat{a}_2y_{T-1} + \hat{\beta}_1\hat{\varepsilon}_T \tag{2.58}$$

and the one-step-ahead forecast error will be

$$e_T(1) = y_{T+1} - (\hat{a}_0 + \hat{a}_1y_T + \hat{a}_2y_{T-1} + \hat{\beta}_1\hat{\varepsilon}_T)$$

Clearly, this forecast will not be identical to that from (2.57). When we forecast using (2.58), the coefficients (and the residuals) are estimated imprecisely. The forecasts made using the estimated model extrapolate this coefficient uncertainty into the future. Since coefficient uncertainty increases as the model becomes more complex, it could be that an estimated AR(1) model forecasts the process given by (2.51) better than an estimated ARMA(2, 1) model. The general point is that large models usually contain in-sample estimation errors that induce forecast errors. As shown in Clark and West (2007), Dimitrios and Guerard (2004), and Liu and Enders (2003), forecasts using overly parsimonious models with little parameter uncertainty can provide better forecasts than models consistent with the actual data-generating process. Moreover, it is very difficult to construct confidence intervals for this type of forecast error. Not only is it necessary to include the effects of the stochastic variation in the future values of $\{y_{T+i}\}$, it is also necessary to incorporate the fact that the coefficients are estimated with error.

How do you know which one of several reasonable models has the best forecasting performance? One way to answer this question is to put the alternative models to a head-to-head test. Since the future values of the series are unknown, you can hold back a portion of the observations from the estimation process. As such, you can estimate the alternative models over the shortened span of data and use these estimates to forecast the observations of the holdback period. You can then compare the properties of the forecast errors from the two models. To take a simple example, suppose that $\{y_t\}$ contains a total of 150 observations and that you are unsure as to whether an AR(1) or an MA(1) model best captures the behavior of the series.

One way to proceed is to use the first 100 observations to estimate both models and use each to forecast the value of y_{101}. Since you know the actual value of y_{101}, you can construct the forecast error obtained from the AR(1) and from the MA(1). These two forecast errors are precisely those that someone would have made if they had been making a one-step-ahead forecast in period 100. Now, reestimate an AR(1) and an MA(1) model using the first 101 observations. Although the estimated coefficients will change somewhat, they are those that someone would have obtained in period 101. Use the two models to forecast the value of y_{102}. Given that you know the actual value of y_{102}, you can construct two more forecast errors. Since you know all values of the $\{y_t\}$ sequence through period 150, you can continue this process so as to obtain two series of one-step-ahead forecast errors, each containing 50 observations. To keep the notation simple, let $\{f_{1i}\}$ and $\{f_{2i}\}$ denote the sequence of forecasts from the AR(1) and the MA(1), respectively. If you understand the notation, it should be clear that $f_{11} = E_{100}y_{101}$ is the first forecast using the AR(1) and $f_{2,50}$ is the last forecast from the MA(1).

Obviously, it is desirable that the forecast errors have a mean near zero and a small variance. A regression-based method to assess the forecasts is to use the 50 forecasts from the AR(1) to estimate an equation of the form

$$y_{100+i} = a_0 + a_1 f_{1i} + v_{1i} \qquad i = 1, \ldots, 50$$

If the forecasts are unbiased, an F-test should allow you to impose the restriction $a_0 = 0$ and $a_1 = 1$. Similarly, the residual series $\{v_{1i}\}$ should act as a white-noise process. It is a good idea to plot $\{v_{1i}\}$ against $\{y_{100+i}\}$ to determine if there are periods in which your forecasts are especially poor. Now repeat the process with the forecasts from the MA(1). In particular, use the 50 forecasts from the MA(1) to estimate

$$y_{100+i} = b_0 + b_1 f_{2i} + v_{2i} \qquad i = 1, \ldots, 50$$

Again, if you use an F-test, you should not be able to reject the joint hypothesis $b_0 = 0$ and $b_1 = 1$. If the significance levels from the two F-tests are similar, you might select the model with the smallest residual variance; that is, select the AR(1) if $\text{var}(v_1) < \text{var}(v_2)$.[5]

More generally, you might want to have a holdback period that differs from 50 observations. If you have a large sample, it is possible to hold back as much as 50 percent of the data set. Also, you might want to use j-step-ahead forecasts instead of one-step-ahead forecasts. For example, if you have quarterly data and want to forecast one year into the future, you can perform the analysis using four-step-ahead forecasts. Once you have the two sequences of forecast errors, you can compare their properties. With a very small sample, it may not be possible to hold back many observations. Small samples are a problem since Ashley (2003) showed that very large samples are often necessary to reveal a significant difference between the out-of-sample forecasting performances of similar models. You need to have enough observations to have well-estimated coefficients for the in-sample period and enough out-of-sample forecasts so that the test has good power.

Instead of focusing on the bias, many researchers would select the model with the smallest mean square prediction error (MSPE). Suppose you construct H one-step-ahead forecasts from two different models. Again, let f_{1i} be the forecasts from Model 1 and f_{2i} be the forecasts from Model 2. Since we are using one-step-ahead forecasts, we can suppress the subscript j and denote the two series of forecasts errors as e_{1i} and e_{2i}. As such, the MSPE of Model 1 can be calculated as

$$MSPE = \frac{1}{H}\sum_{i=1}^{H} e_{1i}^2$$

Several methods have been proposed to determine whether one MSPE is statistically different from the other. If you put the larger of the two MSPEs in the numerator, a standard recommendation is to use the F-statistic

$$F = \sum_{i=1}^{H} e_{1i}^2 \Big/ \sum_{i=1}^{H} e_{2i}^2 \tag{2.59}$$

The intuition is that the value of F will equal unity if the forecast errors from the two models are identical. A very large value of F implies that the forecast errors from the first model are substantially larger than those from the second. Under the null hypothesis of equal forecasting performance, (2.59) has a standard F-distribution with (H, H) degrees of freedom if the following three assumptions hold:

1. The forecast errors have zero mean and are normally distributed.
2. The forecast errors are serially uncorrelated.
3. The forecast errors are contemporaneously uncorrelated with each other.

Although it is common practice to assume that the $\{\varepsilon_t\}$ sequence is normally distributed, it is not necessarily the case that the forecast errors are normally distributed with a mean value of zero. Similarly, the forecasts may be serially correlated; this is particularly true if you use multi-step-ahead forecasts. For example, equation (2.55) indicated that the two-step-ahead forecast error for y_{t+2} is

$$e_t(2) = (a_1 + \beta_1)\varepsilon_{t+1} + \varepsilon_{t+2}$$

and updating by one period yields the two-step-ahead forecast error for y_{t+3}:

$$e_{t+1}(2) = (a_1 + \beta_1)\varepsilon_{t+2} + \varepsilon_{t+3}$$

It should be clear that the two forecast errors are correlated. In particular,

$$E[e_t(2)e_{t+1}(2)] = (a_1 + \beta_1)\sigma^2$$

The point is that predicting y_{t+2} from the perspective of period t and predicting y_{t+3} from the perspective of period $t + 1$ both contain an error due to the presence of ε_{t+2}. However, for $i > 1$, $E[e_t(2)e_{t+i}(2)] = 0$ since there are no overlapping forecasts. Hence, the autocorrelations of the two-step-ahead forecast errors cut to zero after lag 1. You should be able to demonstrate the general result that j-step-ahead forecast errors act as an MA($j - 1$) process.

Finally, the forecast errors from the two alternative models will usually be highly correlated with each other. For example, a negative realization of ε_{t+1} will tend to cause the forecasts from both models to be too high. Unfortunately, the violation of any one of these assumptions means that the ratio of the MSPEs in (2.59) does not have an F-distribution.

THE GRANGER–NEWBOLD TEST Granger and Newbold (1976) showed how to overcome the problem of contemporaneously correlated forecast errors. If you have H one-step-ahead forecast errors from each model, use the two sequences of forecast errors to form

$$x_i = e_{1i} + e_{2i} \quad \text{and} \quad z_i = e_{1i} - e_{2i} \quad i = 1, \ldots, H$$

Given that the first two assumptions above are valid, under the null hypothesis of equal forecast accuracy, x_i and z_i should be uncorrelated:

$$\rho_{xz} = Ex_iz_i = E(e_{1i}^2 - e_{2i}^2)$$

If the models forecast equally well, it follows that $Ee_{1i}^2 = Ee_{2i}^2$. Model 1 has a larger MSPE if ρ_{xz} is positive and Model 2 has a larger MSPE if ρ_{xz} is negative. Let r_{xz} denote the sample correlation coefficient between $\{x_i\}$ and $\{z_i\}$. Granger and Newbold (1976) showed that if assumptions 1 and 2 hold,

$$r_{xz}/\sqrt{(1 - r_{xz}^2)/(H - 1)} \tag{2.60}$$

has a t-distribution with $H - 1$ degrees of freedom. Thus, if r_{xz} is statistically different from zero, Model 1 has a larger MSPE if r_{xz} is positive and Model 2 has a larger MSPE if r_{xz} is negative.

THE DIEBOLD-MARIANO TEST There is a very large body of literature that tries to extend the Granger–Newbold test so as to relax assumptions 1 and 2. Moreover, applied econometricians might be interested in measures of forecasting performance other than the sum of squared errors. Indeed, it should be clear that using the sum of squared errors as a criterion makes sense only if the loss from making an incorrect forecast is quadratic. However, there are many other possibilities. For example, if your loss depends on the size of the forecast error, you should be concerned with the absolute values of the forecast errors. Alternatively, an options trader receives a payoff of zero if the value of the underlying asset lies below the strike price but receives a one-dollar payoff for each dollar the asset price rises above the strike price. In such a circumstance, the loss payoff is asymmetric. Diebold and Mariano (1995) developed a test that relaxes assumptions 1 to 3 and allows for an objective function that is not quadratic.

As before, if we consider only one-step-ahead forecasts, we can eliminate the subscript j. As such, we can let the loss from a forecast error in period i be denoted by $g(e_i)$. In the typical case of mean squared errors, the loss is e_i^2. Nevertheless, to allow the loss function to be general, we can write the differential loss in period i from using Model 1 versus Model 2 as $d_i = g(e_{1i}) - g(e_{2i})$. The mean loss can be obtained as

$$\bar{d} = \frac{1}{H}\sum_{i=1}^{H}[g(e_{1i}) - g(e_{2i})] \tag{2.61}$$

Under the null hypothesis of equal forecast accuracy, the value of $\bar{d}$ is zero. Since $\bar{d}$ is the mean of the individual losses, under fairly weak conditions the central limit theorem implies that $\bar{d}$ should have a normal distribution. Hence, it is not necessary to assume that the individual forecast errors are normally distributed. Thus, if we knew var($\bar{d}$), we could construct the ratio $\bar{d}/\sqrt{\text{var}(\bar{d})}$ and test the null hypothesis of equal forecast accuracy using a standard normal distribution. In practice, the implementation of the test is complicated by the fact that we need to estimate var($\bar{d}$).

If the $\{d_i\}$ series is serially uncorrelated with a sample variance of γ_0, the estimate of var($\bar{d}$) is simply $\gamma_0/(H - 1)$. Since we use the estimated value of the variance, the expression $\bar{d}/\sqrt{\gamma_0/(H - 1)}$ has a t-distribution with $H - 1$ degrees of freedom.

There is a very large body of literature on the best way to estimate the standard deviation of $\bar{d}$ in the presence of serial correlation. Many of the technical details are not appropriate here. Diebold and Mariano let γ_i denote the ith autocovariance of the d_t sequence. Suppose that the first q values of γ_i are different from zero. The variance of $\bar{d}$ can be approximated by $\text{var}(\bar{d}) = [\gamma_0 + 2\gamma_1 + \ldots + 2\gamma_q](H - 1)^{-1}$; the standard deviation is the square root. As such, Harvey, Leybourne, and Newbold (1998) recommended constructing the Diebold–Mariano (DM) statistic as

$$\text{DM} = \bar{d}/\sqrt{(\gamma_0 + 2\gamma_1 + \ldots + 2\gamma_q)/(H - 1)} \tag{2.62}$$

Compare the sample value of (2.62) to a t-statistic with $H - 1$ degrees of freedom. As a practical matter, a simple way to proceed is to regress the d_i on a constant and use a t-test (with robust standard errors) to determine whether the constant is statistically different from zero.[6]

It is also possible to use the method for the j-step-ahead forecasts $e_{1i}(j)$ and $e_{2i}(j)$. Construct each $d_i = g(e_{1i}(j)) - g(e_{2i})$ and construct the mean $\bar{d}$. If you construct H forecast errors, the DM statistic is

$$\text{DM} = \bar{d}/\sqrt{(\gamma_0 + 2\gamma_1 + \ldots + 2\gamma_q)/[H + 1 - 2j + H^{-1}j(j - 1)]}$$

An example showing the appropriate use of the Granger–Newbold and Diebold–Mariano tests is provided in the next section. Nevertheless, before proceeding, a strong word of caution is in order. Clark and McCracken (2001) showed that the Granger–Newbold and Diebold–Mariano tests have a t-distribution only when the underlying forecasting models are not nested. For example, the tests might not work well when comparing forecasts from an AR(1) model to those obtained from an ARMA(2, 1) model. Clearly, the AR(1) can be obtained from the ARMA(2, 1) specification by setting $a_2 = \beta_1 = 0$. The problem with nested models is that that under the null hypothesis of equal MSPEs (so that the data are generated by the small model), the two models should predict equally well. However, the large model will always contain some extra error as it contains unnecessary parameters. Hence, if you want to test whether the data are actually generated from the different models, you need to control for the parameter uncertainty.

Clark and West (2007) developed a simple procedure to adjust the forecast errors from the large model so that a simple variant of the DM statistic can be used with nested models. To continue with the notation developed above, denote the H forecasts from Model 1 as f_{1i} and the forecast errors as e_{1i}. Similarly, the H forecasts and forecast errors from Model 2 are f_{2i} and e_{2i}, respectively. Let Model 1 be nested within Model 2. Given that the models are nested, the sole reason for any discrepancy between f_{1i} and f_{2i} is parameter estimation error. If this estimation error is subtracted from e_{2i}, the adjusted forecast errors can be used as the basis for the modified DM test. Consider the z_i series constructed from the squares of these errors as

$$z_i = (e_{1i})^2 - [(e_{2i})^2 - (f_{1i} - f_{2i})^2] \qquad i = 1, \ldots, H$$

Allowing for parameter uncertainty, under the null hypothesis that the two models predict equally well, z_i should be zero. Under the alternative hypothesis, the data are generated from Model 2. Hence, to perform the test, regress the z_i series on a constant. Since the test is one-sided, if the t-statistic for the constant exceeds 1.645, reject the null hypothesis of equal forecast accuracy at the 5 percent significance level. If you reject the null hypothesis, you can conclude that the data are generated from Model 2. Otherwise, the data are more likely to be generated from Model 1. If the $\{z_i\}$ series is serially correlated, you should perform the test with a robust t-statistic, such as that in Newey and West (1987).

10. A MODEL OF THE INTEREST RATE SPREAD

The term "textbook example" is supposed to connote a very clear-cut illustration. If you are looking for a textbook example of the Box–Jenkins methodology, go back to Section 7 or turn to Question 11 at the end of this chapter. In practice, we rarely find a data series that precisely conforms to a theoretical ACF or PACF. This section is intended to illustrate some of the ambiguities that can be encountered when using the Box–Jenkins technique. These ambiguities may lead two equally skilled econometricians to estimate and forecast the same series using very different ARMA processes. Many view the necessity of relying on the researcher's judgment and experience as a serious weakness of a procedure that is designed to be scientific. Yet, if you make reasonable choices, you will select models that come very close to mimicking the actual data-generating process. As a practical matter, when you find several plausible models, it makes sense to forecast using both. In the time-series literature, it is typical to see several reasonable models reported.

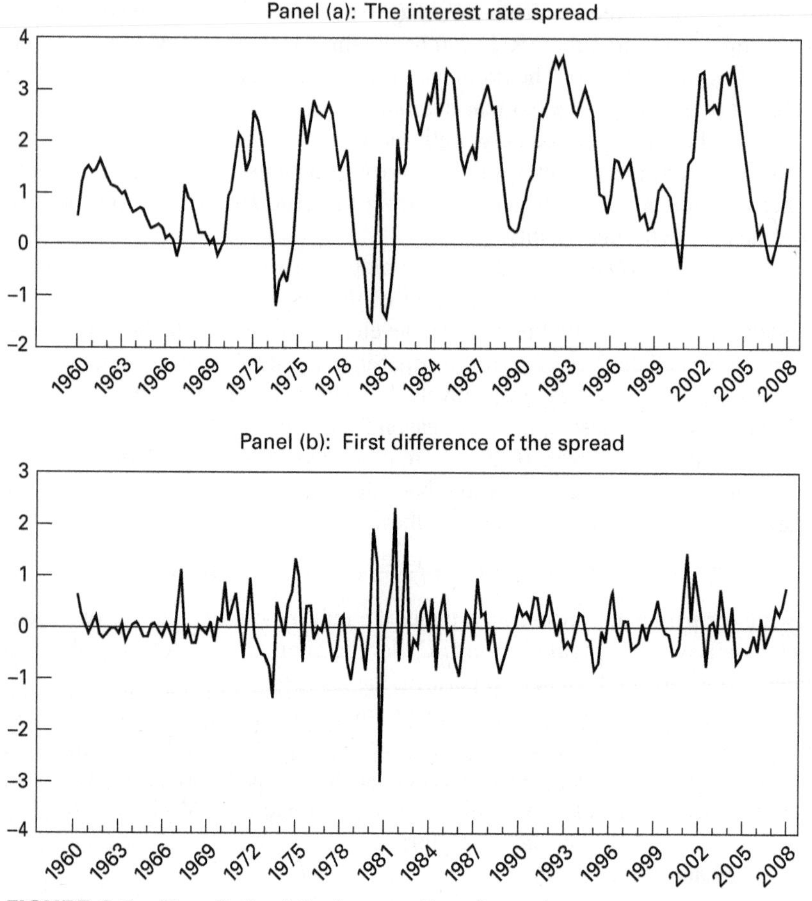

FIGURE 2.5 Time Path of the Interest Rate Spread

It is useful to illustrate the Box–Jenkins modeling procedure by estimating a quarterly model of the spread between a long-term and a short-term interest rate. Specifically, the interest rate spread (s_t) can be formed as the difference between the interest rate on 10-year U.S. government bonds and the rate on three-month treasury bills. The data used in this section are the series labeled R10 and TBILL in the file QUARTERLY.XLS. Exercise 12 at the end of this chapter will help you to reproduce the results reported below.

Panel (a) of Figure 2.5 shows the spread over the period from 1960Q1 to 2008Q1. Although there are a few instances in which the spread is negative, the difference between long- and short-term rates is generally positive (the sample mean is 1.37). Notice that the series shows a fair amount of persistence in that the durations when the spread is above or below the mean can be quite lengthy. Moreover, there do not appear to be any major structural breaks (such as a permanent jump in the mean or variance) in that the dynamic nature of the process seems to be constant over time. As such, it is quite reasonable to suppose that the $\{s_t\}$ sequence is covariance-stationary. In

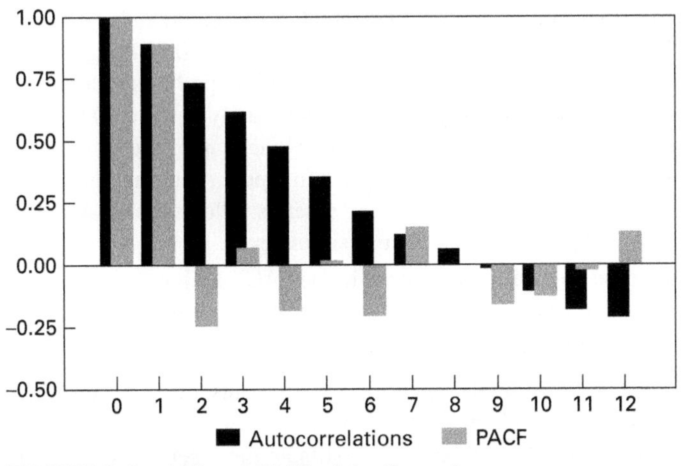

FIGURE 2.6 ACF and PACF of the Spread

contrast, as shown in Panel (b), the first difference of the spread seems to be very erratic. As you will verify in Exercise 12, the Δs_t series has little informational content that can be used to forecast its future values. As such, it seems reasonable to estimate a model of the $\{s_t\}$ sequence without any further transformations. Nevertheless, because there are several large positive and negative jumps in the value of s_t, some researchers might want to transform it so as to diminish its volatility. A reasonable number of such shocks might indicate a departure from the assumption that the errors are normally distributed. Although a logarithmic or a square root transformation is impossible because some realizations of s_t are negative, one could dampen the series using $y_t = \log(s_t + c)$ using a value of c such that $s_t + c$ is always positive. The point is that you should always maintain a healthy skepticism of the accuracy of your model since the behavior of the data-generating process may not fully conform to the underlying assumptions of the methodology.

Before reading on, you should examine the autocorrelation and partial autocorrelation functions of the $\{s_t\}$ sequence shown in Figure 2.6. Try to identify the tentative models that you would want to estimate. Recall that the theoretical ACF of a pure MA(q) process cuts off to zero at lag q and the theoretical ACF of an AR(1) model decays geometrically. Examination of Figure 2.6 suggests that neither of these specifications perfectly describes the sample data. In selecting your set of plausible models, also note the following:

1. The ACF and PACF converge to zero quickly enough that we do not have to worry about a time-varying mean. As suggested above, we do not want to *overdifference* the data and try to model the $\{\Delta s_t\}$ sequence.
2. The ACF does not cut to zero so that we can rule out a pure MA(q) process.
3. The ACF is not really suggestive of a pure AR(1) process in that the decay does not appear to be geometric. The value of ρ_1 is 0.890 and the values of ρ_2, ρ_3, and ρ_4 are 0.741, 0.616, and 0.481, respectively.

4. The estimated values of the PACF are such that $\phi_{11} = 0.890$ and $\phi_{22} = -0.245$. Although ϕ_{33} and ϕ_{55} are close to zero, $\phi_{44} = -0.184$, $\phi_{66} = -0.202$, and $\phi_{77} = 0.150$. Recall that under the null hypothesis of a pure AR(p) model, the variance of $\phi_{p+i,\, p+i}$ is approximately equal to $1/T$. Since there are 193 total observations, the values of ϕ_{22}, ϕ_{44}, ϕ_{66}, and ϕ_{77} are more than two standard deviations from zero. In a pure AR(p) model, the PACF cuts to zero after lag p. Hence, if the s_t series follows a pure AR(p) process, the value of p could be as high as six or seven.
5. There appears to be an oscillating pattern in the PACF in that the first seven values alternate in sign. Oscillating decay of the PACF is characteristic of a positive MA coefficient.

Due to the number of small and marginally significant coefficients, the ACF and PACF of the spread are probably more ambiguous than most of those you will encounter. Hence, suppose you don't know where to start and estimate the s_t series using a pure AR(p) model. To illustrate the point, if you estimate the entire s_t series as an AR(7) process you should obtain the estimates shown in column 2 of Table 2.4. If you examine the table, you will find that all of the t-statistics exceed 1.96 in absolute value (indicating that the coefficients are significant at the 5 percent level). The sum of squared residuals (SSR) is 48.745 and the AIC and SBC are 738.906 and 764.712, respectively. The significance levels of the Q-statistics for lags 4, 8, and 12 indicate no remaining autocorrelation in the residuals.

Although the AR(7) model has some desirable attributes, other pure AR(p) models might perform as well as the AR(7).[7] Suppose that you try a very parsimonious model and estimate an AR(2). As you can see from the third column of the table, the AIC selects the AR(7) model, but SBC selects the AR(2) model. However, the residual autocorrelations from the AR(2) are problematic in that

ρ_1	ρ_2	ρ_3	ρ_4	ρ_5	ρ_6	ρ_7	ρ_8
0.02	−0.10	0.16	−0.03	0.14	−0.09	−0.16	0.13

The Q-statistics from the AR(2) model indicate significant autocorrelation in the residuals. If you examined the AR(7) carefully, you might have noticed that a_3 almost offsets a_4 and that a_5 almost offsets a_6 (since $a_3 + a_4 \approx 0$ and $a_5 + a_6 \approx 0$). If you reestimate the model without s_{t-3}, s_{t-4}, s_{t-5}, and s_{t-6}, you should obtain the results shown in column 4 of Table 2.4. Since the coefficient for s_{t-7} is now statistically insignificant, it might seem preferable to use the AR(2) instead of the model with lags 1, 2, and 7.

Even though the AR(7) model performs extremely well, it is not necessarily the best forecasting model. There are several possible alternatives since the patterns of the ACF and PACF are not immediately clear. Results for a number of models with MA terms are shown in columns 5, 6, and 7 of Table 2.4:

1. From the decaying ACF someone might try to estimate the ARMA(1, 1) model reported in column 5 of the table. The estimated value of a_1 (0.817) is statistically different from zero and is almost four standard deviations from unity. The estimated value of β_1 (0.368) is statistically different from zero and

Table 2.4 Estimates of the Interest Rate Spread

	$p = 1$ to 7 $q = 0$	$p = 1$ and 2 $q = 0$	$p = 1, 2,$ and 7 $q = 0$	$p = 1$ $q = 1$	$p = 1$ and 2 $q = 1$	$p = 1$ and 2 $q = 1$ and 7
a_0	0.210 (3.08)	0.189 (3.11)	0.231 (3.27)	0.248 (2.79)	0.280 (2.79)	0.273 (2.67)
a_1	1.177 (15.84)	1.108 (15.39)	1.088 (14.77)	0.817 (17.44)	0.444 (2.93)	0.312 (3.02)
a_2	−0.466 (−4.17)	−0.244 (−3.39)	−0.210 (−2.73)		0.350 (2.42)	0.488 (4.82)
a_3	0.386 (3.37)					
a_4	−0.339 (−2.931)					
a_5	0.319 (2.78)					
a_6	−0.379 (−3.39)					
a_7	0.150 (2.02)		−0.045 (−1.22)			
β_1				0.368 (4.88)	0.725 (6.20)	0.862 (13.80)
β_7						−0.144 (−3.77)
SSR	48.745	54.207	53.765	53.324	51.848	49.178
AIC	738.906	748.664	749.139	745.608	742.389	734.553
SBC	764.712	758.341	762.042	755.285	755.292	750.682
$Q(4)$	0.230	6.874	6.246	6.598	1.422	1.355
$Q(8)$	4.501	21.286	21.320	20.481	12.509	3.107
$Q(12)$	13.072	29.559	28.081	27.579	21.090	12.231

Notes:

To ensure comparability, each equation was estimated over the 1961Q4–2008Q1 period.

Values in parentheses are the t-statistics for the null hypothesis that the estimated coefficient is equal to zero. SSR is the sum of squared residuals. $Q(n)$ are the Ljung–Box Q-statistics of the residual autocorrelations.

For ARMA models, many software packages do not actually report the intercept term a_0. Instead, they report the estimated mean of process μ_y along with the t-statistic for the null hypothesis that $\mu_y = 0$. The historical reason for this convention is that it was easier to first demean the data and then estimate the ARMA coefficients than to estimate all values in one step. If your software package reports a constant term approximately equal to 1.39, it is reporting the estimated mean.

implies that the process is invertible. Nevertheless, the ARMA(1, 1) specification is inadequate. The Ljung-Box Q-statistic for 4 lags of the residuals is equal to 6.598. As such, we cannot reject the null that $Q(4) = 0$ at any conventional significance level. However, the $Q(8)$ and $Q(12)$ statistics indicate that the residuals from this model exhibit substantial serial autocorrelation. As such, we must eliminate the ARMA(1, 1) model from consideration.

2. Since the ACF decays and the PACF seems to oscillate beginning with lag 2 ($\phi_{22} = -0.245$), it seems plausible to estimate an ARMA(2, 1) model. As shown in column 6 of the table, the model is an improvement over the ARMA(1, 1) specification. The estimated coefficients ($a_1 = 0.444$ and $a_2 = 0.350$) are each significantly different from zero at conventional levels and imply characteristic roots in the unit circle. The AIC selects the ARMA(2, 1) model although the SBC selects the ARMA(1, 1) model. The values for $Q(4)$ and $Q(8)$ indicate that the autocorrelations of the residuals are not statistically significant at the 10 percent level. Since the $Q(12)$ statistic is just significant at the 5 percent level, there is some residual autocorrelation at the longer lags. Consider the ACF of the residuals:

ρ_1	ρ_2	ρ_3	ρ_4	ρ_5	ρ_6	ρ_7	ρ_8	ρ_9	ρ_{10}	ρ_{11}	ρ_{12}
0.01	0.05	0.07	−0.01	0.08	−0.08	−0.17	0.13	−0.00	−0.01	−0.21	−0.02

3. In order to account for the serial correlation at lag 7, it might seem plausible to add an MA term to the model at lag 7. Again, all of the estimated coefficients are of high quality. Note that the coefficient for β_7 has a t-statistic of -3.77. The estimated values of a_1 and a_2 are similar to those of the ARMA(2, 1) model. Now, the Q-statistics indicate that the autocorrelations of the residuals are not significant at conventional level. The sum of squared residuals (SSR) is smaller than that of any of the other models except the AR(7). Moreover, both the AIC and SBC select the ARMA(2, (1, 7)) specification over any of the other models. You can easily verify that the MA coefficient at lag 7 provides a better fit than an AR coefficient at lag 7.

Although the AR(7) and ARMA(2, (1, 7)) appear to be adequate, other researchers might have selected a decidedly different model. Consider some of the alternatives listed below.

1. **Parsimony versus Overfitting:** In Section 7 above, we examined the issue of fitting an MA coefficient at lag 16 to a true AR(2) process. If you reexamine the example, you can understand why some researchers shy away from estimating a model with long lag lengths that are disjointed from those of other periods. In the example of the spread, the problem with the ARMA(2, 1) model is that there was a small amount of residual autocorrelation at lag 7 and the $Q(12)$ statistic indicated the possibility of serial correlation at the longer lags. The addition of the MA coefficient at lag 7 remedied yielded a model with a better fit and remedied the serial correlation problem. However, is it really plausible that ε_{t-7} has a direct effect on the current value of the interest rate spread while lags 3, 4, 5, and 6 have no direct effects? In other words, do the markets for securities work in such a way that what happens seven quarters in the past has a larger effect on today's interest rates than events occurring in the more recent past? Moreover, as you can verify by estimating the ARMA (2, (1, 7)) model, the t-statistic for β_7 over the $1982Q1-2008Q1$ period is equal to 0.994 and is not statistically significant. Notice that Panel (b) of Figure 2.5 suggests that the volatility of the spread in the late 1970s and early 1980s is not

typical of the entire sample. It could be the case that the realizations from this period are anomalies that have large effects on the coefficient estimates and their standard errors. Thus, even though the AIC and SBC select the ARMA (2, (1, 7)) model over the ARMA(2, 1) model, some researchers would prefer the latter.

More generally, *overfitting* refers to a situation in which an equation is fit to some of the idiosyncrasies present in a particular sample that are not actually representative of the data-generating process. In applied work, no data set will perfectly correspond to every assumption required for the Box–Jenkins methodology. Since it is not always clear which characteristics of the sample are actually present in the data-generating process, the attempt to expand a model so as to capture every feature of the data may lead to overfitting.

2. **Volatility:** Given the volatility of the $\{s_t\}$ series during the late 1970s and early 1980s, transforming the spread using some sort of a square root or logarithmic transformation might be appropriate. Moreover, the s_t series has a number of sharp jumps indicating that the assumption of normality might be violated. For a constant c such that $s_t + c$ is always positive, transformations such as $\ln(s_t + c)$ or $(s_t + c)^{0.5}$ yield series with less volatility than the s_t series itself. Alternatively, it is possible to model the difference between the log of the 10-year rate and the log of the 3-month rate.

   A general class of transformations was proposed by Box and Cox (1964). Suppose that all values of $\{y_t\}$ are positive so that it is possible to construct the transformed $\{y_t^*\}$ sequence as

$$
\begin{aligned}
y_t^* &= (y_t^\lambda - 1)/\lambda \qquad && \lambda \neq 0 \\
&= \ln(y_t) && \lambda = 0
\end{aligned}
$$

   The common practice is to transform the data using a preselected value of λ. The selection of a value for λ that is close to zero acts to "smooth" the sequence. An ARMA model can be fitted to the transformed data. Although some software programs have the capacity to simultaneously estimate λ along with the other parameters of the ARMA model, this approach has fallen out of fashion. Instead, it is possible to actually model the variance using the methods discussed in Chapter 3.

3. **Trends:** Suppose that the span of the data had been somewhat different in that the first observation was for 1973*Q*1 and the last was for 2004*Q*4. If you examine Panel (a) of Figure 2.4, you can see that someone might be confused and believe that the data contained an upward trend. Their misinterpretation of the data might be reinforced by the fact that the ACF converges to zero rather slowly. As such, they might have estimated a model of the Δs_t series. Others might have detrended the data using a deterministic time trend.

Out-of-Sample Forecasts

We can assess the forecasting performance of the AR(7) and ARMA(2, (1, 7)) models by examining their bias and mean square prediction errors. Given that the data set contains a total of 186 usable observations, it is possible to use a holdback period of

50 observations. This way, there are at least 136 observations in each of the estimated models and an adequate number of out-of-sample forecasts. First, the two models were estimated using all available observations through 1995*Q*3 and the two one-step-ahead forecasts were obtained. The actual value of $s_{1995Q4} = 0.62333$; the AR(7) predicted a value of 0.80667 and the ARMA(2, (1, 7)) model predicted a value of 1.07524. Thus, the forecast of the ARMA(7) is superior to that of the ARMA(2, (1, 7)) for this first period. An additional 49 forecasts were obtained for periods 1996*Q*1 to 2008*Q*1. Let e_{1t} denote the forecast errors from the AR(7) model and e_{2t} denote the forecast errors from the ARMA(2, (1, 7)) model. The mean of e_{1t} is -0.017824, the mean of e_{2t} is -0.002352, and the estimated variances are $\text{var}(e_1) = 0.212819$ and $\text{var}(e_2) = 0.201455$. As such, there is an advantage in the forecasting performance of the ARMA(2, (1, 7)) in that it has the smallest bias and MSPE.

To ascertain whether these differences are statistically significant, we first check the bias. Let f_{1t} be the 50 forecasts of the AR(7) model and f_{2t} be the 50 forecasts from the ARMA(2, (1, 7)) model. Beginning with $t = 1995Q4$, we can estimate the two regression equations:

$$s_t = 0.0359 + 0.986f_{1t} \quad \text{and} \quad s_t = -0.020 + 1.013f_{2t}$$

For the AR(7) model, the F-statistic for the restriction that the intercept equals zero and the slope equals unity is 0.060 with significance level of 0.942. Clearly, the restriction of unbiased forecasts does not appear to be binding. For the ARMA(2, (1, 7)) model, the F-statistic is 0.021 with a significance level of 0.979. Hence, there is strong evidence that both models have unbiased forecasts.

Next, consider the Granger–Newbold test for equal mean square prediction errors. Form the x_i and z_i series as $x_i = e_{1i} + e_{2i}$ and $z_i = e_{1i} - e_{2i}$. The correlation coefficient between x_i and z_i is $r_{xz} = 0.117181$. Given that there are 50 observations in the holdback period, form the Granger–Newbold statistic

$$r_{xz}/\sqrt{(1 - r_{xz}^2)/(H - 1)} = 0.117181/\sqrt{(1 - (0.117181)^2)/49} = 0.82595$$

With 49 degrees of freedom, a value of $t = 0.82595$ is not statistically significant. We can conclude that the forecasting performance of the AR(7) is not statistically different from that of the ARMA(2, (1, 7)).

Since the e_{1i} and e_{2i} series contain only a low amount of serial correlation, we obtain virtually the same answer using the DM statistic. Oftentimes, forecasters are concerned about the MSPE. However, there are many other possibilities. In Exercise 12 at the end of this chapter, you will be asked to use the mean absolute error. Now, to illustrate the use of the DM test, suppose that the cost of a forecast error rises extremely quickly in the size of the error. In such circumstances, the loss function might be best represented by the forecast error raised to the fourth power. Hence,

$$d_i = (e_{1i})^4 - (e_{2i})^4 \tag{2.63}$$

The mean value of the $\{d_i\}$ sequence from (2.63) (i.e., $\bar{d}$) is 0.025196 and the estimated variance is 0.011913. Since $H = 50$, we can form the DM statistic

$$\text{DM} = 0.025196/(0.011913/49)^{1/2} = 1.61592$$

The null hypothesis is that the models have equal forecasting accuracy and the alternative hypothesis is that the forecast errors from the AR(2, (1, 7)) are smaller. With 49 degrees of freedom, the t-value of 1.61592 is significant at the 5.637% level. Hence, there is only mild evidence in favor of the AR(2, (1, 7)) model. If there is serial correlation in the $\{d_t\}$ series, we need to use the specification in (2.63). Toward this end, we would want to select the statistically significant values of γ_q. The autocorrelations of d_t are

ρ_1	ρ_2	ρ_3	ρ_4	ρ_5	ρ_6	ρ_7	ρ_8	ρ_9	ρ_{10}	ρ_{11}	ρ_{12}
−0.04	−0.00	0.14	−0.05	0.39	0.05	−0.05	0.01	−0.01	−0.07	−0.11	−0.02

$$Q(4) = 1.257;\ Q(8) = 10.411;\ \text{and}\ Q(12) = 11.526$$

Although ρ_5 is large, many applied econometricians would dismiss it as spurious. It does not seem plausible that correlations for ρ_1 and ρ_2 are actually very close to zero while the correlation between d_t and d_{t-5} is very large. Moreover, the Ljung–Box $Q(4)$, $Q(8)$, and $Q(12)$ statistics do not indicate that the autocorrelations are significant. Respectively, the significance levels are 0.869, 0.237, and 0.484. Nevertheless, if you do estimate the long-run variance using (2.63) with 5 lags, you should find that DM = 1.18971 (so that the MSPEs are *not* statistically different from each other). The example underscores the point made earlier that there is no clear answer as to the best way to measure the long-run variance of $\bar{d}$ in the presence of serial correlation. The more general result is that the two models are not substantially different from each other. Both should provide reasonable forecasts.

11. SEASONALITY

Many economic processes exhibit some form of seasonality. The agricultural, construction, and travel sectors have obvious seasonal patterns resulting from their dependence on the weather. Similarly, the Thanksgiving-to-Christmas holiday season has a pronounced influence on the retail trade. In fact, the seasonal variation of a series may account for the preponderance of its total variance. Forecasts that ignore important seasonal patterns will have a high variance.

Too many people fall into the trap of ignoring seasonality if they are working with **deseasonalized** or **seasonally adjusted** data. Suppose you collect a data set that the U.S. Census Bureau has "seasonally adjusted" using its X–11 or X–12 method.[8] In principle, the seasonally adjusted data should have the seasonal pattern removed. However, caution is necessary. Although a standardized procedure may be necessary for a government agency reporting hundreds of series, the procedure might not be best for an individual wanting to model a single series. Even if you use seasonally adjusted data, a seasonal pattern might remain. This is particularly true if you do not use the entire span of data; the portion of the data used in your study can display more (or less) seasonality than the overall span. There is another important reason to be concerned about seasonality when using deseasonalized data. Implicit in any method of seasonal adjustment is a two-step procedure. First, the seasonality is removed, and second, the autoregressive and moving-average coefficients are estimated using

Box–Jenkins techniques. As surveyed in Bell and Hillmer (1984), often the seasonal and the ARMA coefficients are best identified and estimated jointly. In such circumstances, it is wise to avoid using seasonally adjusted data.

Models of Seasonal Data

The Box–Jenkins technique for modeling seasonal data is only a bit different from that of nonseasonal data. The twist introduced by seasonal data of period s is that the seasonal coefficients of the ACF and PACF appear at lags s, $2s$, $3s$, ..., rather than at lags 1, 2, 3, ... For example, two purely seasonal models for quarterly data might be

$$y_t = a_4 y_{t-4} + \varepsilon_t, |a_4| < 1 \tag{2.64}$$

and

$$y_t = \varepsilon_t + \beta_4 \varepsilon_{t-4} \tag{2.65}$$

You can easily convince yourself that the theoretical correlogram for (2.64) is such that $\rho_i = (a_4)^{i/4}$ if $i/4$ is an integer and $\rho_i = 0$ otherwise; thus, the ACF exhibits decay at lags 4, 8, 12, ... For model (2.65), the ACF exhibits a single spike at lag 4 and all other correlations are zero.

In practice, identification will be complicated by the fact that the seasonal pattern will interact with the nonseasonal pattern in the data. The ACF and PACF for a combined seasonal/nonseasonal process will reflect both elements. Note that with quarterly data, a seasonal MA term can have the form

$$y_t = a_1 y_{t-1} + \varepsilon_t + \beta_1 \varepsilon_{t-1} + \beta_4 \varepsilon_{t-4} \tag{2.66}$$

Alternatively, an autoregressive coefficient at lag 4 might have been used to capture the seasonality

$$y_t = a_1 y_{t-1} + a_4 y_{t-4} + \varepsilon_t + \beta_1 \varepsilon_{t-1}$$

Both of these methods treat the seasonal coefficients additively; an AR or an MA coefficient is added at the seasonal period. **Multiplicative seasonality** allows for the interaction of the ARMA and the seasonal effects. Consider the multiplicative specifications

$$(1 - a_1 L) y_t = (1 + \beta_1 L)(1 + \beta_4 L^4) \varepsilon_t \tag{2.67}$$
$$(1 - a_1 L)(1 - a_4 L^4) y_t = (1 + \beta_1 L) \varepsilon_t \tag{2.68}$$

Equation (2.67) differs from (2.66) in that it allows the moving-average term at lag 1 to interact with the seasonal moving-average effect at lag 4. In the same way, (2.68) allows the autoregressive term at lag 1 to interact with the seasonal autoregressive effect at lag 4. Many researchers prefer the multiplicative form since a rich interaction pattern can be captured with a small number of coefficients. Rewrite (2.67) as

$$y_t = a_1 y_{t-1} + \varepsilon_t + \beta_1 \varepsilon_{t-1} + \beta_4 \varepsilon_{t-4} + \beta_1 \beta_4 \varepsilon_{t-5}$$

Estimating only three coefficients (i.e., a_1, β_1, and β_4) allows us to capture the effects of an autoregressive term and the effects of moving-average terms at lags 1, 4, and 5. Of course, you do not really get something for nothing. The estimates of the three moving-average coefficients are interrelated. A researcher estimating the unconstrained model $y_t = a_1 y_{t-1} + \varepsilon_t + \beta_1 \varepsilon_{t-1} + \beta_4 \varepsilon_{t-4} + \beta_5 \varepsilon_{t-5}$ would necessarily obtain a smaller residual sum of squares. However, (2.67) is clearly the more parsimonious model. If the unconstrained value of β_5 approximates the product $\beta_1 \beta_4$, the multiplicative model will be preferable. For this reason, most software packages have routines capable of estimating multiplicative models. Otherwise, there are no theoretical grounds leading us to prefer one form of seasonality over another. As illustrated in the last section, experimentation and diagnostic checks are probably the best way to obtain the most appropriate model.

Seasonal Differencing

The Christmas shopping season is accompanied by an unusually large number of transactions, and the Federal Reserve expands the money supply to accommodate the increased demand for money. As shown by the dashed line in Figure 2.7, the U.S. money supply, as measured by M1, has a decidedly upward trend. The series, called M1NSA, is contained in the file QUARTERLY.XLS. You can use the data to follow along with the discussion below. The logarithmic change, shown by the solid line, appears to be stationary. Nevertheless, there is a clear seasonal pattern in that the value of the fourth quarter for any year is substantially higher than that for the adjacent quarters.

This combination of strong seasonality and nonstationarity is often found in economic data. The ACF for a process with strong seasonality is similar to that for a nonseasonal process; the main difference is that the spikes at lags s, $2s$, $3s$, ... do not exhibit rapid decay. We know that it is necessary to difference (or take the logarithmic change of) a nonstationary process. Similarly, if the autocorrelations at the seasonal

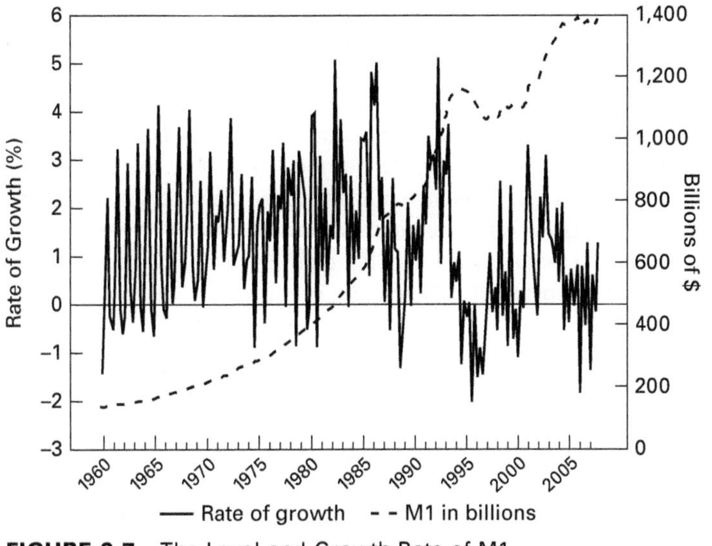

FIGURE 2.7 The Level and Growth Rate of M1

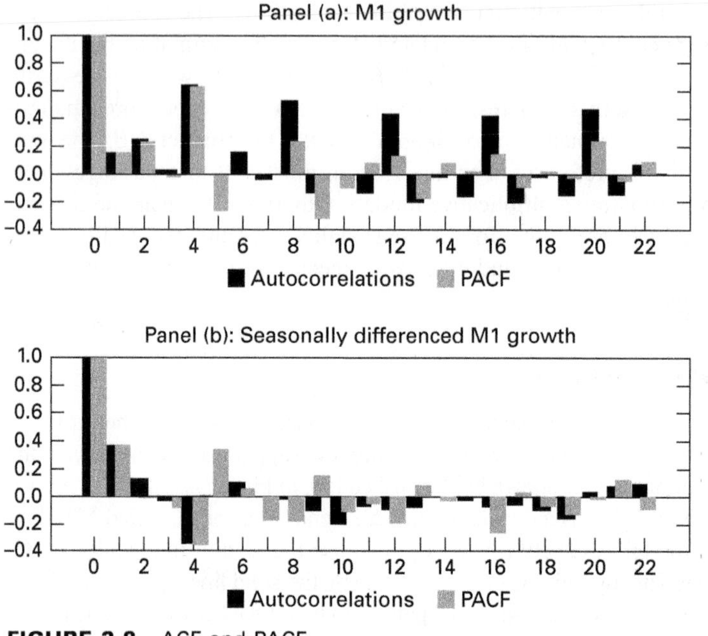

FIGURE 2.8 ACF and PACF

lags do not decay, it is necessary to take the seasonal difference so that the other autocorrelations are not dwarfed by the seasonal effects. The ACF and PACF for the growth rate of M1 are shown in Panel (a) of Figure 2.8. For now, just focus on the autocorrelations at the seasonal lags. All seasonal autocorrelations are large and show no tendency to decay. In particular, $\rho_4 = 0.65$, $\rho_8 = 0.53$, $\rho_{12} = 0.43$, $\rho_{16} = 0.41$, $\rho_{20} = 0.46$, and $\rho_{24} = 0.49$. These large autocorrelations reflect the fact that the change in M1 from one Christmas season to the next is not as pronounced as the change between the fourth quarter and other quarters.

The first step in the Box–Jenkins method is to transform the data so as to make it stationary. As such, a logarithmic transformation is helpful because it can straighten the nonlinear trend in M1. Let y_t denote the log of M1. As mentioned above, the first difference of the $\{y_t\}$ sequence, illustrated by the solid line in Figure 2.7, appears to be stationary. However, to remove the strong seasonal persistence in the data, we need to take the seasonal difference. For quarterly data, the seasonal difference is $y_t - y_{t-4}$. Since the order of differencing is irrelevant, we can form the transformed sequence

$$m_t = (1 - L)(1 - L^4)y_t$$

Thus, we use the seasonal difference of the first difference. The ACF and PACF for the $\{m_t\}$ sequence are shown in Panel (b) of Figure 2.8; the properties of this series are much more amenable to the Box–Jenkins methodology. The autocorrelation and partial autocorrelations for the first few lags are strongly suggestive of an AR(1) process ($\rho_1 = \phi_{11} = 0.37$, $\rho_2 = 0.12$, and $\phi_{22} = 0.01$). Recall that the ACF for an

Table 2.5 Three Models of Money Growth

	Model 1	Model 2	Model 3
a_1	0.541 (8.59)	0.496 (7.66)	
a_4		−0.476 (−7.28)	
β_1			0.453 (6.84)
β_4	−0.759 (−15.11)		−0.751 (−14.87)
SSR	0.0177	0.0214	0.0193
AIC	−735.9	−701.3	−720.1
SBC	−726.2	−691.7	−710.4
$Q(4)$	1.39 (0.845)	3.97 (0.410)	22.19 (0.000)
$Q(8)$	6.34 (0.609)	24.21 (0.002)	30.41 (0.000)
$Q(12)$	14.34 (0.279)	32.75 (0.001)	42.55 (0.000)

To ensure comparability, the three models are estimated over the 1962Q3–2008Q2 period. The estimated intercepts are not reported since all were insignificantly different from zero. The figures in parentheses following the Q-statistics are significance levels.

AR(1) process will decay and the PACF will cut to zero after lag 1. Given that $\rho_4 = -0.34$, $\rho_5 = 0.01$, $\phi_{44} = -0.36$, and $\phi_{55} = 0.34$, there is evidence of remaining seasonality in the $\{m_t\}$ sequence. The seasonal term is most likely to be in the form of an MA coefficient since the autocorrelation cuts to zero whereas the PACF does not. Nevertheless, it is best to estimate several similar models and then select the best. Estimates of the following three models are reported in Table 2.5:

$$m_t = a_0 + a_1 m_{t-1} + \varepsilon_t + \beta_4 \varepsilon_{t-4} \qquad \text{Model 1: AR(1) with Seasonal MA}$$

$$m_t = a_0 + (1 + a_1 L)(1 + a_4 L^4) m_{t-1} + \varepsilon_t \qquad \text{Model 2: Multiplicative Autoregressive}$$

$$m_t = a_0 + (1 + \beta_1 L)(1 + \beta_4 L^4) \varepsilon_t \qquad \text{Model 3: Multiplicative Moving Average}$$

The point estimates of the coefficients all imply stationarity and invertibility. Moreover, except for the intercepts, all are at least six standard deviations from zero. However, the diagnostic statistics all suggest that Model 1 is preferred. Model 1 has the best fit in that it has the lowest sum of squared residuals (SSR), AIC, and SBC. Moreover, the Q-statistics for lags 4, 8, and 12 indicate that the residual autocorrelations are insignificant. In contrast, the residual correlations for Model 2 are significant at the long lags [i.e., $Q(8)$ and $Q(12)$ are significant at the 0.002 and 0.001 levels]. This is because the multiplicative seasonal autoregressive (SAR) term does not adequately capture the seasonal pattern. An SAR term implies autoregressive decay from period s into period $s + 1$. In Panel (b) of Figure 2.8, the value of ρ_4 is -0.34 but ρ_5 is almost zero. As such, a multiplicative seasonal moving-average

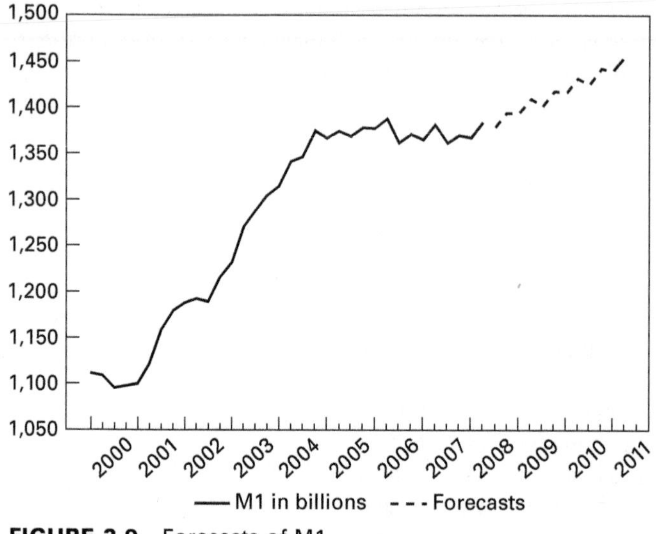

FIGURE 2.9 Forecasts of M1

(SMA) term might be more appropriate. Model 3 properly captures the seasonal pattern, but the MA(1) term does not capture the autoregressive decay present at the short lags. Other diagnostic methods, including splitting the sample, suggest that Model 1 is appropriate.

The out-of-sample forecasts are shown in Figure 2.9. To create the one- through twelve-step-ahead forecasts, Model 1 was estimated over the full sample period 1961Q3–2008Q2. The estimated model is

$$m_t = 0.528m_{t-1} + \varepsilon_t - 0.755\varepsilon_{t-4} \tag{2.69}$$

Given that $m_{2008Q2} = 4.092 \times 10^{-4}$ and the residual for 2007Q3 was -0.01231 (i.e., $\hat{\varepsilon}_{2007Q3} = -0.01231$), the forecast of m_{2008Q3} is -0.00951. Now, use this forecast and the value of $\hat{\varepsilon}_{2007Q4}$ to forecast m_{2008Q4}. You can continue in this fashion so as to obtain the out-of-sample forecasts for the $\{m_t\}$ sequence. Although you do not have the residuals for periods beyond 2008Q2, you can simply use their forecasted values of zero. The trick to forecasting future values of M1 from the $\{m_t\}$ sequence is to sum the changes and the seasonal changes so as to obtain the logarithm of the forecasted values of M1. Since $m_t = (1 - L)(1 - L^4)\ln(\text{M1}_t)$, it follows that the value of $\ln(\text{M1}_t)$ can be obtained from $m_t + \ln(\text{M1}_{t-1}) + \ln(\text{M1}_{t-4}) - \ln(\text{M1}_{t-5})$. The first 12 of the forecasted values are plotted in Figure 2.9.

The procedures illustrated in this example with highly seasonal data are typical of many other series. With highly seasonal data it is necessary to supplement the Box–Jenkins method:

1. In the identification stage, it is usually necessary to seasonally difference the data and to check the ACF of the resultant series. Often, the seasonally differenced data will not be stationary. In such instances, the data may also need to be first differenced.

2. Use the ACF and PACF to identify potential models. Try to estimate models with low-order nonseasonal ARMA coefficients. Consider both additive and multiplicative seasonality. Allow the appropriate form of seasonality to be determined by the various diagnostic statistics.

A compact notation has been developed that allows for the efficient representation of intricate models. As in previous sections, the dth difference of a series is denoted by Δ^d. Hence,

$$\begin{aligned}\Delta^2 y_t &= \Delta(y_t - y_{t-1}) \\ &= y_t - 2y_{t-1} + y_{t-2}\end{aligned}$$

A seasonal difference is denoted by Δ_s where s is the period of the data. The Dth such seasonal difference is Δ_s^D. For example, if we want the second seasonal difference of a monthly series, we can form

$$\begin{aligned}\Delta_{12}^2 y_t &= \Delta_{12}(y_t - y_{t-12}) \\ &= \Delta_{12} y_t - \Delta_{12} y_{t-12} \\ &= y_t - y_{t-12} - (y_{t-12} - y_{t-24}) \\ &= y_t - 2y_{t-12} + y_{t-24}\end{aligned}$$

Combining the two types of differencing yields $\Delta^d \Delta_s^D$. Multiplicative models are written in the form ARIMA$(p, d, q)(P, D, Q)_s$

where: p and q = the nonseasonal ARMA coefficients
d = number of nonseasonal differences
P = number of multiplicative autoregressive coefficients
D = number of seasonal differences
Q = number of multiplicative moving-average coefficients
s = seasonal period

Using this notation, we can say that the fitted equation for $m_t = \Delta\Delta^4 \ln(\text{M1}_t)$ is an ARIMA$(1, 1, 0)(0, 1, 1)_s$ model. In applied work, the ARIMA$(1, 1, 0)(0, 1, 1)_s$ and the ARIMA$(0, 1, 1)(0, 1, 1)_s$ models occurs routinely; the latter is called the *airline model* ever since Box and Jenkins (1976) used this model to analyze airline travel data.

12. PARAMETER INSTABILITY AND STRUCTURAL CHANGE

One key assumption of the Box–Jenkins methodology is that the structure of the data-generating process does not change. As such, the values of the a_i and β_i should be constant from one period to the next. However, in some circumstances, there may be reasons to suspect a structural break in the data-generating process. For example, in a model of GDP growth, it seems natural to inquire whether the oil price shocks of 1973, the events surrounding the tragedy of 9/11, and/or the financial crisis of 2008 had any significant impacts on the coefficients. Of course, parameter instability need not result from a single discrete event. The recent evidence concerning climate change suggests that weather-sensitive series such as crop yields, rainfall, and the number of ski days at Snowmass are most likely to be affected in a sustained, but gradual, way.

Testing for Structural Change

If you have reason to suspect a structural break at a particular date, it is straightforward to use a Chow test. The essence of the Chow test is to fit the same ARMA model to the pre-break data and to the post-break data. If the two models are not sufficiently different, it can be concluded that there has not been any structural change in the data-generating process.

In general, suppose you estimated an ARMA(p, q) model using a sample size of T observations. Denote the sum of the squared residuals as SSR. Also suppose that you have reason to suspect a structural break immediately following date t_m. You can perform a Chow test by dividing the T observations into two subsamples with t_m observations in the first and $t_n = T - t_m$ observations in the second. Use each subsample to estimate the two models:

$$y_t = a_0(1) + a_1(1)y_{t-1} + \ldots + a_p(1)y_{t-p} + \varepsilon_t + \beta_1(1)\varepsilon_{t-1} + \ldots + \beta_q(1)\varepsilon_{t-q}$$
$$\text{using } t_1, \ldots, t_m$$

$$y_t = a_0(2) + a_1(2)y_{t-1} + \ldots + a_p(2)y_{t-p} + \varepsilon_t + \beta_1(2)\varepsilon_{t-1} + \ldots + \beta_q(2)\varepsilon_{t-q}$$
$$\text{using } t_{m+1}, \ldots, t_T$$

Let the sum of the squared residuals from each model be SSR_1 and SSR_2, respectively. To test the restriction that all coefficients are equal [i.e., $a_0(1) = a_0(2)$ and $a_1(1) = a_1(2)$ and ... $a_p(1) = a_p(2)$ and $\beta_1(1) = \beta_1(2)$ and ... $\beta_q(1) = \beta_q(2)$], use an F-test and form:[9]

$$F = \frac{(SSR - SSR_1 - SSR_2)/n}{(SSR_1 + SSR_2)/(T - 2n)} \tag{2.70}$$

where n = number of parameters estimated ($n = p + q + 1$ if an intercept is included and $p + q$ otherwise) and the number of degrees of freedom are (n, $T - 2n$).

Intuitively, if the restriction is not binding (i.e., if the coefficients are equal), the sum $\text{SSR}_1 + \text{SSR}_2$ should equal the sum of the squared residuals from the entire sample estimation. Hence, F should equal zero. The larger the calculated value of F, the more restrictive is the assumption that the coefficients are equal.

Of course, the method requires that there be a reasonable number of observations in each subsample. If either t_m or t_n is very small, the estimated coefficients will have little precision. An alternative type of Chow test is to a use dummy variable to detect a break in one or more of the coefficients. For example, if a break is suspected right after period t_m, you can create a dummy variable, D_t, such that $D_t = 0$ for all $t \leq t_m$ and $D_t = 1$ for $t > t_m$. To test for a break in the intercept of an AR(1) model, for example, check for the significance of D_t in the regression $y_t = a_0 + \alpha_0 D_t + a_1 y_{t-1} + \varepsilon_t$. To allow for a break in both coefficients, also create the variable $D_t y_{t-1}$ and estimate the regression equation $y_t = a_0 + \alpha_0 D_t + a_1 y_{t-1} + \alpha_1 D_t y_{t-1} + \varepsilon_t$. You can test for a break by examining the individual t-statistics of α_0 and α_1 and the F-statistic for the null hypothesis $\alpha_0 = \alpha_1 = 0$.

Return to the example of the interest rate spread examined in Section 10. Suppose that there is reason to believe a break occurred at the end of 1981Q4. Consider the estimates for the two subperiods:

$$s_t = 0.923 + 0.367s_{t-1} + 0.285s_{t-2} + \varepsilon_t + 0.815\varepsilon_{t-1} - 0.153\varepsilon_{t-7} \qquad (1960Q3\text{–}1981Q4)$$

and

$$s_t = 1.799 + 0.800s_{t-1} + 0.053s_{t-2} + \varepsilon_t + 0.354\varepsilon_{t-1} + 0.097\varepsilon_{t-7} \qquad (1982Q1\text{–}2008Q1)$$

Although the coefficients of the models appear to be dissimilar, we can formally test for the equality of coefficients using (2.70). Respectively, the sum of squared residuals for the two equations are $SSR_1 = 27.564$ and $SSR_2 = 21.414$. Estimating the model over the full sample period yields $SSR = 49.692$. Since there are 191 usable observations in the sample and $n = 5$ (the intercept plus the four estimated coefficients), (2.70) becomes

$$F = [(49.692 - 27.564 - 21.414)/5]/[(27.564 + 21.414)/(191 - 10)] = 0.527$$

With 5 degrees of freedom in the numerator and 181 in the denominator, we cannot reject the null of no structural change in the coefficients (i.e., we can accept the hypothesis that there is no structural change in the coefficients).

Alternatively, to test for a break in the intercept only, we can create the dummy variable D_t equal to zero prior to 1982Q1 and equal to unity beginning in 1982Q1. Now, consider the equation for the spread estimated over the entire 1960Q1–2008Q1 period:

$$\underset{(3.55)}{s_t = 1.277} + \underset{(0.82)}{0.312D_t} + \underset{(3.23)}{0.336s_{t-1}} + \underset{(4.43)}{0.435s_{t-2}} + \varepsilon_t + \underset{(13.14)}{0.837\varepsilon_{t-1}} - \underset{(-3.33)}{0.134\varepsilon_{t-7}}$$

Since D_t jumps from 0 to 1 in 1982Q1, the estimate for the intercept is 1.227 prior to 1982Q1 and 1.589 (= 1.277 + 0.312) beginning in 1982Q1. However, since the t-statistic for the null hypothesis $D_t = 0$ cannot be rejected, there is no evidence of a significant intercept break.

Endogenous Breaks

The Chow test asks whether there is a break beginning at some particular known break date t_m. A break occurring at a date not pre-specified by the researcher is called an **endogenous break** to denote the fact that it was not the result of a fixed break date such as 9/11. To determine whether there is a break anywhere in the sample, you could perform a Chow test for every potential break date t_m. It should not be surprising that the break date that results in the largest value of the F-statistic provides a consistent estimate of the actual break date, if any. In order to ensure an adequate number of observations in each of the two subsamples, it is necessary to have a "trimming" such that the break could not occur before the first t_0 observations or after the last $T - t_0$ observations. In applied research, it is common to use a trimming value of 10 percent so that there are at least 10 percent of the observations in each of the two subsamples. In the interest rate spread example, there are 191 usable observations in the 1960*Q*1–2008*Q*1 period (since the first two are lost when estimating the coefficient for s_{t-2}). If you used a 10 percent trimming, you could check for a break everywhere in the interval 1965*Q*1 to 2003*Q*2 (each about 19 observations from the beginning and end of the usable data). Unfortunately, searching for the most likely break date means that the F-statistic for the null hypothesis of no break is inflated. After all, you have just searched for the date that leads to the maximum, or supremum, value of the sample F-statistic. As such, the distribution for the F-statistic is not standard and cannot be obtained from a traditional F-table. Andrews (1993) and Andrews and Ploberger (1994) contain asymptotic critical values for this type of test. As discussed in Chapter 7, Hansen (1997) shows how to obtain the appropriate critical values using bootstrapping methods. Fortunately, a number of software packages can readily perform such tests.[10]

Parameter Instability

Notice that the Chow test and its variants require the researcher to specify a particular break date and to assume that the break fully manifests itself at that date. The intercept, for example, is $a_0(1)$ up to t_m and is precisely $a_0(2)$ beginning at t_{m+1}. However, the assumption that a break occurs exactly at one point in time may not always be appropriate. As mentioned above, there is no particular date at which we can say that significant climate change has occurred. Similarly, it is not clear how we can provide a specific break date to denote the advent "financial deregulation" in the asset markets or to assign a specific date to the development of the microcomputer. These are processes that have been evolving over time. Even if we could date the precise start of financial deregulation or the computer revolution, the full effects of these changes would not occur instantly. As such, it should not be surprising that a number of procedures have been developed that check for parameter stability without the need to identify a particular break date. Probably the simplest method is to estimate the model recursively. For example, if you have 150 observations, you can estimate the model using only the first few, say 10, observations. Plot the individual coefficients and then reestimate the model using the first 11 observations. You can keep repeating this process until you use all 150 observations. In general, the plots of the coefficients will be not be flat since the preliminary values are estimated using a very small number of observations. However, after a "burn-in" period, the time plots of the individual coefficients can provide evidence of coefficient stability. If the magnitude of a coefficient suddenly begins to change, you should suspect a structural change at that point. A sustained change in a coefficient might indicate a model misspecification. One particularly helpful modification of this procedure is to plot each coefficient along with its estimated ± 2 standard deviation band. The bands represent confidence intervals for the estimated coefficients. In this way, it can be seen if the coefficients are always statistically significant and whether the coefficients in the early periods appear to be statistically different from those of the latter periods.

At each step along the way, it is also possible to create the one-step-ahead forecast error. Let $e_t(1)$ be the one-step-ahead forecast error made using all observations through t. In other words, $e_t(1)$ is the difference between y_{t+1} and your conditional forecast of y_{t+1} (i.e., $E_t y_{t+1}$). If you start with the first 10 observations, the value of

$e_{10}(1)$ will be $y_{11} - E_{10}y_{11}$ and the value of $e_{149}(1)$ will be $y_{150} - E_{149}y_{150}$. [Note: If you understand the notation, it should be clear that you cannot create the value $e_{150}(1)$ since you do not have the value of y_{151}.] If your model fits the data well, the forecasts should be unbiased so that the sum of these forecast errors should not be "too far" from zero. In fact, Brown, Durbin, and Evans (1975) calculate whether the cumulated sum of the forecast errors is statistically different from zero. To be a bit more formal, define

$$CUSUM_N = \sum_{i=n}^{N} e_i(1)/\sigma_e \qquad N = n, \ldots, T - 1$$

where n denotes the date of the first forecast error you constructed, T denotes the date of the last observation in the data set, and σ_e is the estimated standard deviation of the forecast errors. With 150 total observations ($T = 150$), if you start the procedure using the first 10 observations ($n = 10$), 140 forecast errors ($T - n$) can be created. Note that σ_e is created using all $T - n$ forecast errors. Starting with $N = n$, to create $CUSUM_{10}$, use the first three observations to create $e_{10}(1)/\sigma_e$. Now let $N = 11$ and create $CUSUM_{11}$ as $[e_{10}(1) + e_{11}(1)]/\sigma_e$. Similarly, $CUSUM_{T-1} = [e_{10}(1) + \ldots + e_{T-1}(1)]/\sigma_e$. If you use the 5 percent significance level, the plot value of each value of $CUSUM_N$ should be within a band of approximately $\pm 0.948\ [(T - n)^{0.5} + 2(N - n)\,(T - n)^{-0.5}]$.

An Example of a Break

In order to illustrate a breaking series, the first panel of Figure 2.10 on the next page shows 150 observations of the simulated series $y_t = 1 + 0.5y_{t-1} + \varepsilon_t$ for $t < 101$ and $y_t = 2.5 + 0.65y_{t-1} + \varepsilon_t$ for $t \geq 101$. The series is contained in the file Y-BREAK.XLS. Of course, in applied work, the break may not be so readily apparent. If you ignore the break and estimate the entire series as an AR(1) process, you should obtain:

$$y_t = \underset{(2.635)}{0.4442} + \underset{(22.764)}{0.8822}y_{t-1}$$

As indicated by the remaining three panels of the figure, the estimated AR(1) model is seriously misspecified. Respectively, Panels (b) and (c) show the estimates of the intercept term and the AR(1) coefficient (along with their (± 2 standard deviation bands) resulting from a recursive estimation. The initial confidence intervals are quite wide since the first few estimations use a very small number of observations. The estimates all seem reasonable until about $t = 100$. At this point, the estimates of the intercept actually decline while the estimates of the AR(1) coefficient rise (the reverse of what happens in the data-generating process). Note that the confidence bands for the latter periods do not even overlap those from the middle periods. The clear suggestion is that there has been a significant structural change. The *CUSUMs*, shown in Panel (d), are clearly within the 90 percent confidence interval for $t < 101$. At this point, they begin to drift upward and depart from the band at $t = 125$. As such, the hypothesis of coefficient stability can be rejected.

Notice that the *CUSUMs* do not actually depart from the band until late in the sample. This is indicative of the problem that the *CUSUM* test may not detect coefficient instability occurring late in the sample period. Moreover, the test may not have much power if there are multiple changes with little overall effect on the *CUSUMs*.[11]

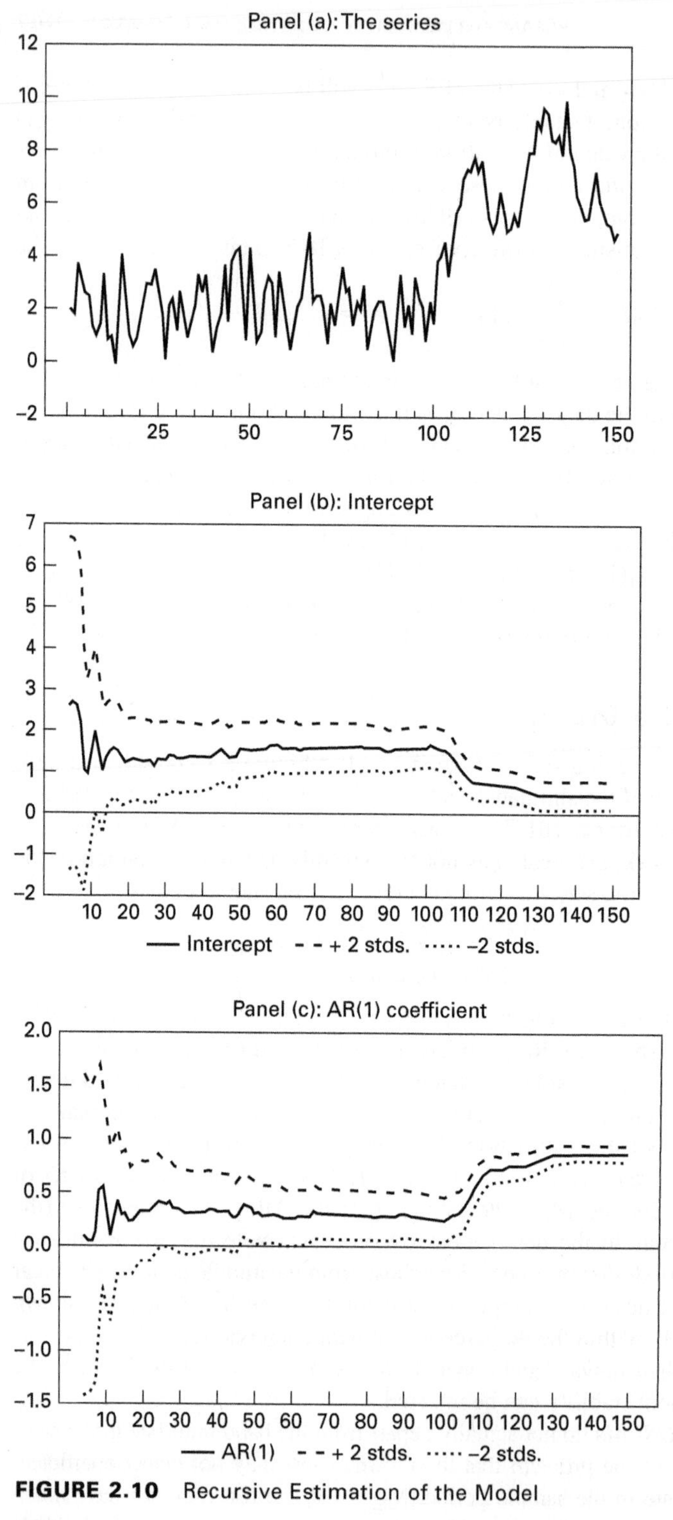

FIGURE 2.10 Recursive Estimation of the Model

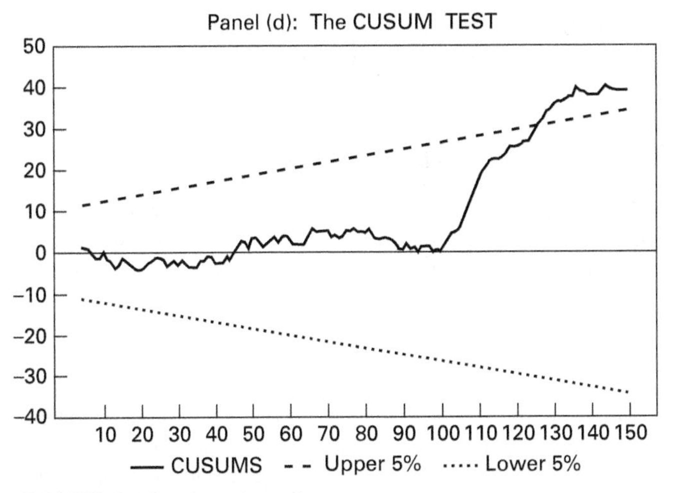

FIGURE 2.10 *(continued)*

Nevertheless, the test is a useful diagnostic tool that does not require the researcher to stipulate the nature of the model's misspecification. It is able to detect model misspecifications from such varied sources including smooth structural breaks, multiple breaks, neglected nonlinearities in the data-generating process, or an overly parsimonious model.

If you had strong reason to believe that the break occurred in period 101, you could form a dummy variable $D_t = 0$ from $t = 1$ to 100 and $D_t = 1$ thereafter. To check for an intercept break, estimate

$$\underset{(5.36)}{y_t = 0.9254} + \underset{(8.91)}{0.5683y_{t-1}} + \underset{(5.88)}{1.936D_t}$$

Since the coefficient for D_t is highly significant, you can conclude that there was a break in the intercept. To check for a break in the intercept and slope coefficient, also form the variable $D_t y_{t-1}$ and estimate:

$$y_t = \underset{(7.22)}{1.6015} + \underset{(2.76)}{0.2545y_{t-1}} - \underset{(-0.391)}{0.2244D_t} + \underset{(4.47)}{0.5433D_t y_{t-1}}$$

In this particular case, the dummy variables indicate that there is a break but do not measure the size of the break very well. (Note: The actual break in the intercept is +1.5 and the actual break in the AR(1) coefficient is 0.15.) The coefficient for the intercept break is not significant while the break in the slope coefficient is highly significant. The F-statistic for the joint hypothesis that the coefficients on D_t and $D_t y_{t-1}$ are equal to zero is 29.568. With 2 degrees of freedom in the numerator and 145 in the denominator, this value is significant at any conventional level. The important point is that you can conclude that the simple AR(1) model is misspecified because of a structural break.

If you wanted to estimate the most likely value for t_m, you could repeat the estimation for every time period in the interval $15 < t_m < 135$. The sum of squared residuals is smallest for $t_m = 100$. Although this consistent estimate of the break date turns out to be exactly correct, you should expect a discrepancy when using actual data.

Also note that the F-test (and the t-statistics of the individual coefficients) for the null hypothesis of no structural change can be tested using Hansen's (1997) bootstrapping method (see Chapter 7).

13. SUMMARY AND CONCLUSIONS

The chapter focuses on the Box–Jenkins (1976) approach to identification, estimation, diagnostic checking, and forecasting a univariate time series. ARMA models can be viewed as a special class of linear stochastic difference equations. By definition, an ARMA model is covariance-stationary in that it has a finite and time-invariant mean and covariances. For an ARMA model to be stationary, the characteristic roots of the difference equation must lie inside the unit circle. Moreover, the process must have started infinitely far in the past or the process must always be in equilibrium.

In the identification stage, the series is plotted and the sample autocorrelations and partial correlations are examined. A slowly decaying autocorrelation function suggests nonstationarity behavior. In such circumstances, Box and Jenkins recommend differencing the data. Formal tests for nonstationarity are presented in Chapter 4. A common practice is to use a logarithmic or Box–Cox transformation if the variance does not appear to be constant. Chapter 3 presents some modern techniques that can be used to model the variance.

The sample autocorrelations and partial correlations of the suitably transformed data are compared to those of various theoretical ARMA processes. All plausible models are estimated and compared using a battery of diagnostic criteria. A well-estimated model: (i) is parsimonious; (ii) has coefficients that imply stationarity and invertibility; (iii) fits the data well; (iv) has residuals that approximate a white-noise process; (v) has coefficients that do not change over the sample period; and (vi) has good out-of-sample forecasts.

A useful check for coefficient instability involves recursive estimation techniques. A sudden change in the recursive estimates of one or more coefficients is indicative of a structural break. The Chow test can be used to test for a break at a known date and the Andrews and Ploberger (1994) test can detect an endogenous break. Bai and Perron (1998, 2003) show how to test for multiple endogenous breaks. More gradual changes can be detected by recursive estimation or by a *CUSUM* test.

In utilizing the Box–Jenkins methodology, you will find yourself making many seemingly *ad hoc* choices. The most parsimonious model may not have the best fit but may have the best out-of-sample forecasts. You will find yourself addressing the following types of questions: What is the most appropriate data transformation? Is an ARMA(2, 1) model more appropriate than an ARMA(1, 2) specification? How can seasonality best be modeled? What should be done about seemingly significant coefficients at reasonably long lags? Given this latitude, many view the Box–Jenkins methodology as an art rather than a science. Nevertheless, the technique is best learned through experience. The exercises at the end of this chapter are designed to guide you through the types of choices you will encounter in your own research.

QUESTIONS AND EXERCISES

1. In the coin-tossing example of Section 1, your average winnings on the last four tosses (w_t) can be denoted by

$$w_t = 1/4\varepsilon_t + 1/4\varepsilon_{t-1} + 1/4\varepsilon_{t-2} + 1/4\varepsilon_{t-3}$$

a. Find the expected value of your winnings. Find the expected value given that $\varepsilon_{t-3} = \varepsilon_{t-2} = 1$.

b. Find var(w_t). Find var(w_t) conditional on $\varepsilon_{t-3} = \varepsilon_{t-2} = 1$.

c. Find cov(w_t, w_{t-1}), cov(w_t, w_{t-2}), and cov(w_t, w_{t-5}).

2. Consider the second-order autoregressive process $y_t = a_0 + a_2y_{t-2} + \varepsilon_t$ where $|a_2| < 1$.

a. Find: **i.** $E_{t-2}y_t$ **ii.** $E_{t-1}y_t$ **iii.** E_ty_{t+2} **iv.** $\text{cov}(y_t, y_{t-1})$
v. $\text{cov}(y_t, y_{t-2})$ **vi.** the partial autocorrelations ϕ_{11} and ϕ_{22}

b. Find the impulse response function. Given y_{t-2}, trace out the effects of an ε_t shock on the $\{y_t\}$ sequence.

c. Determine the forecast function: E_ty_{t+s}. The forecast error $e_t(s)$ is the difference between y_{t+s} and E_ty_{t+s}. Derive the correlogram of the $\{e_t(s)\}$ sequence. (*Hint*: Find $E_te_t(s)$, $\text{var}[e_t(s)]$, and $E_t[e_t(s)e_t(s-j)]$ for $j = 0$ to s.)

3. Two different balls are drawn from a jar containing three balls numbered 1, 2, and 4. Let x denote the number on the first ball drawn and y denote the sum of the two balls drawn.

a. Find the joint probability distribution for x and y; that is, find prob$(x = 1, y = 3)$, prob$(x = 1, y = 5)$, ..., prob $(x = 4, y = 6)$.

b. Find each of the following: $E(x)$, $E(y)$, $E(y \mid x = 1)$, $E(x \mid y = 5)$, $\text{var}(x \mid y = 5)$, and $E(y^2)$.

c. Consider the two functions $w_1 = 3x^2$ and $w_2 = x^{-1}$. Find $E(w_1 + w_2)$ and $E(w_1 + w_2 \mid y = 3)$.

d. How would your answers change if the balls were drawn with replacement?

4. Substitute (2.10) into $y_t = a_0 + a_1y_{t-1} + \varepsilon_t$. Show that the resulting equation is an identity.

a. Find the homogeneous solution to $y_t = a_0 + a_1y_{t-1} + \varepsilon_t$.

b. Find the particular solution given that $| a_1 | < 1$.

c. Show how to obtain (2.10) by combining the homogeneous and particular solutions.

5. The general solution to an nth-order difference equation requires n arbitrary constants. Consider the second-order equation $y_t = a_0 + 0.75y_{t-1} - 0.125y_{t-2} + \varepsilon_t$.

a. Find the homogeneous and particular solutions. Discuss the shape of the impulse response function.

b. Find the values of the initial conditions that ensure the $\{y_t\}$ sequence is stationary.

c. Given your answer to part (b), derive the correlogram for the $\{y_t\}$ sequence.

6. Consider the second-order stochastic difference equation: $y_t = 1.5y_{t-1} - 0.5y_{t-2} + \varepsilon_t$.

a. Find the characteristic roots of the homogeneous equation.

b. Demonstrate that the roots of $1 - 1.5L + 0.5L^2$ are the reciprocals of your answer in part a.

c. Given initial conditions for y_0 and y_1, find the solution for y_t in terms of the current and past values of the $\{\varepsilon_t\}$ sequence.

d. Find the forecast function for y_{T+s} (i.e., find the solution for all values of y_{T+s} given the values of y_T and y_{T-1}).

e. Find Ey_t, Ey_{t+1}, $\text{var}(y_t)$, $\text{var}(y_{t+1})$, and $\text{cov}(y_{t+1}, y_t)$.

7. There are often several representations for the identical time-series process. In the text, the standard equation for an AR(1) model is given by $y_t = a_0 + a_1y_{t-1} + \varepsilon_t$.

a. Show that an equivalent representation is $y_t = a_0/(1 - a_1) + \mu_t$ where $\mu_t = a_1\mu_{t-1} + \varepsilon_t$.

b. In Chapter 1, we considered several models with a deterministic time trend. For example, a modified version of equation (1.62) is $y_t = a_0 + a_1y_{t-1} + a_2t + \varepsilon_t$ where $|a_1| < 1$. Explain why the y_t sequence is not stationary. Also explain why the y_t sequence is stationary about the trend line $a_0 + a_2t$. What does it mean to say that the y_t sequence is **trend-stationary?**

c. Verify that the process generated by $y_t = 16.2 + 0.2t + \mu_t$ where $\mu_t = 0.95\mu_{t-1} + \varepsilon_t$ is identical to the process generated by $y_t = 1 + 0.95y_{t-1} + 0.01t + \varepsilon_t$.

d. Show that the first-order trend-stationary process $y_t = a_0 + a_1y_{t-1} + a_2t + \varepsilon_t$ where $|a_1| < 1$ can be written in the form $y_t = c_0 + c_1t + \mu_t$ where $\mu_t = c_2\mu_{t-1} + \varepsilon_t$. Also, use the method of undetermined coefficients to find the values of c_0, c_1, and c_2.

8. The file entitled SIM_2.XLS contains the simulated data sets used in this chapter. The first series, denoted Y1, contains the 100 values of the simulated AR(1) process used in Section 7. Use this series to perform the following tasks (*Note*: Due to differences in data handling and rounding, your answers need only approximate those presented here):

a. Plot the sequence against time. Does the series appear to be stationary?

b. Use the data to verify the results given in Table 2.2.

c. Estimate the series as an AR(2) process without an intercept. You should obtain:

$$y_t = \underset{(7.01)}{0.710}\, y_{t-1} + \underset{(1.04)}{0.105}\, y_{t-2} + e_t \qquad \text{usable observations: 98}$$

Ljung–Box Q-Statistics: $Q(8) = 5.13$. $Q(16) = 15.86$. $Q(24) = 21.02$

d. Estimate the series as an ARMA(1, 1) process without an intercept. You should obtain:

$$y_t = \underset{(12.16)}{0.844}\, y_{t-1} - \underset{(-1.12)}{0.142}\, e_{t-1} + e_t \qquad \text{usable observations: 99}$$

Verify that the ACF and PACF of the residuals do not indicate any serial correlation.

9. The second column in file SIM_2.XLS contains the 100 values of the simulated ARMA(1, 1) process used in Section 7. This series is entitled Y2. Use this series to perform the following tasks (*Note*: Due to differences in data handling and rounding, your answers need only approximate those presented here):

a. Plot the sequence against time. Does the series appear to be stationary? Plot the ACF.

b. Verify the results in Table 2.3.

c. Estimate the process using a pure MA(2) model. You should obtain:

$$y_t = \underset{(-13.22)}{-1.15}\, e_{t-1} + \underset{(5.98)}{0.522}\, e_{t-2} + e_t \qquad \text{usable observations: 100}$$

Verify that the Ljung–Box Q-statistics are $Q(8) = 28.48$, $Q(16) = 37.47$, and $Q(24) = 38.84$ with significance levels of 0.000, 0.000, and 0.015, respectively.

d. Compare the MA(2) to the ARMA(1, 1).

10. The third column in file SIM_2.XLS contains the 100 values of the simulated AR(2) process used in Section 7. This series is entitled Y3. Use this series to perform the following tasks (*Note*: Due to differences in data handling and rounding, your answers need only approximate those presented here):

a. Plot the sequence against time. Verify the ACF and the PACF coefficients reported in Section 7. Compare the sample ACF and PACF to the those of a theoretical AR(2) process.

b. Estimate the series as an AR(1) process. You should find that the estimated AR(1) coefficient and the t-statistic are

$$y_t = \underset{(5.24)}{0.467}\, y_{t-1} + e_t$$

Show that the standard diagnostic checks indicate that this AR(1) model is inadequate. Be sure to perform a recursive estimation of the *AR*(1) model and to plot the *CUSUMs*.

c. Could an ARMA(1, 1) process generate the type of sample ACF and PACF found in part a? Estimate the series as an ARMA(1, 1) process. You should obtain:

$$y_t = \underset{(1.15)}{0.183}y_{t-1} + \underset{(3.64)}{0.510}e_{t-1} + e_t \qquad \text{usable observations: 99}$$

Use the Ljung–Box Q-statistics to show that the ARMA(1, 1) model is inadequate.

d. Estimate the series as an AR(2) process to verify the results reported in the text.

11. The file QUARTERLY.XLS contains a number of series including the U.S. index of industrial production (indprod), unemployment rate (urate), and producer price index for finished goods (finished). All of the series run from 1960Q1 to 2008Q1.

a. Exercises with indprod.

i. Construct the growth rate of the series as $y_t = \log(\text{indprod}_t) - \log(\text{indprod}_{t-1})$. Since the first few autocorrelations suggest an AR(1), estimate $y_t = 0.0035 + 0.543y_{t-1} + \varepsilon_t$ (the t-statistics are 3.50 and 9.02, respectively).

ii. Show that adding an MA term at lag 8 improves the fit and removes the serial correlation.

b. Exercises with urate.

i. Examine the ACF of the series. It should be clear that the ACF decays slowly and that the PACF has a very negative spike at lag 2.

ii. Temporarily ignore the necessity of differencing the series. Estimate urate as an AR(2) process including a constant. You should find that $a_1 + a_2 = 0.961$.

iii. Form the difference as $y_t = \text{urate}_t - \text{urate}_{t-1}$. Show that the model $y_t = a_0 + a_1 y_{t-1} + \beta_4\varepsilon_{t-4} + \beta_8\varepsilon_{t-8} + \varepsilon_t$ exhibits no remaining serial correlation and has a better fit than $y_t = a_0 + a_1 y_{t-1} + a_4 y_{t-4} + \beta_8\varepsilon_{t-8} + \varepsilon_t$ or $y_t = a_0 + a_1 y_{t-1} + a_4 y_{t-4} + a_8 y_{t-8} + \varepsilon_t$.

c. Exercises with finished.

i. Construct the growth rate of the series as $y_t = \log(\text{finished}_t) - \log(\text{finished}_{t-1})$.

ii. Show that an ARMA(1, 1) model fits the data well but that the residual correlation at lag 3 is problematic.

iii. Show that an ARMA((1, 3), 1) is not appropriate. Show that an ARMA(1, (1, 3)) shows no remaining residual correlation.

iv. Show that an AR(3) has the best fit and has no remaining serial correlation.

12. The file QUARTERLY.XLS contains U.S. interest rate data from 1960Q1 to 2008Q1. As indicated in Section 10, form the spread by subtracting the T-bill rate from the 10-year rate.

a. Use the full sample period to obtain estimates of the AR(7) and the ARMA(1, 1) model reported in Section 10.

b. Estimate the AR(7) and ARMA(1, 1) models over the period 1960Q1 to 1994Q3. Obtain the one-step-ahead forecast and the one-step-ahead forecast error from each. As in Section 10, continue to update the estimation period so as to obtain the fifty one-step-ahead forecast errors from each model. Let $\{f_{1t}\}$ denote the forecasts from the AR(7) and f_{2t} denote the forecasts from the ARMA(1, 1). You should find that the properties of the forecasts are such that

$$y_{1993Q3+t} = 0.0359 + 0.986f_{1t} \text{ and } y_{1993Q3+t} = -0.055 + 1.042f_{2t}$$

Are the forecasts unbiased?

c. Construct the Diebold–Mariano test using the mean absolute error. How do the results compare to those reported in Section 10?

d. Use the Granger–Newbold test to compare the AR(7) model to the ARMA(1, 1).

e. Construct the ACF and PACF of the first-difference of the spread. What type of model is suggested?

f. Show that a model with 2 AR lags and MA lags at 3, 5, and 8 has a better fit than any of the models reported in the text.

13. As you read more of the time-series literature, you will find that different authors and different software packages report the AIC and the SBC in various ways. The purpose of this exercise is to show that, regardless of the method you use, you will always select the same model. The examples in the text use AIC = $T \ln(SSR) + 2n$ and SBC = $T \ln(SSR) + n \ln(T)$ where SSR = sum of squared residuals. However, other common formulas include

$$\text{AIC}^* = -2 \ln(L)/T + 2n/T \text{ and } \text{SBC}^* = -2 \ln(L)/T + n \ln(T)/T$$

and

$$\text{AIC}' = \exp(2n/T) \cdot SSR/T \;\; \text{SBC}' = T^{n/T} \cdot SSR/T$$

where SSR = sum of squared residuals, $\ln(L)$ = maximized value of the log of the likelihood function = $-(T/2)\ln(2\pi) - (T/2)\ln(\sigma^2) - (1/2\sigma^2)(SSR)$, and σ^2 = variance of the residuals.

a. Jennifer estimates two different models over the same time period and assesses their fit using the formula AIC* = $-2 \ln(L)/T + 2n/T$. She denotes the two values AIC*(1) and AIC*(2) and finds that AIC*(1) < AIC*(2). Justin estimates the same two models over the same time period but assesses the fit using the formula AIC = $T \ln(SSR) + 2n$. Show that Justin's results must be such that AIC(1) < AIC(2). *Hint*: Since AIC*(1) < AIC*(2), it must be the case that $\ln(2\pi) + \ln(\sigma_1^2) + T(1/\sigma_1^2)(SSR_1) + 2n_1/T < \ln(2\pi) + \ln(\sigma_2^2) + T(1/\sigma_2^2)(SSR_2) + 2n_2/T$ where n_i, SSR_i, and σ_i^2 are the number of parameters, the sum of squared residuals, and the residual variance of model i. Recall that the estimate of σ^2 is SSR/T. If you simplify the inequality relationship, you should find that it is equivalent to $T \ln(SSR_1) + 2n_1 < T \ln(SSR_2) + 2n_2$.

b. Show that all three methods of calculating the SBC will necessarily select the same model.

c. Select one of the three pairs above. Show that the AIC will never select a more parsimonious model than the SBC.

14. The file labeled Y_BREAK.XLS contains the 150 observations of the series constructed as $y_t = 1 + 0.5y_{t-1} + (1 + 0.1y_{t-1})D_t + \varepsilon_t$ where D_t is a dummy variable equal to 0 for $t < 101$ and equal to 1.5 for $t \geq 101$.

a. Explain how this representation of the model allows the intercept to jump from 1 to 2.5 and the $AR(1)$ coefficient to jump from 0.5 to 0.65.

b. Use the data to verify the results reported in the text.

c. Why do you think that the estimated intercept actually falls beginning with period 101?

d. Estimate the series as an $AR(2)$ process. In what sense does the $AR(2)$ model perform better than the $AR(1)$ model estimated in part a?

e. Perform a recursive estimation of the $AR(2)$ model and plot the *CUSUMs*. Is the $AR(2)$ model adequate?

15. The file QUARTERLY.XLS contains the U.S. money supply as measured by M1 (M1NSA) and as measured by M2 (M2NSA). The series are quarterly averages over the period 1960Q1 to 2008Q2.

a. Reproduce the results for M1 that are reported in Section 11 of the text.

b. How do the three models of M1 reported in the text compare to a model with a seasonal AR(1) term with an additive MA(1) term?

c. Obtain the ACF for the growth rate of the M2NSA series. What type of model is suggested by the ACF?

d. Call the seasonally differenced growth rate $m2_t$. Estimate an AR(1) model with a seasonal MA term over the 1962Q3 to 2008Q2 period. You should obtain: $m2_t = 0.5708m2_{t-1} + \varepsilon_t - 0.8547\varepsilon_{t-4}$. Show that this model is preferable to (i) an AR(1) with a seasonal AR term, (ii) MA(1) with a seasonal AR term, and (iii) an MA(1) with a seasonal MA term.

e. Would you recommend including an MA term at lag 2 to remove any remaining serial correlation in the residuals?

16. The file QUARTERLY.XLS contains the quarterly values of the Consumer Price Index (excluding food and fuel) that have not been seasonally adjusted (CPINSA). Form the inflation rate, as measured by the CPI, as ln(CPINSA$_t$/CPINSA$_{t-1}$).

a. Plot the CPI. Does there appear to be a plausible seasonal pattern in the data? Seasonally difference the CPI using: $\Delta_4$log(CPINSA$_t$) = log(CPINSA)$_t$ − log(CPINSA)$_{t-4}$. You should find the ACF is such that:

ρ_1	ρ_2	ρ_3	ρ_4	ρ_5	ρ_6	ρ_7	ρ_8	ρ_9	ρ_{10}	ρ_{11}	ρ_{12}
0.98	0.93	0.88	0.82	0.76	0.71	0.66	0.61	0.57	0.53	0.51	0.49

Why is it difficult to use the Box–Jenkins methodology on this series?

b. For convenience, let π_t denote the series $(1 - L)(1 - L^4)$ln(CPINSA$_t$). Find the most appropriate model for the $\{\pi_t\}$ series. In particular, compare the following four models for the series:

$\pi_t = a_0 + a_1\pi_{t-1} + \varepsilon_t + \beta_4\varepsilon_{t-4}$ — Model 1: AR(1) with Seasonal MA

$\pi_t = a_0 + (1 + a_1L)(1 + a_4L^4)\pi_{t-1} + \varepsilon_t$ — Model 2: Multiplicative Autoregressive

$\pi_t = a_0 + (1 + \beta_1L)(1 + \beta_4L^4)\varepsilon_t$ — Model 3: Multiplicative Moving Average

$\pi_t = a_0 + (a_1\pi_{t-1} + \beta_1\varepsilon_{t-1})(1 + \beta_4\varepsilon_{t-4}) + \varepsilon_t$ — Model 4: ARMA(1, 1) with Seasonal MA

In what sense is Model 4 the most appropriate?

ENDNOTES

1. Often, the variance is estimated as

$$\hat{\sigma}^2 = [1/(T - 1)]\sum_{t=1}^{T}(y_t - y)^2$$

2. As discussed in Appendix 2.1, the estimation of lagged MA coefficients does not entail a loss of any usable observations. Hence, the two models are estimated over the same sample period.
3. Most software programs will not be able to estimate (2.43) since there is not a unique set of parameter values that minimizes the likelihood function.
4. Some software programs report the Durbin–Watson test statistic as a check for first-order serial correlation. This well-known test statistic is biased toward finding no serial correlation in the presence of lagged dependent variables. Hence, it is usually not used in ARMA models.
5. Some researchers prefer to drop the first observation when adding an additional observation. As such, the model is always estimated using a fixed number of observations. A potential advantage of this "rolling window" method is that structural change occurring early in the sample will not affect all of the forecasts. Of course, the disadvantage is that some of the data is not used when estimating the model.
6. Unfortunately, the construction of (2.62) can be sensitive to the choice of q and, if one or more of the $\gamma_1 < 0$, the estimated variance can be negative. In such circumstances, it is preferable to use robust standard errors—such as those in Newey and West (1987). All professional software packages allow you to directly obtain the Newey–West estimator of the variance. Additional details are included in the *Supplementary Manual* available at www.cba.ua.edu/~wenders.
7. In a pure AR(p) model, one observation is also lost for each lag. Since the data set begins in 1960Q1, estimation of the AR(7) model can begin no earlier than 1961Q4. To ensure comparability, all of the models reported in Table 2.4 were estimated over the same sample period. Note that no usable observations are lost in estimating MA(q) models. Some software programs initialize the values of $\varepsilon_1, \ldots, \varepsilon_q$ to be zero so that no additional usable observations are lost. Others will "backcast" the initial values of $\varepsilon_1, \ldots, \varepsilon_q$.
8. The details of the X-11 and X-12 procedures are not important for our purposes. The technical details along with several versions of the X-12-ARIMA seasonal adjustment procedure can be downloaded from the Bureau of the Census Web page: www.census.gov/srd/www/x12a/.
9. As formulated, the test can also detect a break in the variance of the error process. Estimation of an AR(p) model usually entails a loss of the number of usable observations. Hence, to estimate a model using T usable observations it will be necessary to have a total of $(T + p)$ observations. Also note that the procedure outlined necessitates that the second subsample period incorporate the lagged values $t_m, t_{m-1}, \ldots t_{m-p+1}$.
10. A more detailed discussion of the bootstrapping method used to obtain the critical values is contained in Chapter 7.
11. A variant of the test, often called *CUSUM*(2), is to form the *CUSUM*s using the squared errors. The use of the squared errors can help detect changes in the variance.

APPENDIX 2.1: ESTIMATION OF AN MA(1) PROCESS

How do you estimate an MA or an ARMA process? When you estimate a regression using ordinary least squares (OLS), you have a dependent variable and a set of independent variables. In an AR model, the list of regressors is simply the lagged values of the $\{y_t\}$ series. Estimating an MA process is different because you do not know the values of the $\{\varepsilon_t\}$ sequence. Since you cannot directly estimate a regression equation, maximum-likelihood estimation is used. Suppose that $\{\varepsilon_t\}$ is a white-noise sequence drawn from a normal distribution. The likelihood of any realization ε_t is

$$\frac{1}{\sqrt{2\pi\sigma^2}} \exp\left(\frac{-\varepsilon_t^2}{2\sigma^2}\right)$$

Since the ε_t are independent, the likelihood of the joint realizations $\varepsilon_1, \varepsilon_2, \ldots, \varepsilon_T$ is

$$\prod_{t=1}^{T} \frac{1}{\sqrt{2\pi\sigma^2}} \exp\left(\frac{-\varepsilon_t^2}{2\sigma^2}\right)$$

If you take the log of the likelihood, you obtain

$$\ln L = \frac{-T}{2}\ln(2\pi) - \frac{T}{2}\ln\sigma^2 - \frac{1}{2\sigma^2}\sum_{t=1}^{T}\varepsilon_t^2$$

Now suppose that we observe T values of the MA(1) series $y_t = \beta\varepsilon_{t-1} + \varepsilon_t$. The problem is to construct the $\{\varepsilon_t\}$ sequence from the observed values of $\{y_t\}$. If we knew the true value of β and knew that $\varepsilon_0 = 0$, we could construct $\varepsilon_1, \ldots, \varepsilon_T$ recursively. Given that $\varepsilon_0 = 0$, it follows that

$$\begin{aligned}
\varepsilon_1 &= y_1 \\
\varepsilon_2 &= y_2 - \beta\varepsilon_1 = y_2 - \beta y_1 \\
\varepsilon_3 &= y_3 - \beta\varepsilon_2 = y_3 - \beta(y_2 - \beta y_1) \\
\varepsilon_4 &= y_4 - \beta\varepsilon_3 = y_4 - \beta[y_3 - \beta(y_2 - \beta y_1)]
\end{aligned}$$

In general, $\varepsilon_t = y_t - \beta\varepsilon_{t-1}$ so that if L is the lag operator

$$\varepsilon_t = y_t/(1 + \beta L) = \sum_{i=0}^{t-1}(-\beta)^i y_{t-i}$$

As long as $|\beta| < 1$, the values of ε_t will represent a convergent process. This is the justification for the assumption that the MA process be invertible. If $|\beta| > 1$, we cannot represent the $\{\varepsilon_t\}$ series in terms of the observed $\{y_t\}$ series. If we now substitute the solution for ε_t into the formula for the log likelihood, we obtain

$$\ln L = \frac{-T}{2}\ln(2\pi) - \frac{T}{2}\ln\sigma^2 - \frac{1}{2\sigma^2}\sum_{t=1}^{T}\left(\sum_{i=0}^{t-1}(-\beta)^i y_{t-i}\right)^2$$

Now that we have expressed $\ln L$ in terms of the observable $\{y_t\}$ series, it is possible to select the values of β and σ^2 that maximize the value of $\ln L$. Unlike OLS, if you actually take the partial derivatives of $\ln L$ with respect to β and σ^2 you will not obtain a simple set of first-order conditions. Moreover, the formula becomes much more complicated in higher-order MA(q) processes. Nevertheless, computers can use a number of available search algorithms to find the values of β and σ^2 that maximize $\ln L$. Be aware that numerical optimization routines cannot guarantee exact solutions for the estimated coefficients. Instead, various "hill-climbing" methods are used to find the parameter values that maximize $\ln L$. If the partial derivatives of the likelihood function are close to zero (so that the likelihood function is flat), the algorithms may not be able to find a maximum.

APPENDIX 2.2: MODEL SELECTION CRITERIA

Hypothesis testing is not particularly well suited to testing nonnested models. For example, if you wanted to chose between an AR(1) and an MA(2), you could estimate ARMA(1, 2) and then try to restrict the MA(2) coefficients to equal zero. Alternatively, you could try to restrict the AR(1) coefficient to equal zero. Nevertheless, the method is unsatisfactory because it necessitates estimating the overparameterized ARMA (1, 2) model. Instead, model selection criteria, such as the AIC and the SBC, can be used to choose between alternative models. Such model selection criteria can be viewed as measures of *goodness-of-fit* that include a cost, or penalty, for each parameter estimated.

One reason it is not desirable to have an overparameterized model is that forecast error variance increases as a result of errors arising from parameter estimation. In other words, small models tend to have better out-of-sample performance than large models. Suppose that the actual data-generating process (DGP) is the AR(1) model

$$y_t = ay_{t-1} + \varepsilon_t$$

If a is known, the one-step-ahead forecast of y_{t+1} is $E_t y_{t+1} = ay_t$. Hence, the mean squared forecast error is $E_t(y_{t+1} - ay_t)^2 = E_t \varepsilon_{t+1}^2 = \sigma^2$. However, when a is estimated from the data, the one-step-ahead forecast of y_{t+1} is

$$E_t\, y_{t+1} = \hat{a}y_t$$

where $\hat{a}$ is the estimated value of a.

Hence, the mean squared forecast, or prediction, error is

$$\text{MSPE} = E_t(y_{t+1} - \hat{a}y_t)^2 = E_t[(ay_t - \hat{a}y_t) + \varepsilon_{t+1}]^2$$

Since ε_{t+1} is independent of $\hat{a}$ and y_t, it follows that

$$\begin{aligned} E_t(y_{t+1} - \hat{a}y_t)^2 &= E_t[(a - \hat{a})y_t]^2 + \sigma^2 \\ &\approx E_t[(a - \hat{a})]^2(y_t)^2 + \sigma^2 \end{aligned}$$

Since $E_t[(a - \hat{a})]^2$ is strictly positive, parameter uncertainty contributes to forecast error variance in that the mean squared forecast error exceeds σ^2. The point is that errors in parameter estimation contribute to forecast error variance. Moreover, the more parameters estimated, the greater the parameter uncertainty. It is easy to show that the problem is particularly acute in small samples. Since $\text{var}(y_t) = \sigma^2/(1 - a^2)$ and, in large samples, $\text{var}(\hat{a}) = E_t[(a - \hat{a})]^2 \approx (1 - a^2)/T$, it follows that

$$\begin{aligned} E_t[(a - \hat{a})]^2\,(y_t)^2 + \sigma^2 &\approx [(1 - a^2)/T](1 - a^2)^{-1}\sigma^2 + \sigma^2 \\ &= [1 + (1/T)]\sigma^2 \end{aligned}$$

Thus, as T increases, the MSPE approaches σ^2.

The Finite Prediction Error (FPE) Criterion

The FPE criterion seeks to minimize the one-step-ahead mean squared prediction error. Now consider the AR(p) process:

$$y_t = a_1 y_{t-1} + \ldots + a_p y_{t-p} + \varepsilon_t$$

If you use the argument in the previous section, the MSPE can be shown to be equal to

$$[1 + (p/T)]\sigma^2$$

We do not know the true variance σ^2. However, σ^2 can be replaced by its unbiased estimate $SSR/(T-p)$ to get

$$\text{FPE} = [1 + (p/T)]\,[SSR/(T - p)]$$

Select p so as to minimize FPE. We can use logs and note that $\ln(1 + p/T)$ can be approximated by p/T. Hence, it is possible to select p to minimize

$$p/T + \ln(SSR) - \ln(T - p)$$

which is the same as minimizing

$$p + T\ln(SSR) - T\ln(T - p)$$

Since $\ln(T-p) \cong \ln T - p/T$, the problem can be written in terms of selecting p so as to minimize

$$p + T\ln(SSR) - \ln(T) + p$$

which has the same solution as minimizing

$$T\ln(SSR) + 2p$$

The AIC and the SBC

The more general AIC selects the $(1 + p + q)$ parameters of an ARMA model so as to maximize the log likelihood function including a penalty for each parameter estimated:

$$\text{AIC} = -2 \ln \text{maximized value of log likelihood} + (1 + p + q)/T$$

For a given value of T, selecting the values of p and q so as to minimize AIC is equivalent to selecting p and q so as to minimize the sum:

$$T\ln(SSR) + 2(1 + p + q)$$

Notice that if $q = 0$ and there is no intercept, this is the result obtained using the FPE. Minimizing the value of the AIC implies that each estimated parameter entails a benefit and a cost. Clearly, a benefit of adding another parameter is that the value of SSR is reduced. The cost is that degrees of freedom are reduced and there is added parameter uncertainty. Thus, adding additional parameters will decrease $\ln(SSR)$ but will increase $(1 + p + q)$. The AIC allows you to add parameters until the marginal cost (i.e., the marginal cost is 2 for each parameter estimated) equals the marginal benefit.

The SBC incorporates the larger penalty $(1 + p + q)\ \ln T$. To use the SBC, select the values of p and q so as to minimize

$$T\ln(SSR) + (1 + p + q)\ln(T)$$

For any reasonable sample size, $\ln(T) > 2$ so that the marginal cost of adding parameters using the SBC exceeds that of the AIC. Hence, the SBC will select a more parsimonious model than the AIC. As indicated in the text, the SBC has superior large sample properties. It is possible to prove that the SBC is asymptotically consistent while the AIC is biased toward selecting an overparameterized model. However, Monte Carlo studies have shown that in small samples, the AIC can work better than the SBC.

CHAPTER 3

MODELING VOLATILITY

Many economic time series do not have a constant mean and most exhibit phases of relative tranquility followed by periods of high volatility. Much of the current econometric research is concerned with extending the Box–Jenkins methodology to analyze these types of time-series variables. This chapter has three aims:

1. Examine the so-called *stylized facts* concerning the properties of economic time-series data. Casual inspection of GDP, financial aggregates, interest rates, and exchange rates suggests they do not have a constant mean and variance. Many seem to have a decided upward trend, while others seem to meander and show periods of high and low volatility. A stochastic variable with a constant variance is called **homoskedastic** as opposed to **heteroskedastic.**[1] For series exhibiting volatility, the unconditional variance may be constant even though the variance during some periods is unusually large.
2. Formalize simple models of variables exhibiting heteroskedasticity. Asset holders are interested in the volatility of returns over the holding period, not over some historical period. This forward-looking view of risk means that it is important to be able to estimate and forecast the risk associated with holding a particular asset. As such, this chapter will develop the tools necessary to model and forecast conditional heteroskedasticity.
3. Analyze a number of variants of the basic model for conditional volatility. The simplest model posits that the conditional heteroskedasticity can be estimated as an autoregressive process (ARCH). This chapter examines a number of important generalizations of the basic ARCH model. It concludes with an examination of conditional volatility models in a multivariate setting.

1. ECONOMIC TIME SERIES: THE STYLIZED FACTS

Figures 3.1 through 3.6 illustrate the behavior of some of the more important variables encountered in macroeconomic analysis. Casual inspection does have its perils, and formal testing is necessary to substantiate any first impressions. However, the strong visual pattern is that these series are *not* stationary; the sample means do not appear to be constant and/or there is the strong appearance of heteroskedasticity. We can characterize the key features of the various series with these *stylized facts:*

1. **Most of the series contain a clear trend.** Real GDP and consumption (see Figure 3.1) exhibit a decidedly upward trend. Although other components

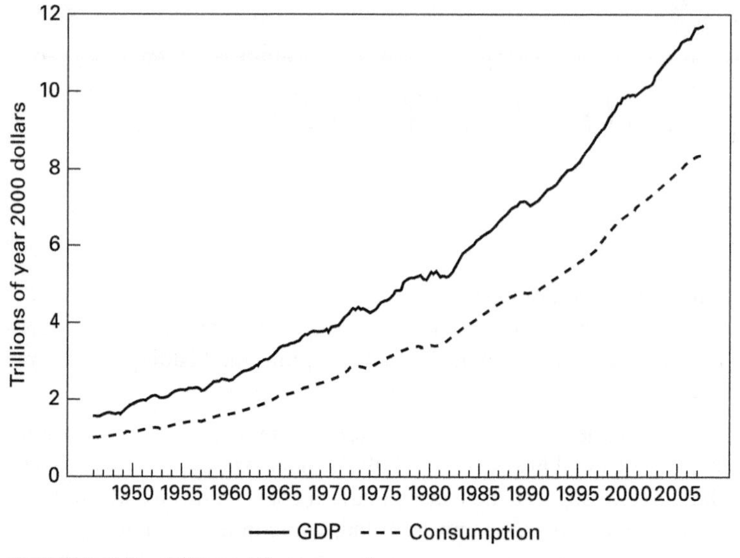

FIGURE 3.1 GDP and Consumption

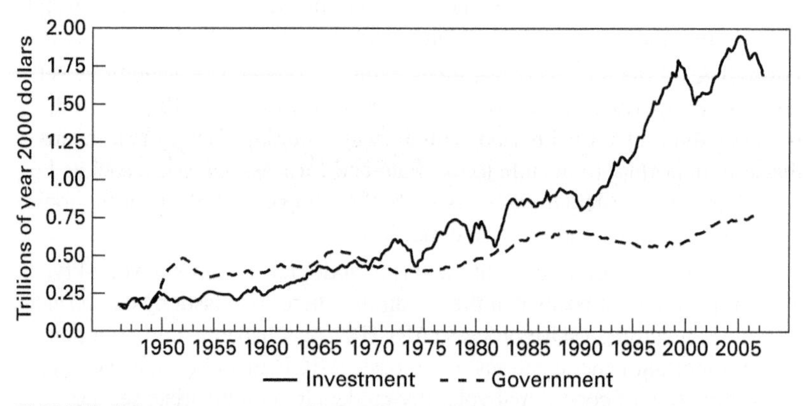

FIGURE 3.2 Government Expenditures and Investment

of real GDP have an upward trend, real investment and government purchases (see Figure 3.2) are more volatile than either real GDP or consumption.

2. **Shocks to a series can display a high degree of persistence.** Neither of the interest rate series shown in Figure 3.3 has a clear upward or downward trend. Nevertheless, both show a high degree of persistence. Notice that the Federal Funds Rate experienced two upward surges in the 1970s and remained at those high levels for several years. Similarly, after a sharp decrease in 1987, the rate never again displayed the levels attained in the early 1980s.

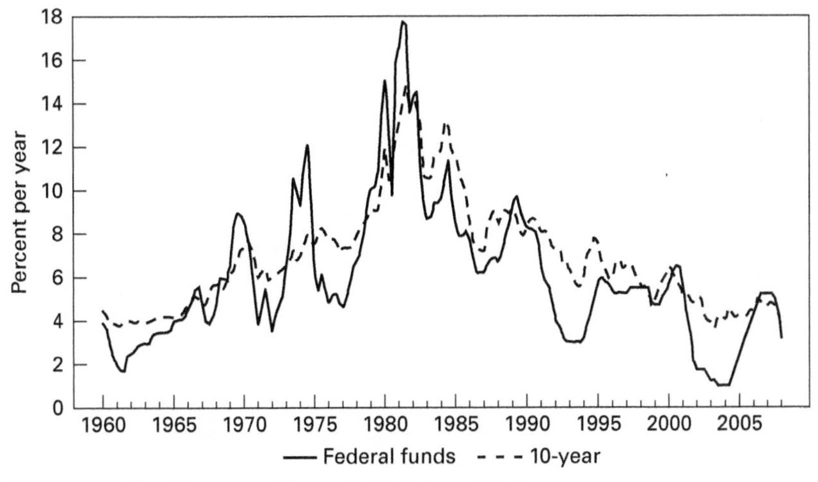

FIGURE 3.3 Short- and Long-Term Interest Rates

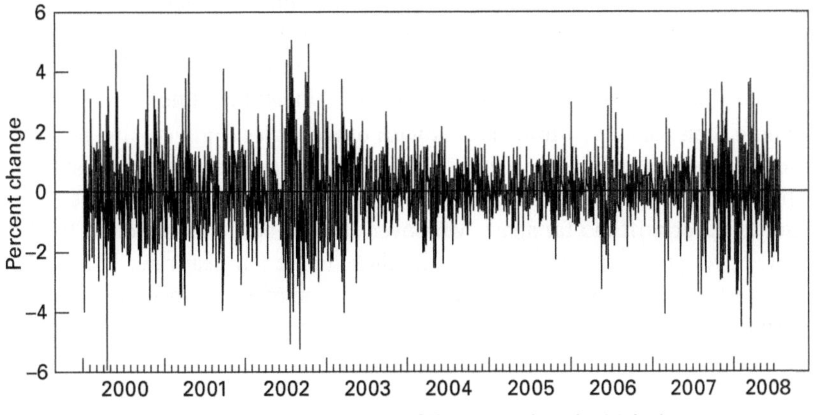

FIGURE 3.4 Daily Changes in the NYSE International 100 Index
Jan 3, 2000–July 30, 2008

3. **The volatility of many series is not constant over time.** Real investment grew smoothly throughout most of the 1960s but also became highly variable in the 1970s. More dramatic are the daily changes in the NYSE index of the 100 leading international stocks. In Figure 3.4, you can see the periods where the stock market seems tranquil alongside periods with large increases and decreases in the market. Such series are called *conditionally heteroskedastic* if the unconditional (or long-run) variance is constant, but there are periods in which the variance is relatively high.
4. **Some series seem to meander.** The real effective exchange rates (see Figure 3.5) show no particular tendency to increase or decrease. The real values of the U.S. and the Canadian dollar seem to go through sustained

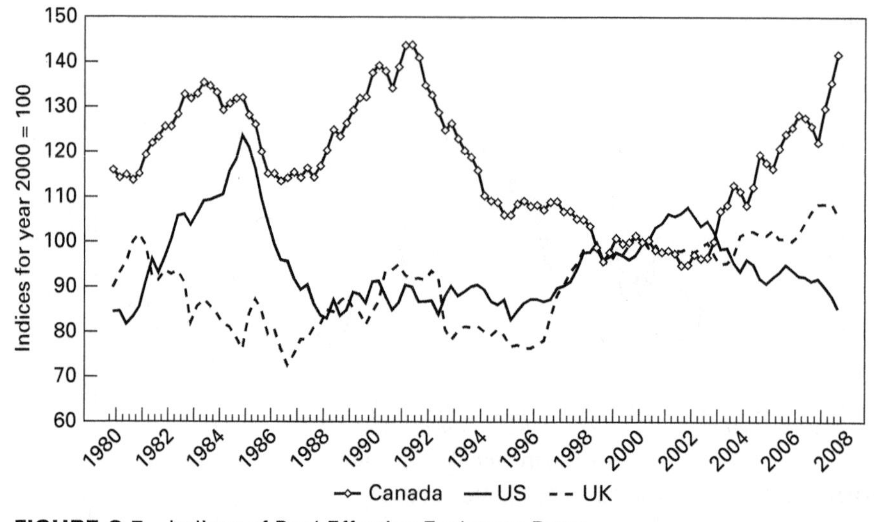

FIGURE 3.5 Indices of Real Effective Exchange Rates

periods of appreciation and then depreciation with no tendency to revert to a long-run mean. This type of "random walk" behavior is typical of nonstationary series. On the other hand, without any formal testing, it is difficult to tell if the real value of the British pound is mean-reverting.

5. **Some series share co-movements with other series.** Individually, the federal funds rate and the ten-year yield on U.S. government securities show no tendency to revert to a long-run mean. However, it is clear that the two series never drift too far apart. Moreover, large shocks to the federal funds rate appear to be timed similarly with those to the ten-year rate. The presence of such co-movements should not be too surprising since the forces driving short-term and long-term rates should be similar. On the other hand, it is not clear whether the various indices of real GDP levels (see Figure 3.6) share the same trend. The movements in the three GDP series are such that all seem to experience recessions and expansions simultaneously. However, it is not clear whether the differences among the trend rates of growth are statistically significant.

Please be aware that "eyeballing" the data is not a substitute for formally testing for the presence of conditional heteroskedasticity or nonstationary behavior.[2] Although most of the variables shown in the figures are neither stationary nor homoskedastic ARMA processes, the issue will not always be so obvious. Fortunately, it is possible to modify the tools developed in the last chapter to help in the identification and estimation of such series. The remainder of this chapter considers the issue of conditional heteroskedasticity. Models and formal tests for the presence of trends (either deterministic and/or stochastic) are contained in the next chapter. The order in which you read Chapters 3 and 4 is immaterial; some instructors may wish to cover the material in Chapter 4 and then the material in Chapter 3.

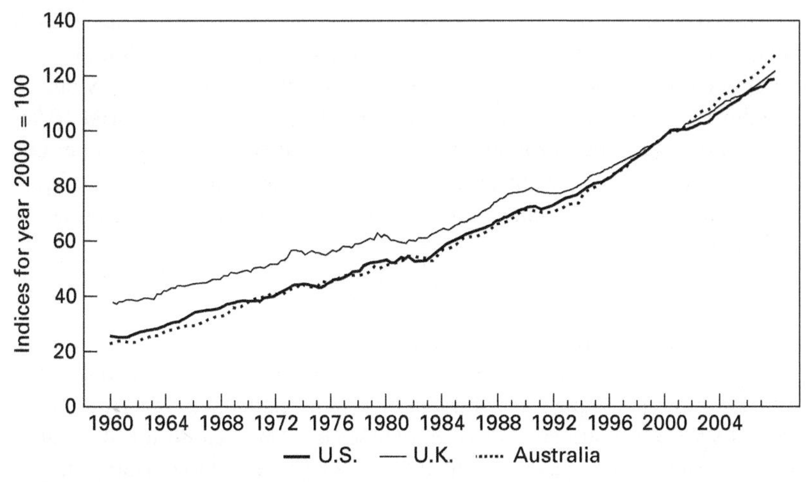

FIGURE 3.6 Real GDP Indices

However, the issue of co-movements in multivariate time series must wait until Chapters 5 and 6.

2. ARCH PROCESSES

In conventional econometric models, the variance of the disturbance term is assumed to be constant. However, Figures 3.1 to 3.6 demonstrate that many economic time series exhibit periods of unusually large volatility, followed by periods of relative tranquility. In such circumstances, the assumption of a constant variance (**homoskedasticity**) is inappropriate. It is easy to imagine instances in which you might want to forecast the conditional variance of a series. As an asset holder, you would be interested in forecasts of the rate of return *and* its variance over the holding period. The unconditional variance (i.e., the long-run forecast of the variance) would be unimportant if you plan to buy the asset at t and sell at $t + 1$.

One approach to forecasting the variance is to explicitly introduce an independent variable that helps to predict the volatility. Consider the simplest case in which

$$y_{t+1} = \varepsilon_{t+1}x_t$$

where: y_{t+1} is the variable of interest
ε_{t+1} is a white-noise disturbance term with variance σ^2
x_t is an independent variable that can be observed at period t

If $x_t = x_{t-1} = x_{t-2} = \dots =$ constant, the $\{y_t\}$ sequence is the familiar white-noise process with a constant variance. However, when the realizations of the $\{x_t\}$ sequence are not all equal, the variance of y_{t+1} conditional on the observable value of x_t is

$$\text{var}(y_{t+1}|x_t) = x_t^2\sigma^2$$

Here the conditional variance of y_{t+1} is dependent on the realized value of x_t. Since you can observe x_t at time period t, you can form the variance of y_{t+1} conditionally on the realized value of x_t. If the magnitude $(x_t)^2$ is large (small), the variance of y_{t+1} will be large (small) as well. Furthermore, if the successive values of $\{x_t\}$ exhibit positive serial correlation (so that a large value of x_t tends to be followed by a large value of x_{t+1}), the conditional variance of the $\{y_t\}$ sequence will exhibit positive serial correlation as well. In this way, the introduction of the $\{x_t\}$ sequence can explain periods of volatility in the $\{y_t\}$ sequence. In practice, you might want to modify the basic model by introducing the coefficients a_0 and a_1 and estimating the regression equation in logarithmic form as

$$\ln(y_t) = a_0 + a_1\ln(x_{t-1}) + e_t$$

where e_t is the error term [formally, $e_t = \ln(\varepsilon_t)$].

This procedure is simple to implement since the logarithmic transformation results in a linear regression equation; OLS can be used to estimate a_0 and a_1 directly. A major difficulty with this strategy is that it assumes a specific cause for the changing variance. Moreover, the methodology also forces $\{x_t\}$ to affect the mean of $\ln(y_t)$. Oftentimes, you might not have a firm theoretical reason for selecting one candidate for the $\{x_t\}$ sequence over other reasonable choices. Was it the oil price shocks, a change in the conduct of monetary policy, and/or the breakdown of the Bretton Woods system that was responsible for the volatility of real investment during the 1970s? Moreover, the technique necessitates a transformation of the data such that the resulting series has a constant variance. In the example at hand, the $\{e_t\}$ sequence is assumed to have a constant variance. If this assumption is violated, some other transformation of the data is necessary.

ARCH Processes

Instead of using ad hoc variable choices for x_t and/or data transformations, Engle (1982) shows that it is possible to simultaneously model the mean *and* the variance of a series. As a preliminary step to understanding Engle's methodology, note that conditional forecasts are vastly superior to unconditional forecasts. To elaborate, suppose you estimate the stationary ARMA model $y_t = a_0 + a_1y_{t-1} + \varepsilon_t$ and want to forecast y_{t+1}. The conditional mean of y_{t+1} is

$$E_ty_{t+1} = a_0 + a_1y_t$$

If we use this conditional mean to forecast y_{t+1}, the forecast error variance is $E_t[(y_{t+1} - a_0 - a_1y_t)^2] = E_t\varepsilon_{t+1}^2 = \sigma^2$. However, if unconditional forecasts are used, the unconditional forecast is always the long-run mean of the $\{y_t\}$ sequence, that is, equal to $a_0/(1 - a_1)$. The unconditional forecast error variance is

$$\begin{aligned} E\{[y_{t+1} - a_0/(1 - a_1)]^2\} &= E[(\varepsilon_{t+1} + a_1\varepsilon_t + a_1^2\,\varepsilon_{t-1} + a_1^3\,\varepsilon_{t-2} + \ldots)^2] \\ &= \sigma^2/(1 - a_1^2) \end{aligned}$$

Since $1/(1 - a_1^2) > 1$, the unconditional forecast has a greater variance than the conditional forecast. Thus, conditional forecasts (since they take into account the known current and past realizations of series) are preferable.

Similarly, if the variance of $\{\varepsilon_t\}$ is not constant, you can estimate any tendency for sustained movements in the variance using an ARMA model. For example, let $\{\hat{\varepsilon}_t\}$

denote the estimated residuals from the model $y_t = a_0 + a_1 y_{t-1} + \varepsilon_t$ so that the conditional variance of y_{t+1} is

$$\begin{aligned} \text{var}(y_{t+1}|y_t) &= E_t[(y_{t+1} - a_0 - a_1 y_t)^2] \\ &= E_t(\varepsilon_{t+1})^2 \end{aligned}$$

To this point, we have set $E_t(\varepsilon_{t+1})^2$ equal to the constant σ^2. Now suppose that the conditional variance is not constant. One simple strategy is to forecast the conditional variance as an AR(q) process using *squares* of the estimated residuals

$$\hat{\varepsilon}_t^2 = \alpha_0 + \alpha_1 \hat{\varepsilon}_{t-1}^2 + \alpha_2 \hat{\varepsilon}_{t-2}^2 + \ldots + \alpha_q \hat{\varepsilon}_{t-q}^2 + v_t \tag{3.1}$$

where v_t is a white-noise process.

If the values of $\alpha_1, \alpha_2, \ldots, \alpha_n$ all equal zero, the estimated variance is simply the constant α_0. Otherwise, the conditional variance of y_t evolves according to the autoregressive process given by (3.1). As such, you can use (3.1) to forecast the conditional variance at $t + 1$ as

$$E_t \hat{\varepsilon}_{t+1}^2 = \alpha_0 + \alpha_1 \hat{\varepsilon}_t^2 + \alpha_2 \hat{\varepsilon}_{t-1}^2 + \ldots + \alpha_q \hat{\varepsilon}_{t+1-q}^2$$

For this reason, an equation like (3.1) is called an **autoregressive conditional heteroskedastic** (ARCH) model. There are many possible applications for ARCH models since the residuals in (3.1) can come from an autoregression, an ARMA model, or a standard regression model.

In actuality, the linear specification of (3.1) is not the most convenient. The reason is that the model for $\{y_t\}$ and the conditional variance are best estimated simultaneously using maximum likelihood techniques. Moreover, instead of the specification given by (3.1), it is more tractable to specify v_t as a multiplicative disturbance.

The simplest example from the class of multiplicative conditionally heteroskedastic models proposed by Engle (1982) is

$$\varepsilon_t = v_t \sqrt{\alpha_0 + \alpha_1 \varepsilon_{t-1}^2} \tag{3.2}$$

where v_t = white-noise process such that $\sigma_v^2 = 1$, v_t and ε_{t-1} are independent of each other, and α_0 and α_1 are constants such that $\alpha_0 > 0$ and $0 \leq \alpha_1 \leq 1$.

Consider the properties of the $\{\varepsilon_t\}$ sequence. Since v_t is white noise and is independent of ε_{t-1}, it is easy to show that the elements of the $\{\varepsilon_t\}$ sequence have a mean of zero and are uncorrelated. The proof is straightforward. Take the unconditional expectation of ε_t. Since $Ev_t = 0$, it follows that

$$\begin{aligned} E\varepsilon_t &= E[v_t(\alpha_0 + \alpha_1 \varepsilon_{t-1}^2)^{1/2}] \\ &= Ev_t E(\alpha_0 + \alpha_1 \varepsilon_{t-1}^2)^{1/2} = 0 \end{aligned} \tag{3.3}$$

Since $Ev_t v_{t-i} = 0$, it also follows that

$$E\varepsilon_t \varepsilon_{t-i} = 0 \qquad i \neq 0 \tag{3.4}$$

The derivation of the unconditional variance of ε_t is also straightforward. Square ε_t and take the unconditional expectation to form

$$\begin{aligned} E\varepsilon_t^2 &= E[v_t^2(\alpha_0 + \alpha_1 \varepsilon_{t-1}^2)] \\ &= Ev_t^2 E(\alpha_0 + \alpha_1 \varepsilon_{t-1}^2) \end{aligned}$$

Since $\sigma_v^2 = 1$ and the unconditional variance of ε_t is identical to that of ε_{t-1} (i.e., $E\varepsilon_t{}^2 = E\varepsilon_{t-1}^2$), the unconditional variance is

$$E\varepsilon_t^2 = \alpha_0/(1 - \alpha_1) \tag{3.5}$$

Thus, the unconditional mean and variance are unaffected by the presence of the error process given by (3.2). Similarly, it is easy to show that the conditional mean of ε_t is equal to zero. Given that v_t and ε_{t-1} are independent and that $Ev_t = 0$, the conditional mean of ε_t is

$$E(\varepsilon_t | \varepsilon_{t-1}, \varepsilon_{t-2}, \ldots) = E_{t-1}v_t E_{t-1}(\alpha_0 + \alpha_1\varepsilon_{t-1}^2)^{1/2} = 0$$

At this point you might be thinking that the properties of the $\{\varepsilon_t\}$ sequence are not affected by (3.2) since the mean is zero, the variance is constant, and all autocovariances are zero. However, the influence of (3.2) falls entirely on the conditional variance. Because $Ev_t^2 = 1$, the variance of ε_t conditioned on the past history of ε_{t-1}, $\varepsilon_{t-2}, \ldots$ is

$$E[\varepsilon_t^2 | \varepsilon_{t-1}, \varepsilon_{t-2}, \ldots] = \alpha_0 + \alpha_1\varepsilon_{t-1}^2 \tag{3.6}$$

In (3.6), the conditional variance of ε_t is dependent on the realized value of ε_{t-1}^2. If the realized value of ε_{t-1}^2 is large, the conditional variance in t will be large as well. In (3.6) the conditional variance follows a first-order autoregressive process denoted by ARCH(1). As opposed to a usual autoregression, the coefficients α_0 and α_1 have to be restricted. In order to ensure that the conditional variance is never negative, it is necessary to assume that both α_0 and α_1 are positive. After all, if α_0 is negative, a sufficiently small realization of ε_{t-1} will mean that (3.6) is negative. Similarly, if α_1 is negative, a sufficiently large realization of ε_{t-1} can render a negative value for the conditional variance. Moreover, to ensure the stability of the process, it is necessary to restrict α_1 such that $0 \leq \alpha_1 \leq 1$.

Equations (3.3), (3.4), (3.5), and (3.6) illustrate the essential features of any ARCH process. In an ARCH model, the conditional and unconditional expectations of the error terms are equal to zero. Moreover, the $\{\varepsilon_t\}$ sequence is serially uncorrelated because for all $s \neq 0$, $E\varepsilon_t\varepsilon_{t-s} = 0$. The key point is that the errors are *not* independent since they are related through their second moment (recall that correlation is a linear relationship). The conditional variance itself is an autoregressive process resulting in conditionally heteroskedastic errors. When the realized value of ε_{t-1} is far from zero—so that $\alpha_1(\varepsilon_{t-1})^2$ is relatively large—the variance of ε_t will tend to be large. As you will see momentarily, the conditional heteroskedasticity in $\{\varepsilon_t\}$ will result in $\{y_t\}$ being heteroskedastic itself. Thus, the ARCH model is able to capture periods of tranquility and volatility in the $\{y_t\}$ series.

The four panels of Figure 3.7 depict two different ARCH models. Panel (a), representing the $\{v_t\}$ sequence, shows 100 serially uncorrelated and normally distributed random deviates. From casual inspection, the $\{v_t\}$ sequence appears to fluctuate around a mean of zero and have a constant variance. Note the moderate increase in volatility between periods 50 and 60. Given the initial condition $\varepsilon_0 = 0$, these realizations of the $\{v_t\}$ sequence were used to construct the next 100 values of the $\{\varepsilon_t\}$ sequence using equation (3.2) and setting $\alpha_0 = 1$ and $\alpha_1 = 0.8$. As illustrated in Panel (b),

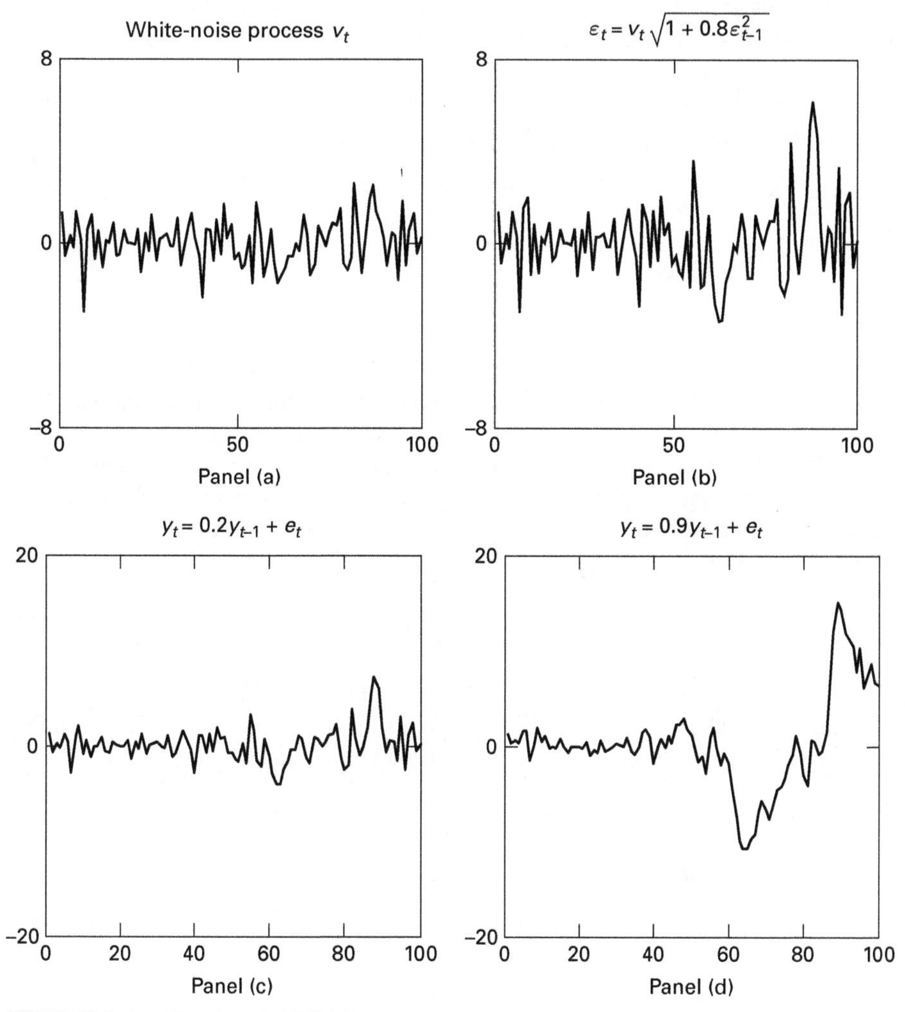

FIGURE 3.7 Simulated ARCH Processes

the $\{\varepsilon_t\}$ sequence also has a mean of zero, but the variance appears to experience an increase in volatility around $t = 50$.

How does the error structure affect the $\{y_t\}$ sequence? Clearly, if the autoregressive parameter a_1 is zero, y_t is nothing more than ε_t. Thus, Panel (b) can be used to depict the time path of the $\{y_t\}$ sequence for the case of $a_1 = 0$. Panels (c) and (d) show the behavior of the $\{y_t\}$ sequence for the cases of $a_1 = 0.2$ and 0.9, respectively. The essential point to note is that the ARCH error structure and the autocorrelation parameters of the $\{y_t\}$ process interact with each other. Comparing Panels (c) and (d) illustrates that the volatility of $\{y_t\}$ is increasing in a_1 and α_1. The explanation is intuitive. Any unusually large (in absolute value) shock in v_t will be associated with a persistently large variance in the $\{\varepsilon_t\}$ sequence; the larger α_1 is, the larger the persistence.

Moreover, the greater the autoregressive parameter a_1, the more persistent is any given change in y_t. The stronger the tendency for $\{y_t\}$ to remain away from its mean, the greater the variance.

To formally examine the properties of the $\{y_t\}$ sequence, the conditional mean and variance are given by

$$E_{t-1}y_t = a_0 + a_1 y_{t-1}$$

and

$$\begin{aligned} \text{var}(y_t | y_{t-1}, y_{t-2}, \ldots) &= E_{t-1}(y_t - a_0 - a_1 y_{t-1})^2 \\ &= E_{t-1}(\varepsilon_t)^2 \\ &= \alpha_0 + \alpha_1(\varepsilon_{t-1})^2 \end{aligned}$$

Since α_1 and ε_{t-1}^2 cannot be negative, the minimum value for the conditional variance is α_0. For any nonzero realization of ε_{t-1}, the conditional variance of y_t is positively related to α_1. The unconditional mean and variance of y_t can be obtained by solving the difference equation for y_t and then taking expectations. If the process began sufficiently far in the past (so that the arbitrary constant A can safely be ignored), the solution for y_t is

$$y_t = \frac{a_0}{1 - a_1} + \sum_{i=0}^{\infty} a_1^i \varepsilon_{t-i} \tag{3.7}$$

Since $E\varepsilon_t = 0$ for all t, the unconditional expectation of (3.7) is $Ey_t = a_0/(1 - a_1)$. The unconditional variance can be obtained in a similar fashion using (3.7). Given that $E\varepsilon_t\varepsilon_{t-i}$ is zero for all $i \neq 0$, the unconditional variance of y_t follows directly from (3.7) as

$$\text{var}(y_t) = \sum_{i=0}^{\infty} a_1^{2i} \text{var}(\varepsilon_{t-i})$$

From the result that the unconditional variance of ε_t is constant [i.e., $\text{var}(\varepsilon_t) = \text{var}(\varepsilon_{t-1}) = \text{var}(\varepsilon_{t-2}) = \ldots = \alpha_0/(1 - \alpha_1)$], it follows that

$$\text{var}(y_t) = \left(\frac{\alpha_0}{1 - \alpha_1}\right)\left(\frac{1}{1 - a_1^2}\right)$$

Clearly, the variance of the $\{y_t\}$ sequence is increasing in α_1 and in the absolute value of a_1. Although the algebra can be a bit tedious, the essential point is that an ARCH error process can be used to model periods of volatility within the univariate framework.

The ARCH process given by (3.2) has been extended in several interesting ways. Engle's (1982) original contribution considered the entire class of higher-order ARCH(q) processes:

$$\varepsilon_t = v_t \sqrt{\alpha_0 + \sum_{i=1}^{q} \alpha_i \varepsilon_{t-i}^2} \tag{3.8}$$

In (3.8), all shocks from ε_{t-1} to ε_{t-q} have a direct effect on ε_t, so that the conditional variance acts like an autoregressive process of order q. It is a good exercise to demonstrate that the forecast for $E_t\varepsilon_{t+1}^2$ arising from (3.1) is precisely the same as that from (3.8).

The GARCH Model

Bollerslev (1986) extended Engle's original work by developing a technique that allows the conditional variance to be an ARMA process. Now let the error process be such that

$$\varepsilon_t = v_t\sqrt{h_t}$$

where $\sigma_v^2 = 1$, and

$$h_t = \alpha_0 + \sum_{i=1}^{q}\alpha_i\varepsilon_{t-i}^2 + \sum_{i=1}^{p}\beta_i h_{t-i} \tag{3.9}$$

Since $\{v_t\}$ is a white-noise process, the conditional and unconditional means of ε_t are equal to zero. Taking the expected value of ε_t, it is easy to verify that

$$E\varepsilon_t = Ev_t(h_t)^{1/2} = 0$$

The important point is that the conditional variance of ε_t is given by $E_{t-1}\varepsilon_t^2 = h_t$. Thus, *the conditional variance of* ε_t *is the ARMA process given by the expression* h_t *in (3.9).*

This **generalized** ARCH(p, q) model—called GARCH(p, q)—allows for both autoregressive and moving-average components in the heteroskedastic variance. If we set $p = 0$ and $q = 1$, it is clear that the first-order ARCH model given by (3.2) is simply a GARCH(0, 1) model. Hence, if all values of β_i equal zero, the GARCH (p, q) model is equivalent to an ARCH(q) model. The benefits of the GARCH model should be clear; a high-order ARCH model may have a more parsimonious GARCH representation that is much easier to identify and estimate. This is particularly true since all coefficients in (3.9) must be positive. Moreover, to ensure that the variance is finite, all characteristic roots of (3.9) must lie inside the unit circle. Clearly, the more parsimonious model will entail fewer coefficient restrictions.[3]

The key feature of GARCH models is that the conditional variance of the *disturbances* of the $\{y_t\}$ sequence constitutes an ARMA process. Hence, it is to be expected that the *squared* residuals from a fitted ARMA model should display this characteristic pattern. To explain, suppose you estimate $\{y_t\}$ as an ARMA process. If your model of $\{y_t\}$ is adequate, the ACF and PACF of the residuals should be indicative of a white-noise process. Equation (3.9) looks very much like a standard ARMA(p, q) process. As such, if there is conditional heteroskedasticity, the correlogram of the squared residuals should be suggestive of such a process. The technique to construct the correlogram of the squared residuals is as follows:

STEP 1: Estimate the $\{y_t\}$ sequence using the "best fitting" ARMA model (or regression model) and obtain the squares of the fitted errors $\{\hat{\varepsilon}_t^2\}$. Also calculate the sample variance of the residuals ($\hat{\sigma}^2$) defined as

$$\hat{\sigma}^2 = \sum_{t=1}^{T}\hat{\varepsilon}_t^2/T$$

where T = number of residuals.

STEP 2: Calculate and plot the sample autocorrelations of the squared residuals as

$$\rho_i = \frac{\sum_{t=i+1}^{T} (\hat{\varepsilon}_t^2 - \hat{\sigma}^2)(\hat{\varepsilon}_{t-i}^2 - \hat{\sigma}^2)}{\sum_{t=1}^{T} (\hat{\varepsilon}_t^2 - \hat{\sigma}^2)^2}$$

STEP 3: In large samples, the standard deviation of ρ_i can be approximated by $T^{-0.5}$. Individual values of ρ_i that are significantly different from zero are indicative of GARCH errors. Ljung–Box Q-statistics can be used to test for groups of significant coefficients. As in Chapter 2, the statistic

$$Q = T(T+2)\sum_{i=1}^{n} \rho_i^2/(T-i)$$

has an asymptotic χ^2 distribution with n degrees of freedom if the $\{\hat{\varepsilon}_t^2\}$ sequence is serially uncorrelated. Rejecting the null hypothesis that the $\{\hat{\varepsilon}_t^2\}$ are serially uncorrelated is equivalent to rejecting the null hypothesis of no ARCH or GARCH errors. In practice, you should consider values of n up to $T/4$.

A more formal Lagrange multiplier test for ARCH errors is the McLeod and Li (1983) test. The methodology involves the following two steps:[4]

STEP 1: Use OLS to estimate the most appropriate regression equation or ARMA model, and let $\{\hat{\varepsilon}_t^2\}$ denote the squares of the fitted errors.

STEP 2: Regress these squared residuals on a constant and on the q lagged values $\hat{\varepsilon}_{t-1}^2, \hat{\varepsilon}_{t-2}^2, \hat{\varepsilon}_{t-3}^2, \ldots, \hat{\varepsilon}_{t-q}^2$; that is, estimate a regression of the form

$$\hat{\varepsilon}_t^2 = \alpha_0 + \alpha_1\hat{\varepsilon}_{t-1}^2 + \alpha_2\hat{\varepsilon}_{t-2}^2 + \ldots + \alpha_q\hat{\varepsilon}_{t-q}^2$$

If there are no ARCH or GARCH effects, the estimated values of α_1 through α_q should be zero. Hence, this regression will have little explanatory power so that the coefficient of determination (i.e., the usual R^2) will be quite low. Using a sample of T residuals, under the null hypothesis of no ARCH errors, the test statistic TR^2 converges to a χ^2 distribution with q degrees of freedom. If TR^2 is sufficiently large, rejection of the null hypothesis that α_1 through α_q are jointly equal to zero is equivalent to rejection of the null hypothesis of no ARCH errors. On the other hand, if TR^2 is sufficiently low, it is possible to conclude that there are no ARCH effects. In the small sample sizes typically used in applied work, an F-test for the null hypothesis $\alpha_1 = \ldots = \alpha_q = 0$ has been shown to be superior to a χ^2 test. Compare the sample value of F to the values in an F-table with q degrees of freedom in the numerator and $T - q$ degrees of freedom in the denominator.

3. ARCH AND GARCH ESTIMATES OF INFLATION

ARCH and GARCH models have become very popular in that they enable the econometrician to estimate the variance of a series at a particular point in time. Clearly, asset pricing models indicate that the risk premium will depend on the expected return and the variance of that return. The relevant measure is the risk over the holding period, not the unconditional risk. Similarly, a portfolio manager who uses **value-at-risk** might be unwilling to hold a portfolio with a 5 percent chance of losing $1 million. The assessment of the risk should be determined using the conditional distribution of asset returns. To use Engle's example of the importance of using the conditional variance rather than the unconditional variance, consider the nature of the wage-bargaining process. Clearly, firms and unions need to forecast the inflation rate over the duration of the labor contract. Economic theory suggests that the terms of the wage contract will depend on the inflation forecasts and the uncertainty concerning the accuracy of these forecasts. Let $E_t\pi_{t+1}$ denote the conditional expected rate of inflation for $t + 1$ and let $\sigma^2_{\pi t}$ denote the conditional variance. If parties to the contract have rational expectations, the terms of the contract will depend on $E_t\pi_{t+1}$ and $\sigma^2_{\pi t}$ as opposed to the unconditional mean or the unconditional variance.

This example illustrates a very important point. The rational expectations hypothesis asserts that agents do not waste useful information. In forecasting any time series, rational agents use the conditional, rather than the unconditional, distribution of the series. Hence, any test of the wage bargaining model above that uses the historical variance of the inflation rate would be inconsistent with the notion that rational agents make use of all available information (i.e., conditional means and variances). A student of the *economics of uncertainty* can immediately see the importance of ARCH and GARCH models. Theoretical models using *variance as a measure of risk (such as mean-variance analysis) can be tested using the conditional variance.* As such, the growth in the use of ARCH/GARCH methods has been nothing short of impressive. In fact, there are so many types of models of conditional volatility that it is common practice to refer to the entire class of models as ARCH or GARCH models.

Engle's Model of U.K. Inflation

Although Section 2 focused on the residuals of a pure ARMA model, it is possible to estimate the residuals of a standard multiple regression model as ARCH or GARCH processes. In fact, Engle's (1982) seminal paper considered the residuals of a simple model of the wage/price spiral for the U.K. over the 1958*Q*2–1977*Q*2 period. Let p_t denote the log of the U.K. consumer price index and w_t denote the log of the index of nominal wage rates. Thus, the rate of inflation is $\pi_t = p_t - p_{t-1}$ and the real wage is $r_t = w_t - p_t$. Engle reports that after some experimentation, he chose the following model of the U.K. inflation rate (standard errors are in parentheses):

$$\begin{aligned} \pi_t &= \underset{(0.006)}{0.0257} + \underset{(0.103)}{0.334}\pi_{t-1} + \underset{(0.110)}{0.408}\pi_{t-4} - \underset{(0.114)}{0.404}\pi_{t-5} + \underset{(0.014)}{0.0559}r_{t-1} + \varepsilon_t \qquad (3.10) \\ h_t &= 8.9 \times 10^{-5} \end{aligned}$$

where h_t is the variance of $\{\varepsilon_t\}$.

The nature of the model is such that increases in the previous period's real wage increase the current inflation rate. Lagged inflation rates at $t - 4$ and $t - 5$ are intended to capture seasonal factors. All coefficients have a t-statistic greater than 3.0, and a battery of diagnostic tests did not indicate the presence of serial correlation. The estimated variance was the constant value 8.9×10^{-5}. In testing for ARCH errors, the

Lagrange multiplier test for ARCH(1) errors was not significant, but the test for an ARCH(4) error process yielded a value of TR^2 equal to 15.2. At the 0.01 significance level, the critical value of χ^2 with four degrees of freedom is 13.28; hence, Engle concludes that there are ARCH errors.

Engle specified an ARCH(4) process forcing the following declining set of weights on the errors:

$$h_t = \alpha_0 + \alpha_1(0.4\,\varepsilon_{t-1}^2 + 0.3\,\varepsilon_{t-2}^2 + 0.2\,\varepsilon_{t-3}^2 + 0.1\,\varepsilon_{t-4}^2) \tag{3.11}$$

The rationale for choosing a two-parameter variance function was to ensure the nonnegativity and stationarity constraints that might not be satisfied using an unrestricted estimating equation. Given this particular set of weights, the necessary and sufficient conditions for the two constraints to be satisfied are $\alpha_0 > 0$ and $0 < \alpha_1 < 1$.

Engle shows that the estimation of the parameters of (3.10) and (3.11) can be considered separately without loss of asymptotic efficiency. One procedure is to estimate (3.10) using OLS and to save the residuals. From these residuals, an estimate of the parameters of (3.11) can be constructed, and based on these estimates, new estimates of (3.10) can be obtained. To estimate both with full efficiency, continued iterations can be checked to determine whether the separate estimates are converging. Now that many statistical software packages contain nonlinear maximum-likelihood estimation routines, the current procedure is to simultaneously estimate both equations using the methodology discussed in Section 8.

Engle's maximum-likelihood estimates of the model are

$$\begin{aligned}
\pi_t &= \underset{(0.005)}{0.0328} + \underset{(0.108)}{0.162}\pi_{t-1} + \underset{(0.089)}{0.264}\pi_{t-4} - \underset{(0.099)}{0.325}\pi_{t-5} + \underset{(0.012)}{0.0707}r_{t-1} + \varepsilon_t \qquad (3.12)\\
h_t &= \underset{(8.5\times10^{-6})}{1.4\times10^{-5}} + \underset{(0.298)}{0.955}(0.4\varepsilon_{t-1}^2 + 0.3\varepsilon_{t-2}^2 + 0.2\varepsilon_{t-3}^2 + 0.1\varepsilon_{t-4}^2)
\end{aligned}$$

The estimated values of h_t are the conditional forecast error variances. All coefficients (except the first lag of the inflation rate) are significant at conventional levels. For a given real wage, the point estimates of (3.12) imply that the inflation rate is a convergent process. Using the calculated values of the $\{h_t\}$ sequence, Engle finds that the standard deviation of inflation forecasts more than doubled as the economy moved from the "predictable sixties into the chaotic seventies." The point estimate of 0.955 indicates an extreme amount of persistence.

Bollerslev's Estimates of U.S. Inflation

Bollerslev's (1986) estimate of U.S. inflation provides an interesting comparison of a standard autoregressive time-series model (which assumes a constant variance), a model with ARCH errors, and a model with GARCH errors. He notes that the ARCH procedure has been useful in modeling different economic phenomena but points out (see pp. 307–308):

> Common to most ... applications, however, is the introduction of a rather arbitrary linear declining lag structure in the conditional variance equation

to take account of the long memory typically found in empirical work, since estimating a totally free lag distribution often will lead to violation of the nonnegativity constraints.

There is no doubt that the lag structure Engle used to model h_t in (3.12) is subject to this criticism. Using quarterly data over the 1948*Q*2 to 1983*Q*4 period, Bollerslev (1986) calculated the inflation rate (π_t) as the logarithmic change in the U.S. GNP deflator. He then estimated the autoregression:

$$\underset{}{\pi_t} = \underset{(0.080)}{0.240} + \underset{(0.083)}{0.552}\pi_{t-1} + \underset{(0.089)}{0.177}\pi_{t-2} + \underset{(0.090)}{0.232}\pi_{t-3} - \underset{(0.080)}{0.209}\pi_{t-4} + \varepsilon_t \qquad (3.13)$$

$$h_t = 0.282$$

Equation (3.13) seems to have all the properties of a well-estimated time-series model. All coefficients are significant at conventional levels (the standard errors are in parentheses) and the estimated values of the autoregressive coefficients imply stationarity. Bollerslev reports that the ACF and PACF do not exhibit any significant correlations at the 5 percent significance level. However, as is typical of ARCH errors, the ACF and PACF of the *squared* residuals (i.e., ε_t^2) show significant correlations. The Lagrange multiplier tests for ARCH(1), ARCH(4), and ARCH(8) errors are all highly significant.

Bollerslev next estimates the restricted ARCH(8) model originally proposed by Engle and Kraft (1983). By way of comparison to (3.13), he finds

$$\pi_t = \underset{(0.059)}{0.138} + \underset{(0.081)}{0.423}\pi_{t-1} + \underset{(0.108)}{0.222}\pi_{t-2} + \underset{(0.078)}{0.377}\pi_{t-3} - \underset{(0.104)}{0.175}\pi_{t-4} + \varepsilon_t \qquad (3.14)$$

$$h_t = \underset{(0.003)}{0.058} + \underset{(0.265)}{0.802}\sum_{i=1}^{8}[(9-i)/36]\varepsilon_{t-i}^2$$

Note that the autoregressive coefficients of (3.13) and (3.14) are similar. The models of the variance, however, are quite different. Equation (3.13) assumes a constant variance, whereas (3.14) assumes that the variance (h_t) is a geometrically declining weighted average of the variance in the previous eight quarters.[5] Hence, the inflation rate predictions of the two models should be similar, but the confidence intervals surrounding the forecasts will differ. Equation (3.13) yields a constant interval of unchanging width. Equation (3.14) yields a confidence interval that expands during periods of inflation volatility and contracts in relatively tranquil periods.

In order to test for the presence of a first-order GARCH term in the conditional variance, it is possible to estimate the equation:

$$h_t = \alpha_0 + \alpha_1\sum_{i=1}^{8}[(9-i)/36]\varepsilon_{t-i}^2 + \beta_1 h_{t-1} \qquad (3.15)$$

The finding that $\beta_1 = 0$ would imply an absence of a GARCH term in the conditional variance. Given the difficulties of estimating (3.15), Bollerslev (1986) uses the simpler Lagrange multiplier (LM) test. Formally, the test involves constructing the residuals of the conditional variance of (3.14). The next step is to regress these residuals on a constant and h_{t-1}; the expression TR^2 has a χ^2 distribution with one degree of freedom. Bollerslev finds that $TR^2 = 4.57$; at the 5 percent significance level, he cannot reject the presence of a first-order GARCH process. He then estimates the following GARCH(1, 1) model:

$$\pi_t = \underset{(0.060)}{0.141} + \underset{(0.081)}{0.433}\pi_{t-1} + \underset{(0.110)}{0.229}\pi_{t-2} + \underset{(0.077)}{0.349}\pi_{t-3} - \underset{(0.104)}{0.162}\pi_{t-4} + \varepsilon_t$$

$$h_t = \underset{(0.006)}{0.007} + \underset{(0.070)}{0.135}\varepsilon_{t-1}^2 + \underset{(0.068)}{0.829}h_{t-1}$$

Diagnostic checks indicate that the ACF and PACF of the squared residuals do not reveal any coefficients exceeding $2T^{-0.5}$. LM tests for the presence of additional lags of ε_t^2 and for the presence of h_{t-2} are not significant at the 5 percent level.

4. TWO EXAMPLES OF GARCH MODELS

GARCH models have found their greatest use in modeling financial data. However, this section and Section 5 are intended to illustrate some other uses of GARCH models. In the first example, the issue is whether there has been a significant reduction in the volatility of real GDP. In the second example, the intent is to obtain reasonable conditional confidence intervals when forecasting. The example also shows that inference in an ARMA (or regression) framework can be improved by accounting for GARCH effects. The example in Section 5 uses a GARCH framework to measure the attitudes and behavior toward risk in the U.S. broiler market.

Volatility Moderation

There is a large body of literature indicating that the volatility of important macroeconomic variables in the industrialized economies decreased in early 1984. For example, Stock and Watson (2002) reported that the standard deviation of real U.S. GDP growth during the 1984–2002 period was 61 percent smaller than that during the 1960–1983 period. As discussed in Romer (1999), some have argued that better monetary policies enabled central bankers to better stabilize economic activity. Others have argued it is a matter of luck that there have not been any major negative supply shocks (such as oil price shocks or widespread failures) since the 1970s. Although this so-called "Great Moderation" probably came to an end with the financial crisis of 2008, we can use the GARCH framework to test whether or not there was a volatility break in 1984Q1.

The file RGDP.XLS contains the four series that were used to construct Figures 3.1 and 3.2. You can use the data in the file to construct the growth rate of real U.S. GDP as $y_t = \log(\text{RGDP}_t/\text{RGDP}_{t-1})$. Without going into detail, if you worked through Chapter 2, it should be clear that a reasonable model for the growth rate of real GDP is

$$y_t = \underset{(7.14)}{0.006} + \underset{(5.47)}{0.331}y_{t-1} + \varepsilon_t$$

The issue is to measure the extent of the volatility break in 1984Q1. As a preliminary test, we can try to determine if there is any conditional volatility. Since we are using quarterly data, it makes sense to use the McLeod–Li (1983) test with a four-quarter lag. Consider

$$\hat{\varepsilon}_t^2 = 5.48 \times 10^{-5} + 0.099\hat{\varepsilon}_{t-1}^2 + 0.131\hat{\varepsilon}_{t-2}^2 - 0.015\hat{\varepsilon}_{t-3}^2 + 0.140\hat{\varepsilon}_{t-4}^2$$

The sample value of the F-statistic for the null hypothesis that the coefficients α_1 through α_4 all equal zero is 3.44. Given that there are four degrees of freedom, this is significant at the 0.009 level. Hence, there is strong evidence that the $\{y_t\}$ series exhibits conditional volatility.

Now create the dummy variable D_t such that D_t is equal to 1 beginning in 1984Q1 and is equal to 0 before 1984Q1. If you estimate the y_t series allowing for ARCH(1) errors and include D_t in the variance equation, you should find

$$\begin{aligned} y_t &= \underset{(8.23)}{0.005} + \underset{(5.10)}{0.321}y_{t-1} + \varepsilon_t \\ h_t &= \underset{(7.55)}{1.16 \times 10^{-4}} + \underset{(1.08)}{0.086}\varepsilon_{t-1}^2 - \underset{(-6.24)}{9.54 \times 10^{-5}}D_t \end{aligned}$$

Given that the coefficient on ε_{t-1}^2 is not statistically different from zero, we can conclude that there is no volatility clustering. Instead, there is a sharp volatility break. Notice that the intercept of the variance equation was 1.16×10^{-4} prior to 1984Q1 and experienced a significant decline to 2.06×10^{-5} $(= 1.16 \times 10^{-4} - 9.54 \times 10^{-5})$ beginning in 1984Q1. The estimated decline is even greater than the 61 percent figure indicated by Stock and Watson (2002).

A GARCH Model of the Spread

To get a better idea of the actual process of fitting a GARCH model, reconsider the example of the interest rate spread used in the last chapter. Recall that the Box–Jenkins approach led us to give serious consideration to the ARMA(2, (1, 7)) model. If you estimate the model for the entire 1960Q3–2008Q3 period, you should obtain

$$s_t = \underset{(2.97)}{0.293} + \underset{(3.49)}{0.371}s_{t-1} + \underset{(4.15)}{0.425}s_{t-2} + \varepsilon_t + \underset{(11.90)}{0.816}\varepsilon_{t-1} - \underset{(-3.13)}{0.132}\varepsilon_{t-7} \qquad (3.16)$$

As shown in Chapter 2, the estimated model performed quite well. All estimated parameters were significant at conventional levels, and both the AIC and SBC selected this specification. The ACF and PACF of the residuals do not indicate any serial correlation. Recall that the MA term at lag 7 is suspect because there it seems very unlikely that events at lag 7 affect the current value of the spread, but that events at lags 3–6 have no direct effects. The Ljung–Box Q-statistics for lags of 4, 8, and 12 quarters are not significant at conventional levels. Moreover, there was no evidence of structural change in the estimated coefficients. Nevertheless, during the very late 1970s and early 1980s, there was a period of unusual volatility that could be indicative of a GARCH process. The aim of this section is to illustrate a step-by-step analysis of a GARCH estimation of the spread. You should be able to follow along using the data in the file labeled QUARTERLY.XLS.

Formal Tests for ARCH Errors

Although (3.16) appears to be quite reasonable, the volatility during the 1970s suggests that it is prudent to examine the ACF and PACF of the squared residuals. The autocorrelations of the *squared* residuals are such that $\rho_1 = 0.040$, $\rho_2 = 0.221$, $\rho_3 = 0.083$, and $\rho_4 = 0.257$. Other values for ρ_i are generally 0.13 or less. The Ljung–Box Q-statistics for the squared residuals are all highly significant; for example, $Q(4) = 24.21$ and $Q(8) = 28.45$, which are both highly significant at any conventional level.

Next, let $\hat{\varepsilon}_t$ denote the residuals of (3.16) and consider the McLeod–Li (1983) test using a lag length of four quarters:

$$\hat{\varepsilon}_t^2 = \underset{(2.46)}{0.145} - \underset{(-0.01)}{0.001}\hat{\varepsilon}_{t-1}^2 + \underset{(2.35)}{0.170}\hat{\varepsilon}_{t-2}^2 + \underset{(0.94)}{0.068}\hat{\varepsilon}_{t-3}^2 + \underset{(3.00)}{0.217}\hat{\varepsilon}_{t-4}^2 \tag{3.17}$$

The value of $TR^2 = 18.51$ so that there is strong evidence of ARCH errors; with four degrees of freedom, the 5 percent critical value of χ^2 is 9.49 and the 1 percent critical value is 13.25. In small samples, it is common to use an F test to determine whether it is possible to reject the restriction $\alpha_1 = \alpha_2 = \ldots = \alpha_q = 0$. In (3.17) with $q = 4$, the sample value of F is 5.00; with four degrees of freedom, this is highly significant.

At this point, you might be tempted to plot the ACF and PACF of the squared residuals and estimate the squared residuals using Box–Jenkins methods. In this way, a parsimonious model of the error process could be obtained. Also, you might be concerned that one of the coefficients in (3.17) is negative and try to reestimate the equation using some other value for q. However, a word of caution is in order. The problem with this strategy is that (3.16) was estimated under the assumption that the conditional variance was constant. As such, it does not make sense to use the residuals of (3.16) to estimate the time-varying conditional variance. Hence, equations such as (3.17) can tell you whether or not there are GARCH errors, but not the precise order of p and/or q.

Alternative Estimates of the Model

The appropriate way to obtain the proper order of the GARCH process is to estimate the model of the spread and the model of the conditional variance simultaneously. As such, GARCH processes are typically estimated by maximum-likelihood techniques so as to obtain estimates that are fully efficient. A low-order ARCH(p) process seems like a reasonable starting place for a model of the conditional variance. For example, you can begin with an ARCH(1) model for the conditional variance

$$s_t = \underset{(2.17)}{0.167} + \underset{(3.51)}{0.461}s_{t-1} + \underset{(3.37)}{0.403}s_{t-2} + \varepsilon_t + \underset{(6.48)}{0.700}\varepsilon_{t-1} - \underset{(-1.83)}{0.095}\varepsilon_{t-7}$$

$$h_t = \underset{(6.17)}{0.156} + \underset{(3.81)}{0.534\varepsilon_{t-1}^2}$$

The model seems to be quite plausible. All of the coefficients are sensible and highly significant. The autoregressive coefficients in the model of the mean imply convergence. The coefficients in the h_t equation are both positive and the ARCH(1) term is less than unity. Now form the standardized errors as the residuals divided by their conditional standard deviations, i.e., form the series $\varepsilon_t/(h_t)^{0.5}$. The estimated standardized residuals are an estimate of the v_t series. If you check for serial correlation in the standardized residuals, you will find that the autocorrelations are such that

ρ_1	ρ_2	ρ_3	ρ_4	ρ_5	ρ_6	ρ_7	ρ_8
0.11	0.03	0.00	0.01	0.00	−0.12	−0.05	0.01

The Q-statistics are $Q(4) = 2.52$ and $Q(8) = 5.82$ so that we can be confident that there is no remaining serial correlation in the standardized residuals. Now, the issue is whether or not the ARCH(1) specification was sufficient to capture all of the dynamics of the conditional variance. To answer this question, form the autocorrelations of the squared standardized residuals:

ρ_1	ρ_2	ρ_3	ρ_4	ρ_5	ρ_6	ρ_7	ρ_8
−0.08	0.31	0.08	0.16	0.01	0.06	0.18	−0.04

The value of ρ_2 is quite large and the Q-statistics are $Q(4) = 25.59$ and $Q(8) = 32.54$. Hence, there is still some remaining conditional volatility so that the ARCH(1) process is inadequate. Next, estimate the ARMA(2, (1, 7)) model allowing for ARCH(2) errors:

$$s_t = \underset{(2.55)}{0.076} + \underset{(10.85)}{1.609s_{t-1}} - \underset{(-4.63)}{0.675s_{t-2}} + \varepsilon_t - \underset{(-1.71)}{0.333\varepsilon_{t-1}} + \underset{(1.22)}{0.071\varepsilon_{t-7}}$$

$$h_t = \underset{(4.73)}{0.096} + \underset{(2.31)}{0.343\varepsilon_{t-1}^2} + \underset{(4.13)}{0.432\varepsilon_{t-2}^2}$$

Interestingly, the problematic MA(7) term is not statistically significant and can be eliminated. If you go on to estimate an ARMA(2, 1) with ARCH(2) errors, you will find

$$s_t = \underset{(2.38)}{0.069} + \underset{(11.56)}{1.586s_{t-1}} - \underset{(-4.75)}{0.648s_{t-2}} - \underset{(-1.64)}{0.303\varepsilon_{t-1}} + \varepsilon_t$$

$$h_t = \underset{(4.62)}{0.096} + \underset{(2.19)}{0.336\varepsilon_{t-1}^2} + \underset{(4.23)}{0.456\varepsilon_{t-2}^2}$$

Although the model of the mean implies that the spread is quite persistent (the sum of the autoregressive coefficients is 0.94), for reasons discussed in Section 10 of Chapter 2, we do not want to use the first difference of the spread. Note that the rather small t-statistic (= −1.64) for the coefficient of ε_{t-1} might lead some researchers to eliminate this term. The properties of the error term are reasonably good. The autocorrelations of the standardized errors and the squared standardized errors are given by:

Correlations	ρ_1	ρ_2	ρ_3	ρ_4	ρ_5	ρ_6	ρ_7	ρ_8
$\varepsilon_t/h_t^{0.5}$	0.02	−0.11	0.08	−0.04	−0.01	−0.14	−0.08	0.07
ε_t^2/h_t	−0.07	0.02	0.05	0.21	−0.04	0.17	0.08	0.03

The ACF for the standardized residuals does not indicate any troublesome serial correlation. The sample values of $Q(4) = 3.95$ and $Q(8) = 10.15$ are not significant at conventional levels. Although the autocorrelations of the squared standardized residuals for lags 4 and 6 are somewhat large, we do not want to simply increase the value of q. As such, many researchers would be content with the ARCH(2) specification for the conditional variance. The solid line in Figure 3.8 shows the one-step-ahead forecast of $E_t s_{t+1}$ using the ARMA(2, 1) model with ARCH(2) errors. Since h_t is an estimate of the conditional variance of s_t, $(h_{t+1})^{0.5}$ is the standard error of the one-step-ahead forecast. The dashed lines in the figure represent a band of $\pm 2\ (h_{t+1})^{0.5}$ surrounding the one-step-ahead forecast of s_{t+1}. In contrast to the assumption of a constant conditional variance, note that the band width increases in the late 1970s through the mid-1980s.

For our purposes, it is instructive to try the GARCH(1, 1) specification as an alternative to the ARCH(2) specification. Unfortunately, if you estimate a GARCH(1, 1) you will find that the sum $\alpha_1 + \beta_1$ exceeds unity. Moreover, you will find the coefficient on ε_{t-1} has a very small t-statistic. However, if you restrict the sum of $\alpha_1 + \beta_1$ to equal unity and eliminate the MA(1) term in the model of the mean, you should find

$$s_t = \underset{(6.39)}{0.128} + \underset{(23.58)}{1.275}s_{t-1} - \underset{(-7.58)}{0.382}s_{t-2} + \varepsilon_t$$

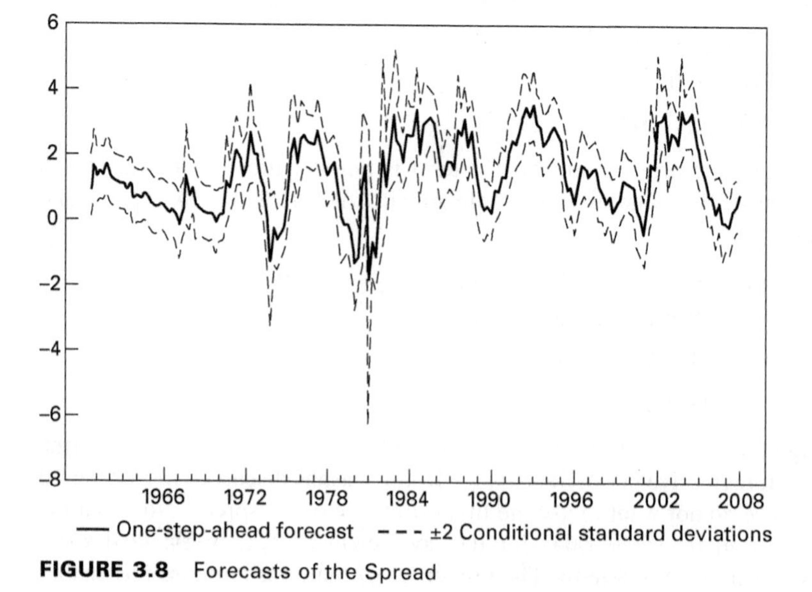

FIGURE 3.8 Forecasts of the Spread

$$h_t = 0.193\varepsilon_{t-1}^2 + 0.807h_{t-1}$$
$$\quad (7.97) \qquad (33.34)$$

The resulting model is very parsimonious and seems to fit the data well. The autocorrelations of the standardized errors and the squared standardized errors are given by:

Correlations	ρ_1	ρ_2	ρ_3	ρ_4	ρ_5	ρ_6	ρ_7	ρ_8
$\varepsilon_t/h_t^{0.5}$	0.08	−0.12	0.09	0.09	0.00	−0.11	0.00	0.13
ε_t^2/h_t	−0.04	0.16	−0.04	0.04	−0.03	−0.02	−0.01	−0.02

Since the model is quite parsimonious, and the autocorrelations are reasonably small, this so-called IGARCH model (with $\alpha_1 + \beta_1 = 1$) is a reasonable alternative to the model with ARCH(2) errors.

5. A GARCH MODEL OF RISK

An interesting application of GARCH modeling is provided by Holt and Aradhyula (1990). Their theoretical framework stands in contrast to the cobweb model (see Section 5 of Chapter 1) in that rational expectations are assumed to prevail in the agricultural sector. The aim of the study is to examine the extent to which producers in the U.S. broiler (i.e., chicken) industry exhibit risk-averse behavior. To this end, the supply function for the U.S. broiler industry takes the form[6]

$$q_t = a_0 + a_1 p_t^e - a_2 h_t - a_3 pfeed_{t-1} + a_4 hatch_{t-1} + a_5 q_{t-4} + \varepsilon_{1t} \qquad (3.18)$$

where: q_t = quantity of broiler production (in millions of pounds) in t

p_t^e = expected real price of broilers at t conditioned on the information at $t-1$ (so that $p_t^e = E_{t-1}p_t$)

h_t = expected variance of the price of broilers in t conditioned on the information at $t-1$

$pfeed_{t-1}$ = real price of broiler feed (in cents per pound) at $t-1$

$hatch_{t-1}$ = hatch of broiler-type chicks in commercial hatcheries (measured in thousands) in period $t-1$

ε_{1t} = supply shock in t

and the length of the time period is one quarter.

The supply function is based on the biological fact that the production cycle of broilers is about two months. Since bimonthly data are unavailable, the model assumes that the supply decision is positively related to the price expectation formed by producers in the previous quarter. Given that feed accounts for the bulk of production costs, real feed prices lagged one quarter are negatively related to broiler production in t. Obviously, the *hatch* available in $t-1$ increases the number of broilers that can be marketed in t. The fourth lag of broiler production is included to account for the possibility that production in any period may not fully adjust to the desired level of production.

For our purposes, the most interesting part of the study is the negative effect of the conditional variance of price on broiler supply. The timing of the production process is such that feed and other production costs must be incurred before output is sold in the market. In the planning stage, producers must forecast the price that will prevail two months hence. The greater p_t^e, the greater the number of chicks that will be fed and brought to market. If price variability is very low, these forecasts can be held with confidence. Increased price variability decreases the accuracy of the forecasts and decreases broiler supply. Risk-averse producers will opt to raise and market fewer broilers when the conditional volatility of price is high.

In the initial stage of the study, broiler prices are estimated as the AR(4) process:

$$(1 - \beta_1 L - \beta_2 L^2 - \beta_3 L^3 - \beta_4 L^4)p_t = \beta_0 + \varepsilon_{2t} \tag{3.19}$$

Ljung–Box Q-statistics for various lag lengths indicate that the residual series appear to be white noise at the 5 percent level. However, the Ljung–Box Q-statistics for the squared residuals—that is, the $\{\varepsilon_{2t}^2\}$—of 32.4 are significant at the 5 percent level. Thus, Holt and Aradhyula conclude that the variance of the price is conditionally heteroskedastic.

In the second stage of the study, several low-order GARCH estimates of (3.19) are compared. Goodness-of-fit statistics and significance tests suggest a GARCH(1, 1) process. In the third stage, the supply equation (3.18) and a GARCH(1, 1) process are simultaneously estimated. The estimated price equation (with standard errors in parentheses) is

$$\underset{}{(1 - \underset{(0.092)}{0.511}L - \underset{(0.098)}{0.129}L^2 - \underset{(0.094)}{0.130}L^3 - \underset{(0.073)}{0.138}L^4)p_t = \underset{(1.347)}{1.632} + \varepsilon_{2t}} \tag{3.20}$$

$$h_t = \underset{(0.747)}{1.353} + \underset{(0.80)}{0.162}\varepsilon_{2t-1}^2 + \underset{(0.175)}{0.591}h_{t-1} \tag{3.21}$$

Equation (3.20) and (3.21) are well-behaved in that (1) all estimated coefficients are significant at conventional significance levels; (2) all coefficients of the conditional variance equation are positive; and (3) the coefficients all imply convergent processes.

Holt and Aradhyula assume that producers use (3.20) and (3.21) to form their price expectations. Combining these estimates with (3.18) yields the supply equation

$$q_t = \underset{(0.585)}{2.767}p_t^e - \underset{(0.344)}{0.521}h_t - \underset{(1.463)}{4.325}pfeed_{t-1} + \underset{(0.205)}{1.887}hatch_{t-1} + \underset{(0.065)}{0.603}q_{t-4} + \varepsilon_{1t}$$

All estimated coefficients are significant at conventional levels and have the appropriate sign. An increase in the expected price increases broiler output. Increased uncertainty, as measured by conditional variance, acts to decrease output. This forward-looking rational expectations formulation is at odds with the more traditional cobweb model discussed in Chapter 1. In order to compare the two formulations, Holt and Aradhyula (1990) also considered an adaptive expectations formulation (see Exercise 2 in Chapter 1). Under adaptive expectations, price expectations are formed according to a weighted average of the previous period's price and the previous period's price expectation:

$$p_t^e = \alpha p_{t-1} + (1 - \alpha)p_{t-1}^e$$

or, solving for p_t^e in terms of the $\{p_t\}$ sequence, we obtain

$$p_t^e = \alpha \sum_{i=0}^{\infty} (1 - \alpha)^i p_{t-1-i}$$

Similarly, the adaptive expectations formulation for conditional risk is given by

$$h_t = \beta \sum_{i=0}^{\infty} (1 - \beta)^i (p_{t-1-i} - p_{t-1-i}^e)^2 \tag{3.22}$$

where $0 < \beta < 1$ and $(p_{t-1-i} - p_{t-1-i}^e)^2$ is the forecast-error variance for period $t - i$.

Note that in (3.22) the expected measure of risk as viewed by producers is not necessarily the actual conditional variance. The estimates of the two models differ concerning the implied long-run elasticities of supply with respect to expected price and conditional variance.[7] Respectively, the estimated long-run elasticities of supply with respect to expected price are 0.587 and 0.399 in the rational expectations and adaptive expectations formulations. Similarly, rational and adaptive expectations formulations yield long-run supply elasticities of conditional variance of −0.030 and −0.013, respectively. Not surprisingly, the adaptive expectations model suggests a more sluggish supply response than does the forward-looking rational expectations model.

6. THE ARCH-M MODEL

Engle, Lilien, and Robins (1987) extended the basic ARCH framework to allow the mean of a sequence to depend on its own conditional variance. This class of model, called the ARCH in mean (ARCH-M) model, is particularly suited to the study of asset markets. The basic insight is that risk-averse agents will require compensation for holding a risky asset. Given that an asset's *riskiness* can be measured by the variance of returns, the risk premium will be an increasing function of the conditional variance of returns. Engle, Lilien, and Robins express this idea by writing the excess return from holding a risky asset as

$$y_t = \mu_t + \varepsilon_t \tag{3.23}$$

where: y_t = excess return from holding a long-term asset relative to a one-period treasury bill

μ_t = risk premium necessary to induce the risk-averse agent to hold the long-term asset rather than the one-period bond

ε_t = unforecastable shock to the excess return on the long-term asset

To explain (3.23), note that the expected excess return from holding the long-term asset must be just equal to the risk premium:

$$E_{t-1}y_t = \mu_t$$

Engle, Lilien, and Robins assume that the risk premium is an increasing function of the conditional variance of ε_t; in other words, the greater the conditional variance of returns, the greater the compensation necessary to induce the agent to hold the long-term asset. Mathematically, if h_t is the conditional variance of ε_t, the risk premium can be expressed as

$$\mu_t = \beta + \delta h_t \qquad \delta > 0 \tag{3.24}$$

where h_t is the ARCH(q) process:

$$h_t = \alpha_0 + \sum_{i=1}^{q} \alpha_i \varepsilon_{t-i}^2 \tag{3.25}$$

As a set, equations (3.23), (3.24), and (3.25) constitute the basic ARCH-M model. From (3.23) and (3.24), the conditional mean of y_t depends on the conditional variance h_t. From (3.25), the conditional variance is an ARCH(q) process. It should be pointed out that if the conditional variance is constant (i.e., if $\alpha_1 = \alpha_2 = \ldots = \alpha_q = 0$) the ARCH-M model degenerates into the more traditional case of a constant risk premium.

Figure 3.9 illustrates two different ARCH-M processes. Panel (a) of the figure shows 60 realizations of a simulated white-noise process denoted by $\{\varepsilon_t\}$. Note the temporary increase in volatility during periods 20 to 30. By initializing $\varepsilon_0 = 0$, the conditional variance was constructed as the first-order ARCH process:

$$h_t = 1 + 0.65\,\varepsilon_{t-1}^2$$

As you can see in Panel (b), the volatility in $\{\varepsilon_t\}$ translates into increases in conditional variance. Note that large positive *and* negative realizations of ε_{t-1} result in a large value of h_t; it is the square of each $\{\varepsilon_t\}$ realization that enters the

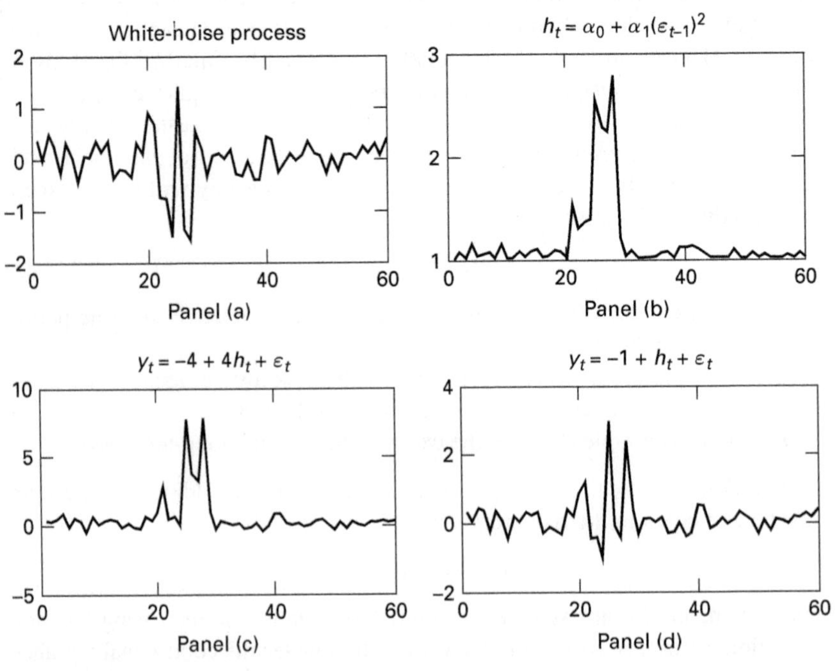

FIGURE 3.9 Simulated ARCH-M Processes

conditional variance. In Panel (c), the values of β and δ are set equal to -4 and $+4$, respectively. As such, the y_t sequence is constructed as $y_t = -4 + 4h_t + \varepsilon_t$. You can clearly see that y_t is above its long-run value during the period of volatility. In the simulation, conditional volatility translates itself into increases in the values of $\{y_t\}$. In the latter portion of the sample, the volatility of $\{\varepsilon_t\}$ diminishes and the values y_{30} through y_{60} fluctuate around their long-run mean.

Panel (d) reduces the influence of ARCH-M effects by reducing the magnitude of δ and β (see Exercise 4 at the end of this chapter). Obviously, if $\delta = 0$, there are no ARCH-M effects at all. As you can see by comparing the two lower graphs, y_t more closely mimics the ε_t sequence when the magnitude of δ is diminished from $\delta = 4$ to $\delta = 1$.[8]

As in any ARCH or GARCH model, a Lagrange multiplier test can be used to detect the presence of conditional volatility. The LM tests are relatively simple to conduct since they do not require estimation of the full model. The statistic TR^2 is asymptotically distributed as χ^2 with degrees of freedom equal to the number of restrictions. Nevertheless, as in the model for the spread, some trial and error is necessary to arrive at the most appropriate model of the conditional variance.

Implementation

Using quarterly data from 1960*Q*1 to 1984*Q*2, Engle, Lilien, and Robins (1987) constructed the excess yield on six-month treasury bills as follows. Let r_t denote the quarterly yield on a three-month treasury bill held from t to $(t + 1)$. Rolling over all proceeds, at the end of two quarters an individual investing \$1 at the beginning of period t will have $(1 + r_t)(1 + r_{t+1})$ dollars. In the same fashion, if R_t denotes the quarterly yield on a six-month treasury bill, buying and holding the six-month bill for the full two quarters will result in $(1 + R_t)^2$ dollars. The excess yield, y_t, due to holding the six-month bill is

$$y_t = (1 + R_t)^2_t - (1 + r_{t+1})(1 + r_t)$$

which is approximately equal to

$$y_t = 2R_t - r_{t+1} - r_t$$

The results from regressing the excess yield on a constant are as follows, with the t-statistic in parentheses:

$$\underset{(4.04)}{y_t = 0.142} + \varepsilon_t \tag{3.26}$$

The excess yield of 0.142 percent per quarter is more than four standard deviations from zero. The problem with this estimation method is that the post-1979 period showed markedly higher volatility than the earlier sample period. To test for the presence of ARCH errors, the squared residuals were regressed on a weighted average of past squared residuals, as in (3.11). The LM test for the restriction $\alpha_1 = 0$ yields a value of $TR^2 = 10.1$, which has a χ^2 distribution with one degree of freedom. At the 1 percent significance level, the critical value of χ^2 with one degree of freedom is 6.635; hence, there is strong evidence of heteroskedasticity. Thus, there appear to be ARCH errors; as a result, (3.26) is misspecified if individuals demand a risk premium.

The maximum likelihood estimates of the ARCH-M model and the associated t-statistics are

$$y_t = -0.0241 + 0.687h_t + \varepsilon_t$$
$$(-1.29) \quad (5.15)$$
$$h_t = 0.0023 + 1.64(0.4\varepsilon_{t-1}^2 + 0.3\varepsilon_{t-2}^2 + 0.2\varepsilon_{t-3}^2 + 0.1\varepsilon_{t-4}^2)$$
$$(1.08) \quad (6.30)$$

The estimated coefficients imply a time-varying risk premium. The estimated parameter of the ARCH equation of 1.64 implies that the unconditional variance is infinite. Although this is troublesome, the conditional variance is finite. Shocks to ε_{t-i} act to increase the conditional variance so that there are periods of tranquility and volatility. During volatile periods, the risk premium rises as risk-averse agents seek assets that are conditionally less risky.

Exercise 6 at the end of this chapter asks you to estimate such an ARCH-M model using simulated data. The questions are designed to guide you through a typical estimation procedure.

7. ADDITIONAL PROPERTIES OF GARCH PROCESSES

Whenever you estimate a GARCH process, you will be estimating the two interrelated equations

$$y_t = a_0 + \beta x_t + \varepsilon_t$$

and

$$\varepsilon_t = v_t(\alpha_0 + \alpha_1\varepsilon_{t-1}^2 + \ldots + \alpha_q\varepsilon_{t-q}^2 + \beta_1 h_{t-1} + \ldots + \beta_p h_{t-p})^{0.5} \tag{3.27}$$

where x_t can be an ARMA process of order (p^m, q^m). Moreover, x_t can contain exogenous variables.

The first equation is a model of the mean and the second yields the model of the variance. The symbols p^m and q^m are used to denote that the order of the ARMA process for the mean need not equal the order of the GARCH(p, q) equation. The two equations are related in that h_t is the conditional variance of ε_t; hence, the GARCH process of (3.27) is the conditional variance of the mean equation. Do not make the mistake of assuming that ε_t^2 is the conditional variance itself. Given that $\varepsilon_t = v_t(h_t)^{0.5}$, it follows that the relationship between h_t and ε_t^2 is

$$\varepsilon_t^2 = v_t^2 h_t$$

and, since $Ev_t^2 = E_{t-1}v_t^2 = 1$,

$$E_{t-1}\varepsilon_t^2 = h_t$$

Thus, h_t is the conditional variance of the $\{\varepsilon_t\}$ sequence.

A GARCH(1, 1) specification is the most popular form of conditional volatility. This is especially true for financial data where volatility shocks are very persistent. As such, it is worthwhile to pay special attention to this form of GARCH process.

Properties of GARCH(1, 1) Error Processes

Given the large number of GARCH(1, 1) models found in the literature, it is desirable to establish the properties of this particular type of error process. In doing so, we can generalize some of the discussion of ARCH(1) models presented in Section 2. If you take the conditional expectation of the GARCH(1, 1) process, you should have no trouble verifying that

$$E_{t-1}\varepsilon_t^2 = \alpha_0 + \alpha_1\varepsilon_{t-1}^2 + \beta_1 h_{t-1}$$

or

$$h_t = \alpha_0 + \alpha_1\varepsilon_{t-1}^2 + \beta_1 h_{t-1} \tag{3.28}$$

The mean of ε_t: The unconditional mean of ε_t is zero. If you take the expected value of (3.27), you obtain

$$E\varepsilon_t = E[v_t(h_t)^{1/2}]$$

Since h_t does not depend on v_t and $Ev_t = 0$, it immediately follows that $E\varepsilon_t = 0$.

The variance of ε_t: Since

$$\varepsilon_t^2 = v_t^2(\alpha_0 + \alpha_1\varepsilon_{t-1}^2 + \beta_1 h_{t-1})$$

it follows that the unconditional variance of a GARCH(1, 1) process is

$$E\varepsilon_t^2 = Ev_t^2(\alpha_0 + \alpha_1 E\varepsilon_{t-1}^2 + \beta_1 Eh_{t-1}) \tag{3.29}$$

We can simplify this expression if we recognize that $Ev_t^2 = 1$ and $E\varepsilon_{t-i}^2 = Eh_{t-i}$. This second relationship follows from the **law of iterated expectations.** The form of the law we need guarantees that $E\varepsilon_t^2 = E(E_{t-1}\varepsilon_t^2)$. In essence, the unconditional expectation of the conditional variance is just the unconditional variance. As such, we can lag the relationship one period and write $E\varepsilon_{t-1}^2 = E(E_{t-2}\varepsilon_{t-1}^2)$, so that $E\varepsilon_{t-1}^2 = E(h_{t-1})$. If we substitute this condition into (3.29), it follows that

$$E\varepsilon_t^2 = \alpha_0 + (\alpha_1 + \beta_1)E\varepsilon_{t-1}^2 \tag{3.30}$$

Thus, (3.30) yields the solution for the unconditional variance. Given that $\alpha_1 + \beta_1 < 0$, the unconditional variance is

$$E\varepsilon_t^2 = \alpha_0/(1 - \alpha_1 - \beta_1)$$

For the more general GARCH(p, q) model, it follows that the variance will be finite if

$$1 - \sum_{i=1}^{q}\alpha_i - \sum_{i=1}^{p}\beta_i > 0$$

The autocorrelation function: The autocorrelations $E\varepsilon_t\varepsilon_{t-j}$ are all equal to zero. Consider

$$E\varepsilon_t\varepsilon_{t-j} = E[v_t(h_t)^{1/2}v_{t-j}(h_{t-j})^{1/2}]$$

Since h_t, v_{t-j}, and h_{t-j} do not depend on the value of v_t and $Ev_t = 0$, it follows that all autocorrelations are zero for $j \neq 0$.

The conditional variance: The conditional variance of the error process is h_t. Consider

$$E_{t-1}\varepsilon_t^2 = E_{t-1}v_t^2 h_t = h_t$$

This simple result is the essential feature of GARCH modeling. The conditional variance of the error process is *not* constant. With the appropriate specification of the parameters of h_t, it is possible to model and forecast the conditional variance of the $\{y_t\}$ process.

Volatility persistence: In a GARCH process, the errors are uncorrelated in that $E\varepsilon_t\varepsilon_{t-j} = 0$. However, as shown in (3.28), the squared errors of a GARCH(1, 1) process are correlated. You should be able to show that the degree of autoregressive decay of the squared residuals is $(\alpha_1 + \beta_1)$. In fact, the ACF of the squared residuals of a GARCH(1, 1) process tends to behave like that of an ARMA(1, 1) process.

Large values of both α_1 and β_1 act to increase the conditional volatility but they do so in different ways. The response of h_t to new information is increasing in the magnitude of α_1; clearly, if α_1 is large, a v_t shock has a sizable effect on ε_t^2 and h_{t+1}. To illustrate the point, two GARCH(1, 1) processes were simulated using the identical set of random numbers for the $\{v_t\}$ sequence. In both cases, h_0 and ε_0 were initialized and the remaining values of the series were constructed using the relationship $\varepsilon_t^2 = v_t^2 h_t$ and

Model 1: $h_t = 1 + 0.6\varepsilon_{t-1}^2 + 0.2h_{t-1}$

Model 2: $h_t = 1 + 0.2\varepsilon_{t-1}^2 + 0.6h_{t-1}$

In order to avoid the effect of selecting the specific values for the initial conditions, the first 100 realizations were eliminated; the remaining 250 realizations are shown in Figure 3.10. Given the value of h_t, a large v_t shock has its immediate effect on ε_t^2. Since Model 1 has a larger value of α_1, the effect of this shock is very pronounced in period $t + 1$. For Model 2, a_1 is equal to only 0.2 so that peaks in the $\{h_t\}$ series are not as large as those from Model 1. However, since Model 2 has the larger value of β_1, its conditional variance displays more autoregressive persistence.

Also note that the value of α_1 must be strictly positive. Hence, the analogy between the ACF of the squared residuals of a GARCH(1, 1) and the ACF of the residuals from an ARMA(1, 1) process is not perfect. If $\alpha_1 = 0$, it is possible to write (3.28) as

$$h_t = \alpha_0 + \beta_1 h_{t-1}$$

so that there is no way for the $\{e_t\}$ series to affect the $\{h_t\}$ series. As such, the model for the conditional variance cannot be identified. The analogy is even less clear in the more general case of a GARCH(p, q) process. Bollerslev (1986) proves that the ACF of the squared residuals resulting from a GARCH(p, q) process acts like that of an ARMA(m, p) process where $m = \max(p, q)$. This makes identification of the most

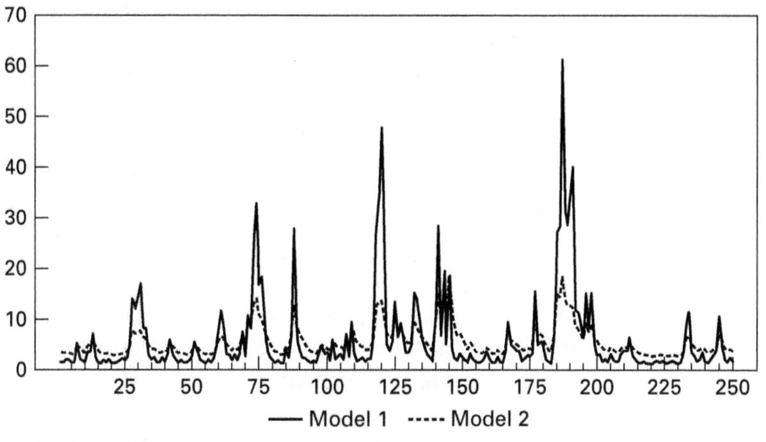

FIGURE 3.10 Persistence in the GARCH (1, 1) Model

appropriate values of p and q somewhat difficult. Question 3 at the end of the chapter guides you through a proof that the ACF of a GARCH(2, 1) has the same properties as the ACF of a GARCH(2, 2) model.

Assessing the Fit

One way to assess the adequacy of a GARCH model is to see how well it fits the data. Some authors use the AIC and SBC discussed in Chapter 2. Moreover, some statistical software packages report these same goodness-of-fit measures for any type of GARCH estimation. However, you need to be careful in interpreting these results since, as stated by Bollerslev, Engle, and Nelson (1994), "... their statistical properties in the ARCH context are largely unknown." Consider the sum of squared residuals (SSR) as a measure of the goodness of fit. Since SSR $= \Sigma\varepsilon_t^2$, the sum of the squared residuals actually measures squared deviations of the model of the mean. Moreover, since $e_t = v_t(h_t)^{1/2}$, the pure innovations in the GARCH model are given by the v_t sequence. Instead of using SSR, in a GARCH model, a reasonable measure of the goodness of fit is

$$\text{SSR}' = \sum_{t=1}^{T} v_t^2$$

Given that $e_t = v_t(h_t)^{1/2}$, you can also write SSR' as

$$\text{SSR}' = \sum_{t=1}^{T} (\varepsilon_t^2/h_t)$$

The point is that SSR' is a measure of the squared errors relative to the fitted values of the conditional variance. Since SSR' will be small if the fitted values of h_t are close to ε_t^2, you can select the model that yields the smallest value of SSR'. Another way to make the same point is to recognize that $e_t/h_t^{0.5}$ is a standardized residual in that the value of e_t is divided by its *conditional* standard error. Hence, SSR' measures the sum of squares of the standardized residuals.

Another goodness-of-fit measure is simply the maximized value of the likelihood function. As explained in more detail in Section 8, if you assume that the error process is normal, the maximized value of the log likelihood function can be written such that

$$2 \ln L = -\sum_{t=1}^{T} [\ln(h_t) + \varepsilon_t^2/h_t] - T\ln(2\pi)$$

$$= -\sum_{t=1}^{T} [\ln(h_t) + v_t^2] - T\ln(2\pi)$$

where L = maximized value of the likelihood function.

Hence, models with a large value of L will tend to have small values of h_t and/or small values of SSR'. Notice that L does not include a penalty for the estimation of additional parameters. However, you can construct the AIC and SBC using

$$\text{AIC} = -2\ln L + 2n$$

$$\text{SBC} = -2\ln L + n\ln(T)$$

where L is defined above and n is the number of estimated parameters. As discussed in Chapter 2, some programmers will not incorporate the expression $-T\ln(2\pi)$ into the calculation of the likelihood function when reporting model-selection criteria.

Diagnostic Checks for Model Adequacy

In addition to providing a good fit, an estimated GARCH model should capture all dynamic aspects of the model of the mean and the model of the variance. The estimated residuals should be serially uncorrelated and should not display any remaining conditional volatility. You can test to ensure that your model has captured these properties by standardizing the residuals as indicated above. Simply divide $\hat{\varepsilon}_t$ by $\hat{h}_t^{1/2}$ in order to obtain an estimate of what we have been calling the $\{v_t\}$ sequence. Since ε_t has a zero mean and a variance of h_t, you can think of $v_t = \varepsilon_t/(h_t)^{1/2}$ as the standardized value of ε_t. The resulting series, which we will call s_t, should have a mean of zero and a variance of unity.

If there is any serial correlation in the $\{s_t\}$ sequence, the model of the mean is not properly specified. To test the model of the mean, form the Ljung–Box Q-statistics for the $\{s_t\}$ sequence. You should not be able to reject the null hypothesis that the various Q-statistics are equal to zero.

To test for remaining GARCH effects, form the Ljung–Box Q-statistics of the squared standardized residuals (i.e., s_t^2). The basic idea is that s_t^2 is an estimate of $\varepsilon_t^2/h_t = v_t^2$. Hence, the properties of the s_t^2 sequence should mimic those of v_t^2. If there are no remaining GARCH effects, you should not be able to reject the null hypothesis that the sample values of the Q-statistics are equal to zero. Otherwise, there is remaining conditional volatility.

Once you have obtained a satisfactory model, you can forecast future values of y_t and its conditional variance. Moreover, you can place confidence bands around the forecast using the estimates of conditional standard deviation. Since $E_t\varepsilon_{t+1}^2 = h_{t+1}$, a two-standard deviation confidence interval for your forecast can be constructed using

$$E_t y_{t+1} \pm 2(h_{t+1})^{0.5}$$

The result is quite general; since the mean of each value of $\{\varepsilon_t\}$ is zero, *the optimal j-step-ahead forecast of* y_{t+j} *does not depend on the presence of GARCH errors.* However, the size of any confidence interval surrounding the forecasts does depend on the conditional volatility. Clearly, in times when there is substantial conditional volatility (i.e., when h_{t+1} is large), the variance of the forecast error will be large. Simply put, we cannot be as confident of our forecasts in periods when conditional volatility is high.

Forecasting the Conditional Variance

The one-step-ahead forecast of the conditional variance is easy to obtain. If we update h_t by one period, we find

$$h_{t+1} = \alpha_0 + \alpha_1 \varepsilon_t^2 + \beta_1 h_t$$

Since ε_t^2 and h_t are known in period t, the one-step-ahead forecast is simply $\alpha_0 + \alpha_1 \varepsilon_t^2 + \beta_1 h_t$. It is only somewhat more difficult to obtain the j-step-ahead forecasts. To begin, use the fact that $\varepsilon_t^2 = v_t^2 h_t$ so that $\varepsilon_{t+j}^2 = v_{t+j}^2 h_{t+j}$. If you update by j periods and take the conditional expectation of each side, it should be clear that

$$E_t \varepsilon_{t+j}^2 = E_t(v_{t+j}^2 h_{t+j})$$

Since v_{t+j} is independent of h_{t+j}, and $E_t v_{t+j}^2 = 1$, it follows that

$$E_t \varepsilon_{t+j}^2 = E_t h_{t+j} \tag{3.31}$$

We can use (3.31) to obtain the forecasts of the conditional variance of the GARCH(1, 1) process. Update (3.28) by j periods to obtain

$$h_{t+j} = \alpha_0 + \alpha_1 \varepsilon_{t+j-1}^2 + \beta_1 h_{t+j-1}$$

and take the conditional expectation

$$E_t h_{t+j} = \alpha_0 + \alpha_1 E_t \varepsilon_{t+j-1}^2 + \beta_1 E_t h_{t+j-1}$$

If you combine this relationship with (3.31), it is easy to verify that

$$E_t h_{t+j} = \alpha_0 + (\alpha_1 + \beta_1) E_t h_{t+j-1} \tag{3.32}$$

Given h_{t+1}, we can use (3.32) to forecast all subsequent values of the conditional variance as

$$E_t h_{t+j} = \alpha_0 [1 + (\alpha_1 + \beta_1) + (\alpha_1 + \beta_1)^2 + \ldots + (\alpha_1 + \beta_1)^{j-1}] + (\alpha_1 + \beta_1)^j h_t$$

If $\alpha_1 + \beta_1 < 1$, the conditional forecasts of h_{t+j} will converge to the long-run value

$$Eh_t = a_0/(1 - \alpha_1 - \beta_1)$$

Similarly, we can forecast the conditional variance of the ARCH(q) process

$$h_t = \alpha_0 + \alpha_1 \varepsilon_{t-1}^2 + \ldots + \alpha_q \varepsilon_{t-q}^2 \tag{3.33}$$

If we update (3.33) by one period, we obtain

$$h_{t+1} = \alpha_0 + \alpha_1 \varepsilon_t^2 + \ldots + \alpha_q \varepsilon_{t-q+1}^2$$

As mentioned above, at period t, we have all of the information necessary to calculate the value of h_{t+1} for any GARCH process. Now, if we update (3.33) by two periods and take the conditional expectation, we obtain

$$E_t h_{t+2} = \alpha_0 + \alpha_1 E_t \varepsilon_{t+1}^2 + \ldots + \alpha_q \varepsilon_{t-q+2}^2$$

Since $E_t \varepsilon_{t+1}^2 = h_{t+1}$, it follows that

$$E_t h_{t+2} = \alpha_0 + \alpha_1 h_{t+1} + \ldots + \alpha_q \varepsilon_{t-q+2}^2$$

The point is that it is possible to obtain the j-step-ahead forecasts of the conditional variance recursively. As the value of $j \to \infty$, the forecasts of h_{t+j} should converge to the unconditional mean

$$E\varepsilon_t^2 = \alpha_0/(1 - \alpha_1 - \alpha_2 - \ldots - \alpha_q)$$

It should be clear that a necessary condition for convergence is for the roots of the inverse characteristic equation $1 - \alpha_1 L - \ldots - \alpha_q L^q$ to lie outside the unit circle. This is a necessary condition for the long-run mean to have the representation $\alpha_0/(1 - \Sigma\alpha_i)$. To ensure that the variance is always positive, we also require that $\alpha_0 > 0$ and $\alpha_i \geq 0$ for $i \geq 1$.

You should have no trouble generalizing these results to the general GARCH (p, q) process. Fortunately, most statistical software packages can perform these calculations automatically.

8. MAXIMUM-LIKELIHOOD ESTIMATION OF GARCH MODELS

Many software packages contain built-in routines that estimate GARCH and ARCH-M models such that the researcher simply specifies the order of the process and the computer does the rest. Even if you have access to an automated routine, it is important to understand the numerical procedures used by your software package. Other packages require user input in the form of a small optimization algorithm. This section explains the maximum-likelihood methods required to understand and write a program for GARCH-type models.

Suppose that values of $\{\varepsilon_t\}$ are drawn from a normal distribution having a mean of zero and a constant variance σ^2. From standard distribution theory, the likelihood of any realization of ε_t is

$$L_t = \left(\frac{1}{\sqrt{2\pi\sigma^2}}\right)\exp\left(\frac{-\varepsilon_t^2}{2\sigma^2}\right)$$

where L_t is the likelihood of ε_t.

Since the realizations of $\{\varepsilon_t\}$ are independent, the likelihood of the joint realizations of $\varepsilon_1, \varepsilon_2, \ldots \varepsilon_T$ is the product in the individual likelihoods. Hence, if all have the same variance, the likelihood of the joint realizations is

$$L = \prod_{t=1}^{T}\left(\frac{1}{\sqrt{2\pi\sigma^2}}\right)\exp\left(\frac{-\varepsilon_t^2}{2\sigma^2}\right)$$

It is far easier to work with a sum than with a product. As such, it is convenient to take the natural log of each side so as to obtain

$$\ln L = -\frac{T}{2}\ln(2\pi) - \frac{T}{2}\ln\sigma^2 - \frac{1}{2\sigma^2}\sum_{t=1}^{T}(\varepsilon_t)^2 \tag{3.34}$$

The procedure used in maximum-likelihood estimation is to select the distributional parameters so as to maximize the likelihood of drawing the observed sample. In Appendix 1 of Chapter 2, we considered the case where the $\{y_t\}$ sequence was an MA(1) process. In the example at hand, suppose that $\{\varepsilon_t\}$ is generated from the model:

$$\varepsilon_t = y_t - \beta x_t \tag{3.35}$$

In the classical regression model, the mean of ε_t is assumed to be zero, the variance is the constant σ^2, and the various realizations of $\{\varepsilon_t\}$ are independent. Using a sample with T observations, we can substitute (3.35) into the log-likelihood function given by (3.34) to obtain

$$\ln L = -\frac{T}{2}\ln(2\pi) - \frac{T}{2}\ln\sigma^2 - \frac{1}{2\sigma^2}\sum_{t=1}^{T}(y_t - \beta x_t)^2 \tag{3.36}$$

Maximizing the log-likelihood function (3.36) with respect to σ^2 and β yields:

$$\frac{\partial \ln L}{\partial \sigma^2} = -\frac{T}{2\sigma^2} + \frac{1}{2\sigma^4}\sum_{t=1}^{T}(y_t - \beta x_t)^2$$

and

$$\frac{\partial \ln \text{L}}{\partial \beta} = \frac{1}{\sigma^2}\sum_{t=1}^{T}(y_t x_t - \beta x_t^2) \tag{3.37}$$

Setting these partial derivatives equal to zero and solving for the values of σ^2 and β that yield the maximum value of ln L results in the familiar OLS estimates of the variance and β (denoted by $\hat{\beta}$ and $\hat{\sigma}^2$). Hence,

$$\hat{\sigma}^2 = \sum \varepsilon_t^2/T \tag{3.38}$$

and

$$\hat{\beta} = \sum x_t y_t / \sum x_t^2 \tag{3.39}$$

All of this should be familiar ground since most econometric texts concerned with regression analysis discuss maximum-likelihood estimation. The point to emphasize here is that the first-order conditions are easily solved since they are all linear. Calculating the appropriate sums may be tedious, but the methodology is straightforward. Unfortunately, this is not the case in estimating an ARCH or GARCH model since the first-order equations are nonlinear. Instead, the solution requires some sort of search algorithm. The simplest way to illustrate the issue is to introduce an ARCH(1) error process into the regression model given by (3.35). Continue to assume that ε_t is the error term in linear equation $y_t - \beta x_t = \varepsilon_t$. Now let ε_t be given by

$$\varepsilon_t = v_t(h_t)^{0.5}$$

Although the conditional variance of ε_t is not constant, the necessary modification of (3.34) is clear. Since each realization of ε_t has the conditional variance h_t, the joint likelihood of realization ε_1 through ε_T is

$$L = \prod_{t=1}^{T}\left(\frac{1}{\sqrt{2\pi h_t}}\right)\exp\left(\frac{-\varepsilon_t^2}{2h_t}\right)$$

so that the log-likelihood function is

$$\ln L = -\frac{T}{2}\ln(2\pi) - 0.5\sum_{t=1}^{T}\ln h_t - 0.5\sum_{t=1}^{T}(\varepsilon_t^2/h_t)$$

Now suppose that $\varepsilon_t = y_t - \beta x_t$ and that the conditional variance is the ARCH(1) process $h_t = \alpha_0 + \alpha_1\varepsilon_{t-1}^2$. Substituting for h_t and y_t yields

$$\ln L = -\frac{T-1}{2}\ln(2\pi) - 0.5\sum_{t=2}^{T}\ln(\alpha_0 + \alpha_1\varepsilon_{t-1}^2)$$

$$-\frac{1}{2}\sum_{t=2}^{T}[(y_t - \beta x_t)^2/(\alpha_0 + \alpha_1\varepsilon_{t-1}^2)]$$

Note that the initial observation is lost since ε_0 is outside the sample. Once you substitute $(y_{t-1} - \beta x_{t-1})^2$ for ε_{t-1}^2, it is possible to maximize $\ln L$ with respect to α_0, α_1, and β. As you can surmise, there are no simple solutions to the first-order conditions for a maximum. Fortunately, computers are able to select the parameter values that maximize this log-likelihood function. In most time-series software packages, the procedure necessary to write such programs is quite simple.

9. OTHER MODELS OF CONDITIONAL VARIANCE

Financial analysts are especially keen to obtain precise estimates of the conditional variance of an asset price. Since GARCH models can forecast conditional volatility, they are able to measure the risk of an asset over the holding period. As such, a number of extensions of the basic GARCH model have been developed that are especially suited to estimating the conditional volatility of financial instruments.

The IGARCH Model

In financial time series, the conditional volatility is persistent. In fact, if you estimate a GARCH(1, 1) model using a long time series of stock returns, you will find that the sum of α_1 and β_1 is very close to unity. Nelson (1990) argued that constraining $\alpha_1 + \beta_1$ to equal unity can yield a very parsimonious representation of the distribution of an asset's return. In some respects, this constraint forces the conditional variance to act like a process with a unit root. This integrated-GARCH (IGARCH) has some very interesting properties. From (3.32), if $\alpha_1 + \beta_1 = 1$, the one-step-ahead forecast of the conditional variance is

$$E_t h_{t+1} = \alpha_0 + h_t$$

and the j-step-ahead forecast is

$$E_t h_{t+j} = j\alpha_0 + h_t$$

Thus, except for the intercept term α_0, the forecast of the conditional variance for the next period is the current value of the conditional variance. Moreover, the unconditional variance is clearly infinite. Nevertheless, Nelson (1990) showed that the analogy between the IGARCH process and an ARIMA process with a unit root is not perfect. Given that $\alpha_1 + \beta_1 = 1$, and that $h_{t-1} = Lh_t$, we can write the conditional variance as

$$h_t = \alpha_0 + (1 - \beta_1)\varepsilon_{t-1}^2 + \beta_1 L h_t$$

and, solving for h_t,

$$h_t = \alpha_0/(1 - \beta_1) + (1 - \beta_1)\sum_{i=0}^{\infty} \beta_1^i \varepsilon_{t-1-i}^2$$

Thus, unlike a true nonstationary process, the conditional variance is a geometrically *decaying* function of the current and past realizations of the $\{\varepsilon_t^2\}$ sequence. As such, an IGARCH model can be estimated like any other GARCH model.

Models with Explanatory Variables

Just as the model of the mean can contain explanatory variables, the specification of h_t also allows for exogenous variables. In Section 4, you saw an example concerning the Great Moderation. Alternatively, suppose that you want to determine whether the terrorist attacks of September 11, 2001, increased the volatility of asset returns. One way to accomplish the task would be to create a dummy variable D_t equal to 0 before 9/11 and equal to 1 thereafter. Consider the following modification of the GARCH(1, 1) specification

$$h_t = \alpha_0 + \alpha_1\varepsilon_{t-1}^2 + \beta_1 h_{t-1} + \gamma D_t$$

If it is found that $\gamma > 0$, it is possible to conclude that the terrorist attacks increased the mean of the conditional volatility.

Models with Asymmetry: TARCH and EGARCH

An interesting feature of asset prices is that "bad" news seems to have a more pronounced effect on volatility than does "good" news. For many stocks, there is a strong negative correlation between the current return and the future volatility. The tendency for volatility to decline when returns rise and to rise when returns fall is often called the **leverage effect.** The idea of the leverage effect is captured in Figure 3.11, where "new information" is measured by the size of ε_t. If $\varepsilon_t = 0$, expected volatility $(E_t h_{t+1})$ is the distance $0a$. Any news increases volatility; however, if the news is "good" (i.e, if ε_t is positive), volatility increases along *ab*. If the news is "bad," volatility increases along *ac*. Since segment *ac* is steeper than *ab*, a positive ε_t shock will have a smaller effect on volatility than a negative shock of the same magnitude.

Glosten, Jaganathan, and Runkle (1994) showed how to allow the effects of good and bad news to have different effects on volatility. In a sense, $\varepsilon_{t-1} = 0$ is a threshold such that shocks greater than the threshold have different effects than shocks below the threshold. Consider the threshold-GARCH (TARCH) process

$$h_t = \alpha_0 + \alpha_1\varepsilon_{t-1}^2 + \lambda_1 d_{t-1}\varepsilon_{t-1}^2 + \beta_1 h_{t-1}$$

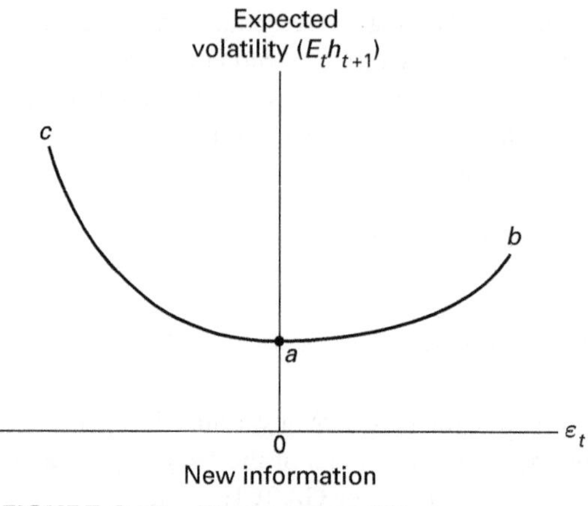

FIGURE 3.11 The Leverage Effect

where d_{t-1} is a dummy variable that is equal to one if $\varepsilon_{t-1} < 0$ and is equal to zero if $\varepsilon_{t-1} \geq 0$.

The intuition behind the TARCH model is that positive values of ε_{t-1} are associated with a zero value of d_{t-1}. Hence, if $\varepsilon_{t-1} \geq 0$, the effect of an ε_{t-1} shock on h_t is $\alpha_1\varepsilon_{t-1}^2$. When $\varepsilon_{t-1} < 0, d_{t-1} = 1$, and the effect of an ε_{t-1} shock on h_t is $(\alpha_1 + \lambda_1)\varepsilon_{t-1}^2$. If $\lambda_1 > 0$, negative shocks will have larger effects on volatility than positive shocks. You can easily create a dummy variable d_t and the product $d_{t-1}\varepsilon_{t-1}^2$. If the coefficient λ_1 is statistically different from zero, you can conclude that your data contain a threshold effect.

Another model that allows for the asymmetric effect of news is the exponential-GARCH (EGARCH) model. One problem with a standard GARCH model is that it is necessary to ensure that all of the estimated coefficients are positive. Nelson (1991) proposed a specification that does not require nonnegativity constraints. Consider

$$\ln(h_t) = \alpha_0 + \alpha_1(\varepsilon_{t-1}/h_{t-1}^{0.5}) + \lambda_1|\varepsilon_{t-1}/h_{t-1}^{0.5}| + \beta_1\ln(h_{t-1}) \tag{3.40}$$

Equation (3.40) is called the exponential-GARCH or EGARCH model. There are three interesting features to notice about the EGARCH model:

1. The equation for the conditional variance is in log-linear form. Regardless of the magnitude of $\ln(h_t)$, the implied value of h_t can never be negative. Hence, it is permissible for the coefficients to be negative.
2. Instead of using the value of ε_{t-1}^2, the EGARCH model uses the level of standardized value of ε_{t-1} [i.e., ε_{t-1} divided by $(h_{t-1})^{0.5}$]. Nelson argues that this standardization allows for a more natural interpretation of the size and persistence of shocks. After all, the standardized value of ε_{t-1} is a unit-free measure.
3. The EGARCH model allows for leverage effects. If $\varepsilon_{t-1}/(h_{t-1})^{0.5}$ is positive, the effect of the shock on the log of the conditional variance is $\alpha_1 + \lambda_1$. If $\varepsilon_{t-1}/(h_{t-1})^{0.5}$ is negative, the effect of the shock on the log of the conditional variance is $-\alpha_1 + \lambda_1$.

Although the EGARCH model has some advantages over the TARCH model, it is difficult to forecast the conditional variance of an EGARCH model. For the TARCH model, it makes sense to assume that $E_t d_{t+j} = 0.5$. If asset returns are symmetric, there is a 50:50 chance that the realized value of ε_{t+j} will be positive.

Testing for Leverage Effects

One way to test for leverage is to estimate the TARCH or EGARCH model and perform a t-test for the null hypothesis $\hat{\lambda}_1 = 0$. However, there is a specific diagnostic test that allows you to determine whether there are any leverage effects in your residuals. After you estimate an ARCH or GARCH model, form the standardized residuals

$$s_t = \hat{\varepsilon}_t / \hat{h}_t^{1/2}$$

Thus, the $\{s_t\}$ sequence consists of each residual divided by its standard deviation. To test for leverage effects, estimate a regression of the form

$$s_t^2 = a_0 + a_1 s_{t-1} + a_2 s_{t-2} + \ldots$$

If there are no leverage effects, the squared errors should be uncorrelated with the level of the error terms. Hence, you can conclude there are leverage effects if the sample value of F for the null hypothesis $a_1 = a_2 = \ldots$ exceeds the critical value obtained from an F table.

Engle and Ng (1993) developed a second way to determine whether positive and negative shocks have different effects on the conditional variance. Again, let d_{t-1} be a dummy variable that is equal to 1 if $\hat{\varepsilon}_{t-1} < 0$ and is equal to zero if $\hat{\varepsilon}_{t-1} \geq 0$. The test is to determine whether the estimated squared residuals can be predicted using the $\{d_{t-1}\}$ sequence. The Sign Bias test uses the regression equation of the form

$$s_t^2 = a_0 + a_1 d_{t-1} + \varepsilon_{1t}$$

where ε_{1t} is a regression residual.

If a t-test indicates that a_1 is statistically different from zero, the sign of the current period shock is helpful in predicting the conditional volatility. To generalize the test, you can estimate the regression

$$s_t^2 = a_0 + a_1 d_{t-1} + a_2 d_{t-1} s_{t-1} + a_3 (1 - d_{t-1}) s_{t-1} + \varepsilon_{1t}$$

The presence of $d_{t-1} s_{t-1}$ and $(1 - d_{t-1}) s_{t-1}$ is designed to determine whether the effects of positive and negative shocks also depend on their size. You can use an F-statistic to test the null hypothesis $a_1 = a_2 = a_3 = 0$. If you conclude that there is a leverage effect, you can estimate a specific form of the TARCH or EGARCH model.[9]

Nonnormal Errors

For most financial assets, the distribution function for the rate of return is **fat-tailed.** A fat-tailed distribution has more weight in the tails than a normal distribution. Suppose that the rate of return on a particular stock has a higher probability of a very large loss (or gain) than indicated by the normal distribution. As such, you might not want to perform a maximum-likelihood estimation using a normal distribution.

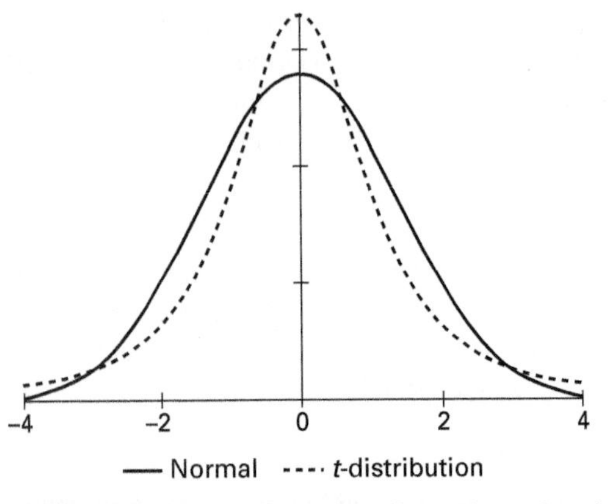

FIGURE 3.12 Comparison of the Normal and *t*-Distributions

Figure 3.12 compares the standardized normal distribution to a *t*-distribution with one degree of freedom. You can see that the *t*-distribution places a greater likelihood on large realizations than does the normal distribution. As such, many computer packages allow you to estimate a GARCH model using a *t*-distribution.[10]

10. ESTIMATING THE NYSE INTERNATIONAL 100 INDEX

We can illustrate the process of fitting a GARCH model to financial data by using the logarithmic change in the NYSE Index of 100 International Stocks shown in Figure 3.4. You can follow along using the series labeled INT100 in the data set NYSE.XLS. The series consists of the daily closing values of the index over the period January 3, 2000, through July 30, 2008.[11] The series is a good candidate to be a GARCH process; you can clearly see periods in which there are only small changes in the series (such as the 2003–2005 period) and others where there are clusters of large increases and decreases in the index.

In Section 4, the main focus of the example of the interest rate spread was to estimate a model of the mean and to estimate the appropriate conditional confidence intervals. Here, the model of the mean is of little interest. Asset prices tend to behave as random walks or as random walks with a drift term. For this reason, there is little informational content in the model of the mean. Instead, our goal is to accurately capture the behavior of the conditional volatility. Accurately modeling the conditional variance requires a large number of observations. Moreover, since financial data are readily available, GARCH models of asset prices typically use large data sets.

The Model of the Mean

The first step in modeling any GARCH process is to estimate the model of the mean. Since the level of the index is clearly nonstationary, we construct the daily rate of return on the index as

$$r_t = 100*\ln(\text{int}\,100_t/\text{int}\,100_{t-1})$$

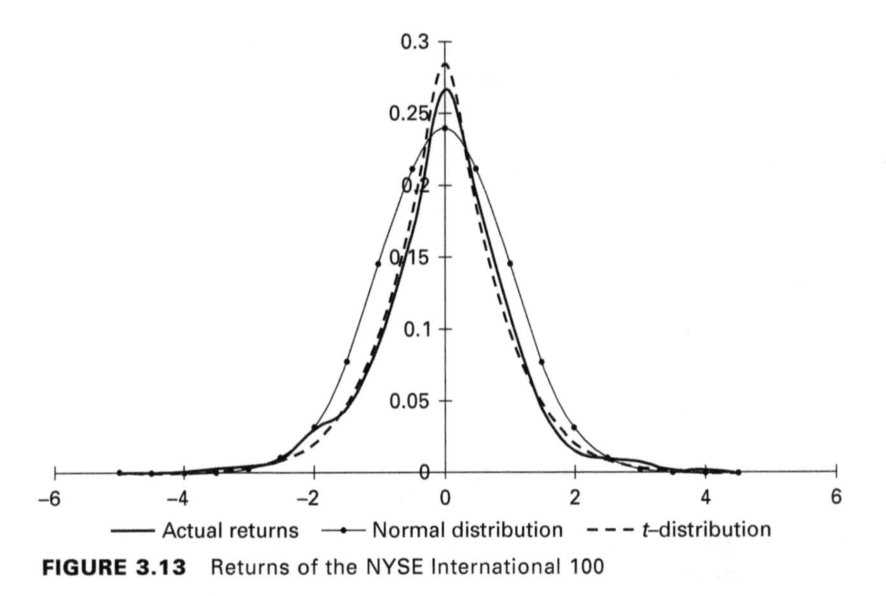

FIGURE 3.13 Returns of the NYSE International 100

The 2,237 observations in the $\{r_t\}$ series have a mean value of 0.0067 and a sample variance of 1.4512. These 2,237 observations are plotted in Figure 3.4, while the distribution of the $\{r_t\}$ sequence is shown as the solid line in Figure 3.13. You can see that the distribution of returns is more peaked than a normal distribution. The dashed line in the figure is a t-distribution with three degrees of freedom. You can see that it makes sense to estimate the $\{r_t\}$ series using a t-distribution. Most professional software packages can estimate a GARCH process using a t-distribution. You do not need to specify the degrees of freedom since it can be estimated along with the other parameters of the model. Since the t-distribution approaches the normal as the the degrees of freedom increase, a large value for the degrees-of-freedom estimate indicates that the series appears to be normally distributed.

Although the autocorrelations of the $\{r_t\}$ sequence are all very small, with such a large number of observations, several appear to be statistically significant. For example, $\rho_1 = 0.061$ and $\rho_6 = -0.052$. Since $2^*(1930)^{-1/2} = 0.0456$, both of these correlations are significant at the 5 percent level. However, it seems foolish to believe that stock returns are directly affected by events six days in the past. As such, it seems prudent to focus only on the first lag. If you form the $\{r_t\}$ series, you can verify that the AIC and the SBC select an MA(1) model over any other low-order specification. Consider the equation over the entire sample period

$$r_t = \underset{(0.28)}{0.0067} + \varepsilon_t - \underset{(-2.90)}{0.0613}\varepsilon_{t-1} \tag{3.41}$$

Notice that it is possible to eliminate the intercept terms from the regressions since the t-statistics are relatively low. Nevertheless, there are advantages to using regressions containing intercept terms. As the t-statistics can change as we posit different models for the conditional variance, intercepts will be included in the model of the mean. Once we have found the most appropriate GARCH representation for h_t, we can consider reestimating a model without an intercept term. If you check the ACF of the residuals, all are insignificant except for the fact that $\rho_6 = -0.054$.

Testing for GARCH Errors

Given that the model of the mean is satisfactory, we can test for GARCH errors by using the squared residuals of (3.41). The ACF of the squared residuals is such that $\rho_1 = 0.12$, $\rho_2 = 0.21$, $\rho_3 = 0.24$, $\rho_4 = 0.12$, and $\rho_5 = 0.16$. The correlations of the squared residuals are significant so that there is strong evidence of GARCH errors. It is also possible to test for the presence of GARCH errors using a Lagrange multiplier test. Let $\hat{\varepsilon}_t^2$ denote the squared residuals from (3.41). If we use five lags of the $\hat{\varepsilon}_t^2$ series (since there are 5 workdays in a week), we obtain

$$\begin{aligned}\hat{\varepsilon}_t^2 = \underset{(9.56)}{0.721} + \underset{(1.75)}{0.037}\hat{\varepsilon}_{t-1}^2 + \underset{(6.90)}{0.145}\hat{\varepsilon}_{t-2}^2 + \underset{(8.84)}{0.185}\hat{\varepsilon}_{t-3}^2 + \underset{(2.55)}{0.054}\hat{\varepsilon}_{t-4}^2 + \underset{(3.43)}{0.072}\hat{\varepsilon}_{t-5}^2\end{aligned}$$

The sample F-statistic for the null hypothesis that all coefficients on the lagged values of $\{\hat{\varepsilon}_t^2\}$ are equal to zero is 46.37; with four degrees of freedom in the numerator and 2,226 in the denominator (we estimate five coefficients and lose five usable observations), the *prob*-value is 0.000. Hence, we can conclude that there are GARCH errors.

Now take a little quiz. How accurately does the lag length need to be estimated to perform the test? The obvious answer ("As accurately as possible!") begs the question. Clearly, you do not want to include lags that have very small t-values; including lags that are insignificant will reduce the power of the test. If your lag length is too short, you could fail to detect the presence of conditional volatility. However, if your lag length is shorter than the true structure, and if you still detect GARCH effects, you can conclude that GARCH effects are present in the data. To take a simple example, if you find GARCH effects using only $\hat{\varepsilon}_{t-1}^2$, you can conclude there is some type of ARCH effect.

Alternative Estimates of the Model

As in the Box–Jenkins method, we want to estimate a parsimonious model. Not only can we alter the lag lengths for the GARCH(p, q) process, we can also allow for ARCH-M effects and for specifications with asymmetry. Given the tremendous number of possible specifications, is it very easy to over-fit the data. Consequently, it is best to start with a simple model and determine whether or not it is adequate. If it fails any of the diagnostic checks, it is possible to use a more complicated model. We begin by estimating (3.41) using a GARCH(1, 1) error process. The results from maximum-likelihood estimation assuming normality are

$$\begin{aligned} r_t &= \underset{(2.50)}{0.050} + \varepsilon_t - \underset{(-2.34)}{0.052}\varepsilon_{t-1} \\ h_t &= \underset{(2.90)}{0.012} + \underset{(6.91)}{0.060}\varepsilon_{t-1}^2 + \underset{(95.04)}{0.932}h_{t-1} \end{aligned}$$

Instead, if we use a t-distribution, we obtain

$$r_t = 0.054 + \varepsilon_t - 0.045\varepsilon_{t-1}$$
$$(3.00) \qquad (-2.06)$$
$$h_t = 0.010 + 0.063\varepsilon_{t-1}^2 + 0.932h_{t-1}$$
$$(2.19) \quad (6.91) \qquad (82.64)$$

where the estimated number of degrees of freedom for the shape of the t-distribution is 9.91. Since the two estimations yield such similar results and the estimated number of degrees of freedom is high, we will proceed assuming normality.

Since the sum of the coefficients on ε_{t-1}^2 and h_{t-1} is almost identically equal to unity, we can estimate the IGARCH(1, 1) model:

$$r_t = 0.050 + \varepsilon_t - 0.052\varepsilon_{t-1} \quad \text{AIC} = 6746.22, \text{SBC} = 6774.78$$
$$(3.00) \qquad (-2.21)$$
$$h_t = 0.007 + 0.064\,\varepsilon_{t-1}^2 + (1 - 0.064)h_{t-1}$$
$$(2.99) \quad (7.99) \qquad (116.66)$$

The fit of the IGARCH model will not be as good as that from the GARCH(1, 1) model since the IGARCH model imposes a constraint on the sum of the coefficients. However, the IGARCH model is more parsimonious than the GARCH(1, 1) model. If you experiment with alternative values of p and/or q in a GARCH(p, q) or an IGARCH(p, q) model, you will find that the other lags are not statistically significant. Moreover, the ARCH-M specification is not favorable to the presumption that the return on the return on the International 100 contains a risk premium. For example, if we use the GARCH(1, 1) specification for h_t, we find that the model for the mean is

$$r_t = 0.065 + \varepsilon_t - 0.052\varepsilon_{t-1} - 0.015h_t$$
$$(1.97) \qquad (-2.29) \qquad (-0.52)$$

Diagnostic Checking

Now we need to know whether the IGARCH(1, 1) model passes the various diagnostic checks for model adequacy. As all diagnostic tests are performed on the standardized residuals, begin by forming the series $s_t = \hat{\varepsilon}_t / \hat{h}_t^{0.5}$.

Remaining serial correlation: The autocorrelations of the $\{s_t\}$ series are all very small; the Ljung–Box $Q(5)$, $Q(10)$, and $Q(15)$ statistics are 2.59, 10.90, and 15.71, respectively. None of these values is significant at conventional levels; hence, we conclude that the standardized residuals are serially uncorrelated.

Remaining GARCH effects: It appears IGARCH(1, 1) is sufficient to capture all of the GARCH effects. The ACF of the squared standardized residuals is such that

ρ_1	ρ_2	ρ_3	ρ_4	ρ_5	ρ_6	ρ_7	ρ_8	ρ_9	ρ_{10}
−0.04	0.03	0.03	0.00	0.01	−0.03	−0.00	−0.27	0.00	0.06

Now use the standardized squared residuals s_t^2 to estimate a regression of the form

$$s_t^2 = a_0 + a_1 s_{t-1}^2 + \dots + a_n s_{t-n}^2$$

If you use various values of n, you will find that none of the a_1 through a_n are statistically significant. More importantly, you cannot reject any test of the null hypothesis $a_1 = a_2 = \dots = a_n = 0$. For example, if you estimate the regression

$$s_t^2 = 0.97 - 0.03s_{t-1}^2 + 0.03s_{t-2}^2$$

you will find that the restriction $a_1 = a_2 = 0$ has an F-value of 2.28 and a *prob*-value of 0.102. Hence, you can reasonably conclude that there are no remaining GARCH effects.

Leverage effects: If there are no leverage effects, s_t^2 should be uncorrelated with the lagged levels of $\{s_t\}$. However, consider the regression equation

$$\underset{(26.34)}{s_t^2 = 0.954} - \underset{(-1.71)}{0.06s_{t-1}} - \underset{(-4.47)}{0.165s_{t-2}}$$

The coefficient on s_{t-2} is highly significant and the F-statistic for the null hypothesis that the coefficients on both lagged values are jointly equal to zero is 11.57 with a *prob*-value of 0.000. Given that the signs are negative, we conclude that negative shocks are associated with large values of the conditional variance. However, we need to be cautious since the delay seems to be two periods. Most financial analysts would argue that asset markets work quickly enough so that there should not be a two-day delay in the market's reaction to news. This result is reinforced by the Engle–Ng sign test. Set $d_{t-1} = 1$ if $s_{t-1} < 0$; otherwise, set $d_{t-1} = 0$. Now if you perform the sign bias test, you will find

$$s_t^2 = \underset{(17.60)}{0.937} + \underset{(1.11)}{0.082d_{t-1}}$$

Since the coefficient of d_{t-1} is insignificant, we should be wary of concluding that negative shocks tend to increase the conditional variance of r_t. If you use the general form of the test, you will find

$$s_t^2 = \underset{(13.92)}{1.07} - \underset{(-0.39)}{0.042d_{t-1}} + \underset{(0.45)}{0.027d_{t-1}s_{t-1}} - \underset{(-2.76)}{0.184(1 - d_{t-1})s_{t-1}}$$

The implication is that there is a leverage effect such that positive shocks are associated with a diminished variance. Since $(1 - d_{t-1})s_{t-1}$, but not d_{t-1}, is significant, the size of the leverage effect depends on the magnitude of the shock (not just the direction).

The Asymmetric Models

A TARCH model is unsatisfactory since the coefficient of ε_{t-1}^2 is negative and insignificant. The estimated equation for the conditional variance is

$$h_t = \underset{(6.46)}{0.020} - \underset{(-0.11)}{0.001\varepsilon_{t-1}^2} + \underset{(6.67)}{0.010d_{t-1}\varepsilon_{t-1}} + \underset{(310.4)}{0.934h_{t-1}}$$

It is not possible to reestimate the model without the variable ε_{t-1}^2. Recall the argument demonstrating that α_1 must be positive for a GARCH(1, 1) model to be identified. You should be able to show that the identical reasoning applies to the TARCH model. One possibility is to constrain the coefficients to be positive. An alternative is to estimate an EGARCH model. Consider

$$r_t = \underset{(1.50)}{0.023} - \underset{(-2.73)}{0.058}\varepsilon_{t-1} + \varepsilon_t$$

$$\ln(h_t) = \underset{(-48.41)}{-0.069} + \underset{(28.30)}{0.092}\varepsilon_{t-1}/h_{t-1}^{0.5} - \underset{(7.32)}{0.072}|\varepsilon_{t-1}/h_{t-1}^{0.5}| + \underset{(303.11)}{0.982}\ln(h_{t-1})$$

All of the coefficients in the equation for $\ln(h_t)$ are highly significant. Given the value of h_{t-1}, a one-unit increase in ε_{t-1} will induce a change in the log of the conditional variance by 0.02 units [$0.092 - 0.072 = 0.02$]. However, if ε_{t-1} falls by one unit, the log of conditional volatility actually declines by -0.164 units ($-0.092 - 0.072 = -0.164$). The implication is that "good news" has a smaller effect on the conditional volatility than "bad news."

The two plausible models seem to be the IGARCH and EGARCH models. Not only does the EGARCH model capture the leverage effect, but it also fits the data better than the IGARCH model. The maximized value of log-likelihood function is larger for the EGARCH model than for the IGARCH model ($-3346.73 > -3368.11$). The AIC and SBC also select the EGARCH model. As a final check on the adequacy of the EGARCH model, the following diagnostic checks were performed:

1. **Checks of the standardized residuals:** The standardized residuals were checked to determine whether they exhibited no serial correlation. Similarly, the squares of the standardized residuals were checked for serial correlation. Any correlation in the $\{s_t^2\}$ series implies that there are neglected GARCH effects in the residuals. Moreover, the sign bias test on s_t^2 did not reveal any remaining leverage effects.
2. **Alternative goodness-of-fit measures:** The value SSR' for the IGARCH model is smaller than that of the EGARCH model ($2168.35 < 2248.77$). However, all of the other goodness-of-fit measures select the EGARCH model. The values AIC' and SBC' from the IGARCH model are 6746.22 and 6774.79, respectively. The same values for the EGARCH model are AIC = 6705.46 and SBC = 6739.74. Hence, the IGARCH model seems to fit the $\{v_t\}$ series better than the EGARCH model, but the EGARCH model has a better overall fit. Nevertheless, it is not clear how to interpret these results since the sums of squares do not differentiate between positive and negative errors.
3. **Normally distributed errors:** In order to determine whether the standardized errors are normally distributed, the fractiles of the $\{s_t\}$ sequence can be plotted against the fractiles of the normal distribution. After all, if $\{s_t\}$ has a standardized normal distribution, 0.5 percent should be below -2.54, 2.5 percent of the values should be below -1.64, 50 percent should be negative, 95 percent should be above 1.64 standard deviations, and 99.5 percent should

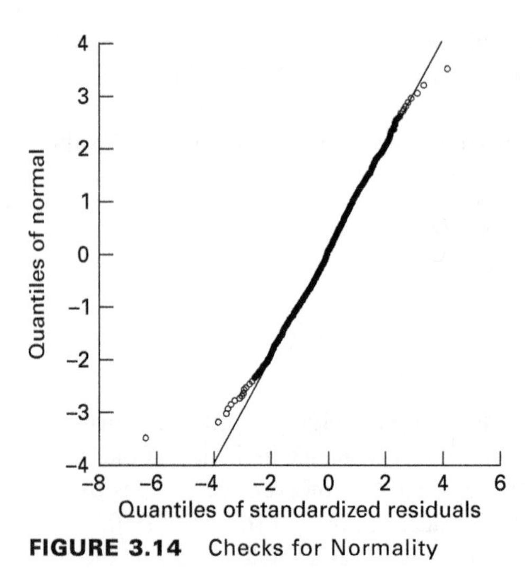

FIGURE 3.14 Checks for Normality

be above 2.54. The point is that if $\{s_t\}$ is truly normally distributed, the fractiles should lie along a straight line when plotted against the fractiles of the normal distribution. As you can see in Figure 3.14, except for some extreme observations (generally below -3 standard deviations from zero), the standardized residuals do appear to be normally distributed. Nevertheless, to ensure the appropriateness of the estimates, when the EGARCH model of the $\{r_t\}$ series is estimated using a t-distribution, the results are very similar to those shown above. The estimated number of degrees of freedom (= 11.13) is large enough that the t-distribution is close to a normal distribution.

Figure 3.15 shows the fitted values of h_t for the period from January 3, 2000, through July 30, 2008. You can see the very large values of the conditional variance beginning in mid-2002, a relatively tranquil period from mid-2004 to mid-2006, and increases in h_t toward the end of the sample.

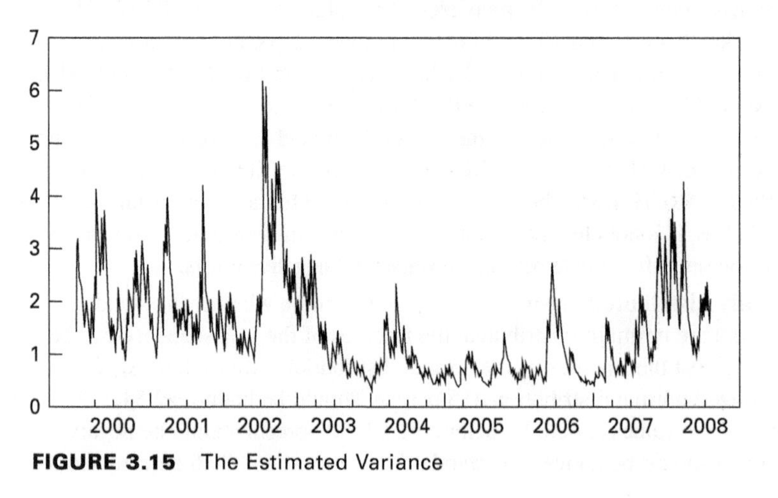

FIGURE 3.15 The Estimated Variance

11. MULTIVARIATE GARCH

If you have a data set with several variables, it often makes sense to estimate the conditional volatilities of the variables simultaneously. Multivariate GARCH models take advantage of the fact that the contemporaneous shocks to variables can be correlated with each other. Moreover, multivariate GARCH models allow for volatility spillovers in that volatility shocks to one variable might affect the volatility of other related variables. For example, instead of simply modeling the NYSE International 100, suppose that we also wanted to model the NYSE Composite Index. Although we could separately model the variance of each index, we might expect the volatilities of the two series to be interrelated. After all, shocks that increase the uncertainty of one index are likely to increase the uncertainty of the other. (If you are comfortable with matrix algebra, you may want to look at the first part of the appendix to this chapter before proceeding.)

To keep the analysis as simple as possible, suppose there are just two variables, y_{1t} and y_{2t}. For now, we are not interested in the means of the series so we can consider only the two error processes

$$\varepsilon_{1t} = v_{1t}(h_{11t})^{0.5}$$
$$\varepsilon_{2t} = v_{2t}(h_{22t})^{0.5}$$

As in the univariate case, if we assume $\text{var}(v_{1t}) = \text{var}(v_{2t}) = 1$, we can think of h_{11t} and h_{22t} as the conditional variances of ε_{1t} and ε_{2t}, respectively. Since we want to allow for the possibility that the shocks are correlated, denote h_{12t} as the conditional covariance between the two shocks. Specifically, let $h_{12t} = E_{t-1}\varepsilon_{1t}\varepsilon_{2t}$.

A natural way to construct a multivariate GARCH(1, 1) process is to allow all of the volatility terms to interact with each other. Consider the so-called *vech* model

$$h_{11t} = c_{10} + \alpha_{11}\varepsilon_{1t-1}^2 + \alpha_{12}\varepsilon_{1t-1}\varepsilon_{2t-1} + \alpha_{13}\varepsilon_{2t-1}^2 + \beta_{11}h_{11t-1} + \beta_{12}h_{12t-1} + \beta_{13}h_{22t-1} \quad (3.42)$$

$$h_{12t} = c_{20} + \alpha_{21}\varepsilon_{1t-1}^2 + \alpha_{22}\varepsilon_{1t-1}\varepsilon_{2t-1} + \alpha_{23}\varepsilon_{2t-1}^2 + \beta_{21}h_{11t-1} + \beta_{22}h_{12t-1} + \beta_{23}h_{22t-1} \quad (3.43)$$

$$h_{22t} = c_{30} + \alpha_{31}\varepsilon_{1t-1}^2 + \alpha_{32}\varepsilon_{1t-1}\varepsilon_{2t-1} + \alpha_{33}\varepsilon_{2t-1}^2 + \beta_{31}h_{11t-1} + \beta_{32}h_{12t-1} + \beta_{33}h_{22t-1} \quad (3.44)$$

Here, the conditional variance of each variable (h_{11t} and h_{22t}) depends on its own past, the conditional covariance between the two variables (h_{12t}), the lagged squared errors (ε_{1t-1}^2 and ε_{2t-1}^2), and the product of lagged errors ($\varepsilon_{1t-1}\varepsilon_{2t-1}$). Notice that the conditional covariance depends on the same set of variables. Clearly, there is a rich interaction between the variables. For example, after one period, a v_{1t} shock affects h_{11t}, h_{12t}, and h_{22t}.

Although simple to conceptualize, multivariate GARCH models in the form of (3.42)–(3.44) can be very difficult to estimate. Notice:

- The number of parameters necessary to estimate can get quite large. In the 2-variable case above, there are 21 parameters. The number grows very quickly as more variables are added to the system and as the order of the

GARCH process increases. If you understand the nature of the multivariate model above, you should be able to show that a GARCH(2, 1) model necessitates the estimation of nine additional parameters. You can also verify that in the 3-variable case, a GARCH(1, 1) model contains six equations (since there are equations for h_{11t}, h_{22t}, h_{33t}, h_{12t}, h_{13t}, and h_{23t}) and that each of the equations entails the estimation of 12 coefficients plus a constant.

Moreover, we have not begun to specify the models of the mean. If we have two variables y_{1t} and y_{2t}, it is possible to estimate the means by specifying $y_{1t} - \mu_1 = \varepsilon_{1t}$ and $y_{2t} - \mu_2 = \varepsilon_{2t}$. Once lagged values of $\{y_{1t}\}$ and $\{y_{2t}\}$ and/or explanatory variables are added to the mean equation, it should be clear that the estimation problem can be quite complicated.

- As in the univariate case, there is not an analytic solution to the maximization problem. As such, it is necessary to use numerical methods to find the parameter values that maximize the function L. Unfortunately, such methods may not be able to find a maximum value if the model is overparameterized. To explain, if a coefficient is small relative to its standard error, it necessarily has a large confidence interval. As such, there is a large range in which the coefficient may lie and slight changes in the coefficient's value will have little influence on the value of L. The numerical "hill climbing" techniques that computers use in their maximization routines will have difficulty pinning down the value of such a coefficient. Hence, when attempting to estimate an overparameterized model, it is typical for a software package to indicate that its search algorithm did not converge.
- Since conditional variances are necessarily positive, the restrictions for the multivariate case are far more complicated than for the univariate case. The results of the maximization problem must be such that every one of the conditional variances is always positive and that the implied correlation coefficients, $\rho_{ij} = h_{ij}/(h_{ii}h_{jj})^{0.5}$, are between -1 and $+1$.

In order to circumvent these problems, much of the recent work involving multivariate GARCH modeling involves finding suitable restrictions on the general model of (3.42)–(3.44). One set of restrictions that became popular in the early literature is the so-called diagonal *vech* model. The idea is to diagonalize the system such that h_{ijt} contains only lags of itself and the cross products of $\varepsilon_{it}\varepsilon_{jt}$. For example, the diagonalized version of (3.42)–(3.44) is

$$h_{11t} = c_{10} + \alpha_{11}\varepsilon_{1t-1}^2 + \beta_{11}h_{11t-1}$$
$$h_{12t} = c_{20} + \alpha_{22}\varepsilon_{1t-1}\varepsilon_{2t-1} + \beta_{22}h_{12t-1}$$
$$h_{22t} = c_{30} + \alpha_{33}\varepsilon_{2t-1}^2 + \beta_{33}h_{22t-1}$$

Given the large number of restrictions, the model is relatively easy to estimate. Each conditional variance is equivalent to that of a univariate GARCH process and the conditional covariance is quite parsimonious as well. The problem is that setting all $\alpha_{ij} = \beta_{ij} = 0$ (for $i \neq j$) means that there are no interactions among the variances. An ε_{1t-1} shock, for example, affects h_{11t} and h_{12t}, but does not affect the conditional variance

h_{2t}. Notice that the systemwide estimation does have the advantage of controlling for the contemporaneous correlation of the residuals across equations.

Engle and Kroner (1995) popularized what is now called the BEK (or BEKK) model that ensures that the conditional variances are positive. The idea is to force all the parameters to enter the model via quadratic forms ensuring that all the variances are positive. Although there are several different variants of the model, consider the specification

$$H_t = C'C + A'\varepsilon_{t-1}\varepsilon_{t-1}'A + B'H_{t-1}B$$

where for the 2-variable case considered in (3.42)−(3.44),

$$H_t = \begin{bmatrix} h_{11t} & h_{12t} \\ h_{12t} & h_{22t} \end{bmatrix}; C = \begin{bmatrix} c_{11} & c_{12} \\ c_{12} & c_{22} \end{bmatrix}; A = \begin{bmatrix} \alpha_{11} & \alpha_{12} \\ \alpha_{21} & \alpha_{22} \end{bmatrix}; B = \begin{bmatrix} \beta_{11} & \beta_{12} \\ \beta_{21} & \beta_{22} \end{bmatrix}$$

For example, if you perform the indicated matrix multiplications you will find

$$h_{11t} = (c_{11}^2 + c_{12}^2) + (\alpha_{11}^2\varepsilon_{1t-1}^2 + 2\alpha_{11}\alpha_{21}\varepsilon_{1t-1}\varepsilon_{2t-1} + \alpha_{21}^2\varepsilon_{2t-1}^2) \\ + (\beta_{11}^2 h_{11t-1} + 2\beta_{11}\beta_{21}h_{12t-1} + \beta_{21}^2 h_{22t-1})$$

In general, h_{ijt} will depend on the squared residuals, cross-products of the residuals, and the conditional variances and covariances of all variables in the system. As such, the model allows for shocks to the variance of one of the variables to "spill over" to the others. The problem is that the BEK formulation can be quite difficult to estimate. The model has a large number of parameters that are not globally identified. Changing the signs of all elements of A, B, or C will have no effect on the value of the likelihood function. As such, convergence can be quite difficult to achieve.

Another popular multivariate GARCH specification is **constant conditional correlation** model. As the name suggests, the constant correlation coefficient (CCC) model restricts the correlation coefficients to be constant. As such, for each $i \neq j$, the CCC model assumes $h_{ijt} = \rho_{ij}(h_{iit}h_{jjt})^{0.5}$. In a sense, the CCC model is a compromise in that the variance terms need not be diagonalized, but the covariance terms are always proportional to $(h_{iit}h_{jjt})^{0.5}$. For example, a CCC model could consist of (3.42), (3.44), and

$$h_{12t} = \rho_{12}(h_{11t}h_{22t})^{0.5}$$

Hence, the covariance equation entails only one parameter instead of the seven parameters appearing in (3.43).

Bollerslev (1990) illustrates the usefulness of the CCC specification by examining the weekly values of the nominal exchange rates for five different countries—the German mark (DM), the French franc (FF), the Italian lira (IL), the Swiss franc (SF), and the British pound (BP)—relative to the U.S. dollar. Note that a five-equation system would be too unwieldy to estimate in an unrestricted form. For the model of the mean, the log of each exchange rate series was modeled as a random walk plus a drift. If y_{it} is the percentage change in the nominal exchange rate for country i, the model of the mean for each country is simply

$$y_{it} = \mu_i + \varepsilon_{it} \tag{3.45}$$

Ljung–Box tests indicated each series of residuals did not contain any serial correlation. This is consistent with the general finding that, when using high-frequency data, nominal exchange rates behave as random walk processes. As a next step, Bollerslev (1990) tested the squared residuals for serial dependence. He reported that the autocorrelations of the squared residuals are strongly indicative of GARCH effects. For example, for the British pound, the $Q(20)$-statistic has a value of 113.020; this is significant at any conventional level. Given the presence of conditional heteroskedasticity, Bollerslev next turned to finding the appropriate orders for the five GARCH(p, q) processes. Individually, each residual series could be well-estimated as a GARCH(1, 1) process. As such, the specification for the full model has the form of (3.45) plus

$$h_{iit} = c_{i0} + \alpha_{ii}\varepsilon_{it-1}^2 + \beta_{ii}h_{iit-1} \qquad (i = 1, \ldots, 5)$$
$$h_{ijt} = \rho_{ij}(h_{iit}h_{jjt})^{0.5} \qquad (i \neq j)$$

Notice that the full model requires that only 30 parameters be estimated (five values of μ_i, the five equations for h_{iit} each have three parameters, and ten values of the ρ_{ij}). He reports that with 333 observations, the required number of matrix inversions is reduced from 10,323 to 31. Also notice that the CCC model has an important advantage over the separate estimation of each equation. As in a seemingly unrelated regression framework, the system-wide estimation provided by the CCC model captures the contemporaneous correlation between the various error terms. As such, the coefficient estimates of the GARCH process are more efficient than those from a set of single equation estimations. The estimated correlations for the period during which the European Monetary System (EMS) prevailed are

	DM	FF	IL	SF
FF	0.932			
IL	0.886	0.876		
SW	0.917	0.866	0.816	
BP	0.674	0.678	0.622	0.635

It is interesting that correlations among continental European currencies were all far greater than those for the pound. Moreover, the correlations were much greater than those of the pre-EMS period. Clearly, EMS acted to keep the exchange rates of Germany, France, Italy, and Switzerland tightly in line prior to the introduction of the Euro.

If you are familiar with matrix algebra, the last part of Appendix 3.1 shows you how to generalize Bollerslev's method so as to estimate time-varying (or dynamic) conditional correlations.

Updating the Study

The file labeled EXRATES(DAILY).XLS contains the 2,342 daily values of the euro, British pound, and Swiss franc over the Jan. 3, 2000–Dec. 23, 2008 period. Denote the U.S. dollar value of each of these nominal exchange rates as e_{it} where i = EU, BP, and SW. If you plot the three currencies, you will see that all generally fell during the early portion of the sample, rose between mid-2001 through 2004, and fell sharply in 2006. As a preliminary step, construct the logarithmic change of each nominal exchange rate as $y_{it} = \log(e_{it}/e_{it-1})$. As in any GARCH estimation, the first step is to properly estimate the model of the mean. If you follow Bollerslev (1990) and estimate equations in the form of (3.45), you should obtain the means as

euro	BP	SF
1.36×10^{-4}	-4.15×10^{-5}	1.36×10^{-4}

Although the residual autocorrelations are all very small in magnitude, a few are statistically significant. For example, the autocorrelations for the euro are

ρ_1	ρ_2	ρ_3	ρ_4	ρ_5	ρ_6
0.036	−0.004	−0.004	0.063	0.001	−0.036

With $T = 2342$, the value of ρ_4 is statistically significant and the value of the Ljung–Box $Q(4)$ statistic is 12.37. Nevertheless, most researchers would not attempt to model this small value of the fourth lag. Moreover, the SBC always selects models with no lagged changes in the mean equation.

For the second step, you should check the squared residuals for the presence of GARCH errors. Since we are using daily data (with a five-day week), it seems reasonable to begin using a model of the form

$$\hat{\varepsilon}_t^2 = \alpha_0 + \sum_{i=1}^{5} \alpha_i \hat{\varepsilon}_{t-5}^2$$

The sample values of the F-statistics for the null hypothesis that $\alpha_1 = \ldots = \alpha_5 = 0$ are 43.36, 89.74, and 20.96 for the euro, BP, and SW, respectively. Since all of these values are highly significant, it is possible to conclude that all three series exhibit GARCH errors.

For the third step, you should try to find the proper form of the GARCH model for each exchange rate series. Although some other GARCH forms (such as the IGARCH model) might seem more appropriate than Bollerslev's specification, proceed as if the GARCH(1, 1) model is appropriate for each series. If you estimate the three series as GARCH(1, 1) processes using the CCC restriction, you should find the results reported in Table 3.1.

Table 3.1 The CCC Model of Exchange Rates

	c	α_1	β_1
Euro	4.48×10^{-7}	0.055	0.934
	(4.48)	(9.43)	(135.36)
Pound	4.78×10^{-7}	0.050	0.934
	(3.28)	(6.97)	(88.18)
Franc	5.90×10^{-7}	0.046	0.942
	(4.24)	(8.90)	(139.89)

Table 3.2 Estimates Using the Diagonal *vech* Model

	h_{11t}	h_{12t}	h_{13t}	h_{22t}	h_{23t}	h_{33t}
c	3.32×10^{-7}	2.54×10^{-7}	3.48×10^{-7}	3.25×10^{-7}	2.21×10^{-7}	4.67×10^{-7}
	(22.24)	(4.74)	(13.77)	(3.12)	(5.24)	(6.32)
α_1	0.041	0.033	0.039	0.039	0.031	0.039
	(26.59)	(9.93)	(108.30)	(8.12)	(10.65)	(23.88)
β_1	0.951	0.957	0.953	0.951	0.961	0.952
	(445.43)	(187.70)	(634.62)	(132.41)	(265.67)	(324.46)

If we let the numbers 1, 2, and 3 represent the euro, pound, and franc, the correlations are $\rho_{12} = 0.71$, $\rho_{13} = 0.92$, and $\rho_{23} = 0.65$. As in Bollerslev's paper, the pound and the franc continue to have the lowest correlation coefficient.

By way of contrast, it is instructive to estimate the model using the diagonal *vech* specification such that each variance and covariance is estimated separately. The estimation results are given in Table 3.2. Now, the correlation coefficients are time-varying. For example, the correlation coefficient between the pound and the franc is given by $h_{23t}/(h_{22t}h_{33t})^{0.5}$. The time path of this correlation coefficient is shown in Figure 3.16. Although the correlation does seem to fluctuate around 0.64 (the value found by the CCC method), there are substantial departures from this average value. Beginning in mid-2006, the correlation between the pound and the franc began a long and steady decline ending in March of 2008. The correlation increased with fears of a U.S. recession and then sharply fell with the onset on the U.S. financial crisis in the fall of 2008.

12. SUMMARY AND CONCLUSIONS

Many economic time series exhibit periods of volatility. Conditionally heteroskedastic models (ARCH or GARCH) allow the conditional variance of a series to depend on the past realizations of the error process. A large realization of the current period's disturbance increases the conditional variance in subsequent periods. For a stable process, the conditional variance will

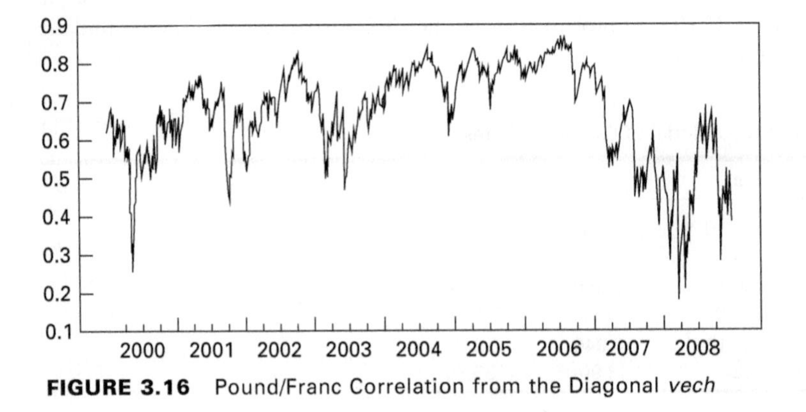

FIGURE 3.16 Pound/Franc Correlation from the Diagonal *vech*

eventually decay to the long-run (unconditional) variance. As such, ARCH and GARCH models can capture periods of turbulence and tranquility.

Conditional variance is a measure of risk. ARCH and GARCH effects have been included in a regression framework to test hypotheses of risk-averse agents. For example, if producers are risk-averse, conditional price variability will affect product supply. Producers may reduce their exposure by withdrawing from the market in periods of substantial risk. Similarly, asset prices should be negatively related to their conditional volatility. Such ARCH effects in the mean of a series (ARCH-M) are a natural implication of asset-pricing models.

The basic ARCH and GARCH models have been extended in a number of interesting ways. The IGARCH model allows volatility shocks to be permanent and the TARCH and EGARCH models allow negative shocks to behave differently than positive shocks. You can also include explanatory variables in the equation for the conditional volatility.

One interesting development is the application of GARCH models in a multivariate setting. The problem is that an unrestricted multivariate GARCH has too many parameters to reasonably estimate. Nevertheless, most software packages now incorporate a number of specifications that restrict the parameters of the multivariate model. For example, EVIEWS and RATS are able to use Engle and Kroner's (1995) method and Bollerslev's (1990) constant conditional correlations.

Estimating any type of GARCH model can be difficult. Here are some suggestions to improve your estimates.

1. Be sure that your model of the mean is appropriate. Any misspecification in the mean equation will carry over into the variance equation. Clearly, the estimated $\{\varepsilon_t\}$ series must be serially uncorrelated in order to obtain a sensible model of the conditional variance.
2. It is very easy to "overfit" the data; you could wind up with a very complicated model when a far more parsimomous model actually captures the nature of the data-generating process. Pretest the squared residuals for the presence of ARCH errors. Similarly, do not simply include leverage effects, ARCH-M effects, or large values of p and or q without good reason.
3. It is very common to find that the sum of the α_i plus the β_i is very close to unity. Such highly persistent volatilities do seem to be the case for financial data. However, Hillebrand (2005) showed that a neglected structural break in the variance series can create the appearance of a highly persistent conditional volatility. After all, if the conditional variance is always small before some date t^* and then always large, the conditional volatility is definitely persistent. However, in such a circumstance, the volatility would be best captured by a dummy variable indicating the break date. Plot the conditional volatility to ensure that there are several periods with high and low volatilities.

   Moreover, as shown by Ma, Nelson, and Startz (2007), the estimated sum of the GARCH coefficients can also be close to unity when the true GARCH effect is very small or absent. To explain, suppose you estimate a GARCH(1, 1) model and find that $\alpha_1 + \beta_1 \approx 1$. As such, the current level of conditional volatility is expected to prevail into the future. However, this could happen because the actual data-generating process is a near-IGARCH process or because the amount of conditional volatility is always constant (so that $h_t = h_{t-1} \ldots = \alpha_0$). Be particularly concerned about this problem if α is very small but $\alpha + \beta$ is close to unity. As such, it is important to examine the ACF of the squared residuals and pretest for conditional heteroskedasticity. You can also compare the GARCH model to a low-order ARCH(q) process check if the persistence is actually large. You do not want to estimate a near-IGARCH process when the amount of conditional volatility is actually quite small.

4. Multivariate GARCH models can be quite difficult to estimate. There are a number of different specifications that ease the estimation problems. If the diagonal *vech* model does not provide sufficient interaction among the conditional variances and covariances, try the BEK specification. The CCC model (or the DCC model described in the appendix) can be especially helpful in a large system.

QUESTIONS AND EXERCISES

1. Suppose that the $\{\varepsilon_t\}$ sequence is the ARCH(q) process

$$\varepsilon_t = v_t(\alpha_0 + \alpha_1\varepsilon_{t-1}^2 + \ldots + \alpha_q\varepsilon_{t-q}^2)^{1/2}$$

Show that the conditional expectation of $E_{t-1}\varepsilon_t^2$ has the same form as the conditional expectation of (3.1).

2. Consider the ARCH-M model represented by equations (3.23) to (3.25). Recall that $\{\varepsilon_t\}$ is a white-noise disturbance; for simplicity, let $E\varepsilon_t^2 = E\varepsilon_{t-1}^2 = \ldots = 1$.
   a. Find the unconditional mean Ey_t. How does a change in δ affect the mean? Using the example of Section 6, show that changing β and δ from $(-4, 4)$ to $(-1, 1)$ preserves the mean of the $\{y_t\}$ sequence.
   b. Show that the unconditional variance of y_t when $h_t = \alpha_0 + \alpha_1\varepsilon_{t-1}^2$ does not depend on β, δ, or α_0.

3. Bollerslev (1986) proved that the ACF of the squared residuals resulting from the GARCH(p, q) process represented by (3.9) act as an ARMA(m, p) process where $m = \max(p, q)$. You are to illustrate this result using the examples below.
   a. Consider the GARCH(1, 2) process: $h_t = \alpha_0 + \alpha_1\varepsilon_{t-1}^2 + \alpha_2\varepsilon_{t-2}^2 + \beta_1 h_{t-1}$. Add the expression $(\varepsilon_t^2 - h_t)$ to each side so that

$$\begin{aligned}\varepsilon_t^2 &= \alpha_0 + \alpha_1\varepsilon_{t-1}^2 + \alpha_2\varepsilon_{t-2}^2 + \beta_1 h_{t-1} + (\varepsilon_t^2 - h_t)\\ &= \alpha_0 + (\alpha_1 + \beta_1)\varepsilon_{t-1}^2 + \alpha_2\varepsilon_{t-2}^2 - \beta_1(\varepsilon_{t-1}^2 - h_{t-1}) + (\varepsilon_t^2 - h_t)\end{aligned}$$

   Define $\eta_t = (\varepsilon_t^2 - h_t)$, so that:

$$\varepsilon_t^2 = \alpha_0 + (\alpha_1 + \beta_1)\varepsilon_{t-1}^2 + \alpha_2\varepsilon_{t-2}^2 - \beta_1\eta_{t-1} + \eta_t$$

   Show that:
      i. η_t is serially uncorrelated.
      ii. The $\{\varepsilon_t^2\}$ sequence acts as an ARMA(2, 1) process.
   b. Consider the GARCH(2, 1) process $h_t = \alpha_0 + \alpha_1\varepsilon_{t-1}^2 + \beta_1 h_{t-1} + \beta_2 h_{t-2}$. Show that it is possible to add η_t to each side so as to obtain:

$$\varepsilon_t^2 = \alpha_0 + \alpha_1\varepsilon_{t-1}^2 + \beta_1 h_{t-1} + \eta_t + \beta_2 h_{t-2}$$

   Show that adding and subtracting the terms $\beta_1\eta_{t-1}$ and $\beta_2\eta_{t-2}$ to the right-hand side of this equation yields an ARMA(2, 2) process.
   c. Provide an intuitive explanation of the statement: "The Lagrange multiplier for the ARCH errors test cannot be used to test the null of white-noise squared residuals against an alternative of a specific GARCH(p, q) process."

d. Sketch the proof of the general statement that the ACF of the squared residuals resulting from the GARCH(p, q) process represented by (3.9) acts as an ARMA(m, p) process where $m = \max(p, q)$.

4. Let $y_0 = 0$, and let the first five realizations of the $\{\varepsilon_t\}$ sequence be $(1, -1, -2, 1, 1)$. Plot each of the following sequences:

$$\text{Model 1:} \quad y_t = 0.5y_{t-1} + \varepsilon_t$$
$$\text{Model 2:} \quad y_t = \varepsilon_t - \varepsilon_{t-1}^2$$
$$\text{Model 3:} \quad y_t = 0.5y_{t-1} + \varepsilon_t - \varepsilon_{t-1}^2$$

a. How does the ARCH-M specification affect the behavior of the $\{y_t\}$ sequence? What is the influence of the autoregressive term in Model 3?

b. For each of the three models, calculate the sample mean and variance of $\{y_t\}$.

5. The file labeled ARCH.XLS contains the 100 realizations of the simulated $\{y_t\}$ sequence used to create the lower-right-hand panel of Figure 3.7. Recall that this series was simulated as $y_t = 0.9y_{t-1} + \varepsilon_t$ where ε_t is the ARCH(1) error process $\varepsilon_t = v_t(1 + 0.8\varepsilon_{t-1})^{1/2}$. You should find the series has a mean of 0.263, a standard deviation of 4.894, and minimum and maximum values of −10.8 and 15.15, respectively.

a. Estimate the series using OLS and save the residuals. You should obtain:

$$y_t = \underset{(26.51)}{0.944}y_{t-1} + \varepsilon_t$$

Note that the estimated value of a_1 differs from the theoretical value of 0.9. This is due to nothing more than sampling error; the simulated values of $\{v_t\}$ do not precisely conform to the theoretical distribution. However, can you provide an intuitive explanation of why positive serial correlation in the $\{v_t\}$ sequence might shift the estimate of a_1 upward in small samples?

b. Obtain the ACF and the PACF of the residuals. Use Ljung–Box Q-statistics to determine whether the residuals approximate white noise. You should find:

	1	2	3	4	5	6	7	8
ACF	0.149	0.004	−0.018	−0.013	0.072	−0.002	−0.110	−0.152
PACF	0.149	−0.018	−0.016	−0.008	0.077	−0.025	−0.109	−0.122

$Q(4) = 2.31$, $Q(8) = 6.39$, $Q(24) = 18.49$.

c. Obtain the ACF and the PACF of the squared residuals. You should find:

	1	2	3	4	5	6	7	8
ACF	0.474	0.128	−0.057	−0.077	0.055	0.245	0.279	0.223
PACF	0.474	−0.125	−0.087	0.005	0.132	0.205	0.074	0.067

Based on the ACF and PACF of the residuals and the squared residuals, what can you conclude about the presence of ARCH errors?

d. Estimate the squared residuals as $\varepsilon_t^2 = \alpha_0 + \alpha_1\varepsilon_{t-1}^2$. You should verify $\alpha_0 = 1.55$ (t-statistic = 2.83) and $\alpha_1 = 0.474$ (t-statistic = 5.28).

Show that the Lagrange multiplier test for ARCH(1) errors is $TR^2 = 22.03$ with a significance level of 0.00000269.

e. Simultaneously estimate the $\{y_t\}$ sequence and the ARCH(1) error process using maximum-likelihood estimation. You should find:

$$y_t = \underset{(32.79)}{0.886}y_{t-1} + \varepsilon_t \qquad h_t = \underset{(4.02)}{1.19} + \underset{(2.89)}{0.663}\,\varepsilon_{t-1}^2$$

6. The second series in the file ARCH.XLS contains 100 observations of a simulated ARCH-M process.

a. Estimate the $\{y_t\}$ sequence using the Box–Jenkins methodology. Try to improve on the model

$$y_t = \underset{(22.32)}{1.07} + \varepsilon_t + \underset{(2.57)}{0.254}\varepsilon_{t-3} - \underset{(-2.64)}{0.262}\varepsilon_{t-6}$$

b. Examine the ACF and the PACF of the residuals from the MA(||3, 6||) model above. Why might someone conclude that the residuals appear to be white noise? Now examine the ACF and PACF of the squared residuals. You should find:

	1	2	3	4	5	6
ACF	0.498	0.251	0.290	0.163	0.043	0.114
PACF	0.498	0.004	0.217	−0.088	−0.041	0.101

Perform the LM test for ARCH errors.

c. Estimate the $\{y_t\}$ sequence as the ARCH-M process:

$$y_t = \underset{(14.05)}{0.908} + \underset{(1.79)}{0.625}h_t + \varepsilon_t$$

$$h_t = \underset{(5.59)}{0.108} + \underset{(2.50)}{0.597}\,\varepsilon_{t-1}^2$$

d. Check ACF and the PACF of the estimated $\{\varepsilon_t\}$ sequence. Do they appear to be satisfactory? Experiment with several other simple formulations of the ARCH-M process.

7. Consider the ARCH(2) process $E_{t-1}\varepsilon_t^2 = \alpha_0 + \alpha_1\varepsilon_{t-1}^2 + \alpha_2\varepsilon_{t-2}^2$.

a. Suppose that the residuals come from the model $y_t = a_0 + a_1 y_{t-1} + \varepsilon_t$. Find the conditional and unconditional variance of $\{y_t\}$ in terms of the parameters a_1, α_0, α_1, and α_2.

b. Suppose that $\{y_t\}$ is an ARCH-M process such that the level of y_t is positively related to its own conditional variance. For simplicity, let $y_t = \alpha_0 + \alpha_1\varepsilon_{t-1}^2 + \alpha_2\varepsilon_{t-2}^2 + \varepsilon_t$. Trace out the impulse response function of $\{y_t\}$ to an $\{\varepsilon_t\}$ shock. You may assume that the system has been in long-run equilibrium ($\varepsilon_{t-2} = \varepsilon_{t-1} = 0$) but now $\varepsilon_1 = 1$. Thus, the issue is to find the values of y_1, y_2, y_3, and y_4 given that $\varepsilon_2 = \varepsilon_3 = \ldots = 0$.

c. Use your answer to part b to explain the following result. A student estimated $\{y_t\}$ as an MA(2) process and found the residuals to be white noise. A second student estimated the same series as the ARCH-M process $y_t = \alpha_0 + \alpha_1\varepsilon_{t-1}^2 + \alpha_2\varepsilon_{t-2}^2 + \varepsilon_t$. Why might both estimates appear reasonable? How would you decide which is the better model?

d. In general, explain why an ARCH-M model might appear to be a moving-average process.

8. The file RGDP.XLS contains the data used to construct Figures 3.1 and 3.2. Use the method presented in Section 4 to show that there were significant volatility reductions in real consumption, investment, and government expenditures in 1984Q1.

9. The file NYSE.XLS contains the daily values of the New York Stock Exchange International 100 Index that was used in Section 10. Reproduce the results of Section 10.

10. Use the data in the file EXRATES(DAILY).XLS to estimate a bivariate model of the pound and euro exchange rates. In particular:

    a. Does a bivariate diagonal *vech* model yield very different results from those shown in Section 11?

    b. Experiment with the convergence criteria and search methods on your software package to determine how they influence the estimates you found in part a. Pay particular attention to the standard errors on the coefficients.

    c. Try to get convergence for a pure *vech* model. Compare the results to those you found in part a.

11. In answering the following, you should consult Appendix 3.1.

    a. Justin finds that a GARCH(2, 1) specification is appropriate for all h_{ijt} in a 2-variable diagonal *vech* model. What is the formula for h_{12t}?

    b. Jennifer finds that a GARCH(1, 2) specification is appropriate for all h_{ijt} in a 2-variable diagonal *vech* model. What is the formula for h_{12t}?

    c. In the 2-variable of the BEK model, it was shown that

$$h_{11t} = (c_{11}^2 + c_{12}^2) + (\alpha_{11}^2\varepsilon_{1t-1}^2 + 2\alpha_{11}\alpha_{21}\varepsilon_{1t-1}\varepsilon_{2t-1} + \alpha_{21}^2\varepsilon_{2t-1}^2) + (\beta_{11}^2 h_{11t-1} + 2\beta_{11}\beta_{21}h_{12t-1} + \beta_{21}^2 h_{22t-1})$$

    Let all of the coefficients be positive. If the ε_{1t-1} and ε_{2t-1} are of opposite signs, can the term $\alpha_{11}^2\varepsilon_{1t-1}^2 + 2\alpha_{11}\alpha_{21}\varepsilon_{1t-1}\varepsilon_{2t-1} + \alpha_{21}^2\varepsilon_{2t-1}^2$ be negative?

    d. Suppose that in period t, $h_{11t} = 2$ and $h_{22t} = 4.5$. If the CCC model indicates $\rho_{12} = -0.5$, find h_{12t}.

12. Chapter 1.6 of the *Programming Manual* that accompanies this text contains a discussion of nonlinear least squares and maximum-likelihood estimation. If you have not already done so, download the manual from the Wiley Web site. Also download the data set MONEYDEM.XLS and the programs. Program 1.4 creates the interest rate spread (s_t) as tb1yr– tb3mo. As indicated in the *Programming Manual*, estimate the spread as an AR(3) process. You should find:

$$s_t = 0.0476 + 0.890s_{t-1} - 0.39s_{t-2} + 0.162s_{t-3} + \varepsilon_t$$

    a. Show that the residuals pass the standard diagnostic tests for serial correlation.

    b. Form the squared residuals and perform the test for ARCH errors. You should find

$$\hat{\varepsilon}_t^2 = 0.046 + 0.217\hat{\varepsilon}_{t-1}^2 - 0.026\hat{\varepsilon}_{t-2}^2 + 0.160\hat{\varepsilon}_{t-3}^2$$

    Determine the value of TR^2 for this equation.

    c. Estimate the spread as an AR(3) process with ARCH(3) errors.

    d. As shown in Program 1.4, estimate the spread as an ARMA(1, 1) process with GARCH(1, 1) errors. Show that this model is preferable to that in part c.

ENDNOTES

1. Some authors prefer the spellings *homoscedastic* and *heteroscedastic;* both forms are correct.
2. If the unconditional variance of a series is not constant, the series is nonstationary. However, *conditional* heteroskedasticy is *not* a source of nonstationarity.
3. Letting $\alpha(L)$ and $\beta(L)$ be polynomials in the lag operator L, we can rewrite h_t in the form $h_t = \alpha_0 + \alpha(L)\varepsilon_{t-1}^2 + \beta(L)h_{t-1}$. The notation $\alpha(1)$ denotes the polynomial $\alpha(L)$ evaluated at $L = 1$; i.e., $\alpha(1) = \alpha_1 + \alpha_2 + \ldots + \alpha_q$. Bollerslev (1986) showed that the GARCH process is stationary with $E\varepsilon_t = 0$, $\text{var}(\varepsilon_t) = \alpha_0/(1-\alpha(1) - \beta(1))$, and $\text{cov}(\varepsilon_t, \varepsilon_{t-s}) = 0$ for $s \neq 0$ if $\alpha(1) + \beta(1) < 1$.
4. Unfortunately, there is no available method to test the null of white-noise errors versus the specific alternative of GARCH(p, q) errors. Bollerslev (1986) proved that the ACF of the squared residuals resulting from (3.9) are an ARMA(m, p) model where $m = \max(p, q)$.
5. Constraining the coefficients of h_t to follow a decaying pattern conserves degrees of freedom and considerably eases the estimation process. Moreover, the lagged coefficients given by $(9 - i)$ are necessarily positive.
6. In addition to the intercept term, three seasonal dummy variables were included in the supply equation.
7. If the underlying data-generating process is autoregressive, adaptive expectations and rational expectations can be perfectly consistent with each other.
8. The unconditional mean of y_t is altered by changing only δ. Changing β and δ commensurately maintains the mean value of the $\{y_t\}$ sequence.
9. A t-test on only the coefficient a_2 is called the negative size bias test, and the test on only the coefficient a_3 is called the positive size bias test.
10. Unless they are specifically interested in the tails of the distribution, many researchers simply ignore the issue of a fat-tailed distribution. Quasi-maximum likelihood estimates use the normal distribution even though the actual distribution of the $\{\varepsilon_t\}$ sequence is fat-tailed. Under fairly weak assumptions, the parameter estimates for the model of the mean and the conditional variance are consistent and normally distributed.
11. In constructing the data set, no attempt was made to account for the fact that the market was closed on holidays and on important key dates such as September 11, 2001. For simplicity, we interpolated to obtain values for non-weekend dates when the market was closed.

APPENDIX 3.1: MULTIVARIATE GARCH MODELS

The Log-Likelihood Function

In the multivariate case, the likelihood function presented in Section 8 needs to be modified. For the 2-variable case, suppose that ε_{1t} and ε_{2t} are zero-mean random variables that are jointly normally distributed. For the time being, we can keep the analysis simple if we assume the variances and the covariance terms are constant. As such, we can drop the time subscripts on the h_{ijt}. In such a circumstance, the log likelihood function for the joint realization of ε_{1t} and ε_{2t} is

$$L_t = \frac{1}{2\pi\sqrt{h_{11}h_{22}(1-\rho_{12}^2)}} \exp\left[-\frac{1}{2(1-\rho_{12}^2)}\left(\frac{\varepsilon_{1t}^2}{h_{11}} + \frac{\varepsilon_{2t}^2}{h_{22}} - \frac{2\,\rho_{12}\varepsilon_{1t}\varepsilon_{2t}}{(h_{11}h_{22})^{0.5}}\right)\right] \quad \text{(A3.1)}$$

where ρ_{12} is the correlation coefficient between ε_{1t} and ε_{2t}; $\rho_{12} = h_{12}/(h_{11}h_{22})^{0.5}$.

Now if we define the matrix H such that

$$H = \begin{bmatrix} h_{11} & h_{12} \\ h_{12} & h_{22} \end{bmatrix}$$

the likelihood function can be written in the compact form

$$L_t = \frac{1}{2\pi\,|H|^{1/2}} \exp\left[-\frac{1}{2}\varepsilon_t' H^{-1} \varepsilon_t \right] \tag{A3.2}$$

where $\varepsilon_t = (\varepsilon_{1t}, \varepsilon_{2t})'$, and $|H|$ is the determinant of H. To see that the two representations given by (A3.1) and (A3.2) are equivalent, note that $|H| = h_{11}h_{22} - (h_{12})^2$. Since $h_{12} = \rho_{12}(h_{11}h_{22})^{0.5}$, it follows that $|H| = (1 - (\rho_{12})^2)h_{11}h_{22}$. Moreover,

$$\varepsilon_t' H^{-1} \varepsilon_t = \frac{\varepsilon_{1t}^2 h_{22} - 2\varepsilon_{1t}\varepsilon_{2t}h_{12} + \varepsilon_{2t}^2 h_{11}}{h_{11}h_{22} - h_{12}^2}$$

Since $h_{12} = \rho_{12}(h_{11}h_{22})^{0.5}$, it follows that

$$\varepsilon_t' H^{-1} \varepsilon_t = \left[\frac{1}{(1 - \rho_{12}^2)} \left(\frac{\varepsilon_{1t}^2}{h_{11}} + \frac{\varepsilon_{2t}^2}{h_{22}} - \frac{2\rho_{12}\varepsilon_{1t}\varepsilon_{2t}}{(h_{11}h_{22})^{0.5}} \right) \right]$$

Since the realizations of $\{\varepsilon_t\}$ are conditionally dependent, the conditional likelihood of the joint realizations of $\varepsilon_1, \varepsilon_2, \ldots \varepsilon_T$ is the product in the individual likelihoods. Hence, if all have the same variance, the conditional likelihood of the joint realizations is

$$L = \prod_{t=1}^{T} \frac{1}{2\pi\,|H|^{1/2}} \exp\left[-\frac{1}{2}\varepsilon_t' H^{-1} \varepsilon_t \right]$$

It is far easier to work with a sum than with a product. As such, it is convenient to take the natural log of each side so as to obtain

$$\ln L = -\frac{T}{2}\ln(2\pi) - \frac{T}{2}\ln|H| - \frac{1}{2}\sum_{t=1}^{T} \varepsilon_t' H^{-1} \varepsilon_t$$

The procedure used in maximum-likelihood estimation is to select the distributional parameters so as to maximize the likelihood of drawing the observed sample. Given the realizations in ε_t, it is possible to select h_{11}, h_{12}, and h_{22} so as to maximize the likelihood function.

For our purposes, we want allow the values of h_{ij} to be time-varying. If you worked through Section 8, it should be clear how to modify this equation if h_{11}, h_{22}, and h_{12} are time-varying. Consider

$$L = \prod_{t=1}^{T} \frac{1}{2\pi|H_t|^{1/2}} \exp\left[-\frac{1}{2}\varepsilon_t' H_t^{-1} \varepsilon_t \right]$$

where

$$H_t = \begin{bmatrix} h_{11t} & h_{12t} \\ h_{12t} & h_{22t} \end{bmatrix}$$

Now, if we take the log of the likelihood function,

$$\ln L = -\frac{T}{2}\ln(2\pi) - \frac{1}{2}\sum_{t=1}^{T} (\ln|H_t| + \varepsilon_t' H_t^{-1} \varepsilon_t) \tag{A3.3}$$

The convenience of working with (A3.2) and (A3.3) is that the form of the likelihood function is identical for models with k variables. In such circumstances, H is a symmetric $k \times k$ matrix, ε_t is a $k \times 1$ column vector, and the constant term (2π) is raised to the power k.

Multivariate GARCH Specifications

Given the log-likelihood function given by (A3.3), it is necessary to specify the functional forms for each of the h_{ijt}. The most familiar specifications are given below:

1. **The *vech* Model:** The *vech* operator transforms the upper (lower) triangle of a symmetric matrix into a column vector. Consider the symmetric covariance matrix

$$H_t = \begin{bmatrix} h_{11t} & h_{12t} \\ h_{12t} & h_{22t} \end{bmatrix}$$

so that

$$vech(H_t) = [h_{11t}, h_{12t}, h_{22t}]'$$

Now consider the vector $\varepsilon_t = [\varepsilon_{1t}, \varepsilon_{2t}]'$. The product $\varepsilon_t\varepsilon_t' = [\varepsilon_{1t}, \varepsilon_{2t}]'[\varepsilon_{1t}, \varepsilon_{2t}]$ is the 2×2 matrix

$$\begin{bmatrix} \varepsilon_{1t}^2 & \varepsilon_{1t}\varepsilon_{2t} \\ \varepsilon_{1t}\varepsilon_{2t} & \varepsilon_{1t}^2 \end{bmatrix}$$

Hence, $vech(\varepsilon_t\varepsilon_t') = [\varepsilon_{1t}^2, \varepsilon_{1t}\varepsilon_{2t}, \varepsilon_{2t}^2]'$. If we now let $C = [c_1, c_2, c_3]'$, A = the 3×3 matrix with elements α_{ij}, and B = the 3×3 matrix with elements β_{ij}, we can write

$$vech(H_t) = C + A\, vech(\varepsilon_{t-1}\varepsilon_{t-1}') + B\, vech(H_{t-1})$$

If you are familiar with matrix operations, it should be clear that this is precisely the system represented by (3.42)–(3.44). The diagonal *vech* uses only the diagonal elements of A and B and sets all values of $\alpha_{ij} = \beta_{ij} = 0$ for $i \neq j$.

2. **The BEK Model:** In a system with k variables, the BEK specification has the form

$$H_t = C'C + A'\varepsilon_{t-1}\varepsilon_{t-1}'A + B'H_{t-1}B$$

where A and B are $k \times k$ matrices. However, C must be a symmetric $k \times k$ matrix in order to ensure that the intercepts of the off-diagonal elements h_{ijt} are identical. As suggested in the text for the 2-variable case,

$$\begin{aligned} h_{11t} = (c_{11}^2 + c_{12}^2) &+ (\alpha_{11}^2\varepsilon_{1t-1}^2 + 2\alpha_{11}\alpha_{21}\varepsilon_{1t-1}\varepsilon_{2t-1} + \alpha_{21}^2\varepsilon_{2t-1}^2) \\ &+ (\beta_{11}^2 h_{11t-1} + 2\beta_{11}\beta_{21}h_{12t-1} + \beta_{21}^2 h_{22t-1}) \end{aligned}$$

$$
\begin{aligned}
h_{12t} = c_{12}(c_{11} + c_{22}) + \alpha_{12}\alpha_{11}\varepsilon_{1t-1}^2 + (\alpha_{11}\alpha_{22} + \alpha_{12}\alpha_{21})\varepsilon_{1t-1}\varepsilon_{2t-1} \\
+ \alpha_{21}\alpha_{22}\varepsilon_{2t-1}^2 + \beta_{11}\beta_{12}h_{11t-1} \\
+ (\beta_{11}\beta_{22} + \beta_{12}\beta_{21})h_{12t-1} + \beta_{21}\beta_{22}h_{22t-1}
\end{aligned}
$$

$$
\begin{aligned}
h_{22t} = (c_{22}^2 + c_{12}^2) + (\alpha_{12}^2\varepsilon_{1t-1}^2 + 2\alpha_{12}\alpha_{22}\varepsilon_{1t-1}\varepsilon_{2t-1} + \alpha_{22}^2\varepsilon_{2t-1}^2) \\
+ (\beta_{12}^2h_{11t-1} + 2\beta_{12}\beta_{22}h_{12t-1} + \beta_{22}^2h_{22t-1})
\end{aligned}
$$

3. **Constant Conditional Correlations:** The CCC formulation is clearly a special case of the more general multivariate GARCH model. In the 2-variable case, we can write H_t as

$$
H_t = \begin{bmatrix} h_{11t} & \rho_{12}(h_{11t}h_{22t})^{0.5} \\ \rho_{12}(h_{11t}h_{22t})^{0.5} & h_{22t} \end{bmatrix}
$$

Now, if h_{11t} and h_{22t} are both GARCH(1, 1) processes, there are seven parameters to estimate (the six values of c_i, α_{ii} and β_{ii}, and ρ_{12}).

4. **Dynamic Conditional Correlations:** Engle (2002) showed how to generalize the CCC model so that the correlations vary over time. Instead of estimating all the parameters simultaneously, the Dynamic Conditional Correlation (DCC) model uses a two-step estimation process. The first step is to use Bollerslev's CCC model to obtain the GARCH estimates of the variances and the standardized residuals. Note that the standardized residuals, $s_{it} = \hat{\varepsilon}_{it}/\hat{h}_{iit}^{0.5}$, are estimates of the v_{it}. The second step uses the standardized residuals to estimate the conditional covariances. Specifically, in the second step you create the correlations by smoothing the series of standardized residuals obtained from the first step. Engle examines several smoothing methods. The simplest is the exponential smoother $q_{ijt} = (1 - \lambda)s_{it}s_{jt} + \lambda q_{ijt-1}$ for $\lambda < 1$. Hence, each $\{q_{iit}\}$ series is an exponentially weighted moving average of the cross-products of the standardized residuals. The dynamic conditional correlations are created from the q_{ijt} as

$$
\rho_{ijt} = q_{ijt}/(q_{iit}q_{jjt})^{0.5} \tag{A3.4}
$$

Engle shows that a two-step procedure yields consistent estimates of the time-varying correlation coefficients. However, the estimates are not as efficient as those from one-step procedures such as the BEK and diagonal *vech* models. Restricting the coefficient on $\bar{s}_{ij}$ to equal $(1 - \alpha - \beta)$ ensures that the q_{ijt} converge to the unconditional covariances.

An alternative smoothing function is to estimate $q_{ijt} = (1 - \alpha - \beta)\bar{s}_{ij} + \alpha s_{it}s_{jt} + \beta q_{ijt-1}$ where $\bar{s}_{ij}$ is the unconditional covariance between s_{it} and s_{jt} by maximum-likelihood estimation. Plug the estimated coefficients from the first step (i.e., from the CCC model) into the likelihood function so that only α and β need to be estimated.

For those of you wanting a formal proof that the two-step procedure is feasible, you should be able to convince yourself that it is possible to write the H_t matrix as

$$
H_t = D_tR_tD_t
$$

where D_t = the diagonal matrix with $(h_{iit})^{0.5}$ on the diagonals and R_t is the matrix of time-varying correlations. This follows from the definition of a correlation coefficient; R_t consists of the elements $r_{ijt} = (h_{ijt})/(h_{iit}h_{jjt})^{0.5}$. For example, in the 2-variable case it is easy to verify $H_t = D_tR_tD_t$ or $R_t = (D_t)^{-1}H_t(D_t)^{-1}$ since

$$R_t = \begin{pmatrix} h_{11t}^{0.5} & 0 \\ 0 & h_{22t}^{0.5} \end{pmatrix}^{-1} \begin{pmatrix} h_{11t} & h_{12t} \\ h_{12t} & h_{22t} \end{pmatrix} \begin{pmatrix} h_{11t}^{0.5} & 0 \\ 0 & h_{22t}^{0.5} \end{pmatrix}^{-1}$$

$$= \begin{pmatrix} 1 & h_{12t}/(h_{11t}h_{22t})^{0.5} \\ h_{12t}/(h_{11t}h_{22t})^{0.5} & 1 \end{pmatrix}$$

Now write the likelihood function (A3.3) by substituting $D_tR_tD_t$ for H_t:

$$\ln L = -\frac{T}{2}\ln(2\pi) - \frac{1}{2}\sum_{t=1}^{T}(\ln|D_tR_tD_t| + \varepsilon_t'(D_tR_tD_t)^{-1}\varepsilon_t)$$

$$= -\frac{T}{2}\ln(2\pi) - \frac{1}{2}\sum_{t=1}^{T}(2\ln|D_t| + \ln|R_t| + \varepsilon_t'(R_t)^{-1}\varepsilon_t) \qquad \text{(A3.5)}$$

Notice that D_t and R_t enter the likelihood separately and that $\varepsilon_t'R_t\varepsilon_t$ represents the squared standardized residuals. The final step is to add and subtract the sum of the squared standardized residuals to (A3.5). If we represent the standardized residuals by v_t, the sum of the squared standardized residuals is $v_t'v_t$. It is also possible to show that $v_t'v_t = \varepsilon_t'D_t^{-1}D_t^{-1}\varepsilon_t$. For example, in the 2-variable case,

$$\varepsilon_t'D_t^{-1}D_t^{-1}\varepsilon_t = \begin{pmatrix} \varepsilon_{1t} \\ \varepsilon_{2t} \end{pmatrix}' \begin{pmatrix} h_{11t}^{0.5} & 0 \\ 0 & h_{11t}^{0.5} \end{pmatrix}^{-1} \begin{pmatrix} h_{11t}^{0.5} & 0 \\ 0 & h_{11t}^{0.5} \end{pmatrix}^{-1} \begin{pmatrix} \varepsilon_{1t} \\ \varepsilon_{2t} \end{pmatrix} = \frac{\varepsilon_{1t}^2}{h_{11t}} + \frac{\varepsilon_{2t}^2}{h_{22t}}$$

Thus, we can write the likelihood function as

$$\ln L = -\frac{T}{2}\ln(2\pi) - \frac{1}{2}\sum_{t=1}^{T}(2\ln|D_t| + \ln|R_t| + \varepsilon_t'(R_t)^{-1}\varepsilon_t - v_t'v_t + \varepsilon_t'D_t^{-1}D_t^{-1}\varepsilon_t)$$

The point of the exercise is to show that the two-step procedure is appropriate. Notice that D_t and R_t enter the equation separately. As such, the parameters of the two matrices can be estimated separately. You can use the CCC model to estimate the parameters of D_t; this can be done without any knowledge of the values of R_t. Use these estimates to construct the values of $|D_t|$ and the standardized residuals. Plug these values into the likelihood function and then select the optimal values of R_t. In essence, in the first stage, you maximize

$$-\frac{1}{2}\sum_{t=1}^{T}(2\ln|D_t| + \varepsilon_t'D_t^{-1}D_t^{-1}\varepsilon_t)$$

and in the second stage you maximize

$$-\frac{1}{2}\sum_{t=1}^{T}(\ln|R_t| + \varepsilon_t'(R_t)^{-1}\varepsilon_t - v_t'v_t)$$

CHAPTER 4

MODELS WITH TREND

Inspection of the autocorrelation function serves as a rough indicator of whether a trend is present in a series. A slowly decaying ACF is indicative of a large characteristic root, a true unit root process, or a trend-stationary process. Formal tests can help determine whether a system contains a trend and whether the trend is deterministic or stochastic. However, the existing tests have difficulty distinguishing between near–unit root and unit root processes. This chapter has five aims:

1. Formalize simple models of variables with a time-dependent mean. A trend can be completely deterministic or may contain stochastic components. It is essential to properly model the trend if you intend to do any hypothesis testing or long-term forecasting.
2. Develop and illustrate the Dickey–Fuller and augmented Dickey–Fuller tests for the presence of a unit root. Several variants of the test are presented, including a test for seasonal unit roots. In order to develop the test statistics it is necessary to understand the nature of Monte Carlo experiments.
3. Consider tests for unit roots in the presence of structural change. Structural change can complicate the tests for trends; a policy regime change can result in a structural break that makes an otherwise stationary series appear to be nonstationary.
4. Illustrate the lack of power of the standard Dickey–Fuller tests. Unit root tests are sensitive to the presence of deterministic regressors, such as an intercept term or a deterministic time trend. The so-called Generalized Least Squares (GLS) detrending methods can enhance the power of the Dickey–Fuller tests. Panel unit root tests can also have good power.
5. Decompose a series with a trend into its stationary and trend components. Several methodologies that can be used to decompose a series into its temporary and permanent components are presented.

1. DETERMINISTIC AND STOCHASTIC TRENDS

It is helpful to represent the general solution to a linear stochastic difference equation as consisting of these three distinct parts:[1]

$$y_t = \text{trend} + \text{stationary component} + \text{noise}$$

Chapter 2 explained how to model the stationary component using the Box–Jenkins methodology. Chapter 3 showed you how to model the variance of the error (i.e., noise)

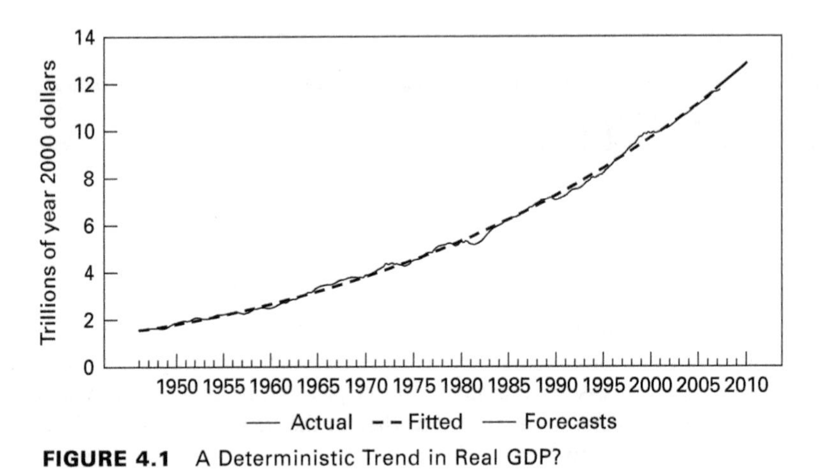

FIGURE 4.1 A Deterministic Trend in Real GDP?

component. A critical task for applied econometricians is to develop simple stochastic difference equation models that mimic the behavior of trending variables. The file RGDP.XLS contains the quarterly values of real U.S. GDP over the 1947Q1–2008Q1 period (in trillions of year 2000 dollars). From the plot of the data shown in Figure 4.1, it is clear that the distinguishing feature of real GDP $\{rgdp_t\}$ is that it increases over time. For such a series, a naïve forecaster might estimate the sustained increase using the following cubic polynomial model for the trend:

$$\underset{}{rgdp_t} = \underset{(44.39)}{1.477} + \underset{(16.59)}{0.019t} + \underset{(1.89)}{2.070*10^{-5}t^2} + \underset{(10.17)}{2.961*10^{-7}t^3} \qquad (4.1)$$

The fitted values are shown as the dashed line in the figure and the forecasted values are shown as the solid line extending past 2008Q1. Regardless of the t-statistics, the use of such a model for the trend of real GDP is problematic. Since there are no stochastic components in the trend, (4.1) implies that there is a deterministic (and accelerating) long-run growth rate of the real economy. The "Real Business Cycle" school argues that technological advances have permanent effects on the trend of the macroeconomy. Since technological innovations are stochastic, the trend should reflect this underlying randomness. Adherents to other schools of macroeconomics would also argue that the trend is not completely deterministic. For example, they might point out that an oil price shock or a targeted tax reduction could affect investment and the economy's long-term growth rate. Moreover, the implications for the behavior of the business cycle are not credible. The deterministic trend implies that whenever real GDP is below trend, in subsequent periods there will be unusually high growth as real GDP returns to the trend. The reaction to the financial crisis beginning in 2008 suggests that most economists and politicians do not take this notion very seriously.

Consider the federal funds rate and the yield on 10-year U.S. federal government securities, shown in Figure 4.2. The two interest rates have no obvious tendency to increase or decrease. Moreover, there are no decided structural breaks that induce

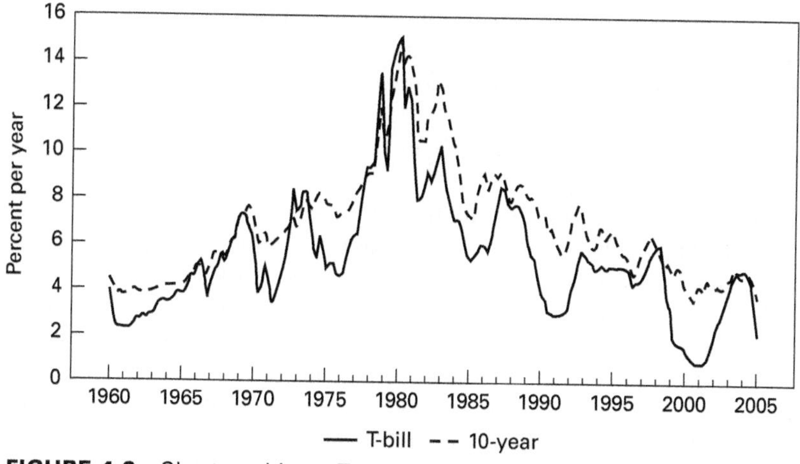

FIGURE 4.2 Short- and Long-Term Interest Rates

one-time shifts in the mean. Nevertheless, there is no pronounced tendency for either series to revert to a long-run mean. The key feature of a trend is that it has a permanent effect on a series. If the trend is defined as the permanent or nondecaying component of a time series, the two interest rates have a trend.

Suppose that a series always changes by the same fixed amount from one period to the next. To be more specific, suppose that

$$\Delta y_t = a_0$$

As you know from Chapter 1, the solution to this linear difference equation is

$$y_t = y_0 + a_0 t$$

where y_0 is the initial condition for period zero.

Hence, the solution for $\Delta y_t = a_0$ turns out to be nothing more than a deterministic linear time trend; the intercept is y_0 and the slope is a_0. Now, if we add the stationary component $A(L)\varepsilon_t$ to the trend, we obtain

$$y_t = y_0 + a_0 t + A(L)\varepsilon_t \tag{4.2}$$

In (4.2), y_t can differ from its trend value by the amount $A(L)\varepsilon_t$. Since this deviation is stationary, the $\{y_t\}$ sequence will exhibit only temporary departures from the trend. As such, the long-term forecast of y_{t+s} will converge to the trend line $y_0 + a_0(t + s)$. In the jargon of the profession, this type of model is called a **trend-stationary** (TS) model.

Now suppose that the *expected* change in y_t is a_0 units. In particular, let Δy_t be equal to a_0 plus a white-noise term:

$$\Delta y_t = a_0 + \varepsilon_t \tag{4.3}$$

Sometimes Δy_t exceeds a_0 and sometimes it falls short of a_0. Since $E_{t-1}\varepsilon_t = 0$, (4.3) implies that y_t is expected to change by a_0 units from one period to the next. The seemingly innocuous modification of (4.2) has profound differences for the trend. If y_0 is the initial condition, it is readily verified that the general solution to the first-order difference equation represented by (4.3) is

$$y_t = y_0 + \sum_{i=1}^{t} \varepsilon_i + a_0 t$$

Here, y_t consists of the deterministic trend component $a_0 t$ and the component $y_0 + \Sigma\varepsilon_i$. We can think of this second component as a stochastic intercept term. In the absence of any shocks, the intercept is y_0. However, each ε_i shock represents a shift in the intercept. Since all values of $\{\varepsilon_i\}$ have a coefficient of unity, the effect of each shock on the intercept term is permanent. In the time-series literature, such a sequence is said to have a **stochastic trend** since each ε_i shock imparts a permanent, albeit random, change in the conditional mean of the series. If $a_0 = 0$, this type of model seems to capture some of the behavior of the interest rates shown in Figure 4.2. The two rates have no particular tendency to increase or decrease over time; neither do they exhibit any tendency to revert to a given mean value.

The Random Walk Model

Equation (4.3) is the basic building block for modeling series containing stochastic trends. Since these models are probably unfamiliar to you, the remainder of this section explores the nature of stochastic trends. We begin by considering the special case of (4.3) when $a_0 = 0$. This model, known as the **random walk** model, has a special place in the economics and finance literature. For example, some formulations of the efficient market hypothesis posit that the change in the price of a stock from one day to the next is completely random. As such, the current price (y_t) should be equal to last period's price plus a white-noise term, so that

$$y_t = y_{t-1} + \varepsilon_t \qquad (\text{or } \Delta y_t = \varepsilon_t) \tag{4.4}$$

Similarly, suppose you were betting on the outcome of a coin toss, and a heads added \$1 to your wealth while a tails cost you \$1. We could let $\varepsilon_t = +\$1$ if a heads appears and $-\$1$ in the event of a tails. Thus, your current wealth (y_t) equals last period's wealth (y_{t-1}) plus the realized value of ε_t. If you play again, your wealth in $t + 1$ is $y_{t+1} = y_t + \varepsilon_{t+1}$.

If y_0 is a given initial condition, it can be readily verified that the general solution to the first-order difference equation represented by the random walk model is

$$y_t = y_0 + \sum_{i=1}^{t} \varepsilon_i$$

Taking expected values, we obtain $Ey_t = Ey_{t-s} = y_0$; thus, the mean of a random walk is a constant. However, all stochastic shocks have nondecaying effects on the $\{y_t\}$ sequence. Given the first t realizations of the $\{\varepsilon_t\}$ process, the conditional mean of y_{t+1} is

$$E_t y_{t+1} = E_t(y_t + \varepsilon_{t+1}) = y_t$$

Similarly, the conditional mean of y_{t+s} (for any $s > 0$) can be obtained from

$$y_{t+s} = y_t + \sum_{i=1}^{s} \varepsilon_{t+i}$$

so that

$$E_t y_{t+s} = y_t + E_t \sum_{i=1}^{s} \varepsilon_{t+i} = y_t$$

For any positive value of s, the conditional means for all values of y_{t+s} are equivalent. Hence, the constant value of y_t is the unbiased estimator of all future values of y_{t+s}. To interpret, note that an ε_t shock has a permanent effect on y_t. This permanence is directly reflected in the forecasts for y_{t+s}.

It is easy to show that the variance is time-dependent. Given the value of y_0, the variance can be constructed as

$$\text{var}(y_t) = \text{var}(\varepsilon_t + \varepsilon_{t-1} + \ldots + \varepsilon_1) = t\sigma^2$$

and

$$\text{var}(y_{t-s}) = \text{var}(\varepsilon_{t-s} + \varepsilon_{t-s-1} + \ldots + \varepsilon_1) = (t - s)\sigma^2$$

Since the variance is not constant [i.e., $\text{var}(y_t) \neq \text{var}(y_{t-s})$], the random walk process is nonstationary. Moreover, as $t \to \infty$ the variance of y_t also approaches infinity. Thus, the random walk meanders without exhibiting any tendency to increase or decrease. It is also instructive to calculate the covariance of y_t and y_{t-s}. Since the mean is constant, we can form the covariance γ_{t-s} as

$$\begin{aligned} E[(y_t - y_0)(y_{t-s} - y_0)] &= E[(\varepsilon_t + \varepsilon_{t-1} + \ldots + \varepsilon_1)(\varepsilon_{t-s} + \varepsilon_{t-s-1} + \ldots + \varepsilon_1)] \\ &= E[(\varepsilon_{t-s})^2 + (\varepsilon_{t-s-1})^2 + \ldots + (\varepsilon_1)^2] \\ &= (t - s)\sigma^2 \end{aligned}$$

To form the correlation coefficient ρ_s, we can divide γ_{t-s} by the product of the standard deviation of y_t multiplied by the standard deviation of y_{t-s}. Thus, the correlation coefficient ρ_s is

$$\begin{aligned} \rho_s &= (t - s)/\sqrt{(t - s)t} \\ &= [(t - s)/t]^{0.5} \end{aligned}$$

This result plays an important role in the detection of nonstationary series. For the first few autocorrelations, the sample size t will be large relative to the number of autocorrelations formed; for small values of s, the ratio $(t - s)/t$ is approximately equal to unity. However, as s increases, the values of ρ_s will decline. Hence, in using sample data, *the autocorrelation function for a random walk process will show a slight tendency to decay.* Thus, it will not be possible to use the autocorrelation function to distinguish between a unit root process and a stationary process with an autoregressive coefficient that is close to unity.

Panel (a) in Figure 4.3 shows the time path of a simulated random walk process. First, 100 normally distributed random deviates were drawn from a theoretical distribution

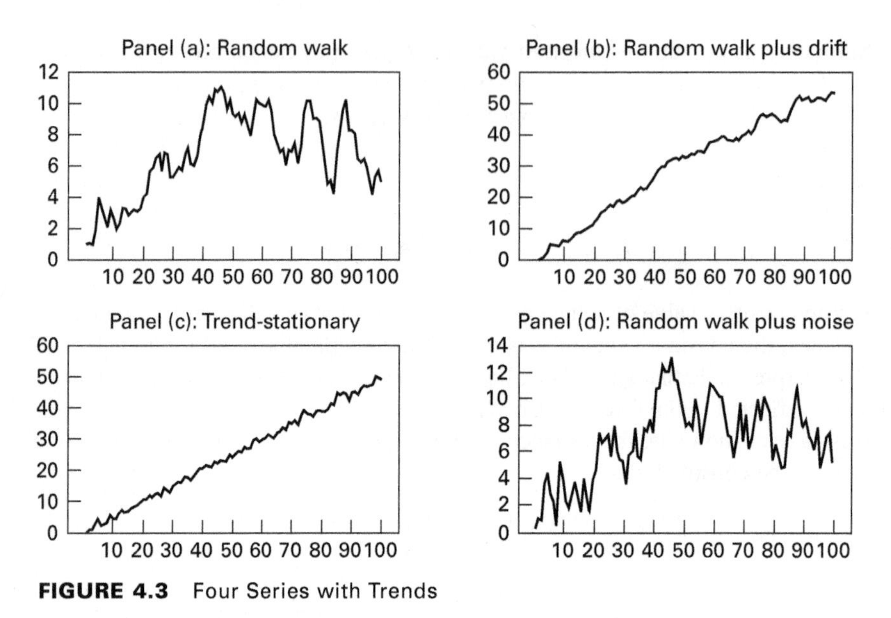

FIGURE 4.3 Four Series with Trends

with a mean of zero and a variance equal to unity. By setting $y_0 = 1$, each value of $y_t(t = 1, ..., 100)$ was constructed by adding the random variable to the value of y_{t-1}. As expected, the series meanders without any tendency to revert to a long-run value. However, there does appear to be a slight positive trend in the simulated data. The reason for the upward trend is that this particular simulation happened to contain more positive values than negative values. The impression of a steadily increasing trend in the true data-generating process is false and serves as a reminder not to rely solely on casual inspection.

The Random Walk Plus Drift Model

Now let the change in y_t be partially deterministic and partially stochastic. The **random walk plus drift model** augments the random walk model by adding a constant term a_0, so that

$$y_t = y_{t-1} + a_0 + \varepsilon_t$$

Hence, you can see that (4.3) is actually a random walk plus drift process. Given the initial condition y_0, the general solution for y_t is given by

$$y_t = y_0 + a_0 t + \sum_{i=1}^{t} \varepsilon_i \tag{4.5}$$

Here, the behavior of y_t is governed by two nonstationary components: a linear deterministic trend and the stochastic trend $\sum \varepsilon_i$. As such, a random walk plus drift is a pure model of a trend; there is no separate stationary component in (4.5).

If we take expectations, the mean of y_t is $y_0 + a_0t$ and the mean of y_{t+s} is $Ey_{t+s} = y_0 + a_0(t + s)$. To explain, the deterministic change in each realization of $\{y_t\}$ is a_0; after t periods the cumulated change is a_0t. In addition, there is the stochastic trend $\sum\varepsilon_i$; each ε_i shock has a permanent effect on the mean of y_t. Notice that the first difference of the series is stationary; taking the first difference yields the stationary sequence $\Delta y_t = a_0 + \varepsilon_t$.

Panel (b) of Figure 4.3 illustrates a simulated random walk plus drift model. The value of a_0 was set equal to 0.5 and (4.5) was simulated using the same 100 deviates used for the random walk model above. Clearly, the deterministic time trend dominates the time path of the series. In a very large sample, asymptotic theory suggests this will always be the case. However, you should not conclude that it is always easy to discern the difference between a random walk model and a model with drift. In a small sample, increasing the variance of $\{\varepsilon_t\}$ or decreasing the absolute value of a_0 could cloud the long-run properties of the sequence. Panel (c) uses the same random numbers to generate the trend-stationary (TS) series $y_t = 0.5t + \varepsilon_t$. The patterns evident in the random walk plus drift model and in the TS series look strikingly similar to each other and to the real GDP series shown in Figure 4.1.

To obtain the s-step-ahead forecast for a random walk plus drift, update (4.5) by s periods to obtain

$$y_{t+s} = y_0 + a_0(t+s) + \sum_{i=1}^{t+s} \varepsilon_i$$

$$= y_t + a_0 s + \sum_{i=1}^{s} \varepsilon_{t+i}$$

Taking the conditional expectation of y_{t+s}, it follows that

$$E_t y_{t+s} = y_t + a_0 s$$

In contrast to the pure random walk model, the forecast function is not flat. The fact that the average change in y_t is always the constant a_0 is reflected in the forecast function. In addition to the given value of y_t, we project this deterministic change s times into the future.

Generalizations of the Stochastic Trend Model

It is not too difficult to generalize the random walk model to allow y_t to be the sum of a stochastic trend and a white-noise component. Formally, this third model—called **random walk plus noise**—is represented by

$$y_t = y_0 + \sum_{i=1}^{t} \varepsilon_i + \eta_t \tag{4.6}$$

where $\{\eta_t\}$ is a white-noise process with variance σ_η^2, and ε_t and η_{t-s} are independently distributed for all t and s [i.e., $E(\varepsilon_t \eta_{t-s}) = 0$].

An alternative representation of the random walk plus noise model is

$$\Delta y_t = \varepsilon_t + \Delta\eta_t \tag{4.7}$$

You can easily verify that (4.6) and (4.7) are equivalent by writing y_{t-1} as

$$y_{t-1} = y_0 + \sum_{i=1}^{t-1} \varepsilon_i + \eta_{t-1}$$

Subtract this expression from (4.6) to obtain (4.7). From (4.6), you can see that the key properties of the random walk plus noise model are:

1. Given the value y_0, the mean of the $\{y_t\}$ sequence is constant ($Ey_t = y_0$), and updating by s periods yields $Ey_{t+s} = y_0$. Notice that the successive ε_t shocks have permanent effects on the $\{y_t\}$ sequence in that there is no decay factor on past values of ε_i. Hence, y_t has the stochastic trend component $\Sigma\varepsilon_i$.

2. The $\{y_t\}$ sequence has a pure noise component in that the $\{\eta_t\}$ sequence has only a temporary effect on the $\{y_t\}$ sequence. The current realization of η_t affects only y_t but not the subsequent values y_{t+s}.

3. The variance of $\{y_t\}$ is not constant: $\text{var}(y_t) = t\sigma^2 + \sigma_\eta^2$ and $\text{var}(y_{t-s}) = (t - s)\sigma^2 + \sigma_\eta^2$. As in the other models with a stochastic trend, the variance of y_t approaches infinity as t increases. The presence of the noise component means that the correlation coefficient between y_t and y_{t-s} is smaller than that for the pure random walk model.

To prove that the sample correlogram will exhibit even faster decay than in the pure random walk model, note that the covariance between y_t and y_{t-s} is

$$\begin{aligned}\text{cov}(y_t, y_{t-s}) &= E[(y_t - y_0)(y_{t-s} - y_0)] \\ &= E[(\varepsilon_1 + \varepsilon_2 + \varepsilon_3 + \dots + \varepsilon_t + \eta_t)(\varepsilon_1 + \varepsilon_2 + \varepsilon_3 + \dots + \varepsilon_{t-s} + \eta_{t-s})]\end{aligned}$$

Since $\{\varepsilon_t\}$ and $\{\eta_t\}$ are independent white-noise sequences,

$$\text{cov}(y_t, y_{t-s}) = (t - s)\sigma^2$$

Thus, the correlation coefficient ρ_s is

$$\rho_s = \frac{(t - s)\sigma^2}{\sqrt{(t\sigma^2 + \sigma_\eta^2)[(t - s)\sigma^2 + \sigma_\eta^2)]}}$$

Comparison of ρ_s with the correlation coefficient for the pure random walk model verifies that the autocorrelations for the random walk plus noise model are always smaller for $\sigma_\eta^2 > 0$. Panel (d) of Figure 4.3 shows a random walk plus noise model. The series was simulated by drawing a second 100 normally distributed random terms to represent the $\{\eta_t\}$ series. For each value of t, y_t was calculated using (4.6). If we compare Panels (a) and (d), it can be seen that the two series track each other quite well. The random walk plus noise model could mimic the same set of macroeconomic variables as the random walk model. The effect of the "noise" component $\{\eta_t\}$ is to increase the variance of $\{y_t\}$ without affecting its long-run behavior. After all, the random walk plus noise series is nothing more than the random walk model with a purely temporary component added.

The random walk plus noise and the random walk plus drift models are the building blocks of more complex time-series models. For example, the noise and drift components can easily be incorporated into a single model by modifying (4.6) such that the trend in y_t contains a deterministic and a stochastic component. Specifically, replace (4.6) with

$$\Delta y_t = a_0 + \varepsilon_t + \Delta \eta_t$$

or

$$y_t = y_0 + a_0 t + \sum_{i=1}^{t} \varepsilon_i + \eta_t \tag{4.8}$$

Equation (4.8) is called the **trend plus noise model;** y_t is the sum of a deterministic trend, a stochastic trend, and a pure white-noise term. Moreover, the noise sequence does not need to be a white-noise process. Let $A(L)$ be a polynomial in the lag operator L; it is possible to augment a random walk plus drift process with the stationary process $A(L)\eta_t$ so that the **general trend plus irregular model** is

$$y_t = y_0 + a_0 t + \sum_{i=1}^{t} \varepsilon_i + A(L)\eta_t \tag{4.9}$$

Thus, (4.9) has a deterministic trend, a stochastic trend, and a stationary component.

2. REMOVING THE TREND

From the previous section, it should be clear that there are important differences between a series with a trend and a stationary series. Shocks to a stationary time series are necessarily temporary; over time, the effects of the shocks will dissipate and the series will revert to its long-run mean level. On the other hand, a series containing a trend will not revert to a long-run level. Note that the trend can have deterministic and stochastic components. These components of the trend have important implications for the appropriate transformation necessary to attain a stationary series. The usual methods for eliminating the trend are **differencing** and **detrending.** Detrending entails regressing a variable on time and saving the residuals.[2] A series containing a unit root can be made stationary by differencing. In fact, we already know that the dth difference of ARIMA (p, d, q) model is stationary. The aim of this section is to compare these two methods of isolating the trend.

Differencing

First consider the solution for the random walk plus drift model:

$$y_t = y_0 + a_0 t + \sum_{i=1}^{t} \varepsilon_i$$

Taking the first difference, we obtain $\Delta y_t = a_0 + \varepsilon_t$. Clearly, the $\{\Delta y_t\}$ sequence—equal to a constant plus a white-noise disturbance—is stationary. Viewing Δy_t as the variable of interest, we have

$$E(\Delta y_t) = E(a_0 + \varepsilon_t) = a_0$$
$$\mathrm{var}(\Delta y_t) \equiv E(\Delta y_t - a_0)^2 = E(\varepsilon_t)^2 = \sigma^2$$

and

$$\text{cov}(\Delta y_t, \Delta y_{t-s}) \equiv E[(\Delta y_t - a_0)(\Delta y_{t-s} - a_0)] = E(\varepsilon_t \varepsilon_{t-s}) = 0$$

Since the mean and variance are constants and the covariance between Δy_t and Δy_{t-s} depends solely on s, the $\{\Delta y_t\}$ sequence is stationary.

The random walk plus noise model is an interesting case study. In first differences, the model can be written as $\Delta y_t = \varepsilon_t + \Delta\eta_t$. In this form, it is easy to show that Δy_t is stationary. Clearly, the mean is zero because

$$E\Delta y_t = E(\varepsilon_t + \Delta\eta_t) = 0$$

Moreover, the variance and all autocovariances are constant and time-invariant because

$$\begin{aligned}
\text{var}(\Delta y_t) &= E[(\Delta y_t)^2] = E[(\varepsilon_t + \Delta\eta_t)^2] \\
&= E[(\varepsilon_t)^2 + 2\varepsilon_t\Delta\eta_t + (\Delta\eta_t)^2] \\
&= \sigma^2 + 2E[\varepsilon_t\Delta\eta_t] + E[(\eta_t)^2 - 2\eta_t\eta_{t-1} + (\eta_{t-1})^2] \\
&= \sigma^2 + 2\sigma_\eta^2 \\
\text{cov}(\Delta y_t, \Delta y_{t-1}) &= E[(\varepsilon_t + \eta_t - \eta_{t-1})(\varepsilon_{t-1} + \eta_{t-1} - \eta_{t-2}) = -\sigma_\eta^2
\end{aligned}$$

and

$$\text{cov}(\Delta y_t, \Delta y_{t-s}) = E[(\varepsilon_t + \eta_t - \eta_{t-1})(\varepsilon_{t-s} + \eta_{t-s} - \eta_{t-s-1}) = 0 \qquad \text{for } s > 1$$

If we set $s = 1$, the correlation coefficient between Δy_t and Δy_{t-1} is

$$\rho_1 = \frac{\text{cov}(\Delta y_t, \Delta y_{t-1})}{\text{var}(\Delta y_t)} = \frac{-\sigma_\eta^2}{\sigma^2 + 2\sigma_\eta^2}$$

Examination reveals $-0.5 < \rho_1 < 0$ and that all other correlation coefficients are zero. Since the first difference of y_t acts exactly as an MA(1) process, the random walk plus noise model is ARIMA(0, 1, 1). Since adding a constant to a series has no effect on the correlogram, it additionally follows that the trend plus noise model of (4.8) also acts as an ARIMA(0, 1, 1) process.

Now consider the general class of ARIMA(p, d, q) models:

$$A(L)y_t = B(L)\varepsilon_t \tag{4.10}$$

where $A(L)$ and $B(L)$ are polynomials of orders p and q in the lag operator L.[3]

First suppose that $A(L)$ has a single unit root and that $B(L)$ has all roots outside the unit circle.[4] We can factor $A(L)$ into two components $(1 - L)A^*(L)$ where $A^*(L)$ is a polynomial of order $p - 1$. Since $A(L)$ has only one unit root, it follows that all roots of $A^*(L)$ are outside the unit circle. Thus, we can write (4.10) as

$$(1 - L)A^*(L)y_t = B(L)\varepsilon_t$$

Now define $y_t^* = \Delta y_t$ so that

$$A^*(L)y_t^* = B(L)\varepsilon_t \tag{4.11}$$

The $\{y_t^*\}$ sequence is stationary since all roots of $A^*(L)$ lie outside the unit circle. The point is that the first difference of a unit root process is stationary. If $A(L)$ has two unit roots, the same argument can be used to show that the second difference of $\{y_t\}$ is stationary. *The general point is that the dth difference of a process with d unit roots is stationary.* Such a sequence is integrated of order d and is denoted by $I(d)$. An ARIMA(p, d, q) model has d unit roots; the dth difference of such a model is a stationary ARMA(p, q) process.

Detrending

We have shown that differencing can sometimes be used to transform a nonstationary model into a stationary model with an ARMA representation. This does not mean that all nonstationary models can be transformed into well-behaved ARMA models by appropriate differencing. Consider, for example, a model that is the sum of a deterministic trend and a pure noise component:

$$y_t = y_0 + a_1 t + \varepsilon_t$$

The first difference of y_t is not well-behaved because

$$\Delta y_t = a_1 + \varepsilon_t - \varepsilon_{t-1}$$

Here Δy_t is not *invertible* in the sense that Δy_t cannot be expressed in the form of an autoregressive process. Recall that invertibility of a stationary process requires that the MA component does not have a unit root.

Instead, an appropriate way to transform this model is to estimate the regression equation $y_t = a_0 + a_1 t + \varepsilon_t$. Subtracting the estimated values of y_t from the observed series yields estimated values of the $\{\varepsilon_t\}$ series. More generally, a time series may have the polynomial trend as

$$y_t = a_0 + a_1 t + a_2 t^2 + a_3 t^3 + \ldots + a_n t^n + e_t$$

where $\{e_t\}$ = a stationary process.

Detrending is accomplished by regressing $\{y_t\}$ on a deterministic polynomial time trend, as in (4.1). The appropriate degree of the polynomial can be determined by standard t-tests, F-tests, and/or using statistics such as the AIC or SBC. The common practice is to estimate the regression equation using the largest value of n deemed reasonable. If the t-statistic for a_n is zero, consider a polynomial trend of order $n - 1$. Continue to pare down the order of the polynomial trend until a nonzero coefficient is found. F-tests can be used to determine whether a group of coefficients, say, a_{n-i} through a_n, is statistically different from zero. The AIC and SBC statistics can be used to reconfirm the appropriate degree of the polynomial.

Simply subtracting the estimated values of the $\{y_t\}$ sequence from the actual values yields an estimate of the stationary sequence $\{e_t\}$. The detrended process can then be modeled using traditional methods (such as ARMA estimation).

Difference versus Trend-Stationary Models

We have encountered two ways to eliminate a trend. A trend-stationary series can be transformed into a stationary series by removing the deterministic trend. A series with a unit root, sometimes called a **difference-stationary** (DS) series, can be transformed into a stationary series by differencing. A serious problem is encountered when an inappropriate method is used to eliminate a trend. We saw an example of the problem in attempting to difference the equation $y_t = y_0 + a_1 t + \varepsilon_t$. Consider a more general trend-stationary process of the form

$$A(L)y_t = a_0 + a_1 t + e_t$$

where the characteristic roots of the polynomial $A(L)$ are all outside the unit circle and the expression e_t is allowed to have the form $e_t = B(L)\varepsilon_t$. Subtracting an estimate of the deterministic time trend yields a stationary and invertible ARMA model. However, if we use the notation of (4.11), the first difference of such a model yields

$$A(L)y_t^* = a_1 + (1 - L)B(L)\varepsilon_t$$

First-differencing the *TS* process has introduced a noninvertible unit root process into the MA component of the model. Of course, the same problem is encountered in a model with a polynomial time trend.

In the same way, subtracting a deterministic time trend from a difference-stationary process is also inappropriate. For example, in the general trend plus irregular model of (4.9), subtracting $(y_0 + a_0 t)$ from each observation does not result in a stationary series since the stochastic portion of the trend is not eliminated.

Are There Business Cycles?

Traditional business cycle research decomposed real macroeconomic variables into a long-run (secular) trend and a cyclical component. The typical decomposition is illustrated by the hypothetical data in Figure 4.4. The secular trend, portrayed by the straight line, was deemed to be in the domain of growth theory. The slope of the trend line was thought to be determined by long-run factors such as technological growth, fertility, immigration, and educational attainment levels.

One source of the deviations from trend occurs because of the wavelike motion of real economic activity called the **business cycle.** Although the actual period of the cycle was never thought to be as regular as that depicted in the figure, the periods of prosperity and recovery were regarded to be as inevitable as the tides. The goal of monetary and fiscal policy was to reduce the amplitude of the cycle (measured by distance *ab*). In terms of our previous discussion, the trend is the nonstationary component and the cyclical and irregular components are stationary.

Although there have been recessions and periods of high prosperity, the post–World War II experience taught us that business cycles do not have a regular period. Even so, there is a widespread belief that over the long run, macroeconomic variables grow at a constant trend rate and that any deviations from trend are eventually eliminated by the *invisible hand.* The belief that trend is unchanging over time leads to the common practice of *detrending* macroeconomic data using a linear (or polynomial) deterministic regression equation. The lower portion of the figure shows the cycle and the noise (or irregular) component after detrending.

Nelson and Plosser (1982) challenged the traditional view by demonstrating that important macroeconomic variables tend to be DS rather than TS processes. They

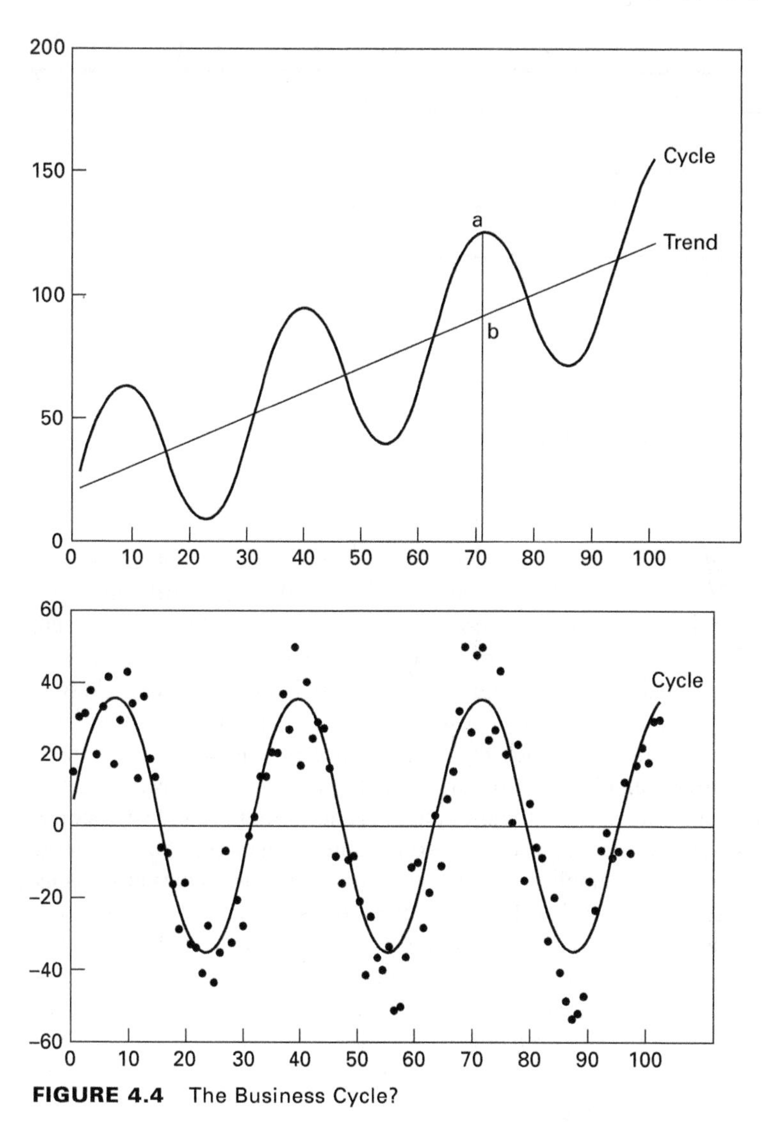

FIGURE 4.4 The Business Cycle?

obtained time-series data for 13 important macroeconomic time series: real GNP, nominal GNP, industrial production, employment, unemployment rate, GNP deflator, consumer prices, wages, real wages, money stock, velocity, bond yields, and an index of common stock prices. The sample began as early as 1860 for consumer prices to as late as 1909 for GNP data and ended in 1970 for all of the series. Some of the findings are reported in Table 4.1. The first two columns report the first- and second-order autocorrelations of real and nominal GNP, industrial production, and the unemployment rate. Notice that the autocorrelations of the first three of the series are strongly indicative of a unit root process. Although ρ_1 for the unemployment rate is 0.75, the second-order autocorrelation is less than 0.5.

Table 4.1 Selected Autocorrelations from Nelson and Plosser

	ρ_1	ρ_2	$r(1)$	$r(2)$	$d(1)$	$d(2)$
Real GNP	.95	.90	.34	.04	.87	.66
Nominal GNP	.95	.89	.44	.08	.93	.79
Industrial production	.97	.94	.03	−.11	.84	.67
Unemployment rate	.75	.47	.09	−.29	.75	.46

Notes:

1. Full details of the correlogram can be obtained from Nelson and Plosser (1982), who report the first six sample autocorrelations.
2. Respectively, ρ_i, $r(i)$, and $d(i)$ refer to the ith-order autocorrelation coefficient for each series, for the first difference of the series, and for the detrended values of the series.

First differences of the series yield the first- and second-order sample autocorrelations $r(1)$ and $r(2)$, respectively. Sample autocorrelations of the first differences are indicative of stationary processes. The evidence supports the claim that the data are generated from DS processes. Nelson and Plosser point out that the positive autocorrelation of differenced real and nominal GNP at lag 1 only is suggestive of an MA(1) process. To further strengthen the argument for DS processes, recall that differencing a TS process yields a noninvertible moving-average process. None of the differenced series reported by Nelson and Plosser appears to have a unit root in the MA terms.

The results from fitting a linear trend to the data and forming sample autocorrelations of the residuals are shown in the last two columns of the table. An interesting feature of the data is that the sample autocorrelations of the detrended data are reasonably high. This is consistent with the fact that detrending a DS series will not eliminate the nonstationarity. Notice that detrending the unemployment rate has *no effect* on the autocorrelations. The implication is that macroeconomic variables do not grow at a smooth long-run rate. Some macroeconomic shocks are of a permanent nature; the effects of such shocks are never eliminated.

The Trend in Real GDP

Another way to make the same point is to note that the real GDP series shown in Figure 4.1 has a clear trend. However, the tight fit of the estimated model might fool a researcher into thinking the series is actually stationary around the cubic trend line shown in Figure 4.1. Our eyes can be deceived because such trend lines are fit so as to make the observed residuals as small as possible. The ACF and PACF of the residuals from (4.1) are shown in Panel (a) of Figure 4.5. You can see that the ACF decays slowly while the PACF cuts to zero after one lag. In fact, this type of slow decay in the ACF might mimic that of a series with a stochastic trend. Thus, detrending the data does not seem to result in a stationary series. Panel (b) shows the ACF and PACF of the logarithmic change in real GDP. The ACF and PACF quickly converge to zero; after two

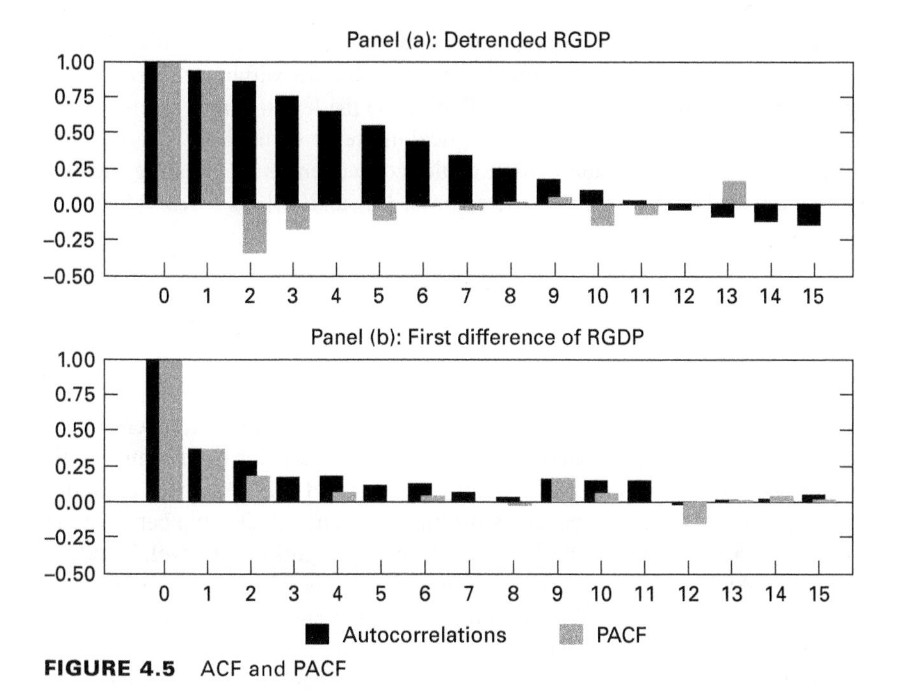

FIGURE 4.5 ACF and PACF

lags, all autocorrelations and partial autocorrelations are not statistically different from zero. The estimated model for logarithmic change in real GDP ($\Delta lrgdp$) is

$$\Delta lrgdp_t = \underset{(7.14)}{0.0055} + \underset{(5.47)}{0.3309}\Delta lrgdp_{t-1}$$

Unlike the model of the deterministic trend, the residuals from this model all appear to be white noise. Thus, differencing is sufficient to remove the trend.

Rather than rely solely on an analysis of correlograms, it is possible to formally test whether a series is stationary. We examine such tests in the next several sections. The testing procedure is not as straightforward as it may seem. We cannot use the usual testing techniques because classical procedures all presume that the data are stationary. For now, it suffices to say that Nelson and Plosser are not able to reject the null hypothesis of a unit root. However, before we examine the tests for a unit root, it is important to note that the issue of nonstationarity also arises quite naturally in the context of the standard regression model.

3. UNIT ROOTS AND REGRESSION RESIDUALS

Consider the regression equation[5]

$$y_t = a_0 + a_1 z_t + e_t \tag{4.12}$$

The assumptions of the classical regression model necessitate that both the $\{y_t\}$ and $\{z_t\}$ sequences be stationary and that the errors have a zero mean and a finite variance. In the presence of nonstationary variables, there might be what Granger and

Newbold (1974) call a **spurious regression.** A spurious regression has a high R^2 and t-statistics that appear to be significant, but the results are without any economic meaning. The regression output "looks good" because the least-squares estimates are not consistent and the customary tests of statistical inference do not hold. Granger and Newbold provided a detailed examination of the consequences of violating the stationarity assumption by generating two sequences, $\{y_t\}$ and $\{z_t\}$, as *independent* random walks using the formulas

$$y_t = y_{t-1} + \varepsilon_{yt} \tag{4.13}$$

and

$$z_t = z_{t-1} + \varepsilon_{zt} \tag{4.14}$$

where ε_{yt} and ε_{zt} are white-noise processes that are independent of each other.

Granger and Newbold generated many such samples, and for each sample estimated a regression in the form of (4.12). Since the $\{y_t\}$ and $\{z_t\}$ sequences are independent of each other, (4.12) is necessarily meaningless; any relationship between the two variables is spurious. Surprisingly, at the 5 percent significance level, standard t-tests rejected the null hypothesis $a_1 = 0$ in approximately 75 percent of the cases. Moreover, the regressions usually had very high R^2 values and the estimated residuals exhibited a high degree of autocorrelation.

To explain the Granger and Newbold findings, note that the regression equation (4.12) is necessarily meaningless if the residual series $\{e_t\}$ is nonstationary. Obviously, if the $\{e_t\}$ sequence has a stochastic trend, any error in period t never decays, so that any deviation from the model is permanent. It is hard to imagine attaching any importance to an economic model having permanent errors. The simplest way to examine the properties of the $\{e_t\}$ sequence is to abstract from the intercept term a_0 and rewrite (4.12) as

$$e_t = y_t - a_1 z_t$$

If y_t and z_t are generated by (4.13) and (4.14), we can impose the initial conditions $y_0 = z_0 = 0$ so that

$$e_t = \sum_{i=1}^{t} \varepsilon_{yi} - a_1 \sum_{i=1}^{t} \varepsilon_{zi} \tag{4.15}$$

Clearly, the variance of the error becomes infinitely large as t increases. Moreover, the error has a permanent component in that $E_t e_{t+i} = e_t$ for all $i \geq 0$. Hence, the assumptions embedded in the usual hypothesis tests are violated so that any t-test, F-test, or R^2 values are unreliable. It is easy to see why the estimated residuals from a spurious regression will exhibit a high degree of autocorrelation. Updating (4.15), you should be able to demonstrate that the theoretical value of the correlation coefficient between e_t and e_{t+1} goes to unity as t increases.

Under the null hypothesis that $a_1 = 0$, the data-generating process in (4.12) is $y_t = a_0 + e_t$. Given that $\{y_t\}$ is integrated of order one [i.e., $I(1)$], it follows that $\{e_t\}$ is $I(1)$ under the null hypothesis. Yet the assumption that the error term is a unit root process is inconsistent with the distributional theory underlying the use of OLS. This problem will not disappear in large samples. In fact, Phillips (1986) proved that the larger the sample, the more likely you are to falsely conclude that $a_1 \neq 0$.

WORKSHEET 4.1

SPURIOUS REGRESSIONS: EXAMPLE 1

Consider the two random walk processes:

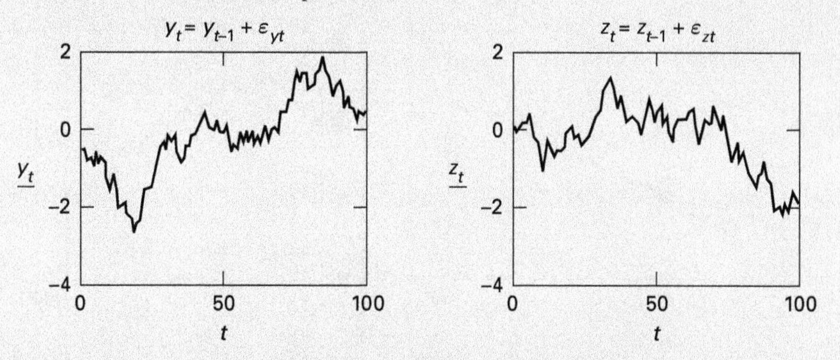

Since the $\{\varepsilon_{yt}\}$ and $\{\varepsilon_{zt}\}$ sequences are independent, the regression of y_t on z_t is spurious. Given the realizations of the random disturbances, it appears as if the two sequences are related. In the scatter plot of y_t against z_t, you can see that y_t tends to rise as z_t decreases. A regression equation of y_t on z_t will capture this tendency. The correlation coefficient between y_t and z_t is -0.372 and a linear regression yields $y_t = -0.46z_t - 0.31$. However, the residuals from the regression equation are nonstationary.

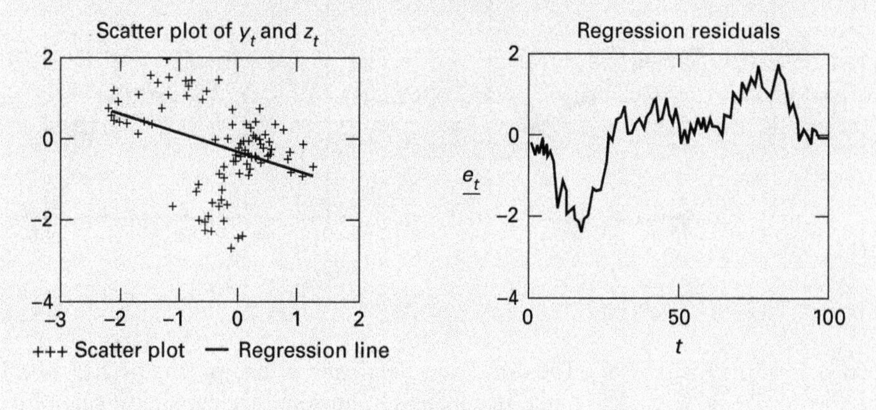

Worksheet 4.1 illustrates the problem of spurious regressions. The top two graphs show 100 realizations of the $\{y_t\}$ and $\{z_t\}$ sequences generated according to (4.13) and (4.14). Although $\{\varepsilon_{yt}\}$ and $\{\varepsilon_{zt}\}$ are drawn from white-noise distributions, the realizations of the two sequences are such that y_{100} is positive and z_{100} is negative.

You can see that the regression of y_t on z_t captures the *within-sample* tendency of the sequences to move in opposite directions. The straight line shown in the scatter plot is the OLS regression line $y_t = -0.31 - 0.46z_t$. The correlation coefficient between $\{y_t\}$ and $\{z_t\}$ is -0.372. The residuals from this regression have a unit root; as such, the coefficients -0.31 and -0.46 are spurious. Worksheet 4.2 illustrates the same problem using two simulated random walk plus drift sequences: $y_t = 0.2 + y_{t-1} + \varepsilon_{yt}$

WORKSHEET 4.2

SPURIOUS REGRESSIONS: EXAMPLE 2

Consider the two random walk plus drift processes:

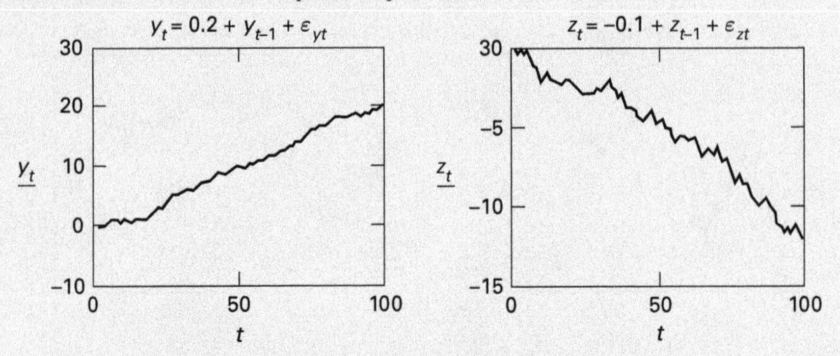

Again, the $\{\varepsilon_{yt}\}$ and $\{\varepsilon_{zt}\}$ sequences are independent so that the regression of y_t on z_t is spurious. The scatter plot of y_t against z_t strongly suggests that the two series are related. It is the deterministic time trend that causes the sustained increase in y_t and the sustained fall in z_t. The residuals from the regression equation $y_t = -2z_t + e_t$ are nonstationary.

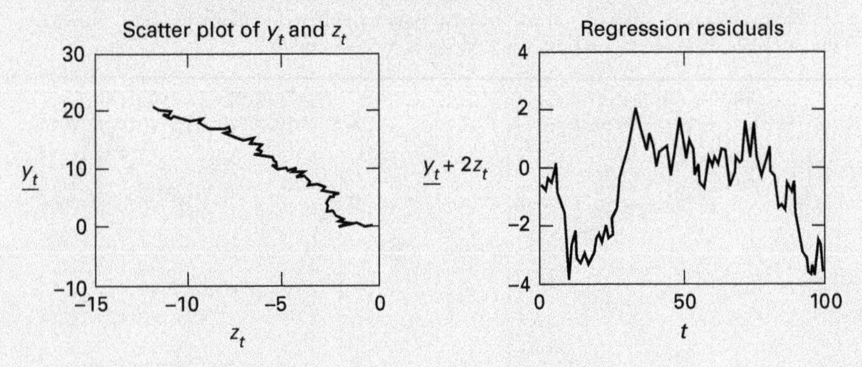

and $z_t = -0.1 + z_{t-1} + \varepsilon_{zt}$. The drift terms dominate so that for small values of t, it appears that $y_t = -2z_t$. As sample size increases, however, the cumulated sum of the errors (i.e., $\sum e_t$) will pull the relationship further and further from -2.0. The scatter plot of the two sequences suggests that the R^2 statistic will be close to unity; in fact, R^2 is almost 0.97. However, as you can see in the last panel of Worksheet 4.2, the residuals from the regression equation are nonstationary. All departures from this relationship are necessarily permanent.

The point is that the econometrician has to be very careful in working with nonstationary variables. In terms of (4.12), there are four cases to consider:

CASE 1

Both $\{y_t\}$ and $\{z_t\}$ are stationary. When both variables are stationary, the classical regression model is appropriate.

CASE 2

The $\{y_t\}$ and $\{z_t\}$ sequences are integrated of different orders. Regression equations using such variables are meaningless. For example, replace (4.14) with the stationary process $z_t = \rho z_{t-1} + \varepsilon_{zt}$ where $|\rho| < 1$ so that $z_t = \sum \rho^i \varepsilon_{zt-i}$. Now (4.15) is replaced by $e_t = \sum \varepsilon_{yi} - a_1 \sum \rho^i \varepsilon_{zt-i}$. Although the expression $\sum \rho^i \varepsilon_{zt-i}$ is convergent, the $\{e_t\}$ sequence still contains a trend component.[6]

CASE 3

The nonstationary $\{y_t\}$ and $\{z_t\}$ sequences are integrated of the same order and the residual sequence contains a stochastic trend. This is the case in which the regression is spurious. The results from such spurious regressions are meaningless in that all errors are permanent. In this case, it is often recommended that the regression equation be estimated in first differences. Consider the first difference of (4.12):

$$\Delta y_t = a_1 \Delta z_t + \Delta e_t$$

Since y_t, z_t, and e_t each contain unit roots, the first difference of each is stationary. Hence, the usual asymptotic results apply. Of course, if one of the trends is deterministic and the other is stochastic, first-differencing each is not appropriate.

CASE 4

The nonstationary $\{y_t\}$ and $\{z_t\}$ sequences are integrated of the same order and the residual sequence is stationary. In this circumstance, $\{y_t\}$ and $\{z_t\}$ are **cointegrated.** A trivial example of a cointegrated system occurs if ε_{zt} and ε_{yt} are perfectly correlated. If $\varepsilon_{zt} = \varepsilon_{yt}$, then (4.15) can be set equal to zero (which is stationary) by setting $a_1 = 1$. To consider a more interesting example, suppose that both z_t and y_t are the random walk plus noise processes:

$$y_t = \mu_t + \varepsilon_{yt}$$

$$z_t = \mu_t + \varepsilon_{zt}$$

where ε_{yt} and ε_{zt} are white-noise processes and μ_t is the random walk process $\mu_t = \mu_{t-1} + \varepsilon_t$. Note that both $\{z_t\}$ and $\{y_t\}$ are $I(1)$ processes, but $y_t - z_t = \varepsilon_{yt} - \varepsilon_{zt}$ is stationary. The subtraction of z_t from y_t serves to nullify the stochastic trend.

All of Chapter 6 is devoted to the issue of cointegrated variables. For now it is sufficient to note that pretesting the variables in a regression for nonstationarity is extremely important. Estimating a regression in the form of (4.12) is meaningless if case 2 or 3 applies. If the variables are cointegrated, the results of Chapter 6 apply. The remainder of this chapter considers the formal test procedures for the presence of unit roots and/or deterministic time trends.

4. THE MONTE CARLO METHOD

As an applied researcher, you need to know whether a data series contains a trend, and also the best way to estimate the trend. You also need to avoid several critical mistakes. Clearly, you do not want to difference or detrend a stationary series. Moreover, you do not want to detrend a unit root process or difference a trend-stationary process. Although the properties of a sample correlogram are useful tools for detecting the possible presence of unit roots or deterministic trends, the method is necessarily imprecise. What may appear as a unit root to one observer may appear as a stationary process to another. The problem is difficult because a near–unit root process will have the same-shaped ACF as that of a process containing a trend. For example, the correlogram of a stationary AR(1) process such that $\rho_1 = 0.99$ will exhibit the type of gradual decay indicative of a nonstationary process. To illustrate some of the issues involved, suppose that we know a series is generated from the following first-order process:[7]

$$y_t = a_1 y_{t-1} + \varepsilon_t \tag{4.16}$$

where $\{\varepsilon_t\}$ is white noise.

Now consider a test of the null hypothesis that $a_1 = 0$. Under the maintained null hypothesis of $a_1 = 0$, we can estimate (4.16) using OLS. The fact that ε_t is a white-noise process and that $|a_1| < 1$ guarantees that the $\{y_t\}$ sequence is stationary and that the estimate of a_1 is efficient. Calculating the standard error of the estimate of a_1, the researcher can use a t-test to determine whether a_1 is significantly different from zero.

The situation is quite different if we want to test the hypothesis $a_1 = 1$. Now, under the null hypothesis, the $\{y_t\}$ sequence is generated by the nonstationary process

$$y_t = y_0 + \sum_{i=1}^{t} \varepsilon_i \tag{4.17}$$

Thus, if $a_1 = 1$, the variance becomes infinitely large as t increases. Under the null hypothesis, it is inappropriate to use classical statistical methods to estimate and perform significance tests on the coefficient a_1. If the $\{y_t\}$ sequence is generated as in (4.17), it is simple to show that the OLS estimate of (4.16) will yield a biased estimate of a_1. In Section 1, it was shown that the first-order autocorrelation coefficient in a random walk model is

$$\rho_1 = [(t-1)/t]^{0.5} < 1$$

Since the estimate of a_1 is directly related to the value of ρ_1, the estimated value of a_1 is biased to be below its true value of unity. The estimated model will mimic that of a stationary AR(1) process with a near–unit root. Hence, the usual t-test cannot be used to test the hypothesis $a_1 = 1$.

Figure 4.6 shows the sample correlogram for a simulated random walk process. One hundred normally distributed random deviates were obtained so as to mimic the $\{\varepsilon_t\}$ sequence. Assuming $y_0 = 0$, the next 100 values in the $\{y_t\}$ sequence were calculated as $y_t = y_{t-1} + \varepsilon_t$. This particular correlogram is characteristic of most sample

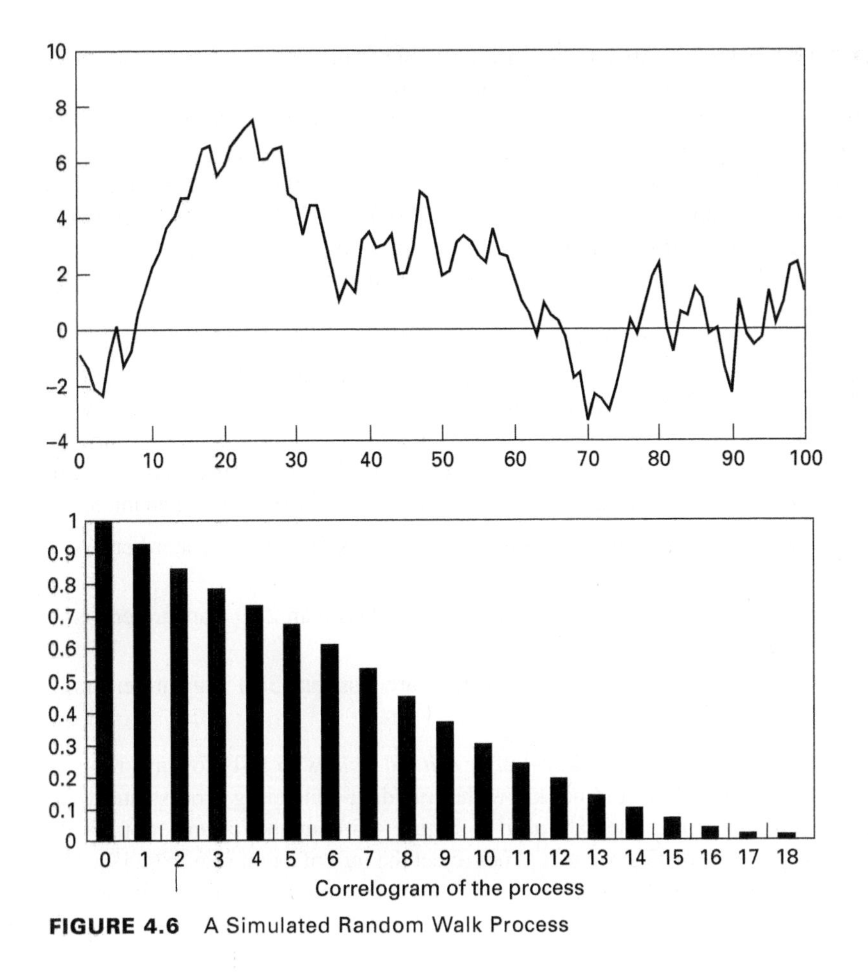

FIGURE 4.6 A Simulated Random Walk Process

correlograms constructed from nonstationary data. The estimated value of ρ_1 is close to unity and the sample autocorrelations die out slowly. If we did not know the way in which the data were generated, inspection of Figure 4.6 might lead us to falsely conclude that the data were generated from a stationary process. With these particular data, estimates of an AR(1) model with and without an intercept yield (standard errors are in parentheses):

$$y_t = \underset{(0.030)}{0.9546}y_{t-1} + \varepsilon_t \qquad R^2 = 0.860 \tag{4.18}$$

$$y_t = 0.164 + \underset{(0.037)}{0.9247}y_{t-1} + \varepsilon_t \qquad R^2 = 0.864 \tag{4.19}$$

Examining (4.18), a careful researcher would not be willing to dismiss the possibility of a unit root since the estimated value of a_1 is only 1.5133 standard deviations from unity: [(1 − 0.9546)/0.30 = 1.5133]. We might correctly recognize that, under the null hypothesis of a unit root, the estimate of a_1 will be biased below unity. If we knew the true distribution of a_1 under the null of a unit root, we could perform such a significance test. Of course, if we did not know the true data-generating process, we might estimate the model with an intercept. In (4.19), the estimate of a_1 is more than two standard deviations from unity: (1 − 0.9247)/0.037 = 2.035. However, it would be wrong to use this information to reject the null of a unit root. After all, the point of this section has been to indicate that such t-tests are inappropriate under the null of a unit root.

Fortunately, Dickey and Fuller (1979, 1981) devised a procedure to formally test for the presence of a unit root. Their methodology is similar to that used in constructing the data reported in Figure 4.6. Suppose that we generated thousands of random walk sequences and that for each we calculated the estimated value of a_1. Although most of the estimates would be close to unity, some would be further from unity than others. In performing this experiment, Dickey and Fuller found that in the presence of an intercept:

- 90 percent of the estimated values of a_1 are less than 2.58 standard errors from unity;
- 95 percent of the estimated values of a_1 are less than 2.89 standard errors from unity;
- 99 percent of the estimated values of a_1 are less than 3.51 standard errors from unity.[8]

The application of these Dickey–Fuller *critical values* to tests for unit roots is straightforward. Suppose we did not know the true data-generating process and were trying to ascertain whether the data used in Figure 4.6 contained a unit root. Using these Dickey–Fuller statistics, we would not reject the null of a unit root in (4.19). The estimated value of a_1 is only 2.035 standard deviations from unity. In fact, if the true value of a_1 does equal unity, we should find the estimated value to be within 2.58 standard deviations from unity 90 percent of the time.

Be aware that stationarity necessitates $-1 < a_1 < 1$ or, equivalently, $a_1^2 < 1$. Thus, if the estimated value of a_1 is close to −1, you should also be concerned about nonstationarity. If we define $\gamma = a_1 - 1$, the equivalent restriction is $-2 < \gamma < 0$. In conducting a Dickey–Fuller test, *it is possible to check that the estimated value of* γ *is greater than* -2.[9]

Monte Carlo Experiments

The procedure that Dickey and Fuller (1979, 1981) used to obtain their critical values is typical of that found in the modern time-series literature. Hypothesis tests concerning the coefficients of nonstationary variables cannot be conducted using traditional t-tests or F-tests. The distributions of the appropriate test statistics are nonstandard and cannot be analytically evaluated. However, given the trivial cost of computer time, the nonstandard distributions can easily be derived using a Monte Carlo simulation.

A Monte Carlo experiment attempts to replicate an actual data-generating process (DGP) on a computer. To be more specific, you simulate a data set with the essential characteristics of the actual data in question. A Monte Carlo experiment generates a random sample of size T, and the parameters and/or sample statistics of interest are calculated. This process is repeated N times (where N is a large number) so that the distribution of the desired parameters and/or sample statistics can be tabulated. These empirical distributions are used as estimates of the actual distributions.

All major statistical software packages have a built-in random number generator. The first step in a Monte Carlo experiment is to computer-generate a set of random numbers (sometimes called pseudo-random numbers) from a given distribution. Of course, the numbers cannot be entirely random since all computer algorithms rely on a deterministic number-generating mechanism. However, the numbers are drawn so as to mimic a random process having some specified distribution. Usually, the numbers are designed to be normally distributed and serially uncorrelated. The idea is to use these numbers to represent one replication of the entire $\{\varepsilon_t\}$ sequence.

The second step is to construct the $\{y_t\}$ sequence using the random numbers and the parameters of the data-generating process. For example, Dickey and Fuller (1979, 1981) obtained 100 values for $\{\varepsilon_t\}$, set $a_1 = 1$ and $y_0 = 0$, and calculated 100 values for $\{y_t\}$ according to (4.16). Once a series has been generated, the third step is to estimate the parameters of interest (such as the estimate of a_1 and the in-sample variance of the $\{y_t\}$ series).

The beauty of this method is that all important attributes of the constructed $\{y_t\}$ sequence are known to the researcher. For this reason, a Monte Carlo simulation is often referred to as an "experiment." The only problem is that the set of random numbers drawn is just one possible outcome. Obviously, the estimates in (4.18) and (4.19) are dependent on the values of the simulated $\{\varepsilon_t\}$ sequence. Different outcomes for $\{\varepsilon_t\}$ will yield different values of the simulated $\{y_t\}$ sequence.

This is why Monte Carlo studies perform many replications of the process outlined above. The fourth step is to replicate steps 1 to 3 thousands of times. The goal is to ensure that the statistical properties of the constructed $\{y_t\}$ sequence are in accord with the true distribution. Thus, for each replication, the parameters of interest are tabulated and critical values (or confidence intervals) obtained. As such, the properties of your data can be compared to the properties of the simulated data so that hypothesis tests can be performed.

For our purposes, it suffices to say that the use of the Monte Carlo method is warranted by the Law of Large Numbers. Consider the simplest case where v_t is an identically and independently distributed (*i.i.d.*) random number with mean μ and variance σ^2 so that

$$v_t \sim (\mu, \sigma^2)$$

The sample mean constructed by using T observations of the $\{v_t\}$ sequence is

$$\bar{v} = (1/T)\sum_{t=1}^{T} v_t$$

By the Law of Large Numbers, as the sample size T grows sufficiently large, $\bar{v}$ converges to the true mean μ. Hence, the sample mean $\bar{v}$ is an unbiased estimate of the population mean. This is the justification for using the Dickey–Fuller critical values to test the hypothesis $a_1 = 1$. Moreover, if the draws are independent and the sample size T grows sufficiently large, the distribution of $\bar{v}$ approaches a normal distribution with mean μ and variance σ^2/T.[10]

An important limitation of a Monte Carlo experiment is that the results are specific to the assumptions used to generate the simulated data. If you change the sample size, include (or delete) an additional parameter in the data-generating process, or use alternative initial conditions, an entirely new simulation needs to be performed. Moreover, the precision of your estimates depends on the number of replications you use. Often you do not need many replications to obtain a good estimate of a population mean. However, it is necessary to use many thousands of replications to obtain good estimates of critical values. Nevertheless, you should be able to envision many applications of Monte Carlo experiments. As discussed in Hendry, Neale, and Ericsson (1990), these experiments are particularly helpful for studying the large- and small-sample properties of time-series data. As you will see shortly, Monte Carlo experiments are the workhorse of many tests used in modern time-series analysis.

Example of the Monte Carlo Method

Suppose you did not know the probability distribution for the sum of the roll of two dice. One way to calculate the probability distribution would be to buy a pair of dice and roll them several thousand times. If the dice were fair, you would find that a sum of your rolls would approximate this result:

sum	2	3	4	5	6	7	8	9	10	11	12
percent	1/36	2/36	3/36	4/36	5/36	6/36	5/36	4/36	3/36	2/36	1/36

Instead of actually rolling the dice, you can easily replicate the experiment on a computer. You could draw a random number from a uniform [0, 1] distribution to replicate the roll of the first die. If the computer-generated number falls within the interval [0, 1/6], set the variable $r_1 = 1$. Similarly, if the number falls within the interval (1/6, 2/6], set $r_1 = 2$, and so on. In this way, r_1 will be some integer 1 through 6, each with a probability 1/6. Next, draw a second number from the same uniform [0, 1] distribution to represent the roll of die 2 (r_2). You complete your first Monte Carlo replication by computing the sum $r_1 + r_2$. If you compute several thousand such sums, the sample distribution of the sums will approximate the true distribution.

Of course, more complicated experiments are possible. It is interesting to note that this method was used to reform a standard recommendation at the blackjack tables. At one time, the recommendation was to "stick" if the dealer shows a 2 or a 3 and you hold a 12. Monte Carlo experiments of a game of blackjack showed that this recommendation was incorrect. Now, a sharp blackjack player will take another card in these circumstances.

Generating the Dickey–Fuller Distribution

We need to modify the procedure above only slightly to obtain the Dickey and Fuller (1979) distribution. To generate the distribution for a sample size of 100, we can perform the following steps:

STEP 1: First, we need a set of random numbers to represent the $\{\varepsilon_t\}$ sequence. If we use the usual set of assumptions, we can draw a set of 100 random numbers from a standard normal distribution. Of course, the Monte Carlo method would allow us to experiment with other distributions.

STEP 2: We need to generate the sequence $y_t = y_{t-1} + \varepsilon_t$. Note that we need to initialize the value of y_0. Once we draw the value of ε_1, we cannot construct y_1 without positing some value for y_0. However, we do not want the results to be sensitive to the initial value chosen for the series. Two slightly different procedures are used to purge the effects of the initial condition from the Monte Carlo results. First, you can initialize the value of y_0 to equal the unconditional mean of the $\{y_t\}$ sequence. Alternatively, suppose you want to generate T values of the $\{y_t\}$ sequence. You can pick an initial condition for y_0 and then generate the next $T + 50$ realizations. Discard the first 50 realizations and use only the last T values. The idea is that the effect of the initial condition will dissipate after 50 periods.

STEP 3: We need to estimate the model under the alternative hypothesis. As such, we estimate an equation of the form $\Delta y_t = a_0 + \gamma y_{t-1} + \varepsilon_t$. Obtain the t-statistic for the null hypothesis $\gamma = 0$. Note that the data are generated under the null hypothesis of a unit root and estimated under the alternative hypothesis.

STEP 4: Repeat Steps 1–3, 10,000 or more times. If you use a sample size such that $T = 100$, you should obtain something very similar to the Dickey–Fuller τ_μ distribution plotted in Figure 4.7. Of course, you will not obtain the exact numbers used in the figure since you will be using a different set of random numbers.

The data used to draw Figure 4.7 contain 10,000 replications. Additional replications would reveal a somewhat smoother probability distribution. As you might expect, the mean of the distribution is far below zero. The mean of the t-statistics shown in the figure is -1.53. The distribution of t-statistics for the null hypothesis

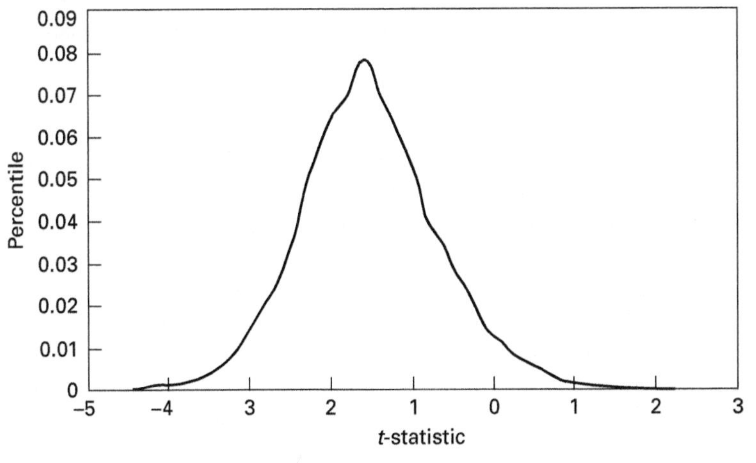

FIGURE 4.7 The Dickey–Fuller Distribution

$\gamma = 0$ is only slightly different from those reported by Dickey and Fuller; about 95 percent are more than -2.89 and 99 percent are more than -3.51. Hence, if you estimate a model in the form $\Delta y_t = a_0 + \gamma y_{t-1} + \varepsilon_t$ and find that the t-statistic for the null hypothesis $\gamma = 0$ is -3.00, you can reject the null hypothesis of a unit root at the 5 percent, but not at the 1 percent, level of significance. We will encounter a number of additional applications of Monte Carlo experiments throughout the text. Additional details of the methodology are discussed in Appendix 4.1 to this chapter and in the *Programming Manual* that accompanies this text.

5. DICKEY–FULLER TESTS

The last section outlined a simple procedure to determine whether $a_1 = 1$ in the model $y_t = a_1 y_{t-1} + \varepsilon_t$. Begin by subtracting y_{t-1} from each side of the equation in order to write the equivalent form: $\Delta y_t = \gamma y_{t-1} + \varepsilon_t$ where $\gamma = a_1 - 1$. Of course, testing the hypothesis $a_1 = 1$ is equivalent to testing the hypothesis $\gamma = 0$. Dickey and Fuller (1979) actually consider three different regression equations that can be used to test for the presence of a unit root:

$$\Delta y_t = \gamma y_{t-1} + \varepsilon_t \tag{4.20}$$

$$\Delta y_t = a_0 + \gamma y_{t-1} + \varepsilon_t \tag{4.21}$$

$$\Delta y_t = a_0 + \gamma y_{t-1} + a_2 t + \varepsilon_t \tag{4.22}$$

The difference between the three regressions concerns the presence of the deterministic elements a_0 and $a_2 t$. Under the null hypothesis $\gamma = 0$, the first is a pure random walk model, the second adds an intercept or *drift* term, and the third includes both a drift and a linear time trend.

The parameter of interest in all the regression equations is γ, if $\gamma = 0$, the $\{y_t\}$ sequence contains a unit root. The test involves estimating one (or more) of the equations above using OLS in order to obtain the estimated value of γ and the associated standard error. Comparing the resulting t-statistic with the appropriate value reported in the Dickey–Fuller tables allows the researcher to determine whether to accept or reject the null hypothesis $\gamma = 0$.

Recall that in (4.18), the estimate of $y_t = a_1 y_{t-1} + \varepsilon_t$ was such that $a_1 = 0.9546$ with a standard error of 0.030. Clearly, the OLS regression in the form $\Delta y_t = \gamma y_{t-1} + \varepsilon_t$ will yield an estimate of γ equal to -0.0454 with the same standard error of 0.030. Hence, the associated t-statistic for the hypothesis $\gamma = 0$ is -1.5133 ($-0.0454/0.03 = -1.5133$).

The methodology is precisely the same regardless of which of the three forms of the equations is estimated. However, be aware that the critical values of the t-statistics do depend on whether an intercept and/or time trend is included in the regression equation. In their Monte Carlo study, Dickey and Fuller (1979) found that the critical values for $\gamma = 0$ depend on the form of the regression and on sample size. The statistics called τ, τ_μ, and τ_τ are the appropriate statistics to use for equations (4.20), (4.21), and (4.22), respectively.

Now, look at Table A at the end of this book. With 100 observations, there are three different critical values for the t-statistic $\gamma = 0$. For a regression without the intercept and trend terms ($a_0 = a_2 = 0$), use the section labeled τ. With 100 observations, the critical values for the t-statistic are -1.61, -1.95, and -2.60 at the 10 percent, 5 percent, and 1 percent significance levels, respectively. Thus, in the hypothetical example with $\gamma = -0.0454$ and a standard error of 0.03 (so that $t = -1.5133$), it is not possible to reject the null of a unit root at conventional significance levels. Note that the appropriate critical values depend on sample size. As in most hypothesis tests, for any given level of significance, the critical values of the t-statistic decrease as sample size increases.

Including an intercept term but not a trend term (only $a_2 = 0$) necessitates the use of the critical values in the section labeled τ_μ. Estimating (4.19) in the form $\Delta y_t = a_0 + \gamma y_{t-1} + \varepsilon_t$ necessarily yields a value of γ equal to $(0.9247 - 1) = -0.0753$ with a standard error of 0.037. The appropriate calculation for the t-statistic yields $-0.0753/0.037 = -2.035$. If we read from the appropriate row of Table A, with the same 100 observations, the critical values are -2.58, -2.89, and -3.51 at the 10 percent, 5 percent, and 1 percent significance levels, respectively. Again, the null of a unit root cannot be rejected at conventional significance levels. Finally, with both intercept and trend, use the critical values in the section labeled τ_τ; now the critical values are -3.45 and -4.04 at the 5 percent and 1 percent significance levels, respectively. The equation was not estimated using a time trend; inspection of Figure 4.5 indicates there is little reason to include a deterministic trend in the estimating equation.

As discussed Section 7, these critical values are unchanged if (4.20), (4.21), and (4.22) are replaced by the autoregressive processes[11]

$$\Delta y_t = \gamma y_{t-1} + \sum_{i=2}^{p} \beta_i \Delta y_{t-i+1} + \varepsilon_t \tag{4.23}$$

$$\Delta y_t = a_0 + \gamma y_{t-1} + \sum_{i=2}^{p} \beta_i \Delta y_{t-i+1} + \varepsilon_t \tag{4.24}$$

$$\Delta y_t = a_0 + \gamma y_{t-1} + a_2 t + \sum_{i=2}^{p} \beta_i \Delta y_{t-i+1} + \varepsilon_t \tag{4.25}$$

The same τ, τ_μ, and τ_τ statistics are all used to test the hypotheses $\gamma = 0$. Dickey and Fuller (1981) provided three additional F-statistics (called ϕ_1, ϕ_2, and ϕ_3) to test **joint** hypotheses on the coefficients. Using (4.21) or (4.24), the null hypothesis $\gamma = a_0 = 0$ is tested using the ϕ_1 statistic. Including a time trend in the regression—so that (4.22) or (4.25) is estimated—the joint hypothesis $a_0 = \gamma = a_2 = 0$ is tested using the ϕ_2 statistic and the joint hypothesis $\gamma = a_2 = 0$ is tested using the ϕ_3 statistic.

The ϕ_1, ϕ_2, and ϕ_3 statistics are constructed in exactly the same way as ordinary F-tests:

$$\phi_i = \frac{[SSR(\text{restricted}) - SSR(\text{unrestricted})]/r}{SSR(\text{unrestricted})/(T - k)}$$

Table 4.2 Summary of the Dickey–Fuller Tests

Model	Hypothesis	Test Statistic	Critical Values for 95% and 99% Confidence Intervals
$\Delta y_t = a_0 + \gamma y_{t-1} + a_2 t + \varepsilon_t$	$\gamma = 0$	τ_τ	−3.45 and −4.04
	$\gamma = a_2 = 0$	ϕ_3	6.49 and 8.73
	$a_0 = \gamma = a_2 = 0$	ϕ_2	4.88 and 6.50
$\Delta y_t = a_0 + \gamma y_{t-1} + \varepsilon_t$	$\gamma = 0$	τ_μ	−2.89 and −3.51
	$a_0 = \gamma = 0$	ϕ_1	4.71 and 6.70
$\Delta y_t = \gamma y_{t-1} + \varepsilon_t$	$\gamma = 0$	τ	−1.95 and −2.60

Note: Critical values are for a sample size of 100.

where: SSR(restricted) and SSR(unrestricted) = the sums of the squared residuals from the restricted and unrestricted models
r = number of restrictions
T = number of usable observations
k = number of parameters estimated in the unrestricted model.

Hence, $T - k$ = degrees of freedom in the unrestricted model.

Comparing the calculated value of ϕ_i to the appropriate value reported in Dickey and Fuller (1981) allows you to determine the significance level at which the restriction is binding. The null hypothesis is that the data are generated by the restricted model and the alternative hypothesis is that the data are generated by the unrestricted model. If the restriction is not binding, SSR(restricted) should be close to SSR(unrestricted) and ϕ_i should be small; hence, large values of ϕ_i suggest a binding restriction and a rejection of the null hypothesis. Thus, if the calculated value of ϕ_i is smaller than that reported by Dickey and Fuller, you can accept the restricted model (i.e., you do not reject the null hypothesis that the restriction is not binding). If the calculated value of ϕ_i is larger than that reported by Dickey and Fuller, you can reject the null hypothesis and conclude that the restriction is binding. The critical values of the three ϕ_i statistics are reported in Table B at the end of this text. The complete set of test statistics and their critical values for a sample size of 100 is summarized in Table 4.2.

An Example

To illustrate the use of the various test statistics, Dickey and Fuller (1981) used quarterly values of the logarithm of the Federal Reserve Board's Production Index over the 1950Q1–1977Q4 period to estimate the following three equations:

$$\underset{}{\Delta y_t} = \underset{(0.15)}{0.52} + \underset{(0.00034)}{0.00120t} - \underset{(0.033)}{0.119y_{t-1}} + \underset{(0.081)}{0.498\Delta y_{t-1}} + \varepsilon_t \quad SSR = 0.056448 \tag{4.26}$$

$$\Delta y_t = \underset{(0.0025)}{0.0054} + \underset{(0.083)}{0.447\Delta y_{t-1}} + \varepsilon_t \quad SSR = 0.063211 \tag{4.27}$$

$$\Delta y_t = \underset{(0.079)}{0.511\Delta y_{t-1}} + \varepsilon_t \quad SSR = 0.065966 \tag{4.28}$$

where SSR = sum of squared residuals, and standard errors are in parentheses.

To test the null hypothesis that the data are generated by (4.28) against the alternative that (4.26) is the "true" model, use the ϕ_2 statistic. Dickey and Fuller tested the null hypothesis $a_0 = a_2 = \gamma = 0$ as follows. Note that the residual sums of squares of the restricted and unrestricted models are 0.065966 and 0.056448 and that the null hypothesis entails three restrictions. With 110 usable observations and four estimated parameters, the unrestricted model contains 106 degrees of freedom. Since $0.056448/106 = 0.000533$, the ϕ_2 statistic is given by:

$$\phi_2 = (0.065966 - 0.056448)/[3(0.000533)] = 5.95$$

With 110 observations, the critical value of ϕ_2 calculated by Dickey and Fuller is 5.59 at the 2.5 percent significance level. Hence, it is possible to reject the null hypothesis of a random walk against the alternative that the data contain an intercept and/or a unit root and/or a deterministic time trend (i.e., rejecting $a_0 = a_2 = \gamma = 0$ means that one or more of these parameters does not equal zero).

Dickey and Fuller also tested the null hypothesis $a_2 = \gamma = 0$ given the alternative of (4.26). If we now view (4.27) as the restricted model and (4.26) as the unrestricted model, the ϕ_3 statistic is calculated as

$$\phi_3 = (0.063211 - 0.056448)/[2(0.000533)] = 6.34$$

With 110 observations, the critical value of ϕ_3 is 6.49 at the 5 percent significance level and 5.47 at the 10 percent significance level.[12] At the 10 percent level, Dickey and Fuller reject the null hypothesis and accept the alternative that the series is trend-stationary. However, at the 5 percent level, the calculated value of ϕ_3 is smaller than the critical value of 6.49; at this significance level, they do not reject the null hypothesis. Hence, at the 5 percent significance level, they maintain the hypothesis that the series contains a unit root and/or a deterministic time trend.

To compare with the τ_τ test (i.e., the hypothesis that only $\gamma = 0$), note that

$$\tau_\tau = -0.119/0.033 = -3.61$$

which rejects the null of a unit root at the 5 percent level.

6. EXAMPLES OF THE DICKEY–FULLER TEST

Section 2 reviewed the evidence reported by Nelson and Plosser (1982) suggesting that macroeconomic variables are difference-stationary rather than trend-stationary. We are now in a position to consider their formal tests of the hypothesis. For each series under study, Nelson and Plosser estimated the regression in the form of (4.25):

$$\Delta y_t = a_0 + \gamma y_{t-1} + a_2 t + \sum_{i=2}^{p} \beta_i \Delta y_{t-i+1} + \varepsilon_t$$

The chosen lag lengths are reported in the column labeled p in Table 4.3. The estimated values a_0, a_2, and γ are reported in columns 3, 4, and 5, respectively.

Recall that the traditional view of business cycles maintains that GNP and production levels are trend-stationary rather than difference-stationary. An adherent to this view must assert that γ is different from zero; if $\gamma = 0$, the series has a unit root and is difference-stationary. Given the sample sizes used by Nelson and Plosser

Table 4.3 Nelson and Plosser's Tests for Unit Roots

	p	a_0	a_2	γ	$\gamma + 1$
Real GNP	2	0.819	0.006	−0.175	0.825
		(3.03)	(3.03)	(−2.99)	
Nominal GNP	2	1.06	0.006	−0.101	0.899
		(2.37)	(2.34)	(−2.32)	
Industrial production	6	0.103	0.007	−0.165	0.835
		(4.32)	(2.44)	(−2.53)	
Unemployment rate	4	0.513	−0.000	−0.294*	0.706
		(2.81)	(−0.23)	(−3.55)	

Notes:

1. p is the chosen lag length. Coefficients divided by their standard errors are in parentheses. Thus, entries in parentheses represent the t-test for the null hypothesis that a coefficient is equal to zero. Under the null of nonstationarity, it is necessary to use the Dickey–Fuller critical values. At the .05 significance level, the critical value for the t-statistic is −3.45.
2. A (*) denotes significance at the 0.05 level. For real and nominal GNP and for industrial production, it is not possible to reject the null hypothesis $\gamma = 0$ at the 0.05 level. Hence, the unemployment rate appears to be stationary.
3. The expression $\gamma + 1$ is the estimate of a_1.

(1982), at the 0.05 level, the critical value of the t-statistic for the null hypothesis $\gamma = 0$ is −3.45. Thus, only if the estimated value of γ is more than 3.45 standard deviations from zero is it possible to reject the hypothesis that $\gamma = 0$. As can be seen from inspection of Table 4.3, the estimated values of γ for real GNP, nominal GNP, and industrial production are not statistically different from zero. Only the unemployment rate has an estimated value of γ that is significantly different from zero at the 0.05 level.

Quarterly Real U.S. GDP

Now use the data on the file RGDP.XLS to estimate the logarithmic change in real GDP as

$$\underset{}{\Delta lrgdp_t} = \underset{(3.52)}{0.0246} + \underset{(2.49)}{0.00028t} - \underset{(-2.59)}{0.0360 lrgdp_{t-1}} + \underset{(5.66)}{0.3426 \Delta lrgdp_{t-1}} \qquad (4.29)$$

The t-statistic on the coefficient for $lrgdp_{t-1}$ is −2.59. Table A indicates that with 244 usable observations, the 10 percent and 5 percent critical values of τ_τ are about −3.13 and −3.43, respectively. As such, we cannot reject the null hypothesis of a unit root. The sample value of ϕ_3 for the null hypothesis $a_2 = \gamma = 0$ is 4.12. Since Table B indicates that the 10 percent critical value is 5.39, we cannot reject the joint hypothesis of a unit root and no deterministic time trend. The sample value of ϕ_2 is 20.20. As such, the growth rate of the real GDP series acts as a random walk plus drift plus the irregular term $0.3426\Delta lrgdp_{t-1}$. Additional details are contained in Appendix 4.2.

Unit Roots and Purchasing Power Parity

Purchasing power parity (PPP) is a simple relationship linking national price levels and exchange rates. In its simplest form, PPP asserts that the rate of currency depreciation is approximately equal to the difference between domestic and foreign inflation rates. If p_t and p_{ft} denote the logarithms of U.S. and foreign price levels and e_t denotes the logarithm of the dollar price of foreign exchange, PPP implies

$$e_t = p_t - p_{ft} + d_t$$

where d_t represents the deviation from PPP in period t.

In applied work, p_t and p_{ft} usually refer to national price indices in t relative to a base year, so that e_t refers to an index of the domestic currency price of foreign exchange relative to a base year. For example, if the U.S. inflation rate is 10 percent while the foreign inflation rate is 15 percent, the dollar price of foreign exchange should fall by approximately 5 percent. The presence of the term d_t allows for short-run deviations from PPP.

Because of its simplicity and intuitive appeal, PPP has been used extensively in theoretical models of exchange rate determination. However, as in the well-known Dornbusch (1976) "overshooting" model, real economic shocks, such as productivity or demand shocks, can cause permanent deviations from PPP. For our purposes, the theory of PPP serves as an excellent vehicle to illustrate many time-series testing procedures. One test of long-run PPP is to determine whether d_t is stationary. After all, if the deviations from PPP are nonstationary (i.e., if the deviations are permanent in nature), we can reject the theory. Note that PPP does allow for persistent deviations; the autocorrelations of the $\{d_t\}$ sequence need not be zero. One popular testing procedure is to define the "real" exchange rate in period t:

$$r_t \equiv e_t + p_{ft} - p_t$$

Long-run PPP is said to hold if the $\{r_t\}$ sequence is stationary. For example, in Enders (1988), I constructed real exchange rates for three major U.S. trading partners: Germany, Canada, and Japan. The data were divided into two periods: January 1960 to April 1971 (representing the fixed-exchange-rate period) and January 1973 to November 1986 (representing the flexible-exchange-rate period). Each nation's wholesale price index (WPI) was multiplied by an index of the U.S. dollar price of the foreign currency and then divided by the U.S. WPI. The log of the constructed series is the $\{r_t\}$ sequence.

A critical first step in any econometric analysis is to visually inspect the data. The plots of the three real-exchange-rate series during the flexible-exchange-rate period are shown in Figure 4.8.[13] Each series seems to meander in a fashion characteristic of a random walk process. Notice that there is little visual evidence of explosive behavior or of a deterministic time trend. The autocorrelation functions for all of the series in the analysis look similar to that in Figure 4.6. In particular, the autocorrelation

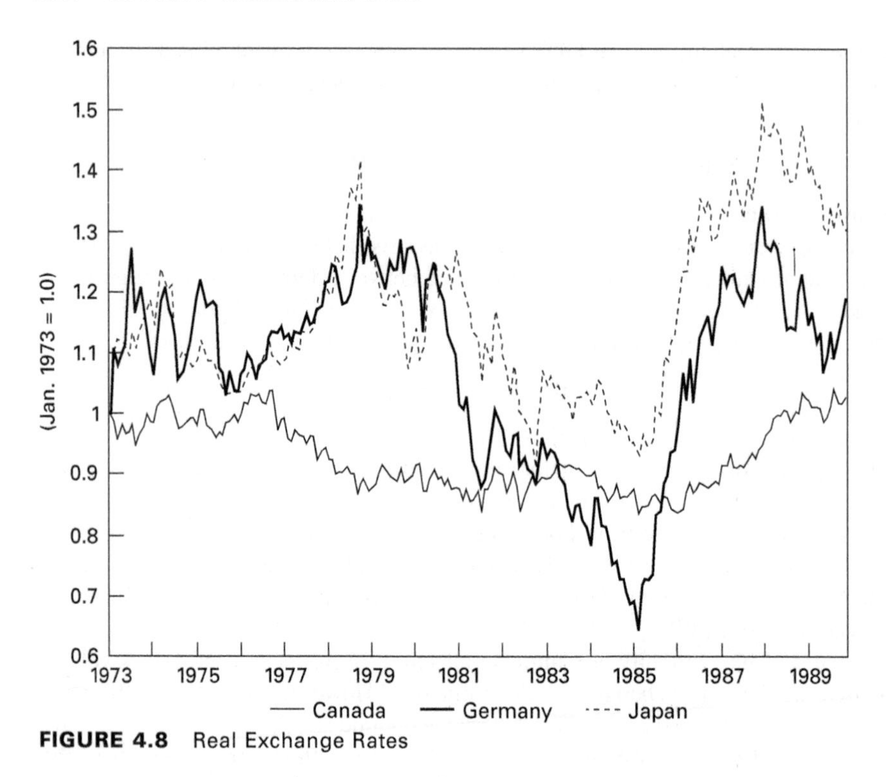

FIGURE 4.8 Real Exchange Rates

functions show little tendency to decay, while the autocorrelations of the first differences display the classic pattern of a stationary series.

To formally test for the presence of a unit root in the real exchange rates, augmented Dickey–Fuller tests of the form given by (4.24) were conducted. The regression $\Delta r_t = a_0 + \gamma r_{t-1} + \beta_2 \Delta r_{t-1} + \beta_3 \Delta r_{t-2} + \ldots + \varepsilon_t$ was estimated based on the following considerations:

1. The theory of PPP does not allow for a deterministic time trend. Any such findings would refute the theory as posited. Given that the series all decline throughout the early 1980s and all rise during the middle to late 1980s, there is no reason to entertain the notion of trend stationarity. As such, the expression $a_2 t$ was not included in the estimating equation.
2. For the fixed-exchange-rate period, various lag length tests indicated that all values of β_i could be set equal to zero for all three countries. However, different lag length tests yielded ambiguous results for the flexible-exchange-rate period. The general-to-specific method indicated that β_{11} was statistically different from zero for all three countries. In contrast, F-tests and the SBC selected two lags for Germany and Japan and no lagged changes for Canada. As such, for the flexible-rate period, the Dickey–Fuller tests were conducted using two different lag lengths for each country.

For the Canadian case during the 1973 to 1986 period, the t-statistic for the null hypothesis that $\gamma = 0$ is -1.42 using no lags and -1.51 using all 11 lags. Given the critical value of the τ_μ statistic, it is not possible to reject the null of a unit root in the Canadian/U.S. real-exchange-rate series. Hence, PPP fails for the Canadian-U.S. case. In the 1960 to 1971 period, the calculated value of the t-statistic is -1.59; again it is possible to conclude that PPP fails.

Table 4.4 reports the results of all six estimations using the short lag lengths suggested by the F-tests and the SBC. Notice the following properties of the estimated models:

1. For all six models, it is not possible to reject the null hypothesis that PPP fails. As can be seen from the third column of Table 4.4, the absolute value of the t-statistic for the null $\gamma = 0$ is never more than 1.59. The economic interpretation is that real productivity and/or demand shocks have had a permanent influence on real exchange rates.
2. As measured by the sample standard deviation (SD), real exchange rates were far more volatile in the 1973 to 1986 period than in the 1960 to 1971 period. Moreover, as measured by the standard error of the estimate (SEE), real-exchange-rate volatility is associated with unpredictability. The SEE during the flexible-exchange-rate period is several hundred times that of the fixed-rate period. It seems reasonable to conclude that the change in the exchange rate regime (i.e., the end of Bretton Woods) affected the volatility of the real exchange rate.
3. Care must be taken to keep the appropriate null hypothesis in mind. Under the null of a unit root, classical test procedures are inappropriate, and we resort to the statistics tabulated by Dickey and Fuller. However, classical test procedures (which assume stationary variables) are appropriate under the null that the real rates are stationary. Thus, the following possibility arises: Suppose that the t-statistic in the Canadian case happened to be -2.16 instead of -1.42. If you used the Dickey–Fuller critical values, you would not reject the null of a unit root. Hence, you could conclude that PPP fails. However, under the null of stationarity (where we can use classical procedures), γ is more than two standard deviations from zero and you would *not reject the null of stationarity.*

   This apparent dilemma commonly occurs when analyzing series with roots close to unity in absolute value. Unit root tests do not have much power in discriminating between characteristic roots close to unity and actual unit roots. The dilemma is only apparent since the two null hypotheses are quite different. It is perfectly consistent to maintain a null that PPP holds and not be able to reject a null that PPP fails! Notice that this dilemma does not arise for any of the series reported in Table 4.4; for each, it is not possible to reject a null of $\gamma = 0$ at conventional significance levels.

   One way to help circumvent this problem is to directly test the null hypothesis of stationarity against the alternative of nonstationarity. Kwiatowski, Phillips, Schmidt, and Shin (1992) showed how to perform this type of test.

Table 4.4 Real Exchange Rate Estimation

	γ[1]	H_0: $\gamma = 0$[2]	Lags	Mean[3]	ρ/DW	F	SD/SEE
1973–1986							
Canada	−0.022	$t = -1.42$	0	1.05	0.059	0.194	5.47
	(0.016)				1.88		1.16
Japan	−0.047	$t = -0.64$	2	1.01	−0.007	0.226	10.44
	(0.074)				2.01		2.81
Germany	−0.027	$t = -0.28$	2	1.11	−0.014	0.858	20.68
	(0.076)				2.04		3.71
1960–1971							
Canada	−0.031	$t = -1.59$	0	1.02	−0.107	0.434	.014
	(0.019)				2.21		.004
Japan	−0.030	$t = -1.04$	0	0.98	0.046	0.330	.017
	(0.028)				1.98		.005
Germany	−0.016	$t = -1.23$	0	1.01	0.038	0.097	.026
	(0.012)				1.93		.004

Notes:

1. Standard errors are in parentheses.
2. Entries are the t-statistic for the hypothesis $\gamma = 0$.
3. Mean is the sample mean of the series. SD is the standard deviation of the real exchange rate. SEE is the estimated standard deviation of the residuals (i.e., the standard error of the estimate). F is the significance level of the test that lags 2 (or 3) through 12 can be excluded. DW is the Durbin-Watson statistic for first-order serial correlation and ρ is the estimated autocorrelation coefficient.

4. Looking at some of the diagnostic statistics, the F-statistics all indicate that it is appropriate to exclude lags 2 (or 3) through 12 from the regression equation. To reinforce the use of short lags, notice that the first-order correlation coefficient of the residuals (ρ) is low and that the Durbin–Watson statistic is close to two. It is interesting that the point estimates of the characteristic roots all indicate that real exchange rates are convergent. To obtain the characteristic roots, rewrite the estimated equations in the autoregressive form $r_t = a_0 + a_1 r_{t-1}$ or $r_t = a_0 + a_1 r_{t-1} + a_2 r_{t-2}$. For the four AR(1) models, the point estimates of the slope coefficients are all less than unity. In the post–Bretton Woods period (1973–1986), the point estimates of the characteristic roots of Japan's second-order process are 0.931 and 0.319; for Germany, the roots are 0.964 and 0.256. Yet, this is precisely what we would expect if PPP fails; under the null of a unit root, we know that γ is biased downward.

To update the study, the file PANEL.XLS contains quarterly values of the real effective exchange rates (CPI-based) for Australia, Canada, France, Germany, Japan, Netherlands, the United Kingdom, and the United States over the 1980Q1–2008Q1 period. These are multilateral (not bilateral) real exchange rates. The rates for the United States, Canada, and Germany are shown in Figure 5 of Chapter 3. As an exercise, you should use these data to verify that very little has changed. You should find that only for France and Germany is it possible to reject a unit root in the real exchange rate at the 5 percent significance level. (Try not to peek; however, for each country, the estimated value of γ and the appropriate lag length are reported in Table 4.8.)

7. EXTENSIONS OF THE DICKEY–FULLER TEST

Not all time-series variables can be well represented by the first-order autoregressive process $\Delta y_t = a_0 + \gamma y_{t-1} + a_2 t + \varepsilon_t$. It is possible to use the Dickey–Fuller tests in higher-order equations such as (4.23), (4.24), and (4.25). Consider the pth order autoregressive process:

$$y_t = a_0 + a_1 y_{t-1} + a_2 y_{t-2} + a_3 y_{t-3} + \dots + a_{p-2} y_{t-p+2} + a_{p-1} y_{t-p+1} + a_p y_{t-p} + \varepsilon_t$$

To best understand the methodology of the **augmented Dickey–Fuller** (ADF) test, add and subtract $a_p y_{t-p+1}$ to obtain

$$\begin{aligned} y_t = {} & a_0 + a_1 y_{t-1} + a_2 y_{t-2} + a_3 y_{t-3} + \dots + a_{p-2} y_{t-p+2} \\ & + (a_{p-1} + a_p) y_{t-p+1} - a_p \Delta y_{t-p+1} + \varepsilon_t \end{aligned}$$

Next, add and subtract $(a_{p-1} + a_p) y_{t-p+2}$ to obtain

$$y_t = a_0 + a_1 y_{t-1} + a_2 y_{t-2} + a_3 y_{t-3} + \dots - (a_{p-1} + a_p) \Delta y_{t-p+2} - a_p \Delta y_{t-p+1} + \varepsilon_t$$

Continuing in this fashion, we obtain

$$\Delta y_t = a_0 + \gamma y_{t-1} + \sum_{i=2}^{p} \beta_i \Delta y_{t-i+1} + \varepsilon_t \tag{4.30}$$

where $\gamma = -\left(1 - \sum_{i=1}^{p} a_i\right)$ and $\beta_i = -\sum_{j=i}^{p} a_j$.

In (4.30), the coefficient of interest is γ; if $\gamma = 0$, the equation is entirely in first differences and so has a unit root. We can test for the presence of a unit root using the same Dickey–Fuller statistics discussed above. Again, the appropriate statistic to use depends on the deterministic components included in the regression equation. Without an intercept or trend, use the τ statistic; with only the intercept, use the τ_μ statistic; and with both intercept and trend, use the τ_τ statistic. It is worthwhile pointing out that the results here are perfectly consistent with our study of difference equations in Chapter 1. If the coefficients of a difference equation sum to 1, *at least* one characteristic root is unity. Here, if $\sum a_i = 1$, $\gamma = 0$ and the system has a unit root.

Note that the Dickey–Fuller tests assume that the errors are independent and have a constant variance. This raises six important problems related to the fact that we do not know the true data-generating process:

1. We cannot properly estimate γ and its standard error unless all of the autoregressive terms are included in the estimating equation. Clearly, the simple regression $\Delta y_t = a_0 + \gamma y_{t-1} + \varepsilon_t$ is inadequate to this task if (4.30)

is the true data-generating process. Since the true order of the autoregressive process is unknown, the problem is to select the appropriate lag length.

2. The DGP may contain both autoregressive and moving-average components. We need to know how to conduct the test if the order of the moving-average terms (if any) is unknown.
3. The Dickey–Fuller test considers only a single unit root. However, a pth order autoregression has p characteristic roots; if there are $d \leq p$ unit roots, the series needs to be differenced d times to achieve stationarity.
4. As we saw in Chapter 2, there may be roots that require first differences and others that necessitate seasonal differencing. We need to develop a method that can distinguish between these two types of unit root processes.
5. There might be structural breaks in the data. As shown in Section 8, such breaks can impart an apparent trend to the data.
6. It might not be known whether an intercept and/or time trend belongs in (4.30). Section 9 and Appendix 4.2 are concerned with the issue of the appropriate deterministic regressors.

Selection of the Lag Length

It is important to use the correct number of lags in conducting a Dickey–Fuller test. Too few lags means that the regression residuals do not behave like white-noise processes. The model will not appropriately capture the actual error process, so γ and its standard error will not be well estimated. Including too many lags reduces the power of the test to reject the null of a unit root since the increased number of lags necessitates the estimation of additional parameters and a loss of degrees of freedom. The degrees of freedom decrease since the number of parameters estimated has increased and the number of usable observations has decreased. (We lose one observation for each additional lag included in the autoregression.) As such, the presence of unnecessary lags will reduce the power of the Dickey–Fuller test to detect a unit root. In fact, an augmented Dickey–Fuller test may indicate a unit root for some lag lengths but not for others.

How does a careful researcher select the appropriate lag length in such circumstances? One approach is the **general-to-specific** methodology. The idea is to start with a relatively long lag length and pare down the model by the usual t-test and/or F-tests. For example, one could estimate equation (4.30) using a lag length of p^*. If the t-statistic on lag p^* is insignificant at some specified critical value, reestimate the regression using a lag length of $p^* - 1$. Repeat the process until the last lag is significantly different from zero. In the pure autoregressive case, such a procedure will yield the true lag length with an asymptotic probability of unity, provided the initial choice of lag length includes the true length. Using seasonal data, the process is a bit different. For example, using quarterly data, one could start with three years of lags ($p = 12$). If the t-statistic on lag 12 is insignificant at some specified critical value, and if an F-test indicates that lags 9–12 are also insignificant, move to lags 1–8. Repeat the process for lag 8 and lags 5–8 until a reasonable lag length has been determined.

Once a tentative lag length has been determined, diagnostic checking should be conducted. As always, plotting the residuals is a most important diagnostic tool. There should not appear to be any strong evidence of structural change or serial correlation. Moreover, the correlogram of the residuals should appear to be white noise. The Ljung–Box Q-statistic should not reveal any significant autocorrelations among the residuals. It is inadvisable to use the alternative procedure of beginning with the most parsimonious model and continuing to add lags until the first insignificant lag is found. Monte Carlo studies show that this procedure is biased toward selecting a value of p that is less than the true value.

As long as the regression equation does not omit a deterministic regressor present in the data-generating process, it is possible to perform lag-length tests using t-tests or F-tests. The rationale follows from an important result proved by Sims, Stock, and Watson (1990). Here is the key finding of interest:

> **Rule 1:** Consider a regression equation containing a mixture of $I(1)$ and $I(0)$ variables such that the residuals are white noise. If the model is such that the coefficient of interest can be written as a coefficient on zero-mean stationary variables, then asymptotically, the OLS estimator converges to a normal distribution. As such, a t-test is appropriate.

Although this rule refers to any regression equation estimated by OLS, it applies directly to unit root tests. As shown above, the pth order autoregressive process:

$$y_t = a_0 + a_1 y_{t-1} + a_2 y_{t-2} + a_3 y_{t-3} + \dots + a_{p-2} y_{t-p+2} + a_{p-1} y_{t-p+1} + a_p y_{t-p} + \varepsilon_t$$

can be written as

$$\Delta y_t = \gamma y_{t-1} + \beta_2 \Delta y_{t-1} + \beta_3 \Delta y_{t-2} + \dots + \beta_p \Delta y_{t-p+1} + \varepsilon_t \qquad (4.31)$$

From Rule 1, all the coefficients on the expressions Δy_{t-i} converge to t-distributions. As such, groups of these coefficients will converge to an F-distribution. Hence, you can perform a test of the form $\beta_i = \beta_{i+1} = \dots = \beta_p = 0$ using an F-test. Nevertheless, under the null hypothesis of a unit root, the value of γ multiplies a nonstationary variable. Thus, a test of $\gamma = 0$ *cannot* be conducted using a standard t-test.

In addition to the use of F-tests and t-tests, it is also possible to determine the lag length using an information criterion such as the AIC or the SBC. Of course, in very large samples with normally distributed errors, the methods should all select the same lag length. In practice, the SBC will select a more parsimonious model than will either the AIC or t-tests. Nevertheless, whichever method is used, the researcher must ensure that residuals act as white-noise processes.

An Example: In order to illustrate the various procedures to select the lag length for an augmented Dickey–Fuller test, 200 realizations of the following unit root process were generated:

$$\Delta y_t = 0.5 + 0.5 \Delta y_{t-1} + 0.2 \Delta y_{t-3} + \varepsilon_t$$

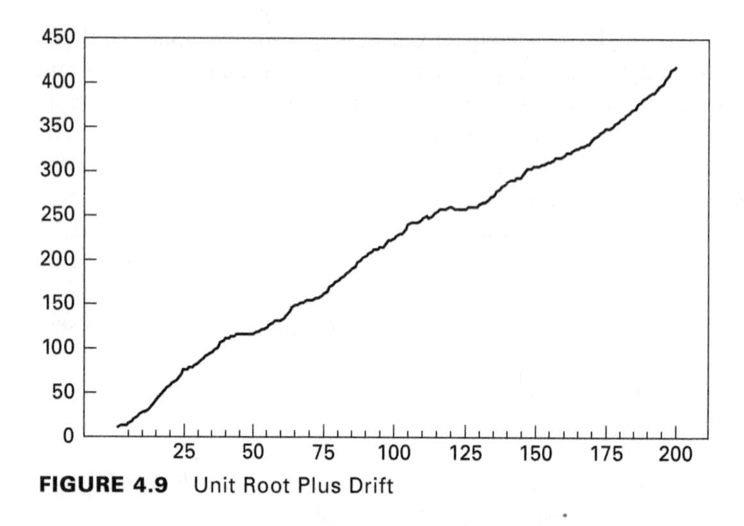

FIGURE 4.9 Unit Root Plus Drift

Notice that the $\{y_t\}$ sequence contains a single unit root and that the appropriate lag length is 3. The drift parameter gives the series the decidedly increasing pattern shown in Figure 4.9. (You can follow along using the data on the file LAGLENGTH.XLS.) Pretend that you do not know the actual DGP. The time path of the sequence allows for two possible DGPs; the series may be trend-stationary or a unit root process containing a drift term. Hence, the null hypothesis is that of a unit root process containing a drift against the alternative of a trend-stationary process. The appropriate way to proceed is to estimate the series under the alternative hypothesis; hence, we estimate a regression equation of the form

$$\Delta y_t = a_0 + \gamma y_{t-1} + a_2 t + \sum_{i=1}^{p} \beta_i \Delta y_{t-i} + \varepsilon_t$$

If it is possible to reject the null hypothesis $\gamma = 0$, the process is trend-stationary. The problem is to determine the appropriate value for p. Toward this end, the equation was estimated for lag lengths of 1 through 4. As shown in Table 4.5, the AIC selects a lag length of 3 and the SBC selects a lag length of 1. Nevertheless, in this instance, the lag length seems not to make a difference; at the 5 percent significance level, the critical value for the null hypothesis $\gamma = 0$ is –3.43. As such, the lag lengths selected by the AIC and the SBC are such that the null hypothesis of a unit root is not rejected. We can conclude that the sequence is not trend-stationary.

Table 4.5 Dickey–Fuller Tests and Lag Length

p	AIC	SBC	γ	*t*-statistic	ϕ_2	ϕ_3
1	1076.211	1089.303	−0.017	−1.776	17.390	1.579
2	1073.076	1089.441	−0.020	−2.049	11.188	2.101
3	1071.817	1091.455	−0.022	−2.285	8.622	2.616
4	1073.799	1096.710	−0.022	−2.276	8.026	2.595

The ϕ_3 allows us to test the null hypothesis $\gamma = a_2 = 0$; at the 5 percent significance level, the critical value is 6.49. As such, for any lag length, we would not reject the null hypothesis and conclude that the sequence has a stochastic trend. However, at the 5 percent significance level, the critical value for the null hypothesis $a_0 = \gamma = a_2 = 0$ (i.e., the critical value of the ϕ_2 statistic) is 4.88. For the lag lengths selected by the AIC and the SBC, this null hypothesis is clearly rejected. The test statistics reflect the fact that the data-generating process does contain the drift term a_0.

It is also possible to use t-tests and F-tests to determine the lag length. Estimating the equation using the lag length $p = 4$ yields

$$\begin{aligned}\Delta y_t = {} & \underset{(4.05)}{1.24} + \underset{(2.28)}{0.042t} - \underset{(-2.28)}{0.022y_{t-1}} + \underset{(5.57)}{0.397\Delta y_{t-1}} + \underset{(1.42)}{0.108\Delta y_{t-2}} \\ & + \underset{(1.64)}{0.125\Delta y_{t-3}} + \underset{(0.13)}{0.009\Delta y_{t-4}} + \varepsilon_t\end{aligned}$$

A t-test for the coefficient on Δy_{t-4} suggests a lag length no greater than 3. Moreover, the F-statistic for the null hypothesis $\beta_3 = \beta_4 = 0$ is 1.59 with a *prob*-value of 0.206. Thus, we can eliminate lags 3 and 4. Moreover, the F-statistic for the null hypothesis $\beta_2 = \beta_3 = \beta_4 = 0$ is 2.76 with a *prob*-value of 0.043. Hence, if we use a 5 percent significance level, the F-tests select a model with two lags. In this instance, the results are not very sensitive to the alternative lag lengths.

The Test with MA Components

Since an invertible MA model can be transformed into an autoregressive model, the procedure can be generalized to allow for moving-average components. Let the $\{y_t\}$ sequence be generated from the mixed autoregressive/moving-average process

$$A(L)y_t = C(L)\varepsilon_t$$

where $A(L)$ and $C(L)$ are polynomials of orders p and q, respectively.

If the roots of $C(L)$ are outside the unit circle, we can write the $\{y_t\}$ sequence as the autoregressive process

$$A(L)y_t/C(L) = \varepsilon_t$$

or, defining $D(L) = A(L)/C(L)$, we can write the process as

$$D(L)y_t = \varepsilon_t$$

Even though $D(L)$ will generally be an infinite-order polynomial, in principle we can use the same technique as used to obtain (4.30) to form the infinite-order autoregressive model

$$\Delta y_t = \gamma y_{t-1} + \sum_{i=2}^{\infty} \beta_i \Delta y_{t-i+1} + \varepsilon_t$$

As it stands, this is an infinite-order autoregression that cannot be estimated using a finite data set. Fortunately, Said and Dickey (1984) have shown that an unknown ARIMA(p, 1, q) process can often be well approximated by an ARIMA(n, 1, 0) autoregression of order n where $n \leq T^{1/3}$. Thus, we can usually solve the problem by using a finite-order approximation of the infinite order autoregression. The test for $\gamma = 0$ can be conducted using the aforementioned Dickey–Fuller τ, τ_μ, or τ_τ test statistics.

LAG LENGTHS AND NEGATIVE MA TERMS Unit root tests generally work poorly if the error process has a strongly negative MA component. While Said and Dickey's (1984) result that an ARIMA(p, 1, q) process can be well approximated by an ARIMA(n, 1, 0) process, the interaction between the unit root and the negative MA component can lead to over rejections of a unit root. To explain the nature of the problem, consider the ARIMA(0, 1, 1) process:

$$y_t = y_{t-1} + \varepsilon_t - \beta_1 \varepsilon_{t-1} \qquad 0 < \beta_1 < 1$$

If we have the initial condition y_0, we can write the general solution for y_t as

$$y_t = y_0 + \varepsilon_t + (1 - \beta_1) \sum_{i=1}^{t-1} \varepsilon_i$$

Clearly, the $\{y_t\}$ sequence is not stationary since the effects of an ε_t shock never decay to zero. However, unlike a random walk process for which $\beta_1 = 0$, the presence of the negative MA term means that ε_t has a one-unit effect on y_t in period t only. Since $\partial y_{t+i}/\partial \varepsilon_t = (1 - \beta_1) < 1$, the magnitude of the effect for all subsequent periods is diminished when compared to that of a pure random walk. For a finite sample with t observations, we can construct the autocovariances as

$$\begin{aligned}
\gamma_0 &= E[(y_t - y_0)^2] = \sigma^2 + (1 - \beta_1)^2 E[(\varepsilon_{t-1})^2 + (\varepsilon_{t-2})^2 + \ldots + (\varepsilon_1)^2] \\
&= [1 + (1 - \beta_1)^2 (t - 1)]\sigma^2 \\
\gamma_s &= E[(y_t - y_0)(y_{t-s} - y_0)] \\
&= E[(\varepsilon_t + (1 - \beta_1)\varepsilon_{t-1} + \ldots + (1 - \beta_1)\varepsilon_1)(\varepsilon_{t-s} + \\
&\qquad (1 - \beta_1)\varepsilon_{t-s-1} + \ldots + (1 - \beta_1)\varepsilon_1) \\
&= (1 - \beta_1)[1 + (1 - \beta_1)(t - s - 1)]\sigma^2
\end{aligned}$$

The autocorrelations are formed from $\rho_s = \gamma_s/(\gamma_s\gamma_0)^{0.5}$. It is easy to verify that all of the autocorrelations ρ_i approach unity as the sample size t becomes infinitely large. However, for the sample sizes usually found in applied work, the autocorrelations can be small. To see the point, let β_1 be close to unity so that terms containing $(1 - \beta_1)^2$ can be safely ignored. In such circumstances, the ACF can be approximated by $\rho_1 = \rho_2 = \ldots = (1 - \beta_1)^{0.5}$. For example, if $\beta_1 = 0.95$, all of the autocorrelations should be close to 0.22. As such, the autocorrelations will be small, appear to be marginally significant, and show little tendency to decay.

From the example, it should not be surprising than that unit root tests do not work well in the presence of a strongly negative MA component. Since many of the autocorrelations are small, the ACF will resemble that of a truly stationary process. In fact, if β_1 is very close to unity, there is a common factor such that $y_t = y_{t-1} + \varepsilon_t - \beta_1\varepsilon_{t-1}$ approximates the white-noise process $y_t = \varepsilon_t$. Any test will have a difficult time distinguishing between the two types of processes and will overreject the null hypothesis of a unit root. Moreover, in conducting the test, it is necessary to use a large number of lags. We can use lag operators to write $\Delta y_t = (1 - \beta_1 L)\varepsilon_t$, so that $\Delta y_t = \beta_1 \Delta y_{t-1} + (\beta_1)^2 \Delta y_{t-2} + (\beta_1)^3 \Delta y_{t-3} + \ldots + \varepsilon_t$. When β_1 is large, many autoregressive lags are needed to properly capture the dynamics of the process. The need to estimate a large number of coefficients can diminish the power of the test.

Nevertheless, there are some precautions to take when testing for a unit root in the presence of a negative MA component. Clearly, you want to use a methodology that properly captures the need to use a large number of lags. Ng and Perron (2001) showed that a modified version of the AIC (MAIC) yields a better estimate of the lag length than either the AIC or the BIC. Consider

$$\text{MAIC} = T\ln(\text{sum of squared residuals}) + 2n + 2\tau(n)$$

where $\tau(n) = \hat{\rho}^2 \Sigma y_{t-1}^2 / \hat{\sigma}^2$, $\hat{\rho}$ is the estimated value of ρ, and $\hat{\sigma}^2$ is the estimated variance.

Notice that the MAIC is equal to the usual expression for the AIC plus an additional penalty term $2\tau(n)$. Given that all models are estimated over the same sample period, the value of Σy_{t-1}^2 is the same for all models. As such, $\tau(n)$ will generally be small for models with a small value of ρ^2 relative to the variance σ^2. Hence, the MAIC will tend to select the lag length resulting in a value of ρ closest to that of a unit root.

Multiple Roots

Dickey and Pantula (1987) suggested a simple extension of the basic procedure if more than one unit root is suspected. In essence, the methodology entails nothing more than performing Dickey–Fuller tests on successive differences of $\{y_t\}$. When exactly one root is suspected, the Dickey–Fuller procedure is to estimate an equation such as $\Delta y_t = a_0 + \gamma y_{t-1} + \varepsilon_t$. In contrast, if two roots are suspected, estimate the equation

$$\Delta^2 y_t = a_0 + \beta_1 \Delta y_{t-1} + \varepsilon_t \tag{4.32}$$

Use the appropriate statistic (i.e., τ, τ_μ, or τ_τ, depending on the deterministic elements actually included in the regression) to determine whether β_1 is significantly different from zero. If you cannot reject the null hypothesis that $\beta_1 = 0$, conclude that the $\{y_t\}$ sequence is $I(2)$. If β_1 does differ from zero, go on to determine whether there is a single unit root by estimating

$$\Delta^2 y_t = a_0 + \beta_1 \Delta y_{t-1} + \beta_2 y_{t-1} + \varepsilon_t \tag{4.33}$$

Since there are not two unit roots, you should find that β_1 and/or β_2 differ from zero. Under the null hypothesis of a single unit root, $\beta_1 < 0$ and $\beta_2 = 0$; under the alternative hypothesis, $\{y_t\}$ is stationary so that β_1 and β_2 are both negative. Thus, estimate (4.33) and use the Dickey–Fuller critical values to test the null hypothesis $\beta_2 = 0$. If you reject this null hypothesis, you can conclude that $\{y_t\}$ is stationary.

As a rule of thumb, economic series do not need to be differenced more than two times. However, in the odd case in which *at most r* unit roots are suspected, the procedure is to first estimate

$$\Delta^r y_t = a_0 + \beta_1 \Delta^{r-1} y_{t-1} + \varepsilon_t$$

If $\Delta^r y_t$ is stationary, you should find that $-2 < \beta_1 < 0$. If the Dickey–Fuller critical values for β_1 are such that it is not possible to reject the null of a unit root, you can accept the hypothesis that $\{y_t\}$ contains r unit roots. If we reject this null of exactly r unit roots, the next step is to test for $r - 1$ roots by estimating

$$\Delta^r y_t = a_0 + \beta_1 \Delta^{r-1} y_{t-1} + \beta_2 \Delta^{r-2} y_{t-1} + \varepsilon_t$$

If both β_1 and β_2 differ from zero, reject the null hypothesis of $r - 1$ roots. You can use the Dickey–Fuller statistics to test the null of exactly $r - 1$ unit roots if the t-statistics for β_1 *and* β_2 are both statistically different from zero. If you can reject this null, the next step is to form

$$\Delta^r y_t = a_0 + \beta_1 \Delta^{r-1} y_{t-1} + \beta_2 \Delta^{r-2} y_{t-1} + \beta_3 \Delta^{r-3} y_{t-1} + \varepsilon_t$$

As long as it is possible to reject the null hypothesis that the various values of the β_i are nonzero, continue toward the equation

$$\Delta^r y_t = a_0 + \beta_1 \Delta^{r-1} y_{t-1} + \beta_2 \Delta^{r-2} y_{t-1} + \beta_3 \Delta^{r-3} y_{t-1} + \ldots + \beta_r y_{t-1} + \varepsilon_t$$

Continue in this fashion until it is not possible to reject the null of a unit root or until the $\{y_t\}$ series is shown to be stationary. Notice that this procedure is quite different from the sequential testing for successively greater numbers of unit roots. It might seem tempting to test for a single unit root and, if the null cannot be rejected, go on to test for the presence of a second root. In repeated samples, this method tends to select too few roots.

Seasonal Unit Roots

You will recall that the best-fitting model for U.S. money supply data used in Chapter 2 had the form

$$(1 - L^4)(1 - L)(1 - a_1 L)y_t = (1 + \beta_4 L^4)\varepsilon_t$$

The specification implies that the money supply has a unit root and a seasonal unit root. Since seasonality is a key feature of many economic series, a sizable body of literature has been developed to test for seasonal unit roots. Before proceeding, note that the first difference of a seasonal unit root process will not be stationary. To keep matters simple, suppose that the quarterly observations of $\{y_t\}$ are generated by

$$y_t = y_{t-4} + \varepsilon_t$$

Here the seasonal difference of $\{y_t\}$ is stationary; using the notation of Chapter 2, we can write $\Delta_4 y_t = \varepsilon_t$. Given the initial condition $y_0 = y_{-1} = \ldots = 0$, the solution for y_t is

$$y_t = \varepsilon_t + \varepsilon_{t-4} + \varepsilon_{t-8} + \ldots$$

so that

$$y_t - y_{t-1} = \sum_{i=0}^{t/4} \varepsilon_{4i} - \sum_{i=0}^{t/4} \varepsilon_{4i-1}$$

Hence, Δy_t equals the difference between two stochastic trends. Since each shock has a permanent effect on the level of Δy_t, the sequence is not mean-reverting. However, the seasonal difference of a unit root process may be stationary. For example, if $\{y_t\}$ is generated by $y_t = y_{t-1} + \varepsilon_t$, the fourth difference (i.e., $\Delta_4 y_t = \varepsilon_t + \varepsilon_{t-1} + \varepsilon_{t-2} + \varepsilon_{t-3}$) is stationary. The point is that the Dickey–Fuller procedure must be modified in order to test for seasonal unit roots and to distinguish between seasonal and nonseasonal roots.

There are several alternative ways to treat seasonality in a nonstationary sequence. The most direct method occurs when the seasonal pattern is purely deterministic. For example, let D_1, D_2, and D_3 represent quarterly seasonal dummy variables such that the value of D_i is unity in season i and zero otherwise. Estimate the regression equation

$$\Delta y_t = a_0 + \alpha_1 D_1 + \alpha_2 D_2 + \alpha_3 D_3 + \gamma y_{t-1} + \sum_{i=2}^{p} \beta_i \Delta y_{t-i+1} + \varepsilon_t \qquad (4.34)$$

The null hypothesis of a unit root (i.e., $\gamma = 0$) can be tested using the Dickey–Fuller τ_μ statistic. (Note that you use the τ_μ statistic since the original data contain an intercept.) Rejecting the null hypothesis is equivalent to accepting the alternative that the $\{y_t\}$ sequence is stationary. The test is possible as Dickey, Bell, and Miller (1986) showed that the limiting distribution for γ is not affected by the removal of the deterministic seasonal components. If you want to include a time trend in (4.34), use the τ_τ statistic.

Notice that the specification in (4.34) makes it difficult to test a hypothesis concerning a_0. Since the mean of each D_i series is 1/4, the presence of the seasonal dummies affects the magnitude of the drift term a_0. To correct for this, it is common to use **centered** seasonal dummy variables. Simply let $D_i = 0.75$ in season i and –0.25 in each of the other three quarters of the year. Hence, the mean of $D_i = 0$ so that the magnitude of a_0 is unchanged.

If you suspect a seasonal unit root, it is necessary to use an alternative procedure. To keep the notation simple, suppose you have quarterly observations on the $\{y_t\}$ sequence and want to test for the presence of a seasonal unit root. To explain the methodology, note that the polynomial $(1 - \gamma L^4)$ can be factored such that there are four distinct characteristic roots:

$$(1 - \gamma L^4) = (1 - \gamma^{1/4} L)(1 + \gamma^{1/4} L)(1 - i\gamma^{1/4} L)(1 + i\gamma^{1/4} L) \qquad (4.35)$$

If y_t has a seasonal unit root, $\gamma = 1$. Equation (4.35) is a bit restrictive in that it only allows for a unit root at an annual frequency. Hylleberg et al. (1990) developed a clever technique that allows you to test for unit roots at various frequencies; you can test for a unit root (i.e., a root at a zero frequency), a unit root at a semiannual frequency, or a seasonal unit root. To understand the HEGY test (named after the four authors of the paper), suppose y_t is generated by

$$A(L) y_t = \varepsilon_t$$

where $A(L)$ is a fourth-order polynomial such that

$$(1 - a_1 L)(1 + a_2 L)(1 - a_3 i L)(1 + a_4 i L) y_t = \varepsilon_t \qquad (4.36)$$

Now, if $a_1 = a_2 = a_3 = a_4 = 1$, (4.36) is equivalent to setting $\gamma = 1$ in (4.35). Hence, if $a_1 = a_2 = a_3 = a_4 = 1$, there is a seasonal unit root. Consider some of the other possible cases:

CASE 1

If $a_1 = 1$, one homogeneous solution to (4.36) is $y_t = y_{t-1}$. As such, the $\{y_t\}$ sequence will act as a random walk in that it tends to repeat itself each and every period. This is the case of a nonseasonal unit root; the appropriate period of differencing is Δy_t.

CASE 2

If $a_2 = 1$, one homogeneous solution to (4.36) is $y_t + y_{t-1} = 0$. In this instance, the sequence tends to replicate itself at six-month intervals so that there is a semiannual unit root. For example, if $y_t = 1$, it follows that $y_{t+1} = -1, y_{t+2} = +1, y_{t+3} = -1, y_{t+4} = 1$, and so forth.

CASE 3

If either a_3 or a_4 is equal to unity, the $\{y_t\}$ sequence has an annual cycle. For example, if $a_3 = 1$, a homogeneous solution to (4.36) is $y_t = iy_{t-1}$. Thus, if $y_t = 1$, it follows that $y_{t+1} = i, y_{t+2} = i^2 = -1, y_{t+3} = -i$, and $y_{t+4} = -i^2 = 1$; hence, the sequence replicates itself every fourth period. The appropriate degree of differencing is $\Delta_4 y_t = (1 - L^4)y_t$.

To develop the test, view (4.36) as a function of a_1, a_2, a_3, and a_4 and take a Taylor series approximation of $A(L)$ around the point $a_1 = a_2 = a_3 = a_4 = 1$. Although the details of the expansion are messy, first take the partial derivative:

$$\begin{aligned} \partial A(L)/\partial a_1 &= \partial(1 - a_1L)(1 + a_2L)(1 - a_3iL)(1 + a_4iL)/\partial a_1 \\ &= -(1 + a_2L)(1 - a_3iL)(1 + a_4iL)L \end{aligned}$$

Evaluating this derivative at the point $a_1 = a_2 = a_3 = a_4 = 1$ yields

$$-L(1 + L)(1 - iL)(1 + iL) = -L(1 + L)(1 + L^2) = -L(1 + L + L^2 + L^3)$$

Next, form

$$\begin{aligned} \partial A(L)/\partial a_2 &= \partial(1 - a_1L)(1 + a_2L)(1 - a_3iL)(1 + a_4iL)/\partial a_2 \\ &= (1 - a_1L)(1 - a_3iL)(1 + a_4iL)L \end{aligned}$$

Evaluating at the point $a_1 = a_2 = a_3 = a_4 = 1$ yields $(1 - L + L^2 - L^3)L$. It should not take too long to convince yourself that evaluating $\partial A(L)/\partial a_3$ and $\partial A(L)/\partial a_4$ at the point $a_1 = a_2 = a_3 = a_4 = 1$ yields

$$\partial A(L)/\partial a_3 = -(1 - L^2)(1 + iL)iL$$

and

$$\partial A(L)/\partial a_4 = (1 - L^2)(1 - iL)iL$$

Since $A(L)$ evaluated at $a_1 = a_2 = a_3 = a_4 = 1$ is $(1 - L^4)$, it is possible to approximate (4.36) by

$$\begin{aligned}[(1 - L^4) - L(1 + L + L^2 + L^3)(a_1 - 1) + (1 - L + L^2 - L^3)L(a_2 - 1)\\ - (1 - L^2)(1 + iL)iL(a_3 - 1) + (1 - L^2)(1 - iL)iL(a_4 - 1)]y_t = \varepsilon_t\end{aligned}$$

Define γ_i such that $\gamma_i = (a_i - 1)$ and note that $(1 + iL)i = i - L$ and $(1 - iL)i = i + L$; hence,

$$\begin{aligned}(1 - L^4)y_t = {}& \gamma_1(1 + L + L^2 + L^3)y_{t-1} - \gamma_2(1 - L + L^2 - L^3)y_{t-1}\\ & + (1 - L^2)[\gamma_3(i - L) - \gamma_4(i + L)]y_{t-1} + \varepsilon_t\end{aligned}$$

so that

$$\begin{aligned}(1 - L^4)y_t = {}& \gamma_1(1 + L + L^2 + L^3)y_{t-1} - \gamma_2(1 - L + L^2 - L^3)y_{t-1}\\ & + (1 - L^2)[(\gamma_3 - \gamma_4)i - (\gamma_3 + \gamma_4)L]y_{t-1} + \varepsilon_t\end{aligned} \tag{4.37}$$

To purge the imaginary numbers from this expression, define γ_5 and γ_6 such that $2\gamma_3 = -\gamma_6 - i\gamma_5$ and $2\gamma_4 = -\gamma_6 + i\gamma_5$. Hence, $(\gamma_3 - \gamma_4)i = \gamma_5$ and $\gamma_3 + \gamma_4 = \gamma_6$. Substituting into (4.37) yields

$$\begin{aligned}(1 - L^4)y_t = {}& \gamma_1(1 + L + L^2 + L^3)y_{t-1} - \gamma_2(1 - L + L^2 - L^3)y_{t-1}\\ & + (1 - L^2)(\gamma_5 - \gamma_6 L)y_{t-1} + \varepsilon_t\end{aligned}$$

Fortunately, many software packages can perform the test directly on quarterly and monthly data. However, to understand the mechanics necessary to implement the procedure, use the following steps:

STEP 1: For quarterly data, form the following variables:

$$\begin{aligned}y_{1t-1} &= (1 + L + L^2 + L^3)y_{t-1} = y_{t-1} + y_{t-2} + y_{t-3} + y_{t-4}\\ y_{2t-1} &= (1 - L + L^2 - L^3)y_{t-1} = y_{t-1} - y_{t-2} + y_{t-3} - y_{t-4}\\ y_{3t-1} &= (1 - L^2)y_{t-1} = y_{t-1} - y_{t-3} \qquad \text{so that } y_{3t-2} = y_{t-2} - y_{t-4}\end{aligned}$$

STEP 2: Estimate the regression

$$(1 - L^4)y_t = \gamma_1 y_{1t-1} - \gamma_2 y_{2t-1} + \gamma_5 y_{3t-1} - \gamma_6 y_{3t-2} + \varepsilon_t$$

You might want to modify the form of the equation by including an intercept, deterministic seasonal dummies, and a linear time trend. As in the augmented form of the Dickey–Fuller test, lagged values of $(1 - L^4)y_{t-i}$ may also be included. Perform the appropriate diagnostic checks to ensure that the residuals from the regression equation approximate a white-noise process.

STEP 3: Form the t-statistic for the null hypothesis $\gamma_1 = 0$; some of the appropriate critical values are reported below. If you do not reject the hypothesis $\gamma_1 = 0$, conclude that $a_1 = 1$ so that there is a nonseasonal unit root. Next form the t-test for the hypothesis $\gamma_2 = 0$. If you do not reject the null hypothesis, conclude that $a_2 = 1$ and that there is a unit root with a semiannual frequency.

Finally, perform the F-test for the hypothesis $\gamma_5 = \gamma_6 = 0$. If the calculated value is less than the critical value reported in Hylleberg et al. (1990), you can conclude that γ_5 and/or γ_6 is zero so that there is a seasonal unit root. Be aware that the three null hypotheses are not alternatives; a series may have nonseasonal, semiannual, and seasonal unit roots.

At the 5 percent significance level, Hylleberg et al. (1990) reported that the critical values for 100 and 200 observations are:

	$T = 100$			$T = 200$		
	$\gamma_1 = 0$	$\gamma_2 = 0$	$\gamma_5 = \gamma_6 = 0$	$\gamma_1 = 0$	$\gamma_2 = 0$	$\gamma_5 = \gamma_6 = 0$
Intercept	−2.88	−1.95	3.08	−2.87	−1.92	3.12
Intercept + Trend	−3.47	−1.95	2.96	−3.44	−1.95	3.07
Intercept + Seasonal Dummies	−2.95	−2.94	6.57	−2.91	−2.89	6.62
Intercept + Seasonal Dummies + Trend	−3.53	−2.94	6.60	−3.49	−2.91	6.57

An Example: In Chapter 2, we took the nonseasonal and the seasonal differences of the U.S. money supply and estimated a model of the form

$$m_t = a_0 + a_1 m_{t-1} + \varepsilon_t + \beta_4 \varepsilon_{t-4}$$

where

$$m_t = (1 - L)(1 - L^4)y_t$$

and y_t is the logarithm of U.S. money supply as measured by M1.

We can use the HEGY test to determine if it is appropriate to use the seasonal and nonseasonal differences. Since it is clear that the money supply series has a sustained upward movement (see Section 11 in Chapter 2), we want to allow for the possibility that the series is trend-stationary. Hence, we include a deterministic trend and an intercept in the regression. Consider the following regression:

$$\underset{}{(1 - L^4)y_t} = \underset{(0.14)}{0.004} - \underset{(-0.16)}{1.45{*}10^{-5}t} - \underset{(-0.10)}{1.52{*}10^{-4}y_{1t-1}} - \underset{(-3.78)}{0.618y_{2t-1}} - \underset{(-3.01)}{0.269y_{3t-1}}$$

$$- \underset{(-1.26)}{0.115y_{3t-2}} + \sum_{i=1}^{3} \alpha_i Di + \sum_{i=1}^{7} \beta_i (1 - L^4) y_{t-i}$$

where the lag length of 7 was chosen by the general-to-specific method beginning with a lag length of 12, the D_i are seasonal dummies, and y_{1t-1}, y_{2t-1}, y_{3t-1}, and y_{3t-2} are defined above.

The coefficient on y_{1t-1} has a t-statistic of −0.10. Given the 5 percent critical value, we cannot reject the null hypothesis of a nonseasonal unit root. The t-statistic for the coefficient on y_{2t-1} is −3.78 so that it is unlikely that there is a seasonal unit root at a semiannual frequency. The sample F-statistic for the null hypothesis that the coefficients on y_{3t-1} and y_{3t-2} jointly equal zero is 5.35. Hence, at the 5 percent significance level, there is a seasonal unit root at the annual frequency ($5.35 < 6.57$). Thus, as in Chapter 2, it was correct to difference and seasonally difference the money supply series. As a group, the seasonal dummies are highly significant; the sample F-statistic for the presence of the seasonal dummies is 5.60. Nevertheless, if you experiment with the model in the form $m_t = (1 - L)(1 - L^4)y_t$ used in Chapter 2, you should find the AR(1) and MA(4) terms forecast better than a model with seasonal dummy variables.

8. STRUCTURAL CHANGE

In performing unit root tests, special care must be taken if it is suspected that structural change has occurred. When there are structural breaks, the various Dickey–Fuller test statistics are biased toward the nonrejection of a unit root. To explain, consider the situation in which there is a one-time change in the mean of an otherwise stationary sequence. In the top graph of Figure 4.10, the $\{y_t\}$ sequence was constructed so as to be stationary around a mean of zero for $t = 0, \ldots, 50$ and then to fluctuate around a mean of 6 for $t = 51, \ldots, 100$. The sequence was formed by drawing 100 normally and independently distributed values for the $\{\varepsilon_t\}$ sequence. Setting $y_0 = 0$, the next 100 values in the sequence were generated using the formula

$$y_t = 0.5y_{t-1} + \varepsilon_t + D_L \tag{4.38}$$

where D_L is a dummy variable such that $D_L = 0$ for $t = 1, \ldots, 50$ and $D_L = 3$ for $t = 51, \ldots, 100$. The subscript L is designed to indicate that the *level* of the dummy changes. At times it will be convenient to refer to the value of the dummy variable in period t as $D_L(t)$; in the example at hand, $D_L(50) = 0$ and $D_L(51) = 3$.

In practice, the structural change may not be as apparent as the break shown in the figure. However, the large simulated break is useful for illustrating the problem of using a Dickey–Fuller test in such circumstances. The straight line shown in the figure highlights the fact that the series appears to have a deterministic trend. In fact, the straight line is the best-fitting OLS equation

$$y_t = a_0 + a_2 t + e_t$$

In the figure, you can see that the fitted value of a_0 is negative and the fitted value of a_2 is positive. The proper way to estimate (4.38) is to fit a simple AR(1) model and allow the intercept to change by including the dummy variable D_L. However, suppose that we unsuspectingly fit the regression equation

$$y_t = a_0 + a_1 y_{t-1} + e_t \tag{4.39}$$

As you can infer from Figure 4.10, the estimated value of a_1 is necessarily biased toward unity. The reason for this upward bias is that the estimated value of a_1 captures the property that "low" values of y_t (i.e., those fluctuating around zero) are followed by other "low" values, and "high" values (i.e., those fluctuating around a mean of six) are followed by other "high" values. For a formal demonstration, note that as a_1 approaches unity, (4.39) approaches a random walk plus drift. We know that the solution to the random walk plus drift model includes a deterministic trend; that is,

$$y_t = y_0 + a_0 t + \sum_{i=1}^{t} \varepsilon_i$$

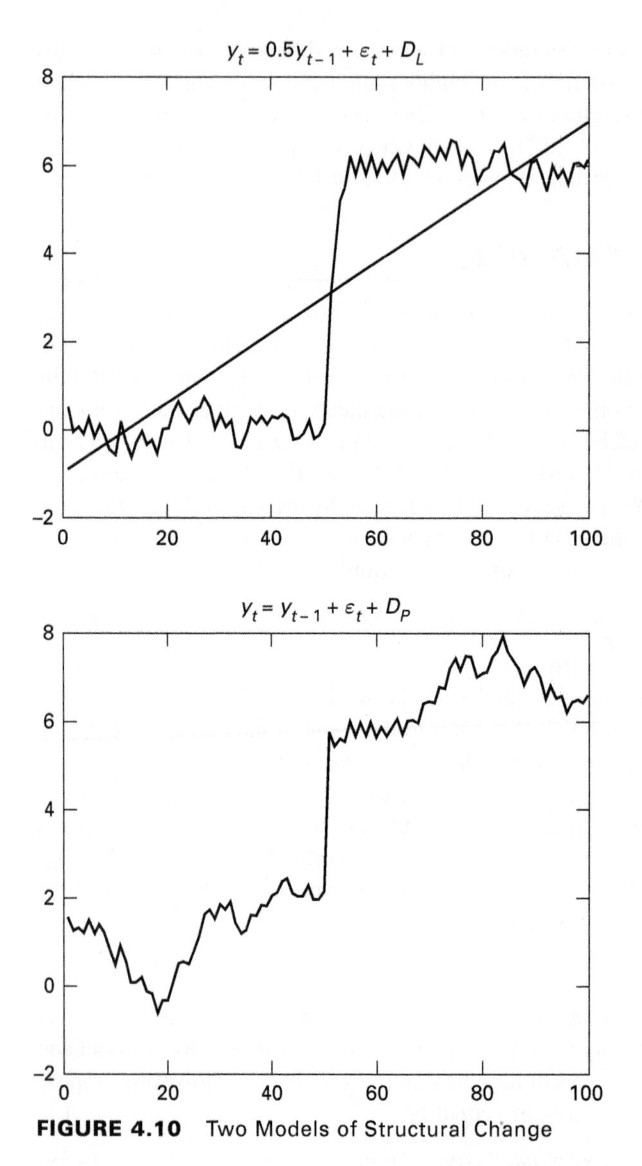

FIGURE 4.10 Two Models of Structural Change

Thus, the misspecified equation (4.39) will tend to mimic the trend line shown in Figure 4.10 by biasing a_1 toward unity. This bias in a_1 means that the Dickey–Fuller test is biased toward accepting the null hypothesis of a unit root *even though the series is stationary within each of the subperiods.*

Of course, a unit root process can also exhibit a structural break. The lower portion of Figure 4.10 simulates a random walk process with a structural change occurring at $t = 51$. This second simulation used the same 100 realizations for the $\{\varepsilon_t\}$ sequence and the initial condition $y_0 = 2$. The 100 realizations of the $\{y_t\}$ sequence were constructed as

$$y_t = y_{t-1} + \varepsilon_t + D_P$$

where $D_P(51) = 4$ and all other values of $D_P = 0$.

Here, the subscript P refers to the fact that there is a single *pulse* in the dummy variable. In a unit root process, a single pulse in the dummy will have a permanent effect on the level of the $\{y_t\}$ sequence. In $t = 51$, the pulse in the dummy is equivalent to an ε_{t+51} shock of four extra units. Hence, the *one-time* shock to $D_P(51)$ has a *permanent* effect on the mean value of the sequence for $t \geq 51$. In the figure, you can see that the level of the process takes a discrete jump in $t = 51$, never exhibiting any tendency to return to the prebreak level.

This bias in the Dickey–Fuller tests was confirmed in a Monte Carlo experiment. Perron (1989) generated 10,000 replications of a process like that of (4.38). Each replication was formed by drawing 100 normally and independently distributed values for the $\{\varepsilon_t\}$ sequence. For each of the 10,000 replicated series, he used OLS to estimate a regression in the form of (4.39).[14] As could be anticipated from our earlier discussion, Perron found that the estimated values of a_1 were biased toward unity. Moreover, the bias became more pronounced as the magnitude of the break increased.

Perron's Test for Structural Change

Returning to the two graphs of Figure 4.10, there may be instances in which the unaided eye cannot easily detect the difference between the alternative types of sequences. One econometric procedure to test for unit roots in the presence of a structural break involves splitting the sample into two parts and using Dickey–Fuller tests on each part. The problem with this procedure is that the degrees of freedom for each of the resulting regressions are diminished. Moreover, you may not know when the break point actually occurs. It is preferable to have a single test based on the full sample.

Perron (1989) went on to develop a formal procedure to test for unit roots in the presence of a structural change at time period $t = \tau + 1$. Consider the null hypothesis of a one-time jump in the level of a unit root process against the alternative of a one-time change in the intercept of a trend-stationary process. Formally, let the null and alternative hypotheses be

$$H_1\colon y_t = a_0 + y_{t-1} + \mu_1 D_P + \varepsilon_t \tag{4.40}$$

$$A_1\colon y_t = a_0 + a_2 t + \mu_2 D_L + \varepsilon_t \tag{4.41}$$

where D_P represents a *pulse* dummy variable such that $D_P = 1$ if $t = \tau + 1$ and zero otherwise, and D_L represents a *level* dummy variable such that $D_L = 1$ if $t > \tau$ and zero otherwise.

Under the null hypothesis, $\{y_t\}$ is a unit root process with a one-time jump in the level of the sequence in period $t = \tau + 1$. Under the alternative hypothesis, $\{y_t\}$ is trend-stationary with a one-time jump in the intercept. Figure 4.11 can help you to visualize the two hypotheses. Simulating (4.40) by setting $a_0 = 1$ and using 100 realizations for the $\{\varepsilon_t\}$ sequence, the erratic line in the figure depicts the time path of $\{y_t\}$ under the null hypothesis. You can see the one-time jump in the level of the process occurring in period 51. Thereafter, the $\{y_t\}$ sequence continues the original random walk plus drift process. The alternative hypothesis posits that the $\{y_t\}$ sequence is stationary around the broken trend line. Up to $t = \tau$, $\{y_t\}$ is stationary around $a_0 + a_2 t$,

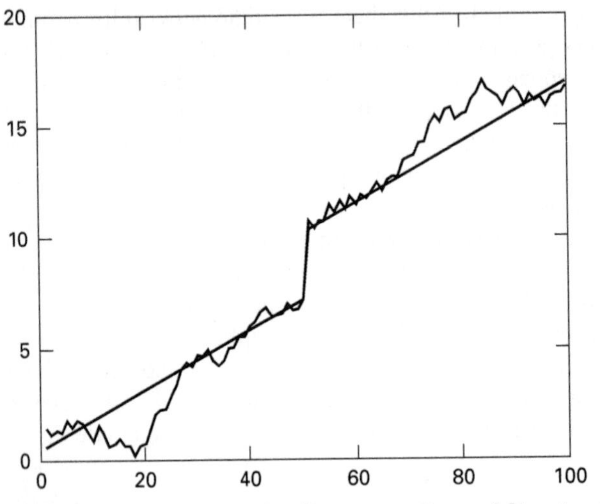

FIGURE 4.11 Alternative Representations of Structural Change

and beginning at $\tau + 1$, y_t is stationary around $a_0 + a_2t + \mu_2$. As illustrated by the broken line, there is a one-time increase in the intercept of the trend if $\mu_2 > 0$.

The econometric problem is to determine whether an observed series is best modeled by (4.40) or (4.41). The implementation of Perron's (1989) technique is straightforward:

STEP 1: Detrend the data by estimating the model under the alternative hypothesis and calling the residuals $\hat{y}_t$. Hence, each value of $\hat{y}_t$ is the residual from the regression $y_t = a_0 + a_2t + \mu_2 D_L + \hat{y}_t$.

STEP 2: Estimate the regression:

$$\hat{y}_t = a_1\hat{y}_{t-1} + \varepsilon_t$$

Under the null hypothesis of a unit root, the theoretical value of a_1 is unity. Perron (1989) showed that when the residuals are identically and independently distributed, the distribution of a_1 depends on the proportion of observations occurring prior to the break. Denote this proportion by $\lambda = \tau/T$ where T = total number of observations.

STEP 3: Perform diagnostic checks to determine if the residuals from Step 2 are serially uncorrelated. If there is serial correlation, use the augmented form of the regression:

$$\hat{y}_t = a_1\hat{y}_{t-1} + \sum_{i=1}^{k} \beta_i \Delta\hat{y}_{t-i} + \varepsilon_t$$

STEP 4: Calculate the t-statistic for the null hypothesis $a_1 = 1$. This statistic can be compared to the critical values calculated by Perron. Perron generated 5,000 series according to H_1 using values of λ ranging from 0 to 1 by increments of 0.1. For each value of λ, he estimated the regressions $\hat{y}_t = a_1\hat{y}_{t-1} + \varepsilon_t$

and calculated the sample distribution of a_1. Naturally the critical values are identical to the Dickey–Fuller statistics when $\lambda = 0$ and $\lambda = 1$; in effect, there is no structural change unless $0 < \lambda < 1$. The maximum difference between the two statistics occurs when $\lambda = 0.5$. For $\lambda = 0.5$, the critical value of the t-statistic at the 5 percent level of significance is –3.76 (which is larger in absolute value than the corresponding Dickey–Fuller statistic of –3.41). If you find a t-statistic greater than the critical value calculated by Perron, it is possible to reject the null hypothesis of a unit root.

Of course, it is possible to incorporate Step 1 directly into Steps 2 or 3. To combine Steps 1 and 3, simply estimate the equation

$$y_t = a_0 + a_1 y_{t-1} + a_2 t + \mu_2 D_L + \sum_{i=1}^{k} \beta_i \Delta y_{t-i} + \varepsilon_t$$

The t-statistic for the null hypothesis $a_1 = 1$ can then be compared to the appropriate critical value calculated by Perron. In addition, the methodology is quite general in that it can also allow for a one-time change in the drift or a one-time change in both the mean and the drift. For example, it is possible to test the null hypothesis of a permanent change in the drift term versus the alternative of a change in the slope of the trend. Here the null hypothesis is

$$H_2\colon y_t = a_0 + y_{t-1} + \mu_2 D_L + \varepsilon_t$$

where $D_L = 1$ if $t > \tau$ and zero otherwise. With this specification, the $\{y_t\}$ sequence is generated by $\Delta y_t = a_0 + \varepsilon_t$ up to period τ and by $\Delta y_t = a_0 + \mu_2 + \varepsilon_t$ thereafter. If $\mu_2 > 0$, the magnitude of the drift increases for $t > \tau$. Similarly, a reduction in the drift occurs if $\mu_2 < 0$.

The alternative hypothesis posits a trend-stationary series with a change in the slope of the trend for $t > \tau$:

$$A_2\colon y_t = a_0 + a_2 t + \mu_3 D_T + \varepsilon_t$$

where $D_T = t - \tau$ for $t > \tau$ and zero otherwise.

For example, suppose that the break occurs in period 51 so that $\tau = 50$. Thus, $D_T(1)$ through $D_T(50)$ are all zero, so that for the first 50 periods, $\{y_t\}$ evolves as $y_t = a_0 + a_2 t + \varepsilon_t$. Beginning with period 51, $D_T(51) = 1$, $D_T(52) = 2$, ... so that for $t > \tau$, $\{y_t\}$ evolves as $y_t = a_0 + a_2 t + \mu_3(t - 50) + \varepsilon_t = a_0 + (a_2 + \mu_3)t - 50\mu_3 + \varepsilon_t$. Hence, D_T changes the slope of the deterministic trend line. The slope of the trend is a_2 for $t \leq \tau$ and $a_2 + \mu_3$ for $t > \tau$.

To be even more general, it is possible to combine the two null hypotheses H_1 and H_2. A change in both the level and drift of a unit root process can be represented by

$$H_3\colon y_t = a_0 + y_{t-1} + \mu_1 D_P + \mu_2 D_L + \varepsilon_t$$

where D_P and D_L are the pulse and level dummies defined above.

The appropriate alternative for this case is

$$A_3\colon y_t = a_0 + a_2 t + \mu_2 D_L + \mu_3 D_T + \varepsilon_t$$

Again, the procedure entails estimating the regression A_2 or A_3. Next, using the residuals $\hat{y}_t$, estimate the regression

$$\hat{y}_t = a_1\hat{y}_{t-1} + e_t$$

If the errors from this second regression equation do not appear to be white noise, estimate the equation in the form of an augmented Dickey–Fuller test. The t-statistic for the null hypothesis $a_1 = 1$ can be compared to the critical values calculated by Perron (1989). For $\lambda = 0.5$, Perron reported the critical value of the t-statistic at the 5 percent significance level to be –3.96 for H_2 and –4.24 for H_3.

Perron's Test and Real Output

Perron (1989) used his analysis of structural change to challenge the findings of Nelson and Plosser (1982). With the same variables, his results indicate that most macroeconomic variables are *not* characterized by unit root processes. Instead, the variables appear to be TS processes coupled with structural breaks. According to Perron (1989), the stock market crash of 1929 and the dramatic oil price increase of 1973 were exogenous shocks having permanent effects on the mean of most macroeconomic variables. The crash induced a one-time fall in the mean. Otherwise, macroeconomic variables appear to be trend-stationary.

All variables in Perron's study (except real wages, stock prices, and the stationary unemployment rate) appeared to have a trend with a constant slope and exhibited a major change in the level around 1929. In order to entertain various hypotheses concerning the effects of the stock market crash, consider the regression equation

$$y_t = a_0 + \mu_1 D_L + \mu_2 D_p + a_2 t + a_1 y_{t-1} + \sum_{i=1}^{k} \beta_i \Delta y_{t-i} + \varepsilon_t$$

where: $D_P(1930) = 1$ and zero otherwise
$D_L = 1$ for all t beginning in 1930 and zero otherwise

Under the presumption of a one-time change in the level of a unit root process, $a_1 = 1$, $a_2 = 0$, and $\mu_2 \neq 0$. Under the alternative hypothesis of a permanent one-time break in the trend-stationary model, $a_1 < 1$ and $\mu_1 \neq 0$. Perron's (1989) results using real GNP, nominal GNP, and industrial production are reported in Table 4.6. Given the length of each series, the 1929 crash means that λ is 1/3 for both real and nominal GNP and equal to 2/3 for industrial production. Lag lengths (i.e., the values of k) were determined using t-tests on the coefficients β_i. The value k was selected if the t-statistic on β_k was greater than 1.60 in absolute value and the t-statistic on β_i for $i > k$ was less than 1.60.

First consider the results for real GNP. When you examine the last column of the table, it is clear that there is little support for the unit root hypothesis; the estimated value of $a_1 = 0.282$ is significantly different from unity at the 1 percent level. Instead, real GNP appears to have a deterministic trend (a_2 is estimated to be over five standard deviations from zero). Also note that the point estimate $\mu_1 = -0.189$ is significantly different from zero at conventional levels. Thus, the stock market crash is estimated to have induced a permanent one-time decline in the intercept of real GNP.

Table 4.6 Retesting Nelson and Plosser's Data for Structural Change

	T	λ	k	a_0	μ_1	μ_2	a_2	a_1
Real GNP	62	0.33	8	3.44	−0.189	−0.018	0.027	0.282
				(5.07)	(−4.28)	(−0.30)	(5.05)	(−5.03)
Nominal GNP	62	0.33	8	5.69	−3.60	0.100	0.036	0.471
				(5.44)	(−4.77)	(1.09)	(5.44)	(−5.42)
Industrial production	111	0.66	8	0.120	−0.298	−0.095	0.032	0.322
				(4.37)	(−4.56)	(−.095)	(5.42)	(−5.47)

Notes:

1. T = number of observations; λ = proportion of observations occurring before the structural change; k = lag length.
2. The appropriate t-statistics are in parentheses. For a_0, μ_1, μ_2, and a_2, the null is that the coefficient is equal to zero. For a_1, the null hypothesis is $a_1 = 1$. Note that all estimated values of a_1 are significantly different from unity at the 1 percent level. For example, 0.282 is 5.03 standard deviations below unity.

These findings receive additional support since the estimated coefficients and their t-statistics are quite similar across the three equations. All values of a_1 are about five standard deviations from unity and the coefficients of the deterministic trends (a_2) are all over five standard deviations from zero. Since all estimated values of μ_1 are significant at the 1 percent level and negative, the data seem to support the contention that real macroeconomic variables are TS, except for a structural break resulting from the stock market crash.

Tests with Simulated Data

To further illustrate the procedure, 100 random numbers were drawn to represent the $\{\varepsilon_t\}$ sequence. By setting $y_0 = 0$, the next 100 values in the $\{y_t\}$ sequence were drawn as

$$y_t = 0.5y_{t-1} + \varepsilon_t + D_L$$

where: $D_L = 0$ for $t = 1, ..., 50$
$D_L = 1$ for $t = 51, ..., 100$

Thus, the simulation is identical to (4.38) except that the magnitude of the structural break is diminished. This simulated series is in the data file labeled BREAK.XLS; you should try to reproduce the following results. If you were to plot the data, you would see the same pattern as in Figure 4.10. However, if you did not plot the data or were otherwise unaware of the break, you might easily conclude that the $\{y_t\}$ sequence had a unit root. The ACF of the $\{y_t\}$ sequence suggests a unit root process; for example, the first 6 autocorrelations are

	ρ_1	ρ_2	ρ_3	ρ_4	ρ_5	ρ_6
Levels	0.945	0.889	0.856	0.835	0.804	0.772
First Differences	−0.002	−0.211	−0.112	0.083	0.007	−0.075

Dickey–Fuller tests yield

$\Delta y_t = 0.032y_{t-1} + \varepsilon_t$ t-statistic for $\gamma = 0$: 1.36
$\Delta y_t = -0.043 + 0.053y_{t-1} + \varepsilon_t$ t-statistic for $\gamma = 0$: 1.61
$\Delta y_t = 0.092 + 0.166y_{t-1} - 0.004t + \varepsilon_t$ t-statistic for $\gamma = 0$: 3.03

Diagnostic tests indicate that longer lags are not needed. Regardless of the presence of the constant or the trend, the $\{y_t\}$ sequence appears to be difference-stationary. Note that the estimated values of γ are all positive. Of course, the problem is that the structural break biases the estimates upward.

Now, using the Perron procedure, the first step is to estimate the model $y_t = a_0 + a_2t + \mu_2D_L + \hat{y}_t$. The residuals from this equation are the detrended $\{\hat{y}_t\}$ sequence. The second step is to test for a unit root in the residuals by estimating $\hat{y}_t = a_1\hat{y}_{t-1} + \varepsilon_t$. The resulting regression is

$$\hat{y}_t = 0.4843\hat{y}_{t-1} + \varepsilon_t$$

In the third step, all of the diagnostic statistics indicate that $\{\varepsilon_t\}$ approximates a white-noise process. Finally, since the standard error of a_1 is 0.0897, the t-statistic for $a_1 = 1$ is $(0.4843-1)/0.0897=-5.749$. Hence, we can reject the null of a unit root and conclude that the simulated data are stationary around a break point at $t = 51$.

Some care must be exercised in using Perron's procedure since it assumes that the date of the structural break is known. In your own work, if the date of the break is uncertain, you should consult Amsler and Lee (1995), Perron (1997), Vogelsang and Perron (1998), Zivot and Andrews (2002), or Lee and Strazicich (2003). The entire issue of the July 1992 *Journal of Business and Economic Statistics* is devoted to break points and unit roots. An interesting application is found in Ben-David and Papell (1995). They considered a long span (of up to 130 years) of GDP data for 16 countries. Allowing for breaks, they rejected the null of a unit root in approximately half of the cases.

9. POWER AND THE DETERMINISTIC REGRESSORS

Tests for unit roots are not especially good at distinguishing between a series with a characteristic root that is close to unity and a true unit root process. Part of the problem concerns the power of the test and the presence of the deterministic regressors in the estimating equations.

Power

Formally, the **power** of a test is equal to the probability of rejecting a false null hypothesis (i.e., one minus the probability of a type II error). A test with good power would correctly reject the null hypothesis of a unit root when the series in question is actually stationary. Monte Carlo simulations have shown that the power of the various Dickey–Fuller tests can be very low. As such, these tests will too often indicate that a series contains a unit root. Moreover, they have little power to distinguish between trend-stationary and drifting processes. In finite samples, any trend-stationary process can be arbitrarily well approximated by a unit root process, and a unit root process can be arbitrarily well approximated by a trend-stationary process. These results should not be too surprising after examining Figure 4.12. The top graph of the figure shows a stationary process and a unit root process that were constructed from the equations

$$y_t = 1.1y_{t-1} - 0.10y_{t-2} + \varepsilon_t$$
$$z_t = 1.1z_{t-1} - 0.15z_{t-2} + \varepsilon_t$$

where $y_0 = y_1 = z_0 = z_1 = 0$ and the same simulated values for $\{\varepsilon_t\}$ were used to construct the $\{y_t\}$ and $\{z_t\}$ sequences.

The $\{y_t\}$ sequence has a unit root; the roots of the $\{z_t\}$ sequence are 0.9405 and 0.1594. Although $\{z_t\}$ is stationary, it can be called a near–unit root process. If you didn't know the actual data-generating processes, it would be difficult to tell that only $\{z_t\}$ is stationary.

Similarly, as illustrated in the lower graph of Figure 4.12, it can be quite difficult to distinguish between a trend-stationary and a unit root plus drift process. Still

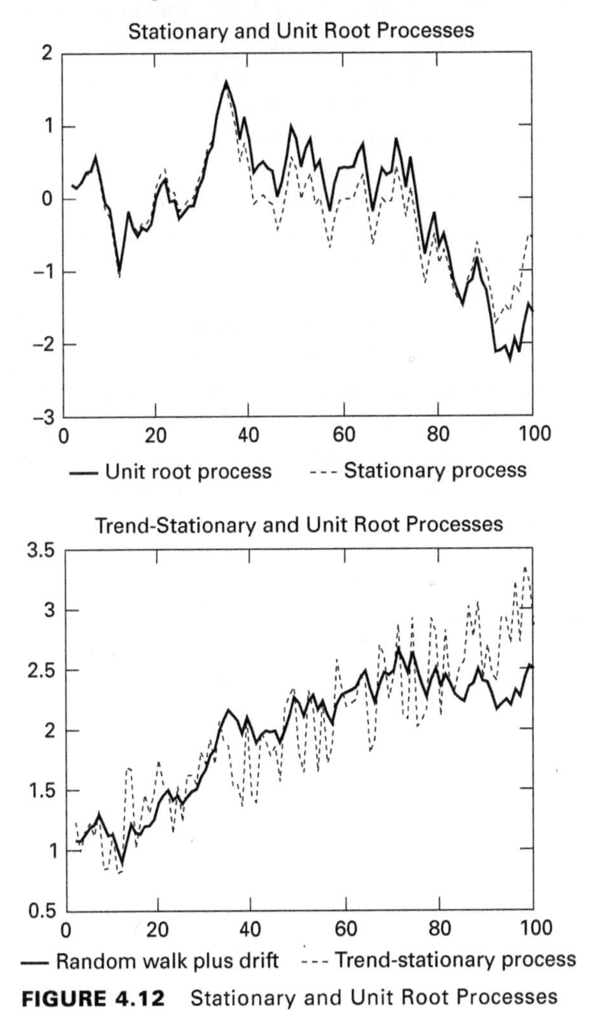

FIGURE 4.12 Stationary and Unit Root Processes

using the same 100 values for $\{\varepsilon_t\}$, two other sequences were constructed using the equations

$$w_t = 1 + 0.02t + \varepsilon_t \tag{4.42}$$

$$x_t = 0.02 + x_{t-1} + (1/3)\varepsilon_t \tag{4.43}$$

where $x_0 = 1$.

Here the trend and the drift terms dominate the time paths of the two sequences. Again, it is very difficult to distinguish between the sequences. This is especially true since dividing each realization of ε_t by 3 acts to smooth out the $\{x_t\}$ sequence. Just as it is difficult for the naked eye to perceive the differences in the sequences, it is also difficult for the Dickey–Fuller test to select the correct specification.

It is simple to conduct a Monte Carlo experiment that determines the power of the Dickey–Fuller test. To be more specific, suppose that the true data-generating process for a series is $y_t = a_0 + a_1 y_{t-1} + \varepsilon_t$ where $|a_1| < 1$. Since you do not know the actual data-generating process, you might test the series for a unit root using a Dickey–Fuller test. Of course, the series is actually stationary so that an ordinary t-statistic is appropriate. The question at hand is: How often will the Dickey–Fuller test fail to detect that the series is actually stationary? Since the confidence intervals for the t-statistics of the Dickey–Fuller test exceed those for the usual t-test, it is to be expected that the power of the Dickey–Fuller test is low. To find out the exact answer to the question, we can generate 10,000 stationary series and apply a Dickey–Fuller test to each. We can then calculate the percentage of the times that the test correctly identifies a truly stationary process.

The ability of the test to properly detect that the series is stationary will depend on the value of a_1. We would expect the test to have the least power when $|a_1|$ is close to unity. Thus, it makes sense to examine how the magnitude of a_1 affects the power of the test. We first construct 100 observations of the series $y_t = a_0 + a_1 y_{t-1} + \varepsilon_t$ using a value of $a_1 = 0.8$ and an $\{\varepsilon_t\}$ sequence drawn from a standardized normal distribution. The magnitude of a_0 is unimportant and so is set equal to zero. The initial value of y_0 is set equal to the unconditional mean of zero. Next, the simulated series is estimated in the form $\Delta y_t = a_0 + \gamma y_{t-1} + \varepsilon_t$. The Dickey–Fuller τ_μ statistics are used to determine whether the null hypothesis that $\gamma = 0$ can be rejected at the 10 percent, 5 percent, and 1 percent significance levels. The experiment is repeated 10,000 times and the proportion of the instances in which the null hypothesis is correctly rejected is recorded. Finally, the entire experiment is repeated for other values of a_1. Consider the following table of proportions:

a_1	10%	5%	1%
0.80	95.9	87.4	51.4
0.90	52.1	33.1	9.0
0.95	23.4	12.7	2.6
0.99	10.5	5.8	1.3

When the true value of $a_1 = 0.8$, the test does reasonably well. For example, at the 5 percent significance level, the false null hypothesis of a unit root is rejected in 87.4

percent of the Monte Carlo replications. However, when $a_1 = 0.95$, the probability of correctly rejecting the null hypothesis of a unit root is estimated to be only 12.7 percent at the 5 percent significance level and 2.6 percent at the 1 percent level. Thus, the test has very low power to detect near–unit root series.

Does it matter that it is often impossible to distinguish between borderline-stationary, trend-stationary, and unit root processes? The realistic answer is that it depends on the question at hand. In borderline cases, the short-run forecasts from the alternative models may have nearly identical forecasting performance. In fact, Monte Carlo studies indicate that when the true data-generating process is stationary but has a root close to unity, the one-step-ahead forecasts from a differenced model are usually superior to the forecasts from a stationary model. However, the long-run forecasts of a model with a deterministic trend will be quite different from those of other models.[15]

Determination of the Deterministic Regressors

Unless the researcher knows the actual data-generating process, there is a question concerning whether it is most appropriate to estimate (4.20), (4.21), or (4.22). It might seem reasonable to test the hypothesis $\gamma = 0$ using the most general of the models:

$$\Delta y_t = a_0 + \gamma y_{t-1} + a_2 t + \sum_{i=2}^{p} \beta_i \Delta y_{t-i+1} + \varepsilon_t \tag{4.44}$$

After all, if the true process is a random walk process, this regression should find that $a_0 = \gamma = a_2 = 0$. One problem with this line of reasoning is that the presence of the additional estimated parameters reduces degrees of freedom and the power of the test. Reduced power means that the researcher will not be able to reject the null of a unit root when, in fact, no unit root is present. The second problem is that the appropriate statistic (i.e., τ, τ_μ, and τ_τ) for testing $\gamma = 0$ depends on which regressors are included in the model. As you can see by examining the three Dickey–Fuller tables, for a given significance level, the confidence intervals around $\gamma = 0$ dramatically expand if a drift and a time trend are included in the model. This is quite different from the case in which $\{y_t\}$ is stationary. When $\{y_t\}$ is stationary, the distribution of the t-statistic does not depend on the presence of the other regressors.

The point is that it is important to use a regression equation that mimics the actual data-generating process. Inappropriately omitting the intercept or time trend can cause the power of the test to go to zero.[16] For example, if as in (4.44) the data-generating process includes a trend, omitting the term $a_2 t$ imparts an upward bias in the estimated value of γ. On the other hand, extra regressors increase the critical values so that you may fail to reject the null of a unit root.

Campbell and Perron (1991) reported the following results concerning unit root tests:

1. If the estimated regression includes deterministic regressors that are not in the actual data-generating process, the power of the unit root test against a stationary alternative decreases as additional deterministic regressors are added. Hence, you do not want to include regressors that are not in the data-generating process.

2. If the estimated regression omits an important deterministic trending variable present in the true data-generating process—such as the expression a_2t in (4.44)—the power of the t-statistic test goes to zero as sample size increases. If the estimated regression omits a nontrending variable (such as an intercept), the t-statistic is consistent, but the finite sample power is adversely affected and decreases as the magnitude of the coefficient on the omitted component increases. Hence, you do not want to omit regressors that are in the data-generating process.

The direct implication of these findings is that the researcher may fail to reject the null hypothesis of a unit root because of a misspecification concerning the deterministic part of the regression. Too few or too many regressors may cause a failure of the test to reject the null of a unit root. How do you know whether to include a drift or time trend in performing the tests? The key problem is that *the tests for unit roots are conditional on the presence of the deterministic regressors and tests for the presence of the deterministic regressors are conditional on the presence of a unit root.* Although we can never be sure that we are including the appropriate deterministic regressors in our econometric model, there are some useful guidelines.

1. *Always plot your data.* Visual inspection can help you determine whether there is a clear trend in the data.
2. *Be clear about the appropriate null hypothesis and the alternative hypothesis.* When you perform a Dickey–Fuller test, always estimate the model under the alternative hypothesis and impose the restriction implied by the null hypothesis. Since the null hypothesis is that the series has a unit root, always estimate the series as if it were stationary or TS. For example, the real GDP series shown in Figure 4.1 moves decidedly upward over time. The issue is whether the series is trend-stationary or contains a unit root plus a drift term. As such, the appropriate model to estimate has the form $\Delta y_t = a_0 + \gamma y_{t-1} + a_2t + \Sigma\beta_i\Delta y_{t-i} + \varepsilon_t$. You then test the restrictions $\gamma = 0$ and/or $\gamma = a_2 = 0$. There is no need to estimate a model without a_2t since the alternative hypothesis is not represented in such a specification.
3. *You do not want to reject the null hypothesis when the series actually has a unit root (a Type I error) or incorrectly accept the null of a unit root when the series is stationary or TS (a Type II error).* Nevertheless, any test involves the possibility of making such errors. As such, you do not want to perform needless tests. In the example of real GDP, there is little point in testing the restriction $a_0 = \gamma = a_2 = 0$ since real GDP clearly increases over time.
4. *Testing a restriction on a model that has already been restricted creates the possibility of compounding your errors.* Suppose that a test for the presence of the time trend allows you to set $a_2 = 0$. A subsequent test of the restriction $a_0 = \gamma = 0$ in the model $\Delta y_t = a_0 + \gamma y_{t-1} + \Sigma\beta_i\Delta y_{t-i} + \varepsilon_t$ is conditional on whether the first test was correct in allowing you to exclude the deterministic trend.

At one time, researchers would apply a battery of tests on the values of a_0 and/or a_2 when the form of the deterministic regressors was completely unknown. One standard procedure is discussed in detail in Appendix 4.2. Now, when power seems to be an issue, it is typical to use variants of the Dickey–Fuller test that have enhanced power.

10. TESTS WITH MORE POWER

If you examine the basic regression used in the Dickey–Fuller test, $\Delta y_t = a_0 + \gamma y_{t-1} + a_2 t + \varepsilon_t$, you will see that that there are two different types of regressors. The intercept and the time trend are purely deterministic while y_{t-1} is a unit root process under the null hypothesis. Notice that the coefficients of the deterministic expressions a_0 and a_2 play very different roles under the null and alternative hypotheses. If we change equation numbers and symbols to match those used in the text, Phillips and Schmidt (1992, p. 258) made the following observation about the parameters in the Dickey–Fuller regressions:

> . . . the parameter a_0 represents trend when $\gamma = 0$ (since the solution for y_t then includes the deterministic trend term $a_0 t$), but it determines level when $\gamma < 0$ (since y_t is then stationary around the level $-a_0/\gamma$). Similarly, in equation (4.44), when $\gamma = 0$ the parameter a_0 represents trend and a_2 represents quadratic trend, while under the alternative a_0 determines level and a_2 determines trend. This confusion over the meanings of the parameters shows up in the properties of the Dickey–Fuller tests.

The essential problem is that the intercept and the slope of the trend are often poorly estimated in the presence of a unit root. In a sense, the least squares principle is unable to properly separate the movements of y_t into those induced by the deterministic trend and those induced by the stochastic trend. Even in the circumstance in which $\{y_t\}$ is stationary, the intercept and trend can be poorly estimated if the $\{y_t\}$ series is quite persistent. Of course, if the estimates of a_0 and a_2 have substantial error, the estimate of γ will have a large standard error too. You can see this effect by comparing the Dickey–Fuller critical values for τ, τ_μ, and τ_τ to those in a standard t-table. The overly wide confidence intervals for γ means that you are less likely to reject the null hypothesis of a unit root even when the true value of γ is not zero.

A number of authors have devised clever methods to improve the estimates of the intercept and trend coefficients. For example, Schmidt and Phillips (1992) proposed a two-step testing procedure that has better power than the Dickey–Fuller test. Although they call their test a **Lagrange Multiplier** (LM) test, the method is actually quite simple. Instead of the Dickey–Fuller specification, under the null hypothesis, the $\{y_t\}$ sequence is a random walk plus a drift so that

$$y_t = a_0 + a_2 t + \sum_{i=0}^{t-1} \varepsilon_{t-i}$$

or

$$\Delta y_t = a_2 + \varepsilon_t$$

The idea is to estimate the trend coefficient, a_2, using the regression $\Delta y_t = a_2 + \varepsilon_t$. As such, the presence of the stochastic trend $\sum \varepsilon_i$ does not interfere with the estimation of a_2. The resulting estimate of a_2 (called $\hat{a}_2$) is an estimate of the slope of the time trend. Use this estimate to form the detrended series as $y_t^d = y_t - (y_1 - \hat{a}_2) - \hat{a}_2 t$ where y_1 is the initial value of the $\{y_t\}$ series. Note that for $t = 1$, the deterministic part of $y_1 = a_0 + a_2$. As such, $(y_1 - \hat{a}_2)$ acts as the intercept of the estimated trend line and $\hat{a}_2$ acts as the slope. The use of $(y_1 - \hat{a}_2)$ in the detrending procedure ensures that the initial value of the detrended series (i.e., y_1^d) is zero. In the second step of the procedure, you estimate a variant of the Dickey–Fuller test using the detrended series in place of the level of y_{t-1}:

$$\Delta y_t = a_0 + \gamma y_{t-1}^d + \varepsilon_t$$

or, if there is any serial correlation in the residuals, estimate

$$\Delta y_t = a_0 + \gamma y_{t-1}^d + \sum_{i=1}^{p} c_i \Delta y_{t-i}^d + \varepsilon_t$$

The null of a unit root can be rejected if it is found that $\gamma \neq 0$. The point is that Schmidt and Phillips (1992) show that it is preferable to estimate the parameters of the trend using a model without the persistent variable y_{t-1}. Once the trend is efficiently estimated, it is possible to detrend the data and perform the unit root test on the detrended data. Some of the critical values for the test are:

Critical Values of the Schmidt–Phillips Unit Root Test

T	1%	2.5%	5%	10%
50	−3.73	−3.39	−3.11	−2.80
100	−3.63	−3.32	−3.06	−2.77
200	−3.61	−3.30	−3.04	−2.76
500	−3.59	−3.29	−3.04	−2.76

Elliott, Rothenberg, and Stock (1996) showed that it is possible to further enhance the power of the test by estimating the model using something close to first differences. The idea is that under the alternative hypothesis that the series is stationary, the Schmidt–Phillips model in first differences is misspecified. Hence, consider the TS model:

$$y_t = a_0 + a_2 t + B(L)\varepsilon_t$$

Instead of creating the first difference of y_t, Elliott, Rothenberg, and Stock (ERS) preselect a constant close to unity, say α, and subtract αy_{t-1} from y_t to obtain

$$\tilde{y}_t = a_0 + a_2 t - \alpha a_0 - \alpha a_2(t-1) + e_t \qquad \text{for } t = 2, \ldots, T$$

where $\tilde{y}_t = y_t - \alpha y_{t-1}$ and e_t is a stationary error term. For $t = 1$, such near-differencing is not possible and the initial value $\tilde{y}_1$ is set equal to y_1. For simplicity, collect terms with a_0 and a_2 to obtain

$$\tilde{y}_t = (1-\alpha)a_0 + a_2[(1-\alpha)t + \alpha)] + e_t$$

Now, it should be clear how to obtain estimates of a_0 and a_2 using OLS. Create the variable $z1_t$ equal to the constant $(1 - \alpha)$ and the variable $z2_t$ equal to the deterministic trend $\alpha + (1 - \alpha)t$. To obtain the desired estimates of a_0 and a_2, simply regress $z1_t$ and $z2_t$ on $\tilde{y}_t$. In other words, use OLS to estimate

$$\tilde{y}_t = a_0 z1_t + a_2 z2_t + e_t$$

Note that the test is conditional on the initial value of the $\{y_t\}$ series in that $y_1 = a_0 + a_2 + \varepsilon_1$. As such, the initial values of $z1_t$ and $z2_t$ should be set equal to unity and the initial value of $\hat{y}_t$ should be set equal to y_1 (i.e., set $z1_1 = 1$, $z2_1 = 1$, and $\tilde{y}_t = y_1$). Since the goal is to obtain the estimated values of a_0 and a_2, at this step it is not especially important if the residual, e_t, is serially correlated. The important point is that the estimates a_0 and a_2 can be used to detrend the $\{y_t\}$ series as

$$y_t^d = y_t - \hat{a}_0 - \hat{a}_2 t$$

In the second step of the procedure, estimate the basic Dickey–Fuller regression using the detrended data.[17] Hence, estimate the regression equation

$$\Delta y_t^d = \gamma y_{t-1}^d + \varepsilon_t$$

If there is serial correlation in the residuals, the augmented form of the test can be estimated as

$$\Delta y_t^d = \gamma y_{t-1}^d + \sum_{i=1}^{p} c_i \Delta y_{t-i}^d + \varepsilon_t$$

Elliott, Rothenberg, and Stock (ERS; 1996) recommended selecting the lag length p using the SBC. As in the Schmidt–Phillips test, the null of a unit root can be rejected if it is found that $\gamma \neq 0$. The critical values of the test depend on whether a trend is included in the test. *If there is an intercept, but not a trend, the critical values are precisely those of the Dickey–Fuller τ test reported in the top portion of Table A.* In essence, you use the Dickey–Fuller critical values as if there is no intercept in the data-generating process. If there is a trend, the critical values depend on the value of α selected to perform the "near-differenced" variable $\hat{y}_t$. ERS reported that the value of α that seems to provide the best overall power is $\alpha = (1 - 7/T)$ for the case of an intercept and $\alpha = (1 - 13.5/T)$ if there is an intercept and trend. The table below reports the critical values for the case of a trend and $\alpha = 1 - 13.5/T$. Notice that as the sample size T increases, α approaches unity so that $\tilde{y}_t$ is approximately equal to Δy_t.

Critical Values of the ERS Test with Trend and $\alpha = 1 - 13.5/T$

T	1%	2.5%	5%	10%
50	−3.77	−3.46	−3.19	−2.89
100	−3.58	−3.29	−3.03	−2.74
200	−3.46	−3.18	−2.93	−2.64
∞	−3.48	−3.15	−2.89	−2.57

One aspect of the ERS test that some researchers might find objectionable is the assumption that the initial value $\tilde{y}_1$ is set equal to y_1. This is equivalent to assuming that the first value of the error term is equal to zero. An alternative assumption is that the initial value of the shock is drawn from its unconditional distribution. Note that relaxing the assumption concerning the initial condition acts to reduce the power of this version of the test. In this circumstance, the first value of $\tilde{y}_1$ is set equal to $y_1(1 - \alpha^2)^{0.5}$, $z1_1 = (1 - \alpha^2)^{0.5}$, and $z2_1 = (1 - \alpha^2)^{0.5}$.[18] Hence, instead of conditioning on the magnitude of y_1, you condition on the number of standard deviations from zero. Note that Elliott (1999) recommends using $\alpha = (1 - 10/T)$ regardless of whether or not a trend is included in the regression. The critical values for this test are different from those reported above. The asymptotic critical values for regressions with an intercept and with an intercept plus trend are:

	1%	2.5%	5%	10%
Intercept	−3.28	−2.98	−2.73	−2.46
Trend	−3.71	−3.41	−3.17	−2.91

An Example

To illustrate the appropriate use of the procedure, the file labeled ERSTEST.XLS contains 200 observations generated from the equation $y_t = 1 + 0.95y_{t-1} + 0.01t + \varepsilon_t$. Although the series is clearly trend-stationary, the point of this exercise is to illustrate the appropriate use of the ERS test and to compare the results to those of a Dickey–Fuller test. If you examine the file, you will see that the first five rows are:

t	*y*	*y_tilde*	*z1*	*z2*	*yd*
1	20.03339	20.03339	1.0000	1.0000	0.036376
2	21.85126	3.170125	0.0675	1.0675	1.692188
3	22.01347	1.637169	0.0675	1.1350	1.692338
4	22.08649	1.558934	0.0675	1.2025	1.603304
5	22.17255	1.576890	0.0675	1.2700	1.527297

The series in column 2, called *y*, contains the 200 realizations representing the y_t series. Since the data contain a trend, the appropriate value of α to use is $1 - 13.5/200 = 0.9325$. This value of α was used to construct the next series (*y_tilde*) as $y_t - 0.9325y_{t-1}$. For example, $\tilde{y}_1 = y_1$, $\tilde{y}_2 = y_2 - \alpha y_1 = 21.85126 - 0.9325(20.03339) = 3.170125$ and $\tilde{y}_3 = y_3 - \alpha y_2 = 1.637169$. Since $\alpha = 0.9325$, $z1_2 = z1_3 = ... = 1 - \alpha = 0.0675$. Similarly, $z2_t = 0.9325 + 0.0675t$ so that $z2_1 = 1.0000$, $z2_2 = 1.0675$, $z2_3 = 1.1350$, ... The regression of $\tilde{y}_t$ on $z1_t$ and $z2_t$ yields

$$\tilde{y}_t = 19.835{*}z1_t + 0.162{*}z2_t$$

These estimates of a_0 and a_2 are used to construct the detrended series as

$$y_t^d = y_t - 19.835 - 0.162t$$

This series is reported in the last column of ERSTEST.XLS. Before proceeding, it is interesting to consider the particular solution for the skeleton of $y_t = 1 + 0.95y_{t-1} + 0.01t + \varepsilon_t$. From your knowledge of Chapter 1 (also see question 7 of Chapter 2), you should have no trouble verifying that the desired solution is $16.2 + 0.2t$. The estimated trend equation, $19.835 + 0.162t$, is reasonably close to the particular solution.

Now that y_t has been detrended, it is straightforward to perform the unit root test. If you use the data in the spreadsheet, you should find:

$$\Delta y_t^d = -0.0975 y_{t-1}^d$$
$$(-3.154)$$

The 2.5 percent and 5 percent critical values for the test are −3.15 and −2.89, respectively. As such, the null hypothesis of a unit root is clearly rejected at the 5 percent level and just barely rejected at the 2.5 percent level. You will find that augmenting this regression with lagged values of Δy_{t-i}^d only acts to increase the value of the SBC. You can perform Elliott's (1999) version of the test in the same way, except that you set $\alpha = 1 - 10/200 = 0.95$, $y_1(1 - \alpha^2)^{0.5} = 6.255$, $z1_1 = (1 - \alpha^2)^{0.5} = 0.3122$, and $z2_1 = (1 - \alpha^2)^{0.5} = 0.3122$. Hence, assuming that the initial value of the series is drawn from its unconditional mean, you should obtain the t-statistic −3.147. The null hypothesis of a unit root is not rejected (although it is very close to being rejected) using the 5 percent critical value of −3.17.

The results of Elliott's (1999) test are very similar to the result found from the Schmidt–Phillips test. To perform the Schmidt–Phillips LM test, you should first regress Δy_t on a constant and obtain $\Delta y_t = 0.1713$. Since $y_1 = 20.03339$, you detrend the y_t series using $y_t^d = 20.03339 - (20.03339 - 0.1713) - 0.1713t$. Now, you should be able to reproduce the regression equation $\Delta y_t = 0.0691 - 0.0903y_t^d$. Since the t-statistic for the coefficient on y_t^d is −3.052, the null hypothesis of a unit root is just rejected at the 5 percent significance level. Very different results are obtained when performing a standard Dickey–Fuller test. Consider the estimated model:

$$\Delta y_t = 2.0809 + 0.0158t - 0.0979y_{t-1} + \varepsilon_t$$
$$(3.265) \quad (3.106) \quad (-3.124)$$

The estimated value of γ is −0.0979 and the t-statistic for the null hypothesis $\hat{\gamma} = 0$ is −3.124. From Table A, the critical values of the τ_τ statistic at the 5 percent and 10 percent significance levels are about −3.45 and −3.15, respectively. Hence, if we use the Dickey–Fuller test, the null hypothesis of a unit root cannot be rejected at conventional significance levels.

Section 9 reported the results of a simple Monte Carlo study of the power of the standard Dickey–Fuller test for the process $y_t = a_0 + a_1y_{t-1} + \varepsilon_t$. Now, if we use the ERS test, the proportions (out of 10,000 replications) in which the null hypothesis of a unit root were correctly rejected are:

a_1	10%	5%	1%
0.80	99.8	99.1	86.6
0.90	93.9	79.0	33.4
0.95	64.3	39.8	10.0
0.99	23.3	11.1	2.3

Although these results are far superior to those of the Dickey–Fuller test, the power of the test for large values of a_1 is still disappointing.

11. PANEL UNIT ROOT TESTS

Section 6 presented some strong evidence that the three real exchange rate series shown in Figure 4.8 are unit root processes. Of course, it is possible that the series are mean-reverting, but the Dickey–Fuller tests have little power to detect the fact that the series are stationary. One way to obtain a more powerful test is to pool the estimates from a number of separate series and then test the pooled value. The theory underlying the test is very simple: If you have n independent and unbiased estimates of a parameter, the mean of the estimates is also unbiased. More importantly, so long as the estimates are independent, the central limit theory suggests that the sample mean will be normally distributed around the true mean.

Im, Pesaran, and Shin (2002) showed how to use this result to construct a test for a unit root when you have a number of similar time-series variables (i.e., a panel). The only complicating factor is that the OLS estimates for γ in the Dickey–Fuller test are biased downward. Suppose you have n series, each containing T observations. For each series, perform an ADF test of the form

$$\Delta y_{it} = a_{i0} + \gamma_i y_{it-1} + a_{i2}t + \sum_{i=1}^{p_i} \beta_{ij}\Delta y_{it-j} + \varepsilon_{it} \qquad i = 1, \ldots, n \qquad (4.45)$$

Because the lag lengths can differ across equations, you should perform separate lag length tests for each equation. Moreover, you may choose to exclude the deterministic time trend. However, if the trend is included in one equation, it should be included in all. Once you have estimated the various γ_i, obtain the t-statistic for the null hypothesis $\gamma_i = 0$. In a traditional Dickey–Fuller test, each of these t-statistics—denoted by t_i—would be compared to the appropriate critical value reported in Table A. However, for the panel unit root test, form the sample mean of the t-statistics as

$$\bar{t} = (1/n)\sum_{i=1}^{n} t_i \qquad (4.46)$$

It is straightforward to construct the statistic Z_{tbar} as

$$Z_{\text{tbar}} = \frac{\sqrt{n}[\bar{t} - E(\bar{t})]}{\sqrt{\text{var}(\bar{t})}}$$

where $E(\bar{t})$ and var $(\bar{t})$ denote the theoretical mean and variance of $\bar{t}$. If the OLS estimates of the individual t_i were unbiased, the value of $E(\bar{t})$ would be zero. However, to correct for the bias, the values $E(\bar{t})$ and var$(\bar{t})$ can be calculated by Monte Carlo simulation. Im, Pesaran, and Shin (IPS) report these values as follows:

T	6	8	10	15	20	50	100	500
$E(\bar{t})$	−1.52	−1.50	−1.50	−1.51	−1.52	−1.53	−1.53	−1.53
var$(\bar{t})$	1.75	1.23	1.07	0.92	0.85	0.76	0.74	0.72

Table 4.7 Selected Critical Values for the IPS Panel Unit Root Test

	25			50			70		
n/T	10%	5%	1%	10%	5%	1%	10%	5%	1%
				No Time Trend					
5	−2.04	−2.18	−2.46	−2.02	−2.15	−2.42	−2.02	−2.15	−2.40
7	−1.95	−2.08	−2.32	−1.95	−2.06	−2.28	−1.95	−2.06	−2.28
10	−1.88	−1.99	−2.19	−1.88	−1.98	−2.16	−1.88	−1.98	−2.16
15	−1.82	−1.90	−2.07	−1.81	−1.89	−2.05	−1.81	−1.89	−2.04
25	−1.75	−1.82	−1.94	−1.75	−1.81	−1.93	−1.75	−1.81	−1.93
50	−1.69	−1.73	−1.82	−1.68	−1.73	−1.81	−1.68	−1.73	−1.73
				Time Trend					
5	−2.65	−2.80	−3.09	−2.62	−2.76	−3.02	−2.62	−2.75	−3.00
7	−2.58	−2.70	−2.94	−2.56	−2.67	−2.88	−2.55	−2.66	−2.67
10	−2.51	−2.62	−2.82	−2.50	−2.59	−2.77	−2.49	−2.58	−2.75
15	−2.45	−2.53	−2.69	−2.44	−2.52	−2.65	−2.44	−2.51	−2.65
25	−2.39	−2.45	−2.58	−2.38	−2.44	−2.55	−2.38	−2.44	−2.54
50	−2.33	−2.37	−2.45	−2.32	−2.36	−2.44	−2.32	−2.36	−2.44

Im, Pesaran, and Shin show that Z_{tbar} has an asymptotic standardized normal distribution.[19] This fact should not be too surprising. If each of the estimated values of the various t_i are independent, the central limit theorem indicates that deviation of the sample average from the true mean will have a normal distribution. Rejecting the null hypothesis $Z_{\text{tbar}} = 0$ is equivalent to accepting the alternative hypothesis that *at least* one value of γ_i differs from zero. After all, if the sample average of the t-statistics is significantly different, at least one of the values of γ_i is statistically different from zero.

The proof that Z_{tbar} has a normal distribution relies on very large samples. For the sample sizes typically used by applied econometricians, it is preferable to use the critical values contained in Table 4.7. Notice that the critical values depend on n, T, and whether a time trend is included in (4.45). For example, if you have seven series, each containing 50 observations, and you include a time trend in (4.45), the 5 percent critical value for $\bar{t}$ is -2.67. If you had used the Dickey–Fuller test, the corresponding critical value for each of the seven values of t_i would be -3.50 (see Table A). Note that it is necessary to have values of T and n that are greater than four. Large values of T are standard in time-series econometrics. However, if n is too small, the calculation of $\bar{t}$ will not be meaningful.

As mentioned in Section 6, the file PANEL.XLS contains quarterly values of the real effective exchange rates (CPI-based) for Australia, Canada, France, Germany, Japan, Netherlands, the United Kingdom, and the United States over the 1980Q1 to 2008Q1 period. Since purchasing power parity (PPP) does not allow for a deterministic time trend, each was estimated in the form of (4.45) but without the trend. The

Table 4.8 The Panel Unit Root Tests for Real Exchange Rates

	Lags	Estimated γ_i	*t*-statistic	Estimated γ_i	*t*-statistic
		Log of the Real Rate		Minus the Common Time Effect	
Australia	5	−0.077	−2.472	−0.079	−2.466
Canada	6	−0.046	−2.198	−0.043	−2.001
France	3	−0.103	−3.064	−0.128	−3.510
Germany	1	−0.093	−3.068	−0.110	−3.423
Japan	3	−0.043	−1.993	−0.037	−1.869
Netherlands	1	−0.053	−1.775	−0.061	−1.950
U.K.	1	−0.061	−2.114	−0.057	−2.066
U.S.	3	−0.069	−2.455	−0.068	−2.379

results of the individual Dickey–Fuller tests for the logarithmic values of the real rates are shown in the first three columns of Table 4.8. For example, the Australian equation used five lags of $\{\Delta y_{it}\}$ and the estimated value of γ_i was -0.077. Notice that the eight *t*-statistics for the null hypothesis $\gamma_i = 0$ have an average value of -2.39. Since each series has a total of 113 observations, the critical values at the 5 percent and 1 percent significance levels are about -2.06 and -2.28, respectively. Hence, it is possible to reject the null hypothesis that all values of $\gamma_i = 0$.

One problem with the results is that the residuals from the individual equation are contemporaneously correlated in that $E\varepsilon_{it}\varepsilon_{jt} \neq 0$. For example, the correlation coefficient between the residuals from the French and German equations is 0.55. The explanation is that the shocks that affect the French real rate are likely to affect the German real rate. In this circumstance, a common strategy is to subtract a common time effect from each observation. At time period t, the mean value of each series is

$$\bar{y}_t = (1/n)\sum_{i=1}^{n} y_{it}$$

The method is to subtract this common mean from each observation (i.e., form $y_{it}^* = y_{it} - \bar{y}_t$) and estimate (4.45) using the values of y_{it}^*. In the example at hand, y_{it} is the logarithm of real rate i at period t; hence, for each time period t, the average of these logarithmic values was subtracted from y_{it}. The last three columns of Table 4.8 show the test results for the $\{y_{it}^*\}$ sequences. Notice that the lag lengths have not changed but the average value of the *t*-statistics is -2.48. As such, it is possible to reject the null hypothesis that the real rates are not stationary.

Limitations of the Panel Unit Root Test

1. The null hypothesis for the IPS test is $\gamma_i = \gamma_2 = \ldots = \gamma_n = 0$. Rejection of the null hypothesis means that *at least* one of the γ_i differs from zero. Thus, it is possible for only one or two values of the γ_i to differ from zero and still reject the null hypothesis. Unfortunately, there is no particular way of knowing which of the γ_i are statistically different from zero. As such, the

results of a panel unit root test may be dependent on the choice of the time-series variables included in the panel.

2. At this point, there is substantial disagreement about the asymptotic theory underlying the test. Sample size can approach infinity by increasing n for a given T, increasing T for a given n, or by simultaneously increasing n and T. Unfortunately, many of the important findings about the various tests are sensitive to this seemingly innocuous choice among the various assumptions. For example, the critical values reported in Table 4.7 are invariant to augmenting (4.45) with lagged changes for large T. However, for small T and large n, the critical values are dependent on the magnitudes of the various β_{ij}.
3. The test requires that the error terms from (4.45) be serially uncorrelated and contemporaneously uncorrelated. You need to determine the values of p_i to ensure that the autocorrelations of $\{\varepsilon_{it}\}$ are zero. Nevertheless, the errors may be contemporaneously correlated in that $E\varepsilon_{it}\varepsilon_{jt} \neq 0$. If the regression residuals are correlated across equations, the critical values in Table 4.7 are not applicable. The example above illustrates a common technique to correct for correlation across equations. As in the example, you can subtract a common time effect from each observation. However, there is no assurance that this correction will completely eliminate the correlation. Moreover, it is quite possible that $\{\bar{y}_t\}$ is nonstationary. Subtracting a nonstationary component from each sequence is clearly at odds with the notion that the variables are stationary. As an alternative, many researchers would generate their own critical values by bootstrapping the value of $\bar{t}$. Some of the details are described in Appendix 4.1.

There are a number of other panel unit root tests in the literature. The Maddala–Wu (1999) test is similar to the IPS test but requires that you bootstrap your own critical values. The Levin–Lin–Chu (2002) test has the more restrictive alternative hypothesis $\gamma_1 = \gamma_2 = \ldots = \gamma_n$. Nevertheless, the cautions listed above are applicable to all of the panel unit root tests. An interesting comparison of the tests can be found in the August 2001 issue of the *Journal of Money Credit and Banking*. Three different articles perform various panel unit roots for a number of real exchange rate series.

12. TRENDS AND UNIVARIATE DECOMPOSITIONS

Nelson and Plosser's (1982) findings suggest that many economic time series have a stochastic trend plus a stationary component. Having observed a series, but not the individual components, is there any way to decompose the series into its constituent parts? Numerous economic theories suggest it is important to distinguish between temporary and permanent movements in a series. A sale (i.e., a temporary price decline) is designed to induce us to purchase now, rather than in the future. Labor economists argue that "hours supplied" is more responsive to a temporary wage increase than to a permanent increase. The idea is that workers will temporarily substitute income for leisure time. Certainly, modern theories of the consumption function that classify an individual's income into permanent and transitory components highlight the importance of such decomposition.

Any such decomposition is straightforward if it is known that the trend in $\{y_t\}$ is purely deterministic. For example, a linear time trend induces a fixed change in each and every period. This deterministic trend can be subtracted from the actual value of y_t to obtain the stationary component.

A difficult conceptual issue arises if the trend is stochastic. For example, suppose you are asked to measure the current phase of the business cycle. If the trend in GDP is stochastic, how is it possible to tell if GDP is above or below trend? The traditional measurement of a recession by consecutive quarterly declines in real GDP is not helpful. After all, if GDP has a deterministic trend component, a negative realization for the stationary component may be outweighed by the positive deterministic trend component.

If it is possible to decompose a sequence into its separate permanent and stationary components, the issue can be solved. To better understand the nature of stochastic trends, note that—in contrast to a deterministic trend—a stochastic trend increases *on average* by a fixed amount each period. For example, consider the random walk plus drift model:

$$y_t = y_{t-1} + a_0 + \varepsilon_t$$

Since $E\varepsilon_t = 0$, the *average* change in y_t is the deterministic constant a_0. Of course, in any period t, the actual change will differ from a_0 by the stochastic quantity ε_t. Yet each sequential change in $\{y_t\}$ adds to its level regardless of whether the change results from the deterministic or the stochastic component. As we saw in (4.5), the random walk plus drift model has no stationary component; hence, it is a model of pure trend.

The idea that a random walk plus drift is a pure trend has proven especially useful in time-series analysis. Beveridge and Nelson (1981) showed how to decompose any ARIMA(p, 1, q) model into the sum of a random walk plus drift and a stationary component (i.e., the general trend plus irregular model). Before considering the general case, begin with the simple example of an ARIMA(0, 1, 2) model:

$$y_t = y_{t-1} + a_0 + \varepsilon_t + \beta_1\varepsilon_{t-1} + \beta_2\varepsilon_{t-2} \tag{4.47}$$

If $\beta_1 = \beta_2 = 0$, (4.47) is nothing more than the pure random walk plus drift model. The introduction of the two moving-average terms adds a stationary component to the $\{y_t\}$ sequence. The first step in understanding the Beveridge and Nelson (1981) procedure is to obtain the forecast function. For now, keep the issue simple by defining $e_t = \varepsilon_t + \beta_1\varepsilon_{t-1} + \beta_2\varepsilon_{t-2}$ so that we can write $y_t = y_{t-1} + a_0 + e_t$. Given an initial condition for y_0, the general solution for y_t is[20]

$$y_t = a_0 t + y_0 + \sum_{i=1}^{t} e_i \tag{4.48}$$

Updating by s periods, we obtain

$$y_{t+s} = a_0(t+s) + y_0 + \sum_{i=1}^{t+s} e_i \tag{4.49}$$

Substituting (4.48) into (4.49) so as to eliminate y_0 yields

$$y_{t+s} = a_0 s + y_t + \sum_{i=1}^{s} e_{t+i} \tag{4.50}$$

To express the solution for y_{t+s} in terms of $\{\varepsilon_t\}$ rather than $\{e_t\}$, note that

$$\sum_{i=1}^{s} e_{t+i} = \sum_{i=1}^{s} \varepsilon_{t+i} + \beta_1 \sum_{i=1}^{s} \varepsilon_{t-1+i} + \beta_2 \sum_{i=1}^{s} \varepsilon_{t-2+i} \tag{4.51}$$

so that the solution for y_{t+s} can be written as

$$y_{t+s} = a_0 s + y_t + \sum_{i=1}^{s} \varepsilon_{t+1} + \beta_1 \sum_{i=1}^{s} \varepsilon_{t-1+i} + \beta_2 \sum_{i=1}^{s} \varepsilon_{t-2+i} \tag{4.52}$$

Now consider the forecast of y_{t+s} for various values of s. Since all values of $E_t\varepsilon_{t+i} = 0$ for $i > 0$, it follows that

$$E_t y_{t+1} = a_0 + y_t + \beta_1\varepsilon_t + \beta_2\varepsilon_{t-1}$$

$$E_t y_{t+2} = 2a_0 + y_t + (\beta_1 + \beta_2)\varepsilon_t + \beta_2\varepsilon_{t-1}$$

$$\ldots$$

$$E_t y_{t+s} = sa_0 + y_t + (\beta_1 + \beta_2)\varepsilon_t + \beta_2\varepsilon_{t-1} \tag{4.53}$$

Here, the forecasts for all $s > 1$ are equal to the expression $sa_0 + y_t + (\beta_1 + \beta_2)\varepsilon_t + \beta_2\varepsilon_{t-1}$. Thus, the forecast function converges to a linear function of the forecast horizon s; the slope of the function equals a_0 and the level equals $y_t + (\beta_1 + \beta_2)\varepsilon_t + \beta_2\varepsilon_{t-1}$. This stochastic level can be called the value of the stochastic trend at t and is denoted by μ_t. This trend plus the deterministic value $a_0 s$ constitutes the forecast $E_t y_{t+s}$. There are several interesting points to note:

1. The trend is defined to be the conditional expectation of the limiting value of the forecast function. In lay terms, the trend is the "long-term" forecast. This forecast will differ at each period t as additional realizations of $\{\varepsilon_t\}$ become available. At any period t, the stationary component of the series is the difference between y_t and the trend μ_t. Hence, the stationary component of the series is

$$y_t - \mu_t = -(\beta_1 + \beta_2)\varepsilon_t - \beta_2\varepsilon_{t-1} \tag{4.54}$$

   At any point in time that y_t is given, the trend and the stationary components are perfectly correlated (the correlation coefficient being -1).

2. By definition, ε_t is the innovation in y_t and the variance of the innovation is σ^2. Since the change in the trend resulting from a change in ε_t is $1 + \beta_1 + \beta_2$, the variance of the innovation in the trend can exceed the variance of y_t itself. If $(1 + \beta_1 + \beta_2)^2 > 1$, the trend is more volatile than y_t since the negative correlation between μ_t and the stationary component act to smooth the $\{y_t\}$ sequence.

3. The trend is a random walk plus drift. Since the trend at t is μ_t, it follows that $\mu_t = y_t + (\beta_1 + \beta_2)\varepsilon_t + \beta_2\varepsilon_{t-1}$. Hence

$$\begin{aligned}\Delta\mu_t &= \Delta y_t + (\beta_1 + \beta_2)\Delta\varepsilon_t + \beta_2\Delta\varepsilon_{t-1} \\ &= (y_t - y_{t-1}) + (\beta_1 + \beta_2)\varepsilon_t - \beta_1\varepsilon_{t-1} - \beta_2\varepsilon_{t-2}\end{aligned}$$

Since $y_t - y_{t-1} = a_0 + \varepsilon_t + \beta_1\varepsilon_{t-1} + \beta_2\varepsilon_{t-2}$,

$$\Delta\mu_t = a_0 + (1 + \beta_1 + \beta_2)\varepsilon_t$$

Thus, $\mu_t = \mu_{t-1} + a_0 + (1 + \beta_1 + \beta_2)\varepsilon_t$, so that the trend at t is composed of the drift term a_0 plus the white-noise innovation $(1 + \beta_1 + \beta_2)\varepsilon_t$.

Beveridge and Nelson show how to recover the trend and stationary components from the data. In the example at hand, estimate the $\{y_t\}$ series using the Box–Jenkins technique. After differencing the data, an appropriately identified and estimated ARMA model will yield high-quality estimates of a_0, β_1, and β_2. Next, obtain ε_t and ε_{t-1} as the one-step-ahead forecast errors of y_t and y_{t-1}, respectively. To obtain these values, use the estimated ARMA model to make in-sample forecasts of each observation of y_{t-1} and y_t. The resulting forecast errors become ε_t and ε_{t-1}. Combining the estimated values of β_1, β_2, ε_t, and ε_{t-1} as in (4.54) yields the irregular component. Repeating for each value of t yields the entire irregular sequence. From (4.54), this irregular component is y_t minus the value of the trend; hence, the permanent component can be obtained directly.

The General ARIMA (*p*, 1, *q*) Model

The first difference of any ARIMA(p, 1, q) series has the stationary infinite-order moving-average representation:

$$y_t - y_{t-1} = a_0 + \varepsilon_t + \beta_1\varepsilon_{t-1} + \beta_2\varepsilon_{t-2} + \ldots$$

As in the earlier example, it is useful to define $e_t = \varepsilon_t + \beta_1\varepsilon_{t-1} + \beta_2\varepsilon_{t-2} + \beta_3\varepsilon_{t-3} + \ldots$, so that it is possible to write the solution for y_{t+s} in the same form as (4.50):

$$y_{t+s} = y_t + a_0 s + \sum_{i=1}^{s} e_{t+i}$$

The next step is to express the $\{e_t\}$ sequence in terms of the various values of the $\{\varepsilon_t\}$ sequence. In this general case, (4.51) becomes

$$\sum_{i=1}^{s} e_{t+i} = \sum_{i=1}^{s} \varepsilon_{t+i} + \beta_1 \sum_{i=1}^{s} \varepsilon_{t-1+i} + \beta_2 \sum_{i=1}^{s} \varepsilon_{t-2+i} + \beta_3 \sum_{i=1}^{s} \varepsilon_{t-3+i} + \ldots \tag{4.55}$$

Since $E_t\varepsilon_{t+i} = 0$, it follows that the forecast function can be written as

$$E_t y_{t+s} = y_t + a_0 s + \left(\sum_{i=1}^{s} \beta_i\right)\varepsilon_t + \left(\sum_{i=2}^{s+1} \beta_i\right)\varepsilon_{t-1} + \left(\sum_{i=3}^{s+2} \beta_i\right)\varepsilon_{t-2} + \ldots \tag{4.56}$$

Now, to find the stochastic trend, take the limiting value of the forecast $E_t(y_{t+s} - a_0 s)$ as s becomes infinitely large. As such, the stochastic trend is[21]

$$y_t + \left(\sum_{i=1}^{\infty} \beta_i\right)\varepsilon_t + \left(\sum_{i=2}^{\infty} \beta_i\right)\varepsilon_{t-1} + \left(\sum_{i=3}^{\infty} \beta_i\right)\varepsilon_{t-2} + \ldots$$

The key to operationalizing the decomposition is to recognize that y_{t+s} can be written as

$$y_{t+s} = \Delta y_{t+s} + \Delta y_{t+s-1} + \Delta y_{t+s-2} + \ldots + \Delta y_{t+1} + y_t$$

As such, the trend can always be written as the current value of y_t plus the sum of all of the forecasted changes in the sequence. Abstracting from $a_0 s$, the stochastic portion of the trend is

$$\begin{aligned}\lim_{s\to\infty} E_t y_{t+s} &= \lim_{s\to\infty} E_t[(y_{t+s} - y_{t+s-1}) + (y_{t+s-1} - y_{t+s-2}) + \ldots \\ &\quad + (y_{t+2} - y_{t+1}) + (y_{t+1} - y_t)] + y_t \\ &= \lim_{s\to\infty} E_t(\Delta y_{t+s} + \Delta y_{t+s-1} + \ldots + \Delta y_{t+2} + \Delta y_{t+1}) + y_t \qquad (4.57)\end{aligned}$$

The useful feature of (4.57) is that the Box–Jenkins method allows you to calculate each value of $E_t\Delta y_{t+s}$. For each observation in your data set, find all s-step-ahead forecasts and construct the sum given by (4.57). Since the irregular component is y_t minus the sum of the deterministic and stochastic trends, the irregular component can be constructed as

$$y_t - \lim_{s\to\infty}(E_t y_{t+s} + a_0 s) = -\lim_{s\to\infty} E_t(\Delta y_{t+s} + \Delta y_{t+s-1} + \ldots + \Delta y_{t+2} + \Delta y_{t+1}) - a_0 s$$

Thus, to use the Beveridge and Nelson (1981) technique:

STEP 1: Estimate the first difference of the series using the Box–Jenkins technique. Select the best-fitting ARMA(p, q) model of the $\{\Delta y_t\}$ sequence.

STEP 2: Using the best-fitting ARMA model, for each time period $t = 1, \ldots T$, find the one-step-ahead, two-step-ahead, … , s-step-ahead forecasts: that is, find $E_t\Delta y_{t+s}$ for each value of t and s. For each value of t, use these forecasted values to construct the sums: $E_t[\Delta y_{t+s} + \Delta y_{t+s-1} + \ldots + \Delta y_{t+1}] + y_t$. In practice, it is necessary to find a reasonable approximation to (4.57); in their own work, Beveridge and Nelson let $s = 100$. For example, for the first usable observation (i.e., $t = 1$), find the sum:

$$\mu_1 = E_1(\Delta y_{101} + \Delta y_{100} + \ldots + \Delta y_2) + y_1$$

The value of y_1 plus the sum of these forecasted changes equals $E_1 y_{101}$; the stochastic portion of trend in period 1 is $E_1 y_{101} - a_0 s$ and the deterministic portion is $a_0 s$. Similarly, for $t = 2$, construct

$$\mu_2 = E_2(\Delta y_{102} + \Delta y_{101} + \ldots + \Delta y_3) + y_2$$

If there are T observations in your data set, the trend component for the last period is

$$\mu_T = E_T(\Delta y_{T+100} + \Delta y_{T+99} + \ldots + \Delta y_{T+1}) + y_T$$

The entire sequence of constructed trends (i.e., $\mu_1, \mu_2, \ldots, \mu_T$) constitutes the $\{\mu_t\}$ sequence.

STEP 3: Form the irregular component at t by subtracting the stochastic portion of the trend at t from the value of y_t. Thus, for each observation t, the irregular component is $-E_t(\Delta y_{t+100} + \Delta y_{t+99} + \ldots + \Delta y_{t+1})$.

Note that for many series the value of s can be quite small. For example, in the ARIMA(0, 1, 2) model of (4.47), the value of s can be set equal to 2 since all forecasts for $s > 2$ are equal to zero. If the ARMA model that is estimated in Step 1 has slowly decaying autoregressive components, the value of s should be large enough that the s-step-ahead forecasts converge to the deterministic change a_0.

Two Examples: The file PANEL.XLS contains quarterly values of the real British pound estimated by the ARIMA(0, 1, 1) process:

$$\underset{}{\Delta y_t} = \underset{(0.24)}{0.0009} + \varepsilon_t + \underset{(3.56)}{0.324\varepsilon_{t-1}}$$

where Δy_t is the logarithmic change in the real British pound.

Although it is often desirable to maintain an insignificant intercept term in a regression, in this case it is clearly undesirable since it imparts a deterministic trend into the real exchange rate. As such, reestimate the model without the intercept to obtain

$$\Delta y_t = \varepsilon_t + 0.324\varepsilon_{t-1}$$

Step 2 requires that for each observation, we form the one-step- through s-step-ahead forecasts. For this model, the mechanics are trivial since for each period t, the one-step-ahead forecast is

$$E_t\Delta y_{t+1} = 0.324\varepsilon_t$$

and all other s-step-ahead forecasts are zero.

Thus, for each observation t, the summation $E_t(\Delta y_{t+100} + \Delta y_{t+99} + \ldots + \Delta y_{t+1})$ is equal to $0.324\varepsilon_t$. As such, for 1980Q2 (the first usable observation in the sample), the *stochastic* portion of the trend is $y_{1980Q2} + 0.324\varepsilon_{1980Q2}$ and the temporary portion of y_{1980Q2} is $-0.324\varepsilon_{1980Q2}$. Repeating for each point in the data set yields the irregular and permanent components of the sequence. The estimated ARIMA(0, 1, 1) model is the special case of (4.47) in which a_0 and β_2 are set equal to zero. As such, you should be able to write the equivalent of (4.48) through (4.54) for the real pound.

We have verified that the real U.S. GDP is the unit root process

$$\Delta lrgdp_t = 0.0055 + 0.3309\Delta lrgdp_{t-1}$$

Now it is more difficult to calculate the sum of the forecasted changes. Nevertheless, it is worthwhile to illustrate the process for the first few values. In 1947Q2 the value of $lrgdp_t$ was 0.45022 and the value of $\Delta lrgdp_t$ was -0.001189. Since we are not interested in the deterministic portion of the trend, based on the information available in 1947Q2 the one-step-ahead forecast for 1947Q3 is

$-3.934*10^{-4} = (0.3309)(-0.001189)]$ and the two-step-ahead forecast is $-1.301*10^{-4} = (0.3309)(-3.934*10^{-4})]$. The forecasts quickly converge to zero after a few periods. Adding up all such forecasted changes, you should obtain $-5.88*10^{-4}$. Thus, abstracting from the deterministic portion of the trend, the log of real GDP is forecasted to change by $-5.88*10^{-4}$ in the very long run. Adding this sum to $lrgdp_{1947Q2}$ yields the stochastic component as $0.45022 - 5.88*10^{-4} = 0.44963$. If you take the anti-logs, you find the actual level of real GDP in 1947Q2 to be \$1.568657 trillion and the permanent component to be \$1.567735 trillion.

Repeating this process for all observations in the data set yields the time path of the trend component of real GDP. If you were to plot the trend along with the actual values, you would find that the two series virtually overlap. Since the autoregressive coefficient is so small, virtually all of the movements in the real GDP series are permanent.

The Unobserved Components Decomposition

The Beveridge and Nelson (1981) decomposition has proven especially useful in that it provides a straightforward method to decompose any ARIMA(p, 1, q) process into a temporary and a permanent component. However, it is important to note that *the Beveridge and Nelson decomposition is not unique*. Equations (4.53) and (4.54) provide an example in which the Beveridge and Nelson decomposition forces the innovation in the trend and the stationary components to have a perfect negative correlation.

In fact, this result applies to the more general ARIMA(p, 1, q) model. Obtaining the irregular component as the difference between y_t and its trend forces the correlation coefficient between the innovations to equal -1. However, there is no reason to constrain the two innovations in the two components to be perfectly correlated. To illustrate the point, consider the random walk plus noise plus drift (i.e., the trend plus noise model) introduced in Section 1:

$$y_t = y_0 + a_0 t + \sum_{i=1}^{t} \varepsilon_i + \eta_t$$

The deterministic portion of the trend is $y_0 + a_0 t$, the stochastic trend is $\sum \varepsilon_i$, and the noise component is η_t. The stochastic trend and the noise components are uncorrelated since we assumed that $E(\varepsilon_i \eta_t) = 0$. Thus, the Beveridge and Nelson methodology would incorrectly identify the trend and the irregular components because it would force the two innovations to have a correlation of -1.

Now consider an alternative identification scheme that relies on an **unobserved components** (UC) model. A UC model posits that a series is composed of several distinct, but unobservable, components (such as the ε_t and η_t shocks in the random walk plus noise model). The goal is to identify the shocks by matching the moments of the UC specification to those from an estimated ARIMA model. In Section 2, the trend plus noise model was shown to have an equivalent ARIMA(0, 1, 1) representation such that

$$E\Delta y_t = 0; \qquad \text{var}(\Delta y_t) = \sigma^2 + 2\sigma_\eta^2, \quad \text{and} \quad \text{cov}(\Delta y_t, \Delta y_{t-1}) = -\sigma_\eta^2 \qquad (4.58)$$

Hence, it is possible to represent the first difference of the trend plus noise model as the MA(1) process:

$$\Delta y_t = a_0 + e_t + \beta_1 e_{t-1} \tag{4.59}$$

where e_t is an independent white-noise disturbance. The notation e_t is designed to indicate that shocks to Δy_t come from two sources: ε_t and η_t. The problem is to decompose the estimated values of $\{e_t\}$ into these two source components.

In this instance, it is possible to recover, or *identify*, the individual $\{\varepsilon_t\}$ and $\{\eta_t\}$ shocks from the estimation of (4.59). The appropriate use of the Box–Jenkins methodology will yield estimates of a_0, β_1, and the elements of the $\{e_t\}$ sequence. Given the coefficient estimates, it is possible to form

$$\text{var}(\Delta y_t) = \text{var}(e_t + \beta_1 e_{t-1}) = (1 + \beta_1)^2 \text{var}(e_t)$$

and

$$\text{cov}(\Delta y_t, \Delta y_{t-1}) = \beta_1 \text{var}(e_t)$$

However, these estimates of the variance and covariance are not arbitrary; for (4.59) to satisfy the restrictions of (4.58) it must be the case that

$$(1 + \beta_1)^2 \text{var}(e_t) = \sigma^2 + 2\sigma_\eta^2$$

and

$$\beta_1 \text{var}(e_t) = -\sigma_\eta^2$$

Now that we have estimated β_1 and var(e_t), it is possible to recover σ^2 and σ_η^2 from the data. The individual values of the $\{\varepsilon_t\}$ and $\{\eta_t\}$ sequences can be recovered as well. From the forecast function, $E_t y_{t+1} = y_t + a_0 - \eta_t$. Hence, it is possible to use one-step-ahead forecasts from (4.59) to find $E_t \Delta y_{t+1} = a_0 + \beta_1 e_t$, so that $E_t y_{t+1} = y_t + a_0 + \beta_1 e_t$. Since the two forecasts must be equivalent, it follows that

$$\beta_1 e_t = -\eta_t$$

Thus, the estimated values of $\beta_1 e_t$ can be used to identify the entire $\{\eta_t\}$ sequence. Given $\{e_t\}$ and $\{\eta_t\}$, the values of $\{\varepsilon_t\}$ can be obtained from $\Delta y_t = a_0 + \varepsilon_t + \Delta \eta_t$. For each value of t, form $\varepsilon_t = \Delta y_t - a_0 - \Delta \eta_t$ using the known values of Δy_t and the estimated values of a_0 and $\Delta \eta_t$.

The point is that it is possible to decompose a series such that the correlation between the trend and irregular components is zero. The example illustrates an especially important point. To decompose a series into a random walk plus drift and a stationary irregular component, it is necessary to specify the correlation coefficient between innovations in the trend and the irregular components. We have seen two ways to decompose an ARIMA(0, 1, 1) model. The Beveridge and Nelson technique assumes that the trend and cycle are perfectly correlated, while the UC decomposition adds the restriction

$$E\varepsilon_t \eta_t = 0$$

In fact, the correlation coefficient between the two components can be any number in the interval -1 to $+1$. Without the extra restriction concerning the correlation between the innovations, the trend and stationary components cannot be identified; in a sense, we are an equation short. This result carries over to more complicated models since it is always necessary to "cleave" or "partition" the contemporaneous movement of a series into its two constituent parts. The problem is important because economic theory does not always provide the relationship between the two innovations. Yet, *without a priori knowledge of the relationship between innovations in the trend and stationary components, the decomposition of a series into a random walk plus drift and a stationary component is not unique.*

Watson (1986) decomposed the logarithm of GNP using the Beveridge and Nelson decomposition and using the UC method. Using a Beveridge and Nelson decomposition, he estimated the following ARIMA(1, 1, 0) model (with standard errors in parentheses):

$$\underset{}{\Delta y_t} = \underset{(0.001)}{0.005} + \underset{(0.077)}{0.406}\Delta y_{t-1} + \varepsilon_t \qquad \text{var}(\varepsilon_t) = 0.0103^2$$

Using the UC method such that the innovations in the trend and irregular components are uncorrelated, Watson estimates GNP as the sum of a trend (τ_t) plus a cyclical term (c_t). The trend is simply a random walk plus drift and the cyclical component is an AR(2) process:[22]

$$\tau_t = \underset{(0.001)}{0.008} + \tau_{t-1} + \varepsilon_t \qquad \text{var}(\varepsilon_t) = 0.0057^2$$

$$c_t = \underset{(0.121)}{1.501}c_{t-1} - \underset{(0.125)}{0.577}c_{t-2} \qquad \text{var}(\eta_t) = 0.0076^2$$

The short-term forecasts of the two models are quite similar. The standard error of the one-step-ahead forecast of UC model is slightly smaller than that from the Beveridge and Nelson decomposition: $(\sigma^2 + \sigma_\eta^2)^{1/2} \cong 0.0095$ is slightly smaller than 0.0103. However, the long-run properties of the two models are quite different. For example, writing $\Delta y_t = (0.005 + \varepsilon_t)/(1 - 0.406L)$ yields the impulse response function using Beveridge and Nelson decomposition. The sum of the coefficients for this impulse response function is 1.68. Hence, a one-unit innovation will eventually increase log(GNP) by a full 1.68 units. Since all coefficients are positive, following the initial shock, y_t steadily increases to its new level. In contrast, the sum of the impulse response coefficients in the UC model is about 0.57. All coefficients beginning with lag 4 are negative. As such, a one-unit innovation in y_t has a larger effect in the short run than in the long run. Most importantly, the Beveridge and Nelson cycle has a small amplitude and is less persistent than the UC cycle.

Morley, Nelson, and Zivot (2003) updated Watson's (1986) study and found similar results using data through 1998*Q*2. They also showed how to use the Kalman filter to estimate the correlation between ε_t and η_t. This is a clear advantage over imposing a particular value for $E\varepsilon_t\eta_t$. It turns out that the estimated correlation is -0.9062 so that the cycle obtained by Beveridge and Nelson is quite reasonable. It is going too far afield to explain the Kalman filter and state-space modeling here. For interested readers, the *Supplemetary Manual* available on my Web site: www.cba.ua.edu/~wenders, contains an introductory discussion of the topic.

The Hodrick–Prescott Decomposition

Another method of decomposing a series into a trend and a stationary component has been developed by Hodrick and Prescott (1997). Suppose you observe the values y_1 through y_T and want to decompose the series into a trend $\{\mu_t\}$ and a stationary component $y_t - \mu_t$. Consider the sum of squares

$$\frac{1}{T}\sum_{t=1}^{T}(y_t - \mu_t)^2 + \frac{\lambda}{T}\sum_{t=2}^{T-1}[(\mu_{t+1} - \mu_t) - (\mu_t - \mu_{t-1})]^2$$

where λ is a constant and T is the number of usable observations.

The problem is to select the $\{\mu_t\}$ sequence so as to minimize this sum of squares. In the minimization problem, λ is an arbitrary constant reflecting the "cost" or penalty of incorporating fluctuations into the trend. In many applications using quarterly data, including Hodrick and Prescott (1984) and Farmer (1993), λ is set equal to 1,600. Increasing the value of λ acts to "smooth out" the trend. If $\lambda = 0$, the sum of squares is minimized when $y_t = \mu_t$; the trend is equal to y_t itself. As $\lambda \to \infty$, the sum of squares is minimized when $(\mu_{t+1} - \mu_t) = (\mu_t - \mu_{t-1})$. As such, as $\lambda \to \infty$, the change in the trend is constant; the result is that there is a linear time trend. Intuitively, for large values of λ, the Hodrick–Prescott (HP) decomposition forces the change in the trend (i.e., $\Delta\mu_{t+1} - \Delta\mu_t$) to be as small as possible. This occurs when the trend is linear.

The benefit of the Hodrick–Prescott decomposition is that it uses the same method to extract the trend from a set of variables. For example, many real business cycle models indicate that all variables will have the same stochastic trend. A Beveridge and Nelson decomposition separately applied to each variable will not yield the same trend for each. Figure 4.13 shows a HP decomposition applied to real

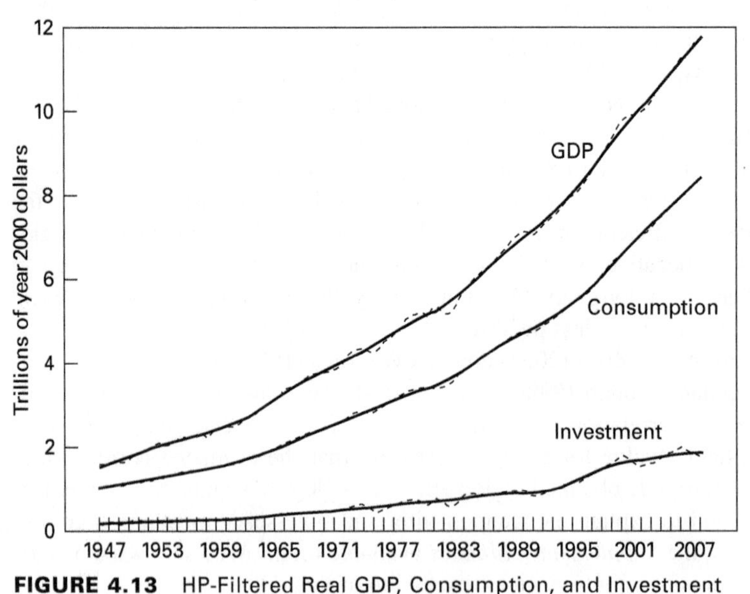

FIGURE 4.13 HP-Filtered Real GDP, Consumption, and Investment

U.S. GDP, consumption, and investment. You can see that the smoothed lines (representing the trends extracted by the HP decomposition) are such that the permanent components of each series account for the majority of the variation. However, a word of warning is in order. Since the HP filter is a function that smooths the trend, it has been shown to introduce spurious fluctuations into the irregular component of a series. The filter forces the stochastic trend to be a smoothed version of $(\mu_{t+1} - \mu_t) - (\mu_t - \mu_{t-1})$. As such, the filter works best if the $\{y_t\}$ series is $I(2)$, so that smoothing the second difference of the stochastic trend is appropriate.

13. SUMMARY AND CONCLUSIONS

The trend in a series can contain both stochastic and deterministic components. Differencing can remove a stochastic trend and detrending can eliminate a deterministic trend. However, it is inappropriate to difference a trend-stationary series and to detrend a series containing a stochastic trend. The resultant irregular component of the series can be estimated using Box–Jenkins techniques.

In contrast to traditional theory, the consensus view is that most macroeconomic time series contain a stochastic trend. In finite samples, the correlogram of a unit root process will decay slowly. As such, a slowly decaying ACF can be indicative of a unit root or a near–unit root process. The issue is especially important since many economic time series appear to have a nonstationary component. When you encounter such a time series, do you detrend, do you first-difference, or do you do nothing since the series might be stationary?

Adherents of the Box–Jenkins methodology recommend differencing a nonstationary variable or a variable with a near–unit root. For very short-term forecasts, the form of the trend is nonessential. Differencing also reveals the pattern of the other autoregressive and moving-average coefficients. However, as the forecast horizon expands, the precise form of the trend becomes increasingly important. Stationarity implies the absence of a trend and long-run mean reversion. A deterministic trend implies steady increases (or decreases) into the infinite future. Forecasts of a series with a stochastic trend converge to a steady level. As illustrated by the distinction between real business cycles and the more traditional formulations, the nature of the trend may have important theoretical implications.

The usual t-statistics and F-statistics are not applicable to determine whether or not a sequence has a unit root. Dickey and Fuller (1979, 1981) provided the appropriate test statistics to determine whether a series contains a unit root, a unit root plus drift, and/or a unit root plus drift plus a time trend. The tests can also be modified to account for seasonal unit roots. Structural breaks will bias the Dickey–Fuller test toward the nonrejection of a unit root. Perron (1989) shows how it is possible to incorporate a known structural change into the tests for unit roots. Caution needs to be exercised because it is always possible to argue that structural change has occurred; each year has something different about it than the previous year.

All the aforementioned tests have very low power to distinguish between a unit root and a near–unit root process. A trend-stationary process can be arbitrarily well approximated by a unit root process, and a unit root process can be arbitrarily well approximated by a trend-stationary process. Moreover, the testing procedure is confounded by the presence of the deterministic regressors (i.e., the intercept and the deterministic trend). The testing regression is misspecified if it omits any of the deterministic regressors in the data-generating process. However, too many regressors reduce the power of the tests. DF-GLS detrending methods generally have much better power than the traditional Dickey–Fuller tests. If a reasonable number of similar series are available (such as the real exchange rates from a number of countries), panel unit root tests can be used.

The fact that macroeconomic variables are not mean-reverting makes it difficult to calculate the trend and cyclical components of GDP and its subcomponents. After all, traditional detrending yields nothing like a stationary cyclical component when a series contains a stochastic trend. Several methods have been devised to decompose real GDP into its permanent and temporary components. The Beveridge and Nelson (1981) decomposition indicates that innovations in the stochastic trend account for a sizable proportion of the period-to-period movements. However, the Beveridge and Nelson decomposition is not unique in that it forces the correlation coefficient between innovations in the trend and irregular components to have a correlation coefficient of -1. The Unobserved Components framework allows the correlation to be estimated along with the other parameters of the model. In contrast, the Hodrick–Prescott filter smooths the trend component of a series. In Chapter 5 you will be shown a multivariate technique that allows for a unique decomposition of a series into its temporary and permanent components.

QUESTIONS AND EXERCISES

1. Given an initial condition for y_0, find the solution for y_t. Also find the s-step-ahead forecast $E_t y_{t+s}$.
   - **a.** $y_t = y_{t-1} + \varepsilon_t + 0.5\varepsilon_{t-1}$
   - **b.** $y_t = 1.1y_{t-1} + \varepsilon_t$
   - **c.** $y_t = y_{t-1} + 1 + \varepsilon_t$
   - **d.** $y_t = y_{t-1} + t + \varepsilon_t$
   - **e.** $y_t = \mu_t + \eta_t + 0.5\eta_{t-1}$, where $\mu_t = \mu_{t-1} + \varepsilon_t$
   - **f.** $y_t = \mu_t + \eta_t + 0.5\eta_{t-1}$, where $\mu_t = 0.5 + \mu_{t-1} + \varepsilon_t$
   - **g.** How can you make the models of parts b and d stationary?
   - **h.** Does model e have an ARIMA$(p, 1, q)$ representation?
2. Given the initial condition y_0, find the general solution and the forecast function (i.e., $E_t y_{t+s}$) for the following variants of the trend plus irregular model:
   - **a.** $y_t = \mu_t + v_t$, where $u_t = u_{t-1} + \varepsilon_t$, $v_t = (1 + \beta_1 L)\eta_t$, and $E\varepsilon_t\eta_t = 0$
   - **b.** $y_t = \mu_t + v_t$, where $u_t = u_{t-1} + \varepsilon_t$, $v_t = (1 + \beta_1 L)\eta_t$, and the correlation between ε_t and η_t equals unity
   - **c.** Find the ARIMA representation of each model.
3. Monte Carlo experiments allow you to replicate a random process so as to calculate the probability of a particular type of outcome. In a properly designed experiment, you can obtain the entire probability distribution. Consider each of the following:
   - **a.** Suppose you toss a coin and a tetrahedron. For the coin, you get 1 point for a tails and 2 points for a heads. The faces of the tetrahedron are labeled 1 through 4. For the tetrahedron, you get the number of points shown on the downward face. Your total score equals the number of points received for the coin and the tetrahedron. Of course, it is impossible to have a score of zero or 1. It is straightforward to calculate that the probabilities of scores 3, 4, and 5 equal 0.25 while the probabilities of scores 2 and 6 equal 0.125.
     - **i.** Describe how you can design a Monte Carlo experiment to calculate these probabilities.
     - **ii.** If your software package allows you to program such an experiment, find out how close the calculated probabilities come to the actual probabilities using 100 replications. How does your answer change if you use 1,000 replications?

b. Replicate the Monte Carlo results that were reported in Section 4 for the t-distribution of a unit root.

c. As discussed in Appendix 4.2, Rule 2 of Sims, Stock, and Watson (1990) states:

> If the data-generating process contains any deterministic regressors (*i.e.*, an intercept or a time trend) and the estimating equation contains these deterministic regressors, inference on all coefficients can be conducted using a t-test or an F-test.

Suppose that a series is generated by the equation $\Delta y_t = a_0 + a_2 t + \varepsilon_t$. A researcher unaware of the actual nature of the series estimates the equation in the form $\Delta y_t = a_0 + \rho y_{t-1} + a_2 t + \varepsilon_t$. Since there is a trend and an intercept in the DGP, Rule 2 indicates that it is appropriate to test the null hypothesis $\rho = 0$ using a normal distribution.

i. Perform the following Monte Carlo experiment to verify this result.

First, select values of a_0 and a_2. If the assertion is correct, you must be able to select any nonzero values for these two coefficients. Generate a series of 100 observations using the equation $y_t = a_0 + y_{t-1} + a_2 t + \varepsilon_t$. Let the initial value of the series be a_0.

Second, estimate the series in the form $\Delta y_t = a_0 + \rho y_{t-1} + a_2 t + \varepsilon_t$ and obtain the t-statistic for the null hypothesis $\rho = 0$.

Third, repeat the first two steps 2,000 times. Obtain the distribution for the calculated t-statistics.

ii. John obtained the following results using 10,000 Monte Carlo replications:

First percentile:	–2.38	Ninetieth percentile:	1.29
Fifth percentile:	–1.67	Ninety-fifth percentile:	1.67
Tenth percentile:	–1.30	Ninety-ninth percentile:	2.33
Twenty-fifth percentile:	–0.66		

Explain how John's findings are consistent with the claim that it is appropriate to test the null hypothesis $\rho = 0$ using a normal distribution.

d. In contrast to part c, suppose that a series is generated by the equation $\Delta y_t = \varepsilon_t$. A researcher unaware of the actual nature of the series estimates the equation in the form $\Delta y_t = a_0 + \rho y_{t-1} + a_2 t + \varepsilon_t$. Repeat steps 1 to 3 in part c(i) above using $a_0 = a_2 = 0$. How close do your results come to the Dickey–Fuller τ_τ-statistic?

4. Use the data sets that come with this text to perform the following:

a. The file PANEL.XLS contains the real exchange rates used to generate the results reported in Table 4.8. Verify the lag lengths, the values of γ, and the t-statistics reported in the left-hand side of the table.

b. Does the ERS test confirm the results you found in part a?

c. The file ERSTEST.XLS contains the data used in Section 10. Reproduce the results reported in the text.

d. The file QUARTERLY.XLS contains the M1NSA series used to illustrate the test for seasonal unit roots. Make the appropriate data transformations and verify the results concerning seasonal unit roots presented in Section 7.

5. The second column in the file BREAK.XLS contains the simulated data used in Section 8.

a. Plot the data to see if you can recognize the effects of the structural break.

b. Verify the results reported in Section 8.

c. The third column in the file BREAK.XLS contains another simulated data series called $\{y2_t\}$ with a structural break at $t = 51$. Plot the series and compare your graph to those of Figures 4.10 and 4.11.

d. Obtain the ACF and PACF of the $\{y2_t\}$ sequence and of the first difference of the sequence. Does the data appear to be difference-stationary?

e. If you perform a Dickey–Fuller test including a constant and a trend, you should obtain:

$$\begin{array}{llll} y2_t = 0.072 & - 1.014{*}10^{-4}t & - 0.022y2_{t-1} \\ \quad (1.01) & \quad (-0.05) & \quad (-0.66) \end{array}$$

In addition to the fact that all t-statistics are small, in what other ways is this regression inadequate? What diagnostic checks would you want to perform?

f. Estimate the equation $y2_t = a_0 + a_2 t + \mu_2 D_L$ and save the residuals. Perform a Dickey–Fuller test on the saved residuals. Perform the appropriate diagnostic tests on this regression to ensure that the residuals approximate white noise. You should conclude that the series is a unit root process with a one-time pulse at $t = 51$.

g. Reestimate the model without the insignificant time trend. How is your answer affected?

6. The file RGDP.XLS contains the real GDP data that were used to estimate (4.29).

   a. Use the series to replicate the results in Appendix 4.2.

   b. It is often argued that the oil price shock in 1973 reduced the trend growth rate of real U.S. GDP. Perform the Perron test to determine whether the series is trend-stationary with a break occurring in mid-1973.

   c. Decompose the real GDP series into the temporary and permanent components using the HP filter and the Beveridge–Nelson decomposition. Plot the transitory component that you obtained from the HP filter and the one you obtained from the Beveridge–Nelson decomposition. In what ways are the two series different?

   d. Suppose that real GDP is trend-stationary with a break occurring in mid-1973. Let the deviations from trend constitute the transitory component of the series. How does this transitory component compare with your answers found in part c?

7. The file PANEL.XLS contains the real exchange rate series used to perform the panel unit root tests reported in Section 11.

   a. Replicate the results of Section 11.

   b. In what way do the results of the test change if Australia, France, Germany, and the United States are excluded from the panel? Why is it inappropriate to include countries based on their t-statistics?

   c. Suppose that you mistakenly included a time trend in the augmented Dickey–Fuller tests. Determine how the results reported in Section 11 change.

8. Section 5 of Chapter 4 in the *Programming Manual* that accompanies this text illustrates a number of Monte Carlo experiments. If you have not already done so, download the manual from my Web site, www.cba.ua.edu/~wenders, the ESTIMA Web site (estima.com), or the Wiley Web site.

   a. Program 4.5 performs the modified coin-tossing problem using a coin and a tetrahedron. How would you modify the program so that you toss two tetrahedra?

   b. Program 4.6 shows the downward bias for the first-order AR coefficient in a sample of 100 observations. Modify the program to calculate the extent of the bias in a smaller sample.

c. Program 4.7 calculates the Dickey–Fuller critical values for τ_μ using a sample size of 100. How would you modify the program to obtain the critical values of the τ_τ statistic?

d. Program 4.7 also calculates the power of the τ_μ statistic. How would you modify the program to obtain the power of the τ_τ statistic?

9. Section 7 of Chapter 4 of the *Programming Manual* that accompanies this text contains a discussion of bootstrapping. In particular, Program 4.14 provides an example of bootstrapping regression coefficients. Use the data in MONEY_DEM.XLS to estimate the logarithmic change in real GDP ($dlrgdp_t$) as an AR(2) process. You should obtain

$$dlrgdp_t = 0.005 + 0.0251dlrgdp_{t-1} + 0.136dlrgdp_{t-2} + \varepsilon_t$$

a. Follow the program to obtain the bootstrap 95 percent confidence interval for the AR(1) and AR(2) coefficients.

b. How would you modify the program so as to obtain the bootstrap 95 percent confidence interval for the presence of a time trend?

10. The file QUARTERLY.XLS contains the U.S. interest rate data used in Section 10 of Chapter 2. Form the spread, s_t, by subtracting the T-bill rate from the 10-year rate. Recall that the spread appeared to be quite persistent in that $\rho_1 = 0.89$ and $\rho_2 = 0.74$.

a. Use a lag length of 8 and perform an augmented Dickey–Fuller test of the spread. You should find:

$$\underset{}{s_t} = \underset{(3.44)}{0.248} - \underset{(-4.06)}{0.177}s_{t-1} + \Sigma\beta_i\Delta s_{t-i}$$

Is the spread stationary?

b. Perform an augmented Dickey–Fuller test of the 10-year rate using five lags. Is the 10-year rate stationary?

c. Perform an augmented Dickey–Fuller test of the T-bill rate using seven lags. Is the T-bill rate stationary?

d. How is it possible that the individual rates act as $I(1)$ processes whereas the spread acts as a stationary process?

e. Suppose that you tested the T-bill rate using five lags. How do you interpret this result?

11. The file QUARTERLY.XLS contains the index of industrial production, the money supply as measured by M1, and the unemployment rate over the 1960Q1 to 2008Q1 period.

a. Show that the results using this data set seem to support Dickey and Fuller's (1981) finding that industrial production (INDPRO) is $I(1)$.

b. Perform an augmented Dickey–Fuller test on the unemployment rate (URATE). If you use eight lagged changes you will find

$$\Delta urate_t = \underset{(2.16)}{0.180} - \underset{(-2.20)}{0.031}urate_{t-1} + \Sigma\beta_i\Delta urate_{t-i}$$

Note the t-statistic on β_8 is -2.68.

c. Now estimate the unemployment rate using only one lagged change. You should find

$$\Delta urate_t = \underset{(3.10)}{0.227} - \underset{(-3.20)}{0.039}urate_{t-1} + \underset{(11.56)}{0.645}\Delta urate_{t-1}$$

The residuals show only mild evidence of serial correlation. Consider:

ρ_1	ρ_2	ρ_3	ρ_4	ρ_5	ρ_6	ρ_7	ρ_8
0.02	−0.02	0.07	−0.13	−0.07	0.09	0.15	−0.15

What do you conclude about the stationarity of the unemployment rate?

d. Regress INDPRO on M1NSA. You should obtain

$$INDPRO_t = \underset{(31.88)}{28.491} + \underset{(48.80)}{0.056} M1NSA_t$$

Examine the ACF of the residuals. Also create a scatter plot of INDPRO_t against M1NSA_t. How do you interpret the fact that $R^2 = 0.99$ and that the t-statistic on the money supply is 48.48?

ENDNOTES

1. Many treatments use the representation y_t = trend + seasonal + cyclical + noise. In this text, the term *cyclical* is avoided because it implies that cyclical economic components are deterministic.
2. A linguist might want to know why detrending entails removing the deterministic trend and not the stochastic trend. The reason is purely historical; originally, trends were viewed as deterministic. Today, subtracting the deterministic time trend is still called *detrending*.
3. If $B(L)$ is of infinite order, it is assumed that $\Sigma\beta_i^2$ is finite.
4. If only $B(L)$ has a unit root, the process is not invertible. The $\{y_t\}$ sequence may be stationary, but the usual estimation techniques are inappropriate. If both $A(L)$ and $B(L)$ have unit roots, there is the *common factor problem* discussed in Chapter 2. The unit root can be factored from $A(L)$ and $B(L)$.
5. Here we use the notation e_t, rather than ε_t, to indicate that the residuals from such a regression will generally not be white noise.
6. For the same reason, it is also inappropriate to use one variable that is trend-stationary and another that is difference-stationary. In such instances, "time" can be included as a so-called *explanatory* variable, or the variable in question can be detrended.
7. Issues concerning higher-order equations, longer lag lengths, serial correlation in the residuals, structural change, and the presence of deterministic components are temporarily ignored.
8. The Dickey–Fuller critical values are all reported in Table A at the end of this text.
9. Suppose that the estimated value of γ is -1.9 (so that the estimate of a_1 is -0.9) with a standard error of 0.04. Since the estimated value of γ is 2.5 standard errors from -2 [$(2 - 1.9)/.04 = 2.5$], the Dickey–Fuller statistics indicate that we cannot reject the null hypothesis $a_1 = -2$ at the 95 percent significance level. Unless stated otherwise, the discussion in this text assumes that a_1 is positive. Also note that if there is no prior information concerning the sign of a_1, a two-tailed test can be conducted.
10. When the distribution for v_t is more complicated, the distribution of the mean may not be normal with variance σ^2/T.
11. Tests using lagged changes in the $\{\Delta y_t\}$ sequence are called augmented Dickey–Fuller tests.
12. In their simulations, Dickey and Fuller (1981) found that 90 percent of the calculated ϕ_3 statistics were 5.47 or less and that 95 percent were 6.49 or less when the actual data were generated according to the null hypothesis.
13. Figure 4.8 runs through 1989.
14. Perron's Monte Carlo study allows for a drift and a deterministic trend. Nonetheless, the value of a_1 is biased toward unity in the presence of the deterministic trend.
15. Moreover, Evans and Savin (1981) found that for an AR(1) model, the limiting distribution of the autoregressive parameter has a normal asymptotic distribution. However, when the parameter is near

one, the unit root distribution is a better finite sample approximation than the asymptotically correct distribution.

16. Campbell and Perron (1991) reported that omitting a variable that is growing at least as fast as any other of the appropriately included regressors causes the power of the tests to approach zero.
17. The ERS procedure is called **Generalized Least Squares** (GLS) detrending because of the way that the near-differencing is performed. Suppose $B(L)$ is the first-order autoregressive process: $\varepsilon_t + \alpha\varepsilon_{t-1}$. Forming $y_t - \alpha y_{t-1}$ yields the serially uncorrelated error structure used in GLS. In the problem at hand, the actual α is unknown. However, if the y_t series is persistent, such differencing should mean that the ACF of $B(L)\varepsilon_t - \alpha B(L)\varepsilon_{t-1}$ is close to that of a white-noise process.
18. To explain, if the error process were such that $B(L) = \sum_{i=0}^{\infty} \alpha^i \varepsilon_{t-i}$, the variance of the error term would be $\sigma^2/(1-\alpha^2)$. Treating $\sigma^2 = 1$ and dividing y_1 by its standard deviation yields $y_1(1 - \alpha^2)^{0.5}$.
19. The result holds for very large values of n and T.
20. Also assume that all values of ε_i are zero for $i < 1$.
21. As an exercise, prove that the first difference of the trend acts as a random walk plus drift. Show that $\mu_t - \mu_{t-1}$ has the intercept a_0 plus a serially uncorrelated error.
22. The assumption that ε_t and η_t are uncorrelated places restrictions on the autoregressive and moving-average coefficients of Δy_t. For example, in the pure random walk plus noise model, β_1 must be negative. To avoid estimating a constrained ARIMA model, Watson estimated the trend and the irregular terms as unobserved components. Many software packages are capable of estimating such equations as time-varying parameter models. Details of the procedure can be obtained in Harvey (1989).

APPENDIX 4.1: THE BOOTSTRAP

Bootstrapping is similar to a Monte Carlo experiment with one essential difference. In a Monte Carlo study, you generate the random variables from a given distribution such as the Normal. The bootstrap takes a different approach—the random variables are drawn from their observed distribution. In essence, the bootstrap uses the **plug-in principle**—the observed distribution of the random variables is the best estimate of their actual distribution.

The idea of the bootstrap was developed in Efron (1979). The key point made by Efron is that the observed data set is a random sample of size T drawn from the actual probability distribution generating the data. In a sense, the empirical distribution of the data is the best estimate of the actual distribution of the data. As such, the empirical distribution function is defined to be the discrete distribution that places a probability of $1/T$ on each of the observed values. It is the empirical distribution function—and not some prespecified distribution such as the Normal—that is used to generate the random variables. The **bootstrap sample** is a random sample of size T drawn *with* replacement from the observed data putting a probability of $1/T$ on each of the observed values.

Example: To be more specific, suppose that we have the following 10 values of x_t:

t	1	2	3	4	5	6	7	8	9	10
x_t	0.8	3.5	0.5	1.7	7.0	0.6	1.3	2.0	1.8	−0.5

The sample mean is 1.87 and the standard deviation is 2.098. The following data show three different bootstrap samples. Each bootstrap sample consists of 10 randomly selected values of x_t drawn with replacement—each of the 10 values listed above is drawn with a probability of 0.1. It might seem that this resampling repeatedly selects the same sample. However, by sampling with replacement, some elements of x_t will appear more than once in the bootstrap sample. The first three bootstrap samples might look like this:

t	1	2	3	4	5	6	7	8	9	10	μ_i^*
x_1^*	3.5	1.7	−0.5	0.5	1.8	2.0	1.7	0.6	0.6	7.0	1.89
x_2^*	−0.5	0.6	0.6	0.8	1.7	7.0	1.8	3.5	1.8	0.8	1.81
x_3^*	0.5	0.6	7.0	1.3	1.3	7.0	1.3	1.8	3.5	0.6	2.49

where x_i^* denotes the bootstrap sample i and μ_i^* is the sample mean.

Notice that 0.6 and 1.7 appear twice in the first bootstrap sample; 0.6, 0.8, and 1.8 appear twice in the second; and 1.3 appears three times in the third bootstrap sample. Unless there is a large outlier, Efron (1979) showed that the moments of the bootstrap samples converge to the population moments as the number of bootstrap samples goes to infinity.

Bootstrapping Regression Coefficients

Suppose you have a data set with T observations and want to estimate the effects of variable x on variable y. Toward this end, you might estimate the linear regression:

$$y_t = a_0 + a_1 x_t + \varepsilon_t$$

Although the properties of the estimators are well known, you might not be confident about using standard t-tests if the estimated residuals do not appear to be normally distributed. You could perform a Monte Carlo study concerning the statistical properties of $\hat{a}_0$ and $\hat{a}_1$. However, instead of selecting various values of ε_t from a normal distribution, you can use the actual regression residuals. The technique is called the method of **bootstrapped residuals.**[1] To use the procedure, perform the following steps:

STEP 1: Estimate the model and calculate the residuals as

$$e_t = y_t - \hat{a}_0 - \hat{a}_1 x_t$$

STEP 2: Generate a bootstrap sample of the error terms containing the elements $e_1^*, e_2^*, \ldots, e_T^*$. Use the bootstrap sample to calculate a bootstrapped y series (called y^*). For each value of t running from 1 to T, calculate y_t^* as

$$y_t^* = \hat{a}_0 + \hat{a}_1 x_t + e_t^*$$

Note that the estimated values of the coefficients are treated as fixed. Moreover, the values of x_t are treated as fixed quantities so that they remain the same across samples.

STEP 3: Use the bootstrap sample of the y_t^* series to estimate new values of a_0 and a_1, calling the resulting values a_0^* and a_1^*.

STEP 4: Repeat Steps 2 and 3 many times and calculate the sample statistics for a_0^* and a_1^*. These should be distributed in the same way as $\hat{a}_0$ and $\hat{a}_1$. For example, you can find the 95 percent confidence interval for $\hat{a}_1$ as the interval containing the lowest 2.5 percent and the highest 97.5 percent of the values of a_1^*.

Step 2 needs to be modified for a time-series model due to the presence of lagged dependent variables. As such, the bootstrap $\{y_t^*\}$ sequence is constructed in a slightly different manner. Consider the simple AR(1) model:

$$y_t = a_0 + a_1 y_{t-1} + \varepsilon_t$$

As in Step 2, we can construct a bootstrap sample of the error terms containing the elements $e_1^*, e_2^*, \ldots, e_T^*$. Now, construct the bootstrap $\{y_t^*\}$ sequence using this sample of error terms. In particular, given the estimates of a_0, and a_1 and an initial condition for y_1^*, the remaining values of the $\{y_t^*\}$ sequence can be constructed using

$$y_t^* = \hat{a}_0 + \hat{a}_1 y_{t-1}^* + e_t^*$$

For an AR(p) model, the initial conditions for y_1^* through y_p^* are usually selected by random draws from the actual $\{y_t\}$ sequence.[2] To avoid problems concerning the selection of the initial condition, it is typical to construct a bootstrap sample with $T + 50$ elements and to discard the first 50.

Example 1: Constructing a Confidence Interval. In Sections 2 and 12, the logarithmic change in real U.S. GDP was estimated as

$$\Delta lrgdp_t = \underset{(7.14)}{0.0055} + \underset{(5.47)}{0.3309}\Delta lrgdp_{t-1}$$

Suppose that you want to obtain a 90 percent confidence interval for the AR(1) coefficient. Given that the t-statistic is 5.47, the standard deviation is 0.0605. Thus, if the estimated value of a_1 does have a t-distribution, a confidence interval that spans ± 1.65 standard deviations on each side of 0.3309 runs from 0.231 to 0.431. However, the residuals do not appear to be normally distributed, so that the t-distribution may not be appropriate.[3] As such, we may want to construct bootstrap confidence intervals for the coefficient 0.3309. For Step 2, we need to generate a bootstrap sample of the regression residuals. This series, denoted by $\{e_t^*\}$, consists of 245 randomly drawn values from the actual residual sequence. We also need an initial condition; we can draw a random value of the $\{\Delta lrgdp_t\}$ sequence to serve as the initial condition. As such, we can create the following bootstrap sequence for $\{y_t^*\}$ containing a total of 246 observations:

$$y_t^* = 0.0055 + 0.3309 y_{t-1}^* + e_t^*$$

For Step 3, we pretend that we do not know the actual data-generating process for $\{y_t^*\}$ and estimate the series as an AR(1) process. Suppose you find that the estimated AR(1) coefficient is 0.3011. This estimate is the first value of a_1^*. Repeat this process several thousand times to obtain the distribution of a_1^*. After performing the 10,000 replications of the experiment, you should find that approximately 5 percent of the estimates lie below 0.2230 and 5 percent lie above 0.4181. These values serve as a 90 percent confidence interval. In this case, it turns out that the bootstrap confidence interval is similar to that obtained using a standard t-distribution.

Example 2: Bootstrapping a Test Statistic. Equation (4.29) was used to perform a Dickey–Fuller test on the real GDP series. Recall that the estimated model was

$$\underset{}{\Delta lrgdp_t} = \underset{(3.52)}{0.0246} + \underset{(2.49)}{0.0003t} - \underset{(-2.59)}{0.0360 lrgdp_{t-1}} + \underset{(5.66)}{0.3426 \Delta lrgdp_{t-1}}$$

The t-statistic for γ (i.e., the coefficient on $lrgdp_{t-1}$) of –2.59 is not significant at conventional levels. However, a concern might be that the residuals from the model are not normally distributed. Moreover, only in large samples do the coefficients of the augmented terms have no bearing on the appropriate critical values. Hence, it seems reasonable to bootstrap the τ_τ-statistic for (4.29). If we bootstrap the series under the null hypothesis of no deterministic time trend and $\gamma = 0$, we need only construct the bootstrap series

$$y_t^* = 0.0055 + 0.3309 y_{t-1}^* + e_t^*$$

Since y_t^* represents a first difference, it is necessary to create the level of the series as $Y_t^* = Y_{t-1}^* + y_t^*$. Now, for Step 3, estimate the bootstrap series in the form

$$y_t^* = a_0^* + \gamma^* Y_{t-1}^* + \lambda^* t + a_1^* y_{t-1}^*$$

Replicate the experiment many times; on each trial, record the t-statistic for the null hypothesis $\gamma^* = 0$. The value of the t-statistic marking the fifth percentile of all of the bootstrapped t-statistics is the 5 percent critical value. In addition, to obtain the bootstrap ϕ_3-statistic, we can obtain the sample values of F for the null hypothesis $\gamma^* = \lambda^* = 0$.

You can select the initial value of Y_1^* as $\log(RGDP_1)$ plus an error term or a randomly drawn value of real GDP. However, as mentioned above, most researchers would actually construct a series with an extra 50 observations and then discard the first 50 of the realizations. When this process was repeated 10,000 times, only 500 (i.e., 5 percent) of the t-statistics for the null hypothesis $\gamma^* = 0$ had values below –3.43. Given the actual value of the t-statistic of –2.59, we cannot reject the null hypothesis $\gamma = 0$ at the 5 percent significance level. Moreover, 95 percent of the F-statistics for the null hypothesis $\gamma^* = \lambda^* = 0$ were less than 6.45. Recall that the sample value ϕ_3 from equation (4.29) was only 4.12. Hence, at the 5 percent significance level, we cannot reject the null hypothesis $\gamma = \lambda = 0$.

It is also possible to use the bootstrap results to test the null hypothesis $\lambda^* = 0$. Of course, if we know that the series is stationary, the t-statistic of 2.49 indicates that the time trend belongs in the regression. However, we cannot use the usual t-distribution if real GDP is not stationary. In total, 5 percent of the bootstrapped t-statistics for the null hypothesis $\lambda^* = 0$ were below 0.895 and 5 percent were above 3.43. Since the actual value of the t-statistic is 2.49, if we use a 90 percent confidence interval we cannot reject the null hypothesis that the time trend does not appear in (4.29).

Example 3: Bootstrapping Correlated Residuals in a Panel. If the residuals from a panel unit root test are highly correlated across equations, you should use bootstrapped critical values. The only modification needed from Example 2 is that in a panel unit root test, you estimate a vector of equations. The essential requirement is that the residuals need to be sampled in such a way that preserves the contemporaneous correlations present in the data. Let $\{e_{it}\}$ denote the regression residual from (4.45) for regression i for time period t. If e_t denotes the vector of residuals for time period t, we can write

$$\begin{aligned} e_1 &= (e_{11}, e_{21}, e_{31}, \ldots, e_{n1}) \\ e_2 &= (e_{12}, e_{22}, e_{32}, \ldots, e_{n2}) \\ e_3 &= (e_{13}, e_{23}, e_{33}, \ldots, e_{n3}) \\ &\ldots \\ e_T &= (e_{1T}, e_{2T}, e_{3T}, \ldots, e_{nT}) \end{aligned}$$

Thus, the first bootstrap sample of the residuals might be

$$\begin{aligned} e_1^* &= (e_{13}, e_{23}, e_{33}, \ldots, e_{n3}) \\ e_2^* &= (e_{15}, e_{25}, e_{35}, \ldots, e_{n5}) \\ &\ldots \\ e_T^* &= (e_{12}, e_{22}, e_{32}, \ldots, e_{n2}) \end{aligned}$$

The point is that you resample in such a way as to maintain the contemporaneous relationships among the regression residuals. As in Example 2, construct a bootstrap series using the resampled values of $\{e_t^*\}$. Once you obtain the average value of the t-statistics for the first bootstrap sample, repeat the entire process several thousand times.

Efron and Tibshirani (1993) presented an extremely accessible treatment of bootstrapping. You may also download a programming manual that illustrates some bootstrapping techniques (at no charge) from my Web site or from the Wiley Web site for this text.

APPENDIX 4.2: DETERMINATION OF THE DETERMINISTIC REGRESSORS

Sometimes the appropriate null and alternative hypotheses are unclear. As indicated in the text, you do not want to lose power by including a superfluous deterministic regressor in a unit root test. However, omitting a regressor that is actually in the data-generating process leads to a misspecification error. Fortunately, Sims, Stock, and Watson (1990) provided a second rule that is helpful in selecting the appropriate set of regressors:

> **Rule 2:** If the data-generating process contains any deterministic regressors (*i.e.*, an intercept or a time trend) and the estimating equation contains these deterministic regressors, inference on all coefficients can be conducted using a t-test or an F-test. This is because a test involving a single restriction across parameters with different rates of convergence is dominated asymptotically by the parameters with the slowest rates of convergence.

While the proof is beyond the scope of this text, the point is that the nonstandard Dickey–Fuller distributions are needed only when you include deterministic regressors not in the actual data-generating process. Hence, if the DGP is known to contain the

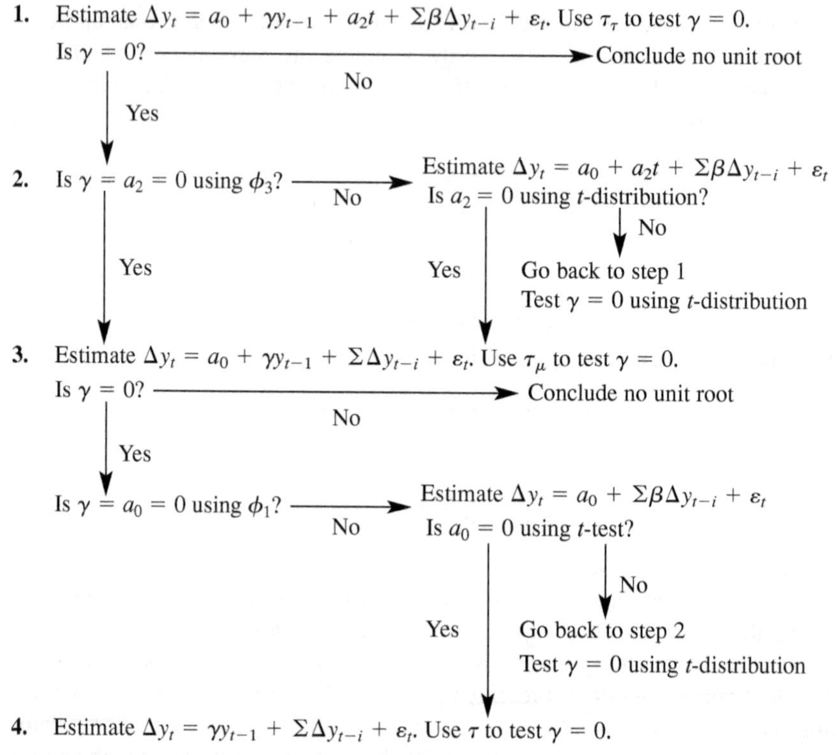

FIGURE A2.1 A Procedure to Test for Unit Roots

deterministic trend term $a_2 t$, the null hypothesis $\gamma = 0$ can be tested using a t-distribution if you estimate the model in the form of (4.25). However, if the superfluous deterministic trend is included, there is a substantial loss of power. As such, papers such as Dolado, Jenkinson, and Sosvilla–Rivero (1990) suggested a procedure to test for a unit root when the form of the data-generating process is *completely* unknown. The following is a straightforward modification of the method:

STEP 1: As shown in Figure A2.1, start with the least restrictive of the plausible models (which will generally include a trend and drift) and use the τ_τ statistic to test the null hypothesis $\gamma = 0$. Thus, in the most general case, you estimate the model in the form of (4.25) so that $\Delta y_t = a_0 + \gamma y_{t-1} + a_2 t + \sum \beta_i \Delta y_{t-i}$. Unit root tests have low power to reject the null hypothesis; hence, if the null hypothesis of a unit root is *rejected*, there is no need to proceed. Conclude that the $\{y_t\}$ sequence does not contain a unit root.

STEP 2: If the null hypothesis is *not rejected*, the next step is to determine whether the trend belongs in the estimating equation. Toward this end, you test the null hypothesis $a_2 = \gamma = 0$ using the ϕ_3 statistic. If you do not reject the null hypothesis, assume the absence of a trend and proceed to Step 3.

If you have reached this point, it is because the τ_τ test indicates that there is a unit root, and the ϕ_3 test indicates that γ and/or a_2 differs from zero.

As such, it would seem that there is a unit root and a trend. You can gain additional support for this result by assuming that there is a unit root and estimating $\Delta y_t = a_0 + a_2 t + \Sigma\beta_i \Delta y_{t-i}$. Since there are no $I(1)$ regressors in this specification, the test for the null hypothesis $a_2 = 0$ can be conducted using a standard t-test. If you conclude $a_2 = 0$, go to Step 3 since it does not appear that there is a trend. If you find that $a_2 \neq 0$, use (4.25) to test the null hypothesis $\gamma = 0$ using a t-distribution. Given that the trend belongs in the regression equation, Rule 2 indicates that the test for $\gamma = 0$ can be conducted using a t-distribution. Go no further; if $\gamma \neq 0$, conclude that the sequence is trend-stationary. If $\gamma = 0$, conclude that there is a unit root (and the $\{y_t\}$ sequence contains a quadratic trend).

STEP 3: Estimate the model without the trend [i.e., estimate a model in the form of (4.24)]. Test for the presence of a unit root using the τ_μ statistic. If the null is rejected, conclude that the model does not contain a unit root. If the null hypothesis of a unit root is not rejected, test for the significance of the intercept by testing the hypothesis $a_0 = \gamma = 0$ using the ϕ_1 statistic. If you do not reject the null hypothesis $a_0 = \gamma = 0$, assume that the intercept is zero and proceed to Step 4. Otherwise, estimate $\Delta y_t = a_0 + \Sigma\beta_i \Delta y_{t-i}$ and test whether $a_0 = 0$ using a t-distribution. If you find $a_0 = 0$, proceed to Step 4. Otherwise, conclude that the process contains an intercept term. In accord with Rule 2, you can use a t-distribution to test whether $\gamma = 0$ in the regression $\Delta y_t = a_0 + \gamma y_{t-1} + \Sigma\beta_i \Delta y_{t-i}$. If the null hypothesis of a unit root is rejected, conclude that the $\{y_t\}$ sequence is stationary around a nonzero mean. Otherwise, conclude that the $\{y_t\}$ sequence contains a unit root and a drift.

STEP 4: Estimate a model without the trend or drift; i.e., estimate a model in the form of (4.23). Use τ to test for the presence of a unit root. If the null hypothesis of a unit root is rejected, conclude that the $\{y_t\}$ sequence does not contain a unit root. Otherwise, conclude that the $\{y_t\}$ sequence contains a unit root.

Remember, no procedure can be expected to work well if it is used in a completely mechanical fashion. Plotting the data is usually an important indicator of the presence of deterministic regressors. The interest rate series shown in Figure 4.2 can hardly be said to contain a deterministic trend. Moreover, theoretical considerations might suggest the appropriate regressors. The efficient market hypothesis is inconsistent with the presence of a deterministic trend in an asset's return. Similarly, in testing for PPP, you should not begin using a deterministic time trend. However, the procedure is a sensible way to test for unit roots when the form of the data-generating process is completely unknown.

GDP and Unit Roots

Although the methodology outlined in Figure A2.1 can be very useful, it does have its problems. Each step in the procedure involves a test that depends on all the previous tests being correct; the significance level of each of the cascading tests is impossible to ascertain.

The procedure and its inherent dangers are nicely illustrated by trying to determine if the real GDP data shown in Figure 4.1 has a unit root. It is a good idea to replicate the results reported below using the data in RGDP.XLS. Using quarterly data over the 1947Q1–2008Q2 period, the correlogram of the logarithm of real GDP exhibits slow decay. At the end of Section 2, the logarithmic first difference of the series was estimated as

$$\underset{}{\Delta lrgdp_t} = \underset{(7.14)}{0.0055} + \underset{(5.47)}{0.3309}\Delta lrgdp_{t-1}$$

The model is well estimated in that the residuals appear to be white noise and all coefficients are of high quality. For our purposes, the interesting point is that the $\{\Delta lrgdp_t\}$ series appears to be a stationary process. Integrating suggests that $\{lrgdp_t\}$ has a stochastic and a deterministic trend. The issue here is to determine whether it was appropriate to difference the log of real GDP. Toward this end, consider the augmented Dickey–Fuller equation with t-statistics in parentheses:

$$\Delta lrgdp_t = \underset{(3.52)}{0.0246} + \underset{(2.49)}{0.00028}t - \underset{(-2.59)}{0.0360}\, lrgdp_{t-1} + \underset{(5.66)}{0.3426}\Delta lrgdp_{t-1} \qquad \text{(A2.1)}$$

As in Step 1, we estimate the least restrictive model; as such, (A2.1) contains an intercept and a deterministic trend. The point estimates of (A2.1) suggest that the real GDP is trend-stationary. However, the issue is to formally test the statistical significance of the null hypothesis $\gamma = 0$. The t-statistic for the null hypothesis $\gamma = 0$ is –2.59. Critical values with exactly 244 usable observations are not reported in the Dickey–Fuller table. However, with 250 observations, the critical value of τ_τ at the 10 percent and 5 percent significance levels are –3.13 and –3.43, respectively. At the 5 and 10 percent levels, we cannot reject the null of a unit root. However, the power of the test may have been reduced due to the presence of an unnecessary time trend and/or drift term. In Step 2, we use the ϕ_3 statistic to test the joint hypothesis $a_2 = \gamma = 0$. The sample value of F for the restriction is 4.12. Since the critical value of ϕ_3 is 6.34 at the 5 percent significance level, it is possible to conclude that the restriction $a_2 = \gamma = 0$ is not binding. Thus, we can proceed to Step 3 and estimate the model without the trend. Consider the following equation:

$$\Delta lrgdp_t = \underset{(4.39)}{0.0077} - \underset{(-1.42)}{0.0014}\, lrgdp_{t-1} + \underset{(5.33)}{0.3233}\Delta lrgdp_{t-1} \qquad \text{(A2.2)}$$

In (A2.2), the t-statistic for the null hypothesis $\gamma = 0$ is –1.42. Since the critical value of the τ_μ statistic is –2.88 at the 5 percent significance level, the null hypothesis of a unit root is not rejected at conventional significance levels. Again, the power of this test will have been reduced if the drift term does not belong in the model. To test for the presence of the drift, use the ϕ_1 statistic. The sample value of F for the restriction $a_0 = \gamma = 0$ is 26.63. Since 26.63 exceeds the critical value of 4.63, we conclude that the restriction is binding. Either $\gamma = 0$ (so there is not a unit root), $a_0 = 0$ (so there is not an intercept term), or both γ and a_0 are zero. In reality, any sensible researcher would stop at this point since it is not plausible to believe that real U.S.

GDP is a unit root process without any deterministic regressors. However, since the point of this section is to illustrate the procedure, test for the presence of the drift using

$$\Delta lrgdp_t = \underset{(7.14)}{0.0055} + \underset{(3.47)}{0.3309}\Delta lrgdp_{t-1}$$

Since we do not reject $a_0 = 0$ using a t-distribution, (A2.2) is our final testing regression. Since we are sure that the intercept belongs in the model, Rule 2 indicates that we can test the null hypothesis $\gamma = 0$ using a t-distribution. Given that the t-statistic is only –1.42, we can conclude that the series has a unit root and a drift term.

ENDNOTES

1. Another possibility discussed by Efron is to resample the paired (y_i, x_i) combinations.
2. A second, although less common, bootstrapping technique used in time-series models is called "moving blocks." For an AR(p) process, select a length L that is longer than p; L is the length of the block. To construct the bootstrap y^* series, randomly select a group of L adjacent data points to represent the first L observations of the bootstrap sample. In total, you need to select T/L of these samples to form a bootstrap sample with T observations. The idea of selecting a block is to preserve the time-dependence of the data. However, observations more than L apart will be nearly independent. Use this bootstrap sample to estimate the bootstrap coefficients a_0^* and a_1^*.
3. The value of the Jarque–Bera test statistic for normality is 15.51; hence, normality can be rejected at any reasonable significance level.

CHAPTER 5

MULTIEQUATION TIME-SERIES MODELS

As we have seen in previous chapters, you can capture many interesting dynamic relationships using single-equation time-series methods. In the recent past, many time-series texts would end with nothing more than a brief discussion of multiequation models. Yet, one of the most fertile areas of contemporary time-series research concerns multiequation models. This chapter has four specific aims:

1. Introduce **intervention analysis** and **transfer function analysis.** These two techniques generalize the univariate methodology by allowing the time path of a *dependent* variable to be influenced by the time path of an *independent* or *exogenous* variable. If it is known that there is no feedback, intervention and transfer function analyses can be very effective tools for forecasting and hypothesis testing.
2. Introduce the concept of a vector autoregression (VAR). The major limitation of intervention and transfer function models is that many economic systems do exhibit feedback. In practice, it is not always known if the time path of a series designated to be the *independent* variable has been unaffected by the time path of the *dependent* variable. The most basic form of a VAR treats all variables symmetrically without making reference to the issue of dependence versus independence.
3. Show that tools employed by a VAR analysis—Granger causality, impulse response analysis, and variance decompositions—can be helpful in understanding the interrelationships among economic variables and in the formulation of a more structured economic model. These tools are illustrated using examples concerning the fight against transnational terrorism.
4. Develop two new techniques, **structural VARs** and **multivariate decompositions,** that blend economic theory and multiple time-series analysis. Economic theories contain behavioral, structural, and/or reduced-form relationships that can be incorporated into a VAR analysis. In a structural VAR, the restrictions of a particular economic model are imposed on the contemporaneous relationship among the variables. The dynamic response of each variable to various economic shocks can be obtained and the restrictions of the model tested. Similarly, long-run neutrality restrictions can aid in decomposing a series into its temporary and permanent components. As

opposed to the class of univariate decompositions considered in Chapter 4, decompositions in a VAR framework can be exactly identified.

1. INTERVENTION ANALYSIS

Although the events of September 11, 2001, brought terrorism to the world's attention, the international community experienced a sharp increase in transnational terrorism beginning in the late 1960s. Terrorists engage in a wide variety of operations, including assassinations, armed attacks, bombings, kidnappings, and skyjackings. Such incidents are particularly heinous because they are often directed at innocent victims who are not part of the decision-making apparatus that the terrorists seek to influence. Figure 5.1 shows the total number of transnational terrorist incidents, along with the number of bombing incidents that have occurred since 1968$Q2$. Note that bombings are a large proportion of the total incident series. It is also interesting to note that the number of incidents began to decline in the mid-1990s and then increased sharply in 2000.

The skyjackings on September 11 and the skyjacking of Pan Am flight 103 over Lockerbie, Scotland, on December 21, 1988, captured the attention of the international community. However, skyjacking incidents have actually been quite numerous. The United States launched a critical response to the rise in skyjackings when it began to install metal detectors in all U.S. airports in January 1973. Other international authorities soon followed suit.

The quarterly totals of all transnational and U.S. domestic skyjackings are shown in Figure 5.2. Although the number of skyjacking incidents appears to take a sizable and permanent decline at this date, we might be interested in actually measuring the effects of installing the metal detectors. If $\{y_t\}$ represents the quarterly total of skyjackings, one might try to take the mean value of $\{y_t\}$ for all $t < 1973Q1$ and compare it to the mean value of $\{y_t\}$ for all $t \geq 1973Q1$. However, such a test is poorly designed because successive values of y_t are serially correlated. As such, some of the effects of the pre–metal detector regime could "carry over" to the post-intervention

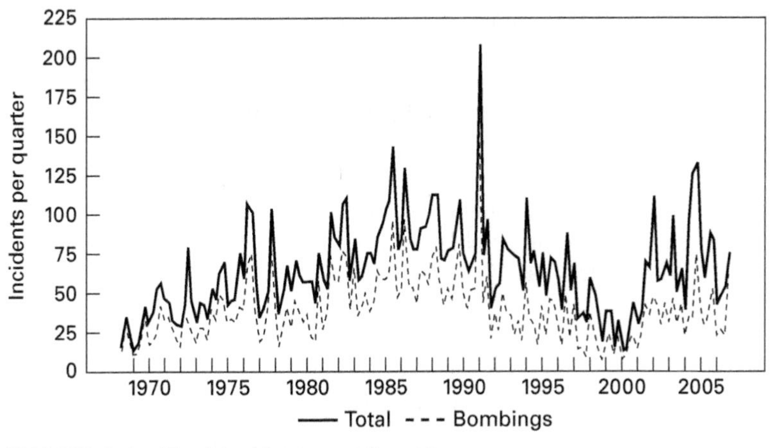

FIGURE 5.1 Total Incidents and Bombings

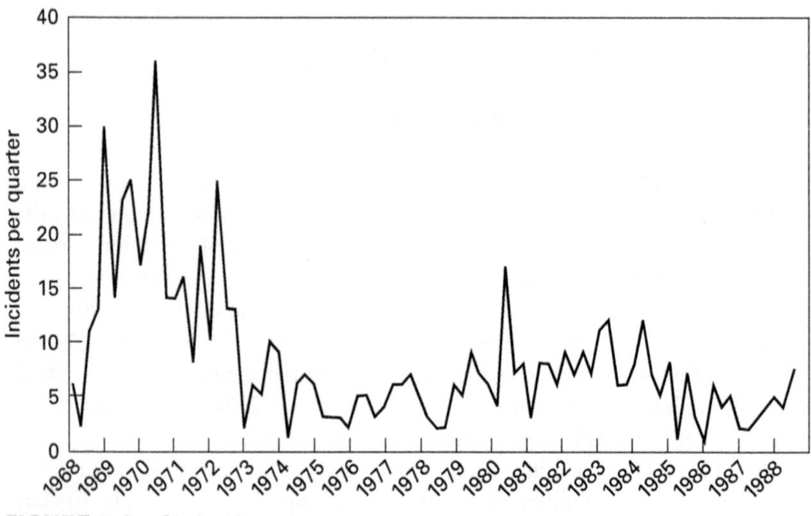

FIGURE 5.2 Skyjackings

date. For example, some planned skyjacking incidents already in the pipeline might not be deterred as readily as others.

Intervention analysis allows for a formal test of a change in the mean of a time series. Consider the model used in Enders, Sandler, and Cauley (1990) to study the impact of metal detector technology on the number of skyjacking incidents:

$$y_t = a_0 + a_1 y_{t-1} + c_0 z_t + \varepsilon_t, \quad |a_1| < 1 \tag{5.1}$$

where: z_t = the intervention (or dummy) variable that takes on the value of zero prior to 1973Q1 and unity beginning in 1973Q1

ε_t = a white-noise disturbance

To explain the nature of the model, notice that for $t < 1973Q1$, the value z_t is zero.[1] As such, the intercept term is a_0 and the long-run mean of the series is $a_0/(1 - a_1)$. Beginning in 1973, the intercept term jumps to $a_0 + c_0$ (since z_{1973Q1} jumps to unity). Thus, the initial or **impact effect** of the metal detectors is given by the magnitude of c_0. The statistical significance of c_0 can be tested using a standard t-test. We would conclude that metal detectors reduced the number of skyjacking incidents if c_0 is negative and statistically different from zero.

The long-run effect of the intervention is given by $c_0/(1 - a_1)$, which is equal to the new long-run mean $(a_0 + c_0)/(1 - a_1)$ minus the value of the original mean $a_0/(1 - a_1)$. The various transitional effects can be obtained from the impulse response function. Using lag operators, rewrite (5.1) as

$$(1 - a_1 L)y_t = a_0 + c_0 z_t + \varepsilon_t$$

so that

$$y_t = a_0/(1 - a_1) + c_0 \sum_{i=0}^{\infty} a_1^i z_{t-i} + \sum_{i=0}^{\infty} a_1^i \varepsilon_{t-i} \tag{5.2}$$

Equation (5.2) is an impulse response function; the interesting twist added by the intervention variable is that we can obtain the responses of the $\{y_t\}$ sequence to the intervention. To trace out the effects of metal detectors on skyjackings, suppose that $t = 1973Q1$ (so that $t + 1 = 1973Q2$, $t + 2 = 1973Q3$, etc.). For time period t, the impact of z_t on y_t is given by the magnitude of the coefficient c_0. The simplest way to derive the remaining impulse responses is to recognize that (1) $dy_t/dz_{t-i} = dy_{t+i}/dz_t$, and (2) $z_{t+i} = z_t = 1$ for all $i > 0$.

Hence, differentiate (5.2) with respect to z_{t-1} and update by one period so that

$$dy_{t+1}/dz_t = c_0 + c_0 a_1$$

The presence of the term c_0 reflects the direct impact of z_{t+1} on y_{t+1}, and the second term $c_0 a_1$ reflects the effect of z_t on y_t $(= c_0)$ multiplied by the effect of y_t on y_{t+1} $(= a_1)$. Continuing in this fashion, we can trace out the entire impulse response function as

$$dy_{t+j}/dz_t = c_0[1 + a_1 + \ldots + (a_1)^j]$$

since $z_{t+1} = z_{t+2} = \ldots = 1$.

Taking limits as $j \to \infty$, we can reaffirm that the long-run impact is given by $c_0/(1 - a_1)$. If it is assumed that $0 < a_1 < 1$, the absolute value of the magnitude of the impacts is an increasing function of j. As we move further away from the date on which the policy was introduced, the absolute value of the magnitude of the policy response becomes greater. If $-1 < a_1 < 0$, the policy has a damped oscillating effect on the $\{y_t\}$ sequence. After the initial jump of c_0, the successive values of $\{y_t\}$ oscillate toward the long-run level of $c_0/(1 - a_1)$.

There are several important extensions to the intervention example provided here. Of course, the model need not be a first-order autoregressive process. A more general ARMA(p, q) intervention model has the form

$$y_t = a_0 + A(L)y_{t-1} + c_0 z_t + B(L)\varepsilon_t$$

where $A(L)$ and $B(L)$ are polynomials in the lag operator L.

Also, the intervention need not be the pure jump illustrated in Panel (a) of Figure 5.3. In our study, the value of the intervention sequence jumps from zero to unity in 1973Q1. However, there are several other possible ways to model the intervention function:

1. *Pulse function.* As shown in Panel (b) of the figure, the function z_t is zero for all periods, except in one particular period in which z_t is unity. This pulse function best characterizes a purely temporary intervention. Of course, the effects of the single impulse may last many periods due to the autoregressive nature of the $\{y_t\}$ series.
2. *Gradually changing function.* An intervention may not reach its full force immediately. Although the U.S. began installing metal detectors in airports in January 1973, it took almost a full year for installations to be completed at some major international airports. Our intervention study of the impact of metal detectors on quarterly skyjackings also modeled the z_t series as 1/4 in 1973Q1, 1/2 in 1973Q2, 3/4 in 1973Q3, and 1.0 in 1973Q4 and all subsequent periods. This type of intervention function is shown in Panel (c) of the figure.

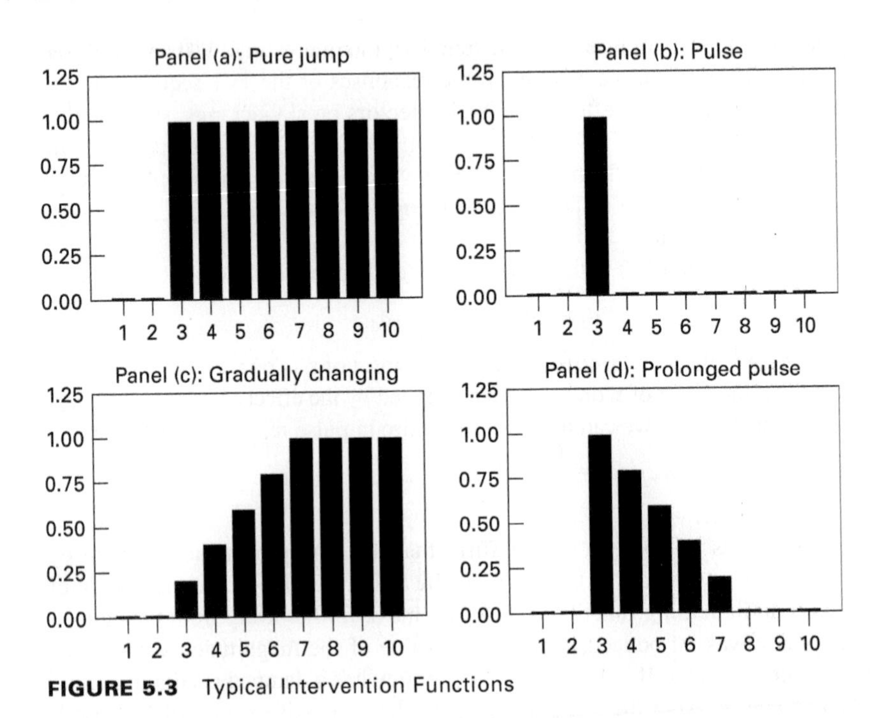

FIGURE 5.3 Typical Intervention Functions

3. *Prolonged impulse function.* Rather than a single pulse, the intervention may remain in place for one or more periods and then begin to decay. For a short time, sky marshals were put on many U.S. flights to deter skyjackings. Since the sky marshal program was allowed to terminate, the $\{z_t\}$ sequence for sky marshals might be represented by the decaying function shown in Panel (d) of Figure 5.3.

Be aware that the effects of these interventions change if $\{y_t\}$ has a unit root. From the discussion of Perron (1989) in Chapter 4, you should recall that a pulse intervention will have a permanent effect on the level of a unit root process. Similarly, if $\{y_t\}$ has a unit root, a pure jump intervention will act as a drift term. As indicated in Question 1 at the end of this chapter, an intervention will have a temporary effect on a unit root process if all values of $\{z_t\}$ sum to zero (e.g., $z_t = 1$, $z_{t+1} = -0.5$, $z_{t+2} = -0.5$, and all other values of the intervention variable equal zero).

Also be aware that the intervention may affect the variable of interest with a delay. Suppose that it takes d periods for z_t to begin to have any effect on the series of interest. It is possible to capture this behavior with a model of the form

$$y_t = a_0 + A(L)y_{t-1} + c_0 z_{t-d} + B(L)\varepsilon_t$$

Often, the shape of the intervention function and the delay factor d are clear from a priori reasoning. When there is an ambiguity, estimate the plausible alternatives and then use the standard Box–Jenkins model selection criteria to choose the most appropriate model. The following two examples illustrate the general estimation procedure.

Estimating the Effect of Metal Detectors on Skyjackings

The linear form of the intervention model $y_t = a_0 + A(L)y_{t-1} + c_0 z_t + B(L)\varepsilon_t$ assumes that the coefficients are invariant to the intervention. A useful check of this assumption is to pretest the data by estimating the most appropriate ARIMA(p, d, q) models for both the pre- and post-intervention periods. If the two ARIMA models are quite different, it is likely that the autoregressive and moving-average coefficients have changed. Usually there are not enough pre-intervention and post-intervention observations to estimate two separate models. In such instances, the researcher must be content to proceed using the best-fitting ARIMA model over the longest data span. The procedure described below is typical of most intervention studies.

STEP 1: Use the longest data span (i.e., either the pre- or the post-intervention observations) to find a plausible set of ARIMA models.

You should be careful to ensure that the $\{y_t\}$ sequence is stationary. If you suspect nonstationarity, you can perform unit root tests on the longest span of data. Alternatively, you can use the Perron (1989) test for structural change discussed in Chapter 4. In the presence of d unit roots, estimate the intervention model using the dth difference of y_t (i.e., $\Delta^d y_t$).

In our study, we were interested in the effects of metal detectors on U.S. domestic skyjackings, transnational skyjackings (including those involving the United States), and all other skyjackings. Call each of these time series $\{DS_t\}$, $\{TS_t\}$, and $\{OS_t\}$, respectively. Since there are only five years of data (i.e., 20 observations) for the pre-intervention period, we estimated the best-fitting ARIMA model over the 1973Q1–1988Q4 period. Using the various criteria discussed in Chapter 2 (including diagnostic checks of the residuals), we selected an AR(1) model for the $\{TS_t\}$ and $\{OS_t\}$ sequences and a pure noise model (i.e., all autoregressive and moving-average coefficients equal to zero) for the $\{DS_t\}$ sequence.

STEP 2: Estimate the various models over the entire sample period, including the effect of the intervention.

The installation of metal detectors was tentatively viewed as an immediate and permanent intervention. As such, we set $z_t = 0$ for $t <$ 1973Q1 and $z_t = 1$ beginning in 1973Q1. The results of the estimations over the entire sample period are reported in Table 5.1. As you can see, the installation of metal detectors reduced each of the three types of skyjacking incidents. The most pronounced effect was that U.S. domestic skyjackings immediately fell by more than 5.6 incidents per quarter. All effects are immediate because the estimate of a_1 is zero. The situation is somewhat different for the $\{TS_t\}$ and $\{OS_t\}$ sequences because the estimated autoregressive coefficients are different from zero. On impact, transnational skyjackings and other types of skyjacking incidents fell by 1.29 and 3.9 incidents per quarter. The long-run effects on $\{TS_t\}$ and $\{OS_t\}$ are estimated to be -1.78 and -5.11 incidents per quarter, respectively.

Table 5.1 Metal Detectors and Skyjackings

	Pre-Intervention Mean	a_1	Impact Effect (c_0)	Long-Run Effect
Transnational $\{TS_t\}$	3.032	0.276	−1.29	−1.78
	(5.96)	(2.51)	(−2.21)	
U.S. Domestic $\{DS_t\}$	6.70		−5.62	−5.62
	(12.02)		(−8.73)	
Other Skyjackings $\{OS_t\}$	6.80	0.237	−3.90	−5.11
	(7.93)	(2.14)	(−3.95)	

Notes:

1. t-statistics are in parentheses.
2. The long-run effect is calculated as $c_0/(1 - a_1)$.

STEP 3: Perform diagnostic checks of the estimated equations.

Diagnostic checking in Step 3 is particularly important since we have merged the observations from the pre- and post-intervention periods. To reiterate the discussion of ARIMA models, a well-estimated intervention model will have the following characteristics:

1. The estimated coefficients should be of "high quality." All coefficients should be statistically significant at conventional levels. As in all ARIMA modeling, we wish to use a parsimonious model. If any coefficient is not significant, an alternative model should be considered. Moreover, the autoregressive coefficients should imply that the $\{y_t\}$ sequence is convergent.
2. The residuals should approximate white noise. If the residuals are serially correlated, the estimated model does not mimic the actual data-generating process. Forecasts from the estimated model cannot possibly make use of all available information. If the residuals do not approximate a normal distribution, the usual tests of statistical inference are not appropriate in small samples. If the errors appear to be ARCH, the entire intervention model can be reestimated as an ARCH process.
3. The tentative model should outperform plausible alternatives. Of course, no one model can be expected to dominate all others in all possible criteria. However, it is good practice to compare the results of the maintained model to those of reasonable rivals. In the skyjacking example, a plausible alternative was to model the intervention as a gradually increasing process. This is particularly true because the impact effect was immediate for U.S. domestic flights and convergent for transnational and other domestic flights. Our conjecture was that metal detectors were gradually installed in non-U.S. airports and, even when installed, the enforcement was sporadic. As a check, we modeled the intervention as

gradually increasing over the year 1973. Although the coefficients were nearly identical to those reported in Table 5.1, the AIC and SBC were slightly lower (indicating a better fit) using the gradually increasing process. Hence, it is reasonable to conclude that metal detector adoption was more gradual outside of the United States.

Estimating the Effect of the Libyan Bombing

We also considered the effects of the U.S. bombing of Libya on the morning of April 15, 1986. The stated reason for the attack was Libya's alleged involvement in the terrorist bombing of the La Belle Discotheque in West Berlin. Since 18 of the F–111 fighter-bombers were deployed from British bases at Lakenheath and Upper Heyford, England, the United Kingdom implicitly assisted in the raid. The remaining U.S. planes were deployed from aircraft carriers in the Mediterranean Sea. Now let y_t denote all transnational terrorist incidents directed against the United States and the United Kingdom during month t. A plot of the $\{y_t\}$ sequence exhibits a large positive spike immediately after the bombing; the immediate effect seemed to be a wave of anti-U.S. and anti-U.K. attacks to protest the retaliatory strike. You can see this spike in each of the two series shown in Figure 5.1.

Preliminary estimates of the monthly data from January 1968 to March 1986 indicated that the $\{y_t\}$ sequence could be estimated as a purely autoregressive model with significant coefficients at lags 1 and 5. We were surprised by a significant coefficient at lag 5, but the AIC and SBC both indicated that the fifth lag is important. Nevertheless, we estimated versions of the model with and without the fifth lag. In addition, we considered two possible patterns for the intervention series. For the first, $\{z_t\}$ was modeled as 0 until April 1986 and 1 in all subsequent periods. Using this specification, we obtained the following estimates (with t-statistics in parentheses):

$$
\begin{array}{ll}
y_t = 5.58 + \underset{(5.56)}{0.336y_{t-1}} + \underset{(3.26)}{0.123y_{t-5}} + \underset{(0.84)}{2.65z_t} \\
\text{AIC} = 1656.03 \qquad \text{SBC} = 1669.95
\end{array}
$$

Note that the coefficient of z_t has a t-statistic of 0.84 (which is not significant at the 0.05 level). Alternatively, when z_t was allowed to be 1 only in the month of the attack, we obtained

$$
\begin{array}{ll}
y_t = 3.79 + \underset{(5.53)}{0.327y_{t-1}} + \underset{(2.59)}{0.157y_{t-5}} + \underset{(6.09)}{38.9z_t} \\
\text{AIC} = 1608.68 \qquad \text{SBC} = 1626.06
\end{array}
$$

In comparing the two estimates, it is clear that magnitudes of the autoregressive coefficients are similar. Although Q-tests indicated that the residuals from both models approximate white noise, the second model is preferable. The coefficient on the pulse term is highly significant, and the AIC and SBC both select the second specification. Our conclusion was that the Libyan bombing did not have the desired effect of reducing terrorist attacks against the United States and the United Kingdom. Instead, the bombing caused an immediate increase of more than 38 attacks. Subsequently, the number of attacks declined; 0.327 of these attacks are estimated to have persisted for one period ($0.327 \cdot 38.9 = 12.7$). Since the autoregressive coefficients imply convergence, the long-run consequences of the raid were estimated to be zero.

You can practice estimating an intervention model with the terrorism data shown in Figure 5.1. Question 2 at the end of this chapter will guide you through the process.

2. TRANSFER FUNCTION MODELS

A natural extension of the intervention model is to allow the $\{z_t\}$ sequence to be something other than a deterministic dummy variable. Consider the following generalization of the intervention model:

$$y_t = a_0 + A(L)y_{t-1} + C(L)z_t + B(L)\varepsilon_t \tag{5.3}$$

where $A(L)$, $B(L)$, and $C(L)$ are polynomials in the lag operator L.

In a typical transfer function analysis, the researcher will collect data on the endogenous variable $\{y_t\}$ and on the exogenous variable $\{z_t\}$. The goal is to estimate the parameter a_0 and the parameters of the polynomials $A(L)$, $B(L)$, and $C(L)$. The major difference between (5.3) and the intervention model is that $\{z_t\}$ is not constrained to have a particular deterministic time path. In a sense, the intervention variable is allowed to be any exogenous process. The polynomial $C(L)$ is called the **transfer function** in that it shows how a movement in the exogenous variable z_t affects the time path of (i.e., is transferred to) the endogenous variable $\{y_t\}$. The coefficients of $C(L)$, denoted by c_i, are called transfer function weights.

It is critical to note that transfer function analysis assumes that $\{z_t\}$ is an exogenous process that evolves independently of the $\{y_t\}$ sequence. Innovations in $\{y_t\}$ are assumed to have no effect on the $\{z_t\}$ sequence so that $Ez_t\varepsilon_{t-s} = 0$ for all values of s and t. Since z_t can be observed and is uncorrelated with the current innovation in y_t (i.e., the disturbance term ε_t), the current and lagged values of z_t are explanatory variables for y_t. Let $C(L)$ be $c_0 + c_1L + c_2L^2 + \ldots$. If $c_0 = 0$, the contemporaneous value of z_t does not directly affect y_t. As such, $\{z_t\}$ is called a **leading indicator** in that the observations z_t, z_{t-1}, z_{t-2}, ... can be used in predicting future values of the $\{y_t\}$ sequence.[2]

It is easy to conceptualize numerous applications for (5.3). After all, a large part of dynamic economic analysis concerns the effects of an "exogenous" or "independent" sequence $\{z_t\}$ on the time path of an endogenous sequence $\{y_t\}$. For example, much of the current research in agricultural economics centers on the effects of the macroeconomy on the agricultural sector. Using (5.3), farm output $\{y_t\}$ is affected by its own past as well as by the current and past state of the macroeconomy $\{z_t\}$. The effects of macroeconomic fluctuations on farm output can be represented by the coefficients of $C(L)$. Here, $B(L)\varepsilon_t$ represents the unexplained portion of farm output. Alternatively, the level of ozone in the atmosphere $\{y_t\}$ is a naturally evolving process; hence, in the absence of other outside influences, we should expect the ozone level to be well represented by an ARIMA model. However, many have argued that the use of fluorocarbons has damaged the ozone layer. Because of a cumulative effect, it is argued that current and past values of fluorocarbon usage affect the value of y_t. Letting z_t denote fluorocarbon usage in t, it is possible to model the effects of fluorocarbon usage on the ozone layer using a model in the form of (5.3). The natural dissipation of ozone is captured through the coefficients of $A(L)$. Stochastic shocks to the ozone layer, possibly due to electrical storms and the presence of measurement errors, are captured by $B(L)\varepsilon_t$. The contemporaneous effect of fluorocarbons on the ozone layer is captured by the coefficient c_0, and the lagged effects are captured by the other transfer function weights (i.e., the values of the various c_i).

In contrast to the pure intervention model, there is no pre- versus post-intervention period so that we cannot estimate a transfer function in the same fashion in which we estimated an intervention model. However, the methods are very similar in that the goal is to estimate a parsimonious model. The procedure involved in fitting a transfer function model is easiest to explain by considering a simple case of (5.3). To begin, suppose $\{z_t\}$ is generated by a white-noise process that is uncorrelated with ε_t at all leads and lags. Also suppose that the realization of z_t affects the $\{y_t\}$ sequence with a lag of unknown duration. Specifically, let

$$y_t = a_1 y_{t-1} + c_d z_{t-d} + \varepsilon_t \tag{5.4}$$

where $\{z_t\}$ and $\{\varepsilon_t\}$ are white-noise processes such that $E(z_t \varepsilon_{t-i}) = 0$; a_1 and c_d are unknown coefficients; and d is the "delay" or lag duration to be determined by the econometrician.

Since $\{z_t\}$ and $\{\varepsilon_t\}$ are assumed to be independent white-noise processes, it is possible to separately model the effects of each type of shock. Since we can observe the various z_t values, the first step is to calculate the **cross-correlations** between y_t and the various z_{t-i}. The cross-correlation between y_t and z_{t-i} is defined to be

$$\rho_{yz}(i) \equiv \text{cov}(y_t, z_{t-i})/(\sigma_y \sigma_z) \tag{5.5}$$

where σ_y and σ_z = the standard deviations of y_t and z_t, respectively. The standard deviation of each sequence is assumed to be time-independent.

Plotting each value of $\rho_{yz}(i)$ yields the cross-correlation function (CCF) or **cross-correlogram.** In practice, we must use the cross-correlations calculated using sample data because we do not know the true covariances or standard deviations. The key point is that the examination of the sample cross-correlations provides the same type of information as the ACF in an ARMA model. To explain, solve (5.4) to obtain

$$y_t = c_d z_{t-d}/(1 - a_1 L) + \varepsilon_t/(1 - a_1 L)$$

Use the properties of lag operators to expand the expression $c_d z_{t-d}/(1 - a_1 L)$:

$$y_t = c_d(z_{t-d} + a_1 z_{t-d-1} + a_1^2 z_{t-d-2} + a_1^3 z_{t-d-3} + \ldots) + \varepsilon_t/(1 - a_1 L)$$

Analogously to our derivation of the Yule–Walker equations, we can obtain the **cross-covariances** by the successive multiplication of y_t by $z_t, z_{t-1}, \ldots$ to form

$$\begin{aligned}
y_t z_t &= c_d(z_t z_{t-d} + a_1 z_t z_{t-d-1} + a_1^2 z_t z_{t-d-2} \\
&\quad + a_1^3 z_t z_{t-d-3} + \ldots) + z_t \varepsilon_t/(1 - a_1 L) \\
y_t z_{t-1} &= c_d(z_{t-1} z_{t-d} + a_1 z_{t-1} z_{t-d-1} + a_1^2 z_{t-1} z_{t-d-2} \\
&\quad + a_1^3 z_{t-1} z_{t-d-3} + \ldots) + z_{t-1} \varepsilon_t/(1 - a_1 L) \\
&\ldots
\end{aligned}$$

$$y_t z_{t-d} = c_d(z_{t-d} z_{t-d} + a_1 z_{t-d} z_{t-d-1} + a_1^2 z_{t-d} z_{t-d-2} + a_1^3 z_{t-d} z_{t-d-3} + \ldots) + z_{t-d}\,\varepsilon_t/(1 - a_1 L)$$

$$y_t z_{t-d-1} = c_d(z_{t-d-1} z_{t-d} + a_1 z_{t-d-1} z_{t-d-1} + a_1^2 z_{t-d-1} z_{t-d-2} + a_1^3 z_{t-d-1} z_{t-d-3} + \ldots) + z_{t-d-1}\,\varepsilon_t/(1 - a_1 L)$$

$$\ldots$$

Now take the expected value of each of the above equations. If we continue to assume that $\{z_t\}$ and $\{\varepsilon_t\}$ are independent white-noise disturbances, it follows that

$$Ey_t z_t = 0$$
$$Ey_t z_{t-1} = 0$$
$$\ldots$$
$$Ey_t z_{t-d} = c_d \sigma_z^2$$
$$Ey_t z_{t-d-1} = c_d a_1 \sigma_z^2$$
$$Ey_t z_{t-d-2} = c_d a_1^2 \sigma_z^2$$
$$\ldots$$

so that in compact form,

$$\begin{aligned} Ey_t z_{t-i} &= 0 \text{ for all } i < d \\ &= c_d a_1^{i-d} \sigma_z^2 \text{ for } i \geq d \end{aligned} \qquad (5.6)$$

Dividing each value of $Ey_t z_{t-i} = \text{cov}(y_t, z_{t-i})$ by $\sigma_y \sigma_z$ yields the CCF. Note that the cross-correlogram consists of zeroes until lag d. The absolute value of the height of the first nonzero cross-correlation is positively related to the magnitudes of c_d and a_1. Thereafter, the cross-correlations decay at the rate a_1. The decay of the correlogram matches the autoregressive patterns of the $\{y_t\}$ sequence.

The pattern exhibited by (5.6) is easily generalized. Suppose we allow both z_{t-d} and z_{t-d-1} to directly affect y_t:

$$y_t = a_1 y_{t-1} + c_d z_{t-d} + c_{d+1} z_{t-d-1} + \varepsilon_t$$

Solving for y_t, we obtain

$$\begin{aligned} y_t &= (c_d z_{t-d} + c_{d+1} z_{t-d-1})/(1 - a_1 L) + \varepsilon_t/(1 - a_1 L) \\ &= c_d(z_{t-d} + a_1 z_{t-d-1} + a_1^2 z_{t-d-2} + a_1^3 z_{t-d-3} + \ldots) \\ &\quad + c_{d+1}(z_{t-d-1} + a_1 z_{t-d-2} + a_1^2 z_{t-d-3} + a_1^3 z_{t-d-4} + \ldots) + \varepsilon_t/(1 - a_1 L) \end{aligned}$$

so that

$$\begin{aligned} y_t = {} & c_d z_{t-d} + (c_d a_1 + c_{d+1}) z_{t-d-1} + a_1(c_d a_1 + c_{d+1}) z_{t-d-2} \\ & + a_1^2(c_d a_1 + c_{d+1}) z_{t-d-3} + \ldots + \varepsilon_t/(1 - a_1 L) \end{aligned}$$

Given that we are assuming $Ez_t = 0$, the cross-covariances are $Ey_t z_{t-i}$ and the cross-correlations are $Ey_t z_{t-i}/\sigma_y \sigma_z$. In the literature, it is also common to work with the standardized cross-covariances denoted by $Ey_t z_{t-i}/\sigma_z^2$. The choice between the two is a matter of indifference since the CCF and the standardized cross-covariance function (CCVF) are proportional to each other. In the example at hand, the CCVF reveals the following pattern:[3]

$$\begin{aligned}\gamma_{yz}(i) &= 0 && \text{for } i < d \\ &= c_d && \text{for } i = d \\ &= c_d a_1 + c_{d+1} && \text{for } i = d + 1 \\ &= a_1^{i-d-1}(c_d a_1 + c_{d+1}) && \text{for } i > d + 1\end{aligned}$$

Panel (a) of Figure 5.4 shows the shape of the standardized cross-covariances for $d = 3$, $c_d = 1$, $c_{d+1} = 1.5$, and $a_1 = 0.8$. Note that there are distinct spikes at lags 3 and 4 corresponding to the nonzero values of c_3 and c_4. Thereafter, the CCVF decays at the rate a_1. Panel (b) of the figure replaces c_4 with the value -1.5. Again, all cross-covariances are zero until lag 3; since $c_3 = 1$, the standardized value of $\gamma_{yz}(3) = 1$. To find the standardized value of $\gamma_{yz}(4)$, form $\gamma_{yz}(4) = 0.8 - 1.5 = -0.7$. The subsequent

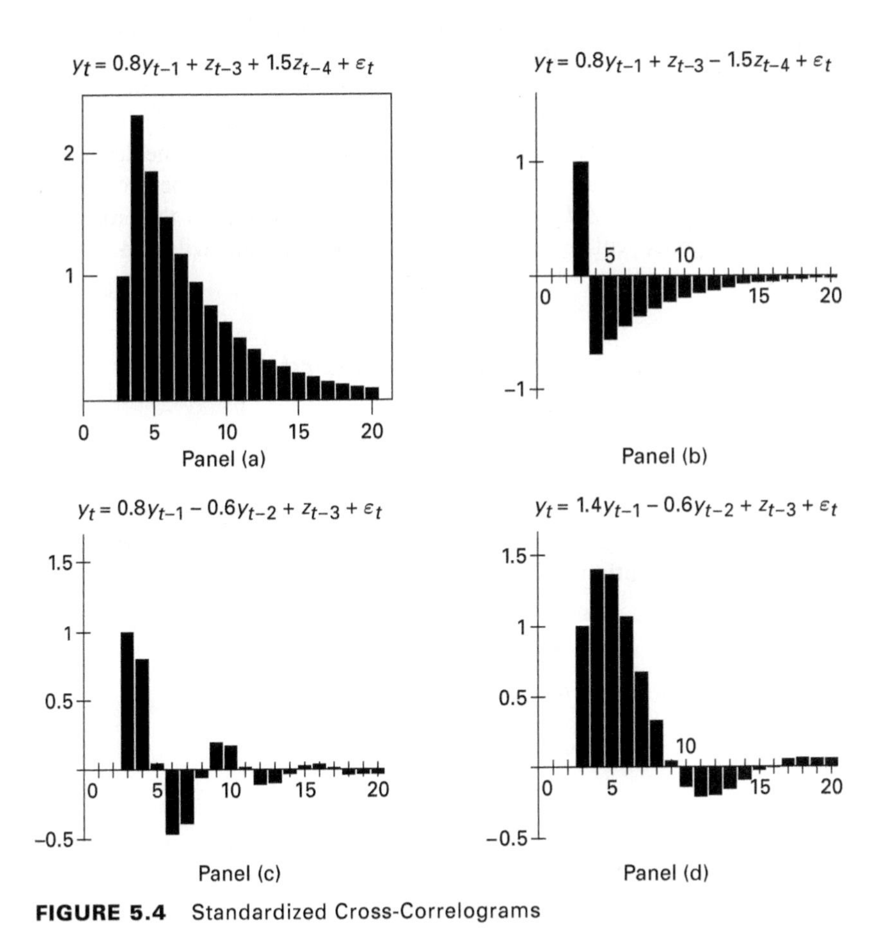

FIGURE 5.4 Standardized Cross-Correlograms

values of $\gamma_{yz}(i)$ decay at the rate 0.8. The pattern illustrated by these two examples generalizes to any intervention model of the form

$$y_t = a_0 + a_1 y_{t-1} + C(L)z_t + B(L)\varepsilon_t \tag{5.7}$$

The theoretical CCVF (and CCF) has a shape with the following characteristics:

1. All $\gamma_{yz}(i)$ will be zero until the first nonzero element of the polynomial $C(L)$.
2. The form of $B(L)$ is immaterial to the *theoretical* CCVF. Since z_t is uncorrelated with ε_t at all leads and lags, the form of the polynomial $B(L)$ will not affect any of the theoretical cross-covariances $\gamma_{yz}(i)$. Obviously, the intercept term a_0 does not affect any of the cross-covariances.
3. A spike in the CCVF indicates a nonzero element of $C(L)$. Thus, a spike at lag d indicates that z_{t-d} directly affects y_t.
4. All spikes decay at the rate a_1; convergence implies that the absolute value of a_1 is less than unity. If $0 < a_1 < 1$, decay in the cross-covariances will be direct, whereas if $-1 < a_1 < 0$, the decay pattern will be oscillatory.

Only the nature of the decay process changes if we generalize equation (5.7) to include additional lags of y_{t-i}. In the general case of (5.3), the *decay pattern in the cross-covariances* is determined by the characteristic roots of the polynomial $A(L)$; the shape is precisely that suggested by the autocorrelations of a pure ARMA model. This should not come as a surprise; in the examples of (5.4) and (5.7), the decay factor was simply the first-order coefficient a_1. We know that there will be decay since all characteristic roots of $1 - A(L)$ must be outside the unit circle for the process to be stationary. Convergence will be direct if the roots are positive and will tend to oscillate if a root is negative.

The Cross-Covariances of a Second-Order Process

To use another example, consider the transfer function

$$y_t = a_1 y_{t-1} + a_2 y_{t-2} + c_d z_{t-d} + \varepsilon_t$$

Using lag operators to solve for y_t is inconvenient since we do not know the numerical values of a_1 and a_2. Instead, use the method of undetermined coefficients and form the **challenge solution:**

$$y_t = \sum_{i=0}^{\infty} W_i z_{t-i} + \sum_{i=0}^{\infty} V_i \varepsilon_{t-i}$$

You should be able to verify that the values of the W_i are given by

$$\begin{aligned} W_0 &= 0 \\ &\dots \\ W_d &= c_d \\ W_{d+1} &= c_d a_1 \\ W_{d+2} &= c_d(a_1^2 + a_2) \end{aligned}$$

$$W_{d+3} = a_1 W_{d+2} + a_2 W_{d+1}$$
$$W_{d+4} = a_1 W_{d+3} + a_2 W_{d+2}$$
$$\dots$$

Thus, for all $i > d + 1$, the successive coefficients satisfy the difference equation $W_i = a_1 W_{i-1} + a_2 W_{i-2}$. At this stage we are not interested in the values of the various V_i, so it is sufficient to write the solution for y_t as

$$y_t = c_d z_{t-d} + c_d a_1 z_{t-d-1} + c_d (a_1{}^2 + a_2) z_{t-d-2} + \dots + \Sigma V_i \varepsilon_{t-i}$$

Next, use this solution for y_t to form all covariances using the Yule–Walker equations. Forming the expressions for $\gamma_{yz}(i)$ as $Ey_t z_{t-i}/\sigma_z^2$,

$$\gamma_{yz}(i) = 0 \text{ for } i < d \quad [\text{since } Ez_t z_{t-i} = 0 \text{ for } i < d]$$
$$\gamma_{yz}(d) = c_d$$
$$\gamma_{yz}(d - 1) = a_1 c_d$$
$$\gamma_{yz}(d - 2) = c_d (a_1{}^2 + a_2)$$
$$\dots$$

Thus, there is an initial spike at lag d reflecting the nonzero value of c_d. After one period, a_1 percent of c_d remains. After two periods, the decay pattern in the standardized cross-covariances begins to satisfy the difference equation

$$\gamma_{yz}(i) = a_1 \gamma_{yz}(i - 1) + a_2 \gamma_{yz}(i - 2)$$

Panel (c) of Figure 5.4 shows the shape of the CCVF for the case of $d = 3$, $c_d = 1$, $a_1 = 0.8$, and $a_2 = -0.6$. The oscillatory pattern reflects the fact that the characteristic roots of the process are imaginary. For purposes of comparison, Panel (d) shows the CCVF of another second-order process with imaginary roots.[4]

Higher-Order Input Processes

The econometrician will rarely be so fortunate as to work with a $\{z_t\}$ series that is white noise. We need to further generalize our discussion of transfer functions to consider the case in which the $\{z_t\}$ sequence is a stationary ARMA process. As discussed below, the estimation of the transfer function becomes more difficult in this case. However, the extra difficulty is worthwhile because a rich set of interactions between the variables is possible. For a moment, we can abstract from the estimation problem and consider the system of equations represented by (5.3)—reproduced for your convenience—and the $\{z_t\}$ process:

$$y_t = a_0 + A(L) y_{t-1} + C(L) z_t + B(L) \varepsilon_t$$
$$D(L) z_t = E(L) \varepsilon_{zt} \tag{5.8}$$

where $D(L)$ and $E(L)$ are polynomials in the lag operator L and ε_{zt} is white noise. The roots of $D(L)$ and $E(L)$ are such that the $\{z_t\}$ sequence is a stationary and invertible ARMA process. Since $\{z_t\}$ is independent of $\{y_t\}$, shocks to the $\{y_t\}$ sequence cannot influence $\{z_t\}$. As such, it must be the case that $E\varepsilon_t \varepsilon_{zt} = 0$.

Once the coefficients of the two equations have been properly estimated, it is possible to trace out two separate impulse response functions. As in a standard

Box–Jenkins model, it is possible to trace out the effects of a one-unit shock to ε_t on the entire y_t sequence. Holding all values of z_t constant, the impulse responses are given by the coefficients of the polynomial $[B(L)/(1 - A(L))]$. More importantly, it is possible to show how shocks to the input series $\{z_t\}$ are transferred to the output series $\{y_t\}$. A one-unit shock to ε_{zt} directly affects z_t by one unit and y_t by $c_0\varepsilon_{zt}$ units. It is relatively straightforward for a computer to trace out the effects of the ε_{zt} shock on the entire $\{z_t\}$ and $\{y_t\}$ sequences. Formally, the impulse responses of ε_{zt} shocks on the $\{y_t\}$ sequence are given by combining (5.3) and (5.8) such that

$$y_t = a_0 + A(L)y_{t-1} + C(L)E(L)\varepsilon_{zt}/D(L) + B(L)\varepsilon_t$$

If you solve for y_t, it will be clear that the impulse responses are the coefficients of $C(L)E(L)/[(D(L)(1 - A(L)L)]$. Similarly, transfer functions are useful because they are conducive to multi-step-ahead forecasting. Since $\{z_t\}$ is an independent process, you can use (5.8) to forecast using the techniques developed in Chapter 2. As such, if you have T observations, you can use (5.8) to form the forecasts $E_T z_{T+1}$, $E_T z_{T+2}$, ... These forecasts are used in the multi-step-ahead forecasts for y_{T+i}. For example, suppose that $z_t = d_1 z_{t-1} + \varepsilon_{zt}$ and $y_t = a_1 y_{t-1} + c_1 z_t + \varepsilon_t$. Since the j-step-ahead forecasts for z_{T+j} are $(d_1)^j z_T$, the multi-step-ahead forecasts for y_{T+j} are

$$\begin{aligned} E_T y_{T+1} &= a_1 y_T + c_1 E_T z_{T+1} = a_1 y_T + c_1 d_1 z_T \\ E_T y_{T+2} &= (a_1)^2 y_T + c_1 d_1 (a_1 + d_1) z_T \\ &\ldots \end{aligned}$$

Identification and Estimation

Since $\{z_t\}$ evolves independently of $\{y_t\}$, we can use the methodology developed in Chapter 2 to estimate $\{z_t\}$ as the ARMA process given by (5.8). The residuals from such a model, denoted by $\{\hat{\varepsilon}_{zt}\}$, should be white noise. The idea is to estimate the *innovations* in the $\{z_t\}$ sequence even though the sequence itself is not a white-noise process.

Once (5.8) has been estimated, you can choose between two techniques to estimate the transfer function. The first is to simply estimate the regression equation

$$y_t = a_0 + \sum_{i=1}^{p} a_i y_{t-i} + \sum_{i=0}^{n} c_i z_{t-i} + \varepsilon_t \tag{5.9}$$

Instead of estimating (5.3), the $\{y_t\}$ series represented by (5.9) is estimated as an AR(p) process that is affected by current and lagged values of $\{z_t\}$. This specification is called an autoregressive distributed lag (ADL) model. Unlike the standard Box–Jenkins approach, begin estimating an ADL using the largest values of p and n deemed feasible. Then, F-tests and t-tests can be used to pare down the model by eliminating unnecessary coefficients. You could also use the AIC or the SBC to find the lag lengths yielding the best fit. As in any time-series estimation, it is crucial to perform the appropriate diagnostic checks to ensure that the residuals are white noise. The benefit of this method is that it is simple to perform. However, you can easily end up with an overly parameterized model. As you know, there might be an ARMA model for $\{y_t\}$ that is more parsimonious than an AR representation. Moreover, since z_t and z_{t-i} are correlated, it is not always straightforward to use t-tests to pare down the coefficients of the transfer function. Nevertheless, the method is quite common and is consistent with the vector-autoregressive methodology discussed in Sections 5 to 13. A fairly detailed example of estimating an ADL is presented in Chapter 6.

The second method estimates (5.3) directly. Although identifying the coefficients of $B(L)$ can be difficult, it can yield a parsimonious model. Just as an ARMA process can approximate a high-order AR process, (5.3) can provide a parsimonious representation of (5.9). As in the case where $\{z_t\}$ is white noise, the idea is to use the cross-correlations to obtain the pattern of the coefficients as they appear in the transfer function. It is tempting to think that we should form the cross-correlations between the $\{y_t\}$ sequence and $\{\hat{\varepsilon}_{zt-i}\}$. However, this procedure would be inconsistent with the maintained hypothesis that the structure of the transfer function is given by (5.3). The reason is that $z_t, z_{t-1}, z_{t-2}, \ldots$ (and not simply the innovations) directly affect the value of y_t. Cross-correlations between y_t and the various ε_{zt-i} would not reveal the pattern of the coefficients in $C(L)$.

The appropriate methodology is to **filter** the $\{y_t\}$ sequence by multiplying (5.3) by the previously estimated polynomial $D(L)/E(L)$. As such, the filtered value of y_t is $D(L)y_t/E(L)$ and is denoted by y_{ft}. The cross-correlations of $\hat{\varepsilon}_{zt}$ and y_{ft} reveal the form of the transfer function. To explain, multiply (5.3) by $D(L)/E(L)$ to obtain

$$D(L)y_t/E(L) = D(L)a_0/E(L) + D(L)A(L)y_{t-1}/E(L) + C(L)D(L)z_t/E(L) + B(L)D(L)\varepsilon_t/E(L)$$

Given that $D(L)y_t/E(L) = y_{ft}$, $D(L)y_{t-1}/E(L) = y_{ft-1}$, and $D(L)z_t/E(L) = \varepsilon_{zt}$, this is equivalent to

$$y_{ft} = D(L)a_0/E(L) + A(L)y_{ft-1} + C(L)\varepsilon_{zt} + B(L)D(L)\varepsilon_t/E(L) \qquad (5.10)$$

Although you could construct the sequence $D(L)y_t/E(L)$, most software packages can make the appropriate transformations automatically. Now compare (5.3) and (5.10). You can see that y_t and z_t in (5.3) will have the same cross-covariances as y_{ft} and ε_{zt} in (5.10). As in the previous case in which $\{z_t\}$ is white noise, we can examine the CCVF between y_{ft} and ε_{zt} to determine the spikes and the decay pattern.

To explain why filtering is important, consider the example where $z_t = d_1 z_{t-1} + \varepsilon_{zt}$ and $y_t = a_1 y_{t-1} + c_1 z_t + \varepsilon_t$. Given that you can never actually observe the form of the transfer function, you might not be able to deduce that only z_t has a direct effect on y_t. In fact, substitution for z_t yields $y_t = a_1 y_{t-1} + c_1(d_1 z_{t-1} + \varepsilon_{zt}) + \varepsilon_t$. As such, you might be fooled into estimating an equation of the form

$$y_t = a_1 y_{t-1} + a_2 z_{t-1} + \varepsilon_{1t}$$

Although there is nothing "wrong" with this equation, the interpretation is such that z_t affects the $\{y_t\}$ sequence with a one-period lag. It should also be clear that $\text{var}(\varepsilon_{1t}) = \text{var}(c_1\varepsilon_{zt} + \varepsilon_t)$. Hence, the estimated transfer function will have a larger variance than that from $y_t = a_1 y_{t-1} + c_1 z_t + \varepsilon_t$. The proper way to identify the form of the transfer function is to filter the values of y_t such that

$$y_{ft} = (1 - d_1 L)y_t = y_t - d_1 y_{t-1}$$

Now multiply each side of the transfer function by $(1 - d_1L)$ to obtain

$$(1 - d_1L)y_t = a_1(1 - d_1L)y_{t-1} + c_1(1 - d_1L)z_t + (1 - d_1L)\varepsilon_t$$

or

$$y_{ft} = a_1y_{ft-1} + c_1\varepsilon_{zt} + \varepsilon_t - d_1\varepsilon_{t-1}$$

Clearly, the covariances between y_{ft} and ε_{zt} will have the same pattern as those between y_t and z_t. In summary, the full procedure for fitting a transfer function entails:

STEP 1: Fit an ARMA model to the z_t sequence. The technique used at this stage is precisely that for estimating any ARMA model. A properly estimated ARMA model should approximate the data-generating process for the $\{z_t\}$ sequence. The calculated residuals $\{\hat{\varepsilon}_{zt}\}$ are called the *filtered* values of the $\{z_t\}$ series. These filtered values can be interpreted as the pure innovations in the $\{z_t\}$ sequence. Calculate and store the $\{\hat{\varepsilon}_{zt}\}$ sequence.

STEP 2: Obtain the filtered $\{y_t\}$ sequence by applying the filter $D(L)/E(L)$ to each value of $\{y_t\}$; that is, use the results of Step 1 to obtain $D(L)/E(L)y_t \equiv y_{ft}$. Form the CCF (or CCVF) between y_{ft} and $\hat{\varepsilon}_{zt-i}$. Of course, the sample cross-covariances will not precisely conform to their theoretical values. Under the null hypothesis that the cross-correlations are all zero, the *sample* variance of cross-correlation coefficient i asymptotically converges to $(T - i)^{-1}$ where T = number of usable observations. Let $r_{yz}(i)$ denote the sample cross-correlation coefficient between y_t and z_{t-i}. Under the null hypothesis that the true values of $\rho_{yz}(i)$ all equal zero, the variance of $r_{yz}(i)$ converges to

$$\text{var}[r_{yz}(i)] = (T - i)^{-1}$$

For example, with 100 usable observations, the standard deviation of the cross-correlation coefficient between y_t and z_{t-1} is approximately equal to 0.10. If the calculated value of $r_{yz}(1)$ exceeds 0.2 (or is less than -0.2), the null hypothesis can be rejected. Significant cross-correlations at lag i indicate that an innovation in z_t affects the value of y_{t+i}. To test the significance of the first k cross-correlations, use the statistic

$$Q = T(T + 2)\sum_{i=0}^{k} r_{yz}^2(i)/(T - k)$$

Asymptotically, Q has a χ^2 distribution with $(k - p_1 - p_2)$ degrees of freedom where p_1 and p_2 denote the number of nonzero coefficients in $A(L)$ and $C(L)$, respectively.

STEP 3: Examine the pattern of the cross-correlogram (or CCVF). Just as the ACF can be used as a guide in identifying an ARMA model, the cross-correlogram can help identify the form of $A(L)$ and $C(L)$. This can be difficult since you try to infer the form of both functions from the cross-correlations. Remember that spikes in the cross-correlogram indicate nonzero values of c_i. The decay pattern of the cross-correlations suggests plausible candidates for coefficients of $A(L)$. For example, geometric decay after a single spike at lag d suggests the model $y_t = a_0 + a_1y_{t-1} + c_dz_{t-d} + \varepsilon_t$. As illustrated by the CCVF in Figure 5.4, this decay pattern is perfectly analogous to the ACF in a traditional ARMA model. In practice, examination of the cross-correlogram will suggest several plausible transfer functions. Estimate each of these plausible models and select the "best-fitting" model. Regress y_t (not y_{ft}) on its own lags and the selected values of $\{z_t\}$. At this point, you will have estimated a model of the form

$$[1 - A(L)L]\, y_t = C(L)z_t + e_t$$

where e_t denotes the error term, which is not necessarily white noise.

STEP 4: The $\{e_t\}$ sequence obtained in Step 3 is an approximation of $B(L)\varepsilon_t$. As such, the ACF of these residuals can suggest the appropriate form for the $B(L)$ function. If the $\{e_t\}$ sequence appears to be white noise, your task is complete. However, the correlogram of the $\{e_t\}$ sequence will usually suggest a number of plausible forms for $B(L)$. Use the $\{e_t\}$ sequence to estimate the various forms of $B(L)$ and select the "best" model for the $B(L)e_t$. As in any model with MA terms, you need to use maximum-likelihood estimates.

STEP 5: Combine the results of Steps 3 and 4 to estimate the full equation. At this stage, you will estimate $A(L)$, $B(L)$, and $C(L)$ simultaneously. The properties of a well-estimated model are such that the coefficients are of high quality, the model is parsimonious, the residuals conform to a white-noise process, and the forecast errors are small. You should compare your estimated model to the other plausible candidates from Steps 3 and 4.

There is no doubt that estimating a transfer function involves judgment on the part of the researcher. Experienced econometricians would agree that the procedure is a blend of skill, art, and perseverance that is developed through practice. Keep in mind that the goal is to find a parsimonious representation of a potentially complicated interaction among the variables. As in an ARMA process, different models can have similar economic implications and yield similar forecasts. Nevertheless, there are some hints that can be quite helpful.

1. After estimating the full model in Step 5, any remaining autocorrelation in the residuals probably means that $B(L)$ is misspecified. Return to Step 4 and reformulate the form of $B(L)$ so as to capture the remaining explanatory power of the residuals.
2. After estimating the full model in Step 5, if the residuals are correlated with $\{z_t\}$, the $C(L)$ function is probably misspecified. Return to Step 3 and reformulate the specifications of $A(L)$ and $C(L)$.
3. Instead of estimating $\{e_t\}$ as a pure autoregressive process in Step 4, you can estimate $B(L)$ as an ARMA process. Thus, $e_t = B(L)\varepsilon_t$ is allowed to have the form $e_t = G(L)\varepsilon_t/H(L)$. Here, $G(L)$ and $H(L)$ are low-order polynomials in the lag operator L. The benefit is that a high-order autoregressive process can often be approximated by a low-order ARMA model.
4. The sample cross-correlations are not meaningful if $\{y_t\}$ and/or $\{z_t\}$ are not stationary. You can test each for a unit root using the procedures discussed in Chapter 4. In the presence of unit roots, Box and Jenkins (1976) recommend differencing each variable until it is stationary. The next chapter

considers unit roots in a multivariate context. For now, it is sufficient to note that this recommendation can lead to overdifferencing.

The interpretation of the transfer function depends on the type of differencing performed. Consider the following three specifications and assume that $|a_1| < 1$:

$$y_t = a_1 y_{t-1} + c_0 z_t + \varepsilon_t \tag{5.11}$$

$$\Delta y_t = a_1 \Delta y_{t-1} + c_0 z_t + \varepsilon_t \tag{5.12}$$

$$y_t = a_1 y_{t-1} + c_0 \Delta z_t + \varepsilon_t \tag{5.13}$$

In (5.11), a one-unit shock in z_t has the initial effect of increasing y_t by c_0 units. This initial effect decays at the rate a_1. In (5.12), a one-unit shock in z_t has the initial effect of increasing *the change in* y_t by c_0 units. The effect on the *change* decays at the rate a_1, but the effect on the *level* of the $\{y_t\}$ sequence never decays. In (5.13), only the change in z_t affects y_t. Here, a pulse in the $\{z_t\}$ sequence will have a temporary effect on the level of $\{y_t\}$. Questions 3 and 4 at the end of this chapter are intended to help you gain familiarity with the different specifications.

3. ESTIMATING A TRANSFER FUNCTION

The clustering of high-profile terrorist events (e.g., the hijacking of TWA flight 847 on June 14, 1985; the hijacking of the *Achille Lauro* cruise ship on October 7, 1985; and the Abu Nidal attacks on the Vienna and Rome airports on December 27, 1985) caused much speculation in the press about tourists changing their travel plans. Similarly, the tourism industry was especially hard-hit after the attacks on September 11, 2001. Although opinion polls of prospective tourists suggest that terrorism affects tourism, the true impact, if any, can best be discovered through the application of statistical techniques. Polls conducted in the aftermath of significant incidents cannot indicate whether respondents actually rebooked trips. Moreover, polls cannot account for tourists not surveyed who may have been induced to take advantage of offers designed to entice tourists back to a troubled spot.

To measure the impact of terrorism on tourism, Enders, Sandler, and Parise (1992) constructed the quarterly values of total receipts from tourism for twelve countries.[5] The logarithmic share of each nation's revenues was treated as the dependent variable $\{y_t\}$, and the number of transnational terrorist incidents occurring within each nation was treated as the independent variable $\{z_t\}$. The crucial assumption for the use of intervention analysis is that there is no feedback from tourism to terrorism. This assumption would be violated if changes in tourism induced terrorists to change their activities.

Consider a transfer function in the form of (5.3):

$$y_t = a_0 + A(L)y_{t-1} + C(L)z_t + B(L)\varepsilon_t$$

where: y_t = deseasonalized values of the logarithmic share of a nation's tourism revenues in quarter t

z_t = the number of transnational terrorist incidents within that country during quarter t.[6]

Using the methodology developed in the previous section, the first step in fitting a transfer function is to fit an ARMA model to the $\{z_t\}$ sequence. For illustrative purposes, it is helpful to consider the Italian case since terrorism in Italy appeared to be white noise (with a constant mean of 4.20 incidents per quarter). Let $\rho_z(i)$ denote the autocorrelations between z_t and z_{t-i}. If you are following along with the data in the file labeled ITALY.XLS, be sure to set the sample for 1971Q1–1988Q4. The correlogram for terrorist attacks in Italy is

Correlogram for Terrorist Attacks in Italy

$\rho_z(0)$	$\rho_z(1)$	$\rho_z(2)$	$\rho_z(3)$	$\rho_z(4)$	$\rho_z(5)$	$\rho_z(6)$	$\rho_z(7)$	$\rho_z(8)$
1	0.14	0.05	−0.06	−0.04	0.13	−0.00	0.01	−0.12

Each value of $\rho_z(i)$ is less than two standard deviations from unity, and the Ljung–Box Q-statistics indicate that no groupings are significant. Since terrorist incidents appear to be a white-noise process, we can skip Step 1; there is no need to fit an ARMA model to the series or to filter the $\{y_t\}$ sequence for Italy. At this point, we conclude that terrorists randomize their acts so that the number of incidents in quarter t is uncorrelated with the number of incidents in previous periods.

Step 2 calls for obtaining the cross-correlogram between tourism and terrorism. The cross-correlogram is

Cross-Correlogram Between Tourism and Terrorism in Italy

$\rho_{yz}(0)$	$\rho_{yz}(1)$	$\rho_{yz}(2)$	$\rho_{yz}(3)$	$\rho_{yz}(4)$	$\rho_{yz}(5)$	$\rho_{yz}(6)$	$\rho_{yz}(7)$	$\rho_{yz}(8)$
−0.18	−0.23	−0.24	−0.05	0.04	0.13	0.04	0.00	0.10

There are several interesting features of the cross-correlogram:

1. With T observations and i lags, the theoretical value of the standard deviation of each value of $\rho_{yz}(i)$ is $(T-i)^{-1/2}$. With 73 observations, $T^{-1/2}$ is approximately equal to 0.117. At the 5 percent significance level (i.e., two standard deviations), the sample value of $\rho_{yz}(0)$ is not significantly different from zero, and $\rho_{yz}(1)$ and $\rho_{yz}(2)$ are just on the margin. However, the Q-statistic for $\rho_{yz}(0) = \rho_{yz}(1) = \rho_{yz}(2) = 0$ is significant at the 0.01 level. Thus, there appears to be a strong negative relationship between terrorism and tourism beginning at lag 1 or at lag 2.
2. It is good practice to examine the cross-correlations between y_t and leading values of z_{t+i}. If the current value of y_t tends to be correlated with *future* values of z_{t+i}, it might be that the assumption of no feedback is violated. The presence of a significant cross-correlation between y_t and leads of z_t might be due to the effect of the current realization of y_t on future values of the $\{z_t\}$ sequence.
3. Since $\rho_{yz}(0)$ is not significantly different from zero at the 5 percent level, it is likely that the delay factor is one quarter; it takes at least one quarter for tourists to significantly revise their travel plans. However, there is no obvious pattern to the cross-correlation function. It is wise to entertain the possibility of several plausible models at this point in the process.

Table 5.2 Terrorism and Tourism in Italy (Estimates from Step 3)

	a_0	a_1	a_2	c_0	c_1	c_2	AIC/ SBC
Model 1	0.0249	0.795	−0.469		−0.0046		−5.09/
	(1.25)	(2.74)	(−1.63)		(−2.34)		4.01
Model 2		0.868	−0.696		−0.0030		−5.54/
		(4.52)	(−3.44)		(−2.23)		1.28
Model 3		1.09	−0.683	−0.0025			−4.94/
		(4.51)	(−2.96)	(−2.10)			1.89
Model 4					−0.0025	−0.0019	−4.84/
					(−1.15)	(−0.945)	3.27
Model 5		−0.217			−.0025	−.0027	−2.93/
		(−0.221)			(−1.16)	(−.080)	3.89

Note: The numbers in parentheses are the t-statistics for the null hypothesis of a zero coefficient.

Step 3 entails examining the cross-correlogram and estimating each of the plausible models. Based on the ambiguous evidence of the cross-correlogram, several different models for the transfer function were estimated. We experimented using delay factors of 0, 1, and 2 quarters. Since the decay pattern of the cross-correlogram is also ambiguous, we allowed $A(L)$ to have the form a_1y_{t-1} and $a_1y_{t-1} + a_2y_{t-2.}$ Some of our estimates are reported in Table 5.2.

Model 1 has the form $y_t = a_0 + a_1y_{t-1} + a_2y_{t-2} + c_1z_{t-1} + e_t$. The problem with this specification is that the intercept term a_0 is not significantly different from zero. Eliminating this coefficient yields Model 2. Notice that all coefficients in Model 2 are significant at conventional levels and that the magnitude of each is quite reasonable. The estimated value of c_1 is such that a terrorist incident reduces the logarithmic share of Italy's tourism by 0.003 in the following period. The point estimates of the autoregressive coefficients imply imaginary characteristic roots (the roots are $0.434 \pm 0.69i$). Since these roots lie inside the unit circle, the effect of any incident decays in a sine-wave pattern.

Model 3 changes the delay so as to allow a contemporaneous effect of z_t on y_t. The point estimates of the coefficients are reasonable and all are more than two standard deviations from zero. However, the AIC and SBC both select Model 2 over Model 3. The appropriate delay seems to be one quarter.

Since the cross-correlogram seems to have two spikes and exhibits little decay, we allowed both z_{t-1} and z_{t-2} to directly affect y_t. You can see from the table that Models 4 and 5 are inadequate in nearly all respects. Thus, we tentatively selected Model 2 as the "best" model.

For Step 4, we obtained the $\{e_t\}$ sequence from the residuals of Model 2. Hence,

$$e_t = y_t - [-0.003z_{t-1}/(1 - 0.868L + 0.696L^2)]$$

The correlogram of the residuals is

$\rho(0)$	$\rho(1)$	$\rho(2)$	$\rho(3)$	$\rho(4)$	$\rho(5)$	$\rho(6)$	$\rho(7)$	$\rho(8)$
1.0	0.621	0.554	0.431	0.419	0.150	0.066	0.021	−0.00

The residuals were then estimated as an ARMA process using standard Box–Jenkins methods. Without going into details, the best-fitting ARMA model of the residuals is

$$e_t = 0.485e_{t-1} + 0.295e_{t-2} + (1 + 0.238L^4)\varepsilon_t$$

where the t-statistics for the coefficients are 4.08, 2.33, and 1.83 (significant at the 0.000, 0.023, and 0.071 levels), respectively.

At this point, our tentative transfer function is

$$y_t = -\frac{0.003z_{t-1}}{1 - 0.868L + 0.696L^2} + \frac{(1 + 0.238L^4)\varepsilon_t}{1 - 0.485L - 0.295L^2} \tag{5.14}$$

The problem with (5.14) is that the coefficients in the first expression were estimated separately from the coefficients in the second expression. In Step 5, we estimated all coefficients simultaneously and obtained

$$y_t = \frac{-0.0022z_{t-1}}{1 - 0.876L + 0.749L^2} + \frac{(1 + 0.293L^4)\varepsilon_t}{1 - 0.504L - 0.245L^2} \tag{5.15}$$

Note that the coefficients of (5.15) are similar to those of (5.14). The t-statistics for the two numerator coefficients are -2.17 and 2.27, and the t-statistics for the four denominator coefficients are -7.78, 5.20, -4.31, and -1.94, respectively. The roots of the inverse characteristic equation for z_{t-1} are imaginary and outside the unit circle (the inverse characteristic roots are $0.585 \pm 0.996i$ so that the characteristic roots are $0.438 \pm 0.246i$). As in Model 2, the effects of a terrorist incident decay in a sine-wave pattern. The roots of the inverse characteristic equation for ε_t are -3.29 and 1.238 so that the characteristic roots are -0.303 and 0.807. As an aside, note that you can obtain the original form of the model given in (5.3) by multiplying (5.15) by the two denominator coefficients.

The Ljung–Box Q-statistics indicate that the residuals of (5.15) appear to be white noise. For example, $Q(8) = 6.54$ and $Q(16) = 12.67$ with significance levels of 0.587 and 0.696, respectively. Additional diagnostic checking included excluding the MA(4) term in the numerator (since the significance level was 5.5 percent) and estimating other plausible forms of the transfer functions. All other models had insignificant coefficients and/or larger values of the AIC and SBC and/or Q-statistics indicating significant correlation in the estimated residuals. Hence, we concluded that (5.15) best captures the effects of terrorism on tourism in Italy.

Our ultimate aim was to use the estimated transfer function to simulate the effects of a typical terrorist incident. Initializing the system such that all values of $y_0 = y_1 = y_2 = y_3 = 0$ and setting all $\{\varepsilon_t\} = 0$, we let the value of $z_t = 1$. Figure 5.5 shows the *impulse response function* for this one-unit change in the $\{z_t\}$ sequence. As you can see from the figure, after a one-period delay, tourism in Italy declines sharply. After a sustained decline, tourism returns to its initial value in approximately one year. The system has a memory, and tourism again falls; notice the oscillating decay pattern.

Integrating over time and over all incidents allowed us to estimate Italy's total losses to tourism. The undiscounted losses exceeded 600 million SDR; using a 5 percent real

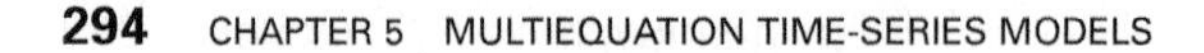

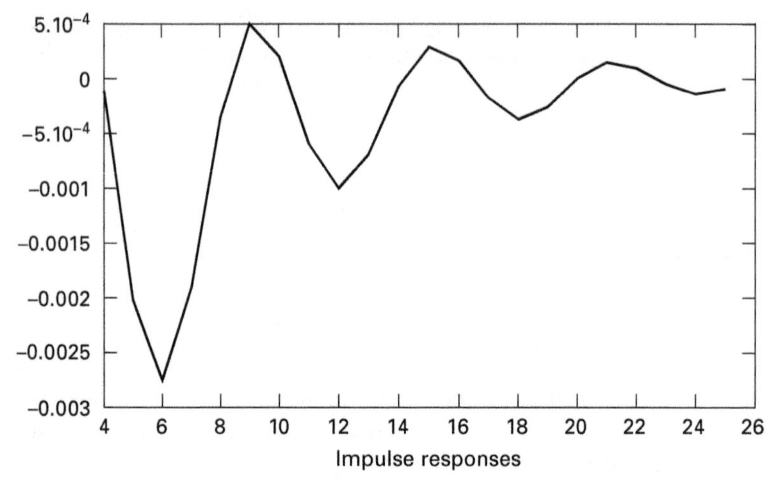

FIGURE 5.5 Italy's Share of Tourism

interest rate, the total value of the losses exceeded 861 million 1988 SDRs (equal to 6 percent of Italy's annual tourism revenues). The data set ITALY.XLS also contains the level of terrorism and the share of tourism for Greece. You can practice the technique by using the data to estimate a transfer function for Greece.

4. LIMITS TO STRUCTURAL MULTIVARIATE ESTIMATION

There are two important difficulties involved in fitting a multivariate equation such as a transfer function. The first concerns the goal of fitting a parsimonious model. Obviously, a parsimonious model is preferable to an overparameterized model. In the relatively small samples usually encountered in economic data, estimating an unrestricted model may so severely limit degrees of freedom as to render forecasts useless. Moreover, the possible inclusion of large but insignificant coefficients will add variability to the model's forecasts. However, in paring down the form of the model, two equally skilled researchers will likely arrive at two different transfer functions. Although one model may have a better "fit" (in terms of the AIC or SBC), the residuals of the other may have better diagnostic properties. There is substantial truth to the consensus opinion that fitting a transfer function model has many characteristics of an "art form." There is a potential cost to using a parsimonious model. Suppose you simply estimate the equation $y_t = A(L)y_{t-1} + C(L)z_t + B(L)\varepsilon_t$ using long lags for $A(L)$, $B(L)$, and $C(L)$. As long as $\{z_t\}$ is exogenous, the estimated coefficients and forecasts are unbiased even though the model is overparameterized. Such is *not* the case if the researcher improperly imposes zero restrictions on any of the polynomials in the model.

The second problem concerns the assumption of no feedback from the $\{y_t\}$ sequence to the $\{z_t\}$ sequence. For the coefficients of $C(L)$ to be unbiased estimates of the impact effects of $\{z_t\}$ on the $\{y_t\}$ sequence, z_t must be uncorrelated with $\{\varepsilon_t\}$ at all leads and lags.

Although certain economic models may assert that policy variables (such as money supply or government spending) are exogenous, there may be feedback such that the policy variables are set with specific reference to the state of other variables in the system. To understand the problem of feedback, suppose that you were trying to keep a constant 70°F temperature inside your apartment by turning the thermostat up or down. Of course, the "true" model is that turning up the heat (the intervention variable z_t) warms up your apartment (the $\{y_t\}$ sequence). However, intervention analysis cannot adequately capture the true relationship in the presence of feedback. Clearly, if you perfectly controlled the inside temperature, there would be no correlation between the *constant* value of the inside temperature and the movement of the thermostat. Alternatively, you might listen to the weather forecast and turn up the thermostat whenever you expected it to be cold. If you underreacted by not turning the heat high enough, the cross-correlogram between the two variables would tend to show a negative spike reflecting the drop in room temperature with the upward movement in the thermostat setting. Instead, if you overreacted by greatly increasing the thermostat setting, both room temperature and the thermostat setting would rise together. Only if you moved the thermostat setting without reference to room temperature would we expect to uncover the actual model.

The need to restrict the form of the transfer function and the problem of feedback or "reverse causality" led Sims (1980) to propose a nonstructural estimation strategy. To best understand the vector-autoregression approach, it is useful to consider the state of macroeconometric modeling that led Sims to his then-radical ideas.

Multivariate Macroeconometric Models: Some Historical Background

Traditionally, macroeconometric hypothesis tests and forecasts were conducted using large-scale macroeconometric models. Usually, a complete set of structural equations was estimated, one equation at a time. Then all equations were aggregated in order to form overall macroeconomic forecasts. Consider two of the equations from the *Brookings Quarterly Econometric Model of the United States*, as reported by Suits and Sparks (p. 208, 1965):

$$\underset{}{C_{\text{NF}}} = \underset{(0.0165)}{0.0656Y_{\text{D}}} - \underset{(2.49)}{10.93(P_{\text{CNF}}/P_{\text{C}})_{t-1}} + \underset{(0.0522)}{0.1889(N + N_{\text{ML}})_{t-1}}$$

$$C_{\text{NEF}} = 4.2712 + \underset{(0.0127)}{0.1691Y_{\text{D}}} - \underset{(0.0213)}{0.0743(ALQD_{\text{HH}}/P_{\text{C}})_{t-1}}$$

where:
C_{NF} = personal consumption expenditures on food
Y_{D} = disposable personal income
P_{CNF} = implicit price deflator for personal consumption expenditures on food
P_{C} = implicit price deflator for personal consumption expenditures
N = civilian population
N_{ML} = military population including armed forces overseas
C_{NEF} = personal consumption expenditures for nondurables other than food
$ALQD_{\text{HH}}$ = end-of-quarter stock of liquid assets held by households

and standard errors are in parentheses.

The remaining portions of the model contain estimates for the other components of aggregate consumption, investment spending, government spending, exports, imports, the financial sector, various price determination equations, and so on. Note that food expenditures, but not expenditures on other nondurables, are assumed to depend on relative price and population. However, expenditures for other nondurables are assumed to depend on real liquid assets held by households in the previous quarter.

Are such ad hoc behavioral assumptions consistent with economic theory? Sims (p. 3, 1980) considers such multiequation models and argues that

> ... what "economic theory" tells us about them is mainly that any variable that appears on the right-hand side of one of these equations belongs in principle on the right-hand side of all of them. To the extent that models end up with very different sets of variables on the right-hand side of these equations, they do so not by invoking economic theory, but (in the case of demand equations) by invoking an intuitive econometrician's version of psychological and sociological theory, since constraining utility functions is what is involved here. Furthermore, unless these sets of equations are considered as a system in the process of specification, the behavioral implications of the restrictions on all equations taken together may be less reasonable than the restrictions on any one equation taken by itself.

On the other hand, many of the monetarists used **reduced-form** equations to ascertain the effects of government policy on the macroeconomy. As an example, consider the following form of the *St. Louis model* estimated by Anderson and Jordan (1968). Using U.S. quarterly data from 1952 to 1968, they estimated the following reduced-form GNP determination equation:

$$\Delta Y_t = 2.28 + 1.54\Delta M_t + 1.56\Delta M_{t-1} + 1.44\Delta M_{t-2} + 1.29\Delta M_{t-3} + 0.40\Delta E_t + 0.54\Delta E_{t-1} - 0.03\Delta E_{t-2} - 0.74\Delta E_{t-3} \quad (5.16)$$

where: ΔY_t = change in nominal GNP
ΔM_t = change in the monetary base
ΔE_t = change in "high employment" budget deficit

In their analysis, Anderson and Jordan used base money and the high employment budget deficit because these are the variables under the control of the monetary and fiscal authorities, respectively. The St. Louis model was an attempt to demonstrate the monetarist policy recommendations that changes in the money supply, but not changes in government spending or taxation, affected GNP. The t-tests for the individual coefficients are misleading because of the substantial multicolinearity between each variable and its lags. However, testing whether the sum of the monetary base coefficients (i.e., $1.54 + 1.56 + 1.44 + 1.29 = 5.83$) differs from zero yields a t-value of 7.25. Hence, they concluded that changes in the money base translate into changes in nominal GNP. Since all the coefficients are positive, the effects of monetary policy are cumulative. On the other hand, the test that the sum of the fiscal coefficients ($0.40 + 0.54 - 0.03 - 0.74 = 0.17$) equals zero yields a t-value of 0.54. According to Anderson and Jordan, the results support "lagged crowding out" in the sense that an increase in the budget deficit initially stimulates the economy. Over time, however, changes in interest rates and other macroeconomic variables lead to reductions in private-sector expenditures. The cumulative effects of the fiscal stimulus are not statistically different from zero.

Sims (1980) also pointed out several problems with this type of analysis. Sims's criticisms are easily understood by recognizing that (5.16) is a transfer function with two independent variables $\{M_t\}$ and $\{E_t\}$ and no lags of the dependent variable. As with any type of transfer function analysis, we must be concerned with two things:

1. *Ensuring that lag lengths are appropriate.* Serially correlated residuals in the presence of lagged dependent variables lead to biased coefficient estimates.
2. *Ensuring that there is no feedback between GNP and the money base or the budget deficit.* However, the assumption of no feedback is unreasonable if the monetary or fiscal authorities deliberately attempt to alter nominal GNP. As in the thermostat example, if the monetary authority attempts to control the economy by changing the money base, we cannot identify the "true" model. In the jargon of time-series econometrics, changes in GNP would "cause" changes in the money supply. One appropriate strategy would be to simultaneously estimate the GNP determination equation *and* the money supply feedback rule.

Comparing the two types of models, Sims (pp. 14–15, 1980) states:

> Because existing large models contain too many incredible restrictions, empirical research aimed at testing competing macroeconomic theories too often proceeds in a single- or few-equation framework. For this reason alone, it appears worthwhile to investigate the possibility of building large models in a style which does not tend to accumulate restrictions so haphazardly. ... It should be feasible to estimate large-scale macromodels as unrestricted reduced forms, treating all variables as endogenous.

5. INTRODUCTION TO VAR ANALYSIS

When we are not confident that a variable is actually exogenous, a natural extension of transfer function analysis is to treat each variable symmetrically. In the two-variable case, we can let the time path of $\{y_t\}$ be affected by current and past realizations of the $\{z_t\}$ sequence *and* let the time path of the $\{z_t\}$ sequence be affected by current and past realizations of the $\{y_t\}$ sequence. Consider the simple bivariate system:

$$y_t = b_{10} - b_{12}z_t + \gamma_{11}y_{t-1} + \gamma_{12}z_{t-1} + \varepsilon_{yt} \tag{5.17}$$

$$z_t = b_{20} - b_{21}y_t + \gamma_{21}y_{t-1} + \gamma_{22}z_{t-1} + \varepsilon_{zt} \tag{5.18}$$

where it is assumed that (1) both y_t and z_t are stationary; (2) ε_{yt} and ε_{zt} are white-noise disturbances with standard deviations of σ_y and σ_z, respectively; and (3) $\{\varepsilon_{yt}\}$ and $\{\varepsilon_{zt}\}$ are uncorrelated white-noise disturbances.

Equations (5.17) and (5.18) constitute a *first-order* vector autoregression (VAR) because the longest lag length is unity. This simple two-variable, first-order VAR is useful for illustrating the multivariate higher-order systems that are introduced in

Section 8. The structure of the system incorporates feedback because y_t and z_t are allowed to affect each other. For example, $-b_{12}$ is the contemporaneous effect of a unit change of z_t on y_t, and γ_{12} is the effect of a unit change in z_{t-1} on y_t. Note that the terms ε_{yt} and ε_{zt} are pure innovations (or shocks) in y_t and z_t, respectively. Of course, if b_{21} is not equal to zero, ε_{yt} has an indirect contemporaneous effect on z_t, and if b_{12} is not equal to zero, ε_{zt} has an indirect contemporaneous effect on y_t. Such a system could be used to capture the feedback effects in our temperature-thermostat example. The first equation allows current and past values of the thermostat setting to affect the time path of the temperature; the second allows for feedback between current and past values of the temperature and the thermostat setting.[7]

Equations (5.17) and (5.18) are not reduced-form equations since y_t has a contemporaneous effect on z_t and z_t has a contemporaneous effect on y_t. Fortunately, it is possible to transform the system of equations into a more usable form. Using matrix algebra, we can write the system in the compact form

$$\begin{bmatrix} 1 & b_{12} \\ b_{21} & 1 \end{bmatrix}\begin{bmatrix} y_t \\ z_t \end{bmatrix} = \begin{bmatrix} b_{10} \\ b_{20} \end{bmatrix} + \begin{bmatrix} \gamma_{11} & \gamma_{12} \\ \gamma_{21} & \gamma_{22} \end{bmatrix}\begin{bmatrix} y_{t-1} \\ z_{t-1} \end{bmatrix} + \begin{bmatrix} \varepsilon_{yt} \\ \varepsilon_{zt} \end{bmatrix}$$

or

$$Bx_t = \Gamma_0 + \Gamma_1 x_{t-1} + \varepsilon_t$$

where

$$B = \begin{bmatrix} 1 & b_{12} \\ b_{21} & 1 \end{bmatrix}, \quad x_t = \begin{bmatrix} y_t \\ z_t \end{bmatrix}, \quad \Gamma_0 = \begin{bmatrix} b_{10} \\ b_{20} \end{bmatrix},$$

$$\Gamma_1 = \begin{bmatrix} \gamma_{11} & \gamma_{12} \\ \gamma_{21} & \gamma_{22} \end{bmatrix}, \quad \varepsilon_t = \begin{bmatrix} \varepsilon_{yt} \\ \varepsilon_{zt} \end{bmatrix}$$

Premultiplication by B^{-1} allows us to obtain the VAR model in *standard* form:

$$x_t = A_0 + A_1 x_{t-1} + e_t \tag{5.19}$$

where $A_0 = B^{-1}\Gamma_0$, $A_1 = B^{-1}\Gamma_1$, and $e_t = B^{-1}\varepsilon_t$.

For notational purposes, we can define a_{i0} as element i of the vector A_0, a_{ij} as the element in row i and column j of the matrix A_1, and e_{it} as the element i of the vector e_t. Using this new notation, we can rewrite (5.19) in the equivalent form:

$$y_t = a_{10} + a_{11}y_{t-1} + a_{12}z_{t-1} + e_{1t} \tag{5.20}$$

$$z_t = a_{20} + a_{21}y_{t-1} + a_{22}z_{t-1} + e_{2t} \tag{5.21}$$

To distinguish between the systems represented by (5.17) and (5.18) versus (5.20) and (5.21), the first is called a structural VAR or the primitive system and the second is called a VAR in standard form. It is important to note that the error terms (i.e., e_{1t} and e_{2t}) are composites of the two shocks ε_{yt} and ε_{zt}. Since $e_t = B^{-1}\varepsilon_t$, we can compute e_{1t} and e_{2t} as

$$e_{1t} = (\varepsilon_{yt} - b_{12}\varepsilon_{zt})/(1 - b_{12}b_{21}) \tag{5.22}$$

$$e_{2t} = (\varepsilon_{zt} - b_{21}\varepsilon_{yt})/(1 - b_{12}b_{21}) \tag{5.23}$$

Since ε_{yt} and ε_{zt} are white-noise processes, it follows that both e_{1t} and e_{2t} have zero means, have constant variances, and are individually serially uncorrelated. To find the properties of $\{e_{1t}\}$, first take the expected value of (5.22):

$$Ee_{1t} = E(\varepsilon_{yt} - b_{12}\varepsilon_{zt})/(1 - b_{12}b_{21}) = 0$$

The variance of e_{1t} is given by

$$\begin{aligned} Ee_{1t}^2 &= E[(\varepsilon_{yt} - b_{12}\varepsilon_{zt})/(1 - b_{12}b_{21})]^2 \\ &= (\sigma_y^2 + b_{12}^2\sigma_z^2)/(1 - b_{12}b_{21})^2 \end{aligned} \tag{5.24}$$

Thus, the variance of e_{1t} is time-independent. The autocorrelations of e_{1t} and e_{1t-i} are

$$Ee_{1t}e_{1t-i} = E[(\varepsilon_{yt} - b_{12}\varepsilon_{zt})(\varepsilon_{yt-i} - b_{12}\varepsilon_{zt-i})]/(1 - b_{12}b_{21})^2 = 0 \quad \text{for } i \neq 0$$

Similarly, (5.23) can be used to demonstrate that e_{2t} is a stationary process with zero mean, constant variance, and all autocovariances equal to zero. A critical point to note is that e_{1t} and e_{2t} are correlated. The covariance of the two terms is

$$\begin{aligned} Ee_{1t}e_{2t} &= E[(\varepsilon_{yt} - b_{12}\varepsilon_{zt})(\varepsilon_{zt} - b_{21}\varepsilon_{yt})]/(1 - b_{12}b_{21})^2 \\ &= -(b_{21}\sigma_y^2 + b_{12}\sigma_z^2)/(1 - b_{12}b_{21})^2 \end{aligned} \tag{5.25}$$

In general, (5.25) will not be zero so that the two shocks will be correlated. In the special case where $b_{12} = b_{21} = 0$ (i.e., if there are no contemporaneous effects of y_t on z_t and z_t on y_t), the shocks will be uncorrelated. It is useful to define the variance/covariance matrix of the e_{1t} and e_{2t} shocks as

$$\Sigma = \begin{bmatrix} \text{var}(e_{1t}) & \text{cov}(e_{1t}, e_{2t}) \\ \text{cov}(e_{1t}, e_{2t}) & \text{var}(e_{2t}) \end{bmatrix}$$

Since all elements of Σ are time-independent, we can use the more compact form

$$\Sigma = \begin{bmatrix} \sigma_1^2 & \sigma_{12} \\ \sigma_{21} & \sigma_2^2 \end{bmatrix} \tag{5.26}$$

where $\text{var}(e_{it}) = \sigma_i^2$ and $\text{cov}(e_{1t}, e_{2t}) = \sigma_{12} = \sigma_{21}$.

Stability and Stationarity

In the first-order autoregressive model $y_t = a_0 + a_1 y_{t-1} + \varepsilon_t$, the stability condition is that a_1 be less than unity in absolute value. There is a direct analogue between this stability condition and the matrix A_1 in the first-order VAR model of (5.19). Using the brute force method to solve the system, iterate (5.19) backward to obtain

$$\begin{aligned} x_t &= A_0 + A_1(A_0 + A_1 x_{t-2} + e_{t-1}) + e_t \\ &= (I + A_1)A_0 + A_1^2 x_{t-2} + A_1 e_{t-1} + e_t \end{aligned}$$

where $I = 2 \cdot 2$ identity matrix.

After n iterations,

$$x_t = (I + A_1 + \ldots + A_1^n)A_0 + \sum_{i=0}^{n} A_1^i e_{t-i} + A_1^{n+1} x_{t-n-1}$$

If we continue to iterate backward, it is clear that convergence requires that the expression A_1^n vanish as n approaches infinity. As shown below, stability requires that the roots of $(1 - a_{11}L)(1 - a_{22}L) - (a_{12}a_{21}L^2)$ lie outside the unit circle (the stability condition for higher-order systems is derived in Appendix 6.2 of the next chapter). For the time being, assuming the stability condition is met, we can write the particular solution for x_t as

$$x_t = \mu + \sum_{i=0}^{\infty} A_1^i e_{t-i} \tag{5.27}$$

where $\mu = [\,\bar{y}\ \bar{z}\,]'$
and

$$\bar{y} = [a_{10}(1 - a_{22}) + a_{12}a_{20}]/\Delta; \quad \bar{z} = [a_{20}(1 - a_{11}) + a_{21}a_{10}]/\Delta$$
$$\Delta = (1 - a_{11})(1 - a_{22}) - a_{12}a_{21}$$

If we take the expected value of (5.27), the unconditional mean of x_t is μ; hence, the unconditional means of y_t and z_t are $\bar{y}$ and $\bar{z}$, respectively. The variances and covariances of y_t and z_t can be obtained as follows. First, form the variance/covariance matrix as

$$E(x_t - \mu)^2 = E\left[\sum_{i=0}^{\infty} A_1^i e_{t-i}\right]^2$$

Next, using (5.26), note that

$$Ee_t^2 = E\begin{bmatrix} e_{1t} \\ e_{2t} \end{bmatrix}[e_{1t}, e_{2t}]$$
$$= \Sigma$$

Since $Ee_t e_{t-i} = 0$ for $i \neq 0$, it follows that

$$E(x_t - \mu)^2 = (I + A_1^2 + A_1^4 + A_1^6 + \ldots)\Sigma$$
$$= [I - A_1^2]^{-1}\Sigma$$

where it is assumed that the stability condition holds, so A_1^n approaches zero as n approaches infinity.

If we can abstract from an initial condition, the $\{y_t\}$ and $\{z_t\}$ sequences will be jointly covariance-stationary if the stability condition holds. Each sequence has a finite and time-invariant mean, and a finite and time-invariant variance.

In order to get another perspective on the stability condition, use lag operators to rewrite the VAR model of (5.20) and (5.21) as

$$y_t = a_{10} + a_{11}Ly_t + a_{12}Lz_t + e_{1t}$$
$$z_t = a_{20} + a_{21}Ly_t + a_{22}Lz_t + e_{2t}$$

or

$$(1 - a_{11}L)y_t = a_{10} + a_{12}Lz_t + e_{1t}$$
$$(1 - a_{22}L)z_t = a_{20} + a_{21}Ly_t + e_{2t}$$

If we use this last equation to solve for z_t, it follows that Lz_t is

$$Lz_t = L(a_{20} + a_{21}Ly_t + e_{2t})/(1 - a_{22}L)$$

so that

$$(1 - a_{11}L)y_t = a_{10} + a_{12}L[(a_{20} + a_{21}Ly_t + e_{2t})/(1 - a_{22}L)] + e_{1t}$$

Notice that we have transformed the first-order VAR in the $\{y_t\}$ and $\{z_t\}$ sequences into a stochastic difference equation in the $\{y_t\}$ sequence. Explicitly solving for y_t, we get

$$y_t = \frac{a_{10}(1 - a_{22}) + a_{12}a_{20} + (1 - a_{22}L)e_{1t} + a_{12}e_{2t-1}}{(1 - a_{11}L)(1 - a_{22}L) - a_{12}a_{21}L^2} \tag{5.28}$$

In the same fashion, you should be able to demonstrate that the solution for z_t is

$$z_t = \frac{a_{20}(1 - a_{11}) + a_{21}a_{10} + (1 - a_{11}L)e_{2t} + a_{21}e_{1t-1}}{(1 - a_{11}L)(1 - a_{22}L) - a_{12}a_{21}L^2} \tag{5.29}$$

Equations (5.28) and (5.29) both have the same characteristic equation; convergence requires that the roots of the polynomial $(1 - a_{11}L)(1 - a_{22}L) - a_{12}a_{21}L^2$ must lie outside the unit circle. (If you have forgotten the stability conditions for second-order difference equations, you might want to refresh your memory by reexamining Chapter 1.) Just as in any second-order difference equation, the roots may be real or complex and may be convergent or divergent. Notice that both y_t and z_t have the same characteristic equation; as long as a_{12} and a_{21} do not both equal zero, the solutions for the two sequences have the same characteristic roots. Hence, both will exhibit similar time paths.

Dynamics of a VAR Model

Figure 5.6 shows the time paths of four simple systems. For each system, 100 sets of normally distributed random numbers representing the $\{e_{1t}\}$ and $\{e_{2t}\}$ sequences were drawn. The initial values of y_0 and z_0 were set equal to zero and the $\{y_t\}$ and $\{z_t\}$ sequences were constructed as in (5.20) and (5.21). Panel (a) uses the values

$$a_{10} = a_{20} = 0; a_{11} = a_{22} = 0.7; \text{and } a_{12} = a_{21} = 0.2$$

When we substitute these values into (5.27), it is clear that the mean of each series is zero. From the quadratic formula, the two roots of the inverse characteristic equation $(1 - a_{11}L)(1 - a_{22}L) - a_{12}a_{21}L^2$ are 1.111 and 2.0. Since both are outside the unit circle, the system is stationary; the two characteristic roots of the solution for $\{y_t\}$ and $\{z_t\}$ are 0.9 and 0.5. Since these roots are positive, real, and less than unity, convergence will be direct. As you can see in the figure, there is a tendency for the

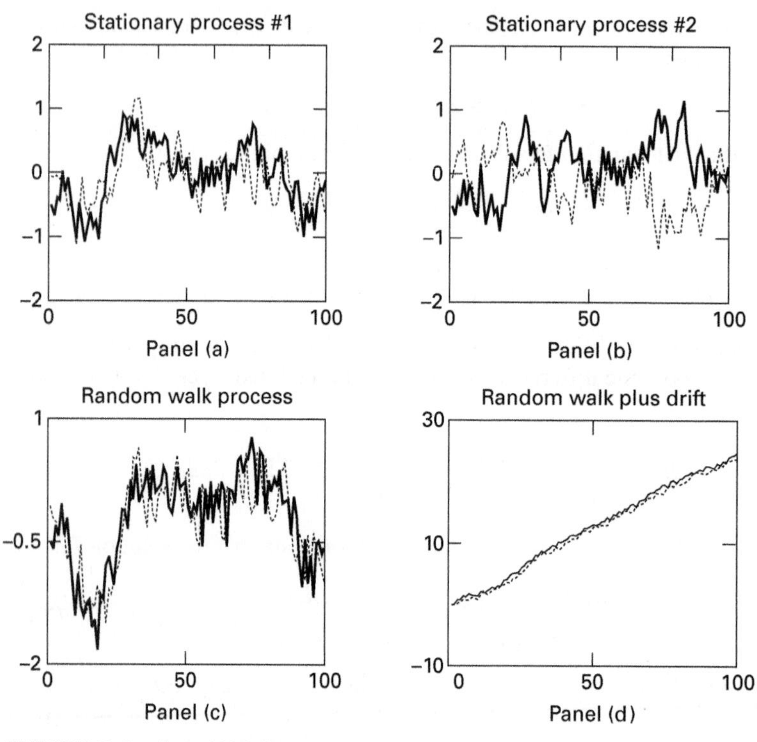

FIGURE 5.6 Four VAR Processes

sequences to move together. Since a_{21} is positive, a large realization in y_t induces a large realization of z_{t+1}; since a_{12} is positive, a large realization of z_t induces a large realization of y_{t+1}. The cross-correlations between the two series are positive.

Panel (b) illustrates a stationary process with $a_{10} = a_{20} = 0$; $a_{11} = a_{22} = 0.5$; and $a_{12} = a_{21} = -0.2$. Again, the mean of each series is zero and the characteristic roots are 0.7 and 0.3. However, in contrast to the previous case, a_{21} and a_{12} are both negative so that positive realizations of y_t can be associated with negative realizations of z_{t+1} and vice versa. As can be seen from comparing the second panel, the two series appear to be negatively correlated.

In contrast, Panel (c) shows a process possessing a unit root; here, $a_{11} = a_{22} = a_{12} = a_{21} = 0.5$. You should take a moment to find the characteristic roots. Undoubtedly, there is little tendency for either of the series to revert to a constant long-run value. Here, the intercept terms a_{10} and a_{20} are equal to zero so that Panel (c) represents a multivariate generalization of the random walk model. You can see how the series seem to meander together. In Panel (d), the VAR process of Panel (c) also contains a nonzero intercept term ($a_{10} = .5$ and $a_{20} = 0$) that takes the role of a "drift." As you can see from Panel (d), the two series appear to move closely together. The drift term adds a deterministic time trend to the nonstationary behavior of the two series. Combined with the unit characteristic root, the $\{y_t\}$ and $\{z_t\}$ sequences are joint random walk plus drift processes. Notice that the presence of the drift dominates the long-run behavior of the series.

6. ESTIMATION AND IDENTIFICATION

One explicit aim of the Box–Jenkins approach is to provide a methodology that leads to parsimonious models. The ultimate objective of making accurate short-term forecasts is best served by purging insignificant parameter estimates from the model. Sims's (1980) criticisms of the "incredible identification restrictions" inherent in structural models argue for an alternative estimation strategy. Consider the following multivariate generalization of an autoregressive process:

$$x_t = A_0 + A_1 x_{t-1} + A_2 x_{t-2} + \ldots + A_p x_{t-p} + e_t \qquad (5.30)$$

where: x_t = an $(n \cdot 1)$ vector containing each of the n variables included in the VAR
A_0 = an $(n \cdot 1)$ vector of intercept terms
A_i = $(n \cdot n)$ matrices of coefficients
e_t = an $(n \cdot 1)$ vector of error terms

Sims's methodology entails little more than a determination of the appropriate variables to include in the VAR and a determination of the appropriate lag length. The variables to be included in the VAR are selected according to the relevant economic model. Lag length tests (to be discussed below) select the appropriate lag length. Otherwise, no explicit attempt is made to "pare down" the number of parameter estimates. The matrix A_0 contains n parameters and each matrix A_i contains n^2 parameters; hence, $n + pn^2$ coefficients need to be estimated. Unquestionably, a VAR will be *overparameterized* in that many of these coefficient estimates will be insignificant. However, the goal is to find the important interrelationships among the variables. Improperly imposing zero restrictions may waste important information. Moreover, the regressors are likely to be highly colinear so that the t-tests on individual coefficients are not reliable guides for paring down the model.

Note that the right-hand side of (5.30) contains only predetermined variables and that the error terms are assumed to be serially uncorrelated with constant variance. Hence, *each equation in the system can be estimated using OLS*. Moreover, OLS estimates are consistent and asymptotically efficient. Even though the errors are correlated across equations, seemingly unrelated regressions (SUR) do not add to the efficiency of the estimation procedure since all regressions have identical right-hand-side variables.

There is an issue of whether the variables in a VAR need to be stationary. Sims (1980) and Sims, Stock, and Watson (1990) recommended against differencing *even if the variables contain a unit root*. They argued that the goal of a VAR analysis is to determine the interrelationships among the variables, *not* to determine the parameter estimates. The main argument against differencing is that it "throws away" information concerning the comovements in the data (such as the possibility of cointegrating relationships). Similarly, it is argued that the data need not be detrended. In a VAR, a trending variable will be well approximated by a unit root plus drift. However, the majority view is that the form of the variables in the VAR should mimic the true data-generating process. This is particularly true if the aim is to estimate a structural model. We return to these issues in the next chapter; for now, it is assumed that all variables are stationary. Questions 8 and 9 at the end of this chapter ask you to compare a VAR in levels to a VAR in first differences.

Forecasting

Once the VAR has been estimated, it can be used as a multiequation forecasting model. Suppose you estimate the first-order model $x_t = A_0 + A_1 x_{t-1} + e_t$ so as to obtain the values of the coefficients in A_0 and A_1. If your data run through period T, it is straightforward to obtain the one-step-ahead forecasts of your variables using the relationship $E_T x_{T+1} = A_0 + A_1 x_T$. Similarly, a two-step-ahead forecast can be obtained recursively from $E_T x_{T+2} = A_0 + A_1 E_T x_{T+1} = A_0 + A_1[A_0 + A_1 x_T]$. Nevertheless, in a higher-order model there can be a large number of coefficient estimates. Since unrestricted VARs are overparameterized, the forecasts may be unreliable. In order to obtain a parsimonious model, many forecasters would purge the insignificant coefficients from the VAR. After reestimating the so-called **near-VAR** model using SUR, it could be used for forecasting purposes. Others might use a Bayesian approach by combining a set of prior beliefs with the traditional VAR methods presented in this text. West and Harrison (1989) provided an approachable introduction to the Bayesian approach. Litterman (1980) proposed a sensible set of Bayesian priors that have become the standard in Bayesian VAR models.

An interesting use of forecasting with a VAR is provided by Eckstein and Tsiddon's (2004) four-equation VAR. The aim of the study was to investigate the effects of terrorism (T) on the growth rates of Israeli real per capita GDP (ΔGDP_t), investment (ΔI_t), exports (ΔEXP_t), and nondurable consumption (ΔNDC_t). The authors use quarterly data running from 1980Q1 to 2003Q3 so that there are 95 total observations. The measure of terrorism is a weighted average of the number of Israeli fatalities, injuries, and noncasualty incidents due to both domestic and transnational attacks occurring in Israel. Consider a simplified version of their VAR model:

$$\begin{bmatrix} \Delta GDP_t \\ \Delta I_t \\ \Delta EXP_t \\ \Delta NDC_t \end{bmatrix} = \begin{bmatrix} A_{11}(L) & \cdots & A_{14}(L) \\ \vdots & \ddots & \vdots \\ A_{41}(L) & \cdots & A_{44}(L) \end{bmatrix} \begin{bmatrix} \Delta GDP_{t-1} \\ \Delta I_{t-1} \\ \Delta EXP_{t-1} \\ \Delta NDC_{t-1} \end{bmatrix} + \begin{bmatrix} c_1 T_{t-1} \\ c_2 T_{t-1} \\ c_3 T_{t-1} \\ c_3 T_{t-1} \end{bmatrix} + \dots + \begin{bmatrix} e_{1t} \\ e_{2t} \\ e_{3t} \\ e_{4t} \end{bmatrix}$$

where the expressions $A_{ij}(L)$ are polynomials in the lag operator L, the c_i measure the influence of lagged terrorism on variable i, and the e_i are the regression errors. The other right-hand-side variables (not shown) are the first difference of the real interest rate, three quarterly seasonal dummies, and an intercept.

The nature of the VAR is such that ΔGDP_t, ΔI_t, ΔEXP_t, and ΔNDC_t are all jointly determined. In contrast, the terrorism variable acts as an independent variable in the system. Notice that the magnitude of T_{t-1} is allowed to affect the four macroeconomic variables, but there is no feedback from these variables to the level of terrorism. The authors report that lagging the terrorism variable for a single period provided a better fit than the use of other lag lengths.

The four equations of the model were estimated through 2003*Q*3 and used to obtain one- through twelve-step-ahead forecasts of ΔGDP_t, ΔI_t, ΔEXP_t, and ΔNDC_t. Unlike forecasting with a pure VAR (in which all variables are jointly determined), it was necessary for Eckstein and Tsiddon (2004) to specify the time path of the terrorism variable. Consider the VAR representation of their model $x_t = A_0 + A_1 x_{t-1} + cT_{t-1} + e_t$ where c is the 4 × 1 vector $[c_1, c_2, c_3, c_4]'$. The one-step-ahead forecast is $E_T x_{T+1} = A_0 + A_1 x_T + cT_T$ and the two-step-ahead forecast is $E_T x_{T+2} = A_0 + A_1 E_T [x_{T+1} + cT_{T+1}]$. Hence, in order to forecast the values of x_{T+2} and beyond, it is necessary to know the magnitude of the terrorism variable over the forecast period. Toward this end, Eckstein and Tsiddon supposed that all terrorism actually ended in 2003*Q*4 (so that all values of $T_j = 0$ for $j >$ 2003*Q*4). Under this assumption, the annual growth rate of GDP was estimated to be 2.5 percent through 2005*Q*3. Instead, when they set the values of T_j at the 2000*Q*4 to 2003*Q*4 period average, the growth rate of GDP was estimated to be zero. Thus, a steady level of terrorism would have cost the Israeli economy all of its real output gains. In actuality, the largest influence of terrorism was found to be on investment. The impact of terrorism on investment was twice as large as the impact on real GDP.

Identification

Suppose that you want to recover the structural VAR from your estimate of the model in standard form. To illustrate the identification procedure, return to the two-variable/first-order VAR of the previous section. Due to the feedback inherent in a VAR process, the primitive equations (5.17) and (5.18) cannot be estimated directly. The reason is that z_t is correlated with the error term ε_{yt} *and* that y_t is correlated with the error term ε_{zt}. Standard estimation techniques require that the regressors be uncorrelated with the error term. Note that there is no such problem in estimating the VAR system in the form of (5.20) and (5.21). OLS can provide estimates of the two elements of A_0 and of the four elements of A_1. Moreover, obtaining the residuals from the two regressions, it is possible to calculate estimates of the variance of e_{1t} and e_{2t}, and of the covariance between e_{1t} with e_{2t}. The issue is whether it is possible to recover all of the information present in the primitive system given by (5.17) and (5.18). In other words, is the primitive system identifiable given the OLS estimates of the VAR model in the form of (5.20) and (5.21)?

The answer to this question is, "No, unless we are willing to appropriately restrict the primitive system." The reason is clear if we compare the number of parameters of the primitive system with the number of parameters recovered from the estimated VAR model. Estimating (5.20) and (5.21) yields six coefficient estimates (a_{10}, a_{20}, a_{11}, a_{12}, a_{21}, and a_{22}) and the calculated values of var(e_{1t}), var(e_{2t}), and cov(e_{1t}, e_{2t}). However, the primitive system (5.17) and (5.18) contains ten parameters. In addition to the two intercept coefficients b_{10} and b_{20}, the four autoregressive coefficients γ_{11}, γ_{12}, γ_{21}, and γ_{22}, and the two feedback coefficients b_{12} and b_{21}, there are the two standard deviations σ_y and σ_z. In all, the primitive system contains 10 parameters whereas the VAR estimation yields only nine parameters. Unless one is willing to restrict one of the parameters, it is not possible to identify the primitive system; equations (5.17) and (5.18) are underidentified.

One way to identify the model is to use the type of **recursive** system proposed by Sims (1980). Suppose that you are willing to impose a restriction on the primitive system such that the coefficient b_{21} is equal to zero. Writing (5.17) and (5.18) with the constraint imposed yields

$$y_t = b_{10} - b_{12}z_t + \gamma_{11}y_{t-1} + \gamma_{12}z_{t-1} + \varepsilon_{yt} \tag{5.31}$$

$$z_t = b_{20} + \gamma_{21}y_{t-1} + \gamma_{22}z_{t-1} + \varepsilon_{zt} \tag{5.32}$$

Similarly, we can rewrite the relationship between the pure shocks and the regression residuals given by (5.22) and (5.23) as

$$e_{1t} = \varepsilon_{yt} - b_{12}\varepsilon_{zt}$$

$$e_{2t} = \varepsilon_{zt}$$

The point is that forcing $b_{21} = 0$ means that z_t has a contemporaneous effect on y_t but y_t affects the $\{z_t\}$ sequence with a one-period lag. Nevertheless, it should be clear that this restriction (which might be suggested by a particular economic model) results in an exactly identified system. Imposing the restriction $b_{21} = 0$ means that B^{-1} is given by

$$B^{-1} = \begin{bmatrix} 1 & -b_{12} \\ 0 & 1 \end{bmatrix}$$

Now, premultiplication of the primitive system by B^{-1} yields

$$\begin{bmatrix} y_t \\ z_t \end{bmatrix} = \begin{bmatrix} 1 & -b_{12} \\ 0 & 1 \end{bmatrix}\begin{bmatrix} b_{10} \\ b_{20} \end{bmatrix} + \begin{bmatrix} 1 & -b_{12} \\ 0 & 1 \end{bmatrix}\begin{bmatrix} \gamma_{11} & \gamma_{12} \\ \gamma_{21} & \gamma_{22} \end{bmatrix}\begin{bmatrix} y_{t-1} \\ z_{t-1} \end{bmatrix} + \begin{bmatrix} 1 & -b_{12} \\ 0 & 1 \end{bmatrix}\begin{bmatrix} \varepsilon_{yt} \\ \varepsilon_{zt} \end{bmatrix}$$

or

$$\begin{bmatrix} y_t \\ z_t \end{bmatrix} = \begin{bmatrix} b_{10} - b_{12}b_{20} \\ b_{20} \end{bmatrix} + \begin{bmatrix} \gamma_{11} - b_{12}\gamma_{21} & \gamma_{12} - b_{12}\gamma_{22} \\ \gamma_{21} & \gamma_{22} \end{bmatrix}\begin{bmatrix} y_{t-1} \\ z_{t-1} \end{bmatrix} + \begin{bmatrix} \varepsilon_{yt} - b_{12}\varepsilon_{zt} \\ \varepsilon_{zt} \end{bmatrix} \tag{5.33}$$

Estimating the system using OLS yields the parameter estimates from

$$y_t = a_{10} + a_{11}y_{t-1} + a_{12}z_{t-1} + e_{1t}$$

$$z_t = a_{20} + a_{21}y_{t-1} + a_{22}z_{t-1} + e_{2t}$$

where $a_{10} = b_{10} - b_{12}b_{20}$, $a_{11} = \gamma_{11} - b_{12}\gamma_{21}$, $a_{12} = \gamma_{12} - b_{12}\gamma_{22}$, $a_{20} = b_{20}$, $a_{21} = \gamma_{21}$, and $a_{22} = \gamma_{22}$.

Since $b_{21} = 0$, it follows that $e_{1t} = \varepsilon_{yt} - b_{12}\varepsilon_{zt}$ and $e_{2t} = \varepsilon_{zt}$. Hence,

$$\text{var}(e_1) = \sigma_y^2 + b_{12}^2\,\sigma_z^2$$

$$\text{var}(e_2) = \sigma_z^2$$

$$\text{cov}(e_1, e_2) = -b_{12}\sigma_z^2$$

Thus, we have nine parameter estimates a_{10}, a_{11}, a_{12}, a_{20}, a_{21}, a_{22}, var(e_1), var(e_2), and cov(e_1, e_2) that can be substituted into the nine equations above in order to simultaneously solve for b_{10}, b_{12}, γ_{11}, γ_{12}, b_{20}, γ_{21}, γ_{22}, σ_y^2, and σ_z^2.

Note also that the estimates of the $\{\varepsilon_{yt}\}$ and $\{\varepsilon_{zt}\}$ sequences can be recovered. The residuals from the second equation (i.e., the $\{e_{2t}\}$ sequence) are estimates of the $\{\varepsilon_{zt}\}$ sequence. Combining these estimates along with the solution for b_{12} allows us to calculate the estimates of the $\{\varepsilon_{yt}\}$ sequence using the relationship $e_{1t} = \varepsilon_{yt} - b_{12}\varepsilon_{zt}$.

Take a minute to examine the restriction. In (5.32), the assumption $b_{21} = 0$ means that y_t does not have a contemporaneous effect on z_t. In (5.33), the restriction manifests itself such that both ε_{yt} and ε_{zt} shocks affect the contemporaneous value of y_t but only ε_{zt} shocks affect the contemporaneous value of z_t. The observed values of e_{2t} are completely attributed to pure shocks to the $\{z_t\}$ sequence. Decomposing the residuals in this triangular fashion is called a **Choleski** decomposition.

In fact, the result is quite general. In an n-variable VAR, B is an $n \cdot n$ matrix since there are n regression residuals and n structural shocks. As shown in Section 10, exact identification requires that $(n^2 - n)/2$ restrictions be placed on the relationship between the regression residuals and the structural innovations. Since the Choleski decomposition is triangular, it forces exactly $(n^2 - n)/2$ values of the B matrix to equal zero.

7. THE IMPULSE RESPONSE FUNCTION

Just as an autoregression has a moving-average representation, a vector autoregression can be written as a vector moving average (VMA). In fact, equation (5.27) is the VMA representation of (5.19) in that the variables (i.e., y_t and z_t) are expressed in terms of the current and past values of the two types of shocks (i.e., e_{1t} and e_{2t}). The VMA representation is an essential feature of Sims's (1980) methodology in that it allows you to trace out the time path of the various shocks on the variables contained in the VAR system. For illustrative purposes, continue to use the two-variable/first-order model analyzed in the previous two sections. Writing the two-variable VAR in matrix form,

$$\begin{bmatrix} y_t \\ z_t \end{bmatrix} = \begin{bmatrix} a_{10} \\ a_{20} \end{bmatrix} + \begin{bmatrix} a_{11} & a_{12} \\ a_{21} & a_{22} \end{bmatrix} \begin{bmatrix} y_{t-1} \\ z_{t-1} \end{bmatrix} + \begin{bmatrix} e_{1t} \\ e_{2t} \end{bmatrix} \tag{5.34}$$

or, using (5.27), we get

$$\begin{bmatrix} y_t \\ z_t \end{bmatrix} = \begin{bmatrix} \bar{y} \\ \bar{z} \end{bmatrix} + \sum_{i=0}^{\infty} \begin{bmatrix} a_{11} & a_{12} \\ a_{21} & a_{22} \end{bmatrix}^i \begin{bmatrix} e_{1t-i} \\ e_{2t-i} \end{bmatrix} \tag{5.35}$$

Equation (5.35) expresses y_t and z_t in terms of the $\{e_{1t}\}$ and $\{e_{2t}\}$ sequences. However, it is insightful to rewrite (5.35) in terms of the $\{\varepsilon_{yt}\}$ and $\{\varepsilon_{zt}\}$ sequences. From (5.22) and (5.23), the vector of errors can be written as

$$\begin{bmatrix} e_{1t} \\ e_{2t} \end{bmatrix} = \frac{1}{1 - b_{12}b_{21}} \begin{bmatrix} 1 & -b_{12} \\ -b_{21} & 1 \end{bmatrix} \begin{bmatrix} \varepsilon_{yt} \\ \varepsilon_{zt} \end{bmatrix} \tag{5.36}$$

so that (5.35) and (5.36) can be combined to form

$$\begin{bmatrix} y_t \\ z_t \end{bmatrix} = \begin{bmatrix} \bar{y} \\ \bar{z} \end{bmatrix} + \frac{1}{1 - b_{12}b_{21}} \sum_{i=0}^{\infty} \begin{bmatrix} a_{11} & a_{12} \\ a_{21} & a_{22} \end{bmatrix}^i \begin{bmatrix} 1 & -b_{12} \\ -b_{21} & 1 \end{bmatrix} \begin{bmatrix} \varepsilon_{yt-i} \\ \varepsilon_{zt-i} \end{bmatrix}$$

Since the notation is getting unwieldy, we can simplify by defining the $2 \cdot 2$ matrix ϕ_i with elements $\phi_{jk}(i)$:

$$\phi_i = \frac{A_1^i}{1 - b_{12}b_{21}} \begin{bmatrix} 1 & -b_{12} \\ -b_{21} & 1 \end{bmatrix}$$

Hence, the moving-average representation of (5.35) and (5.36) can be written in terms of the $\{\varepsilon_{yt}\}$ and $\{\varepsilon_{zt}\}$ sequences:

$$\begin{bmatrix} y_t \\ z_t \end{bmatrix} = \begin{bmatrix} \bar{y} \\ \bar{z} \end{bmatrix} + \sum_{i=0}^{\infty} \begin{bmatrix} \phi_{11}(i) & \phi_{12}(i) \\ \phi_{21}(i) & \phi_{22}(i) \end{bmatrix} \begin{bmatrix} \varepsilon_{yt-i} \\ \varepsilon_{zt-i} \end{bmatrix}$$

or, more compactly,

$$x_t = \mu + \sum_{i=0}^{\infty} \phi_i \varepsilon_{t-i} \tag{5.37}$$

The moving-average representation is an especially useful tool to examine the interaction between the $\{y_t\}$ and $\{z_t\}$ sequences. The coefficients of ϕ_i can be used to generate the effects of ε_{yt} and ε_{zt} shocks on the entire time paths of the $\{y_t\}$ and $\{z_t\}$ sequences. If you understand the notation, it should be clear that the four elements $\phi_{jk}(0)$ are **impact multipliers.** For example, the coefficient $\phi_{12}(0)$ is the instantaneous impact of a one-unit change in ε_{zt} on y_t. In the same way, the elements $\phi_{11}(1)$ and $\phi_{12}(1)$ are the one-period responses of unit changes in ε_{yt-1} and ε_{zt-1} on y_t, respectively. Updating by one period indicates that $\phi_{11}(1)$ and $\phi_{12}(1)$ also represent effects of unit changes in ε_{yt} and ε_{zt} on y_{t+1}.

The accumulated effects of unit impulses in ε_{yt} and/or ε_{zt} can be obtained by the appropriate summation of the coefficients of the impulse response functions. For example, note that after n periods, the effect of ε_{zt} on the value of y_{t+n} is $\phi_{12}(n)$. Thus, after n periods, the cumulated sum of the effects of ε_{zt} on the $\{y_t\}$ sequence is

$$\sum_{i=0}^{n} \phi_{12}(i)$$

Letting n approach infinity yields the total cumulated effect. Since the $\{y_t\}$ and $\{z_t\}$ sequences are assumed to be stationary, it must be the case that for all j and k,

$$\sum_{i=0}^{\infty} \phi_{jk}^2(i) \text{ is finite}$$

The four sets of coefficients $\phi_{11}(i)$, $\phi_{12}(i)$, $\phi_{21}(i)$, and $\phi_{22}(i)$ are called the **impulse response functions.** Plotting the impulse response functions [i.e., plotting the coefficients of $\phi_{jk}(i)$ against i] is a practical way to visually represent the behavior of the $\{y_t\}$ and $\{z_t\}$ series in response to the various shocks. In principle, it might be possible to know all of the parameters of the primitive system (5.17) and (5.18). With such knowledge, it would be possible to trace out the time paths of the effects of pure ε_{yt} or ε_{zt} shocks. However, this methodology is not available to the researcher since an estimated VAR is underidentified. As explained in the previous section, knowledge of the various a_{ij} and the variance/covariance matrix Σ is not sufficient to identify the primitive system. Hence, the econometrician must impose an additional restriction on the two-variable VAR system in order to identify the impulse responses.

One possible identification restriction is to use the Choleski decomposition such that y_t does not have a contemporaneous effect on z_t. Formally, this restriction is represented by setting $b_{21} = 0$ in the primitive system. In terms of (5.36), the error terms can be decomposed as follows:

$$e_{1t} = \varepsilon_{yt} - b_{12}\varepsilon_{zt} \tag{5.38}$$

$$e_{2t} = \varepsilon_{zt} \tag{5.39}$$

As already noted, if we use (5.39), all of the observed errors from the $\{e_{2t}\}$ sequence are attributed to ε_{zt} shocks. Given the calculated $\{\varepsilon_{zt}\}$ sequence, knowledge of the values of the $\{e_{1t}\}$ sequence and the correlation coefficient between e_{1t} and e_{2t} allows for the calculation of the $\{\varepsilon_{yt}\}$ sequence using (5.38). Although this Choleski decomposition constrains the system such that an ε_{yt} shock has no direct effect on z_t, there is an indirect effect in that lagged values of y_t affect the contemporaneous value of z_t. The key point is that the decomposition forces a potentially important asymmetry on the system since an ε_{zt} shock has contemporaneous effects on both y_t and z_t. For this reason, (5.38) and (5.39) are said to be an **ordering** of the variables. An ε_{zt} shock directly affects e_{1t} and e_{2t}, but an ε_{yt} shock does not affect e_{2t}. Hence, z_t is said to be "causally prior" to y_t.

Suppose that estimates of equations (5.20) and (5.21) yield the values $a_{10} = a_{20} = 0$, $a_{11} = a_{22} = 0.7$, and $a_{12} = a_{21} = 0.2$. You will recall that this is precisely the model used in the simulation reported in Panel (a) of Figure 5.6. Also suppose that the elements of the Σ matrix are such that $\sigma_1^2 = \sigma_2^2$ and that cov(e_{1t}, e_{2t}) is such that the correlation coefficient between e_{1t} and e_{2t} (denoted by ρ_{12}) is 0.8. Hence, the decomposed errors can be represented by[8]

$$e_{1t} = \varepsilon_{yt} + 0.8\varepsilon_{zt} \tag{5.40}$$

$$e_{2t} = \varepsilon_{zt} \tag{5.41}$$

Panels (a) and (b) of Figure 5.7 trace out the effects of one-unit shocks to ε_{zt} and ε_{yt} on the time paths of the $\{y_t\}$ and $\{z_t\}$ sequences. As shown in Panel (a), a one-unit shock in ε_{zt} causes z_t to jump by one unit and y_t to jump by 0.8 units. [From (5.40), 80 percent of the ε_{zt} shock has a contemporaneous effect on e_{1t}.] In the next period, ε_{zt+1} returns to zero, but the autoregressive nature of the system is such that y_{t+1} and z_{t+1} do not immediately return to their long-run values. Since $z_{t+1} = 0.2y_t + 0.7z_t + \varepsilon_{zt+1}$, it follows that $z_{t+1} = 0.86$ [$0.2(0.8) + 0.7(1) = 0.86$]. Similarly, $y_{t+1} = 0.7y_t + 0.2z_t = 0.76$. As you can see from the figure, the subsequent values of the $\{y_t\}$ and $\{z_t\}$ sequences converge to their long-run levels. This convergence is assured by the stability of the system; as found earlier, the two characteristic roots are 0.5 and 0.9.

The effects of a one-unit shock in ε_{yt} are shown in Panel (b) of the figure. You can see the asymmetry of the decomposition immediately by comparing the two upper graphs. A one-unit shock in ε_{yt} causes the value of y_t to increase by one unit; however, *there is no contemporaneous effect on the value of* z_t so that $y_t = 1$ and $z_t = 0$. In the subsequent period, ε_{yt+1} returns to zero. The autoregressive nature of the system is

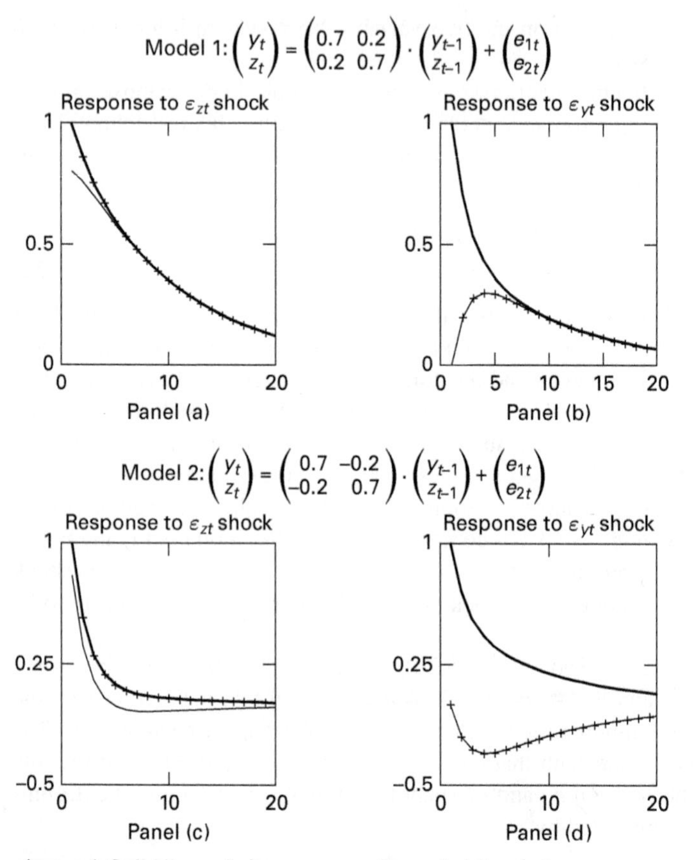

FIGURE 5.7 Two Impulse Response Functions

such that $y_{t+1} = 0.7y_t + 0.2z_t = 0.7$ and $z_{t+1} = 0.2y_t + 0.7z_t = 0.2$. The remaining points in the figure are the impulse responses for periods $(t + 2)$ through $(t + 20)$. Since the system is stationary, the impulse responses ultimately decay.

Can you figure out the consequences of reversing the Choleski decomposition in such a way that b_{12}, rather than b_{21}, is constrained to equal zero? Since matrix A_1 is symmetric (i.e., $a_{11} = a_{22}$ and $a_{12} = a_{21}$), the impulse responses of an ε_{yt} shock would be similar to those in Panel (a) and the impulse responses of an ε_{zt} would be similar to those in Panel (b). The only difference would be that the solid line would represent the time path of the $\{z_t\}$ sequence and the hatched line would represent the time path of the $\{y_t\}$ sequence.

As a practical matter, how does the researcher decide which of the alternative decompositions is most appropriate? In some instances, there might be a theoretical reason to suppose that one variable has no contemporaneous effect on the other. In the terrorism/tourism example, knowledge that terrorist incidents affect tourism with a lag suggests that terrorism does not have a contemporaneous effect on tourism. Usually, there is no such a priori knowledge. Moreover, the very idea of imposing a structure on a VAR system seems contrary to the spirit of Sims's argument against "incredible identifying restrictions." Unfortunately, there is no simple way to circumvent the problem; identification necessitates imposing *some* structure on the system. The Choleski decomposition provides a minimal set of assumptions that can be used to identify the structural model.[9]

It is crucial to note that *the importance of the ordering depends on the magnitude of the correlation coefficient between e_{1t} and e_{2t}*. Let this correlation coefficient be denoted by ρ_{12} so that $\rho_{12} = \sigma_{12}/(\sigma_1\sigma_2)$. Now suppose that the estimated model yields a value of Σ such that ρ_{12} is found to be equal to zero. In this circumstance, the ordering would be immaterial. Formally, (5.38) and (5.39) become $e_{1t} = \varepsilon_{yt}$ and $e_{2t} = \varepsilon_{zt}$. Since there is no correlation across equations, the residuals from the y_t and z_t equations are necessarily equivalent to the ε_{yt} and ε_{zt} shocks, respectively. The point is that if $Ee_{1t}e_{2t} = 0$, b_{12} and b_{21} can both be set equal to zero. At the other extreme, if ρ_{12} is found to be unity, there is a single shock that contemporarily affects both variables. Under the assumption $b_{21} = 0$, (5.38) and (5.39) become $e_{1t} = \varepsilon_{zt}$ and $e_{2t} = \varepsilon_{zt}$, and under the alternative assumption $b_{12} = 0$, it follows that $e_{1t} = \varepsilon_{yt}$ and $e_{2t} = \varepsilon_{yt}$. Usually the researcher will want to test the significance of ρ_{12}. As in a univariate model, you can test the null hypothesis $\rho_{12} = 0$ using a normal distribution with a mean of zero and a standard deviation of $T^{-0.5}$. As such, with 100 usable observations, if $|\rho_{12}| > 0.2$, the correlation is deemed to be significant at conventional levels. If ρ_{12} is significant, the usual procedure is to obtain the impulse response function using a particular ordering. Compare the results to the impulse response function obtained by reversing the ordering. If the implications are quite different, additional investigation into the relationships between the variables is necessary.

The lower half of Figure 5.7, Panels (c) and (d), presents the impulse response functions for a second model; the sole difference between models 1 and 2 is the change in the values of a_{12} and a_{21} to -0.2. Notice that this model was used in the simulation reported in Panel (b) of Figure 5.6. The negative off-diagonal elements of A_1 weaken the tendency for the two series to move together. Panel (c) traces out the effect of a one-unit ε_{zt} shock using ordering represented by (5.40) and (5.41). In period t, z_t rises by one unit and y_t rises by 0.8 units. In period $(t + 1)$, ε_{zt+1} returns to zero but the value of y_{t+1} is $0.7y_t - 0.2z_t = 0.36$ and the value of z_{t+1} is $-0.2y_t + 0.7z_t = 0.54$. The points represented by $t = 2$ through 20 show that the impulse responses converge to zero. Panel (d) traces the responses of a one-unit ε_{yt} shock. Since the value of z_t is unaffected by the shock, in period $(t + 1)$, $y_{t+1} = 0.7y_t - 0.2z_t = 0.7$ and $z_{t+1} = -0.2y_t + 0.7z_t = -0.2$. In the same way, $y_{t+2} = 0.7*0.7 - 0.2*(-0.2) = 0.53$ and $z_{t+2} = -0.2*(0.7) + 0.7*(-0.2) = -0.28$. Since the system is stable, both sequences eventually converge to zero.

Confidence Intervals and Impulse Responses

One key issue concerning the impulse response functions is that they are constructed using the estimated coefficients. Since each coefficient is estimated imprecisely, the impulse responses also contain error. The issue is to construct confidence intervals around the impulse responses that allow for the parameter uncertainty inherent in the estimation process. To illustrate the methodology, consider the following estimate of an *AR*(1) model:

$$y_t = 0.60y_{t-1} + \varepsilon_t$$
$$(4.00)$$

Given the t-statistic of 4.00, the AR(1) coefficient seems to be well estimated. It is easy to form the impulse response function: For any given level of y_{t-1}, a one-unit shock to ε_t will increase y_t by one unit. In subsequent periods, y_{t+1} will be 0.60 and y_{t+2} will be $(0.60)^2$. As you can easily verify, the impulse response function can be written as $\phi(i) = (0.60)^i$.

Notice that the AR(1) coefficient has an estimated mean of 0.6 and standard deviation of 0.15 (0.15 = 0.60/4.00). If we are willing to assume that the coefficient is normally distributed, there is a 95 percent chance that the actual value lies within the two-standard deviation interval 0.3 to 0.9. The problem is much more complicated in higher-order systems since the estimated coefficients will be correlated. Moreover, you may not want to assume normality. A better way to obtain the desired confidence intervals from the AR(p) process $y_t = a_0 + a_1y_{t-1} + \dots + a_py_{t-p} + \varepsilon_t$ is to perform the following Monte Carlo study:

1. Estimate the coefficients a_0 through a_p using OLS and save the residuals. Let $\hat{a}_i$ denote the estimated value of a_i and let $\{\hat{\varepsilon}_i\}$ denote the estimated residuals.
2. For a sample size of T, draw T random numbers so as to represent the $\{\varepsilon_t\}$ sequence. Most software packages will draw the numbers using randomly selected values of $\hat{\varepsilon}_i$ (with replacement). In this way, they actually generate bootstrap confidence intervals. Thus, you will have a simulated series of length T, called ε_t^s, that should have the same properties as the true error process. Use these random numbers to construct the simulated $\{y_t^s\}$ sequence as

$$y_t^s = \hat{a}_0 + \hat{a}_1y_{t-1}^s + \dots + \hat{a}_py_{t-p}^s + \varepsilon_t^s$$

   Be sure that you appropriately initialize the series so as to eliminate the effects of the initial conditions.
3. Now act as if you did not know the coefficient values used to generate the y_t^s series. Estimate y_t^s as an AR(p) process and obtain the impulse response function. If you repeat the process several thousand times, you can generate several thousand impulse response functions. You use these impulse response functions to construct the confidence intervals. For example, you can construct the interval that excludes the lowest 2.5 percent and highest 2.5 percent of the responses to obtain a 95 percent confidence interval.

The benefit of this method is that you do not need to make any special assumptions concerning the distribution of the autoregressive coefficients. The actual calculation of confidence intervals is only a bit more complicated in a VAR. Consider the two-variable system:

$$y_t = a_{11}y_{t-1} + a_{12}z_{t-1} + e_{1t}$$
$$z_t = a_{21}y_{t-1} + a_{22}z_{t-1} + e_{2t}$$

The complicating issue is that the regression residuals are correlated. As such, you need to draw e_{1t} and e_{2t} so as to maintain the appropriate error structure. A simple method is to draw e_{1t} and use the value of e_{2t} that corresponds to that same period. If you use a Choleski decomposition such that $b_{21} = 0$, construct ε_{1t} and ε_{2t} using (5.38) and (5.39). Question 10 at the end of this chapter reports the confidence intervals from a three-variable VAR.

Variance Decomposition

Although an unrestricted VAR is likely to be overparameterized, understanding the properties of the forecast errors is exceedingly helpful in uncovering interrelationships among the variables in the system. Suppose that we knew the coefficients of A_0 and A_1 and wanted to forecast the various values of x_{t+i} conditional on the observed value of x_t. Updating (5.19) one period (i.e., $x_{t+1} = A_0 + A_1x_t + e_{t+1}$) and taking the conditional expectation of x_{t+1}, we obtain

$$E_tx_{t+1} = A_0 + A_1x_t$$

Note that the one-step-ahead forecast error is $x_{t+1} - E_tx_{t+1} = e_{t+1}$. Similarly, updating two periods, we get

$$\begin{aligned} x_{t+2} &= A_0 + A_1x_{t+1} + e_{t+2} \\ &= A_0 + A_1(A_0 + A_1x_t + e_{t+1}) + e_{t+2} \end{aligned}$$

If we take conditional expectations, the two-step-ahead forecast of x_{t+2} is

$$E_tx_{t+2} = (I + A_1)A_0 + A_1^2x_t$$

The two-step-ahead forecast error (i.e., the difference between the realization of x_{t+2} and the forecast) is $e_{t+2} + A_1e_{t+1}$. More generally, it is easily verified that the n-step-ahead forecast is

$$E_tx_{t+n} = (I + A_1 + A_1^2 + \dots + A_1^{n-1})A_0 + A_1^nx_t$$

and that the associated forecast error is

$$e_{t+n} + A_1e_{t+n-1} + A_1^2e_{t+n-2} + \dots + A_1^{n-1}e_{t+1} \tag{5.42}$$

We can also consider these forecast errors in terms of (5.37) (i.e., the VMA form of the structural model). Of course, the VMA and the VAR models contain exactly the same information but it is convenient (and a good exercise) to describe the properties of the forecast errors in terms of the $\{\varepsilon_t\}$ sequence. If we use (5.37) to conditionally forecast x_{t+1}, one step ahead the forecast error is $\phi_0\varepsilon_{t+1}$. In general,

$$x_{t+n} = \mu + \sum_{i=0}^{\infty}\phi_i\varepsilon_{t+n-i}$$

so that the n-period forecast error $x_{t+n} - E_t x_{t+n}$ is

$$x_{t+n} - E_t x_{t+n} = \sum_{i=0}^{n-1} \phi_i \varepsilon_{t+n-i}$$

Focusing solely on the $\{y_t\}$ sequence, we see that the n-step-ahead forecast error is

$$\begin{aligned} y_{t+n} - E_t y_{t+n} &= \phi_{11}(0)\varepsilon_{yt+n} + \phi_{11}(1)\varepsilon_{yt+n-1} + \ldots + \phi_{11}(n-1)\varepsilon_{yt+1} \\ &\quad + \phi_{12}(0)\varepsilon_{zt+n} + \phi_{12}(1)\varepsilon_{zt+n-1} + \ldots + \phi_{12}(n-1)\varepsilon_{zt+1} \end{aligned}$$

Denote the n-step-ahead forecast error variance of y_{t+n} as $\sigma_y(n)^2$:

$$\begin{aligned} \sigma_y(n)^2 &= \sigma_y^2[\phi_{11}(0)^2 + \phi_{11}(1)^2 + \ldots + \phi_{11}(n-1)^2] \\ &\quad + \sigma_z^2[\phi_{12}(0)^2 + \phi_{12}(1)^2 + \ldots + \phi_{12}(n-1)^2] \end{aligned}$$

Because all values of $\phi_{jk}(i)^2$ are necessarily nonnegative, the variance of the forecast error increases as the forecast horizon n increases. Note that it is possible to decompose the n-step-ahead forecast error variance into the proportions due to each shock. Respectively, the proportions of $\sigma_y(n)^2$ due to shocks in the $\{\varepsilon_{yt}\}$ and $\{\varepsilon_{zt}\}$ sequences are

$$\frac{\sigma_y^2[\phi_{11}(0)^2 + \phi_{11}(1)^2 + \ldots + \phi_{11}(n-1)^2]}{\sigma_y(n)^2}$$

and

$$\frac{\sigma_z^2[\phi_{12}(0)^2 + \phi_{12}(1)^2 + \ldots + \phi_{12}(n-1)^2]}{\sigma_y(n)^2}$$

The **forecast error variance decomposition** tells us the proportion of the movements in a sequence due to its "own" shocks versus shocks to the other variable. If ε_{zt} shocks explain none of the forecast error variance of $\{y_t\}$ at all forecast horizons, we can say that the $\{y_t\}$ sequence is exogenous. In this circumstance, $\{y_t\}$ evolves independently of the ε_{zt} shocks and of the $\{z_t\}$ sequence. At the other extreme, ε_{zt} shocks could explain all of the forecast error variance in the $\{y_t\}$ sequence at all forecast horizons, so that $\{y_t\}$ would be entirely endogenous. In applied research it is typical for a variable to explain almost all of its forecast error variance at short horizons and smaller proportions at longer horizons. We would expect this pattern if ε_{zt} shocks had little contemporaneous effect on y_t but acted to affect the $\{y_t\}$ sequence with a lag.

Note that the variance decomposition contains the same problem inherent in impulse response function analysis. In order to identify the $\{\varepsilon_{yt}\}$ and $\{\varepsilon_{zt}\}$ sequences, it is necessary to restrict the B matrix. The Choleski decomposition used in (5.38) and (5.39) necessitates that all of the one-period forecast error variance of z_t is due to ε_{zt}. If we use the alternative ordering, all of the one-period forecast error variance of y_t would be due to ε_{yt}. The dramatic effects of these alternative assumptions are reduced at longer forecasting horizons. In practice, it is useful to examine the variance decompositions at various forecast horizons. As n increases, the variance decompositions should converge. Moreover, if the correlation coefficient ρ_{12} is significantly different from zero, it is customary to obtain the variance decompositions under various orderings.

Nevertheless, impulse analysis and variance decompositions (together called **innovation accounting**) can be useful tools to examine the relationships among economic variables. If the correlations among the various innovations are small, the identification problem is not likely to be especially important. The alternative orderings should yield similar impulse responses and variance decompositions. Of course, the contemporaneous movements of many economic variables are highly correlated. Sections 10 through 13 consider two attractive methods that can be used to identify the structural innovations. Before examining these techniques, we consider hypothesis testing in a VAR framework and reexamine the interrelationships between terrorism and tourism.

8. TESTING HYPOTHESES

In principle, there is nothing to prevent you from incorporating a large number of variables in the VAR. It is possible to construct an n-equation VAR with each equation containing p lags of all n variables in the system. You will want to include those variables that have important economic effects on each other. As a practical matter, degrees of freedom are quickly eroded as more variables are included. For example, using monthly data with 12 lags, the inclusion of one additional variable uses an additional 12 degrees of freedom in each equation. A careful examination of the relevant theoretical model will help you to select the set of variables to include in your VAR model.

An n-equation VAR can be represented by

$$\begin{bmatrix} x_{1t} \\ x_{2t} \\ \cdot \\ x_{nt} \end{bmatrix} = \begin{bmatrix} A_{10} \\ A_{20} \\ \cdot \\ A_{n0} \end{bmatrix} + \begin{bmatrix} A_{11}(L) & A_{12}(L) & \cdot & A_{1n}(L) \\ A_{21}(L) & A_{22}(L) & \cdot & A_{2n}(L) \\ \cdot & \cdot & \cdot & \cdot \\ A_{n1}(L) & A_{n2}(L) & \cdot & A_{nn}(L) \end{bmatrix} \begin{bmatrix} x_{1t-1} \\ x_{2t-1} \\ \cdot \\ x_{nt-1} \end{bmatrix} + \begin{bmatrix} e_{1t} \\ e_{2t} \\ \cdot \\ e_{nt} \end{bmatrix} \quad (5.43)$$

where: A_{i0} = the parameters representing intercept terms
$A_{ij}(L)$ = the polynomials in the lag operator L

The individual coefficients of $A_{ij}(L)$ are denoted by $a_{ij}(1)$, $a_{ij}(2)$, ... Since all equations have the same lag length, the polynomials $A_{ij}(L)$ are all of the same degree. The terms e_{it} are white-noise disturbances that may be correlated. Again, designate the variance/covariance matrix by Σ, where the dimension of Σ is $(n \cdot n)$.

In addition to the determination of the set of variables to include in the VAR, it is important to determine the appropriate lag length. One possible procedure is to allow for different lag lengths for each variable in each equation. However, in order to preserve the symmetry of the system (and to be able to use OLS efficiently) it is common to use the same lag length for all equations. As indicated in Section 6, as long as there are identical regressors in each equation, OLS estimates are consistent and asymptotically efficient. If some of the VAR equations have regressors not included in the others, seemingly unrelated regressions (SUR) provide efficient estimates of the VAR coefficients. Hence, when there is a good reason to let lag lengths differ across equations, estimate the so-called **near-VAR** using SUR.

In a VAR, long lag lengths quickly consume degrees of freedom. If lag length is p, each of the n equations contains np coefficients plus the intercept term. Appropriate lag length selection can be critical. If p is too small, the model is misspecified; if p is too large, degrees of freedom are wasted. To check lag length, begin with the longest plausible length or the longest feasible length given degrees-of-freedom considerations. Estimate the VAR and form the variance/covariance matrix of the residuals. Using quarterly data, you might start with a lag length of 12 quarters based on the a priori notion that three years is sufficiently long to capture the system's dynamics. Call the variance/covariance matrix of the residuals from the 12-lag model Σ_{12}. Now suppose you want to determine whether eight lags are appropriate. After all, restricting the model from 12 to eight lags would reduce the number of estimated parameters by $4n$ in each equation.

Since the goal is to determine whether lag 8 is appropriate for all equations, an equation by equation F-test on lags 9 through 12 is not appropriate. Instead, the proper test for this **cross-equation** restriction is a likelihood ratio test. Reestimate the VAR *over the same sample period* using eight lags and obtain the variance/covariance matrix of the residuals Σ_8. Note that Σ_8 pertains to a system of n equations with $4n$ restrictions in each equation, for a total of $4n^2$ restrictions. The likelihood ratio statistic is

$$(T)(\ln|\Sigma_8| - \ln|\Sigma_{12}|)$$

However, given the sample sizes usually found in economic analysis, Sims (1980) recommended using

$$(T - c)(\ln|\Sigma_8| - \ln|\Sigma_{12}|)$$

where: T = number of usable observations

c = number of parameters estimated in each equation of the unrestricted system

$\ln|\Sigma_n|$ = the natural logarithm of the determinant of Σ_n

In the example at hand, $c = 1 + 12n$ since each equation of the unrestricted model has 12 lags for each variable plus an intercept.

This statistic has an asymptotic χ^2 distribution with degrees of freedom equal to the number of restrictions *in the system*. In the example under consideration, there are $4n$ restrictions in each equation, for a total of $4n^2$ restrictions in the system. Clearly, if the restriction of a reduced number of lags is not binding, we would expect $\ln|\Sigma_8|$ to be equal to $\ln|\Sigma_{12}|$. Large values of this sample statistic indicate that having only eight lags is a binding restriction; hence, we can reject the null hypothesis that lag length = 8. If the calculated value of the statistic is less than χ^2 at a prespecified significance level, we will not be able to reject the null of only eight lags. At that point we could seek to determine whether four lags were appropriate by constructing

$$(T - c)(\ln|\Sigma_4| - \ln|\Sigma_8|)$$

Considerable care should be taken in paring down lag length in this fashion. Often this procedure will not reject the null hypotheses of eight versus twelve lags and four versus eight lags, although it will reject a null of four versus twelve lags. The problem with paring down the model is that you may lose a small amount of explanatory power at each stage. Overall, the total loss in explanatory power can be significant. In such circumstances, it is better to use the longer lag lengths.

This type of likelihood ratio test is applicable to any type of cross-equation restriction. Let Σ_u and Σ_r be the variance/covariance matrices of the unrestricted and restricted systems, respectively. If the equations of the unrestricted model contain different regressors, let c denote the maximum number of regressors contained in the longest equation. Sims's recommendation is to compare the test statistic

$$(T - c)(\ln|\Sigma_r| - \ln|\Sigma_u|) \tag{5.44}$$

to a χ^2 distribution with degrees of freedom equal to the number of restrictions in the system.

To take another example, suppose you wanted to capture seasonal effects by including three seasonal dummies in each of the n equations of a VAR. Estimate the unrestricted model by including the dummy variables and estimate the restricted model by excluding the dummies. The total number of restrictions in the system is $3n$. If lag length is p, the equations of the unrestricted model have $np + 4$ parameters (np lagged variables, the intercept, and the three seasonals). For T usable observations, set $c = np + 4$ and calculate the value of (5.44). If for some prespecified significance level this calculated value χ^2 (with $3n$ degrees of freedom) exceeds the critical value, the restriction of no seasonal effects can be rejected.

The likelihood ratio test is based on asymptotic theory, which may not be very useful in the small samples available to time-series econometricians. Moreover, the likelihood ratio test is only applicable when one model is a restricted version of the other. Alternative test criteria are the multivariate generalizations of the AIC and SBC:

$$\text{AIC} = T\ln|\Sigma| + 2N$$
$$\text{SBC} = T\ln|\Sigma| + N\ln(T)$$

where: $|\Sigma|$ = determinant of the variance/covariance matrix of the residuals
N = total number of parameters estimated *in all equations*

Thus, if each equation in an n-variable VAR has p lags and an intercept, $N = n^2p + n$; each of the n equations has np lagged regressors and an intercept.

Adding additional regressors will reduce $\ln|\Sigma|$ at the expense of increasing N. As in the univariate case, select the model with the lowest AIC or SBC value. Make sure that you adequately compare the models by using the same number of observations in each. Note that the multivariate AIC and SBC cannot be used to *test* the statistical significance of alternative models. Instead, they are measures of the overall fit of the alternatives. As in the univariate case, there are a number of ways that researchers and software packages use to report the multivariate generalizations of the AIC and SBC. Often, these values will be reported as

$$\text{AIC}^* = -2\ln(L)/T + 2N/T$$
$$\text{SBC}^* = -2\ln(L)/T + N\ln(T)/T$$

where L = maximized value of the multivariate log likelihood function.

Granger Causality

One test of causality is whether the lags of one variable enter into the equation for another variable. In a two-equation model with p lags, $\{y_t\}$ does not **Granger cause** $\{z_t\}$ if and only if all of the coefficients of $A_{21}(L)$ are equal to zero. Thus, if $\{y_t\}$ does not improve the forecasting performance of $\{z_t\}$, then $\{y_t\}$ does not Granger cause $\{z_t\}$. *If all variables in the VAR are stationary*, the direct way to test Granger causality is to use a standard F-test of the restriction

$$a_{21}(1) = a_{21}(2) = a_{21}(3) = \ldots = a_{21}(p) = 0$$

It is straightforward to generalize this notion to the n-variable case of (5.43). Since $A_{ij}(L)$ represents the coefficients of lagged values of variable j on variable i, variable j does not Granger cause variable i if all coefficients of the polynomial $A_{ij}(L)$ can be set equal to zero.

Note that Granger causality is something quite different from a test for exogeneity. For z_t to be exogenous, we would require that it not be affected by the contemporaneous value of y_t. However, Granger causality refers only to the effects of past values of $\{y_t\}$ on the current value of z_t. Hence, Granger causality actually measures whether current and past values of $\{y_t\}$ help to forecast future values of $\{z_t\}$. To illustrate the distinction in terms of a VMA model, consider the following equation for z_t:

$$z_t = \bar{z} + \phi_{21}(0)\varepsilon_{yt} + \sum_{i=0}^{\infty} \phi_{22}(i)\varepsilon_{zt-i}$$

If we forecast z_{t+1} conditional on the value of z_t, we obtain the forecast error $\phi_{21}(0)\varepsilon_{yt+1} + \phi_{22}(0)\varepsilon_{zt+1}$. Given the value of z_t, information concerning y_t does not aid in reducing the forecast error for z_{t+1}. In other words, for the model under consideration, $E_t(z_{t+1}|z_t) = E_t(z_{t+1}|z_t, y_t)$. The point is that the only additional information contained in y_t is the past values of $\{\varepsilon_{yt}\}$. However, such values do not affect z_t and so cannot improve on the forecasting performance of the z_t sequence. Thus, $\{y_t\}$ does not Granger cause $\{z_t\}$. On the other hand, since we are assuming that $\phi_{21}(0)$ is not zero, $\{z_t\}$ is not exogenous. Clearly, if $\phi_{21}(0)$ is not zero, pure shocks to y_{t+1} (i.e., ε_{yt+1}) affect the value of z_{t+1} even though the $\{y_t\}$ sequence does not Granger cause the $\{z_t\}$ sequence.

A **block-exogeneity** test is useful for detecting whether to incorporate an additional variable into a VAR. Given the aforementioned distinction between causality and exogeneity, this multivariate generalization of the Granger causality test should actually be called a *block-causality* test. In any event, the issue is to determine whether lags of one variable—say, w_t—Granger cause any other of the variables in the system. In the three-variable case with w_t, y_t, and z_t, the test is whether lags of w_t Granger cause either y_t or z_t. In essence, the block exogeneity restricts all lags of w_t in the y_t and z_t to be equal to zero. This cross-equation restriction is properly tested using the likelihood ratio test given by (5.44). Estimate the y_t and z_t equations using lagged values of $\{y_t\}$, $\{z_t\}$, and $\{w_t\}$ and calculate Σ_u. Reestimate excluding the lagged values of $\{w_t\}$ and calculate Σ_r. Next, find the likelihood ratio statistic:

$$(T - c)(\ln|\Sigma_r| - \ln|\Sigma_u|)$$

As in (5.44), this statistic has a χ^2 distribution with degrees of freedom equal to $2p$ (since p lagged values of $\{w_t\}$ are excluded from each equation). Here $c = 3p + 1$ because the unrestricted y_t and z_t equations contain p lags of $\{y_t\}$, $\{z_t\}$, and $\{w_t)$ plus a constant.

Granger Causality and Money Supply Changes

The usefulness of Granger causality tests can be illustrated by a reconsideration of the type of time-series equation used in the St. Louis model. Through the late 1970s, the conventional wisdom was that fluctuations in money contained useful information about the future values of real income and prices. In fact, the argument in favor of conducting an active monetary policy is that there exists a systematic relationship between current values of the money supply and future values of the price level and/or real income. However, there is a large body of literature indicating that this relationship broke down in the late 1970s. In an influential article, Friedman and Kuttner (1992) argued that the issue is whether fluctuations in money help predict future fluctuations in income that are not already predicted on the basis of income itself or other readily observable variables. Consider the VAR equation

$$\Delta y_t = \alpha + \sum_{i=1}^{4} \beta_i \Delta m_{t-i} + \sum_{i=1}^{4} \gamma_i \Delta g_{t-i} + \sum_{i=1}^{4} \delta_i \Delta y_{t-i} + \varepsilon_t$$

Notice how this equation differs from the St. Louis model given by (5.16). Here, the logarithmic change in nominal income (Δy_t) depends on its own past values, and on past values of the logarithmic changes in the nominal money supply (Δm_t) and federal government expenditures (Δg_t).

The issue is simple; in the presence of past values of $\{\Delta y_t\}$ and $\{\Delta g_t\}$, does knowledge of the money supply series provide any information about the future value of nominal income? Toward this end, Friedman and Kuttner (1992) used several measures of the money supply (e.g., the money base, $M1$, $M2$, and various short-term interest rates) and estimated a three-variable VAR over various sample periods. For the 1960Q2 to 1979Q2 period, the F-statistic for the null hypothesis that the money base does not Granger cause Δy_t is 3.68. At the 1 percent significance level, it is possible to conclude that money Granger causes $\{\Delta y_t\}$. However, for the 1970Q3 to 1990Q4 period, the F-statistic is only 0.82; hence, at any conventional significance level, money does not Granger cause income. The findings are quite robust to the other measures of the monetary variable. Until 1979Q2, all of the monetary aggregates Granger cause nominal income at the 1 percent significance level. None of these aggregates Granger causes nominal income in the latter period.

To provide a better understanding of the interrelationships among the three variables, Friedman and Kuttner also reported the results of the variance decompositions. For the 1960Q2 to 1979Q2 period, $M1$ explained 27 percent of the forecast error variance in $\{\Delta y_t\}$ at both the four- and eight-quarter forecast horizons. In contrast, for the 1970Q3 to 1990Q4 period, $M1$ explained about 10 percent of the forecast error variance in $\{\Delta y_t\}$ at both the four- and eight-quarter forecast horizons. These results are in striking contrast to those of the St. Louis equation. Undoubtedly, money supply changes have become less useful in predicting the future path of nominal income.

Tests with Nonstationary Variables

In Chapter 4, we saw that it is possible to perform hypothesis tests on an individual equation when some of the regressors are stationary and others are nonstationary. In particular, Rule 1 of Sims, Stock, and Watson (1990) was used to select the appropriate lag length in an augmented Dickey–Fuller test. The issue is particularly relevant to VARs since many of the regressors are likely to be nonstationary. Recall that a key finding of Sims, Stock, and Watson (1990) is: *If the coefficient of interest can be written as a coefficient on a zero-mean stationary variable, then a t-test is appropriate.* If the sample size is large, you can use the normal approximation for the t-test. To take a specific example, consider the following equation from a two-variable VAR:

$$y_t = a_{11}y_{t-1} + a_{12}y_{t-2} + b_{11}z_{t-1} + b_{12}z_{t-2} + \varepsilon_t \tag{5.45}$$

First consider the case in which $\{y_t\}$ is $I(1)$ and $\{z_t\}$ is $I(0)$. Since b_{11} and b_{12} are coefficients on stationary variables, it is possible to use a t-test to test the hypothesis $b_{11} = 0$ or $b_{12} = 0$ and an F-test to test the hypothesis $b_{11} = b_{12} = 0$. Hence, lag lengths involving $\{z_t\}$ and the test to determine whether $\{z_t\}$ Granger causes $\{y_t\}$ can be performed using the t- or F-distributions.

Notice that it is possible to use a t-test for the restriction $a_{11} = 0$ or $a_{12} = 0$. You can perform both of these tests even though $\{y_t\}$ is not stationary. However, you cannot test the restriction $a_{11} = a_{12} = 0$ using an F-test. To make the point, add and subtract $a_{12}y_{t-1}$ to the right-hand side of (5.45) to obtain

$$y_t = a_{11}y_{t-1} + a_{12}y_{t-1} - a_{12}(y_{t-1} - y_{t-2}) + b_{11}z_{t-1} + b_{12}z_{t-2} + \varepsilon_t$$

and if we define $a_{11} + a_{12} = \gamma$, we can write

$$y_t = \gamma y_{t-1} - a_{12}\Delta y_{t-1} + b_{11}z_{t-1} + b_{12}z_{t-2} + \varepsilon_t$$

The coefficient a_{12} multiplies the stationary variable (y_{t-1} so that it is permissible to test the null hypothesis $a_{12} = 0$ using a t-test. Alternatively, add and subtract $a_{11}y_{t-2}$ to the right-hand side of (5.45) to obtain

$$y_t = a_{11}\Delta y_{t-1} + \gamma y_{t-2} + b_{11}z_{t-1} + b_{12}z_{t-2} + \varepsilon_t$$

Thus, the null hypothesis $a_{11} = 0$ can similarly be tested using a t-statistic. It is important to recognize that the individual coefficients may have normal distributions, but the sum $a_{11} + a_{12} = \gamma$ does not have a normal distribution. It is impossible to isolate γ as the coefficient on a stationary variable.

Now suppose that $\{y_t\}$ and $\{z_t\}$ are both $I(1)$. It is easy to show that the coefficients a_{12} and b_{12} can be written as coefficients on stationary variables. Add and subtract both $a_{12}y_{t-1}$ and $b_{12}z_{t-1}$ to the right-hand side of (5.45) so that the equation becomes

$$y_t = (a_{11} + a_{12})y_{t-1} - a_{12}(y_{t-1} - y_{t-2}) + (b_{11} + b_{12})z_{t-1} - b_{12}(z_{t-1} - z_{t-2}) + \varepsilon_t$$

or

$$y_t = \gamma_1 y_{t-1} - a_{12}\Delta y_{t-1} + \gamma_2 z_{t-1} - b_{12}\Delta z_{t-1} + \varepsilon_t \tag{5.46}$$

where $\gamma_1 = a_{11} + a_{12}$ and $\gamma_2 = b_{11} + b_{12}$.

Thus, it is possible to perform the lag length test $a_{12} = b_{12} = 0$ using an F-distribution. Equation (5.46) shows that it is possible to rewrite (5.45) in such a way that both coefficients multiply stationary variables. As such, an F-test can be used to test the joint restriction $a_{12} = b_{12} = 0$. However, the restriction that $\{z_t\}$ does not Granger cause $\{y_t\}$ involves the setting $\gamma_2 = b_{12} = 0$. Since γ_2 is a coefficient on a nonstationary variable, the test is nonstandard—a standard F-statistic is not appropriate. Only if you know that $\gamma_2 = 0$ can you perform a test to determine whether $\{z_t\}$ Granger causes $\{y_t\}$. Given that $\gamma_2 = 0$, (5.46) becomes

$$y_t = \gamma_1 y_{t-1} + a_{12}\Delta y_{t-1} + b_{12}\Delta z_{t-1} + \varepsilon_t$$

Now, it is possible to perform the causality test since only b_{12} needs to be restricted. In the same way, if it is known that $\gamma_1 = 1$, we can write[10]

$$\Delta y_t = a_{12}\Delta y_{t-1} + b_{12}\Delta z_{t-1} + \varepsilon_t$$

Now the VAR is entirely in first differences. As such, all coefficients multiply stationary variables. These results are quite general and hold for higher-order systems containing any number of lags. To summarize, in a VAR with stationary and nonstationary variables,

1. You can use t-tests or F-tests on the stationary variables.
2. You can perform a lag length test on any variable or any set of variables. This is true regardless of whether the variable in question is stationary.
3. You *may* be able to use an F-test to determine whether a nonstationary variable Granger causes another nonstationary variable. If the causal variable can be made to appear only in first differences, the test is permissible. For example, suppose that y_t, z_t, and x_t are all $I(1)$ and that it is possible to write the equation for $\{y_t\}$ as $y_t = \gamma_1 y_{t-1} + a_{12}\Delta y_{t-1} + a_{13}\Delta y_{t-2} + b_{12}\Delta z_{t-1} + b_{13}\Delta z_{t-2} + \gamma_3 x_{t-1} + c_{12}\Delta x_{t-1} + c_{13}\Delta x_{t-2} + \varepsilon_t$. It is possible to determine whether z_t Granger causes $\{y_t\}$ but not whether x_t Granger causes $\{y_t\}$. Similarly, you *cannot* test the joint restriction $\gamma_1 = a_{12} = 0$.
4. The issue of differencing is important. If the VAR can be written entirely in first differences, hypothesis tests can be performed on any equation or any set of equations using t-tests or F-tests. This follows because all of the variables are stationary. As you will see in the next chapter, it is possible to write the VAR in first differences if the variables are $I(1)$ and are *not* cointegrated. If the variables in question are cointegrated, the VAR cannot be written in first differences; hence, causality tests *cannot* be performed using t-tests or F-tests.

9. EXAMPLE OF A SIMPLE VAR: TERRORISM AND TOURISM IN SPAIN

In Enders and Sandler (1991), we used the VAR methodology to estimate the impact of terrorism on tourism in Spain during the period from 1970 to 1988. Most transnational terrorist incidents in Spain during this period were perpetrated by left-wing groups, which included the Anti-Fascist Resistance Group of October First (GRAPO), the ETA, the now-defunct International Revolutionary Armed Front (FRAP), and Iraultza. Most incidents are attributed to the ETA (Basque Fatherland and Liberty) and its splinter groups, such as the Autonomous Anti-Capitalist Commandos (CAA). Right-wing terrorist groups included the Anti-Terrorist Liberation Group (GAL), the Anti-Terrorism ETA, and the Warriors of Christ the King. Catalan independence groups, such as Free Land (Terra Lliure) and the Catalan Socialist Party for National Liberation, were active in the late 1980s and often targeted U.S. businesses.

The transfer function model of Section 3 may not be appropriate because of feedback between terrorism and tourism. If high levels of tourism induce terrorist activities, the basic assumption of the transfer function methodology is violated. In fact, there is some evidence that the terrorist organizations in Spain target tourist hotels in the summer season. Since increases in tourism may generate terrorist acts, the VAR methodology allows us to examine the reactions of tourists to terrorism *and* the reactions of terrorists to tourism. We can gain some additional insights into the interrelation between the two series by performing Granger causality tests both of terrorism on tourism and of tourism on terrorism. Impulse-response analysis can quantify and graphically depict the time path of the effects of a typical terrorist incident on tourism.

We assembled a time series of all publicly available transnational terrorist incidents that took place in Spain from 1970 through 1988. In total, there are 228 months of observation in the time series; each observation is the number of terrorist incidents occurring in that month. The tourism data are taken from various issues of the National Statistics Institute's (Instituto Nacional de Estadística) quarterly reports. In particular, we assemble a time series of the number of foreign tourists per month in Spain for the 1970–1988 period.

Empirical Methodology

Our basic methodology involves estimating tourism and terrorism in a vector autoregression (VAR) framework. Consider the following system of equations:

$$n_t = \alpha_{10} + A_{11}(L)n_{t-1} + A_{12}(L)i_{t-1} + e_{1t} \tag{5.47}$$

$$i_t = \alpha_{20} + A_{21}(L)n_{t-1} + A_{22}(L)i_{t-1} + e_{2t} \tag{5.48}$$

where:

n_t = the number of tourists visiting Spain during time period t

i_t = the number of transnational terrorist incidents in Spain during t

α_{i0} = the vectors containing a constant, eleven seasonal (monthly) dummy variables, and a time trend

$A_{ij}(L)$ = the polynomials in the lag operator L

e_{it} = independent and identically distributed disturbance terms such that $E(e_{1t}e_{2t})$ is not necessarily zero

Although Sims (1980) recommended against the use of a deterministic time trend, we decided not to heed this advice. We experimented with several alternative ways to model the series; the model including the time trend had the best diagnostic statistics. Other variants included differencing (5.47) and (5.48) and simply eliminating the trend and letting the random walk plus drift terms capture any nonstationary behavior. We were also concerned that the number of incidents had a large number of zeroes (and could not be negative), so that the normality assumption was violated.

The polynomials $A_{12}(L)$ and $A_{21}(L)$ in (5.47) and (5.48) are of particular interest. If all of the coefficients of $A_{21}(L)$ are zero, then knowledge of the tourism series does not reduce the forecast error variance of terrorist incidents. Formally, tourism would not Granger cause terrorism. Unless there is a contemporaneous response of terrorism to tourism, the terrorism series evolves independently of tourism. In the same way, if all of the coefficients of $A_{12}(L)$ are zero, then terrorism does not Granger cause tourism. The absence of a statistically significant contemporaneous correlation of the error terms would then imply that terrorism cannot affect tourism. If, instead, any of the coefficients in these polynomials differ from zero, there are interactions between the two series. In the case of negative coefficients of $A_{12}(L)$, terrorism would have a negative effect on the number of foreign tourist visits to Spain.

Each equation was estimated using lag lengths of 24, 12, 6, and 3 months (i.e., for four estimations, we set $p = 24$, 12, 6, and 3). Because each equation has identical right-hand-side variables, ordinary least squares (OLS) is an efficient estimation technique. Using χ^2 tests, we determined that a lag length of 12 months was most appropriate (reducing lag length from 24 to 12 months had a χ^2 value that was significant at the 0.56 level, whereas reducing the lag length from 24 to 6 months had a χ^2 value that was significant at the 0.049 level). The AIC indicated that 12 lags were appropriate, whereas the SBC suggested we could use only 6 lags. Since we were using monthly data, we decided to use the 12 lags.

To ascertain the importance of the interactions between the two series, we obtained the variance decompositions. The moving-average representations of Equations (5.47) and (5.48) express n_t and i_t as dependent on the current and past values of both $\{e_{1t}\}$ and $\{e_{2t}\}$ sequences:

$$n_t = c_0 + \sum_{j=1}^{\infty} (c_{1j}e_{1t-j} + c_{2j}e_{2t-j}) + e_{1t} \tag{5.49}$$

$$i_t = d_0 + \sum_{j=1}^{\infty} (d_{1j}e_{1t-j} + d_{2j}e_{2t-j}) + e_{2t} \tag{5.50}$$

where c_0 and d_0 are vectors containing constants, the 11 seasonal dummies, and a trend; and c_{1j}, c_{2j}, d_{1j}, and d_{2j} are parameters.

Because we cannot estimate (5.49) and (5.50) directly, we used the residuals of (5.47) and (5.48) and then decomposed the variances of n_t and i_t into the percentages attributable to each type of innovation. We used the orthogonalized innovations obtained from a Choleski decomposition; the order of the variables in the factorization had no qualitative effects on our results (the contemporaneous correlation between e_{1t} and e_{2t} was -0.0176).

Empirical Results

The variance decompositions for a 24-month forecasting horizon, with significance levels in parentheses, are reported in Table 5.3. As expected, each time series explains the preponderance of its own past values; n_t explains over 91 percent of its forecast error

Table 5.3 Variance Decomposition Percentage of 24-Month Error Variance

Percent of Forecast Error Variance in	Typical shock in n_t	Typical shock in i_t
n_t	91.3	8.7
	(3 × E − 15)	(0.006)
i_t	2.2	97.8
	(17.2)	(93.9)

Note: The numbers in parentheses indicate the significance level for the joint hypothesis that all lagged coefficients of the variable in question can be set equal to zero.

variance, while i_t explains nearly 98 percent of its forecast error variance. It is interesting that terrorist incidents explain 8.7 percent of the forecast error variance of Spain's tourism, while tourism explains only 2.2 percent of the forecast error variance of terrorist incidents. More important, Granger causality tests indicate that the effects of terrorism on tourism are significant at the 0.006 level, whereas the effects of tourism on terrorism are not significant at conventional levels. Thus, causality is unidirectional: Terrorism affects tourism but not the reverse. We also note that the terrorism series appears to be autonomous in the sense that neither series significantly explains the forecast error variance of i_t. This result is consistent with the notion that terrorists randomize their incidents so that any one incident is not predictable on a month-to-month basis.

Forecasts from an unrestricted VAR are known to suffer from overparameterization. Given the results of the variance decompositions and the Granger causality tests, we reestimated (5.47) and (5.48) restricting all of the coefficients of $A_{21}(L)$ to equal zero. Because the right-hand variables were no longer identical, we reestimated the equations with seemingly unrelated regressions (SUR). From the resulting coefficients from the SUR estimates, the effects of a typical terrorist incident on Spain's tourism can be depicted. In terms of the restricted version of (5.50), we set all e_{1t-j} and e_{2t-j} equal to zero for $j > 0$. We then simulated the time paths resulting from the effects of a one-unit shock to e_{2t}. The time path is shown in Figure 5.8, where the

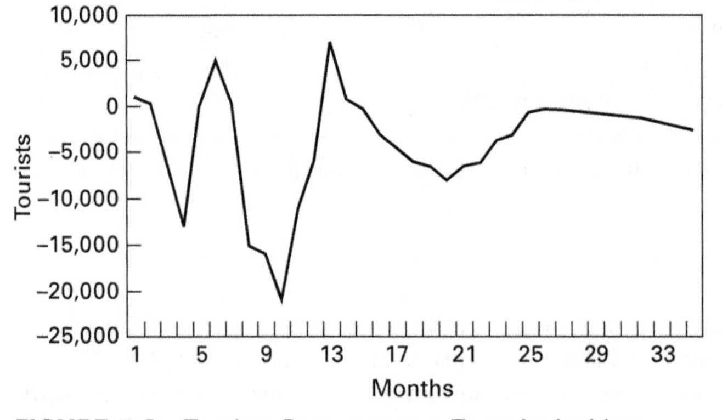

FIGURE 5.8 Tourism Response to a Terrorist Incident

vertical axis measures the monthly impact on the number of foreign tourists and the horizontal axis shows the months following the shock. To smooth out the series, we present the time path of a three-month moving average of the simulated tourism response function.

After a "typical" terrorist incident, tourism to Spain begins to decline in the third month. After the sixth month, tourism begins to revert to its original level. There does appear to be a rebound in months eight and nine. There follows another drop in tourism in month nine, reaching the maximum decline about one year after the original incident. Obviously, some of this pattern is due to the seasonality in the series. However, tourism slowly recovers and generally remains below its preincident level for a substantial period of time. Aggregating all 36 monthly impacts, we estimate that the combined effects of a typical transnational terrorist incident in Spain are to decrease the total number of foreign visits by 140,847 people. By comparison, a total of 5,392,000 tourists visited Spain in 1988 alone.

10. STRUCTURAL VARs

Sims's (1980) VAR approach has the desirable property that all variables are treated symmetrically so that the econometrician does not rely on any "incredible identification restrictions." A VAR can be quite helpful in examining the relationship among a set of economic variables. Moreover, the resulting estimates can be used for forecasting purposes. Consider a first-order VAR system of the type represented by (5.19):

$$x_t = A_0 + A_1 x_{t-1} + e_t$$

Although the VAR approach yields only estimated values of A_0 and A_1, for exposition purposes it is useful to treat each as being known. As we saw in (5.42), the n-step-ahead forecast error is

$$x_{t+n} - E_t x_{t+n} = e_{t+n} + A_1 e_{t+n-1} + A_1^2 e_{t+n-2} + \dots + A_1^{n-1} e_{t+1} \qquad (5.51)$$

Even though econometric analysis will never reveal the actual values of A_0 and A_1, an appropriately specified model will have forecasts that are unbiased and have minimum variance. Of course, if we had a priori information concerning any of the coefficients, it would be possible to improve the precision of the estimates and reduce the forecast-error variance. A researcher interested only in forecasting might want to trim down the overparameterized VAR model. Nonetheless, it should be clear that forecasting with a VAR is a multivariate extension of forecasting using a simple autoregression.

The VAR approach has been criticized as being devoid of any economic content. The sole role of the economist is to suggest the appropriate variables to include in the VAR. From that point on, the procedure is almost mechanical. Since there is so little economic input in a VAR, it should not be surprising that there is little economic content in the results. Of course, innovation accounting does require an ordering of the variables, but the selection of the ordering is generally ad hoc.

Unless the underlying structural model can be identified from the reduced-form VAR model, the innovations in a Choleski decomposition do not have a direct economic interpretation. Reconsider the two-variable VAR of (5.17) and (5.18):

$$y_t + b_{12}z_t = b_{10} + \gamma_{11}y_{t-1} + \gamma_{12}z_{t-1} + \varepsilon_{yt}$$
$$b_{21}y_t + z_t = b_{20} + \gamma_{21}y_{t-1} + \gamma_{22}z_{t-1} + \varepsilon_{zt}$$

so that it is possible to write the model in the form of (5.20) and (5.21):

$$y_t = a_{10} + a_{11}y_{t-1} + a_{12}z_{t-1} + e_{1t}$$
$$z_t = a_{20} + a_{21}y_{t-1} + a_{22}z_{t-1} + e_{2t}$$

where the various a_{ij} are defined as in (5.19). For our purposes, the important point to note is that the two error terms e_{1t} and e_{2t} are actually composites of the underlying shocks ε_{yt} and ε_{zt}. From (5.22) and (5.23),

$$\begin{bmatrix} e_{1t} \\ e_{2t} \end{bmatrix} = \frac{1}{1-b_{12}b_{21}} \begin{bmatrix} 1 & -b_{12} \\ -b_{21} & 1 \end{bmatrix} \begin{bmatrix} \varepsilon_{yt} \\ \varepsilon_{zt} \end{bmatrix}$$

Although these composite shocks are the one-step-ahead forecast errors in y_t and z_t, they do not have a structural interpretation. Hence, there is an important difference between using VARs for forecasting and using VARs for economic analysis. In (5.51), e_{1t} and e_{2t} are forecast errors. If we are interested only in forecasting, the components of the forecast errors are unimportant. Given the economic model of (5.17) and (5.18), ε_{yt} and ε_{zt} are the autonomous changes in y_t and z_t in period t, respectively. If we want to obtain the impulse response functions or the variance decompositions, it is necessary to use the structural shocks (i.e., ε_{yt} and ε_{zt}), not the forecast errors. The aim of a structural VAR is to use economic theory (rather than the Choleski decomposition) to recover the structural innovations from the residuals $\{e_{1t}\}$ and $\{e_{2t}\}$.

The Choleski decomposition actually makes a strong assumption about the underlying structural errors. Suppose, as in (5.32), we select an ordering such that $b_{21} = 0$. With this assumption, the two pure innovations can be recovered as

$$\varepsilon_{zt} = e_{2t}$$

and

$$\varepsilon_{yt} = e_{1t} + b_{12}e_{2t}$$

Forcing $b_{21} = 0$ is equivalent to assuming that an innovation in y_t does not have a contemporaneous effect on z_t. Unless there is a theoretical foundation for this assumption, the underlying shocks are improperly identified. As such, the impulse responses and variance decompositions resulting from this improper identification can be quite misleading.

If the correlation coefficient between e_{1t} and e_{2t} is low, the ordering is not likely to be important. However, in a VAR with several variables, it is improbable that all correlations will be small. After all, in selecting the variables to include in a model, you are likely to choose variables that exhibit strong comovements. When the residuals of a VAR are correlated, it is not practical to try all alternative orderings. With a four-variable model, there are 24 (i.e., 4!) possible orderings. Sims (1986) and Bernanke (1986) proposed modeling the innovations using economic analysis. The basic idea is to estimate the relationships among the structural shocks using an economic model. To understand the procedure, it is useful to examine the relationship between the forecast errors and the structural innovations in an n-variable VAR. Since this relationship is invariant to lag length, consider the first-order model with n variables:

$$\begin{bmatrix} 1 & b_{12} & b_{13} & \cdots & b_{1n} \\ b_{21} & 1 & b_{23} & \cdots & b_{2n} \\ \cdot & \cdot & \cdot & \cdots & \cdot \\ b_{n1} & b_{n2} & b_{n3} & \cdots & b_{1nn} \end{bmatrix} \begin{bmatrix} x_{1t} \\ x_{2t} \\ \cdots \\ x_{nt} \end{bmatrix}$$

$$= \begin{bmatrix} b_{10} \\ b_{20} \\ \cdots \\ b_{n0} \end{bmatrix} + \begin{bmatrix} \gamma_{11} & \gamma_{12} & \gamma_{13} & \cdots & \gamma_{1n} \\ \gamma_{21} & \gamma_{22} & \gamma_{23} & \cdots & \gamma_{2n} \\ \cdot & \cdot & \cdot & \cdots & \cdot \\ \gamma_{n1} & \gamma_{n2} & \gamma_{n3} & \cdots & \gamma_{nn} \end{bmatrix} \begin{bmatrix} x_{1t-1} \\ x_{2t-1} \\ \cdots \\ x_{nt-1} \end{bmatrix} + \begin{bmatrix} \varepsilon_{1t} \\ \varepsilon_{2t} \\ \cdots \\ \varepsilon_{nt} \end{bmatrix}$$

or, in compact form:

$$Bx_t = \Gamma_0 + \Gamma_1 x_{t-1} + \varepsilon_t$$

The multivariate generalization of (5.19) is obtained by premultiplying by B^{-1} so that

$$x_t = B^{-1}\Gamma_0 + B^{-1}\Gamma_1 x_{t-1} + B^{-1}\varepsilon_t$$

Defining $A_0 = B^{-1}\Gamma_0, A_1 = B^{-1}\Gamma_1$, and $e_t = B^{-1}\varepsilon_t$ yields (5.19). The problem, then, is to take the observed values of e_t and to restrict the system so as to recover ε_t as $\varepsilon_t = Be_t$. However, the selection of the various b_{ij} cannot be completely arbitrary. The issue is to restrict the system so as to (1) recover the various $\{\varepsilon_{it}\}$, and (2) preserve the assumed error structure concerning the independence of the various $\{\varepsilon_{it}\}$ shocks. To solve this identification problem, simply count equations and unknowns. Using OLS, we can obtain the variance/covariance matrix Σ:

$$\Sigma = \begin{bmatrix} \sigma_1^2 & \sigma_{12} & \cdots & \sigma_{1n} \\ \sigma_{21} & \sigma_2^2 & \cdots & \sigma_{2n} \\ \cdots & \cdots & \cdots & \cdots \\ \sigma_{n1} & \sigma_{n2} & \cdots & \sigma_n^2 \end{bmatrix}$$

where each element of Σ is constructed as the sum

$$\sigma_{ij} = (1/T)\sum_{t=1}^{T} e_{it}e_{jt}$$

Since Σ is symmetric, it contains only $(n^2 + n)/2$ distinct elements. There are n elements along the principal diagonal, $(n - 1)$ along the first off-diagonal, $(n - 2)$ along the next off-diagonal, ..., and one corner element for a total of $(n^2 + n)/2$ free elements.

Given that the diagonal elements of B are all unity, B contains $(n^2 - n)$ unknown values. In addition, there are the n unknown values var(ε_{it}), for a total of n^2 unknown values in the structural model [i.e., the $(n^2 - n)$ values of B plus the n values var(ε_{it})]. Now the answer to the identification problem is simple; in order to identify the n^2 unknowns from the known $(n^2 + n)/2$ independent elements of Σ, it is necessary to impose additional $n^2 - [(n^2 + n)/2] = (n^2 - n)/2$ restrictions on the system. This result generalizes to a model with p lags: *To identify the structural model from an estimated VAR, it is necessary to impose $(n^2 - n)/2$ restrictions on the structural model.*

Take a moment to count the number of restrictions in a Choleski decomposition. In the system above, the Choleski decomposition requires all elements above the principal diagonal to be zero:

$$
\begin{aligned}
b_{12} = b_{13} = b_{14} = \ldots &= b_{1n} = 0 \\
b_{23} = b_{24} = \ldots &= b_{2n} = 0 \\
b_{34} = \ldots &= b_{3n} = 0 \\
&\ldots \\
b_{n-1n} &= 0
\end{aligned}
$$

Hence, there are a total of $(n^2 - n)/2$ restrictions; the system is exactly identified. To take a specific example, consider the following Choleski decomposition in a three-variable VAR:

$$
\begin{aligned}
e_{1t} &= \varepsilon_{1t} \\
e_{2t} &= c_{21}\varepsilon_{1t} + \varepsilon_{2t} \\
e_{3t} &= c_{31}\varepsilon_{1t} + c_{32}\varepsilon_{2t} + \varepsilon_{3t}
\end{aligned}
$$

From the previous discussion, you should be able to demonstrate that ε_{1t}, ε_{2t}, and ε_{3t} can be identified from the estimates of e_{1t}, e_{2t}, e_{3t}, and variance/covariance matrix Σ. In terms of our previous notation, define matrix $C = B^{-1}$ with elements c_{ij}. Hence, $e_t = C\varepsilon_t$. An alternative way to model the relationship between the forecast errors and the structural innovations is

$$
\begin{aligned}
e_{1t} &= \varepsilon_{1t} + c_{13}\varepsilon_{3t} \\
e_{2t} &= c_{21}\varepsilon_{1t} + \varepsilon_{2t} \\
e_{3t} &= c_{31}\varepsilon_{2t} + \varepsilon_{3t}
\end{aligned}
$$

Notice the absence of a triangular structure. Here, the forecast error of each variable is affected by its own structural innovation and the structural innovation in one other variable. Given the $(9 - 3)/2 = 3$ restrictions on C, the *necessary* condition for the exact identification of B and ε_t is satisfied. However, as illustrated in the next section, imposing $(n^2 - n)/2$ restrictions is not a sufficient condition for exact identification. Unfortunately, the presence of nonlinearities means there are no simple rules that guarantee exact identification.

For those wanting a bit more formality, write the variance/covariance matrix of the regression residuals as

$$
Ee' = \Sigma = \begin{pmatrix} \sigma_1^2 & \sigma_{12} \\ \sigma_{21} & \sigma_2^2 \end{pmatrix}
$$

Given that $e_t = B^{-1}\varepsilon_t$, it must be the case that

$$Ee_te_t' = EB^{-1}\varepsilon_t\varepsilon_t'(B^{-1})' = B^{-1}E(\varepsilon_t\varepsilon_t')(B^{-1})' \tag{5.52}$$

Note that $E(\varepsilon_t\varepsilon_t')$ is the variance/covariance matrix of the structural innovations (Σ_ε). Since the covariance between the structural shocks is zero, we can write Σ_ε as

$$\Sigma_\varepsilon = \begin{bmatrix} \text{var}(\varepsilon_1) & 0 \\ 0 & \text{var}(\varepsilon_2) \end{bmatrix}$$

To find the relationship between the structural innovations and the regression residuals, substitute Σ and Σ_ε into (5.52) to obtain

$$\begin{pmatrix} \sigma_1^2 & \sigma_{12} \\ \sigma_{21} & \sigma_2^2 \end{pmatrix} = B^{-1}\begin{bmatrix} \text{var}(\varepsilon_1) & 0 \\ 0 & \text{var}(\varepsilon_2) \end{bmatrix}(B^{-1})'$$

or

$$\begin{pmatrix} \sigma_1^2 & \sigma_{12} \\ \sigma_{21} & \sigma_2^2 \end{pmatrix} = \begin{pmatrix} 1 & b_{12} \\ b_{21} & 1 \end{pmatrix}^{-1}\begin{bmatrix} \text{var}(\varepsilon_1) & 0 \\ 0 & \text{var}(\varepsilon_2) \end{bmatrix}\left[\begin{pmatrix} 1 & b_{12} \\ b_{21} & 1 \end{pmatrix}^{-1}\right]'$$

Since the four values of Σ are known, it would appear that there are four equations to determine the four unknown values b_{12}, b_{21}, var(ε_1), and var(ε_2). However, the symmetry of the system is such that $\sigma_{21} = \sigma_{12}$ so that there are only three independent equations to determine the four unknown values. To generalize the argument to an nth-order VAR system, we have

$$\Sigma = B^{-1}\Sigma_\varepsilon(B^{-1})'$$

where Σ, B^{-1}, and Σ_ε are $n \cdot n$ matrices. Using the same logic, it is possible to show that it is necessary to impose $(n^2 - n)/2$ additional restrictions on B^{-1} to completely identify the system. Some specific examples are considered in the next section.

11. EXAMPLES OF STRUCTURAL DECOMPOSITIONS

To illustrate a Sims–Bernanke decomposition, suppose there are five residuals for e_{1t} and e_{2t}. Although a usable sample size of five is unacceptable for estimation purposes, it does allow us to do the necessary calculations in a simple fashion. Thus, suppose that the five error terms are

t	1	2	3	4	5
e_{1t}	1.0	−0.5	0.0	−1.0	0.5
e_{2t}	0.5	−1.0	0.0	−0.5	1.0

Since $\{e_{1t}\}$ and $\{e_{2t}\}$ are regression residuals, their sums are zero. It is simple to verify that $\sigma_1^2 = 0.5$, $\sigma_{12} = \sigma_{21} = 0.4$, and $\sigma_2^2 = 0.5$; hence, the variance/covariance matrix Σ is

$$\Sigma = \begin{bmatrix} 0.5 & 0.4 \\ 0.4 & 0.5 \end{bmatrix}$$

Although the covariance between ε_{1t} and ε_{2t} is zero, the variances of ε_{1t} and ε_{2t} are presumably unknown. As in the previous section, let the variance/covariance matrix of these structural shocks be denoted by Σ_ε so that

$$\Sigma_\varepsilon = \begin{bmatrix} \text{var}(\varepsilon_1) & 0 \\ 0 & \text{var}(\varepsilon_2) \end{bmatrix}$$

The reason that the covariance terms are equal to zero is that ε_{1t} and ε_{2t} are deemed to be pure structural shocks. Moreover, the variance of each shock is time-invariant. For notational convenience, the time subscript can be dropped; for example, $\text{var}(\varepsilon_{1t}) = \text{var}(\varepsilon_{1t-1}) = \ldots = \text{var}(\varepsilon_1)$. The relationship between the variance/covariance matrix of the forecast errors (i.e., Σ) and the variance/covariance matrix of the pure shocks (i.e., Σ_ε) is such that $\Sigma_\varepsilon = B\Sigma B'$. Recall that e_t and ε_t are the column vectors $(e_{1t}, e_{2t})'$ and $(\varepsilon_{1t}, \varepsilon_{2t})'$, respectively. Hence,

$$e_t e_t' = \begin{bmatrix} e_{1t}^2 & e_{1t}e_{2t} \\ e_{1t}e_{2t} & e_{2t}^2 \end{bmatrix}$$

so that

$$\Sigma = \frac{1}{T}\sum_{t=1}^{T} e_t e_t' \tag{5.53}$$

Similarly, Σ_ε is

$$\Sigma_\varepsilon = \frac{1}{T}\sum_{t=1}^{T} \varepsilon_t \varepsilon_t' \tag{5.54}$$

To link the two variance/covariance matrices, note that the relationship between ε_t and e_t is such that $\varepsilon_t = Be_t$. Substitute this relationship into (5.54) and recall that the transpose of a product is the product of the transposes [i.e., $(Be_t)' = e_t'B'$], so that

$$\Sigma_\varepsilon = \frac{1}{T}\sum_{t=1}^{T} Be_t e_t' B'$$

Thus, using (5.53), we get

$$\Sigma_\varepsilon = B\Sigma B'$$

Using the specific numbers in the example, it follows that

$$\begin{bmatrix} \text{var}(\varepsilon_1) & 0 \\ 0 & \text{var}(\varepsilon_2) \end{bmatrix} = \begin{bmatrix} 1 & b_{12} \\ b_{21} & 1 \end{bmatrix} \begin{bmatrix} 0.5 & 0.4 \\ 0.4 & 0.5 \end{bmatrix} \begin{bmatrix} 1 & b_{21} \\ b_{12} & 1 \end{bmatrix}$$

Since both sides of this equation are equivalent, they must be the same element by element. Carry out the indicated multiplication of $B\Sigma B'$ to obtain

$$\text{var}(\varepsilon_1) = 0.5 + 0.8b_{12} + 0.5b_{12}^2 \tag{5.55}$$

$$0 = 0.5b_{21} + 0.4b_{21}b_{12} + 0.4 + 0.5b_{12} \tag{5.56}$$

$$0 = 0.5b_{21} + 0.4b_{12}b_{21} + 0.4 + 0.5b_{12} \tag{5.57}$$

$$\text{var}(\varepsilon_2) = 0.5b_{21}^2 + 0.8b_{21} + 0.5 \tag{5.58}$$

As you can see, equations (5.56) and (5.57) are identical. There are three independent equations to solve for the four unknowns b_{12}, b_{21}, var(ε_1), and var(ε_2). As we saw in the last section, in a two-variable system, one restriction needs to be imposed if the structural model is to be identified. Now consider the Choleski decomposition one more time. If $b_{12} = 0$, we find

$\text{var}(\varepsilon_1) = 0.5$

$0 = 0.5b_{21} + 0.4$ so that $b_{21} = -0.8$

$0 = 0.5b_{21} + 0.4$ so that again we find $b_{21} = -0.8$

$\text{var}(\varepsilon_2) = 0.5(b_{21})^2 + 0.8b_{21} + 0.5$ so that $\text{var}(\varepsilon_2) = 0.5(0.64) - 0.64 + 0.5 = 0.18$

Using this decomposition, we can recover each $\{\varepsilon_{1t}\}$ and $\{\varepsilon_{2t}\}$ as $\varepsilon_t = Be_t$:

$$\varepsilon_{1t} = e_{1t}$$

and

$$\varepsilon_{2t} = -0.8e_{1t} + e_{2t}$$

Thus, the identified structural shocks are

t	1	2	3	4	5
ε_{1t}	1.0	−0.5	0.0	−1.0	0.5
ε_{2t}	−0.3	−0.6	0.0	0.3	0.6

If you want to take the time, you can verify that $\text{var}(\varepsilon_1) = \Sigma(\varepsilon_{1t})^2/5 = 0.5$, $\text{var}(\varepsilon_{2t}) = \Sigma(\varepsilon_{2t})^2/5 = 0.18$, and $\text{cov}(\varepsilon_{1t}, \varepsilon_{2t}) = \Sigma\varepsilon_{1t}\varepsilon_{2t}/5 = 0$. Instead, if we impose the alternative restriction of a Choleski decomposition and set $b_{21} = 0$, from (5.55) through (5.58) we obtain

$\text{var}(\varepsilon_1) = 0.5 + 0.8b_{12} + 0.5b_{12}{}^2$

$0 = 0.4 + 0.5b_{12}$ so that $b_{12} = -0.8$

$0 = 0.4 + 0.5b_{12}$ so again $b_{12} = -0.8$

$\text{var}(\varepsilon_2) = 0.5$

Since $b_{12} = -0.8$, $\text{var}(\varepsilon_1) = 0.5 + 0.8(-0.8) + 0.5(0.64) = 0.18$. Now B is identified as

$$B = \begin{bmatrix} 1 & -0.8 \\ 0 & 1 \end{bmatrix}$$

If we use the identified values of B, the structural innovations are such that $\varepsilon_{1t} = e_{1t} - 0.8e_{2t}$ and $\varepsilon_{2t} = e_{2t}$. Hence, we have the structural innovations

t	1	2	3	4	5
ε_{1t}	0.6	0.3	0.0	−0.6	−0.3
ε_{2t}	0.5	−1.0	0.0	−0.5	1.0

In this example, the ordering used in the Choleski decomposition is very important. This should not be too surprising since the correlation coefficient between e_{1t} and e_{2t} is 0.8. The point is that the ordering will have important implications for the resulting variance decompositions and impulse response functions. Selecting the first ordering (i.e., setting $b_{12} = 0$) gives more importance to innovations in ε_{1t}. The assumed timing is such that ε_{1t} can have a contemporaneous effect on x_{1t} and x_{2t} while ε_{2t} shocks can affect x_{1t} only with a one-period lag. Moreover, the amplitude of the impulse responses attributable to ε_{1t} shocks will be increased since the ordering affects the magnitude of a "typical" (i.e., one standard deviation) shock in ε_{1t} and decreases the magnitude of a "typical" ε_{2t} shock.

The important point to note is that *the Choleski decomposition is only one type of identification restriction.* With three independent equations among the four unknowns b_{12}, b_{21}, var(ε_{1t}), and var(ε_{2t}), any other linearly independent restriction will allow for the identification of the structural model. Consider some of the other alternatives:

1. *A coefficient restriction.* Suppose that we know that a one-unit innovation ε_{2t} has a one-unit effect on x_{1t}; hence, suppose we know that $b_{12} = 1$. By using the other three independent equations, it follows that var(ε_{1t}) = 1.8, $b_{21} = -1$, and var(ε_{2t}) = 0.2.

   Given that $\varepsilon_t = Be_t$, we obtain

$$\begin{bmatrix} \varepsilon_{1t} \\ \varepsilon_{2t} \end{bmatrix} = \begin{bmatrix} 1 & 1 \\ -1 & 1 \end{bmatrix} \begin{bmatrix} e_{1t} \\ e_{2t} \end{bmatrix}$$

   so that $\varepsilon_{1t} = e_{1t} + e_{2t}$ and $\varepsilon_{2t} = -e_{1t} + e_{2t}$. If we use the five hypothetical regression residuals, the decomposed innovations become

t	1	2	3	4	5
ε_{1t}	1.5	−1.5	0.0	−1.5	1.5
ε_{2t}	−0.5	−0.5	0.0	0.5	0.5

2. *A variance restriction.* Except for the special case of var(ε_{it}) = 1, variance restrictions are not usually imposed in structural VARs since theory usually gives little guidance about the variances of the shocks. Nevertheless, to illustrate the decomposition, suppose that we know var(ε_{1t}) = 1.8. Given the relationship between Σ_ε and Σ (i.e., $\Sigma_\varepsilon = B\Sigma B'$), a restriction on the variances contained within Σ_ε will always imply multiple solutions for the coefficients of B. The first equation yields two possible solutions for $b_{12} = 1$ and $b_{12} = -2.6$; unless we have a theoretical reason to discard one of these magnitudes, there are two solutions to the model. If $b_{12} = 1$, the remaining solutions are $b_{21} = -1$ and var(ε_{2t}) = 0.2. If $b_{12} = -2.6$, the solutions are $b_{21} = -5/3$ and var(ε_{2t}) = 5/9.

The two solutions can be used to identify two different $\{\varepsilon_{1t}\}$ and $\{\varepsilon_{2t}\}$ sequences, and innovation accounting can be performed using both solutions. Even though there are two solutions, both satisfy the theoretical restriction concerning var(ε_{1t}).

3. *Symmetry restrictions.* A linear combination of the coefficients and variances can be used for identification purposes. For example, the symmetry restriction $b_{12} = b_{21}$ can be used for identification. If we use equation (5.56), there are two solutions: $b_{12} = b_{21} = -0.5$ or $b_{12} = b_{21} = -2.0$. For the first solution, we find var(ε_{1t}) $= 0.225$ and for the second solution var(ε_{1t}) $= 0.9$.[11] From the first solution,

$$\begin{bmatrix} \varepsilon_{1t} \\ \varepsilon_{2t} \end{bmatrix} = \begin{bmatrix} 1 & -0.5 \\ -0.5 & 1 \end{bmatrix} \begin{bmatrix} e_{1t} \\ e_{2t} \end{bmatrix}$$

so that

t	1	2	3	4	5
ε_{1t}	0.75	0.0	0.0	−0.75	0.0
ε_{2t}	0.0	−0.75	0.0	0.0	0.75

An Example

A common assumption in the open-economy macroeconomics literature is that global shocks have little influence on current account balances, relative output levels, and real exchange rates. The notion underlying this assumption is that global shocks affect all nations equally; in a sense, global shocks are like the tides that "cause all boats to rise and fall together." However, this might not be true if nations have different technologies, preferences, and/or factor supplies. In Souki and Enders (2008) we use a four-variable structural VAR to obtain a global shock and three country-specific shocks. The nature of the VAR is that we allow the global shock to have asymmetric effects on the U.S., Japanese, and German economies.

As a first step, we performed unit root tests on the logarithmic levels and on the logarithmic first differences of the variables. All six variables contain a unit root but are stationary in first differences. We then estimated a three-country model represented by the appropriately differenced four-variable VAR:

$$\begin{bmatrix} \Delta rgus_t \\ \Delta rjus_t \\ \Delta ygus_t \\ \Delta yjus_t \end{bmatrix} = \begin{bmatrix} A_{11}(L) & A_{12}(L) & A_{13}(L) & A_{14}(L) \\ A_{21}(L) & A_{22}(L) & A_{23}(L) & A_{24}(L) \\ A_{31}(L) & A_{32}(L) & A_{33}(L) & A_{34}(L) \\ A_{41}(L) & A_{42}(L) & A_{43}(L) & A_{44}(L) \end{bmatrix} \begin{bmatrix} \Delta rgus_{t-1} \\ \Delta rjus_{t-1} \\ \Delta ygus_{t-1} \\ \Delta yjus_{t-1} \end{bmatrix} + \begin{bmatrix} e_{1t} \\ e_{2t} \\ e_{3t} \\ e_{4t} \end{bmatrix}$$

where $rgus_t$ is the log of the real exchange rate between Germany and the U.S., $rjus_t$ is the log of the real exchange rate between Japan and the U.S., $ygus_t$ is the log of German/U.S. output, $yjus_t$ is the log of Japanese/U.S. output, Δ is the difference operator, the $A_{ij}(L)$ are polynomials in the lag operator L, and the e_{it} are the regression residuals. Note that the responses of the German/Japanese real exchange rate and relative income levels can be obtained from $\Delta rgus_t - \Delta rjus_t$ and $\Delta ygus_t - \Delta yjus_t$, respectively. The estimated VAR also includes a constant and four lags of the first difference of each variable (the lag length selection is based on a likelihood ratio test). The estimation period runs from March 1973 to June 2004.

We classify the shocks by their consequences, not by their source. After all, almost any shock emanates from some particular country. In our classification system, shocks—such as 9/11 or the financial crisis beginning in the U.S. housing market—with immediate worldwide consequences are global, not country-specific, shocks. The shock is global because of its immediate worldwide consequences, not because of its source. The sharp rise and then fall in the price of oil is a global shock. In order to ensure that country-specific shocks do not have any immediate worldwide consequences, it is necessary to assume that country-specific shocks are orthogonal to each other and to the global shock. The discussion implies that it makes sense to decompose the regression residuals using the six restrictions $\alpha_{12} = \alpha_{21} = \alpha_{32} = \alpha_{41} = 0$, $\alpha_{13} = \alpha_{23}$, and $\alpha_{33} = \alpha_{34}$. Hence:

$$\begin{bmatrix} e_{1t} \\ e_{2t} \\ e_{3t} \\ e_{4t} \end{bmatrix} = \begin{bmatrix} \alpha_{11} & 0 & \alpha_{13} & \alpha_{14} \\ 0 & \alpha_{22} & \alpha_{13} & \alpha_{24} \\ \alpha_{31} & 0 & \alpha_{33} & \alpha_{34} \\ 0 & \alpha_{42} & \alpha_{33} & \alpha_{44} \end{bmatrix} \begin{bmatrix} \varepsilon_{gt} \\ \varepsilon_{jt} \\ \varepsilon_{ut} \\ \varepsilon_{wt} \end{bmatrix}$$

where ε_{wt} is the global (or worldwide) shock in period t, and ε_{it} is the country-specific shock for i in period t. The nature of the four ε_{it} shocks is that they are all *i.i.d.* zero-mean random variables that are mutually uncorrelated in the sense that $E_{t-1}\varepsilon_{it}\varepsilon_{kt} = 0$ for $i \neq k$. Moreover, we normalize units so that the variance of each structural shock is unity. As such, in this four-variable VAR, we have imposed six additional restrictions to obtain an exactly identified system.

To explain, note that an ε_{jt} shock has no contemporaneous effect on $\Delta rgus_t$ if $\alpha_{12} = 0$ and has no contemporaneous effect on $\Delta ygus_t$ if $\alpha_{32} = 0$. In the same way, an ε_{gt} shock has no contemporaneous effect on $\Delta rjus_t$ if $\alpha_{21} = 0$ and has no contemporaneous effect on $\Delta yjus_t$ if $\alpha_{41} = 0$. To explain the last two restrictions, notice that the log of the real exchange rate between Japan and Germany is $rgus_t - rjus_t$ and the log of the German/Japanese output is $ygus_t - yjus_t$. Hence, if $\alpha_{13} = \alpha_{23}$, the U.S. shock will have no contemporaneous effect on the German/Japanese real exchange rate, and if $\alpha_{33} = \alpha_{34}$, the U.S. shock will have no contemporaneous effect on the German/Japanese output.

Since we do not restrict α_{14}, α_{24}, α_{34}, or α_{44} to zero, our identification scheme allows global shocks to change relative output levels and real exchange rates. Nevertheless, we do not *force* global shocks to have asymmetric effects. If the standard assumption is correct (so that global shocks have only symmetric effects), we should find that all values of α_{i4} are equal to zero. Moreover, the lag structure should be such that global shocks explain none of the forecast error variance of real exchange rates and relative outputs. Hence, any findings that our identified global shocks affect relative output levels and/or real exchange rates are necessarily due to nonproportional effects of global shocks.

The variance decompositions are shown in Table 5.4. The key points are:

- We find little evidence that third-country effects are important. The maximal impact is that the U.S. shock explains 12 percent of the forecast error variance of the Japanese/German industrial production ratio.
- Global shocks have little effect on relative output levels. As such, the conventional wisdom is correct in that global shocks do tend to affect industrial production levels proportionately.
- Global shocks explain almost all of the movements in the DM/dollar real exchange rate and sizable portions of the movements in the other two real rates. As such, our identified global shocks alter relative prices but not relative outputs.

A natural interpretation is that preferences differ across nations. Even if global productivity shocks cause output levels to move together, differences in preferences can induce relative price changes. After all, residents of different nations will use their altered income levels to buy different baskets of goods and services. For our purposes,

Table 5.4 Variance Decompositions Using Structural Shocks

			Percent of Forecast Error Variances Due to German Shock			
Horizon	**$\Delta rgus$**	**$\Delta rjus$**	**Δrgj**	**$\Delta ygus$**	**$\Delta yjus$**	**Δygj**
1-quarter	1.977	0.000	1.002	63.495	0.000	99.486
4-quarter	4.427	0.630	7.459	61.496	0.589	90.711
8-quarter	5.287	4.851	9.742	63.795	1.563	87.010
			Percent of Forecast Error Variances Due to Japanese Shock			
Horizon	**$\Delta rgus$**	**$\Delta rjus$**	**Δrgj**	**$\Delta ygus$**	**$\Delta yjus$**	**Δygj**
1-quarter	0.000	72.366	81.365	0.000	1.483	0.420
4-quarter	2.648	68.731	75.597	2.125	4.539	0.473
8-quarter	3.875	64.193	73.175	2.196	4.603	0.833
			Percent of Forecast Error Variances Due to U.S. Shock			
Horizon	**$\Delta rgus$**	**$\Delta rjus$**	**Δrgj**	**$\Delta ygus$**	**$\Delta yjus$**	**Δygj**
1-quarter	1.948	1.645	0.000	35.600	95.024	0.000
4-quarter	2.713	2.588	2.029	34.709	92.028	7.711
8-quarter	2.782	3.294	2.741	32.388	90.914	11.161
			Percent of Forecast Error Variances Due to Global Shock			
Horizon	**$\Delta rgus$**	**$\Delta rjus$**	**Δrgj**	**$\Delta ygus$**	**$\Delta yjus$**	**Δygj**
1-quarter	96.076	25.990	17.633	0.905	3.492	0.094
4-quarter	90.212	28.052	14.915	1.669	2.844	1.106
8-quarter	88.056	27.661	14.342	1.621	2.920	0.996

the main point is that a combination of coefficient restrictions and symmetry restrictions can be used to identify structural shocks.

Overidentified Systems

It may be that economic theory suggests more than $(n^2 - n)/2$ restrictions. If so, it is necessary to modify the method above. The procedure for identifying an overidentified system entails the following steps:

STEP 1: The restrictions on B or var(ε_{it}) do not affect the estimation of VAR coefficients. Hence, estimate the unrestricted VAR, $x_t = A_0 + A_1x_{t-1} + \ldots + A_px_{t-p} + e_t$. Use the standard lag length and block-causality tests to help determine the form of the VAR.

STEP 2: Obtain the unrestricted variance/covariance matrix Σ. The determinant of this matrix is an indicator of the overall fit of the model.

STEP 3: Restricting B and/or Σ_ε will affect the estimate of Σ. Select the appropriate restrictions and maximize the likelihood function with respect to the free parameters of B and Σ_ε. This will lead to an estimate of the restricted variance/covariance matrix. Denote this second estimate by Σ_R.

For those wanting a more technical explanation, note that the log likelihood function is

$$-\frac{T}{2}\ln|\Sigma| - \frac{1}{2}\sum_{t=1}^{T}(e_t'\Sigma^{-1}e_t)$$

Fix each element of e_t (and e_t') at the level obtained using OLS; call these estimated OLS residuals $\hat{e}_t$. Now use the relationship $\Sigma_\varepsilon = B\Sigma B'$ so that the log likelihood function can be written as

$$-\frac{T}{2}\ln|B^{-1}\Sigma_\varepsilon(B')^{-1}| - \frac{1}{2}\sum_{t=1}^{T}(\hat{e}_t'B'\Sigma_\varepsilon^{-1}B\,\hat{e}_t)$$

Now select the restrictions on B and Σ_ε and maximize with respect to the remaining free elements of these two matrices. The resulting estimates of B and Σ_ε imply a value of Σ that we have dubbed Σ_R. A number of popular software packages can perform this type of estimation using the Generalized Method of Moments.

STEP 4: If the restrictions are not binding, Σ and Σ_R will be equivalent. Let R = the number of overidentifying restrictions; i.e., R = number of restrictions exceeding $(n^2 - n)/2$. Then the χ^2 test statistic

$$\chi^2 = |\Sigma_R| - |\Sigma|$$

with R degrees of freedom can be used to test the restricted system.[12] If the calculated value of χ^2 exceeds that in a χ^2 table, the restrictions can be rejected. Now allow for two sets of overidentifying restrictions such that the number of restrictions in R_2 exceeds that in R_1. In fact, if $R_2 > R_1 \geq (n^2 - n)/2$, the significance of the extra $R_2 - R_1$ restrictions can be tested as

$$\chi^2 = |\Sigma_{R2}| - |\Sigma_{R1}| \text{ with } R_2 - R_1 \text{ degrees of freedom}$$

Similarly, in an overidentified system, the t-statistic for the individual coefficients can be obtained. Sims warned that the calculated standard errors may not be very accurate. Waggoner and Zha (1997) pointed out a problem with this type of structural decomposition. Because of the restrictions on the signs of the $c_{ij}(0)$, they argued that the normalization can have important effects on statistical inference. The essence of the argument relies on the fact that the contemporaneous impulse responses are obtained as the solutions to

$$\begin{bmatrix} e_{1t} \\ e_{2t} \end{bmatrix} = \begin{bmatrix} c_{11}(0) & c_{12}(0) \\ c_{21}(0) & c_{22}(0) \end{bmatrix} \begin{bmatrix} \varepsilon_{1t} \\ \varepsilon_{2t} \end{bmatrix}$$

Suppose that $c_{11}(0)$ is small and is estimated with a large standard error. Restricting $c_{11}(0)$ to be positive (so that ε_{1t} shocks are estimated to increase output with probability one) can artificially inflate the standard errors of the remaining c_{ij}.

Sims's Structural VAR

Sims (1986) used a six-variable VAR of quarterly data over the period 1948Q1 to 1979Q3. The variables included in the study are real GNP (y), real business fixed investment (i), the GNP deflator (p), the money supply as measured by M1 (m), unemployment (u), and the treasury bill rate (r). An unrestricted VAR was estimated with four lags of each variable and a constant term. Sims obtained the 36 impulse response functions using a Choleski decomposition with the ordering $y \rightarrow i \rightarrow p \rightarrow m \rightarrow u \rightarrow r$. Some of the impulse response functions had reasonable interpretations. However, the response of real variables to a money supply shock seemed unreasonable. The impulse responses suggested that a money supply shock had little effect on prices, output, or the interest rate. Given a standard money demand function, it is hard to explain why the public would be willing to hold the expanded money supply. Sims proposes an alternative to the Choleski decomposition that is consistent with money market equilibrium. Sims restricts the B matrix such that

$$\begin{bmatrix} 1 & b_{11} & 0 & 0 & 0 & 0 \\ b_{21} & 1 & b_{23} & b_{24} & 0 & 0 \\ b_{31} & 0 & 1 & 0 & 0 & b_{36} \\ b_{41} & 0 & b_{43} & 1 & 0 & b_{46} \\ b_{51} & 0 & b_{53} & b_{54} & 1 & b_{56} \\ 0 & 0 & 0 & 0 & 0 & 1 \end{bmatrix} \begin{bmatrix} r_t \\ m_t \\ y_t \\ p_t \\ u_t \\ i_t \end{bmatrix} = \begin{bmatrix} \varepsilon_{rt} \\ \varepsilon_{mt} \\ \varepsilon_{yt} \\ \varepsilon_{pt} \\ \varepsilon_{ut} \\ \varepsilon_{it} \end{bmatrix}$$

Notice there are 17 zero restrictions on the b_{ij}. The system is overidentified; with six variables, exact identification requires only $(6^2 - 6)/2 = 15$ restrictions. Imposing these restrictions, Sims identifies the following six relationships among the contemporaneous innovations:

$$r_t = 71.20m_t + \varepsilon_{rt} \tag{5.59}$$

$$m_t = 0.283y_t + 0.224p_t - 0.0081r_t + \varepsilon_{mt} \quad (5.60)$$

$$y_t = -0.00135r_t + 0.132i_t + \varepsilon_{yt} \quad (5.61)$$

$$p_t = -0.0010r_t + 0.045y_t - 0.00364i_t + \varepsilon_{pt} \quad (5.62)$$

$$u_t = -0.116r_t - 20.1y_t - 1.48i_t - 8.98p_t + \varepsilon_{ut} \quad (5.63)$$

$$i_t = \varepsilon_{it} \quad (5.64)$$

Sims views (5.59) and (5.60) as money supply and demand functions, respectively. In (5.59), the money supply rises as the interest rate increases. The demand for money in (5.60) is positively related to income and the price level and negatively related to the interest rate. Investment innovations in (5.64) are completely autonomous. Otherwise, Sims sees no reason to restrict the other equations in any particular fashion. For simplicity, he chooses a Choleski-type block structure for GNP, the price level, and the unemployment rate. The impulse response functions appear to be consistent with the notion that money supply shocks affect prices, income, and the interest rate.

12. THE BLANCHARD–QUAH DECOMPOSITION

Blanchard and Quah (1989) provide an alternative way to obtain a structural VAR. Their aim is to reconsider the Beveridge and Nelson (1981) decomposition of real GNP into its temporary and permanent components. Toward this end, they developed a macroeconomic model such that real GNP is affected by demand-side and supply-side disturbances. In accord with the natural rate hypothesis, demand-side disturbances have no long-run effect on real GNP. On the supply side, productivity shocks are assumed to have permanent effects on output. In a univariate model, there is no unique way to decompose a variable into its temporary and permanent components. However, using a bivariate VAR, Blanchard and Quah show how to decompose real GNP and recover the two pure shocks.

To take a general example, suppose we are interested in decomposing an $I(1)$ sequence, say $\{y_t\}$, into its temporary and permanent components. Let there be a second variable $\{z_t\}$ that is affected by the same two shocks. For the time being, suppose that $\{z_t\}$ is stationary. If we ignore the intercept terms, the bivariate moving-average (BMA) representation of the $\{y_t\}$ and $\{z_t\}$ sequences will have the form

$$\Delta y_t = \sum_{k=0}^{\infty} c_{11}(k)\varepsilon_{1t-k} + \sum_{k=0}^{\infty} c_{12}(k)\varepsilon_{2t-k} \quad (5.65)$$

$$z_t = \sum_{k=0}^{\infty} c_{21}(k)\varepsilon_{1t-k} + \sum_{k=0}^{\infty} c_{22}(k)\varepsilon_{2t-k} \quad (5.66)$$

or, in a more compact form,

$$\begin{bmatrix} \Delta y_t \\ z_t \end{bmatrix} = \begin{bmatrix} C_{11}(L) & C_{12}(L) \\ C_{21}(L) & C_{22}(L) \end{bmatrix} \begin{bmatrix} \varepsilon_{1t} \\ \varepsilon_{2t} \end{bmatrix}$$

where ε_{1t} and ε_{2t} are independent white-noise disturbances, each having a constant variance, and the $C_{ij}(L)$ are polynomials in the lag operator L such that the individual coefficients of $C_{ij}(L)$ are denoted by $c_{ij}(k)$. For example, the third coefficient of $C_{21}(L)$ is $c_{21}(3)$. For convenience, the time subscripts on the variances and the covariance terms are dropped and the shocks are normalized so that $\text{var}(\varepsilon_1) = 1$ and $\text{var}(\varepsilon_2) = 1$. If we call Σ_ε the variance/covariance matrix of the innovations, it follows that

$$\Sigma_\varepsilon = \begin{bmatrix} \text{var}(\varepsilon_1) & \text{cov}(\varepsilon_1, \varepsilon_2) \\ \text{cov}(\varepsilon_1, \varepsilon_2) & \text{var}(\varepsilon_2) \end{bmatrix}$$
$$= \begin{bmatrix} 1 & 0 \\ 0 & 1 \end{bmatrix}$$

In order to use the Blanchard and Quah (BQ) technique, at least one of the variables must be nonstationary since $I(0)$ variables do not have a permanent component. However, to use the method, both variables must be in a stationary form. Since $\{y_t\}$ is $I(1)$, (5.65) uses the first difference of the series. Note that (5.66) implies that the $\{z_t\}$ sequence is $I(0)$; if in your own work you find $\{z_t\}$ is also $I(1)$, use its first difference.

In contrast to the Sims–Bernanke procedure, Blanchard and Quah do not directly associate the $\{\varepsilon_{1t}\}$ and $\{\varepsilon_{2t}\}$ shocks with the $\{y_t\}$ and $\{z_t\}$ sequences. Instead, the $\{y_t\}$ and $\{z_t\}$ sequences are the endogenous variables, and the $\{\varepsilon_{1t}\}$ and $\{\varepsilon_{2t}\}$ sequences represent what an economic theorist would call the exogenous variables. In their example, y_t is the logarithm of real GNP, z_t is unemployment, ε_{1t} is an aggregate demand shock, and ε_{2t} is an aggregate supply shock. The coefficients of $C_{11}(L)$, for example, represent the impulse responses of an aggregate demand shock on the time path of change in the log of real GNP.[13]

The key to decomposing the $\{y_t\}$ sequence into its permanent and stationary components is to assume that one of the shocks has a temporary effect on the $\{y_t\}$ sequence. It is this dichotomy between temporary and permanent effects that allows for the complete identification of the structural innovations from an estimated VAR. For example, Blanchard and Quah assume that an aggregate demand shock has no long-run effect on real GNP. In the long run, if real GNP is to be unaffected by the demand shock, it must be the case that the cumulative effect of an ε_{1t} shock on the Δy_t sequence must be equal to zero. Hence, the coefficients $c_{11}(k)$ in (5.65) must be such that

$$\sum_{k=0}^{\infty} c_{11}(k)\, \varepsilon_{1t-k} = 0$$

Since this must hold for any possible realization of the $\{\varepsilon_{1t}\}$ sequence, it must be the case that

$$\sum_{k=0}^{\infty} c_{11}(k) = 0 \tag{5.67}$$

Since the demand-side and supply-side shocks are not observed, the problem is to recover them from a VAR estimation. Given that the variables are stationary, we know there exists a VAR representation of the form

$$\begin{bmatrix} \Delta y_t \\ z_t \end{bmatrix} = \begin{bmatrix} A_{11}(L) & A_{12}(L) \\ A_{21}(L) & A_{22}(L) \end{bmatrix} \begin{bmatrix} \Delta y_{t-1} \\ z_{t-1} \end{bmatrix} + \begin{bmatrix} e_{1t} \\ e_{2t} \end{bmatrix} \tag{5.68}$$

or, to use a more compact notation,

$$x_t = A(L)x_{t-1} + e_t$$

where: x_t = the column vector $(\Delta y_t, z_t)'$
e_t = the column vector $(e_{1t}, e_{2t})'$
$A(L)$ = the $2 \cdot 2$ matrix with elements equal to the polynomials $A_{ij}(L)$
and the coefficients of $A_{ij}(L)$ are denoted by $a_{ij}(k)$.[14]

The critical insight is that the VAR residuals are composites of the pure innovations ε_{1t} and ε_{2t}. For example, e_{1t} is the one-step-ahead forecast error of y_t; i.e., $e_{1t} = \Delta y_t - E_{t-1}\Delta y_t$. From the BMA, the one-step-ahead forecast error is $c_{11}(0)\varepsilon_{1t}$ + $c_{12}(0)\varepsilon_{2t}$. Since the two representations are equivalent, it must be the case that

$$e_{1t} = c_{11}(0)\varepsilon_{1t} + c_{12}(0)\varepsilon_{2t} \tag{5.69}$$

Similarly, since e_{2t} is the one-step-ahead forecast error of z_t,

$$e_{2t} = c_{21}(0)\varepsilon_{1t} + c_{22}(0)\varepsilon_{2t} \tag{5.70}$$

or, combining (5.69) and (5.70), we get

$$\begin{bmatrix} e_{1t} \\ e_{2t} \end{bmatrix} = \begin{bmatrix} c_{11}(0) & c_{12}(0) \\ c_{21}(0) & c_{22}(0) \end{bmatrix} \begin{bmatrix} \varepsilon_{1t} \\ \varepsilon_{2t} \end{bmatrix}$$

If $c_{11}(0)$, $c_{12}(0)$, $c_{21}(0)$, and $c_{22}(0)$ were known, it would be possible to recover ε_{1t} and ε_{2t} from the regression residuals e_{1t} and e_{2t}. Blanchard and Quah show that the relationship between (5.68) and the BMA model plus the long-run restriction of (5.67) provide exactly four restrictions that can be used to identify these four coefficients. The VAR residuals can be used to construct estimates of $\text{var}(e_1)$, $\text{var}(e_2)$, and $\text{cov}(e_1, e_2)$.[15] Hence, there are the following four restrictions:

RESTRICTION 1

Given (5.69) and noting that $E\varepsilon_{1t}\varepsilon_{2t} = 0$, the normalization $\text{var}(\varepsilon_1) = \text{var}(\varepsilon_2) = 1$ means that the variance of e_{1t} is

$$\text{var}(e_1) = c_{11}(0)^2 + c_{12}(0)^2 \tag{5.71}$$

RESTRICTION 2

Similarly, using (5.70), the variance of e_{2t} is related to $c_{21}(0)$ and $c_{22}(0)$ as

$$\text{var}(e_2) = c_{21}(0)^2 + c_{22}(0)^2 \tag{5.72}$$

RESTRICTION 3

The product of e_{1t} and e_{2t} is

$$e_{1t}e_{2t} = [c_{11}(0)\varepsilon_{1t} + c_{12}(0)\varepsilon_{2t}][c_{21}(0)\varepsilon_{1t} + c_{22}(0)\varepsilon_{2t}]$$

If we take the expectation, the covariance of the VAR residuals is

$$Ee_{1t}e_{2t} = c_{11}(0)c_{21}(0) + c_{12}(0)c_{22}(0) \tag{5.73}$$

Thus, equations (5.71), (5.72), and (5.73) can be viewed as three equations in the four unknowns $c_{11}(0)$, $c_{12}(0)$, $c_{21}(0)$, and $c_{22}(0)$. The fourth restriction is embedded in the assumption that the $\{\varepsilon_{1t}\}$ has no long-run effect on the $\{y_t\}$ sequence. The problem is to transform the restriction (5.67) into its VAR representation. Since the algebra is a bit messy, it is helpful to rewrite (5.68) as

$$x_t = A(L)Lx_t + e_t$$

so that

$$[I - A(L)L]x_t = e_t$$

and, by premultiplying by $[I - A(L)L]^{-1}$, we obtain

$$x_t = [I - A(L)L]^{-1}e_t \tag{5.74}$$

Denote the determinant of $[I - A(L)L]$ by the expression D. It should not take too long to convince yourself that (5.74) can be written as

$$\begin{bmatrix} \Delta y_t \\ z_t \end{bmatrix} = \frac{1}{D}\begin{bmatrix} 1-A_{22}(L)L & A_{12}(L)L \\ A_{21}(L)L & 1-A_{11}(L)L \end{bmatrix}\begin{bmatrix} e_{1t} \\ e_{2t} \end{bmatrix}$$

or, using the definitions of the $A_{ij}(L)$, we get

$$\begin{bmatrix} \Delta y_t \\ z_t \end{bmatrix} = \frac{1}{D}\begin{bmatrix} 1-\Sigma a_{22}(k)L^{k+1} & \Sigma a_{12}(k)L^{k+1} \\ \Sigma a_{21}(k)L^{k+1} & 1-\Sigma a_{11}(k)L^{k+1} \end{bmatrix}\begin{bmatrix} e_{1t} \\ e_{2t} \end{bmatrix}$$

where the summations run from $k = 0$ to infinity.

Thus, the solution for Δy_t in terms of the current and lagged values of $\{e_{1t}\}$ and $\{e_{2t}\}$ is

$$\Delta y_t = \frac{1}{D}\left\{\left[1-\sum_{k=0}^{\infty} a_{22}(k)L^{k+1}\right]e_{1t} + \sum_{k=0}^{\infty} a_{12}(k)L^{k+1}e_{2t}\right\} \tag{5.75}$$

Now e_{1t} and e_{2t} can be replaced by (5.69) and (5.70). Making these substitutions, the restriction that the $\{\varepsilon_{1t}\}$ sequence has no long-run effect on y_t is

$$\left[1-\sum_{k=0}^{\infty} a_{22}(k)L^{k+1}\right]c_{11}(0)\varepsilon_{1t} + \sum_{k=0}^{\infty} a_{12}(k)L^{k+1}c_{21}(0)\varepsilon_{1t} = 0$$

RESTRICTION 4

For all possible realizations of the $\{\varepsilon_{1t}\}$ sequence, ε_{1t} shocks will have only temporary effects on the Δy_t sequence (and on y_t itself) if

$$\left[1-\sum_{k=0}^{\infty} a_{22}(k)\right]c_{11}(0) + \sum_{k=0}^{\infty} a_{12}(k)c_{21}(0) = 0$$

With this fourth restriction, there are four equations that can be used to identify the unknown values $c_{11}(0)$, $c_{12}(0)$, $c_{21}(0)$, and $c_{22}(0)$. To summarize, the steps in the procedure are as follows:

STEP 1: Begin by pretesting the two variables for time trends and for unit roots. If $\{y_t\}$ does not have a unit root, there is no reason to proceed with the decomposition. Appropriately transform the two variables so that the resulting sequences are both $I(0)$. Perform lag-length tests to find a reasonable specification for the VAR. The residuals of the estimated VAR should pass the standard diagnostic checks for white-noise processes (of course, e_{1t} and e_{2t} can be correlated with each other).

STEP 2: Using the residuals of the estimated VAR, calculate the variance/covariance matrix; i.e., calculate var(e_1), var(e_2), cov(e_1, e_2). Also calculate the sums

$$1 - \sum_{k=0}^{p} a_{22}(k) \text{ and } \sum_{k=0}^{p} a_{12}(k)$$

where p = lag length used to estimate the VAR.

Use these values to solve the following four equations for $c_{11}(0)$, $c_{12}(0)$, $c_{21}(0)$, and $c_{22}(0)$:

$$\begin{aligned}
\text{var}(e_1) &= c_{11}(0)^2 + c_{12}(0)^2 \\
\text{var}(e_2) &= c_{21}(0)^2 + c_{22}(0)^2 \\
\text{cov}(e_1, e_2) &= c_{11}(0)c_{21}(0) + c_{12}(0)c_{22}(0) \\
0 &= c_{11}(0)[1 - \Sigma a_{22}(k)] + c_{21}(0)\Sigma a_{12}(k)
\end{aligned}$$

Given these four values $c_{ij}(0)$ and the residuals of the VAR, the entire $\{\varepsilon_{1t}\}$ and $\{\varepsilon_{2t}\}$ sequences can be identified using the formulas[16]

$$e_{1t-i} = c_{11}(0)\varepsilon_{1t-i} + c_{12}(0)\varepsilon_{2t-i}$$

and

$$e_{2t-i} = c_{21}(0)\varepsilon_{1t-i} + c_{22}(0)\varepsilon_{2t-i}$$

STEP 3: As in a traditional VAR, the identified $\{\varepsilon_{1t}\}$ and $\{\varepsilon_{2t}\}$ sequences can be used to obtain impulse response functions and variance decompositions. The difference is that the interpretation of the impulses is straightforward. For example, Blanchard and Quah are able to obtain the impulse responses of the change in the log of real GNP to a typical supply-side shock. Moreover, it is possible to obtain the historical decomposition of each series. For example, set all $\{\varepsilon_{1t}\}$ shocks equal to zero and use the actual $\{\varepsilon_{2t}\}$ series (i.e., use the identified values of ε_{2t}) to obtain the permanent changes in $\{y_t\}$ as[17]

$$\Delta y_t = \sum_{k=0}^{\infty} c_{12}(k)\varepsilon_{2t-k}$$

The Blanchard and Quah Results

In their study, Blanchard and Quah (1989) used the first difference of the logarithm of real GNP and the level of unemployment. They noted that unemployment exhibits an apparent time trend and that there is a slowdown in real growth beginning in the mid-1970s. Since there is no obvious way to address these difficult issues, they estimated four different VARs. Two include a dummy allowing for the change in the rate of growth in output, and two include a deterministic time trend in unemployment. Using quarterly GNP and unemployment data over the period 1950*Q*2 through 1987*Q*4, they estimated a VAR with eight lags.

Imposing the restriction that demand-side shocks have no long-run effect on real GNP, Blanchard and Quah identified the two types of shocks. The impulse response functions for the four VARs are quite similar:

- The time paths of demand-side disturbances on output and unemployment are hump-shaped. The impulse responses are mirror images of each other; initially output increases while unemployment decreases. The effects peak after four quarters; afterward they converge to their original levels.
- Supply-side disturbances have a cumulative effect on output. A supply disturbance having a positive effect on output has a small positive initial effect on unemployment. After this initial increase, unemployment steadily decreases and the cumulated change becomes negative after four quarters. Unemployment remains below its long-run level for nearly five years.

Blanchard and Quah found that the alternative methods of treating the slowdown in output growth and the trend in unemployment affect the variance decompositions. Since the goal here is to illustrate the technique, consider only the variance decomposition using a dummy variable for the decline in output growth and detrended unemployment.

Percent of Forecast Error Variance Due to Demand-Side Shocks

Forecasting Horizon (Quarters)	Output	Unemployment
1	99.0	51.9
4	97.9	80.2
12	67.6	86.2
40	39.3	85.6

At short-run horizons, the huge preponderance of the variation in output is due to demand-side innovations. Demand shocks account for almost all of the movement in GNP at short horizons. Since demand shock effects are necessarily temporary, the findings contradict those of Beveridge and Nelson. The proportion of the forecast error variance falls steadily as the forecast horizon increases; the proportion converges to zero since these effects are temporary. Consequently, the contribution of supply-side innovations to real GNP movements increases at longer forecasting horizons. On the other hand, demand-side shocks generally account for increasing proportions of the variation in unemployment at longer forecasting horizons.

13. DECOMPOSING REAL AND NOMINAL EXCHANGE RATES: AN EXAMPLE

In Enders and Lee (1997) we decomposed real and nominal exchange rate movements into the components induced by real and nominal factors. This section presents a small portion of the paper in order to further illustrate the methodolny of the Blanchard and Quah technique. The results reported below are updated through 2008Q2 using the data in the file labeled REALRATES.XLS. One aim of the study is to explain the deviations from purchasing power parity. As in Chapter 4, the real exchange rate (r_t) can be defined as[18]

$$r_t = e_t + p_t^* - p_t$$

where p_t^* and p_t refer to the logarithms of U.S. and U.K. wholesale price indices and e_t is the logarithm of the pound/dollar nominal exchange rate.

To explain the deviations from PPP, we suppose that there are two types of shocks: a real shock and a nominal shock. The theory suggests that real shocks can cause permanent changes in the real exchange rate but that nominal shocks can cause only temporary movements in the real rate. For example, in the long run, if the U.K. doubles its nominal money supply, the U.K. price level and the exchange rate will both double (i.e., p_t and e_t will double). Hence, in the long run, the real exchange rate remains invariant to a money supply shock.

For Step 1, we perform various unit root tests on the quarterly pound/dollar real and nominal exchange rates over the 1973Q1 to 2008Q2 period. Consistent with other studies focusing on the post–Bretton Woods period, it is clear that real and nominal rates can be characterized by nonstationary processes. If you follow the general-to-specific approach and use five lags of Δr_t in an augmented Dickey–Fuller test, you should find that the coefficient on r_{t-1} is -0.0456 with a t-statistic of -1.799. Although the AIC and SBC select one lag, it still follows that the real exchange rate is $I(1)$. Rejecting the null hypothesis of a unit root is important; if the $\{r_t\}$ series is stationary, it has no permanent component. Finding the order of integration of the nominal exchange rate is a bit trickier. Although many researchers argue that nominal exchange rates should act as $I(1)$ processes, it is worthwhile to formally test this claim using an ADF test. The AIC, SBC, and t-tests all suggest that one lagged change is appropriate. If you estimate the model you should find

$$\Delta e_t = \underset{(-2.62)}{-0.0357} - \underset{(-2.83)}{0.0677} e_{t-1} + \underset{(2.92)}{0.2377} \Delta e_{t-1}$$

The value of -2.83 is not quite significant at the 5 percent level (from Table A, the critical value is -2.89). Given the ambiguity, we might be able to clarify the issue using the Elliott, Rothenberg, and Stock (1996) unit root test discussed in Chapter 4. If e_t^d denotes the detrended nominal exchange rate, you should find

$$\Delta e_t^d = \underset{(-1.16)}{-0.015} e_{t-1}^d + \underset{(2.71)}{0.224} \Delta e_{t-1}^d$$

If you compare the t-statistic of -1.16 to the appropriate critical values in the top portion of Table A, the null hypothesis of a unit root is not rejected. As such, it seems reasonable to proceed treating the $\{e_t\}$ series as an $I(1)$ process. The BMA model has the form

$$\begin{bmatrix} \Delta r_t \\ \Delta e_t \end{bmatrix} = \begin{bmatrix} C_{11}(L) & C_{12}(L) \\ C_{21}(L) & C_{22}(L) \end{bmatrix} \begin{bmatrix} \varepsilon_{rt} \\ \varepsilon_{nt} \end{bmatrix}$$

where ε_{rt} and ε_{nt} represent the zero-mean mutually uncorrelated real and nominal shocks, respectively.

The restriction that the nominal shocks have no long-run effect on the real exchange rate is represented by the restriction that the coefficients in $C_{12}(L)$ sum to zero; thus, if $c_{ij}(k)$ is the kth coefficient in $C_{ij}(L)$, as in (5.67), the restriction is

$$\sum_{k=0}^{\infty} c_{12}(k) = 0 \tag{5.76}$$

The restriction in (5.76) implies that the cumulative effect of ε_{nt} on Δr_t is zero, and consequently, that the long-run effect of ε_{nt} on the level of r_t itself is zero. Put another way, the nominal shock ε_{tn} has only short-run effects on the real exchange rate. Note that there is no restriction on the effects of a real shock on the real rate or on the effects of either real or nominal shocks on the nominal exchange rate.

For Step 2, we estimate a bivariate VAR model for several lag lengths. Likelihood ratio tests indicate that a VAR model with five lags is appropriate. For example, if you compare the five-lag and one-lag models you should find that $\ln(|\Sigma_5|) = -15.113$, $\ln(|\Sigma_1|) = -14.879$, the number of coefficients in each equation of the five-lag model is 11, and the number of usable observations is 136. Using these values, equation (5.44) becomes

$$(136 - 11)*[-14.879 - (-15.113)] = 29.178$$

If you compare 29.178 to a χ^2 distribution with 16 degrees of freedom, you will find that the restriction is binding at the 0.0227 significance level. In contrast, the multivariate AIC and SBC indicate that one lag of each variable is sufficient. The AIC and SBC with five lags are -2011.40 and -1947.32 while the AIC and SBC with one lag are -2011.66 and -1994.18.

Since the lag length selection methods give conflicting answers, a careful researcher might want to perform the analysis using both lag lengths. For ease of exposition, the text reports results using only one lag. You can use the data in the file EXRATES.XLS to see if the key results are dependent on the lag length.

The variance decompositions using a standard Choleski decomposition are shown in the second and third columns of the table below. The ordering is such that the nominal exchange rate has no contemporaneous effect on the real rate. The decompositions using the Blanchard–Quah decomposition are shown in the fourth and fifth columns. The table shows the percentages of the forecast error variances accounted for by the ε_{rt} shock.

Comparison of Choleski and BQ Decompositions

	Choleski		Blanchard–Quah	
Horizon	Δr_t	Δe_t	Δr_t	Δe_t
1 quarter	100.0	80.62	97.08	92.14
4 quarters	97.85	77.57	95.52	98.40
8 quarters	97.85	77.57	95.52	89.40

If we use the Choleski decomposition, it is immediately evident that real shocks explain almost all of the forecast error variance of the real exchange rate at any forecast horizon. Nominal shocks accounted for approximately 20 percent of the forecast error variance of the nominal exchange rate. Our interpretation is that real shocks are responsible for movements in real *and* nominal exchange rates. Hence, we should expect them to display sizeable comovements. The effect of using the BQ decomposition makes little difference for the behavior of the real exchange rate. However, the supply shock accounts for a larger percentage of the forecast error variance of the nominal exchange rate.

Figure 5.9 shows the impulse response functions of the real and nominal exchange rates to both types of shocks. For clarity, the results are shown for the levels of exchange rates (as opposed to first differences) measured in terms of standard deviations. Moreover, each series is divided by its standard error so that the values are standardized.

1. The effect of a "real" shock is to cause an immediate increase in the real and nominal exchange rates. It is interesting to note that the movements in the real value of the dollar are nearly the same as those of the nominal dollar. Moreover, these changes are all of a permanent nature. Real and

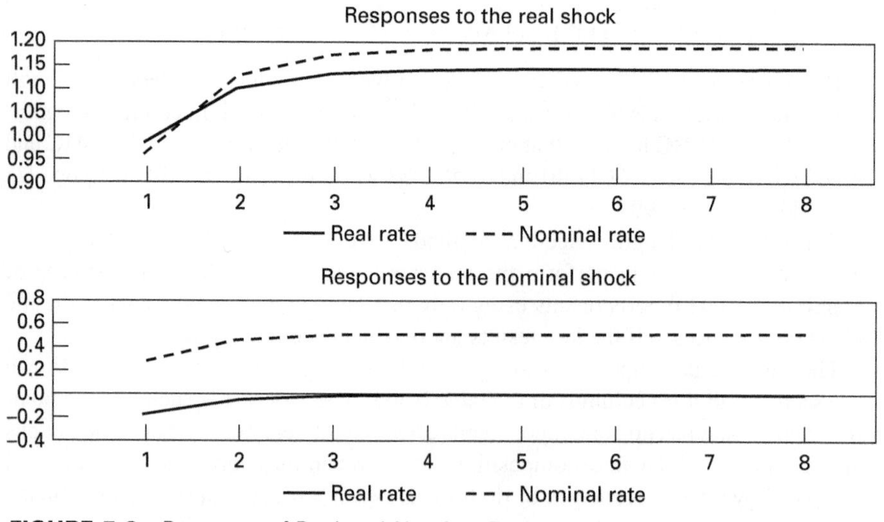

FIGURE 5.9 Response of Real and Nominal Exchange Rates

nominal rates converge to their new long-run levels in about three quarters. Since the movements in r_t are e_t are nearly identical, the implication is that the price ratio $p_t - p_t^*$ shows very little response to a real shock.

2. In response to the nominal shock, the movement in the nominal exchange rate to its long-run level is almost immediate. There is little evidence of exchange rate overshooting. As required by our identification restriction, the effect of a nominal shock on the real exchange rate is necessarily temporary. Nevertheless, even the short-run changes in the real rate show very little response to a nominal shock. The implication is that $p_t - p_t^*$ adjusts to offset the change in the nominal exchange rate.

Limitations of the Technique

A problem with this type of decomposition is that there are many types of shocks. As recognized by Blanchard and Quah (1989), the approach is limited by its ability to identify *at most* only as many types of distinct shocks as there are variables. Blanchard and Quah proved several propositions that are somewhat helpful when the presence of three or more structural shocks is suspected. Suppose there are several disturbances having permanent effects but only one having a temporary effect on $\{y_t\}$. If the variance of one type of permanent disturbance grows "arbitrarily" small relative to the other, then the decomposition scheme approaches the correct decomposition. The second proposition they prove is that if there are multiple permanent disturbances (temporary disturbances), the correct decomposition is possible if and only if the individual distributed lag responses in the real and nominal exchange rate are sufficiently similar across equations. By "sufficiently similar," Blanchard and Quah mean that the coefficients may differ up to a scalar lag distribution. Yet both propositions essentially imply that there are only two types of disturbances. For the first proposition, the third disturbance must be arbitrarily small. For the second proposition, the third disturbance must have a sufficiently similar path to one of the others. It is wise to avoid such a decomposition when the presence of three or more important disturbances is suspected. Alternatively, as in Clarida and Gali (1994), you might be able to develop a model implying three long-run restrictions among three variables.

A second problem is that the Blanchard–Quah restrictions produce a system of quadratic equations so that the signs of the $c_{ij}(0)$ are not identified. Moreover, in a system with many variables, there can be many solutions to the nonlinear system of equations. In these circumstances, Taylor (2003) recommends the use of overidentifying restrictions or those normalizations that are consistent with an underlying economic model.

14. SUMMARY AND CONCLUSIONS

Intervention analysis was used to determine the effects of installing metal detectors in airports. More generally, intervention analysis can be used to ascertain how any *deterministic* function affects an economic time series. Usually the shape of the intervention function is clear, as in the metal detector example. However, there are a great many possible intervention functions. If there is an ambiguity, the shape of the intervention function can be determined using the standard Box–Jenkins criteria for model selection. The crucial assumption in intervention analysis is that the intervention function has only deterministic components.

Transfer function analysis is appropriate if the "intervention" sequence is stochastic. If $\{y_t\}$ is endogenous and $\{z_t\}$ is exogenous, a transfer function can be fit using the five-step procedure discussed in Section 2. The procedure is a straightforward modification of the standard Box–Jenkins methodology. The resulting impulse response function traces out the time path of $\{z_t\}$ realizations on the $\{y_t\}$ sequence. This technique was illustrated by a study showing that terrorist attacks caused Italy's tourism revenues to decline by a total 600 million SDR.

With economic data, it is not always clear that one variable is dependent and the others are independent. In the presence of feedback, intervention and transfer function analyses are inappropriate. Instead, use a vector autoregression, which treats all variables as jointly endogenous. Each variable is allowed to depend on its past realizations and on the past realizations of all other variables in the system. There is no special attention paid to parsimony since the imposition of the "incredible identification restrictions" may be inconsistent with economic theory. Granger causality tests, block exogeneity, and lag length tests can help select a more parsimonious model.

Ordinary least squares yield efficient estimates of the VAR coefficients. One difficulty with VAR analysis is that the underlying structural model cannot be recovered from estimated VAR. An arbitrary Choleski decomposition provides an extra equation necessary for identification of the structural model. For each variable in the system, innovation accounting techniques can be used to ascertain: (i) the percentage of the forecast error variance attributable to each of the other variables, and (ii) the impulse responses to the various innovations. This technique was illustrated by examining the relationship between terrorism and tourism in Spain.

An important development is the convergence of traditional economic theory and the VAR framework. Structural VARs impose an economic model on the contemporaneous movements of the variables. As such, they allow for the identification of the parameters of the economic model and the structural shocks. The Sims–Bernanke procedure can be used to identify (or overidentify) the structural innovations. The Blanchard and Quah methodology imposes long-run restrictions on the impulse response functions to exactly identify the structural innovations. An especially useful feature of the technique is that it provides a unique decomposition of an economic time series into its temporary and permanent components.

Nevertheless, as summarized in an interesting paper by Todd (1990), VAR results may not be robust to reasonable changes in the model's specification. Sometimes, the addition of a time trend, changing the lag length, eliminating a variable from the model, or changing the frequency of the data from monthly to quarterly can alter the results of a VAR. Similarly, using several plausible ways to measure a variable (e.g., using one short-term interest rate instead of another) might lead to different impulse responses or variance decompositions. As such, you need to be careful in estimating a VAR. Some suggestions are:

1. Select your variables carefully. Use the variables that most accurately measure the phenomena of interest. Moreover, incorporating extraneous variables will quickly consume degrees of freedom. Omitting important variables will prevent you from interpreting your impulse responses and variance decompositions properly.
2. You should have some idea as to whether or not the variables in question are stationary, trend-stationary, or difference-stationary. Granger causality tests can be meaningless if they involve nonstationary variables. You do not want to include a time trend unless the variables actually contain a deterministic trend. Moreover, the impulse response functions involving nonstationary variables can have very large standard errors.
3. Be sure to perform robustness checks. Todd (1990), for example, checked the robustness of Sims's results using three different measures of the money supply and

two different interest rate series. He also obtained results with and without a trend. The point is to try a number of reasonable specifications. Compare several different performance measures of the alternative specifications (such as fit, impulse responses, and variance decompositions). Maintain a healthy skepticism of any conclusions if the results from the alternative estimations are very different.

QUESTIONS AND EXERCISES

1. Consider three forms of the intervention variable:

pulse: $z_1 = 1$ and all other $z_i = 0$
pure jump: $z_1 = z_2 = \ldots = 1$ and all other $z_i = 0$ for $i > 10$
prolonged impulse: $z_1 = 1$; $z_2 = 0.75$; $z_3 = 0.5$; $z_4 = 0.25$; and all other values of $z_i = 0$

a. Show how each of the following $\{y_t\}$ sequences responds to the three types of interventions:

i. $y_t = 0.5y_{t-1} + z_t + \varepsilon_t$
ii. $y_t = -0.5y_{t-1} + z_t + \varepsilon_t$
iii. $y_t = 1.25y_{t-1} - 0.5y_{t-2} + z_t + \varepsilon_t$
iv. $y_t = y_{t-1} + z_t + \varepsilon_t$
v. $y_t = 0.75y_{t-1} + 0.25y_{t-2} + z_t + \varepsilon_t$

b. Notice that the intervention models in iv and v have unit roots. Show that the intervention variable $z_1 = 1$, $z_2 = -1$, and all other values of $z_i = 0$ has only a temporary effect on these two sequences.

c. Show that an intervention variable will not have a permanent effect on a unit root process if all values of z_i sum to zero.

d. Discuss the plausible models you might choose if the $\{y_t\}$ sequence is:

i. stationary and you suspect that the intervention has a growing and then a diminishing effect.

ii. nonstationary and you suspect that the intervention has a permanent effect on the level of $\{y_t\}$.

iii. nonstationary and you suspect that the intervention has a temporary effect on the level of the $\{y_t\}$.

iv. nonstationary and you suspect that the intervention increases the trend growth of $\{y_t\}$.

2. Former KGB General Sakharovsky has been quoted as saying, "In today's world, when nuclear arms have made military force obsolete, terrorism should become our main weapon." Now, most analysts believe that the end of the Cold War brought about a dramatic decline in state-sponsored terrorism. If you examine Figure 5.1, you can see that the total number of incidents begins to fall in the early 1990s. If you examine the figure closely, you will also see that the proportion of bombing incidents also seems to have declined at the same time that the Soviet Union fell. The data set INCIDENTS.XLS contains the quarterly values of various types of transnational terrorist incidents over the 1968*Q*2–2006*Q*4 period. The precise definition of the variables is discussed in Liu, Enders, and Prodan (2009).

a. Create the series $\{y_t\}$, the proportion of bombing incidents, by dividing the series labeled BOMBINGS by the series labeled TOTAL. The first step in estimating an intervention model is to examine the ACF and PACF of the $\{y_t\}$ series for the

1968Q2–1991Q4 period and try to identify a plausible set of models. Since 1991Q4 occurs near the middle of the data set, it is also reasonable to examine the ACF and PACF for the 1992Q1–2006Q4 period. What models for $\{y_t\}$ seem most promising?

b. Jennifer created the dummy variable z_t to represent the decline in the state sponsorship of terrorism. Specifically, she let $z_t = 1$ after 1991Q4 and $z_t = 0$ for $t \leq$ 1991Q4. She then estimated the two models

$$y_t = \underset{(60.53)}{0.659} - \underset{(-7.53)}{0.132}z_t \quad \text{and} \quad y_t = \underset{(10.28)}{0.567} + \underset{(1.66)}{0.136}y_{t-1} - \underset{(-5.37)}{0.111}z_t$$

Estimate the two models and determine which seems to be the most satisfactory.

c. Justin, who never liked to take advice, ignored step 1 of the methodology used to estimate transfer functions and looked at the ACF and PACF for the entire sample period. Why might Justin conclude that the y_t series is very persistent?

d. Justin thought that an ARMA(1, 1) model could adequately capture the apparent persistence of the $\{y_t\}$ series. He estimated

$$y_t = \underset{(2.50)}{0.226} + \underset{(5.41)}{0.731}y_{t-1} - \underset{(-5.59)}{0.120}z_t - \underset{(-4.17)}{0.667}\varepsilon_{t-1} + \varepsilon_t$$

In what important ways are Jennifer's and Justin's findings for the long-run effects of z_t on y_t quite different? How does the common factor problem (see Chapter 2) explain the error in Justin's equation?

3. Let the realized value of the $\{z_t\}$ sequence be such that $z_1 = 1$ and all other values of $z_i = 0$.

a. Use equation (5.11) to trace out the effects of the $\{z_t\}$ sequence on the time path of y_t.

b. Use equation (5.12) to trace out the effects of the $\{z_t\}$ sequence on the time paths of y_t and Δy_t.

c. Use equation (5.13) to trace out the effects of the $\{z_t\}$ sequence on the time paths of y_t and Δy_t.

d. Would your answers to parts a through c change if $\{z_t\}$ was assumed to be a white-noise process and you were asked to trace out the effects of a z_t shock on $\{y_t\}$ and $\{\Delta y_t\}$?

e. Assume that $\{z_t\}$ is a white-noise process with a variance equal to unity.

i. Use (5.11) to derive the cross-correlogram between $\{z_t\}$ and $\{y_t\}$.

ii. Use (5.12) to derive the cross-correlogram between $\{z_t\}$ and $\{\Delta y_t\}$.

iii. Use (5.13) to derive the cross-correlogram between $\{z_t\}$ and $\{\Delta y_t\}$.

iv. Now suppose that z_t is the random walk process $z_t = z_{t-1} + \varepsilon_{zt}$. Trace out the effects of an ε_{zt} shock in (5.11), (5.12), and (5.13).

4. Consider the transfer function model $y_t = 0.5y_{t-1} + z_t + \varepsilon_t$ where z_t is the autoregressive process $z_t = 0.5z_{t-1} + \varepsilon_{zt}$.

a. Derive the cross-correlations between the filtered $\{y_t\}$ sequence and the $\{\varepsilon_{zt}\}$ sequence.

b. Now suppose $y_t = 0.5y_{t-1} + z_t + 0.5z_{t-1} + \varepsilon_t$ and $z_t = 0.5z_{t-1} + \varepsilon_{zt}$. Derive the standardized cross-covariances between the filtered $\{y_t\}$ sequence and ε_{zt}. Show that the first and second cross-covariances are proportional to the cross-correlations. Show that the cross-covariances decay at the rate 0.5.

5. Use (5.28) to find the appropriate second-order stochastic difference equation for y_t.

$$\begin{bmatrix} y_t \\ z_t \end{bmatrix} = \begin{bmatrix} 0.8 & 0.2 \\ 0.2 & 0.8 \end{bmatrix} \begin{bmatrix} y_{t-1} \\ z_{t-1} \end{bmatrix} + \begin{bmatrix} e_{1t} \\ e_{2t} \end{bmatrix}$$

a. Determine whether the $\{y_t\}$ sequence is stationary.

b. Discuss the shape of the impulse response function of y_t to a one-unit shock in e_{1t} and to a one-unit shock in e_{2t}.

c. Suppose $e_{1t} = \varepsilon_{yt} + 0.5\varepsilon_{zt}$ and that $e_{2t} = \varepsilon_{zy}$. Discuss the shape of the impulse response function of y_t to a one-unit shock in ε_{yt}. Repeat for a one-unit shock in ε_{zt}.

d. Suppose that $e_{1t} = \varepsilon_{yt}$ and $e_{2t} = 0.5\varepsilon_{zy} + \varepsilon_{zt}$. Discuss the shape of the impulse response function of y_t to a one-unit shock in ε_{yt}. Repeat for a one-unit shock in ε_{zt}.

e. Use your answers to c and d to explain why the ordering in a Choleski decomposition is important.

f. Using the notation in (5.27), find A_1^2 and A_1^3. Does A_1^n appear to approach zero (i.e., the null matrix)?

6. Using the notation of (5.20) and (5.21), suppose $a_{10} = 0$, $a_{20} = 0$, $a_{11} = 0.8$, $a_{12} = 0.2$, $a_{21} = 0.4$, and $a_{22} = 0.1$.

a. Find the appropriate second-order stochastic difference equation for y_t. Determine whether the $\{y_t\}$ sequence is stationary.

b. Answer parts b through f of Question 5 using these new values of the a_{ij}.

c. How would the solution for y_t change if $a_{10} = 0.2$?

7. Suppose the residuals of a VAR are such that $\text{var}(e_1) = 0.75$, $\text{var}(e_2) = 0.5$, and $\text{cov}(e_{1t}, e_{2t}) = 0.25$.

a. Using (5.55) through (5.58) as guides, show that it is not possible to identify the structural VAR.

b. Using Choleski decomposition such that $b_{12} = 0$, find the identified values of b_{21}, $\text{var}(\varepsilon_1)$, and $\text{var}(\varepsilon_2)$.

c. Using Choleski decomposition such that $b_{21} = 0$, find the identified values of b_{12}, $\text{var}(\varepsilon_1)$, and $\text{var}(\varepsilon_2)$.

d. Using a Sims–Bernanke decomposition such that $b_{12} = 0.5$, find the identified values of b_{21}, $\text{var}(\varepsilon_1)$, and $\text{var}(\varepsilon_2)$.

e. Using a Sims–Bernanke decomposition such that $b_{21} = 0.5$, find the identified values of b_{12}, $\text{var}(\varepsilon_1)$, and $\text{var}(\varepsilon_2)$.

f. Suppose that the first three values of e_{1t} are estimated to be $1, 0, -1$ and that the first three values of e_{2t} are estimated to be $-1, 0, 1$. Find the first three values of ε_{1t} and ε_{2t} using each of the decompositions in parts b through e.

8. This set of exercises uses data from the file entitled QUARTELY.XLS in order to estimate the dynamic interrelationships among the level of industrial production, the unemployment rate, and interest rates. In Chapter 2, you created the interest rate spread (s_t) as the difference between the 10-year rate and the T-bill rate. Now create the logarithmic change in the index of industrial production (ip) as $\Delta lip_t = \ln(ip_t) - \ln(ip_{t-1})$ and the seasonal difference of the unemployment rate as $\Delta_4 ur_t = ur_t - ur_{t-4}$.

a. Estimate the three-variable VAR using eight lags of each variable and a constant and save the residuals. Explain why the estimation cannot be begin earlier than

1963Q1. What are the potential advantages of using the variables Δlip_t and $\Delta_4 ur_t$ instead of ip_t and ur_t?

b. Verify that $\ln(|\Sigma_8|) = -13.968$ and (assuming normality) that the log of the likelihood function is 493.647. Calculate the multivariate AIC and SBC using the formulas AIC $= T\ln(|\Sigma|) + 2N$ and SBC $= T\ln(|\Sigma|) + N\ln(T)$. Calculate the multivariate AIC and SBC using the formulas AIC$^* = -2\ln(L)/T + 2n/T$ and SBC$^* = -2\ln(L)/T + n\ln(T)/T$.

c. Estimate the model using three lags of each variable and save the residuals. Show that the AIC selects the eight-lag model and that the SBC selects the three-lag model. Show that the same ambiguity applies to the AIC* and SBC*. Why is it important to estimate the three-variable VAR beginning with 1963Q1?

d. Construct the likelihood ratio test for the null hypothesis of eight lags against the alternative of three lags. How many restrictions are there in the system? How many regressors are there in each of the unrestricted equations? If you answer correctly, you should find that the calculated value χ^2 with 45 degrees of freedom is 95.20, with a significance level smaller than 0.0001. Hence, the restriction of three lags is binding.

e. Now estimate the model with six lags. You should find that the likelihood ratio test selects the eight-lag model, the AIC selects the six-lag model, and the SBC selects the three-lag model.

9. Question 8 indicates that a three-lag VAR seems reasonable for the variables Δlip_t, $\Delta_4 ur_t$, and y_t. Estimate the three-VAR beginning in 1961Q4 and use the ordering such that Δlip_t is causally prior to $\Delta_4 ur_t$ and that $\Delta_4 ur_t$ is causally prior to y_t.

a. If you perform a test to determine whether y_t Granger causes Δlip_t, you should find that the F-statistic is 3.09 with a *prob*-value of 0.0098. How do you interpret this result? (Note: Some software packages will report that the sample χ^2 statistic is 11.71.)

b. Verify that $\Delta_4 ur_t$ does not Granger cause Δlip_t. You should find that the F-statistic is 1.7018 with a *prob*-value of 0.168.

c. It turns out that the correlation coefficient between e_{1t} and e_{2t} is -0.597. The correlation between e_{1t} and e_{3t} is -0.231 and between e_{2t} and e_{3t} is 0.223. Explain why the ordering in a Choleski decomposition is likely to be important for obtaining the impulse responses.

d. Verify that the forecast error variance decompositions are:

	Response to Δlip_t shock			Response to $\Delta_4 ur_t$ shock			Response to y_t shock		
Horizon	Δlip_t	$\Delta_4 ur_t$	y_t	Δlip_t	$\Delta_4 ur_t$	y_t	Δlip_t	$\Delta_4 ur_t$	y_t
1	100.00%	35.63%	5.37%	0.00%	64.36%	1.12%	0.00%	0.00%	93.51%
4	93.60	71.53	20.77	1.66	27.46	1.35%	4.78	1.01	77.88
8	89.15	58.05	37.57	1.86	20.31	1.25%	8.99	21.64	61.18

e. Obtain the impulse response functions. Show that a positive shock to industrial production induces a decline in the unemployment rate and in the long-term interest rate relative to the short-term rate.

10. The data set MONEY_DEM.XLS contains real U.S. GDP (RGDP), nominal GDP, the money supply as measured by M2, and the three-month rate on U.S. Treasury bills. As described in Chapter 2 of the *Programming Manual,* construct the following four variables:

$$dlrgdp_t = \ln(\text{RGDP}_t) - \ln(\text{RGDP}_{t-1})$$
$$price_t = \text{GDP}_t/\text{RGDP}_t$$
$$dlrm2_t = \ln(\text{M2}_t/price_t) - \ln(\text{M2}_{t-1}/price_{t-1})$$
$$drs = tb3mo_t - tb3mo_{t-1}$$

Hence, *dlrgdp* is the logarithmic change in real GDP, *price* is the GDP deflator, *dlrm2* is the logarithmic change in the real money supply, and *drs* is the change in the short-term interest rate.

a. Estimate a three-variable VAR with 12 lags of $dlrdgp_t$, $dlrm2_t$, and drs_t. Include a constant but do not use any seasonal dummy variables.

b. Calculate the multivariate AIC and SBC. Note that there are 37 coefficients (12 lags of three variables plus a constant) in each equation, for a total of 111 estimated coefficients in the system.

c. Reestimate the model using eight lags of each variable. Be sure to estimate the system over the same period used in part a. Note that the multivariate AIC selects the 12-lag model and the SBC selects the eight-lag model. Test for the appropriate lag length using the likelihood ratio test. Given the calculated value of $\chi^2 = 55.33$, the restriction is binding at the 5 percent (but not the 1 percent) significance level.

d. Using the 12-lag model, show that drs_t is not block exogenous for the other two variables in the system.

e. Perform the Granger causality tests.

f. Linda obtained the impulse response functions (for 12 steps) with two standard deviation confidence intervals surrounding each. The results are shown in Figure 5.10 on the next page.

  i. What was the ordering Linda used for the Choleski decomposition?

  ii. The vertical axis contains the magnitude of the effects. Why do you suppose that many econometricians prefer to scale the responses to a variable by the standard deviation of its own shock? What is the standard deviation of the $dlrgdp_t$ shock?

  iii. What variables have significant short-term responses for the other variables in the system?

g. Set up a near-VAR such that the lags of $dlrgdp_t$ are excluded from the equation for $dlrgdp_t$ and $dlrm2_t$. How does the near-VAR perform relative to the unrestricted VAR?

11. Question 10 suggests that it is appropriate to estimate three-variable, 12-lag VAR for $dlrdgp_t$, $dlrm2_t$, and drs_t. Now suppose we want the contemporaneous relationships among the variables to be

$$\begin{bmatrix} e_{yt} \\ e_{mt} \\ e_{rt} \end{bmatrix} = \begin{bmatrix} 1 & 0 & 0 \\ g_{21} & 1 & g_{23} \\ 0 & 0 & 1 \end{bmatrix} \begin{bmatrix} \varepsilon_{yt} \\ \varepsilon_{mt} \\ \varepsilon_{rt} \end{bmatrix}$$

where e_{yt}, e_{mt}, and e_{rt} are the regression residual from the $dlrgdp_t$, $dlrm2_t$, and drs_t equations, and ε_{yt}, ε_{mt}, and ε_{rt} are the pure shocks (i.e., the structural innovations) to $dlrgdp_t$, $dlrm2_t$, and drs_t, respectively.

a. Provide a plausible economic interpretation of this set of restrictions.

b. Is the system identified?

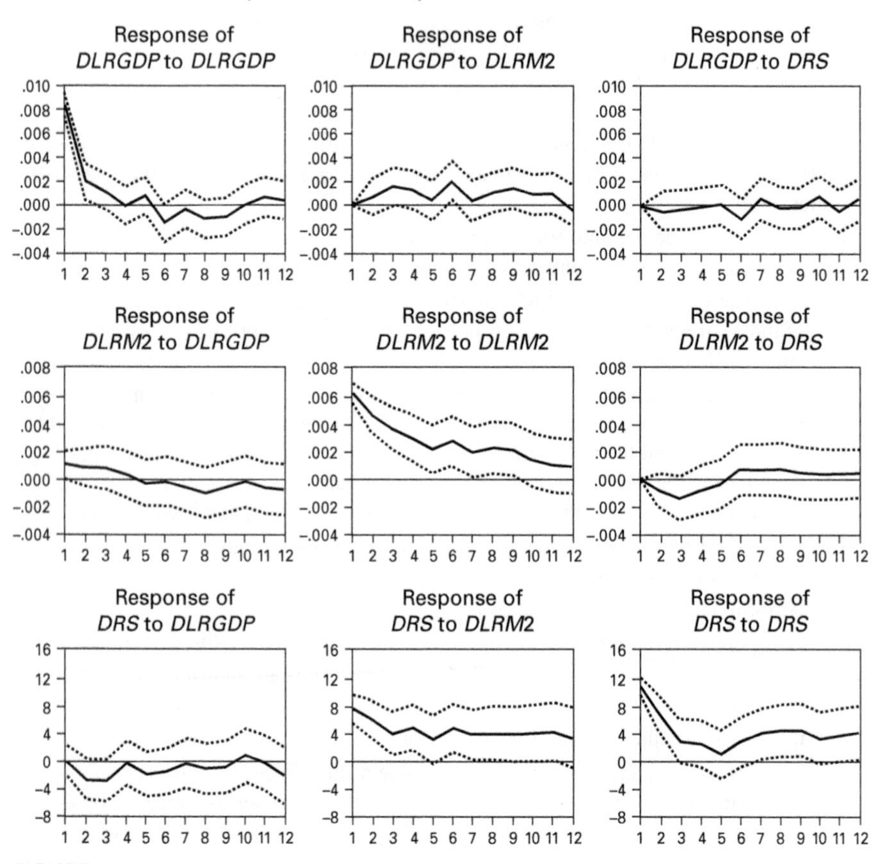

FIGURE 5.10 Impulse Responses for the Three-Variable VAR

c. Verify that the impulse responses of an interest rate shock are those provided in the *Programming Manual*. Compare these results to Linda's results shown in Figure 5.10.

ENDNOTES

1. In terms of the notation of Chapter 4, z_t is equivalent to the level dummy variable D_L.
2. In other words, if $c_0 \neq 0$, predicting y_{t+1} necessitates predicting the value of z_{t+1}.
3. In the identification process, we are primarily interested in the shape, not the height, of the cross-correlation function. It is useful to standardize the covariance by dividing through by σ_z^2; the shape of the correlogram is proportional to the standardized covariance. The text follows the procedure used by most software packages by plotting the standardized cross-covariances.
4. The discussion in the text assumes that $\{y_t\}$ and $\{z_t\}$ are both stationary processes. In other circumstances, Box and Jenkins (1976) recommend differencing y_t and/or z_t so that the resulting series are both stationary. The modern view cautions against this approach; as shown in the next chapter, a

linear combination of nonstationary variables may be stationary. In such circumstances, the Box–Jenkins recommendation leads to overdifferencing.

5. We were able to obtain quarterly data from 1970*Q*1 to 1988*Q*4 for Austria, Canada, Denmark, Finland, France, West Germany, Greece, Italy, the Netherlands, Norway, the United Kingdom, and the United States. The International Monetary Fund's *Balance of Payments Statistics* reports all data in Special Drawing Rights (SDR). The dependent variables were the logarithms of each nation's revenues divided by the sum of the revenues for all twelve countries.
6. Tourism is highly seasonal; we tried several alternative deseasonalization techniques. The results reported here were obtained using seasonal dummy variables. Hence, y_t represents the deseasonalized logarithmic share of tourism receipts. The published paper reports results using quarterly differencing. Using either type of deseasonalization, the final results were similar.
7. Expectations of the future can also be included in this framework. If the temperature $\{y_t\}$ is an autoregressive process, the expected value of the next period's temperature (i.e., y_{t+1}) will depend on the current and past values. In (5.18), the presence of the terms y_t and y_{t-1} can represent how predictions of the next period's temperature affect the current thermostat setting.
8. It is easily verified that this representation implies that $\rho_{12} = 0.8$. By definition, the correlation coefficient ρ_{12} is defined to be $\sigma_{12}/(\sigma_1\sigma_2)$ and the covariance is $Ee_{1t}e_{2t} = \sigma_{12}$. If we use the numbers in the example, $Ee_1e_{2t} = E[\varepsilon_{zt}(\varepsilon_{yt} + 0.8\varepsilon_{zt})] = 0.8\sigma_z^2$. Since the decomposition equates var(e_{2t}) with σ_z^2, it follows that $\rho_{12} = 0.8$ if $\sigma_1^2 = \sigma_2^2$.
9. Other types of identification restrictions are discussed in Sections 10 through 13.
10. Note that γ_1 cannot equal zero if $\{y_t\}$ is $I(1)$. If $\gamma_1 = 0$ and $y_t = a_{12}\Delta y_{t-1} + b_{12}\Delta z_{t-1} + \varepsilon$, the equation is unbalanced. The left-hand side contains the $I(1)$ variable y_t while the right-hand side contains only the three stationary variables Δy_t, Δz_t, and ε_t.
11. In the example under consideration, the symmetry restriction on the coefficients means that var(ε_{1t}) is equal to var(ε_{2t}). This result does not generalize; it holds in the example because of the assumed equality var(e_{1t}) = var(e_{2t}).
12. The value $|\Sigma_R| - |\Sigma|$ is asymptotically distributed as a χ^2 distribution with R degrees of freedom.
13. Since a key assumption of the model of the technique is that $E(\varepsilon_{1t}\varepsilon_{2t}) = 0$, you might wonder how it is possible to assume that aggregate demand and supply shocks are independent. After all, if the stabilization authorities follow a feedback rule, aggregate demand will change in response to aggregate supply shocks. The key to understanding this apparent contradiction is that ε_{1t} is intended to be the orthogonalized portion of the demand shock, i.e., the portion of the demand shock that does not change in response to aggregate supply. Cover, Enders, and Hueng (2006) and Enders and Hurn (2007) show how this assumption can be relaxed.
14. For example, $A_{11}(L) = a_{11}(0) + a_{11}(1)L + a_{11}(2)L^2 + \ldots$.
15. The VAR residuals also have a constant variance/covariance matrix. Hence, the time subscripts can be dropped.
16. Since two of the restrictions contain squared terms, there will be a positive value and an equal but opposite negative value for some of the coefficients. In Blanchard and Quah's example, if $c_{11}(0)$ is positive, positive demand shocks have a positive effect on output, and if $c_{11}(0)$ is negative, the positive shock has a negative effect on output. Taylor (2003) considers the problem of selecting among the alternative sets of solutions.
17. In doing so, it will be necessary to treat all $\varepsilon_{2t-i} = 0$ for $t - i < 1$.
18. Here, the U.K. is treated as the home country so that e_t is the pound price of U.S. dollars and p_t^* refers to the U.S. price level.

CHAPTER **6**

COINTEGRATION AND ERROR-CORRECTION MODELS

This chapter explores an exciting development in econometrics: the estimation of a structural equation or a VAR containing nonstationary variables. In univariate models, we have seen that a stochastic trend can be removed by differencing. The resulting stationary series can be estimated using univariate Box–Jenkins techniques. At one time, the conventional wisdom was to generalize this idea and difference all nonstationary variables used in a regression analysis. However, the appropriate way to treat nonstationary variables is not so straightforward in a multivariate context. It is quite possible for there to be a linear combination of integrated variables that is stationary; such variables are said to be **cointegrated.** Many economic models entail such cointegrating relationships. This chapter has three aims:

1. Introduce the basic concept of cointegration and show that it applies in a variety of economic models. Any equilibrium relationship among a set of nonstationary variables implies that their stochastic trends must be linked. After all, the equilibrium relationship means that the variables cannot move independently of each other. This linkage among the stochastic trends necessitates that the variables be cointegrated.
2. Consider the dynamic paths of cointegrated variables. Since the trends of cointegrated variables are linked, the dynamic paths of such variables must bear some relation to the current deviation from the equilibrium relationship. This connection between the change in a variable and the deviation from equilibrium is examined in detail. It is shown that the dynamics of a cointegrated system are such that the conventional wisdom was incorrect. After all, if the linear relationship is already stationary, differencing the relationship entails a misspecification error.
3. Study the alternative ways to test for cointegration. The econometric methods underlying the test procedures stem from the theory of simultaneous difference equations. The theory is explained and used to develop the three most popular cointegration tests. Although the mathematics can get quite difficult, Appendix 6.1 contains the necessary details.
4. Illustrate the various ways to estimate a system of cointegrated variables. Several examples of each testing methodology are provided. Moreover, the various methods are compared by applying each to the same data set.

1. LINEAR COMBINATIONS OF INTEGRATED VARIABLES

Since money demand studies have stimulated much of the cointegration literature, we begin by considering a simple model of money demand. Theory suggests that individuals want to hold a real quantity of money balances, so that the demand for nominal money holdings should be proportional to the price level. Moreover, as real income and the associated number of transactions increase, individuals will want to hold increased money balances. Finally, since the interest rate is the opportunity cost of holding money, money demand should be negatively related to the interest rate. In logarithms, an econometric specification for such an equation can be written as

$$m_t = \beta_0 + \beta_1 p_t + \beta_2 y_t + \beta_3 r_t + e_t \qquad (6.1)$$

where: m_t = demand for money
p_t = price level
y_t = real income
r_t = interest rate
e_t = *stationary* disturbance term
β_i = parameters to be estimated

and all variables but the interest rate are expressed in logarithms.

The hypothesis that the money market is in equilibrium allows the researcher to collect time-series data of the money supply (= money demand if the money market always clears), the price level, real income (possibly measured using real GDP), and an appropriate short-term interest rate. The behavioral assumptions require that $\beta_1 = 1$, $\beta_2 > 0$, and $\beta_3 < 0$; a researcher conducting such a study would certainly want to test these parameter restrictions. Be aware that the properties of the unexplained portion of the demand for money (i.e., the $\{e_t\}$ sequence) are an integral part of the theory. If the theory is to make any sense at all, any deviation in the demand for money must necessarily be temporary in nature. Clearly, if e_t has a stochastic trend, the errors in the model will be cumulative so that deviations from money market equilibrium will not be eliminated. Hence, a key assumption of the theory is that the $\{e_t\}$ sequence is stationary.

The problem confronting the researcher is that real GDP, the money supply, price level, and interest rate can all be characterized as nonstationary $I(1)$ variables. As such, each variable can meander without any tendency to return to a long-run level. However, the theory expressed in (6.1) asserts that there exists a linear combination of these nonstationary variables that is stationary! Solving for the error term, we can rewrite (6.1) as

$$e_t = m_t - \beta_0 - \beta_1 p_t - \beta_2 y_t - \beta_3 r_t \qquad (6.2)$$

Since $\{e_t\}$ must be stationary, it follows that the linear combination of integrated variables given by the right-hand side of (6.2) must also be stationary. Thus, the theory necessitates that the time paths of the four nonstationary variables $\{m_t\}$, $\{p_t\}$, $\{y_t\}$, and $\{r_t\}$ be linked. This example illustrates the crucial insight that has dominated much of the macroeconometric literature in recent years: *Equilibrium theories involving nonstationary variables require the existence of a combination of the variables that is stationary.*

The money demand function is just one example of a stationary combination of nonstationary variables. Within any equilibrium framework, the deviations from equilibrium must be temporary. Other important economic examples involving stationary combinations of nonstationary variables include:

1. *Consumption function theory.* A simple version of the permanent income hypothesis maintains that total consumption (c_t) is the sum of permanent consumption (c_t^p) and transitory consumption (c_t^t). Since permanent consumption is proportional to permanent income (y_t^p), we can let β be the constant of proportionality and write $c_t = \beta y_t^p + c_t^t$. Transitory consumption is necessarily a stationary variable, and both consumption and permanent income are reasonably characterized as $I(1)$ variables. As such, the permanent income hypothesis requires that the linear combination of two $I(1)$ variables given by $c_t - \beta y_t^p$ be stationary.
2. *Unbiased forward rate hypothesis.* One form of the efficient market hypothesis asserts that the forward (or futures) price of an asset should equal the expected value of that asset's spot price in the future. Foreign exchange market efficiency requires that the one-period forward exchange rate equal the expectation of the spot rate in the next period. Letting f_t denote the log of the one-period price of forward exchange in t and s_t the log of the spot price of foreign exchange in t, the theory asserts that $E_t s_{t+1} = f_t$. If this relationship fails, speculators can expect to make a pure profit on their trades in the foreign exchange market. If agents' expectations are rational, the forecast error for the spot rate in $t + 1$ will have a conditional mean equal to zero, so that $s_{t+1} - E_t s_{t+1} = \varepsilon_{t+1}$ where $E_t \varepsilon_{t+1} = 0$. Combining the two equations yields $s_{t+1} = f_t + \varepsilon_{t+1}$. Since $\{s_t\}$ and $\{f_t\}$ are $I(1)$ variables, the **unbiased forward rate hypothesis** necessitates that there be a linear combination of nonstationary spot and forward exchange rates that is stationary.
3. *Commodity market arbitrage and purchasing power parity.* Theories of spatial competition suggest that in the short run, prices of similar products in varied markets might differ. However, arbiters will prevent the various prices from moving too far apart even if the prices are nonstationary. Similarly, the prices of PCs and Apple computers have exhibited sustained declines. Economic theory suggests that these simultaneous declines are related to each other since a price discrepancy between these similar products cannot continually widen. Also, as we saw in Chapter 4, purchasing power parity places restrictions on the movements of nonstationary price levels and exchange rates. If e_t denotes the log of the price of foreign exchange and p_t and p_{ft} denote, respectively, the logs of domestic and foreign price levels, long-run PPP requires that the linear combination $e_t + p_{ft} - p_t$ be stationary.

All of these examples illustrate the concept of **cointegration** as introduced by Engle and Granger (1987). Their formal analysis begins by considering a set of economic variables in long-run equilibrium when

$$\beta_1 x_{1t} + \beta_2 x_{2t} + \ldots + \beta_n x_{nt} = 0$$

Letting β and x_t denote the vectors $(\beta_1, \beta_2, \ldots, \beta_n)$ and $(x_{1t}, x_{2t}, \ldots, x_{nt})'$, the system is in long-run equilibrium when $\beta x_t = 0$. The deviation from long-run equilibrium—called the **equilibrium error**—is e_t, so that

$$e_t = \beta x_t$$

If the equilibrium is meaningful, it must be the case that the equilibrium error process is stationary. In a sense, the use of the term *equilibrium* is unfortunate because economic theorists and econometricians use the term in different ways. Economic theorists usually use the term to refer to an equality between desired and actual transactions. The econometric use of the term makes reference to any long-run relationship among nonstationary variables. Cointegration does not require that the long-run relationship be generated by market forces or by the behavioral rules of individuals. In Engle and Granger's use of the term, the equilibrium relationship may be causal, behavioral, or simply a reduced-form relationship among similarly trending variables. Engle and Granger (1987) provided the following definition of cointegration:

The components of the vector $x_t = (x_{1t}, x_{2t}, \ldots, x_{nt})'$ are said to be *cointegrated of order d, b*, denoted by $x_t \sim CI(d, b)$ if

1. All components of x_t are integrated of order d.
2. There exists a vector $\beta = (\beta_1, \beta_2, \ldots, \beta_n)$ such that the linear combination $\beta x_t = \beta_1 x_{1t} + \beta_2 x_{2t} + \ldots + \beta_n x_{nt}$ is integrated of order $(d - b)$ where $b > 0$.

The vector β is called the **cointegrating vector.**[1]

In terms of equation (6.1), if the money supply, price level, real income, and interest rate are all $I(1)$ and the linear combination $m_t - \beta_0 - \beta_1 p_t - \beta_2 y_t - \beta_3 r_t = e_t$ is stationary, then the variables are cointegrated of order (1, 1). The vector x_t is $(m_t, 1, p_t, y_t, r_t)'$ and the cointegrating vector β is $(1, -\beta_0, -\beta_1, -\beta_2, -\beta_3)$. The deviation from long-run money market equilibrium is e_t; since $\{e_t\}$ is stationary, this deviation is temporary in nature.

There are four important points to note about the definition:

1. Cointegration typically refers to a *linear* combination of nonstationary variables. Theoretically, it is quite possible that nonlinear long-run relationships exist among a set of integrated variables. However, as discussed in Chapter 7, the current state of econometric practice is just beginning to allow for tests of nonlinear cointegrating relationships. Also note that the cointegrating vector is not unique. If $(\beta_1, \beta_2, \ldots, \beta_n)$ is a cointegrating vector, then for any nonzero value of λ, $(\lambda\beta_1, \lambda\beta_2, \ldots, \lambda\beta_n)$ is also a cointegrating vector. Typically, one of the variables is used to *normalize* the cointegrating vector by fixing its coefficient at unity. To normalize the cointegrating vector with respect to x_{1t}, simply select $\lambda = 1/\beta_1$.

2. From Engle and Granger's original definition, cointegration refers to variables that are integrated of the same order. Of course, this does not imply that all integrated variables are cointegrated; usually, a set of $I(d)$ variables is *not* cointegrated. Such a lack of cointegration implies no long-run equilibrium among the variables, so that they can wander arbitrarily far from each other. If two variables are integrated of different orders, they cannot be cointegrated. Suppose x_{1t} is $I(d_1)$ and x_{2t} is $I(d_2)$ where $d_2 > d_1$. Question 6 at the end of this chapter asks you to prove that any linear combination of x_{1t} and x_{2t} is $I(d_2)$.

   Nevertheless, it is possible to find equilibrium relationships among groups of variables that are integrated of different orders. Suppose that x_{1t} and x_{2t} are $I(2)$ and that the other variables under consideration are $I(1)$. As such, there cannot be a cointegrating relationship between x_{1t} (or x_{2t}) and x_{3t}. However, if x_{1t} and x_{2t} are $CI(2, 1)$, there exists a linear combination of the form $\beta_1 x_{1t} + \beta_2 x_{2t}$ which is $I(1)$. It is possible that *this* combination of x_{1t} and x_{2t} is cointegrated with the $I(1)$ variables. Lee and Granger (1990) use the term **multicointegration** to refer to this type of circumstance.

3. If x_t has n nonstationary components, there may be as many as $n - 1$ linearly independent cointegrating vectors. Clearly, if x_t contains only two variables, there can be *at most* one independent cointegrating vector. The number of cointegrating vectors is called the **cointegrating rank** of x_t. For example, suppose that the monetary authorities followed a feedback rule such that they decreased the money supply when nominal GDP was high and increased the nominal money supply when nominal GDP was low. This feedback rule might be represented by

$$\begin{aligned} m_t &= \gamma_0 - \gamma_1(y_t + p_t) + e_{1t} \\ &= \gamma_0 - \gamma_1 y_t - \gamma_1 p_t + e_{1t} \end{aligned} \tag{6.3}$$

   where $\{e_{1t}\}$ = a stationary error in the money supply feedback rule.

   Given the money demand function in (6.1), there are two cointegrating vectors for the money supply, price level, real income, and interest rate. Let β be the $(2 \cdot 5)$ matrix:

$$\beta = \begin{bmatrix} 1 & -\beta_0 & -\beta_1 & -\beta_2 & -\beta_3 \\ 1 & -\gamma_0 & \gamma_1 & \gamma_1 & 0 \end{bmatrix}$$

   The two linear combinations given by βx_t are stationary. As such, the cointegrating rank of x_t is 2. As a practical matter, if multiple cointegrating vectors are found, it may not be possible to identify the behavioral relationships from what may be reduced-form relationships.

4. Most of the cointegration literature focuses on the case in which each variable contains a single unit root. The reason is that traditional regression or time-series analysis applies when variables are $I(0)$ and few economic variables are integrated of an order higher than unity.[2] When it is unambiguous, many authors use the term *cointegration* to refer to the case in which variables are $CI(1, 1)$.

WORKSHEET 6.1

ILLUSTRATING COINTEGRATED SYSTEMS

CASE 1: The series $\{\mu_t\}$ is a random walk process and $\{\varepsilon_{yt}\}$ and $\{\varepsilon_{zt}\}$ are white noise. Hence, the $\{y_t\}$ and $\{z_t\}$ sequences are both random walk plus noise processes. Although each is nonstationary, the two sequences have the same stochastic trend; hence they are cointegrated such that the linear combination $(y_t - z_t)$ is stationary. The equilibrium error term $(\varepsilon_{yt} - \varepsilon_{zt})$ is an $I(0)$ process.

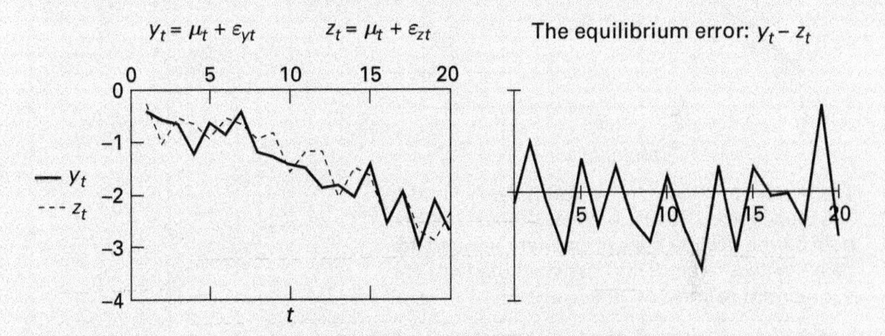

CASE 2: All three sequences are random walk plus noise processes. As constructed, no two are cointegrated. However, the linear combination $(y_t + z_t - w_t)$ is stationary; hence, the three variables are cointegrated. The equilibrium error is an $I(0)$ process.

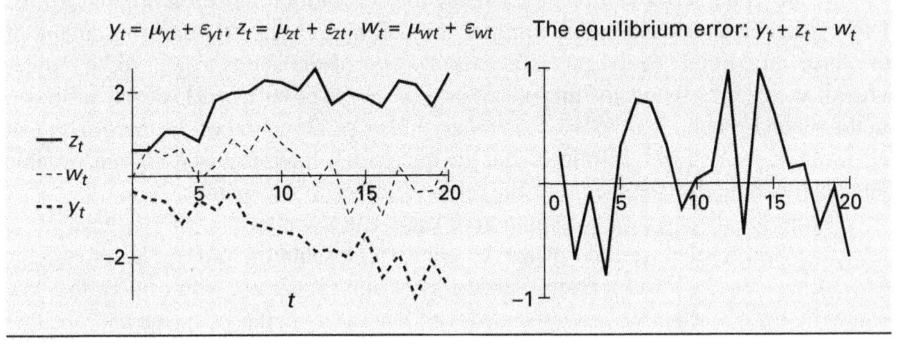

Worksheet 6.1 illustrates some of the important properties of cointegration relationships. In Case 1, both the $\{y_t\}$ and $\{z_t\}$ sequences were constructed so as to be random walk plus noise processes. Although the 20 realizations shown generally decline, extending the sample would eliminate this tendency. In any event, neither series shows any tendency to return to a long-run level, and formal Dickey–Fuller tests are not able to reject the null hypothesis of a unit root in either series. Although each series is nonstationary, you can see that they do move together. In fact, the difference between y_t and z_t (i.e., $y_t - z_t$)—shown in the second graph—is stationary; the *equilibrium error* term $e_t = (y_t - z_t)$ has a zero mean and a constant variance.

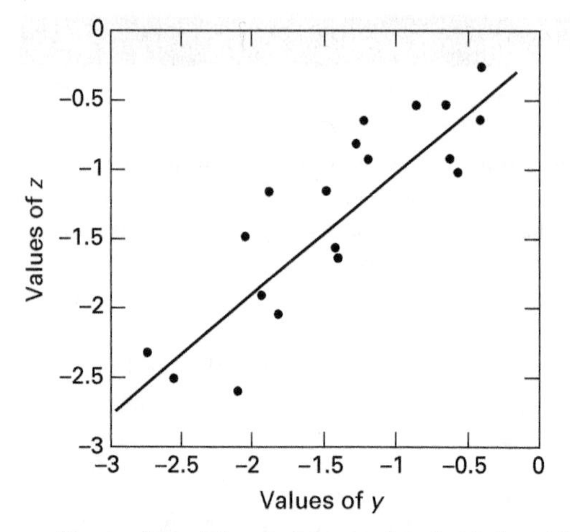

The scatter plot was drawn using the $\{y_t\}$ and $\{z_t\}$ sequences from Case 1 of Worksheet 6.1. Since both series decline over time, there appears to be a positive relationship between the two. The equilibrium regression line is shown.

FIGURE 6.1 Scatter Plot of Cointegrated Variables

Case 2 illustrates cointegration among three random walk plus noise processes. As in Case 1, no series exhibits a tendency to return to a long-run level, and formal Dickey–Fuller tests are not able to reject the null hypothesis of a unit root in any of the three. In contrast to the previous case, no two of the series appear to be cointegrated; each series seems to "meander" away from the other two. However, as shown in the second graph, there exists a stationary linear combination of the three such that $e_t = y_t + z_t - w_t$. Thus, it follows that the dynamic behavior of *at least* one variable must be restricted by the values of the other variables in the system.

Figure 6.1 displays the information of Case 1 in a scatter plot of $\{y_t\}$ against the associated value of $\{z_t\}$; each of the 20 points represents the ordered pairs (y_1, z_1), (y_2, z_2), ..., (y_{20}, z_{20}). Comparing Worksheet 6.1 and Figure 6.1, you can see that low values in the $\{y_t\}$ sequence are associated with low values in the $\{z_t\}$ sequence, and that values near zero in one series are associated with values near zero in the other. Since both series move together over time, there is a positive relationship between the two. The least-squares line in the scatter plot reveals this to be a strong positive association. In fact, this line is the "long-run" equilibrium relationship between the series, and the deviations from the line are the stationary deviations from long-run equilibrium.

For comparison purposes, Panel (a) in Worksheet 6.2 shows the time paths of two random walk plus noise processes that are not cointegrated. Each seems to meander without any tendency to approach the other. The scatter plot shown in Panel (b) confirms the impression of no long-run relationship between the variables. The deviations from the straight line showing the regression of z_t on y_t are substantial. Plotting the regression residuals against time [see Panel (c)] suggests that the regression residuals are not stationary.

WORKSHEET 6.2

NONCOINTEGRATED VARIABLES

The $\{y_t\}$ and $\{z_t\}$ sequences are constructed as independent random walk plus noise processes. There is **no** cointegrating relationship between the two variables. As shown in Panel (a), both seem to meander without any tendency to come together. Panel (b) shows the scatter plot of the two sequences and the regression line $z_t = \beta_0 + \beta_1 y_t$. However, this regression line is spurious. As shown in Panel (c), the regression residuals are nonstationary.

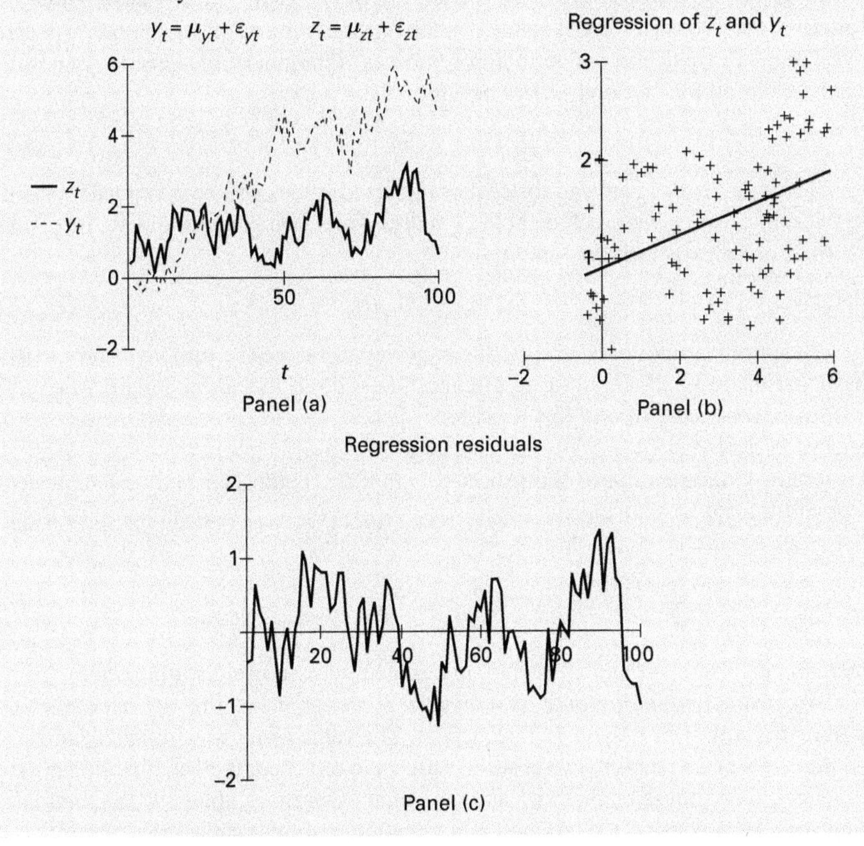

2. COINTEGRATION AND COMMON TRENDS

Stock and Watson's (1988) observation that cointegrated variables share common stochastic trends provides a very useful way to understand cointegration relationships.[3] For ease of exposition, return to the case in which the vector x_t contains only two variables so that $x_t = (y_t, z_t)'$. Ignoring cyclical and seasonal terms, we can write each variable as a random walk plus an irregular (but not necessarily a white-noise) component:

$$y_t = \mu_{yt} + e_{yt} \tag{6.4}$$

$$z_t = \mu_{zt} + e_{zt} \tag{6.5}$$

where: μ_{it} = a random walk process representing the stochastic trend in variable i
e_{it} = the stationary (irregular) component of variable i

If $\{y_t\}$ and $\{z_t\}$ are cointegrated of order (1, 1), there must be nonzero values of β_1 and β_2 for which the linear combination $\beta_1 y_t + \beta_2 z_t$ is stationary. Consider the sum

$$\begin{aligned}\beta_1 y_t + \beta_2 z_t &= \beta_1(\mu_{yt} + e_{yt}) + \beta_2(\mu_{zt} + e_{zt}) \\ &= (\beta_1 \mu_{yt} + \beta_2 \mu_{zt}) + (\beta_1 e_{yt} + \beta_2 e_{zt})\end{aligned} \tag{6.6}$$

For $\beta_1 y_t + \beta_2 z_t$ to be stationary, the term $(\beta_1 \mu_{yt} + \beta_2 \mu_{zt})$ must vanish. After all, if either of the two trends appears in (6.6), the linear combination $\beta_1 y_t + \beta_2 z_t$ will also have a trend. Since the second term in parentheses is stationary, the necessary and sufficient condition for $\{y_t\}$ and $\{z_t\}$ to be $CI(1, 1)$ is

$$\beta_1 \mu_{yt} + \beta_2 \mu_{zt} = 0 \tag{6.7}$$

Clearly, μ_{yt} and μ_{zt} are variables whose realized values will be continually changing over time. Since we preclude both β_1 and β_2 from being equal to zero, it follows that (6.7) holds for all t if and only if

$$\mu_{yt} = -\beta_2 \mu_{zt} / \beta_1$$

For nonzero values of β_1 and β_2, the only way to ensure the equality is for the stochastic trends to be *identical* up to a scalar. Thus, up to the scalar $-\beta_2/\beta_1$, *two I(1) stochastic processes* $\{y_t\}$ *and* $\{z_t\}$ *must have the same stochastic trend if they are cointegrated of order* (1, 1).

Return your attention to Worksheet 6.1. In Case 1, the $\{y_t\}$ and $\{z_t\}$ sequences were constructed so as to satisfy

$$\begin{aligned} y_t &= \mu_t + \varepsilon_{yt} \\ z_t &= \mu_t + \varepsilon_{zt} \\ \mu_t &= \mu_{t-1} + \varepsilon_t \end{aligned}$$

where ε_{yt}, ε_{zt}, and ε_t are independently distributed white-noise disturbances.

By construction, μ_t is a pure random walk process representing the same stochastic trend for both the $\{y_t\}$ and $\{z_t\}$ sequences. The value of μ_0 was initialized to zero, and three sets of 20 random numbers were drawn to represent the $\{\varepsilon_{yt}\}$, $\{\varepsilon_{zt}\}$, and $\{\varepsilon_t\}$ sequences. Using these realizations and the initial value of μ_0, the $\{y_t\}$, $\{z_t\}$, and $\{\mu_t\}$ sequences were constructed. As you can clearly determine, subtracting the realized value of z_t from y_t results in a stationary sequence:

$$y_t - z_t = (\mu_t + \varepsilon_{yt}) - (\mu_t + \varepsilon_{zt}) = \varepsilon_{yt} - \varepsilon_{zt}$$

To state the point using Engle and Granger's terminology, multiplying $\beta = (1, -1)$ by the vector $x_t = (y_t, z_t)'$ yields the stationary sequence $\varepsilon_t = \varepsilon_{yt} - \varepsilon_{zt}$. Indeed, the equilibrium error term shown in the second graph of Worksheet 6.1 has all the hallmarks of a stationary process. The essential insight of Stock and Watson (1988) is that the parameters of the cointegrating vector must be such that they purge the trend from the linear combination. Any other linear combination of the two variables contains a trend so that the cointegrating vector is unique up to a normalizing scalar. Hence, $\beta_3 y_t + \beta_4 z_t$ cannot be stationary unless $\beta_3/\beta_4 = \beta_1/\beta_2$.

Recall that Case 2 illustrates cointegration between three random walk plus noise processes. Each process is $I(1)$, and Dickey–Fuller unit root tests would not be able to reject the null hypothesis that each contains a unit root. As you can see in the lower portion of Worksheet 6.1, no pairwise combination of the series appears to be cointegrated. Each series seems to meander, but, as opposed to Case 1, no one single series appears to remain close to any other series. However, by construction, the trend in w_t is the simple summation of the trends in y_t and z_t:

$$\mu_{wt} = \mu_{yt} + \mu_{zt}$$

Here, the vector $x_t = (y_t, z_t, w_t)'$ has the cointegrating vector $(1, 1, -1)$, so that the linear combination $y_t + z_t - w_t$ is stationary. Consider:

$$y_t + z_t - w_t = (\mu_{yt} + \varepsilon_{yt}) + (\mu_{zt} + \varepsilon_{zt}) - (\mu_{wt} + \varepsilon_{wt}) = \varepsilon_{yt} + \varepsilon_{zt} - \varepsilon_{wt}$$

This example illustrates the general point that cointegration will occur whenever the trend in one variable can be expressed as a linear combination of the trends in the other variable(s). In such circumstances it is always possible to find a vector β such that the linear combination $\beta_1 y_t + \beta_2 z_t + \beta_3 w_t$ does not contain a trend. The result easily generalizes to the case of n variables. Consider the vector representation

$$x_t = \mu_t + e_t \tag{6.8}$$

where: x_t = the vector $(x_{1t}, x_{2t}, \ldots, x_{nt})'$
μ_t = the vector of stochastic trends $(\mu_{1t}, \mu_{2t}, \ldots, \mu_{nt})'$
e_t = an $n \cdot 1$ vector of stationary components

If one trend can be expressed as a linear combination of the other trends in the system, it means that there exists a vector β such that

$$\beta_1\mu_{1t} + \beta_2\mu_{2t} + \ldots + \beta_n\mu_{nt} = 0$$

Premultiply (6.8) by this set of β_is to obtain

$$\beta x_t = \beta\mu_t + \beta e_t$$

Since $\beta\mu_t = 0$, it follows that $\beta x_t = \beta e_t$. Hence, the linear combination βx_t is stationary. As shown in Section 8, this argument easily generalizes to the case of multiple cointegrating vectors.

3. COINTEGRATION AND ERROR CORRECTION

A principal feature of cointegrated variables is that their time paths are influenced by the extent of any deviation from long-run equilibrium. After all, if the system is to return to long-run equilibrium, the movements of at least some of the variables must respond to the magnitude of the disequilibrium. Before proceeding further, be aware that we will be examining the time paths of multiple nonstationary time-series variables. To do so in a tractable way, we will need to rely on the relationship between the rank of a matrix and its characteristic roots. The required mathematics are provided in Appendix 6.1.

The relationship between long-term and short-term interest rates illustrates how variables might adjust to any discrepancies from the long-run equilibrium relationship. Clearly, the theory of the term structure of interest rates implies a long-run relationship between long- and short-term rates. If the gap between the long- and short-term rates is "large" relative to the long-run relationship, the short-term rate must ultimately rise relative to the long-term rate. Of course, the gap can be closed by (1) an increase in the short-term rate and/or a decrease in the long-term rate, (2) an increase in the long-term rate but a commensurately larger rise in the short-term rate, or (3) a fall in the long-term rate but a smaller fall in the short-term rate. Without a full dynamic specification of the model, it is not possible to determine which of the possibilities will occur. Nevertheless, the short-run dynamics must be influenced by the deviation from the long-run relationship.

The dynamic model implied by this discussion is one of **error correction.** In an error-correction model, the short-term dynamics of the variables in the system are influenced by the deviation from equilibrium. If we assume that both interest rates are $I(1)$, a simple error-correction model that could apply to the term structure of interest rates is[4]

$$\Delta r_{St} = \alpha_S(r_{Lt-1} - \beta r_{St-1}) + \varepsilon_{St} \qquad \alpha_S > 0 \tag{6.9}$$

$$\Delta r_{Lt} = -\alpha_L(r_{Lt-1} - \beta r_{St-1}) + \varepsilon_{Lt} \qquad \alpha_L > 0 \tag{6.10}$$

where ε_{St} and ε_{Lt} are white-noise disturbance terms which may be correlated, r_{Lt} and r_{St} are the long- and short-term interest rates, and α_S, α_L, and β are parameters.

As specified, the short- and long-term interest rates change in response to stochastic shocks (represented by ε_{St} and ε_{Lt}) *and* in response to the previous period's deviation from long-run equilibrium. Everything else being equal, if this deviation happened to be positive (so that $r_{Lt-1} - \beta r_{St-1} > 0$), the short-term interest rate would rise and the long-term rate would fall. Long-run equilibrium is attained when $r_{Lt} = \beta r_{St}$.

Here you can see the relationship between error-correcting models and cointegrated variables. By assumption, Δr_{St} is stationary so that the left-hand side of (6.9) is $I(0)$. For (6.9) to be sensible, the right-hand side must be $I(0)$ as well. Given that ε_{St} is stationary, it follows that the linear combination $r_{Lt-1} - \beta r_{St-1}$ must also be stationary; hence, the two interest rates must be cointegrated with the cointegrating vector $(1, -\beta)$. Of course, the identical argument applies to (6.10). The essential point to note is that the error-correction representation necessitates that the two variables be cointegrated of order $CI(1, 1)$. This result is unaltered if we formulate a more general model by introducing the lagged changes of each rate into both equations:[5]

$$\Delta r_{St} = a_{10} + \alpha_S(r_{Lt-1} - \beta r_{St-1}) + \Sigma a_{11}(i)\Delta r_{St-i} + \Sigma a_{12}(i)\Delta r_{Lt-i} + \varepsilon_{St} \tag{6.11}$$

$$\Delta r_{Lt} = a_{20} - \alpha_L(r_{Lt-1} - \beta r_{St-1}) + \Sigma a_{21}(i)\Delta r_{St-i} + \Sigma a_{22}(i)\Delta r_{Lt-i} + \varepsilon_{Lt} \tag{6.12}$$

Again, ε_{St}, ε_{Lt}, and all terms involving Δr_{St-i} and Δr_{Lt-i} are stationary. Thus, the linear combination of interest rates $r_{Lt-1} - \beta r_{St-1}$ must also be stationary.

Inspection of (6.11) and (6.12) reveals a striking similarity to the VAR models of the previous chapter. This two-variable error-correction model is a bivariate VAR in first differences augmented by the error-correction terms $\alpha_S(r_{Lt-1} - \beta r_{St-1})$ and $-\alpha_L(r_{Lt-1} - \beta r_{St-1})$. Notice that α_S and α_L have the interpretation of *speed-of-adjustment* parameters. The larger α_S is, the greater the response of r_{St} to the previous period's deviation from long-run equilibrium. At the opposite extreme, very small values of α_S imply that the short-term interest rate is unresponsive to last period's equilibrium error. For the $\{\Delta r_{St}\}$ sequence to be unaffected by the long-term interest rate sequence, α_S *and* all the $a_{12}(i)$ coefficients must be equal to zero. Of course, at least one of the speed-of-adjustment terms in (6.11) and (6.12) must be nonzero. If both α_S and α_L are equal to zero, the long-run equilibrium relationship does not appear and the model is not one of error correction or cointegration.

The result can easily be generalized to the n-variable model. Formally, the $(n \cdot 1)$ vector $x_t = (x_{1t}, x_{2t}, \ldots, x_{nt})'$ has an error-correction representation if it can be expressed in the form

$$\Delta x_t = \pi_0 + \pi x_{t-1} + \pi_1 \Delta x_{t-1} + \pi_2 \Delta x_{t-2} + \ldots + \pi_p \Delta x_{t-p} + \varepsilon_t \quad (6.13)$$

where: π_0 = an $(n \cdot 1)$ vector of intercept terms with elements π_{i0}
π_i = $(n \cdot n)$ coefficient matrices with elements $\pi_{jk}(i)$
π = a matrix with elements π_{jk} such that one or more of the $\pi_{jk} \neq 0$
ε_t = an $(n \cdot 1)$ vector with elements ε_{it}

Note that the disturbance terms are such that ε_{it} may be correlated with ε_{jt}.

Let all variables in x_t be $I(1)$. Now, if there is an error-correction representation of these variables as in (6.13), there is necessarily a linear combination of the $I(1)$ variables that is stationary. Solving (6.13) for πx_{t-1} yields

$$\pi x_{t-1} = \Delta x_t - \pi_0 - \Sigma \pi_i \Delta x_{t-i} - \varepsilon_t$$

Since each expression on the right-hand side is stationary, πx_{t-1} must also be stationary. Since π contains only constants, each row of π is a cointegrating vector of x_t. For example, the first row can be written as $(\pi_{11}x_{1t-1} + \pi_{12}x_{2t-1} + \ldots + \pi_{1n}x_{nt-1})$. Since each series is $I(1)$, $(\pi_{11}, \pi_{12}, \ldots, \pi_{1n})$ must be a cointegrating vector for x_t.

The key feature in (6.13) is the presence of the matrix π. There are two important points to note:

1. If all elements of π equal zero, (6.13) is a traditional VAR in first differences. In such circumstances there is no error-correction representation since Δx_t does not respond to the previous period's deviation from long-run equilibrium.
2. If one or more of the π_{jk} differs from zero, Δx_t responds to the previous period's deviation from long-run equilibrium. Hence, *estimating x_t as a VAR in first differences is inappropriate if x_t has an error-correction representation.* The omission of the expression πx_{t-1} entails a misspecification error if x_t has an error-correction representation as in (6.13).

A good way to examine the relationship between cointegration and error correction is to study the properties of the simple VAR model

$$y_t = a_{11}y_{t-1} + a_{12}z_{t-1} + \varepsilon_{yt} \quad (6.14)$$

$$z_t = a_{21}y_{t-1} + a_{22}z_{t-1} + \varepsilon_{zt} \quad (6.15)$$

where ε_{yt} and ε_{zt} are white-noise disturbances that may be correlated with each other and, for simplicity, intercept terms have been ignored. Using lag operators, we can write (6.14) and (6.15) as

$$(1 - a_{11}L)y_t - a_{12}Lz_t = \varepsilon_{yt}$$
$$-a_{21}Ly_t + (1 - a_{22}L)z_t = \varepsilon_{zt}$$

The next step is to solve for y_t and z_t. Writing the system in matrix form, we obtain

$$\begin{bmatrix} (1 - a_{11}L) & -a_{12}L \\ -a_{21}L & (1 - a_{22}L) \end{bmatrix} \begin{bmatrix} y_t \\ z_t \end{bmatrix} = \begin{bmatrix} \varepsilon_{yt} \\ \varepsilon_{zt} \end{bmatrix}$$

Using Cramer's Rule or matrix inversion, we can obtain the solutions for y_t and z_t as

$$y_t = \frac{(1 - a_{22}L)\varepsilon_{yt} + a_{12}L\varepsilon_{zt}}{(1 - a_{11}L)(1 - a_{22}L) - a_{12}a_{21}L^2} \tag{6.16}$$

$$z_t = \frac{a_{21}L\varepsilon_{yt} + (1 - a_{11}L)\varepsilon_{zt}}{(1 - a_{11}L)(1 - a_{22}L) - a_{12}a_{21}L^2} \tag{6.17}$$

We have converted the two-variable first-order system represented by (6.14) and (6.15) into two univariate second-order difference equations of the type examined in Chapter 2. Note that both variables have the same inverse characteristic equation: $(1 - a_{11}L)(1 - a_{22}L) - a_{12}a_{21}L^2$. Setting $(1 - a_{11}L)(1 - a_{22}L) - a_{12}a_{21}L^2 = 0$ and solving for L yields the two roots of the inverse characteristic equation. In order to work with the characteristic roots (as opposed to the inverse characteristic roots), define $\lambda = 1/L$ and write the characteristic equation as

$$\lambda^2 - (a_{11} + a_{22})\lambda + (a_{11}a_{22} - a_{12}a_{21}) = 0 \tag{6.18}$$

Since the two variables have the same characteristic equation, the characteristic roots of (6.18) determine the time paths of both variables. The following remarks summarize the time paths of $\{y_t\}$ and $\{z_t\}$:

1. If both characteristic roots (λ_1, λ_2) lie inside the unit circle, (6.16) and (6.17) yield stable solutions for $\{y_t\}$ and $\{z_t\}$. If t is sufficiently large or if the initial conditions are such that the homogeneous solution is zero, the stability condition guarantees that the variables are stationary. The variables cannot be cointegrated of order (1, 1) since each is stationary.
2. If either root lies outside the unit circle, the solutions are explosive. Neither variable is difference-stationary, so they cannot be $CI(1, 1)$. In the same way, if both characteristic roots are unity, the second difference of each variable will be stationary. Since each is $I(2)$, the variables cannot be $CI(1, 1)$.
3. As you can see from (6.14) and (6.15), if $a_{12} = a_{21} = 0$, the solution is trivial. For $\{y_t\}$ and $\{z_t\}$ to be unit root processes, it is necessary for $a_{11} = a_{22} = 1$. It follows that $\lambda_1 = \lambda_2 = 1$ and that the two variables evolve without any long-run equilibrium relationship; hence, the variables cannot be cointegrated.

4. For $\{y_t\}$ and $\{z_t\}$ to be $CI(1, 1)$, it is necessary for one characteristic root to be unity and the other to be less than unity in absolute value. In this instance, each variable will have the same stochastic trend and the first difference of each variable will be stationary. For example, if $\lambda_1 = 1$, (6.16) will have the form

$$y_t = [(1 - a_{22}L)\varepsilon_{yt} + a_{12}L\varepsilon_{zt}]/[(1 - L)(1 - \lambda_2 L)]$$

or, multiplying by $(1 - L)$, we get

$$(1 - L)y_t = \Delta y_t = [(1 - a_{22}L)\varepsilon_{yt} + a_{12}L\varepsilon_{zt}]/(1 - \lambda_2 L)$$

which is stationary if $|\lambda_2| < 1$.

Thus, to ensure that the variables are $CI(1, 1)$, we must set one of the characteristic roots equal to unity and the other to a value that is less than unity in absolute value. For the larger of the two roots to equal unity, it must be the case that

$$0.5(a_{11} + a_{22}) + 0.5\sqrt{(a_{11}^2 + a_{22}^2) - 2a_{11}a_{22} + 4a_{12}a_{21}} = 1$$

so that after some simplification, the coefficients are seen to satisfy[6]

$$a_{11} = [(1 - a_{22}) - a_{12}a_{21}]/(1 - a_{22}) \tag{6.19}$$

Now consider the second characteristic root. Since a_{12} and/or a_{21} must differ from zero if the variables are cointegrated, the condition $|\lambda_2| < 1$ requires

$$a_{22} > -1 \tag{6.20}$$

and

$$a_{12}a_{21} + (a_{22})^2 < 1 \tag{6.21}$$

Equations (6.19), (6.20), and (6.21) are restrictions we must place on the coefficients of (6.14) and (6.15) if we want to ensure that the variables are cointegrated of order (1, 1). To see how these coefficient restrictions bear on the nature of the solution, write (6.14) and (6.15) as

$$\begin{bmatrix} \Delta y_t \\ \Delta z_t \end{bmatrix} = \begin{bmatrix} a_{11} - 1 & a_{12} \\ a_{21} & a_{22} - 1 \end{bmatrix} \begin{bmatrix} y_{t-1} \\ z_{t-1} \end{bmatrix} + \begin{bmatrix} \varepsilon_{yt} \\ \varepsilon_{zt} \end{bmatrix} \tag{6.22}$$

Now, (6.19) implies that $a_{11} - 1 = -a_{12}a_{21}/(1 - a_{22})$ so that after a bit of manipulation, (6.22) can be written in the form

$$\Delta y_t = -[a_{12}a_{21}/(1 - a_{22})]y_{t-1} + a_{12}z_{t-1} + \varepsilon_{yt} \tag{6.23}$$

$$\Delta z_t = a_{21}y_{t-1} - (1 - a_{22})z_{t-1} + \varepsilon_{zt} \tag{6.24}$$

Equations (6.23) and (6.24) form an error-correction model. If both a_{12} and a_{21} differ from zero, we can normalize the cointegrating vector with respect to either variable. Normalizing with respect to y_t, we get

$$\Delta y_t = \alpha_y(y_{t-1} - \beta z_{t-1}) + \varepsilon_{yt}$$
$$\Delta z_t = \alpha_z(y_{t-1} - \beta z_{t-1}) + \varepsilon_{zt}$$

where: $\alpha_y = -a_{12}a_{21}/(1 - a_{22})$
$\beta = (1 - a_{22})/a_{21}$
$\alpha_z = a_{21}$

You can see that y_t and z_t change in response to the previous period's deviation from the long-run equilibrium $y_{t-1} - \beta z_{t-1}$. If $y_{t-1} = \beta z_{t-1}$, y_t and z_t change only in response to ε_{yt} and ε_{zt} shocks. Moreover, if $\alpha_y < 0$ and $\alpha_z > 0$, y_t decreases and z_t increases in response to a positive deviation from long-run equilibrium.

You can easily convince yourself that conditions (6.20) and (6.21) ensure that $\beta \neq 0$ and that at least one of the speed-of-adjustment parameters (i.e., α_y and α_z) is not equal to zero. Now, refer to (6.9) and (6.10); you can see this model is in exactly the same form as the interest rate example presented in the beginning of this section.

Although a_{12} and a_{21} cannot both equal zero, an interesting special case arises if one of these coefficients is zero. For example, if we set $a_{12} = 0$, the speed-of-adjustment coefficient $\alpha_y = 0$. In this case, y_t changes only in response to ε_{yt} as $\Delta y_t = \varepsilon_{yt}$.[7] The $\{z_t\}$ sequence does all of the correction to eliminate any deviation from long-run equilibrium.

To highlight some of the important implications of this simple model, we have shown:

1. *The restrictions necessary to ensure that the variables are CI*(1, 1) *guarantee that an error-correction model exists.* In our example, both $\{y_t\}$ and $\{z_t\}$ are unit root processes but the linear combination $y_t - \beta z_t$ is stationary; the normalized cointegrating vector is $[1, -(1 - a_{22})/a_{21}]$. The variables have an error-correction representation with speed-of-adjustment coefficients $\alpha_y = -a_{12}a_{21}/(1 - a_{22})$ and $\alpha_z = a_{21}$. It was also shown that an error-correction model for $I(1)$ variables necessarily implies cointegration. This finding illustrates the **Granger representation theorem** stating that for any set of $I(1)$ variables, error correction and cointegration are equivalent representations.
2. *A cointegration necessitates coefficient restrictions in a VAR model.* It is important to realize that a cointegrated system can be viewed as a restricted form of a general VAR model. Let $x_t = (y_t, z_t)'$ and $\varepsilon_t = (\varepsilon_{yt}, \varepsilon_{zt})'$ so that we can write (6.22) in the form

$$\Delta x_t = \pi x_{t-1} + \varepsilon_t \tag{6.25}$$

Clearly, it is inappropriate to estimate a VAR of cointegrated variables using only first differences. Estimating (6.25) without the expression πx_{t-1} would eliminate the error-correction portion of the model. It is also important to note that the rows of π are *not* linearly independent if the variables are cointegrated. Multiplying each element in row 1 by $-(1 - a_{22})/a_{12}$ yields the corresponding element in row 2. Thus, the determinant of π is equal to zero, and y_t and z_t have the error-correction representation given by (6.23) and (6.24).

This two-variable example illustrates the very important insights of Johansen (1988) and Stock and Watson (1988) that *we can use the rank of*

π to determine whether or not two variables $\{y_t\}$ and $\{z_t\}$ are cointegrated. Compare the determinant of π to the characteristic equation given by (6.18). If the largest characteristic root equals unity, ($\lambda_1 = 1$), it follows that the determinant of π is zero and that π has a rank equal to unity. If π were to have a rank of zero, it would be necessary for $a_{11} = 1$, $a_{22} = 1$, and $a_{12} = a_{21} = 0$. The VAR represented by (6.14) and (6.15) would be nothing more than $\Delta y_t = \varepsilon_{yt}$ and $\Delta z_t = \varepsilon_{zt}$. In this case, both the $\{y_t\}$ and $\{z_t\}$ sequences are unit root processes without any cointegrating vector. Finally, if the rank of π is full, then neither characteristic root can be unity, so the $\{y_t\}$ and $\{z_t\}$ sequences are jointly stationary.

3. In general, both variables in a cointegrated system will respond to a deviation from long-run equilibrium. However, it is possible that one (but not both) of the speed-of-adjustment parameters is zero. For example, if $\alpha_y = 0$, $\{y_t\}$ does not respond to the discrepancy from long-run equilibrium and $\{z_t\}$ does all of the adjustment. In this circumstance, $\{y_t\}$ is said to be **weakly exogenous.** As such, an econometric model for $\{z_t\}$ can be estimated and hypothesis testing can be conducted without reference to a specific model for $\{y_t\}$. Section 10 and Appendix 6.2 consider modeling in a cointegrated system when a variable is weakly exogenous.

Also, *it is necessary to reinterpret Granger causality in a cointegrated system.* In a cointegrated system, $\{y_t\}$ does not Granger cause $\{z_t\}$ if lagged values Δy_{t-i} do not enter the Δz_t equation *and* if z_t does not respond to the deviation from long-run equilibrium. Hence, $\{z_t\}$ must be weakly exogenous. If $a_{21} = 0$ in (6.24), $\{z_t\}$ is weakly exogenous and is not Granger caused by $\{y_t\}$. Similarly, in the cointegrated system of (6.11) and (6.12), $\{r_{Lt}\}$ does not Granger cause $\{r_{St}\}$ if all $a_{12}(i) = 0$ and if $\alpha_S = 0$.

The *n*-Variable Case

Little is altered in the n-variable case. The relationship between cointegration, error correction, and the rank of the matrix π is invariant to adding additional variables to the system. The interesting feature introduced in the n-variable case is the possibility of multiple cointegrating vectors. Now consider a more general version of (6.25):

$$x_t = A_1 x_{t-1} + \varepsilon_t \tag{6.26}$$

where: x_t = the $(n \cdot 1)$ vector $(x_{1t}, x_{2t}, \ldots, x_{nt})'$
ε_t = the $(n \cdot 1)$ vector $(\varepsilon_{1t}, \varepsilon_{2t}, \ldots, \varepsilon_{nt})'$
A_1 = an $(n \cdot n)$ matrix of parameters

Subtracting x_{t-1} from each side of (6.26) and letting I be an $(n \cdot n)$ identity matrix, we get

$$\begin{aligned} \Delta x_t &= -(I - A_1)x_{t-1} + \varepsilon_t \\ &= \pi x_{t-1} + \varepsilon_t \end{aligned} \tag{6.27}$$

where π is the $(n \cdot n)$ matrix $-(I - A_1)$ and π_{ij} denotes the element in row i and column j of π. As you can see, (6.27) is a special case of (6.13) such that all $\pi_i = 0$.

Again, the crucial issue for cointegration concerns the rank of the $(n \cdot n)$ matrix π. If the rank of this matrix is zero, each element of π must equal zero. In this instance, (6.27) is equivalent to an n-variable VAR in first differences:

$$\Delta x_t = \varepsilon_t$$

Here, each $\Delta x_{it} = \varepsilon_{it}$ so that the first difference of each variable in the vector x_t is $I(0)$. Since each $x_{it} = x_{it-1} + \varepsilon_{it}$, all the $\{x_{it}\}$ sequences are unit root processes and there is no linear combination of the variables that is stationary.

At the other extreme, suppose that π is of full rank. The long-run solution to (6.27) is given by the n independent equations

$$\begin{aligned}
\pi_{11}x_{1t} + \pi_{12}x_{2t} + \pi_{12}x_{3t} + \ldots + \pi_{1n}x_{nt} &= 0 \\
\pi_{21}x_{1t} + \pi_{22}x_{2t} + \pi_{23}x_{3t} + \ldots + \pi_{2n}x_{nt} &= 0 \\
&\cdot \\
&\cdot \\
&\cdot \\
\pi_{n1}x_{1t} + \pi_{n2}x_{2t} + \pi_{n3}x_{3t} + \ldots + \pi_{nn}x_{nt} &= 0
\end{aligned} \tag{6.28}$$

Each of these n equations is an independent restriction on the long-run solution of the variables; the n variables in the system face n long-run constraints. In this case, each of the n variables contained in the vector x_t must be stationary with the long-run values given by (6.28). The variables cannot be $CI(1, 1)$ since all are stationary.

In intermediate cases, in which the rank of π is equal to $r < n$, there are r cointegrating vectors. With r independent equations and n variables, there are $n - r$ stochastic trends in the system. If $r = 1$, there is a single cointegrating vector given by any row of the matrix π. Each $\{\Delta x_{it}\}$ sequence can be written in error-correction form. For example, we can write Δx_{1t} as

$$\Delta x_{1t} = \pi_{11}x_{1t-1} + \pi_{12}x_{2t-1} + \ldots + \pi_{1n}x_{nt-1} + \varepsilon_{1t}$$

or, normalizing with respect to x_{1t-1}, we can set $\alpha_1 = \pi_{11}$ and $\beta_{1j} = \pi_{1j}/\pi_{11}$ to obtain

$$\Delta x_{1t} = \alpha_1(x_{1t-1} + \beta_{12}x_{2t-1} + \ldots + \beta_{1n}x_{nt-1}) + \varepsilon_{1t} \tag{6.29}$$

In the long run, the $\{x_{it}\}$ will satisfy the relationship

$$x_{1t} + \beta_{12}x_{2t} + \ldots + \beta_{1n}x_{nt} = 0$$

Hence, the normalized cointegrating vector is $(1, \beta_{12}, \beta_{13}, \ldots, \beta_{1n})$ and the speed-of-adjustment parameter is α_1. In the same way, with two cointegration vectors the long-run values of the variables will satisfy the two relationships

$$\begin{aligned}
\pi_{11}x_{1t} + \pi_{12}x_{2t} + \ldots + \pi_{1n}x_{nt} &= 0 \\
\pi_{21}x_{1t} + \pi_{22}x_{2t} + \ldots + \pi_{2n}x_{nt} &= 0
\end{aligned}$$

which can be appropriately normalized.

The main point here is that there are three important ways to test for cointegration. The Engle–Granger methodology seeks to determine whether the residuals of the equilibrium relationship are stationary. The Johansen (1988) methodology determines the rank of π and the error-correction method examines the speed-of-adjustment coefficients. The Engle–Granger approach is the subject of the next three sections. Sections 7 through 9 examine the Johansen (1988) methodology, and cointegration using an error-correction framework is examined in Section 10.

4. TESTING FOR COINTEGRATION: THE ENGLE–GRANGER METHODOLOGY

To explain the Engle–Granger testing procedure, let's begin with the type of problem likely to be encountered in applied studies. Suppose that two variables—say, y_t and z_t—are believed to be integrated of order 1 and we want to determine whether there exists an equilibrium relationship between the two. Engle and Granger (1987) proposed a four-step procedure to determine if two $I(1)$ variables are cointegrated of order $CI(1, 1)$.

STEP 1: Pretest the variables for their order of integration. By definition, cointegration necessitates that two variables be integrated of the same order. Thus, the first step in the analysis is to pretest each variable to determine its order of integration. The augmented Dickey–Fuller tests discussed in Chapter 4 can be used to infer the number of unit roots (if any) in each of the variables. If both variables are stationary, it is not necessary to proceed since standard time-series methods apply to stationary variables. If the variables are integrated of different orders, it is possible to conclude that they are *not* cointegrated. However, as detailed in Section 5, if you have more than two variables such that some are $I(1)$ and some are $I(2)$, you may want to determine whether the variables are multicointegrated.

STEP 2: Estimate the long-run equilibrium relationship. If the results of Step 1 indicate that both $\{y_t\}$ and $\{z_t\}$ are $I(1)$, the next step is to estimate the long-run equilibrium relationship in the form

$$y_t = \beta_0 + \beta_1 z_t + e_t \tag{6.30}$$

If the variables are cointegrated, an OLS regression yields a "super-consistent" estimator of the cointegrating parameters β_0 and β_1. Stock (1987) proved that the OLS estimates of β_0 and β_1 converge faster than in OLS models using stationary variables. To explain, reexamine the scatter plot shown in Figure 6.1. You can see that the effect of the common trend dominates the effect of the stationary component; both variables seem to rise and fall in tandem. Hence, there is a strong linear relationship as shown by the regression line drawn in the figure.

In order to determine if the variables are actually cointegrated, denote the residual sequence from this equation by $\{\hat{e}_t\}$. Thus, the $\{\hat{e}_t\}$ series contains the estimated values of the deviations from the long-run relationship. If these deviations are found to be stationary, the $\{y_t\}$ and $\{z_t\}$ sequences are cointegrated of order (1, 1). It would be convenient if we could perform a Dickey–Fuller test on these residuals to determine their order of integration. Consider the autoregression of the residuals:

$$\Delta\hat{e}_t = a_1\hat{e}_{t-1} + \varepsilon_t \tag{6.31}$$

Since the $\{\hat{e}_t\}$ sequence is a residual from a regression equation, there is no need to include an intercept term; the parameter of interest in (6.31) is a_1. If we cannot reject the null hypothesis $a_1 = 0$, we can conclude that the residual series contains a unit root. Hence, we conclude that the $\{y_t\}$ and $\{z_t\}$ sequences are *not* cointegrated. The more precise wording is awkward because of a triple negative, but to be technically correct, *if it is not possible to reject the null hypothesis* $|a_1| = 0$, *we cannot reject the hypothesis that the variables are not cointegrated.* Instead, the rejection of the null hypothesis implies that the residual sequence is stationary.[8] Given that $\{y_t\}$ and $\{z_t\}$ were both found to be $I(1)$ and that the residuals are stationary, we can conclude that the series are cointegrated of order (1, 1).

In most applied studies it is not possible to use the Dickey–Fuller tables themselves. The problem is that the $\{\hat{e}_t\}$ sequence is generated from a regression equation; the researcher does not know the actual error e_t, only the estimate of the error $\hat{e}_t$. The methodology of fitting the regression in (6.30) selects values of β_0 and β_1 that minimize the sum of squared residuals. Since the residual variance is made as small as possible, the procedure is prejudiced toward finding a stationary error process in (6.31). Hence, the test statistic used to test the magnitude of a_1 must reflect this fact. Only if β_0 and β_1 were known in advance and used to construct the true $\{e_t\}$ sequence would an ordinary Dickey–Fuller table be appropriate. When you estimate the cointegrating vector, use the critical values provided in Table C at the end of the text. These critical values depend on sample size and the number of variables used in the analysis. For example, to test for cointegration between two variables using a sample size of 100, the critical value at the 5 percent significance level is −3.398.

If the residuals of (6.31) do not appear to be white noise, an augmented form of the test can be used instead of (6.31). Suppose that diagnostic checks indicate that the $\{\varepsilon_t\}$ sequence of (6.31) exhibits serial correlation. Instead of using the results from (6.31), estimate the autoregression

$$\Delta\hat{e}_t = a_1\hat{e}_{t-1} + \sum_{i=1}^{n} a_{i+1}\Delta\hat{e}_{t-i} + \varepsilon_t \tag{6.32}$$

Again, if we reject the null hypothesis $a_1 = 0$, we can conclude that the residual sequence is stationary and that the variables are cointegrated.

STEP 3: Estimate the error-correction model. If the variables are cointegrated (i.e., if the null hypothesis of no cointegration is rejected), the residuals from the equilibrium regression can be used to estimate the error-correction model. If $\{y_t\}$ and $\{z_t\}$ are $CI(1, 1)$, the variables have the error-correction form

$$\Delta y_t = \alpha_1 + \alpha_y[y_{t-1} - \beta_1 z_{t-1}] + \sum_{i=1} \alpha_{11}(i)\Delta y_{t-i} + \sum_{i=1} \alpha_{12}(i)\Delta z_{t-i} + \varepsilon_{yt} \tag{6.33}$$

$$\Delta z_t = \alpha_2 + \alpha_z[y_{t-1} - \beta_1 z_{t-1}] + \sum_{i=1} \alpha_{21}(i)\Delta y_{t-i} + \sum_{i=1} \alpha_{22}(i)\Delta z_{t-i} + \varepsilon_{zt} \tag{6.34}$$

where β_1 = the parameter of the cointegrating vector given by (6.30), ε_{yt} and ε_{zt} = white-noise disturbances (which may be correlated with each other), and α_1, α_2, α_y, α_z, $\alpha_{11}(i)$, $\alpha_{12}(i)$, $\alpha_{21}(i)$, and $\alpha_{22}(i)$ are all parameters.

Engle and Granger (1987) proposed a clever way to circumvent the cross-equation restrictions involved in the direct estimation of (6.33) and (6.34). The magnitude of the residual $\hat{e}_{t-1}$ is the deviation from long-run equilibrium in period $(t-1)$. Hence, it is possible to use the saved residuals $\{\hat{e}_{t-1}\}$ obtained in Step 2 as an estimate of the expression $y_{t-1} - \beta_1 z_{t-1}$ in (6.33) and (6.34). Thus, using the saved residuals from the estimation of the long-run equilibrium relationship, estimate the error-correcting model as

$$\Delta y_t = \alpha_1 + \alpha_y \hat{e}_{t-1} + \sum \alpha_{11}(i)\Delta y_{t-i} + \sum \alpha_{12}(i)\Delta z_{t-i} + \varepsilon_{yt} \tag{6.35}$$

$$\Delta z_t = \alpha_2 + \alpha_z \hat{e}_{t-1} + \sum \alpha_{21}(i)\Delta y_{t-i} + \sum \alpha_{22}(i)\Delta z_{t-i} + \varepsilon_{zt} \tag{6.36}$$

Other than the error-correction term $\hat{e}_{t-1}$, (6.35) and (6.36) constitute a VAR in first differences. This VAR can be estimated using the same methodology developed in Chapter 5. All of the procedures developed for a VAR apply to the system represented by the error-correction equations. Notably:

1. OLS is an efficient estimation strategy since each equation contains the same set of regressors.
2. Since all terms in (6.35) and (6.36) are stationary [i.e., Δy_t and its lags, Δz_t and its lags, and $\hat{e}_{t-1}$ are $I(0)$], the test statistics used in traditional VAR analysis are appropriate for (6.35) and (6.36). For example, lag lengths can be determined using a χ^2-test, and the restriction that all $\alpha_{jk}(i) = 0$ can be checked using an F-test. If there is a single cointegrating vector, restrictions concerning α_y or α_z can be conducted using a t-test.

STEP 4: Assess model adequacy. There are several procedures that can help determine whether the error-correction estimated model is appropriate.

1. You should be careful to assess the adequacy of the model by performing diagnostic checks to determine whether the residuals of the error-correction equations approximate white noise. If the residuals are serially correlated, lag lengths may be too short. Reestimate the model using lag lengths that yield serially uncorrelated errors. It may be that you need to allow longer lags of some variables than of others. If so, you can gain efficiency by estimating the near-VAR using the seemingly unrelated regressions (SUR) method.
2. The *speed-of-adjustment* coefficients α_y and α_z are of particular interest in that they have important implications for the dynamics of the system.[9] If we focus on (6.36) it is clear that for any given value of $\hat{e}_{t-1}$, a large value of α_z is associated with a large value of Δz_t. If α_z is zero, the change in z_t does not at all respond to the deviation from long-run

equilibrium in $(t - 1)$. If α_z is zero and if all $\alpha_{21}(i) = 0$, then it can be said that $\{\Delta y_t\}$ does not Granger cause $\{\Delta z_t\}$. We know that α_y and/or α_z should be significantly different from zero if the variables are cointegrated. After all, if both α_y and α_z are zero, there is no error correction and (6.35) and (6.36) comprise nothing more than a VAR in first differences. Moreover, the absolute values of these speed-of-adjustment coefficients must not be too large. The point estimates should imply that Δy_t and Δz_t converge to the long-run equilibrium relationship.[10]

If all but one variable are weakly exogenous, you may want to estimate that variable using the error-correction technique described in Section 10.

3. As in a traditional VAR analysis, Lutkepohl and Reimers (1992) showed that innovation accounting (i.e., impulse responses and variance decomposition analysis) can be used to obtain information concerning the interactions among the variables. As a practical matter, the two innovations ε_{yt} and ε_{zt} may be contemporaneously correlated if y_t has a contemporaneous effect on z_t and/or if z_t has a contemporaneous effect on y_t. In obtaining impulse response functions and variance decompositions, some method—such as a Choleski decomposition—must be used to orthogonalize the innovations.

The shape of the impulse response functions and the results of the variance decompositions can indicate whether the dynamic responses of the variables conform to theory. Since all variables in (6.35) and (6.36) are $I(0)$, the impulse responses of Δy_t and Δz_t should converge to zero. You should reexamine your results from each step if you obtain a nondecaying or explosive impulse response function.

Before closing this section, a word of warning is in order. It is very tempting to use t-statistics to perform significance tests on the cointegrating vector. However, you must avoid this temptation since the coefficients have an asymptotic t-distribution only in one special circumstance. Suppose that the cointegration relationship between $\{y_t\}$ and $\{z_t\}$ is such that

$$y_t = \beta_0 + \beta_1 z_t + \varepsilon_{1t}$$
$$\Delta z_t = \varepsilon_{2t}$$

where $E\varepsilon_{1t}\varepsilon_{2t} = 0$.

The notation is designed to illustrate the point that the residuals from both equations are uncorrelated white-noise disturbances. The set of assumptions is fairly restrictive in that the residuals from both equations must be serially uncorrelated and the cross-correlations must be zero. If these conditions hold, the OLS estimates of β_0 and β_1 can be tested using t-tests and F-tests. If the disturbances are not normally distributed, the asymptotic results are such that t-tests and F-tests are appropriate. Be aware that both conditions are necessary to perform such tests. If $E\varepsilon_{1t}\varepsilon_{2t} \neq 0$, $\{z_t\}$ is not exogenous since shocks to ε_{1t} affect z_t. Hence, inference on β_1 is not appropriate since y_t and z_t are jointly determined. Moreover, as in a standard regression, if the residuals of the cointegrating vector are serially correlated, inference concerning the coefficients is inappropriate. Phillips and Hansen (1990) developed a procedure you can use in such circumstances. The details are outlined in Appendix 6.2 of this chapter.

5. ILLUSTRATING THE ENGLE–GRANGER METHODOLOGY

Figure 6.2 shows three simulated variables that can be used to illustrate the Engle–Granger procedure. Inspection of the figure suggests that each is nonstationary, and there is no visual evidence that any pair is cointegrated. As detailed in Table 6.1, each series is constructed as the sum of a stochastic trend component plus an autoregressive irregular component.

The first column of the table contains the formulas used to construct the $\{y_t\}$ sequence. First, 150 realizations of a white-noise process were drawn to represent the $\{\varepsilon_{yt}\}$ sequence. Initializing $\mu_{y0} = 0$, 150 values of the random walk process $\{\mu_{yt}\}$ were constructed using the formula $\mu_{yt} = \mu_{yt-1} + \varepsilon_{yt}$ (see the first cell of the table). Another 150 realizations of a white-noise process were drawn to represent the $\{\eta_{yt}\}$ sequence; given the initial condition $\delta_{y0} = 0$, these realizations were used to construct $\{\delta_{yt}\}$ as $\delta_{yt} = 0.5\delta_{yt-1} + \eta_{yt}$ (see the next lower cell). Adding the two constructed

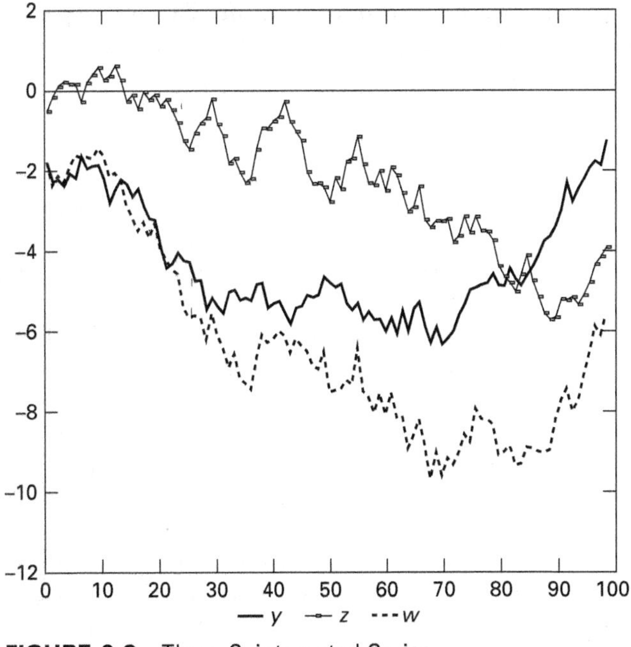

FIGURE 6.2 Three Cointegrated Series

Table 6.1 The Simulated Series

	$\{y_t\}$	$\{z_t\}$	$\{w_t\}$
Trend	$\mu_{yt} = \mu_{yt-1} + \varepsilon_{yt}$	$\mu_{zt} = \mu_{zt-1} + \varepsilon_{zt}$	$\mu_{wt} = \mu_{yt} + \mu_{zt}$
Pure irregular	$\delta_{yt} = 0.5\delta_{yt-1} + \eta_{yt}$	$\delta_{zt} = 0.5\delta_{zt-1} + \eta_{zt}$	$\delta_{wt} = 0.5\delta_{wt-1} + \eta_{wt}$
Series	$y_t = \mu_{yt} + \delta_{yt}$	$z_t = \mu_{zt} + \delta_{zt} + 0.5\delta_{yt}$	$w_t = \mu_{wt} + \delta_{wt} + 0.5\delta_{yt} + 0.5\delta_{zt}$

series yields 150 realizations for $\{y_t\}$. To help ensure randomness, only the last 100 observations are used in the simulated study. Hence, $\{y_t\}$ is the sum of a stochastic trend and a stationary (i.e., irregular) component.

The $\{z_t\}$ sequence was constructed in a similar fashion; the $\{\varepsilon_{zt}\}$ and $\{\eta_{zt}\}$ sequences are each represented by two different sets of 150 random numbers. The trend $\{\mu_{zt}\}$ and the autoregressive irregular term $\{\delta_{zt}\}$ were constructed as shown in the second column of the table. The $\{\delta_{zt}\}$ sequence can be thought of as a pure irregular component in the $\{z_t\}$ sequence. In order to introduce correlation between the $\{y_t\}$ and $\{z_t\}$ sequences, the irregular component in $\{z_t\}$ was constructed as the sum $\delta_{zt} + 0.5\delta_{yt}$. In the third column you can see that the trend in $\{w_t\}$ is the simple summation of the trends in the other two series. As such, the three series have the cointegrating vector $(1, 1, -1)$. The irregular component in $\{w_t\}$ is the sum of pure innovation δ_{wt} and 50 percent of the innovations δ_{yt} and δ_{zt}.

Now pretend that we do not know the data-generating process. The issue is whether the Engle–Granger methodology can uncover the essential details of the process. Use the data in the file COINT6.XLS to follow along. The first step is to pretest the variables in order to determine their order of integration. Consider the augmented Dickey–Fuller regression equation for $\{y_t\}$:

$$\Delta y_t = \alpha_0 + \alpha_1 y_{t-1} + \sum_{i=1}^{n} \alpha_{i+1} \Delta y_{t-i} + \varepsilon_t$$

If the data happened to be quarterly, it would be natural to perform the augmented Dickey–Fuller tests using lag lengths that are multiples of 4 (i.e., $n = 4, 8, \ldots$). For each series, the results of the Dickey–Fuller test and the augmented test using four lags are reported in Table 6.2.

With 100 observations and a constant, the 5 percent critical value for the Dickey–Fuller test is -2.89. Since the absolute values of all t-statistics are well below this critical value, we cannot reject the null hypothesis of a unit root in any of the series. Of course, if there were any serious doubt about the presence of a unit root, we could use the procedures in Chapter 4 to test for the presence of the drift term. If various lag lengths yield different results, we would want to test for the most appropriate lag length.

The luxury of using simulated data is that we can avoid these potentially sticky problems and move on to Step 2. Since all three variables are presumed to be jointly determined, the long-run equilibrium regression can be estimated using either y_t, z_t, or

Table 6.2 Estimated α_1 and the Associated t-statistic

	No Lags	4 Lags
Δy_t	−0.01995 (−0.74157)	−0.02691 (−1.0465)
Δz_t	−0.02069 (−0.99213)	−0.25841 (−1.1437)
Δw_t	−0.03501 (−1.9078)	−0.03747 (−1.9335)

Table 6.3 Estimated a_1 and the Associated t-statistic

	No Lags	4 Lags
Δe_{yt}	−0.44301 (−5.17489)	−0.59525 (−4.0741)
Δe_{zt}	−0.45195 (−5.37882)	−0.59344 (−4.2263)
Δe_{wt}	−0.45525 (−5.3896)	−0.60711 (−4.2247)

w_t as the "left-hand-side" variable. The three estimates of the long-run relationship (with t-values in parentheses) are

$$y_t = \underset{(-0.575)}{-0.4843} - \underset{(-38.095)}{0.9273z_t} + \underset{(53.462)}{0.9769w_t} + e_{yt}$$

$$z_t = \underset{(0.671)}{0.0589} - \underset{(-38.095)}{1.0108y_t} + \underset{(65.323)}{1.0255w_t} + e_{zt}$$

$$w_t = \underset{(-1.009)}{-0.0852} + \underset{(52.462)}{0.9901y_t} + \underset{(65.462)}{0.9535z_t} + e_{wt}$$

where e_{yt}, e_{zt}, and e_{wt} = the residuals from the three equilibrium regressions.

The essence of the test is to determine whether the residuals from the equilibrium regression are stationary. Again, in performing the test, there is no presumption that any one of the three residual series is preferable to any of the others. If we use each of the three series to estimate an equation in the form of (6.31) [or (6.32)], the estimated values of a_1 are given in Table 6.3.

From Table C, you can see that the critical value of the t-statistic is −3.828 Hence, using any one of the three equilibrium regressions, we can conclude that the series are cointegrated of order (1, 1). Fortunately, all three equilibrium regressions yield this same conclusion. We should be very wary of a result indicating that the variables are cointegrated using one variable for the normalization but are not cointegrated using another variable for the normalization. In such circumstances, it is possible that only a subset of the variables are cointegrated. Suppose that x_{1t}, x_{2t}, and x_{3t} are three $I(1)$ variables and that x_{1t} and x_{2t} are cointegrated such that $x_{1t} - \beta_2 x_{2t}$ is stationary. A regression of x_{1t} on the other two variables should yield the stationary relationship $x_{1t} = \beta_2 x_{2t} + 0x_{3t}$. Similarly, a regression of x_{2t} on the other variables should yield the stationary relationship $x_{2t} = (1/\beta_2)x_{1t} + 0x_{3t}$. However, a regression of x_{3t} on x_{1t} and x_{2t} cannot reveal the cointegrating relationship. Nevertheless, the possibility of a contradictory result is a weakness of the test; other methods can be tried if mixed results are found.

You must be careful in conducting significance tests on the estimated equilibrium regressions. As mentioned above, the coefficients *do not* have an asymptotic

t-distribution unless the right-hand-side variables are actually independent and the error terms are serially uncorrelated.

Step 3 entails estimating the error-correction model. Consider the first-order system shown with t-statistics in parentheses:

$$\begin{aligned}\Delta y_t = {} & 0.006 + 0.418 e_{wt-1} + 0.178 \Delta y_{t-1} + 0.313 \Delta z_{t-1} - 0.368 \Delta w_{t-1} + \varepsilon_{yt} \\ & (0.19) \quad (2.79) \qquad (1.08) \qquad (1.94) \qquad (-2.27)\end{aligned} \tag{6.37}$$

$$\begin{aligned}\Delta z_t = {} & -0.042 + 0.074 e_{wt-1} + 0.145 \Delta y_{t-1} + 0.262 \Delta z_{t-1} - 0.313 \Delta w_{t-1} + \varepsilon_{zt} \\ & (-1.12) \quad (0.42) \qquad (0.75) \qquad (1.38) \qquad (-1.63)\end{aligned} \tag{6.38}$$

$$\begin{aligned}\Delta w_t = {} & -0.040 - 0.069 e_{wt-1} + 0.156 \Delta y_{t-1} + 0.301 \Delta z_{t-1} - 0.420 \Delta w_{t-1} + \varepsilon_{wt} \\ & (-0.90) \quad (-0.33) \qquad (0.68) \qquad (1.35) \qquad (-1.87)\end{aligned} \tag{6.39}$$

where $e_{wt-1} = w_{t-1} + 0.0852 - 0.9901 y_{t-1} - 0.9535 z_{t-1}$ so that e_{wt-1} is the lagged value of the residual from the equilibrium relationship using w_t as the dependent variable.

Equations (6.37) through (6.39) comprise a first-order VAR augmented with the single error-correction term e_{wt-1}. Again, there is an area of ambiguity since the residuals from any of the "equilibrium" relationships could have been used in the estimation. The signs of the speed-of-adjustment coefficients are in accord with convergence toward the long-run equilibrium. In response to a positive discrepancy in e_{wt-1}, both y_t and z_t tend to increase while w_t tends to decrease. The error-correction term, however, is significant only in (6.37).

Finally, the diagnostic methods discussed in the last section should be applied to (6.37) through (6.39) in order to assess the model's adequacy. Using actual data, lag length tests and the properties of the residuals need to be considered. Moreover, innovation accounting could help determine whether the model is adequate. These methods are not performed here since we know the actual lag length and there is no economic theory associated with the simulated data.

The Engle–Granger Procedure with $I(2)$ Variables

Multicointegration refers to a situation in which a linear combination of $I(2)$ and $I(1)$ variables is integrated of order zero. For example, suppose that x_{1t} and x_{2t} are $I(2)$ and that z_t is $I(1)$. It is possible that a linear combination of x_{1t} and x_{2t} is $I(1)$ and that this combination is cointegrated with z_t. Hence, it is possible to have a long-run equilibrium relationship of the form[11]

$$x_{1t} = \beta_2 x_{2t} + \alpha_1 z_t$$

However, a richer set of possibilities is given by the stationary relationship

$$x_{1t} = \beta_2 x_{2t} + \gamma_1 \Delta x_{2t} + \alpha_1 z_t$$

This specification allows for the possibility that the linear combination $x_{1t} - \beta_2 x_{2t}$ is $I(1)$ and cointegrated with the other $I(1)$ independent variables in the system, Δx_{2t} and z_t. To make sure you understand the issue, ask yourself if it is possible for β_2 to be zero. The answer is a resounding no. If $\beta_2 = 0$, the $I(2)$ variable x_{1t} cannot, by itself, be cointegrated with the $I(1)$ variables.

In principle, it is possible to check for multicointegration using a two-step procedure. First, search for a cointegrating relationship among the $I(2)$ variables and then use this relationship to check for a possible cointegrating relationship with the remaining $I(1)$ variables. Engsted, Gonzalo, and Haldrup (1997) showed that this procedure is effective only if the cointegrating vector for the first step is known. Otherwise, the second step is contaminated with the errors generated in the first step. In the most general form of the one-step procedure, you estimate an equation in the form

$$x_{1t} = a_0 + a_1 t + a_2 t^2 + \beta_2 x_{2t} + \beta_3 x_{3t} + \gamma_1 \Delta x_{2t} + \gamma_2 \Delta x_{3t} + \alpha_1 z_t + e_t \quad (6.40)$$

where x_{1t}, x_{2t}, and x_{3t} are $I(2)$ variables, z_t is a vector of $I(1)$ variables, and the deterministic regressors can include a quadratic time trend.

Hence, the test allows you to include up to two $I(2)$ variables and an unrestricted number of $I(1)$ variables as regressors. You might want to include the quadratic time trend if $\Delta^2 x_{1t}$ contains a drift. Since the key issue is the stationarity of the $\{e_t\}$ series, estimate a regression of the form

$$\Delta \hat{e}_t = \rho \hat{e}_{t-1} + \sum_{i=1}^{p} \rho_i \Delta \hat{e}_{t-i} + v_t$$

where $\{\hat{e}_t\}$ are the regression residuals from (6.40).

If it is possible to reject the null hypothesis $\rho = 0$, it is possible to conclude that there is multicointegration. In addition to sample size, the critical values of the t-statistic for the null hypothesis $\rho = 0$ depend on the number of $I(2)$ regressors ($m_2 = 1$ or 2), the number of $I(1)$ regressors ($m_1 = 0$ to 4), and the form of the deterministic regressors. The critical values are shown in Table D at the end of the text. Consider the U.K. money demand equations for the sample period 1963Q1 to 1989Q2 estimated by Haldrup (1994):

$$m_t = a_0 + 0.68p_t + 1.57y_t - 2.67r_t - 2.55\Delta p_t \quad (6.41)$$

and

$$m_t = a_0 + a_1 t + 0.89p_t + 2.39y_t - 2.69r_t - 3.25\Delta p_t \quad (6.42)$$

Pretesting the variables indicated that m_t (as measured by $M1$) and p_t (the implicit price deflator) were $I(2)$ and that y_t (total final expenditure) and r_t (a measure of the interest rate differential) were $I(1)$. The only variable needing explanation is the presence of Δp_t in the money demand function. The idea is to allow for the demand for money to depend on the inflation rate (i.e., change in the log of the price level) since high inflation should reduce the desire to hold money balances. Since there is a total of 105 observations, one $I(2)$ regressor (so that $m_2 = 1$), and three $I(1)$ regressors, the 5 percent critical values for models without and with the linear trend are -4.56 and -4.91, respectively. Using the residuals from the money demand equations given by (6.41) and (6.42), Haldrup found that the t-statistics for the null hypothesis $\rho = 0$ were -2.35 and -2.66, respectively. Hence, it is possible to conclude that the two regressions are spurious (i.e., it is not possible to reject the null hypothesis of no multicointegration).

Even though multicointegration fails, Haldrup goes on to experiment with various estimates of the error-correction mechanism. One interesting model (with standard errors in parentheses) is

$$\Delta^2 m_t = \underset{(0.02)}{-0.04\hat{e}_{t-1}} + \text{stationary regressors}$$

where the stationary regressors can include lagged values of $\Delta^2 m_t$ as well as current and lagged values of $\Delta^2 p_t$, Δy_t, Δp_t, and Δr_t. The point estimate is such that $\Delta^2 m_t$ is expected to decline in response to a positive discrepancy from the long-run relationship. The t-statistic of $-0.04/0.02 = 2$ suggests that the effect is just significant at the 5 percent level.

6. COINTEGRATION AND PURCHASING POWER PARITY

To illustrate the Engle–Granger methodology using "real world" data, reconsider the theory of *purchasing power parity (PPP)*. Respectively, if e_t, p_{ft}, and p_t denote the logarithms of the price of foreign exchange, the foreign price level, and the domestic price level, long-run PPP requires that $e_t + p_{ft} - p_t$ be stationary. The unit root tests reported in Chapter 4 indicate that real exchange rates (defined as $r_t = e_t + p_{ft} - p_t$) appear to be nonstationary. Cointegration offers an alternative method to check the theory; if PPP holds, the sequence formed by the sum $\{e_t + p_{ft}\}$ should be cointegrated with the $\{p_t\}$ sequence. Call the constructed dollar value of the foreign price level f_t; that is, $f_t = e_t + p_{ft}$. Long-run PPP asserts that there exists a linear combination of the form $f_t = \beta_0 + \beta_1 p_t + \mu_t$ such that $\{\mu_t\}$ is stationary *and* the cointegrating vector is such that $\beta_1 = 1$.

As reported in Chapter 4, in Enders (1988), I used price and exchange rate data for Germany, Japan, Canada, and the United States for both the Bretton Woods (1960–1971) and post–Bretton Woods (1973–1988) periods.[12] Pretesting the data indicated that for each period, the U.S. price level $\{p_t\}$ and the dollar values of the foreign price levels $\{e_t + p_{ft}\}$ both contained a single unit root. With differing orders of integration, it would have been possible to immediately conclude that long-run PPP had failed.

The next step was to estimate the long-run equilibrium relation by regressing each $f_t = e_t + p_{ft}$ on p_t:

$$f_t = \beta_0 + \beta_1 p_t + \mu_t \tag{6.43}$$

Absolute PPP asserts $f_t = p_t$, so this version of the theory requires that $\beta_0 = 0$ and $\beta_1 = 1$. The intercept β_0 is consistent with the relative version of PPP, requiring only that domestic and foreign price levels are proportional to each other. Unless there are compelling reasons to omit the constant, the recommended practice is to include an intercept term in the equilibrium regression. In fact, Engle and Granger's (1987) original Monte Carlo simulations all include intercept terms.

The estimated values of β_1 and their associated standard errors are reported in Table 6.4. Note that five of the six values are estimated to be quite a bit below unity.

Table 6.4 The Equilibrium Regressions

	Germany	Japan	Canada
1973–1986			
Estimated β_1	0.5374	0.8938	0.7749
Standard error	(0.0415)	(0.0316)	(0.0077)
1960–1971			
Estimated β_1	0.6660	0.7361	1.0809
Standard error	(0.0262)	(0.0154)	(0.0200)

Be especially careful not to make too much of these findings. It is *not* appropriate to conclude that each value of β_1 is significantly different from unity simply because the values of $(1 - \beta_1)$ exceed two or three standard deviations. It is hard to overstate the point that the assumptions underlying this type of t-test are not applicable because there is no presumption that p_t is the exogenous variable while f_t is the dependent variable, or that $\{\mu_t\}$ is white noise.[13]

The residuals from each regression equation, called $\{\hat{\mu}_t\}$, were checked for unit roots. The unit root tests are straightforward because the residuals from a regression equation have a zero mean and do not have a time trend. The following two equations were estimated using the residuals from each long-run equilibrium relationship:

$$\Delta\hat{\mu}_t = a_1\hat{\mu}_{t-1} + \varepsilon_t \tag{6.44}$$

and

$$\Delta\hat{\mu}_t = a_1\hat{\mu}_{t-1} + \Sigma a_{i+1}\Delta\hat{\mu}_{t-i} + \varepsilon_t \tag{6.45}$$

Table 6.5 reports the estimated values of a_1 from (6.44) and from (6.45) using a lag length of four. It bears repeating that failure to reject the null hypothesis $a_1 = 0$ means we cannot reject the null of no cointegration. Alternatively, if $-2 < a_1 < 0$, it is possible to conclude that the $\{\hat{\mu}_t\}$ sequence does not have a unit root and that the $\{f_t\}$ and $\{p_t\}$ sequences are cointegrated. Also note that it is not appropriate to use the confidence intervals reported in Dickey and Fuller. The Dickey–Fuller statistics are inappropriate because the residuals used in (6.44) and (6.45) are not the actual error terms. Rather, these residuals are estimated error terms that are obtained from the estimate of the equilibrium regression. If we knew the magnitudes of the actual errors in each period, we could use the Dickey–Fuller tables.

Under the null hypothesis $a_1 = 0$, the critical values for the t-statistic depend on sample size. Comparing the results reported in Table 6.5 with the critical values provided by Table C indicates that only for Japan during the fixed exchange rate period is it possible to reject the null hypothesis of no cointegration. At the 5 percent significance level, the critical value of t is -3.398 for two variables and $T = 100$. Hence, at the 5 percent significance level we can reject the null of no cointegration (i.e., we accept the alternative that the variables are cointegrated) and find in favor of PPP. For

Table 6.5 Dickey–Fuller Tests of the Residuals

	Germany	Japan	Canada
1973–1986			
No lags			
Estimated a_1	−0.0225	−0.0151	−0.1001
Standard error	(0.0169)	(0.0236)	(0.0360)
t-statistic for $a_1 = 0$	−1.331	−0.640	−2.781
4 lags			
Estimated a_1	−0.0316	−0.0522	−0.0983
Standard error	(0.0170)	(0.0236)	(0.0388)
t-statistic for $a_1 = 0$	−1.859	−2.212	−2.533
1960–1971			
No lags			
Estimated a_1	−0.0189	−0.1137	−0.0528
Standard error	(0.0196)	(0.0449)	(0.0286)
t-statistic for $a_1 = 0$	−0.966	−2.535	−1.846
4 lags			
Estimated a_1	−0.0294	−0.1821	−0.0509
Standard error	(0.0198)	(0.0530)	(0.0306)
t-statistic for $a_1 = 0$	−1.468	−3.437	−1.663

the other countries in each time period, we cannot reject the null hypothesis of no cointegration and must conclude that PPP generally failed.

The third step in the methodology entails estimation of the error-correction model. Only the Japan/U.S. model needs estimation since it is the sole case for which cointegration holds. The final error-correction models for Japanese and U.S. price levels during the 1960 to 1971 period were estimated to be

$$\begin{aligned}\Delta f_t &= 0.00119 - 0.10548\hat{\mu}_{t-1} \\ &\quad\ (0.00044)\quad (0.04184)\end{aligned} \tag{6.46}$$

$$\begin{aligned}\Delta p_t &= 0.00156 + 0.01114\hat{\mu}_{t-1} \\ &\quad\ (0.00033)\quad (0.03175)\end{aligned} \tag{6.47}$$

where $\hat{\mu}_{t-1}$ is the lagged residual from the long-run equilibrium regression. Note that $\hat{\mu}_{t-1}$ is the estimated value of $f_{t-1} - \beta_0 - \beta_1 p_{t-1}$ and that standard errors are in parentheses.

Lag length tests (see the discussion of χ^2 and F-tests for lag length in Chapter 5) indicated that lagged values of Δf_{t-i} or Δp_{t-i} did not need to be included in the error-correction equations. Note that the point estimates in (6.46) and (6.47) indicate a direct convergence to long-run equilibrium. For example, in the presence of a one-unit deviation from long-run PPP in period $t - 1$, the Japanese price level (converted into dollars) falls by 0.10548 units and the U.S. price level rises by 0.01114 units. Both of these price changes in period t act to eliminate the positive discrepancy from long-run PPP present in period $t - 1$.

Notice the discrepancy between the magnitudes of the two speed-of-adjustment coefficients; in absolute value, the Japanese coefficient is approximately ten times that of the U.S. coefficient. As compared to the Japanese price level, the U.S. price level responded only slightly to a deviation from PPP. Moreover, the error-correction term is about 1/3 of a standard deviation from zero for the U.S. (0.01114/0.03175 = 0.3509) and approximately 2.5 standard deviations from zero for Japan (0.10548/0.04184 = 2.5210). Hence, at the 5 percent significance level, we can conclude that the speed-of-adjustment term is insignificantly different from zero for the United States but not for Japan. This result is consistent with the idea that the United States was a large country relative to Japan—movements in U.S. prices evolved independently of events in Japan, but movements in exchange rate–adjusted Japanese prices responded to events in the United States.

You can update the study using the data contained in the file COINT_PPP.XLS. The file contains quarterly values of the U.K., Japanese, and Canadian wholesale prices and bilateral exchange rates with the United States. Germany is not included because the preunification data for Germany are not compatible with the more recent data. The file also contains the U.S. wholesale price level. Questions 9 and 10 at the end of the chapter guide you through the process.

7. CHARACTERISTIC ROOTS, RANK, AND COINTEGRATION

Although the Engle and Granger (1987) procedure is easily implemented, it does have several important defects. The estimation of the long-run equilibrium regression requires that the researcher place one variable on the left-hand side and use the others as regressors. For example, in the case of two variables, it is possible to run the Engle–Granger test for cointegration by using the residuals from either of the following two "equilibrium" regressions:

$$y_t = \beta_{10} + \beta_{11}z_t + e_{1t} \tag{6.48}$$

or

$$z_t = \beta_{20} + \beta_{21}y_t + e_{2t} \tag{6.49}$$

As the sample size grows infinitely large, asymptotic theory indicates that the test for a unit root in the $\{e_{1t}\}$ sequence becomes equivalent to the test for a unit root in the $\{e_{2t}\}$ sequence. Unfortunately, the large-sample properties from which this result is derived may not be applicable to the sample sizes usually available to economists. In practice, it is possible to find that one regression indicates that the variables are cointegrated, whereas reversing the order indicates no cointegration. This is a very undesirable feature of the procedure because the test for cointegration should be invariant to the choice of the variable selected for normalization. The problem is obviously compounded using three or more variables since any of the variables can be selected as the left-hand-side variable. Moreover, in tests using three or more variables, we know that there may be more than one cointegrating vector. The method has no systematic procedure for the separate estimation of the multiple cointegrating vectors.

Another defect of the Engle–Granger procedure is that it relies on a *two-step* estimator. The first step is to generate the residual series $\{\hat{e}_t\}$, and the second step uses these generated errors to estimate a regression of the form $\Delta\hat{e}_t = a_1\hat{e}_{t-1} + \ldots$. Thus, the coefficient a_1 is obtained by estimating a regression using the residuals from another regression. Hence, any error introduced by the researcher in Step 1 is carried into Step 2. Fortunately, several methods have been developed that avoid these problems. The Johansen (1988) and the Stock and Watson (1988) maximum likelihood estimators circumvent the use of two-step estimators *and* can estimate and test for the presence of multiple cointegrating vectors. Moreover, these tests allow the researcher to test restricted versions of the cointegrating vector(s) and the speed-of-adjustment parameters. Often, we want to determine whether it is possible to verify a theory by testing restrictions on the magnitudes of the estimated coefficients.

The Johansen (1988) procedure relies heavily on the relationship between the rank of a matrix and its characteristic roots. Appendix 6.1 to this chapter reviews the essentials of these concepts; those of you wanting more details should review this appendix. For those wanting an intuitive explanation, notice that the Johansen procedure is nothing more than a multivariate generalization of the Dickey–Fuller test. In the univariate case, it is possible to view the stationarity of $\{y_t\}$ as being dependent on the magnitude $(a_1 - 1)$; that is,

$$y_t = a_1 y_{t-1} + \varepsilon_t$$

or

$$\Delta y_t = (a_1 - 1)y_{t-1} + \varepsilon_t$$

If $(a_1 - 1) = 0$, the $\{y_t\}$ process has a unit root. Ruling out the case in which $\{y_t\}$ is explosive, if $(a_1 - 1) \neq 0$ we can conclude that the $\{y_t\}$ sequence is stationary. The Dickey–Fuller tables provide the appropriate statistics to formally test the null hypothesis $(a_1 - 1) = 0$. Now consider the simple generalization to n variables; as in (6.26), let

$$x_t = A_1 x_{t-1} + \varepsilon_t$$

so that

$$\begin{aligned}\Delta x_t &= A_1 x_{t-1} - x_{t-1} + \varepsilon_t \\ &= (A_1 - I)x_{t-1} + \varepsilon_t \\ &= \pi x_{t-1} + \varepsilon_t \end{aligned} \tag{6.50}$$

where: x_t and ε_t are $(n \cdot 1)$ vectors
A_1 = an $(n \cdot n)$ matrix of parameters
I = an $(n \cdot n)$ identity matrix
and π is defined to be $(A_1 - I)$

As indicated in the discussion surrounding (6.27), the rank of $(A_1 - I)$ equals the number of cointegrating vectors. By analogy to the univariate case, if $(A_1 - I)$ consists of all zeroes—so that rank(π) = 0—all of the $\{x_{it}\}$ sequences are unit root processes. Since there is no linear combination of the $\{x_{it}\}$ processes that is stationary, the variables are not cointegrated. If we rule out characteristic roots that are greater than unity and if rank(π) = n, (6.50) represents a convergent system of difference equations, so that all variables are stationary.

There are several ways to generalize (6.50). The equation is easily modified to allow for the presence of a drift term; simply let

$$\Delta x_t = A_0 + \pi x_{t-1} + \varepsilon_t \tag{6.51}$$

where A_0 = the $(n \cdot 1)$ vector of constants $(a_{10}, a_{20}, \ldots, a_{n0})'$.

The effect of including the various a_{i0} is to allow for the possibility of a linear time trend in the data-generating process. You would want to include the drift term if the variables exhibited a decided tendency to increase or decrease. Here, the rank of π can be viewed as the number of cointegrating relationships existing in the "detrended" data. In the long run, $\pi x_{t-1} = 0$ so that each $\{\Delta x_{it}\}$ sequence has an expected value of a_{i0}. Aggregating all such changes over t yields the deterministic expression $a_{i0}t$.

Figure 6.3 illustrates the effects of including a drift in the data-generating process. Two random sequences with 100 observations each were generated; denote these sequences as $\{\varepsilon_{yt}\}$ and $\{\varepsilon_{zt}\}$. Initializing $y_0 = z_0 = 0$, we constructed the next 100 values of the $\{y_t\}$ and $\{z_t\}$ sequences as

$$\begin{bmatrix} \Delta y_t \\ \Delta z_t \end{bmatrix} = \begin{bmatrix} -0.2 & 0.2 \\ 0.2 & -0.2 \end{bmatrix} \begin{bmatrix} y_{t-1} \\ z_{t-1} \end{bmatrix} + \begin{bmatrix} \varepsilon_{yt} \\ \varepsilon_{zt} \end{bmatrix}$$

so that the cointegrating relationship is

$$-0.2y_{t-1} + 0.2z_{t-1} = 0$$

or

$$y_t = z_t$$

In Panel (a) of Figure 6.3, you can see that each sequence resembles a random walk process and that neither wanders too far from the other. Panel (b) adds drift coefficients such that $a_{10} = a_{20} = 0.1$; now each series tends to increase by 0.1 units in each period. In addition to the fact that each sequence shares the same stochastic trend, note that each also has the same deterministic time trend. The fact that each has the same deterministic trend is *not* a result of the equivalence between a_{10} and a_{20}; since y_t and z_t are cointegrated, the general solution to (6.51) necessitates that each have the same linear trend. For verification, Panel (c) sets $a_{10} = 0.1$ and $a_{20} = 0.4$. Again, the sequences have the same stochastic and deterministic trends. As an aside, note that increasing a_{20} and decreasing a_{10} would have an ambiguous effect on the slope of the deterministic trend. This point will be important in a moment; by appropriately manipulating the elements of A_0 it is possible to include a constant in the cointegrating vector(s) without imparting a deterministic time trend to the system.

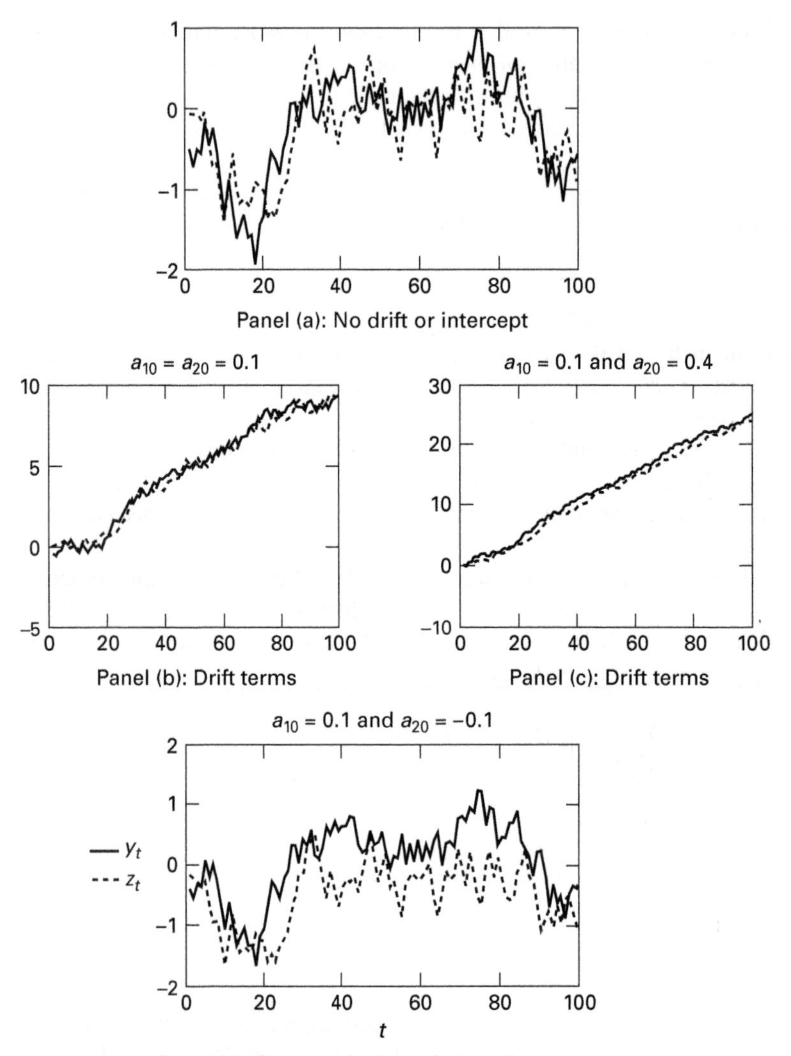

FIGURE 6.3 Drifts and Intercepts in Cointegrating Relationships

One way to include a constant in the cointegrating relationships is to restrict the values of the various a_{i0}. For example, if rank(π) = 1, the rows of π can differ by only a scalar, so that it is possible to write each $\{\Delta x_{it}\}$ sequence in (6.51) as

$$\begin{aligned}
\Delta x_{1t} &= \pi_{11}x_{1t-1} + \pi_{12}x_{2t-1} + \ldots + \pi_{1n}x_{nt-1} + a_{10} + \varepsilon_{1t} \\
\Delta x_{2t} &= s_2(\pi_{11}x_{1t-1} + \pi_{12}x_{2t-1} + \ldots + \pi_{1n}x_{nt-1}) + a_{20} + \varepsilon_{2t} \\
&\ldots \\
\Delta x_{nt} &= s_n(\pi_{11}x_{1t-1} + \pi_{12}x_{2t-1} + \ldots + \pi_{1n}x_{nt-1}) + a_{n0} + \varepsilon_{nt}
\end{aligned}$$

where s_i = scalars such that $s_i\pi_{1j} = \pi_{ij}$.

If the a_{i0} can be restricted such that $a_{i0} = s_i a_{10}$, it follows that all of the $\{\Delta x_{it}\}$ sequences can be written with the constant included in the cointegrating vector:

$$\Delta x_{1t} = (\pi_{11}x_{1t-1} + \pi_{12}x_{2t-1} + \ldots + \pi_{1n}x_{nt-1} + a_{10}) + \varepsilon_{1t}$$
$$\Delta x_{2t} = s_2(\pi_{11}x_{1t-1} + \pi_{12}x_{2t-1} + \ldots + \pi_{1n}x_{nt-1} + a_{10}) + \varepsilon_{2t}$$
$$\ldots$$
$$\Delta x_{nt} = s_n(\pi_{11}x_{1t-1} + \pi_{12}x_{2t-1} + \ldots + \pi_{1n}x_{nt-1} + a_{10}) + \varepsilon_{nt}$$

or, in compact form,

$$\Delta x_t = \pi^* x^*_{t-1} + \varepsilon_t \tag{6.52}$$

where:

$$x_t = (x_t, x_{2t}, \ldots, x_{nt})'$$
$$x^*_{t-1} = (x_{1t-1}, x_{2t-1}, \ldots, x_{nt-1}, 1)'$$
$$\pi^* = \begin{bmatrix} \pi_{11} & \pi_{12} & \cdots & \pi_{1n} & a_{10} \\ \pi_{21} & \pi_{22} & \cdots & \pi_{2n} & a_{20} \\ \cdot & \cdot & \cdots & \cdot & \cdot \\ \pi_{n1} & \pi_{n2} & \cdots & \pi_{nn} & a_{n0} \end{bmatrix}$$

The interesting feature of (6.52) is that the linear trend is purged from the system. In essence, the various a_{i0} have been altered in such a way that the general solution for each $\{x_{it}\}$ does not contain a time trend. The solution to the set of difference equations represented by (6.52) is such that all Δx_{it} are expected to equal zero when $\pi_{11}x_{1t-1} + \pi_{12}x_{2t-1} + \ldots + \pi_{1n}x_{nt-1} + a_{10} = 0$.

To highlight the difference between (6.51) and (6.52), Panel (d) of Figure 6.3 illustrates the consequences of setting $a_{10} = 0.1$ and $a_{20} = -0.1$. You can see that neither sequence contains a deterministic trend. In fact, for the data shown in the figure, the trend will vanish so long as we select values of the drift terms maintaining the relationship $a_{10} = -a_{20}$. (Question 1 at the end of this chapter will help you to demonstrate this result.)

Some econometricians prefer to include an intercept term in the cointegrating vector along with a drift term. This makes sense if the variables contain a drift and if economic theory suggests that the cointegrating vector contains an intercept. However, it should be clear that the intercept in the cointegrating vector is not identified in the presence of a drift term. After all, some portion of the unrestricted drift can always be included in the cointegration vector. In terms of the example above, the system can always be written as

$$\Delta x_{1t} = (\pi_{11}x_{1t-1} + \pi_{12}x_{2t-1} + \ldots + \pi_{1n}x_{nt-1} + b_{10}) + b_{11} + \varepsilon_{1t}$$
$$\ldots$$
$$\Delta x_{nt} = s_n(\pi_{11}x_{1t-1} + \pi_{12}x_{2t-1} + \ldots + \pi_{1n}x_{nt-1} + b_{10}) + b_{n1} + \varepsilon_{nt}$$

where b_{i1} is defined to be the value that satisfies $s_i b_{10} + b_{i1} = a_{10}$.

All that was done is to divide a_{10} into two parts and to place one part inside the cointegrating relationship. As such, some identification strategy is necessary since the proportion of the drift to include in the cointegrating vector is arbitrary. The popular software package E-Views, for example, identifies the portion belonging in the cointegrating vector as the amount necessary to force the error-correction term to have a sample mean of zero. Nevertheless, as you can see from Figure 6.3, a drift term outside of the cointegrating relationship is necessary to capture the effects of a sustained tendency for the variables to increase (or decrease). Most researchers include drift terms if the data match Panels (b) or (c) of Figure 6.3. Otherwise, they include intercepts in the cointegrating vector or exclude the deterministic regressors altogether. If you are unsure, you can use the methods described in the next section to test whether the drifts can be appropriately restricted. Some software packages allow you to include a deterministic time trend in the model. However, it is best to avoid the use of a trend as an explanatory variable unless you have a good reason to include it in the model. Johansen (1994) discusses the role of the deterministic regressors in a cointegrating relationship.

As with the augmented Dickey–Fuller test, the multivariate model can also be generalized to allow for a higher-order autoregressive process. Consider:

$$x_t = A_1 x_{t-1} + A_2 x_{t-2} + \dots + A_p x_{t-p} + \varepsilon_t \tag{6.53}$$

where: x_t = the $(n \cdot 1)$ vector $(x_{1t}, x_{2t}, \dots, x_{nt})'$
ε_t = an independently and identically distributed n-dimensional vector with zero mean and variance matrix Σ_ε

Equation (6.53) can be put in a more usable form by adding and subtracting $A_p x_{t-p+1}$ to the right-hand side to obtain

$$x_t = A_1 x_{t-1} + A_2 x_{t-2} + A_3 x_{t-3} + \dots + A_{p-2} x_{t-p+2} + (A_{p-1} + A_p) x_{t-p+1} - A_p \Delta x_{t-p+1} + \varepsilon_t$$

Next, add and subtract $(A_{p-1} + A_p) x_{t-p+2}$ to obtain

$$x_t = A_1 x_{t-1} + A_2 x_{t-2} + A_3 x_{t-3} + \dots - (A_{p-1} + A_p) \Delta x_{t-p+2} - A_p \Delta x_{t-p+1} + \varepsilon_t$$

Just as in the augmented Dickey–Fuller test developed in Chapter 4, we can continue in this fashion to obtain

$$\Delta x_t = \pi x_{t-1} + \sum_{i=1}^{p-1} \pi_i \Delta x_{t-i} + \varepsilon_t \tag{6.54}$$

where $\pi = -\left(I - \sum_{i=1}^{p} A_i\right)$ and $\pi_i = -\sum_{j=i+1}^{p} A_j$.

Again, the key feature to note in (6.54) is rank of the matrix π, the rank of π is equal to the number of independent cointegrating vectors. Clearly, if rank$(\pi) = 0$, the matrix is null and (6.54) is the usual VAR model in first differences. Instead, if π is of rank n, the vector process is stationary. In intermediate cases, if rank$(\pi) = 1$, there is a single cointegrating vector and the expression πx_{t-1} is the error-correction term. For other cases in which $1 <$ rank$(\pi) < n$, there are multiple cointegrating vectors.

As detailed in Appendix 6.1, the number of distinct cointegrating vectors can be obtained by checking the significance of the characteristic roots of π. We know that the rank of a matrix is equal to the number of its characteristic roots that differ from zero. Suppose we obtained the matrix π and ordered the n characteristic roots such that $\lambda_1 > \lambda_2 > \ldots > \lambda_n$. If the variables in x_t are *not* cointegrated, the rank of π is zero and all of these characteristic roots will equal zero. Since $\ln(1) = 0$, each of the expressions $\ln(1 - \lambda_i)$ will equal zero if the variables are not cointegrated. Similarly, if the rank of π is unity, $0 < \lambda_1 < 1$ so the first expression $\ln(1 - \lambda_1)$ will be negative and all the other $\lambda_i = 0$ so that $\ln(1 - \lambda_2) = \ln(1 - \lambda_3) = \ldots = \ln(1 - \lambda_n) = 0$.

In practice, we can obtain only estimates of π and its characteristic roots. The test for the number of characteristic roots that are insignificantly different from unity can be conducted using the following two test statistics:

$$\lambda_{trace}(r) = -T \sum_{i=r+1}^{n} \ln(1 - \hat{\lambda}_i) \tag{6.55}$$

$$\lambda_{max}(r, r+1) = -T \ln(1 - \hat{\lambda}_{r+1}) \tag{6.56}$$

where: $\hat{\lambda}_i$ = the estimated values of the characteristic roots (also called eigenvalues) obtained from the estimated π matrix

T = the number of usable observations

When the appropriate values of r are clear, these statistics are simply referred to as λ_{trace} and λ_{max}.

The first statistic tests the null hypothesis that the number of distinct cointegrating vectors is less than or equal to r against a general alternative. From the previous discussion, it should be clear that λ_{trace} equals zero when all $\lambda_i = 0$. The further the estimated characteristic roots are from zero, the more negative is $\ln(1 - \hat{\lambda}_i)$ and the larger is the λ_{trace} statistic. The second statistic tests the null that the number of cointegrating vectors is r against the alternative of $r + 1$ cointegrating vectors. Again, if the estimated value of the characteristic root is close to zero, λ_{max} will be small.

Critical values of the λ_{trace} and the λ_{max} statistics are obtained using the Monte Carlo approach. The critical values are reproduced in Table E at the end of this text. The distribution of these statistics depends on two things:

1. The number of nonstationary components under the null hypothesis (i.e., $n - r$).
2. The form of the vector A_0. Use the top portion of Table E if you do not include either a constant in the cointegrating vector or a drift term. Use the middle portion of the table if you include a drift term A_0. Use the bottom portion of the table if you include a constant in the cointegrating vector.

Using quarterly data for Denmark over the sample period 1974Q1 to 1987Q3, Johansen and Juselius (1990) let the x_t vector be represented by

$$x_t = (m2_t, y_t, i_t^d, i_t^b)'$$

where: $m2$ = log of the real money supply as measured by $M2$ deflated by a price index

y = log of real income

i^d = deposit rate on money representing a direct return on money holding

i^b = bond rate representing the opportunity cost of holding money

Including a constant in the cointegrating relationship (i.e., augmenting x_{t-1} with a constant), they report that the residuals from (6.54) appear to be serially uncorrelated.

The four characteristic roots of the estimated π matrix are given in the first column of the following table:[14]

	λ_{max} $-T\ln(1-\hat{\lambda}_{r+1})$	λ_{trace} $-T\Sigma\ln(1-\hat{\lambda}_i)$
$\hat{\lambda}_1 = 0.4332$	30.09	49.14
$\hat{\lambda}_2 = 0.1776$	10.36	19.05
$\hat{\lambda}_3 = 0.1128$	6.34	8.69
$\hat{\lambda}_4 = 0.0434$	2.35	2.35

The second column reports the various λ_{max} statistics as the number of usable observations ($T = 53$) multiplied by $\ln(1 - \hat{\lambda}_{r+1})$. For example, $-53\ln(1 - 0.0434) = 2.35$ and $-53\ln(1 - 0.1128) = 6.34$. The last column reports the λ_{trace} statistics as the summation of the λ_{max} statistics. Simple arithmetic reveals that $8.69 = 2.35 + 6.34$ and $19.05 = 2.35 + 6.34 + 10.36$.

To test the null hypothesis $r = 0$ against the general alternative $r = 1, 2, 3$, or 4, use the λ_{trace} statistic. Since the null hypothesis is $r = 0$ and there are four variables (i.e., $n = 4$), the summation in (6.55) runs from 1 to 4. If we sum over the four values, the calculated value of λ_{trace} is 49.14. Since Johansen and Juselius (1990) include the constant in the cointegrating vector, this calculated value of 49.14 is compared to the critical values reported in the bottom portion of Table E. For $n - r = 4$, the critical values of λ_{trace} are 49.65, 53.12, and 60.16 at the 10, 5, and 1 percent significance levels, respectively. Thus, at the 10 percent level, the restriction is *not* binding, so that the variables are *not* cointegrated using this test.

To make a point and to give you practice in using the table, suppose you want to test the null hypothesis $r \leq 1$ against the alternative $r = 2, 3$, or 4. Under this null hypothesis, the summation in (6.55) runs from 2 to 4 so that the calculated value of λ_{trace} is 19.05. For $n - r = 3$, the critical values of λ_{trace} are 32.00, 34.91, and 41.07 at the 10, 5, and 1 percent significance levels, respectively. The restriction $r = 0$ or $r = 1$ is not binding.

In contrast to the λ_{trace} statistic, the λ_{max} statistic has a specific alternative hypothesis. To test the null hypothesis $r = 0$ against the specific alternative $r = 1$, use equation (6.56). The calculated value of the $\lambda_{max}(0, 1)$ statistic is $-53\ln(1 - 0.4332) = 30.09$. For $n - r = 4$, the critical values of λ_{max} are 25.56, 28.14, 30.32, and 33.24 at the 10, 5, 2.5, and 1 percent significance levels, respectively. Hence, it is possible to reject the null hypothesis $r = 0$ at the 5 percent significance level (but not the 2.5 percent level) and conclude that there is only one cointegrating vector (i.e., $r = 1$). Before reading on, you should take a moment to examine the data and convince yourself that the null hypothesis $r = 1$ against the alternative $r = 2$ cannot be rejected at conventional levels. You should find that the calculated value of the λ_{max} statistic for $r = 1$ is 10.36 and that the critical value at the 10 percent level is 19.77. Hence, there is no significant evidence of more than one cointegrating vector.

The example illustrates the important point that the results of the λ_{trace} and λ_{max} tests can conflict. The λ_{max} test has the sharper alternative hypothesis. It is usually preferred for trying to pin down the number of cointegrating vectors.

8. HYPOTHESIS TESTING

In the Dickey–Fuller tests discussed in Chapter 4, it was important to correctly ascertain the form of the deterministic regressors. A similar situation applies in the Johansen procedure. As you can see in Table E, the critical values of the λ_{trace} and λ_{max} statistics tend to be smallest without any deterministic regressors and largest with an intercept term included in the cointegrating vector. Instead of cavalierly positing the form of A_0, it is possible to test restricted forms of the vector.

One of the most interesting aspects of the Johansen procedure is that it allows for testing restricted forms of the cointegrating vector(s). In a money demand study, you might want to test restrictions concerning the long-run proportionality between money and prices, or the sizes of the income and interest rate elasticities of demand for money. In terms of equation (6.1) (i.e., $m_t = \beta_0 + \beta_1 p_t + \beta_2 y_t + \beta_3 r_t + e_t$), the restrictions of interest are $\beta_1 = 1$, $\beta_2 > 0$, and $\beta_3 < 0$.

The key insight to all such hypothesis tests is that *if there are r cointegrating vectors, only these r linear combinations of the variables are stationary*. All other linear combinations are nonstationary. Thus, suppose you reestimate the model restricting the parameters of π. If the restrictions are not binding, you should find that the number of cointegrating vectors has *not* diminished.

To test for the presence of an intercept in the cointegrating vector as opposed to the unrestricted drift A_0, estimate the two forms of the model. Denote the ordered characteristic roots of the unrestricted π matrix by $\hat{\lambda}_1, \hat{\lambda}_2, \ldots, \hat{\lambda}_n$ and the characteristic roots of the model with the intercept(s) in the cointegrating vector(s) by $\hat{\lambda}_1^*, \hat{\lambda}_2^*, \ldots, \hat{\lambda}_n^*$. Suppose that the unrestricted form of the model has r nonzero characteristic roots. Asymptotically, the statistic

$$-T \sum_{i=r+1}^{n} [\ln(1 - \hat{\lambda}_i^*) - \ln(1 - \hat{\lambda}_i)] \tag{6.57}$$

has a χ^2 distribution with $(n - r)$ degrees of freedom.

The intuition behind the test is that all values of $\ln(1 - \hat{\lambda}_i^*)$ and $\ln(1 - \hat{\lambda}_i)$ should be equivalent if the restriction is not binding. Hence, small values for the test statistic imply that it is permissible to include the intercept in the cointegrating vector. However, the likelihood of finding a stationary linear combination of the n variables is greater with the intercept in the cointegrating vector than if the intercept is absent from the cointegrating vector. Thus, a large value of $\hat{\lambda}_{r+1}^*$ [and a corresponding large value of $-T\ln(1 - \hat{\lambda}_{r+1}^*)$] implies that the restriction artificially inflates the number of cointegrating vectors. Thus, as proven by Johansen (1991), if the test statistic is sufficiently large, it is possible to reject the null hypothesis of an intercept in the cointegrating vector(s) and conclude that there is a linear trend in the variables. This is precisely the case represented by the middle portion of Figure 6.3.

Johansen and Juselius (1990) tested the restriction that their estimated Danish money demand function does not have a drift. Since they found only one cointegrating vector among $m2$, y, i^d, and i^b, set $n = 4$ and $r = 1$. The calculated value of the χ^2 statistic in (6.57) is 1.99. With three degrees of freedom, this is insignificant at conventional levels; they conclude that the variables do not have a linear time trend and find it appropriate to include the constant in the cointegrating vector.

In order to test other restrictions on the cointegrating vector, Johansen defines the two matrices α and β, both of dimension $(n \cdot r)$ where r is the rank of π. The properties of α and β are such that

$$\pi = \alpha\beta'$$

Note that β is the matrix of cointegrating parameters and α is the matrix of weights with which each cointegrating vector enters the n equations of the VAR. In a sense, α can be viewed as the matrix of the speed-of-adjustment parameters. Due to the cross-equation restrictions, it is not possible to estimate α and β using OLS.[15] However, using maximum-likelihood estimation, it is possible to (1) estimate (6.54) as an error-correction model; (2) determine the rank of π; (3) use the r most significant cointegrating vectors to form β'; and (4) select α such that $\pi = \alpha\beta'$. Question 5 at the end of this chapter asks you to find several such α and β' matrices.

It is easy to understand the process in the case of a single cointegrating vector. Given that rank$(\pi) = 1$, the rows of π are all linear multiples of each other. Hence, the equations in (6.54) have the form:

$$\begin{aligned}
\Delta x_{1t} &= \pi_{11}x_{1t-1} + \pi_{12}x_{2t-1} + \ldots + \pi_{1n}x_{nt-1} + \ldots + \varepsilon_{1t} \\
\Delta x_{2t} &= s_2(\pi_{11}x_{1t-1} + \pi_{12}x_{2t-1} + \ldots + \pi_{1n}x_{nt-1}) + \ldots + \varepsilon_{2t} \\
&\ldots \\
\Delta x_{nt} &= s_n(\pi_{11}x_{1t-1} + \pi_{12}x_{2t-1} + \ldots + \pi_{1n}x_{nt-1}) + \ldots + \varepsilon_{nt}
\end{aligned}$$

where the s_i are scalars and, for notational simplicity, the matrices $\pi_i\Delta x_{t-i}$ have not been written out.

Now define $\alpha_i = s_i\pi_{11}$ and $\beta_i = \pi_{1i}/\pi_{11}$ so that each equation can be written as[16]

$$\Delta x_{it} = \alpha_i(x_{1t-1} + \beta_2 x_{2t-1} + \ldots + \beta_n x_{nt-1}) + \ldots + \varepsilon_{it} \qquad (i = 1, \ldots, n)$$

or, in matrix form,

$$\Delta x_t = \sum_{i=1}^{p-1} \pi_i \Delta x_{t-i} + \alpha\beta' x_{t-1} + \varepsilon_t \tag{6.58}$$

where the single cointegrating vector is $\beta = (1, \beta_2, \beta_3, \ldots, \beta_n)'$ and the speed-of-adjustment parameters are given by $\alpha = (\alpha_1, \alpha_2, \ldots, \alpha_n)'$.

Once α and β' are determined, testing various restrictions on α and β' is straightforward if you remember the fundamental point that if there are r cointegrating vectors, only these r linear combinations of the variables are stationary. Thus, the test statistics involve comparing the number of cointegrating vectors under the null and alternative hypotheses. Again, let $\hat{\lambda}_1, \hat{\lambda}_2, \ldots, \hat{\lambda}_n$ and $\hat{\lambda}_1^*, \hat{\lambda}_2^*, \ldots, \hat{\lambda}_n^*$ denote the ordered characteristic roots of the unrestricted and restricted models, respectively. To test restrictions on β, form the test statistic

$$T\sum_{i=1}^{r} [\ln(1 - \hat{\lambda}_i^*) - \ln(1 - \hat{\lambda}_i)] \tag{6.59}$$

Asymptotically, this statistic has a χ^2 distribution with degrees of freedom equal to the number of restrictions placed on β. Small values of $\hat{\lambda}_i^*$ relative to $\hat{\lambda}_i$ (for $i \leq r$) imply a reduced number of cointegrating vectors. Hence, the restriction embedded in the null hypothesis is binding if the calculated value of the test statistic exceeds that in a χ^2 table. For example, Johansen and Juselius test the restriction that money and income move proportionally. Their estimated long-run equilibrium relationship is

$$m2_t = 1.03y_t - 5.21i_t^b + 4.22i_t^d + 6.06$$

They restrict the coefficient of income to be unity and find the restricted values of the $\hat{\lambda}_i^*$ to be such that

	$\hat{\lambda}_i^*$	$T \ln(1 - \hat{\lambda}_i^*)$
$i = 1$	0.433	−30.04
$i = 2$	0.172	−10.01
$i = 3$	0.044	−2.36
$i = 4$	0.006	−0.32

Given that the unrestricted model has $r = 1$ and $-T\ln(1 - \hat{\lambda}_1) = 30.09$, (6.59) becomes $-30.04 + 30.09 = 0.05$. Since there is only one restriction imposed on β, the test statistic has a χ^2 distribution with one degree of freedom. A χ^2 table indicates that 0.05 is not significant; hence, they conclude that the restriction is not binding.

Restrictions on α can be tested in the same way. The procedure is to restrict α and compare the r most significant characteristic roots for the restricted and unrestricted models using (6.59). If the calculated value of (6.59) exceeds that from a χ^2 table, with degrees of freedom equal to the number of restrictions placed on α, the restrictions can be rejected. For example, Johansen and Juselius (1990) test the restriction that only money demand (i.e., $m2_t$) responds to the deviation from long-run equilibrium. Formally, they test the restriction that $\alpha_2 = \alpha_3 = \alpha_4 = 0$. Restricting the three values of α_i to equal zero, they find the largest characteristic root in the restricted model is such that $T\ln(1 - \hat{\lambda}_1^*) = -23.42$. Since the unrestricted model is such that $T\ln(1 - \hat{\lambda}_1) = -30.09$, equation (6.59) becomes $-23.42 - (-30.09) = 7.67$. The χ^2 statistic with three degrees of freedom is 7.81 at the 5 percent significance level. Hence, they find mild support for the hypothesis that the restriction is not binding.

If there is a single cointegrating vector, the Engle–Granger and Johansen methods have the same asymptotic distribution. If it can be determined that only one cointegrating vector exists, it is common to rely on the estimated error-correction model to test restrictions on α. If $r = 1$, and a single value of α is being tested, the usual t-statistic is asymptotically equivalent to the Johansen test.

Lag Length and Causality Tests

The simplest way to understand lag length tests is to consider the system in the form of (6.54):

$$\Delta x_t = \pi x_{t-1} + \sum_{i=1}^{p-1} \pi_i \Delta x_{t-i} + \varepsilon_t$$

Regardless of the rank of π, all of the Δx_{t-i} are stationary variables. Hence, we can use Rule 1 of Sims, Stock, and Watson (1990). Recall that the rule implies that the coefficients of interest on zero-mean stationary variables can be using a normal distribution. Since lag length depends solely on the values of π_i, a χ^2 distribution is appropriate to test any restriction concerning lag length. As in the case of any VAR, let Σ_u and Σ_r be the variance/covariance matrices of the unrestricted and restricted systems, respectively. As in Chapter 5, let c denote the maximum number of regressors contained in the longest equation. The test statistic

$$(T - c)(\log|\Sigma_r| - \log|\Sigma_u|)$$

can be compared to a χ^2 distribution with degrees of freedom equal to the number of restrictions in the system. Alternatively, you can use the multivariate AIC or SBC to determine the lag length. If you want to test the lag lengths for a single equation, an F-test is appropriate.

The rule also means that you cannot perform Granger causality tests in a cointegrated system using a standard F-test. First, suppose that rank $(\pi) = 0$ so that

$$\Delta x_t = \sum_{i=1}^{p-1} \pi_i \Delta x_{t-i} + \varepsilon_t$$

As such, Granger causality involves only stationary variables. Yet, this was precisely the case discussed in Chapter 5 when the variables in a VAR are not cointegrated. Hence, Granger causality tests can be conducted using a standard F-distribution. However, if the variables are cointegrated, a Granger causality test involves the coefficients of π. Since these coefficients multiply nonstationary variables, it is not appropriate to use an F-statistic to test for Granger causality. After all, if rank $(\pi) \neq 0$, it is impossible to write the restrictions of the test as restrictions on a set of $I(0)$ variables. Block-exogeneity tests are also ruled out. If w_t is cointegrated with y_t or z_t, you cannot use a standard χ^2 test to determine whether w_t belongs in the equations for y_t and z_t.

To Difference or Not to Difference

We have reached a point where it is possible to address the issue of differencing the nonstationary variables in an unrestricted VAR. There is no question that differencing leads to a misspecification error if the variables are cointegrated. Suppose that the actual data-generating process is given by the cointegrated system of (6.54) but you estimate the following VAR in first differences:

$$\Delta x_t = \sum_{i=1}^{p-1} \pi_i \Delta x_{t-i} + \varepsilon_t$$

The system is misspecified since it excludes the long-run equilibrium relationships among the variables that are contained in πx_{t-1}. Given the misspecification error, all of the coefficient estimates, t-tests, F-tests, tests of cross-equation restrictions, impulse responses, and variance decompositions are not representative of the true process. Hence, there is a substantial penalty to pay if you estimate a VAR in first differences when the data are actually cointegrated; differencing "throws away" information contained in the cointegrating relationship(s).

Why not simply estimate all VARs in levels? The answer is that it is preferable to use the first differences if the $I(1)$ variables are not cointegrated. There are three consequences if the $I(1)$ variables are not cointegrated and you estimate the VAR in levels:

1. Tests lose power because you estimate n^2 more parameters (one extra lag of each variable in each equation).
2. For a VAR in levels, tests for Granger causality conducted on the $I(1)$ variables do not have a standard F-distribution. If you use first differences, you can use the standard F-distribution to test for Granger causality.
3. When the VAR has $I(1)$ variables, the impulse responses at long forecast horizons are inconsistent estimates of the true responses. Since the impulse responses need not decay, any imprecision in the coefficient estimates will have a permanent effect on the impulse responses. If the VAR is estimated in first differences, the impulse responses decay to zero and so the estimated responses are consistent.

The suggestion is that it is important to properly determine whether the $I(1)$ variables are cointegrated. You can perform lag length tests regardless of whether the variables are cointegrated. As such, the suggested methodology is to estimate an unrestricted VAR. Most researchers would begin with a lag length of approximately $T^{1/3}$. You may want to extend the number of lags if you suspect a substantial amount of seasonality. For example, with 100 observations of two variables using quarterly data, you might want to begin with 12 lags even though $T^{1/3}$ is approximately five. Select the most appropriate lag length and then perform a cointegration test. If the variables are not cointegrated, estimate the system in first differences. If the variables are cointegrated, you can work with the error-correction model. Since the error-correction term and all values of Δx_{t-i} are stationary, you can conduct inference on any variable (except those appearing within the cointegrating vectors) using the usual test statistics. Impulse responses and variance decompositions will yield consistent estimates of the actual values.

Tests on Multiple Cointegrating Vectors

If the rank of π exceeds one, it is not straightforward to interpret the cointegrating vectors. When there are multiple cointegrating vectors, any linear combination of these vectors is also a cointegrating vector. Fortunately, it is often possible to identify separate behavioral relationships by appropriately restricting the individual cointegrating vectors. The only complication is that you need to be clear about the number of restrictions you impose on the system. It is important to note that *if there are r cointegration relationships in an n-variable system, there exists a cointegrating vector for each subset of $(n - r + 1)$ variables.* For example, if there are two cointegrating vectors in a three-variable system, there is a cointegrating vector for each bilateral pair of the variables ($2 = n - r + 1$). To demonstrate the point, let $x_t = (x_{1t}, x_{2t}, x_{3t}, x_{4t})'$ and suppose there are two cointegrating vectors for these four variables. If we normalize each vector with respect to x_{1t}, we can write the condition $\beta' x_t = 0$ as

$$\begin{bmatrix} 1 & -\beta_{12} & -\beta_{13} & -\beta_{14} \\ 1 & -\beta_{22} & -\beta_{23} & -\beta_{24} \end{bmatrix} \begin{bmatrix} x_{1t} \\ x_{2t} \\ x_{3t} \\ x_{4t} \end{bmatrix} = \begin{bmatrix} 0 \\ 0 \end{bmatrix}$$

Consider the $2 \cdot n$ matrix β' consisting of the cointegrating parameters. Subtract row 1 from row 2 to obtain

$$\begin{bmatrix} 1 & -\beta_{12} & -\beta_{13} & -\beta_{14} \\ 1 & -\beta_{22} + \beta_{12} & -\beta_{23} + \beta_{13} & -\beta_{24} + \beta_{14} \end{bmatrix}$$

Now, renormalize row 2 by dividing each of its elements by $(\beta_{12} - \beta_{22})$ to obtain

$$\begin{bmatrix} 1 & -\beta_{12} & -\beta_{13} & -\beta_{14} \\ 0 & 1 & -\beta_{23}^{*} & -\beta_{24}^{*} \end{bmatrix}$$

where $-\beta_{23}^{*} = (\beta_{13} - \beta_{23})/(\beta_{12} - \beta_{22})$ and $-\beta_{24}^{*} = (\beta_{14} - \beta_{24})/(\beta_{12} - \beta_{22})$. Hence, x_{2t}, x_{3t}, and x_{4t} are cointegrated such that $x_{2t} = \beta_{23}^{*} x_{3t} + \beta_{24}^{*} x_{4t}$. Similarly, add β_{12} times row 2 to row 1 to obtain

$$\begin{bmatrix} 1 & 0 & -\beta_{13}^{*} & -\beta_{14}^{*} \\ 0 & 1 & -\beta_{23}^{*} & -\beta_{24}^{*} \end{bmatrix}$$

where $\beta_{1j}^{*} = \beta_{1j} + \beta_{12}\beta_{2j}^{*}$.

Thus, x_1, x_3, and x_4 are cointegrated such that $x_{1t} = \beta_{13}^{*} x_{3t} + \beta_{14}^{*} x_{4t}$. Since the labeling of the variables is irrelevant, it follows that there exists a cointegrating vector for each subset of three variables. More generally, β' will be an $r \cdot n$ matrix of cointegrating parameters, and each subset of $n - r + 1$ variables will be cointegrated. From the preceding discussion, it should be clear that standard row and column operations on β' *do not* entail restrictions on the cointegrating vectors. Such operations merely result in additional cointegrating vectors that are linear combinations of the original vectors.

EXAMPLE 1: VARIABLE EXCLUSION WITHIN AN EQUATION With multiple cointegrating vectors, you cannot test whether any one particular $\beta_{ij} = 0$ since this assumption does not restrict the cointegrating space. In the general case where β' is an $r \cdot n$ matrix, a testable exclusion restriction entails the exclusion of r or more variables from a cointegrating vector. Hence, excluding r variables from a cointegrating vector entails only one restriction. If the sample value of the χ^2 statistic with one degree of freedom (since there is only one restriction involved) exceeds a critical value, reject the null hypothesis that this set of variables contains a cointegrating relationship.

EXAMPLE 2: VARIABLE EXCLUSION ACROSS EQUATIONS Next, suppose that you want to test whether x_{4t} can be excluded from the set of cointegrating relationships. The restriction $\beta_{14} = \beta_{24} = 0$ entails only one restriction on the cointegrating space. In the general case where β' is an $r \cdot n$ matrix, the test $\beta_{1j} = \beta_{2j} = \ldots = \beta_{rj} = 0$ still involves only one restriction. This follows since x_{it} can be eliminated from $r - 1$ equations using simple row and column operations.

EXAMPLE 3: CONDITIONAL RESTRICTIONS It is also possible to restrict one cointegrating vector conditional on the values of all other cointegrating vectors. For example, you might want to determine if $(1, 0, \beta_{23}, \beta_{24})'$ is a cointegrating vector for the given normalized values of β_{12}, β_{13}, and β_{14}. Thus, you fix the values of β_{12}, β_{13}, and β_{14} and determine whether you can exclude x_{2t} from the second vector. Cutler, Davis, and Smith (1999) considered the identification issue in considerable detail. They examined the following four behavioral relationships in a seven-variable system:

$$\begin{aligned} m_t &= d_0 + d_1 y_t + d_2 r_t + d_3 p_t + e_{1t} \\ c_t &= a_0 + a_1 y_t + a_2 r_t + e_{2t} \\ i_t &= b_0 + b_1 y_t + b_2 r_t + e_{3t} \\ im_t &= g_0 + g_1 y_t + g_2 r_t + e_{4t} \end{aligned}$$

where: m_t = log of nominal money holdings
y_t = log of real income
r_t = real interest rate
c_t = log of real consumption
i_t = log of real investment
p_t = log of the price level
im_t = log of real imports
e_{1t}, e_{2t}, e_{3t}, and e_{4t} are stationary error terms.

The first equation is the money demand equation. The next three equations are a simple consumption function, an investment function, and an import demand function, respectively. Consumption, investment, and imports are each assumed to be functions of only income and the interest rate. The issue is to determine whether it is possible to identify these four equations from a seven-variable system. Toward this end, they obtained estimates of a 7×7 π matrix over a number of sample periods. There were at least four cointegrating vectors in every case considered. Over the entire sample, 1960*Q*2 to 1990*Q*4, Cutler, Davis, and Smith (1999) found that they could not reject the restrictions at conventional significance levels (the *prob*-value was 16 percent).

The Test in the Presence of *I*(2) Variables

It is also possible to test for multicointegration using Johansen's methodology. Consider the VAR system:

$$\Delta^2 x_t = \pi x_{t-1} + \Gamma \Delta x_{t-1} + \sum_{i=1}^{p-2} \pi_i \Delta^2 x_{t-i} + \varepsilon_t \qquad (6.60)$$

The issue of multicointegration concerns the ranks of both π and Γ. In principle, it is possible to consider all possible orders of cointegration for the variables in the system. However, to illustrate the procedure, it is useful to begin with a three-variable system consisting of the three $I(2)$ variables x_{1t}, x_{2t}, and x_{3t} that are multicointegrated such that

$$\pi_{11}x_{1t} + \pi_{12}x_{2t} + \pi_{13}x_{3t} + \Gamma_{11}\Delta x_{1t} + \Gamma_{12}\Delta x_{2t} + \Gamma_{13}\Delta x_{3t} = 0$$

Let r denote the rank of π and r_1 denote the rank of Γ so that (6.60) is such that $r = r_1 = 1$. Clearly, if $r = 0$, multicointegration fails since there is no linear combination of the three $I(2)$ variables that forms an equilibrium relationship. If $r = 1$ and $r_1 = 0$, the equilibrium relationship has the form $\pi_{11}x_{1t} + \pi_{12}x_{2t} + \pi_{13}x_{3t} = 0$. As such, $\Delta^2 x_t = \pi x_{t-1} + I(0)$ variables so that $\pi_{11}x_{1t} + \pi_{12}x_{2t} + \pi_{13}x_{3t}$ is necessarily a stationary relationship—the variables are $CI(2, 2)$. All of this may seem straightforward, but there is a complicating factor when the ranks of π and Γ have to be estimated. To illustrate the point, suppose that the $I(2)$ variables are cointegrated such that

$$\pi_{11}x_{1t} + \pi_{12}x_{2t} + \pi_{13}x_{3t} \sim I(1)$$

where $\sim I(d)$ indicates the order of integration.

If you take the first difference, it follows that $\pi_{11}\Delta x_{1t} + \pi_{12}\Delta x_{2t} + \pi_{13}\Delta x_{3t} \sim I(0)$. You should be able to figure out the problem. For any cointegrating vector in π, it is possible to estimate an identical cointegration vector for the first differences of the variables. Yet a linear combination of the two relationships is not stationary. Consider the result obtained by subtracting the $I(0)$ relationship from the $I(1)$ relationship:

$$\begin{aligned}&\pi_{11}x_{1t} + \pi_{12}x_{2t} + \pi_{13}x_{3t} - \pi_{11}\Delta x_{1t} - \pi_{12}\Delta x_{2t} - \pi_{13}\Delta x_{3t}\\ &= \pi_{11}x_{1t-1} + \pi_{12}x_{2t-1} + \pi_{13}x_{3t-1}\end{aligned}$$

Since $\pi_{11}x_{1t-1} + \pi_{12}x_{2t-1} + \pi_{13}x_{3t-1}$ is $I(1)$, all that has been done is to change the time subscript for the variables in the cointegrating relationship. The point is that it is necessary to find cointegrating vectors in Γ that are not linear combinations of those in π.[17]

If we take the more general case considered by Johansen (1995), let rank(π) $= r$ and let s denote the number of cointegrating vectors in Γ that are orthogonal to those in π. In an n-variable system such that some of the variables are $I(2)$, you should be able to verify that:

1. If $r = 0$, there is no relationship among the variables that is stationary.
2. In an n-variable system, if $r + s = n - 1$, there is a unique multicointegrating vector. The number of $I(2)$ stochastic trends in an n-variable system is given by $n - r - s$.
3. The value of s must be such that $s < n - r$. For the analysis of $I(2)$ variables to be appropriate, the values of r and s must be such that $s + r < n$. If $s = n - r$, then x_t contains no $I(2)$ variables.

Johansen's cointegration test with $I(2)$ variables is actually a two-step procedure. In the first step, you estimate a model as in (6.60) to determine the rank of π. Determine the value of r using the λ_{trace} and λ_{max} statistics in the usual way. In the second step, you determine the value of s conditional on the value of r.[18] Let the null hypothesis be $s = s_0$ and consider:

$$Q^*_{r,s} = -T \sum_{i=s_0+1}^{n} \ln(1 - \hat{\lambda}_i) \tag{6.61}$$

Hence, $Q^*_{r,s}$ is constructed in the same fashion as a λ_{trace} statistic. The principal differences are that you test the rank of Γ conditional on the value of r and that you obtain the number of cointegrating vectors orthogonal to those in π.[19] As such, the critical values needed to determine the value of s have to be modified. Given the value of r, if the sample value of $Q^*_{r,s}$ exceeds the critical value calculated by Johansen, reject the null hypothesis $s = s_0$ in favor of the alternative $s > s_0$. For $r = 1$, the critical values at the 10 percent, 5 percent, and 1 percent significance levels are:

Critical Values for $Q^*_{1,s}$

	$s = 0$	$s = 1$
10%	31.88	17.79
5%	34.80	19.99
1%	40.84	24.74

For example, let $r = 1$ and suppose that the sample value of $Q^*_{1,s}$ is found to be 35.00. As such, the null hypothesis $s = 0$ can be rejected at the 5 percent significance level.

9. ILLUSTRATING THE JOHANSEN METHODOLOGY

An interesting way to illustrate the Johansen methodology is to use exactly the same data shown in Figure 6.2. Recall that the data are contained in the file COINT6.XLS. Although the Engle–Granger technique did find that the simulated data were cointegrated, a comparison of the two procedures is useful. Use the following four steps when implementing the Johansen procedure.

STEP 1: It is good practice to pretest all variables to assess their order of integration. Plot the data to see if a linear time trend is likely to be present in the data-generating process. In most instances, you will have variables that are integrated of the same order. In other cases, you can check for multicointegration.

The results of the test can be quite sensitive to the lag length, so it is important to be careful. The most common procedure is to estimate a vector autoregression using the *undifferenced* data. Then use the same lag length tests as in a traditional VAR. Begin with the longest lag length deemed reasonable and test whether it can be shortened. For example, if we want to test whether lags 2 through 4 are important, we can estimate the following two VARs:

$$x_t = A_0 + A_1x_{t-1} + A_2x_{t-2} + A_3x_{t-3} + A_4x_{t-4} + e_{1t}$$
$$x_t = A_0 + A_1x_{t-1} + e_{2t}$$

where: x_t = the $(n \cdot 1)$ vector of variables
$A_0 = (n \cdot 1)$ matrix of intercept terms
$A_i = (n \cdot n)$ matrices of coefficients
e_{1t} and $e_{2t} = (n \cdot 1)$ vectors of error terms

Estimate the first system with four lags of each variable in each equation and call the variance/covariance matrix of residuals Σ_4. Now estimate the second equation using only one lag of each variable in each equation and call the variance/covariance matrix of residuals Σ_1. Even though we are working with nonstationary variables, we can perform lag length tests using the likelihood ratio test statistic recommended by Sims (1980):

$$(T - c)(\log|\Sigma_1| - \log|\Sigma_4|)$$

where: T = number of observations
c = number of parameters in the unrestricted system
$\log|\Sigma_i|$ = natural logarithm of the determinant of Σ_i

Following Sims, use the χ^2 distribution with degrees of freedom equal to the number of coefficient restrictions. Since each A_i has n^2 coefficients, constraining $A_2 = A_3 = A_4 = 0$ entails $3n^2$ restrictions. Alternatively, you can select lag length p using the multivariate generalizations of the AIC or SBC.

STEP 2: Estimate the model and determine the rank of π. Many time-series statistical software packages contain a routine to estimate the model. Here, it suffices to say that OLS is not appropriate because it is necessary to impose cross-equation restrictions on the π matrix. In most circumstances, you may choose to estimate the model in three forms: (1) with all elements of A_0 set equal to zero, (2) with a drift, or (3) with a constant term in the cointegrating vector.

For example, we can use the simulated data shown in Figure 6.2 so that $x_t = (y_t, z_t, w_t)'$. If we pretend that we do not know the form of the data-generating process, we might want to include an intercept term in the cointegrating vector(s). As we saw in the last section, it is possible to test for the presence of the intercept. Lag length tests indicate setting $p = 2$ so that the estimated form of the model is

$$\Delta x_t = A_0 + \pi x_{t-1} + \pi_1 \Delta x_{t-1} + \varepsilon_t \tag{6.62}$$

where A_0 was constrained so as to force the intercept to appear in the cointegrating vector.

As always, carefully analyze the properties of the residuals of the estimated model. Any evidence that the errors are not white noise usually means that lag lengths are too short. Figure 6.4 shows deviations of y_t from the long-run relationship ($\mu_t = -0.01331 - y_t - 1.0350z_t + 1.0162w_t$) and one of the

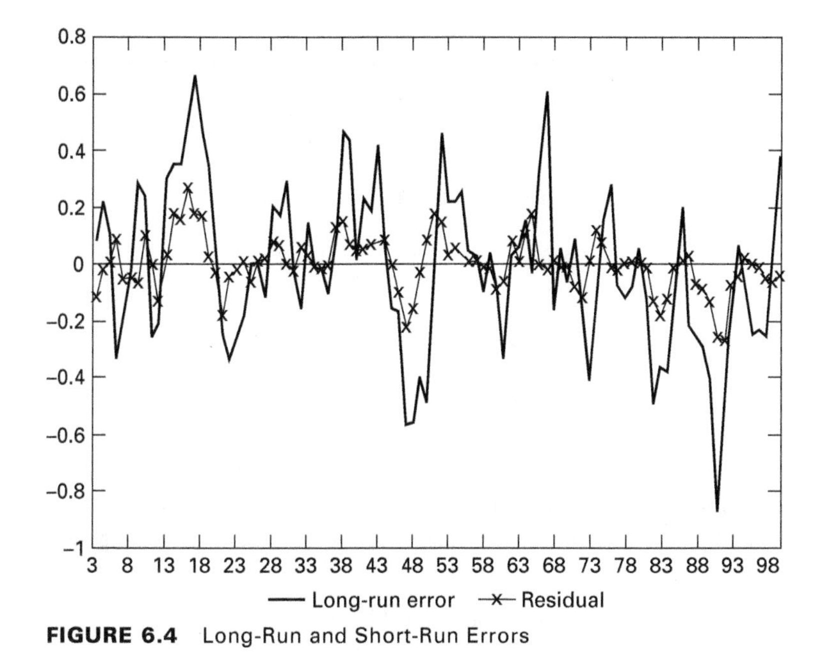

FIGURE 6.4 Long-Run and Short-Run Errors

error sequences [i.e., the $\{\varepsilon_{yt}\}$ sequence that equals the residuals from the y_t equation in (6.62)]. Both sequences conform to their theoretical properties in that the residuals from the long-run equilibrium appear to be stationary and the estimated values of the $\{\varepsilon_{yt}\}$ series approximate a white-noise process.

The estimated values of the characteristic roots of the π matrix in (6.62) are

$$\lambda_1 = 0.32600; \lambda_2 = 0.14032; \text{and } \lambda_3 = 0.033168$$

Since $T = 98$ (100 observations less the two lost as a result of using two lags), the calculated values of $\lambda_{\max}$ and λ_{trace} for the various possible values of r are reported in the center column of Table 6.6.

Consider the hypothesis that the variables are not cointegrated (so that the rank $\pi = 0$). Depending on the alternative hypothesis, we have a choice of two possible test statistics. If we are interested in the hypothesis that the variables are not cointegrated ($r = 0$) against the alternative of one or more cointegrating vectors ($r > 0$), we can calculate the $\lambda_{\text{trace}}(0)$ statistic:

$$\begin{aligned}\lambda_{\text{trace}}(0) &= -T\,[\ln(1-\lambda_1) + \ln(1-\lambda_2) + \ln(1-\lambda_3)]\\ &= -98\,[\ln(1-0.326) + \ln(1-0.14032) + \ln(1-0.033168)]\\ &= 56.786\end{aligned}$$

Since 56.786 exceeds the 5 percent critical value of the λ_{trace} statistic (in the bottom panel of Table E, the critical value is 34.91), it is possible to reject the null hypothesis of no cointegrating vectors and accept the alternative of

Table 6.6 The λ_{max} and λ_{trace} Tests

Null Hypothesis	Alternative Hypothesis		95% Critical Value	90% Critical Value
λ_{trace} tests		λ_{trace} value		
$r = 0$	$r > 0$	56.786	34.91	32.00
$r \leq 1$	$r > 1$	18.123	19.96	17.85
$r \leq 2$	$r > 2$	3.306	9.24	7.52
λ_{max} tests		λ_{max} value		
$r = 0$	$r = 1$	38.663	22.00	19.77
$r = 1$	$r = 2$	14.817	15.67	13.75
$r = 2$	$r = 3$	3.306	9.24	7.52

one or more cointegrating vectors. Next, we can use the $\lambda_{trace}(1)$ statistic to test the null of $r \leq 1$ against the alternative of two or three cointegrating vectors. In this case, the $\lambda_{trace}(1)$ statistic is

$$\begin{aligned}\lambda_{trace}(1) &= -T\,[\ln(1 - \lambda_2) + \ln(1 - \lambda_3)] \\ &= -98\,[\ln(1 - 0.14032) + \ln(1 - 0.033168)] \\ &= 18.123\end{aligned}$$

Since 18.123 is less than the 5 percent critical value of 19.96, we cannot reject the null hypothesis at this significance level. However, 18.123 does exceed the 10 percent critical value of 17.85; some researchers might reject the null and accept the alternative of two or three cointegrating vectors. The $\lambda_{trace}(2)$ statistic indicates no more than two cointegrating vectors at the 10 percent significance level.

The λ_{max} statistic does not help to clarify the issue. The null hypothesis of no cointegrating vectors ($r = 0$) against the specific alternative $r = 1$ is clearly rejected. The calculated value $\lambda_{max}(0, 1) = -98 \ln(1 - 0.326) = 38.663$ exceeds the 5 percent critical value of 22.00. Note that the test of the null hypothesis $r = 1$ against the specific alternative $r = 2$ cannot be rejected at the 5 percent, but can be rejected at the 10 percent, significance level. The calculated value of $\lambda_{max}(1, 2)$ is $-98 \ln(1 - 0.14032) = 14.817$, whereas the critical values at the 5 and 10 percent significance levels are 15.67 and 13.75, respectively. Even though the actual data-generating process contains only one cointegrating vector, the realizations are such that researchers willing to use the 10 percent significance level would incorrectly conclude that there are two cointegrating vectors. Failing to reject an incorrect null hypothesis is always an inherent danger of using wide confidence intervals.

STEP 3: Analyze the normalized cointegrating vector(s) and speed-of-adjustment coefficients. If we select $r = 1$, the estimated cointegrating vector $(\beta_0\, \beta_1\, \beta_2\, \beta_3)$ is

$$\beta = (0.00553, 0.41532, 0.42988, -0.42207)$$

If we normalize with respect to β_1, the normalized cointegrating vector and the speed-of-adjustment parameters are

$$\beta = (-0.01331, -1.0000, -1.0350, 1.0162)$$
$$\alpha_y = 0.54627$$
$$\alpha_z = 0.16578$$
$$\alpha_w = 0.21895$$

Recall that the data were constructed imposing the long-run relationship $w_t = y_t + z_t$; hence, the estimated coefficients of the normalized β vector are close to their theoretical values of (0 −1 −1 1). Consider the following tests:

1. The test that $\beta_0 = 0$ entails one restriction on one cointegrating vector; hence, the likelihood ratio test has a χ^2 distribution with one degree of freedom. The calculated value of $\chi^2 = 0.011234$ is not significant at conventional levels. Hence, we cannot reject the null hypothesis that $\beta_0 = 0$. Thus, it is possible to use the form of the model in which there is neither a drift nor an intercept in the cointegrating vector. Thus, to clarify the issue concerning the number of cointegrating vectors, it would be wise to reestimate the model excluding the constant from the cointegrating vector.
2. To restrict the normalized cointegrating vector such that $\beta_2 = -1$ and $\beta_3 = 1$ entails two restrictions on one cointegrating vector; hence, the likelihood ratio test has a χ^2 distribution with two degrees of freedom. The calculated value of $\chi^2 = 0.55350$ is not significant at conventional levels. Hence, we cannot reject the null hypothesis that $\beta_2 = -1$ and $\beta_3 = 1$.
3. To test the joint restriction $\beta = (0, -1, -1, 1)$ entails the three restrictions $\beta_0 = 0$, $\beta_2 = -1$, and $\beta_3 = 1$. The calculated value of χ^2 with three degrees of freedom is 1.8128 so that the significance level is 0.612. Hence, we cannot reject the null hypothesis that the cointegrating vector is $(0, -1, -1, 1)$.

STEP 4: Finally, innovation accounting and causality tests on the error-correction model of (6.62) could help to identify a structural model and determine whether the estimated model appears to be reasonable. Since the simulated data have no economic meaning, innovation accounting is not performed here.

10. ERROR-CORRECTION AND ADL TESTS

In the Engle–Granger method, it is possible to estimate the long-run equilibrium relationship from a regression of z_t on y_t or from a regression of y_t on z_t. In the Johansen method, all variables are treated symmetrically. Hence, either method can be used in circumstances when you do not want to explicitly specify a "dependent" variable and a set of "independent" variables. This can be especially advantageous if the variables are jointly determined and you are not sure how to disentangle the interdependence among them. For example, in a test for purchasing power parity, it is likely that the exchange rate and the two price levels all have strong effects on each other. In other circumstances, the selection of a dependent variable and the set of independent variables might be clear. As discussed in this section, there are potential benefits to be had by incorporating such information into a cointegration model. The starting point is to be precise about the econometric meaning of the term "exogenous." To begin with the simplest case, suppose that y_t and z_t are cointegrated of order (1, 1) and that the error-correcting model (ECM) is represented by

$$\Delta y_t = \alpha_1(y_{t-1} - \beta z_{t-1}) + e_{1t} \tag{6.63}$$

$$\Delta z_t = \alpha_2(y_{t-1} - \beta z_{t-1}) + e_{2t} \tag{6.64}$$

Notice that (6.63) and (6.64) are in reduced form and not in structural form. In order to allow for the possibility that the error terms are correlated, we can let the relationship between the error terms and the structural shocks be given by

$$\begin{bmatrix} e_{1t} \\ e_{2t} \end{bmatrix} = \begin{bmatrix} c_{11} & c_{12} \\ c_{21} & c_{22} \end{bmatrix} \begin{bmatrix} \varepsilon_{yt} \\ \varepsilon_{zt} \end{bmatrix}$$

where ε_{yt} and ε_{zt} are the structural innovations in Δy_t and Δz_t, and the c_{ij} are coefficients. As in the discussion of structural VARs in Section 10 of Chapter 5, the structural shocks are uncorrelated in that $E\varepsilon_{yt}\varepsilon_{zt} = 0$. Even though $E\varepsilon_{yt}\varepsilon_{zt} = 0$, e_{1t} and e_{2t} will generally be correlated if c_{12} and/or c_{21} differ from zero.

For now, suppose that the values of the c_{ij} are unknown. Nevertheless, it is always possible to use a Choleski decomposition to orthogonalize the two errors such that

$$e_{1t} = \rho e_{2t} + v_t \tag{6.65}$$

where ρ is the regression coefficient of e_{1t} on e_{2t} and v_t is the innovation in e_{1t} that is not correlated with e_{2t}. If we substitute (6.64) and (6.65) into (6.63), we obtain

$$\begin{aligned}\Delta y_t &= \alpha_1(y_{t-1} - \beta z_{t-1}) + \rho e_{2t} + v_t \\ &= \alpha_1(y_{t-1} - \beta z_{t-1}) + \rho[\Delta z_t - \alpha_2(y_{t-1} - \beta z_{t-1})] + v_t \\ &= (\alpha_1 - \rho\alpha_2)\,(y_{t-1} - \beta z_{t-1}) + \rho\Delta z_t + v_t\end{aligned}$$

Now, if we let $\alpha = \alpha_1 - \rho\alpha_2$, we can write

$$\Delta y_t = \alpha(y_{t-1} - \beta z_{t-1}) + \rho\Delta z_t + v_t \tag{6.66}$$

In general, it is not appropriate to estimate (6.66) directly since it contains the jointly determined variables Δy_t and Δz_t. The general problem is that Δz_t will be correlated with the error term v_t so that there is a simultaneity problem. As such, OLS cannot be used to recover meaningful estimates of the parameters of the model. Even if the simultaneity problem is rectified, there is an identification problem since α_1 and α_2 cannot be separately identified from the OLS estimate of α. However, it is possible to specify conditions such that the simultaneity and identification problems disappear and that OLS is an efficient estimation and testing strategy. As will be shown below, the two conditions are $\alpha_2 = 0$ (so that z_t does not respond to the discrepancy from the long-run equilibrium relationship) and $c_{21} = 0$ (so that z_t does not respond to ε_{yt}). Thus, the two required assumptions are that z_t is weakly exogenous and that it is causally prior to y_t.

Cointegration with Weak Exogeneity

In a cointegrated system, if a variable does not respond to the discrepancy from the long-run equilibrium relationship, it is **weakly exogenous.**[20] Hence, if the speed of adjustment parameter α_i is zero, the variable in question is weakly exogenous. In the example used by Johansen and Juselius (1990), it might be possible to argue that real income should be weakly exogenous. After all, in a full-employment environment, discrepancies between long-run money demand and supply would not be expected to change real income. For our purposes, the practical importance is that a weakly exogeneous variable does not experience the type of feedback that necessitates the use of a VAR.

To explain, suppose that you try to estimate an equation like (6.66) using OLS. You could use a two-step method, such as that employed in the Engle–Granger procedure, and regress y_t on z_t to obtain an estimate of β, and then form the variable $y_{t-1} - \beta z_{t-1}$. However, at this point in time, the preference in the literature is to estimate the unrestricted equation

$$\Delta y_t = \beta_1 y_{t-1} + \beta_2 z_{t-1} + \beta_3 \Delta z_t + v_t \tag{6.67}$$

where from (6.66) the estimated coefficients are such that $\beta_1 = \alpha_1 - \rho\alpha_2$, $\beta_2 = (\alpha_1 - \rho\alpha_2)\beta$, and $\beta_3 = \rho$.

Since the coefficients of (6.67) are unrestricted, this form of the model is often called an **autoregressive distributed lag** to distinguish it from an ECM in the form of (6.66). Notice that the value of α_2 appears in the estimates for β_1 and β_2. However, if z_t is weakly exogenous (i.e., if $\alpha_2 = 0$), your coefficient estimates should be such that $\beta_1 = \alpha_1$, $\beta_2 = \alpha_1\beta$, and $\beta_3 = \rho$. Thus, you can identify α_1, β, and ρ from β_1, β_2, and β_3 since the OLS estimation of (6.67) is equivalent to estimating the equation

$$\Delta y_t = \alpha_1 y_{t-1} - \alpha_1\beta z_{t-1} + \rho\Delta z_t + v_t \tag{6.68}$$

Although weak exogeneity allows the model to be identified, there is still the issue of properly testing (6.68) for cointegration. Since $\{y_t\}$ and $\{z_t\}$ are $I(1)$, the test statistics of the null hypothesis $\beta_1 = 0$ and $\beta_2 = 0$ in (6.67) are nonstandard and need to be tabulated. The usual way to test for cointegration is to use the t-statistic for the null hypothesis $\beta_1 = 0$ in (6.67).[21] After all, if $\beta_1 = 0$, there is no error-correction so that y_t is not cointegrated with z_t. Table F uses the work of Ericsson and MacKinnon (2002) to calculate the appropriate critical values necessary to determine whether $\beta_1 < 0$. The critical values depend on the number of $I(1)$ regressors in the model (denoted by k), the adjusted sample size T^a, and the form of the deterministic regressors. For example, if you have an adjusted sample size with 100 observations and estimate a model with an intercept ($d = 1$) and two weakly exogenous variables ($k = 3$), Table F indicates that the appropriate critical values to test the null hypothesis $\beta_1 = 0$ are -4.181, -3.538, and -3.205 at the 1 percent, 5 percent, and 10 percent significance levels, respectively.

If you compare (6.67) with (6.63), you can see the benefit of employing weak exogeneity. Since the two representations are equivalent, e_{1t} is composed of Δz_t and v_t. Since (6.67) will have a smaller variance than the error term in (6.63), the coefficients of (6.67) can be estimated with more precision than the coefficient of (6.63). A second benefit ascribed to estimating such a model is that the coefficients of y_{t-1} and z_{t-1} are unrestricted. As such, the short-run dynamics for Δy_t are not dictated by the long-run equilibrium relationship $y_{t-1} = \beta z_{t-1}$. In the Engle–Granger and Johansen approaches, the so-called **Common Factor Restriction** forces the short-run changes in Δy_t to be a constant proportion of the previous period's deviation from long-run equilibrium.

Inference of the Cointegrating Vector

Suppose you assume that weak exogeneity holds and conclude that the variables are cointegrated (so that $\alpha_1 < 0$ and $\alpha_2 = 0$). As such, it is possible to write (6.64) and (6.67) as

$$\Delta y_t = \alpha_1(y_{t-1} - \beta z_{t-1}) + \rho \Delta z_t + v_t \tag{6.69}$$

and

$$\Delta z_t = e_{2t} \tag{6.70}$$

Now the question becomes: Can you conduct inference on α_1 and β in (6.69) using standard t-tests and F-tests? The answer, quite possibly, is yes! Since all variables in (6.69) are stationary, we are really operating within a standard OLS regression framework. A simultaneity problem exists if the regressors appearing in (6.69) depend on the error term v_t. Clearly, the $I(0)$ variable $y_{t-1} - \beta z_{t-1}$ is predetermined so that there is no need to worry about the influence of v_t on the error-correction term. Hence, the key issue concerns the contemporaneous relationship between Δy_t and Δz_t. If Δz_t is unaffected by innovations in Δy_t, it is appropriate to conduct inference on (6.69) using standard t-tests and F-tests.

Recall that the particular orthogonalization used in (6.65) is such that $e_{1t} = \rho e_{2t} + v_t$ where e_{2t} and v_t are uncorrelated. This is equivalent to a Choleski decomposition in that Δz_t does not respond to innovations in Δy_t but Δy_t responds to innovations in Δz_t. It should be clear that actual error structure has this Choleski form only if $c_{21} = 0$. In other words, if $c_{21} = 0$, (6.65) is equivalent to $e_{1t} = \rho e_{2t} + \varepsilon_{yt}$ and $e_{2t} = \varepsilon_{zt}$. Given that $\Delta z_t = e_{2t}$ does not depend on ε_{yt}, there is no feedback from Δy_t to Δz_t so that it is possible to use standard inference on (6.68) or (6.69).

Thus, testing restrictions on α_1 is straightforward since it is the coefficient on the $I(0)$ variable $(y_{t-1} - \beta z_{t-1})$. As such, given that $\alpha_1 \neq 0$, it is appropriate to form confidence intervals on α_1 using a standard t-distribution. Similarly, given that $\beta \neq 0$, β can be written as the coefficient on the $I(0)$ variable $(y_{t-1}/\beta - \alpha_1 z_{t-1})$. Inference on β can also be conducted using a t-distribution. Finally, note that ρ is the coefficient on the stationary variable Δz_t. Hence, it is appropriate to construct confidence intervals for ρ using a t-distribution.

It is straightforward to generalize these results. Since z_t can actually be a vector of $I(1)$ variables, you can estimate (6.67) for y_t and a set of weakly exogenous variables z_t. For example, with two weakly exogenous variables, z_{1t} and z_{2t}, the error-correction model generalizes to

$$\Delta y_t = \alpha_1(y_{t-1} - \beta_1 z_{1t-1} - \beta_2 z_{2t-1}) + \beta_3 \Delta z_{1t} + \beta_4 \Delta z_{2t} + v_t$$

so that you estimate a model of the form

$$\Delta y_t = \alpha_1 y_{t-1} + b_1 z_{1t-1} + b_2 z_{2t-1} + \beta_3 \Delta z_{1t} + \beta_4 \Delta z_{2t} + v_t$$

where $b_1 = -\beta_1/\alpha_1$ and $b_2 = -\beta_2/\alpha_1$.

To test for cointegration use the t-statistic for the null hypothesis $\alpha_1 = 0$. Since you have three $I(1)$ variables in the model, obtain the critical values from Table F such that $k = 3$. Of course, if we start from a higher-order process, additional lags of Δy_{t-i}, Δz_{1t-i}, and Δz_{2t-i} should be added to the equation. As in the two-variable case, you need to assume that Δy_t has no contemporaneous effects on any values of Δz_{it}.

11. COMPARING THE THREE METHODS

In this section, we compare the Engle–Granger, Johansen, and ADL tests for cointegration using the three-month Treasury bill and 10-year interest rates already analyzed in other chapters. You can follow along using the data in QUARTERLY.XLS. Although we know that the spread acts as a stationary variable, the point of this section is to illustrate the use of the three testing methodologies. Since we have already verified that each rate acts as an $I(1)$ process, we can skip the preliminary step of pretesting for unit roots. To keep the discussion on point, reasonable leg lengths for each test are simply reported. You are asked to verify them in the exercises at the end of the chapter.

The Engle–Granger Methodology

Given that each rate acts as a unit root process, we can begin by estimating the long-run equilibrium relationship

$$\underset{}{r_{Lt}} = \underset{(12.36)}{2.269} + \underset{(28.10)}{0.837} r_{St} \tag{6.71}$$

Next, we test the residuals from (6.71) for stationarity by estimating an equation in the form of (6.32). If you experiment with various lag lengths, you will find that one lagged change seems reasonable. Consider the equation

$$\Delta \hat{e}_t = \underset{(-3.68)}{-0.123} \hat{e}_{t-1} + \underset{(3.15)}{0.224} \Delta \hat{e}_{t-1}$$

In a model with two variables and almost 200 usable observations, the 5 percent critical value shown in Table C is -3.368. As such, we can reject the null hypothesis of no cointegration. Since we are making no assumption concerning weak exogeneity, it is clearly possible to carry out the analysis using r_{St} as the left-hand-side variable. Reversing the variables in (6.71) yields

$$r_{St} = \underset{(-4.40)}{-1.106} + \underset{(28.11)}{0.961} r_{St}$$

In this form, the Engle–Granger test is somewhat more amenable to the finding of cointegration since the regression of the residuals yields

$$\Delta \hat{e}_t = \underset{(-4.17)}{-0.182} \hat{e}_{t-1} + \sum_{t=1}^{8} a_i \Delta \hat{e}_{t-i}$$

From Table C, the 1 percent critical value is about −3.945 so that that the null of no cointegration is strongly rejected. Notice that the two estimates of the long-run equilibrium relationship are quite different from each other. Nevertheless, it is not possible to conduct inference on either of these cointegrating vectors unless you use the methods discussed in Appendix 6.2.

The Johansen Methodology

Let x_t denote the vector $[r_{Lt}, r_{St}]'$. If you estimate the unrestricted VAR in the form of (6.53) (i.e., if you estimate the VAR $x_t = A_0 + \Sigma A_i x_{t-i}$), you should find that a reasonable lag length is $i = 8$. Given this lag length, it is possible to estimate the model in the form of (6.54). Since the interest rates do not continually increase or decrease over time, it seems reasonable to constrain the drift terms so that a constant appears in the cointegrating relationship. The estimated value of the π^* matrix is such that

$$\pi^* x^*_{t-1} = \begin{bmatrix} -1.18 & 1.31 & 0.92 \\ 0.57 & -0.16 & -3.03 \end{bmatrix} \begin{bmatrix} r_{Lt} \\ r_{St} \\ 1 \end{bmatrix}$$

The characteristic roots are such that $\lambda_1 = 0.1282$ and $\lambda_2 = 0.0084$ so that $-T\ln(1 - \lambda_1) = 25.384$ and $-T\ln(1 - \lambda_2) = 1.556$. To test the null hypothesis of no cointegration against the general alternative of 1 or 2 cointegrating vectors, compare the sum $25.364 + 1.556 = 26.92$ to the 5 percent critical value of the λ_{trace} statistic shown in Table E. Since 26.92 exceeds the critical value of 19.96, reject the null and conclude that there is at least one cointegrating vector. To test the null of one cointegrating vector against the alternative of two cointegrating vectors, compare the sample value of 1.556 to the 5 percent critical value of 9.24. As such, we can conclude that there is only one cointegrating vector.

Normalizing the cointegrating vector with respect to r_{Lt} yields

$$\begin{aligned} r_{Lt} = {} & 0.779 + 1.11 r_{St} \\ & (1.76) \quad (15.28) \end{aligned}$$

A key difference between this estimate of the long-run equilibrium relationship and those from the Engle–Granger test is that standard inference can be performed on the coefficients of the cointegrating vector. For example, the value of χ^2 for the test for the null hypothesis that the coefficient on the short-term rate equals unity is 2.21. With one degree of freedom, the 5 percent and 10 percent critical values of χ^2 are 3.841 and 2.706, respectively. As such, we can conclude that the restriction is not binding. Hence, in the long run, the 10-year rate tends to move 1:1 with the short-term rate. If you reestimate the model imposing the restriction, you should find:

$$\begin{aligned} \Delta r_{Lt} &= -0.113(r_{Lt-1} - 1.39 - r_{St-1}) + A_{11}(L)\Delta r_{Lt-1} + A_{12}(L)\Delta r_{St-1} \\ &\quad\ (-2.59) \qquad\quad (-8.58) \\ \Delta r_{St} &= 0.068(r_{Lt-1} - 1.39 - r_{St-1}) + A_{21}(L)\Delta r_{Lt-1} + A_{22}(L)\Delta r_{St-1} \\ &\quad\ (1.09) \qquad\quad (-8.58) \end{aligned}$$

The t-statistic of 8.58 on the constant term in the cointegrating vector is highly significant. The important point is that the t-statistics on the error-correcting terms imply that the long-term rate adjusts to the discrepancy from the long-run equilibrium relationship, but the short rate does not. In other words, r_{St} is weakly exogenous. Consider the dynamic adjustment mechanism if there is a positive one-unit discrepancy from the long-run equilibrium relationship. The estimates imply that the long-term rate falls by -0.113 units and that the short-term rate does none of the adjusting. As such, the deviations from the long-run relationship are quite long-lived.

The Error-Correction/ADL Methodology

In contrast to the Engle–Granger and Johansen methodologies, to use the error-correction test it is necessary to assume that one of the variables is weakly exogenous. Suppose that we were certain that the short-term interest rate did none of the adjustment necessary to restore the long-run equilibrium relationship. Given that the short-term rate is weakly exogenous, we can estimate an equation in the form:

$$\Delta r_{Lt} = \beta_0 + \beta_1 r_{Lt-1} + \beta_2 r_{St-1} + \beta_3 \Delta r_{St} + A_1(L)\Delta r_{Lt-1} + A_2(L)\Delta r_{St-1} + v_t \tag{6.72}$$

Equation (6.72) looks very much like (6.67) except that we have included an intercept term β_0 and included lagged changes of the two interest rates to capture the stationary dynamics of the adjustment process. Since we are not treating all variables symmetrically, there is no need to constrain the lag length represented by the polynomial $A_1(L)$ to be the same as that from $A_2(L)$. However, for this case, it turns out that a lag length of 6 seems appropriate for each variable. Consider the estimated equation

$$\begin{aligned} \Delta r_{Lt} = \underset{(1.33)}{0.109} - \underset{(-4.10)}{0.124} r_{Lt-1} + \underset{(4.47)}{0.135} r_{St-1} + \underset{(13.01)}{0.489}\Delta r_{St} \\ + A_1(L)\Delta r_{Lt-1} + A_2(L)\Delta r_{St-1} + v_t \end{aligned} \tag{6.73}$$

The key point to note is that the t-statistic for the null hypothesis $\beta_1 = 0$ is -4.10. Given the presence of an intercept ($d = 1$), two $I(1)$ variables ($k = 2$), and that the estimation begins in 1961Q4 ($T = 186$), the adjusted sample size is $T^a = 186 - (2{*}2 - 1) - 1 = 182$. From Table F, the critical values at the 1 percent, 5 percent, and 10 percent significance levels are approximately -3.834, -3.231, and -2.916, respectively. Hence, we can reject the null hypothesis of no cointegration and conclude that the variables are cointegrated.

We can reparameterize (6.73) such that

$$\begin{aligned} \Delta r_{Lt} = -0.124(r_{Lt-1} - 1.09 r_{St-1} - 0.879) + 0.489\Delta r_{St} \\ + A_1(L)\Delta r_{Lt-1} + A_2(L)\Delta r_{St-1} + v_t \end{aligned}$$

In this particular example, all three approaches find that the variables are cointegrated. The Engle–Granger approach indicates that the speed of the adjustment parameter is -0.123 (or -0.182), but does not indicate which variable (or variables) does the adjustment. In response to a one-unit deviation from the long-run equilibrium, the Johansen approach indicates that the long-term rate adjusts by -0.113 units, while the ADL approach indicates that it adjusts by -0.124 units. The Engle–Granger approach does not allow us to readily perform inference of the cointegrating vector, but the Johansen approach allows us to conclude that two rates move 1:1 in the long run.

So long as we are willing to assume $\beta_2 \neq 0$, it is possible to perform inference on the coefficient on r_{St-1} in the long-run equilibrium relationship. Clearly, it would have been possible to reparameterize (6.72) such that

$$\Delta r_{Lt} = -0.135(0.919 r_{Lt-1} - r_{St-1} - 0.807) + 0.489\Delta r_{St} + A_1(L)\Delta r_{Lt-1} + A_2(L)\Delta r_{St-1} + v_t \quad (6.74)$$

Hence, β_2 is the coefficient on a stationary variable so that it has a standard t-distribution. The sample t-value for the test is -1.105, so that we can accept the null hypothesis. Alternatively, we could have performed an F-test for the null hypothesis $\beta_1 = \beta_2$ in (6.73). A traditional F-test is appropriate since each coefficient has a t-distribution. With one degree of freedom in the numerator and 170 in the denominator, the sample value of $F = 1.105$ is significant at the 0.294 level. If you reestimate the model such that $\beta_1 = \beta_2$, you should find

$$\Delta r_{Lt} = -0.130(r_{Lt-1} - r_{St-1} - 1.38) + 0.482\Delta r_{St} + A_1(L)\Delta r_{Lt-1} + A_2(L)\Delta r_{St-1} + v_t$$

If you are willing to abstract from the stationary dynamics, it is clear how to trace out the effects of a one-unit shock in Δr_{St}. All else equal, if $\Delta r_{St} = 1$, it follows that $\Delta r_{Lt} = 0.482$. In period $t + 1$, it follows that the discrepancy from the long-run equilibrium is -0.518 ($= 0.482 - 1$) and the change in the long-rate is $(-0.518)(-0.130) = 0.067$. In subsequent periods, the long rate keeps rising by 13 percent of the discrepancy from the long-run equilibrium. At this point, you could go on to perform the innovation accounting by estimating an equation of the form $\Delta r_{St} = A_3(L)\Delta r_{Lt} + A_4(L)\Delta r_{St} + e_{2t}$. Note that the equation is in first differences since the Δr_{St} equation does not contain an error-correction term. Also note that the assumption that Δr_{St} is weakly exogenous implies a causal ordering of the innovations in that a v_t shock has no contemporaneous effect on Δr_{St}, but an e_{2t} shock directly affects Δr_{Lt}.

12. SUMMARY AND CONCLUSIONS

Many economic theories imply that a linear combination of certain nonstationary variables must be stationary. For example, if the variables $\{x_{1t}\}$, $\{x_{2t}\}$, and $\{x_{3t}\}$ are $I(1)$ and the linear combination $e_t = \beta_0 + \beta_1 x_{1t} + \beta_2 x_{2t} + \beta_3 x_{3t}$ is stationary, the variables are said to be cointegrated of order $(1, 1)$. The vector $(\beta_0, \beta_1, \beta_2, \beta_3)$ is called the cointegrating vector. Cointegrated variables share the same stochastic trends and so cannot drift too far apart. Cointegrated variables have an error-correction representation such that each responds to the deviation from "long-run equilibrium."

One way to check for cointegration is to examine the residuals from the long-run equilibrium relationship. If these residuals have a unit root, the variables cannot be cointegrated of order (1, 1). Another way to check for cointegration among $I(1)$ variables is to estimate a VAR in first differences and include the lagged level of the variables. The Johansen methodology uses the λ_{trace} and λ_{max} test statistics to determine if the variables are cointegrated and the number of cointegrating vectors. These tests are sensitive to the presence of the deterministic regressors included in the cointegrating vector(s). Restrictions on the cointegrating vector(s) and/or the speed-of-adjustment parameters can be tested using χ^2 statistics. You should be aware of the role of the deterministic regressors in a cointegration framework. Johansen (1994) shows how to test to determine whether there is a deterministic trend, drift terms that occur outside of the cointegrating vector, or constants that appear in the cointegrating vector. A third way to test for cointegration is to estimate the error-correction model. If only one variable adjusts to the discrepancy from the long-run equilibrium relationship, it can be preferable to estimate an autoregressive distributed lag model. It is straightforward to estimate the model using OLS and to perform hypothesis tests on the coefficients of the cointegrating vector. For more complicated situations, Appendix 6.2 discusses the Phillips–Hansen (1990) method of modeling in a single-equation framework.

QUESTIONS AND EXERCISES

1. Let equations (6.14) and (6.15) contain intercept terms such that

$$y_t = a_{10} + a_{11}y_{t-1} + a_{12}z_{t-1} + \varepsilon_{yt} \quad \text{and} \quad z_t = a_{20} + a_{21}y_{t-1} + a_{22}z_{t-1} + \varepsilon_{zt}$$

a. Show that the solution for y_t can be written as

$$y_t = [(1 - a_{22}L)\varepsilon_{yt} + (1 - a_{22})a_{10} + a_{12}L\varepsilon_{zt} + a_{12}a_{20}]/[(1 - a_{11}L)(1 - a_{22}L) - a_{12}a_{21}L^2]$$

b. Find the solution for z_t.

c. Suppose that y_t and z_t are $CI(1, 1)$. Use the conditions in (6.19), (6.20), and (6.21) to write the error-correcting model. Compare your answer to (6.22) and (6.23). Show that the error-correction model contains an intercept term.

d. Show that $\{y_t\}$ and $\{z_t\}$ have the same deterministic time trend (i.e., show that the slope coefficients of the time trends are identical).

e. What is the condition such that the slope of the trend is zero? Show that this condition is such that the constant can be included in the cointegrating vector.

2. The data file COINT6.PRN contains the three simulated series used in Sections 5 and 9.

a. Use the data to reproduce the results in Section 5.

b. Use the data to reproduce the results in Section 9.

c. Examine Table 6.1. Show that y_t and z_t, but not w_t, are weakly exogenous.

d. Use the data to compare the ECM test to the Engle–Granger and Johansen tests, treating y_t and z_t as weakly exogenous.

3. In Question 11 in Chapter 4, you were asked to use the data in QUARTERLY.XLS to estimate the regression equation

$$\begin{aligned} INDPRO_t &= \underset{(31.88)}{28.491} + \underset{(48.80)}{0.056}\, M1NSA_t \qquad R^2 = 0.99 \end{aligned}$$

a. Use the Engle–Granger test to show that the regression is spurious.

b. Examine the scatter plot of $INDPRO_t$ against $M1NSA_t$. How do you explain the fact that $R^2 = 0.99$ and that the t-statistic on the money supply is 48.48?

c. Use the data in the file labeled RGDP.XLS. Denote the natural logs of real GDP and consumption by ly_t and lc_t, respectively. Estimate the regression

$$lc_t = -0.507 + 1.05ly_t \qquad R^2 = 0.999$$
$$(-130.49) \quad (444.83)$$

If you perform the Engle–Granger test using four lags, you should find

$$\Delta\hat{e}_t = -0.093\hat{e}_{t-1} + \sum_{i=1}^{4}\Delta\beta_i\hat{e}_{t-i}$$

The t-statistic for $\hat{e}_{t-1}$ is -3.37. How do you interpret the consumption-income relationship?

4. The file labeled QUARTERLY.XLS contains the interest rates paid on U.S. three-month, three-year, and 10-year U.S. government securities. The data run from 1960Q1 to 2008Q1. The variables are labeled TBILL, R3, and R10, respectively.

a. Pretest the variables to show that the rates all act as unit root processes. Specifically, perform augmented Dickey–Fuller tests using the lag length selected by deleting lags until the t-statistic on the last lag is significant at the 5 percent level. If you include an intercept (but no time trend) you should obtain:

Series	Lags	Estimated a_1	t-statistic
TBILL	7	−0.04029	−1.99014
R3	7	−0.02468	−1.36935
R10	5	−0.02011	−1.40849

b. Estimate the cointegrating relationships using the Engle–Granger procedure. Perform augmented Dickey–Fuller tests on the residuals. Using TBILL as the "dependent" variable, you should find

$$\text{TBILL}_t = 0.291 + 1.85\text{R3}_t - 0.997\text{R10}_t$$
$$(2.26) \quad (26.18) \quad (-12.09)$$

where t-statistics are in parenthesis.

Perform the Engle–Granger test on the residuals from the equation above. Why is it appropriate to use eight lags in the augmented form of the test? If you use eight lags, you should find that the coefficient on the lagged residual (i.e., e_{t-1}) is -0.328 with a t-statistic of -4.02.

The 5 percent critical value is about -3.76. Based on these data, do you conclude that the variables are cointegrated?

c. Repeat part b using R10 as the dependent variable. If you use three lags in the augmented form of the Engle–Granger test (i.e., estimate $\Delta e_t = \alpha_1 e_{t-1} + \ldots$), you should find $a_1 = -0.167$ and the t-statistic is -3.74.

Using R10 as the dependent variable, are the three interest rates cointegrated?

d. Estimate an error-correcting model using seven lagged changes of each variable. Use the residuals from part b as the error-correction term and do not include intercepts. You should find that the error-corrections are such that:

$\Delta\text{TBILL}_t = 0.070e_{t-1} + \ldots$ t–statistic for the error-correction term: 0.427

$\Delta\text{R3}_t = -0.272e_{t-1} + \ldots$ t–statistic for the error-correction term: −1.83

$\Delta\text{R10}_t = -0.271e_{t-1} + \ldots$ t–statistic for the error-correction term: −2.41

where e_{t-1} is the lagged residual from your estimate in part b.

i. Perform the appropriate diagnostic tests on the system. In particular, determine whether the three residual series appear to be white noise. Are the lag lengths unnecessarily long?

ii. Discuss the nature of the adjustment. Are any of the rates weakly exogenous? In response to a deviation from the long-run relationship, how are the three rates predicted to change?

e. Estimate the model using the Johansen procedure. Use seven lags and include an intercept in the cointegrating vector. You should find the following:

Trace Tests				**Maximum Eigenvalue Tests**			
Null	Alternative	λ_{trace}	5% Value	Null	Alternative	λ_{max}	5% Value
$r = 0$	$r \geq 1$	52.60	34.91	$r = 0$	$r = 1$	37.33	22.00
$r \leq 1$	$r \geq 2$	15.26	19.96	$r = 1$	$r = 2$	13.32	15.67
$r \leq 2$	$r = 3$	1.94	9.24	$r = 2$	$r = 3$	1.94	9.24

i. Explain why the λ_{trace} test strongly suggests there is exactly one cointegrating vector.

ii. To what extent is this result reinforced by the λ_{max} test?

Verify that the normalized cointegrating vector is

$$\text{TBILL}_t = -0.333 + 0.9723\text{R3}_t - 0.066\text{R10}_t$$

Compare this result to your answer in part b.

f. Why might you be wary about testing for cointegration using the error-correction (i.e., ADL) test developed in Section 10?

5. Suppose you estimate π to be:

$$\pi = \begin{bmatrix} 0.6 & -0.5 & 0.2 \\ 0.3 & -0.25 & 0.1 \\ 1.2 & -1.0 & 0.4 \end{bmatrix}$$

a. Show that the determinant of π is zero.

b. Show that two of the characteristic roots are zero and that the third is 0.75.

c. Let $\beta' = (3, -2.5, 1)$ be the single cointegrating vector normalized with respect to x_{3t}. Find the $(3 \cdot 1)$ vector α such that $\pi = \alpha\beta'$. How would α change if you normalized β with respect to x_{1t}?

d. Describe how you could test the restriction $\beta_1 + \beta_2 = 0$.

Now suppose you estimate π to be

$$\pi = \begin{bmatrix} 0.8 & 0.4 & 0.0 \\ 0.2 & 0.1 & 0.0 \\ 0.75 & 0.25 & 0.5 \end{bmatrix}$$

e. Show that the three characteristic roots are 0.0, 0.5, and 0.9.

f. Select β such that

$$\beta = \begin{bmatrix} 0.8 & 0.75 \\ 0.4 & 0.25 \\ 0.0 & 0.5 \end{bmatrix}$$

Find the $(3 \cdot 2)$ matrix α such that $\pi = \alpha\beta'$.

6. Suppose that x_{1t} and x_{2t} are integrated of orders 1 and 2, respectively. You are to sketch the proof that any linear combination of x_{1t} and x_{2t} is integrated of order 2. Toward this end:

a. Allow x_{1t} and x_{2t} to be the random walk processes $x_{1t} = x_{1t-1} + \varepsilon_{1t}$ and $x_{2t} = x_{2t-1} + \varepsilon_{2t}$.

i. Given the initial conditions x_{10} and x_{20}, show that the solutions for x_{1t} and x_{2t} have the form $x_{1t} = x_{10} + \Sigma\varepsilon_{1t-i}$ and $x_{2t} = x_{20} + \Sigma\varepsilon_{2t-i}$.

ii. Show that the linear combination $\beta_1 x_{1t} + \beta_2 x_{2t}$ will generally contain a stochastic trend.

iii. What assumption is necessary to ensure that x_{1t} and x_{2t} are $CI(1, 1)$?

b. Now let x_{2t} be integrated of order 2. Specifically, let $\Delta x_{2t} = \Delta x_{2t-1} + \varepsilon_{2t}$. Given the initial condition for x_{20} and x_{21}, find the solution for x_{2t}. (You may allow ε_{1t} and ε_{2t} to be perfectly correlated.)

i. Is there any linear combination of x_{1t} and x_{2t} that contains only a stochastic trend?

ii. Is there any linear combination of x_{1t} and x_{2t} that does not contain a stochastic trend?

c. Provide an intuitive explanation for the statement: If x_{1t} and x_{2t} are integrated of orders d_1 and d_2 where $d_2 > d_1$, any linear combination of x_{1t} and x_{2t} is integrated of order d_2.

7. The *Programming Manual* that accompanies this text contains a discussion of nonlinear least squares and maximum-likelihood estimation. If you have not already done so, download the manual and programs from the Wiley Web site.

a. Section 5.5 in Chapter 4 contains a discussion of the problem of conducting inference on the parameters of a cointegrating vector. Execute Program 4.10. Why is it a problem that only 16.8 percent of the true values of β_1 lie within a 95 percent confidence interval?

b. How would you modify the program so as to generate the Engle–Granger critical values?

8. Section 3 of Chapter 2 of the *Programming Manual* that comes with this text estimates the relationship between the long-term and short-term interest rate as

$$tb1yr_t = 0.698 + 0.916tb3mo_t$$

a. Use the data set MONEYDEM.XLS to estimate the error-correction model. Use five lags of each variable.

b. As shown in the manual, you can use the error-correction model to obtain the impulse response functions with a Choleski decomposition. Your responses should look like those on page 68 of the manual.

c. In the equation for $\Delta tb1yr_t$, the coefficient on the error-correction term is -0.098 with a t-statistic equal to -0.427. Why is it possible to argue that $tb1yr_t$ is weakly exogenous? How can you model the long-term rate using the error-correction approach?

9. The file COINT_PPP.XLS contains quarterly values of the U.K., Japanese, and Canadian wholesale price levels and the bilateral exchange rates with the United States. The file also contains the U.S. wholesale price level. The names on the individual series should be self-evident. For example, *p_us* is the U.S. price level and *ex_ja* is the Japanese exchange rate with the U.S. All variables run from 1973*Q*1 to 2008*Q*2 and all have been normalized to equal 100 in 1973*Q*1.

a. Form the log of each variable. Estimate the long-run relationship between Japan and the U.S. as

$$\log(ex\_ja) = 4.80 + 1.108\log(p\_ja) - 1.166\log(p\_us)$$

i. Do the point estimates of the slope coefficients seem to be consistent with long-run PPP?

b. Let v_t denote the residuals from the long-run relationship. Use these residuals to perform the Engle–Granger test for cointegration. If you use three lagged changes, you should find

$$\Delta v_t = -0.052v_{t-1} + 0.319\Delta v_{t-1} - 0.189\Delta v_{t-2} + 0.227\Delta v_{t-3} + \varepsilon_t$$

The t-statistic on the coefficient for v_{t-1} is -2.097. Explain why long-run PPP fails.

c. Repeat parts i and ii using the U.K. and Canada.

10. The file COINT_PPP.XLS also contains the Swiss and French nominal effective exchange rates, the Swiss wholesale price level, and the French consumer price level.

a. Show that the logarithms of the Swiss exchange rate, Swiss price level, and U.S. price level all appear to be unit root processes.

b. The nominal effective exchange rate is a multilateral exchange rate constructed by the International Monetary Fund. Let the U.S. represent the "rest of the world" and estimate the long-run equilibrium relationship between the three variables as

$$\log(ex\_sw) = \underset{(0.06)}{0.225} + \underset{(3.07)}{0.411}\log(lp\_sw) + \underset{(9.11)}{0.438}\log(lp\_us)$$

The theory of PPP indicates that the sign of the coefficient on log(*p_us*) should be negative. Why could it be a mistake to conclude that PPP fails because of the sign of the coefficient on log(*p_us*)?

c. Perform the Engle–Granger test using one lagged change of the estimated residual from the equilibrium relationship. You should find

$$\Delta\hat{e}_t = -0.156\hat{e}_{t-1} + 0.382\Delta\hat{e}_{t-1}$$

Given that the t-statistic for is -4.16, do you conclude that long-run PPP holds?

d. Perform lag length tests on the error-correction model. You should find that the multivariate AJC and SBC select a model with one lagged change in favor of a model with four lagged changes (be sure to compare the models over the same sample period). A likelihood ratio test, however, seems to indicate that a model with four lagged changes is best.

e. Estimate the error correction with one lagged change of each variable. You should find that price levels are weakly exogenous. The exchange rate equation has the form:

$$\Delta\log(ex\_sw_t) = 0.004 + 0.165\hat{e}_{t-1} + \text{ lagged changes}$$

The t-statistic for $\hat{e}_{t-1}$ is 4.37.

f. Obtain the variance decompositions using a Choleski decomposition. Let the ordering be such that the U.S. price level is causally prior to the Swiss price level, and the Swiss price level is causally prior to the exchange rate. You should find that with a 24-quarter horizon, the exchange rate, Swiss price level, and U.S. price level explain

54 percent, 12 percent, and 34 percent of the forecast error variance of the exchange rate, respectively.

g. Obtain the impulse responses using the Choleski decomposition from part f. Show that the responses of the exchange rate to the shocks are as shown in Figure 6.5. Note that a U.S. price level shock causes the franc to appreciate while a Swiss price level shock causes the franc to depreciate.

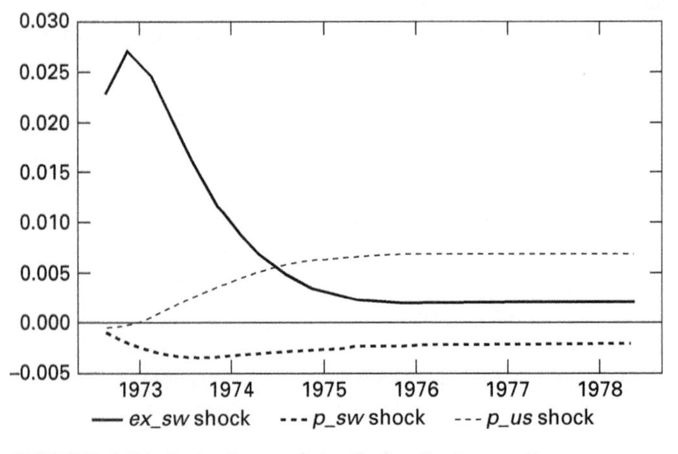

FIGURE 6.5 Responses of the Swiss Exchange Rate

h. Show that long-run PPP fails between the U.S. and France.

ENDNOTES

1. To include an intercept term, simply set all realizations of one $\{x_{it}\}$ sequence equal to unity. In the text, the long-run relationship with an intercept will be denoted by $\beta_0 + \beta_1 x_{1t} + \ldots + \beta_n x_{nt} = 0$. Also note that the definition rules out the trivial case in which all elements of β equal zero. Obviously, if all the $\beta_i = 0$, $\beta x_t' = 0$.
2. As a technical point, note that if all elements of x_t are $I(0)$, it is possible for e_t to be integrated of order -1. However, this case is of little interest for economic analysis. Also note that if $\{x_t\}$ is stationary, $\Delta^d x_t$ is stationary for all $d > 0$.
3. The issue is trivial if both trends are deterministic. Simply detrend each of the variables using a deterministic polynomial time trend of the form $\alpha_0 + \alpha_1 t + \alpha_2 t^2 + \ldots$.
4. The error-correction term could have been written in the form $\alpha_S' (\beta_1 r_{Lt-1} - \beta_2 r_{St-1})$. Normalization with respect to the long-term rate yields (6.9), where $\alpha_S = \alpha_S' \beta_1$ and $\beta = \beta_2/\beta_1$. Here, the cointegrating vector is $(1, -\beta)$.
5. Note that (6.11) and (6.12) represent a system of difference equations. The stability conditions place restrictions on the magnitudes of α_S, α_L, and the various values of $a_{ij}(k)$.
6. Equation (6.18) can be written as $\lambda^2 = a_1\lambda + a_2$ where $a_1 = (a_{11} + a_{22})$ and $a_2 = (a_{12}a_{21} - a_{11}a_{22})$. Now refer all the way back to Figure 1.5 in Chapter 1. For $\lambda_1 = 1$, the coefficients of (6.18) must lie along line segment BC. Hence, $a_1 + a_2 = 1$, or $a_{11} + a_{22} + a_{12}a_{21} - a_{11}a_{22} = 1$. Solving for a_{11} yields (6.19). For $|\lambda_2| < 1$, the coefficients must lie inside region A0BC. Given (6.19), the condition $a_2 - a_1 = 1$ is equivalent to that in (6.21).
7. Another interesting way to obtain this result is to refer back to (6.14). If $a_{12} = 0$, $y_t = a_{11}y_{t-1} + \varepsilon_{yt}$. Imposing the condition that $\{y_t\}$ is a unit root process is equivalent to setting $a_{11} = 1$ so that $\Delta y_t = \varepsilon_{yt}$.
8. The stability condition is that $-2 < a_1 < 0$. Hence, if a_1 is found to be sufficiently negative, we need to be able to reject the null hypothesis $a_1 = -2$.

9. As shown in Section 3, the values of α_y and α_z are directly related to the characteristic roots of the difference equation system. Direct convergence necessitates that α_y be negative and α_z be positive.
10. Engle and Granger (1987) did provide a statistic to test the joint hypothesis $\alpha_y = \alpha_z = 0$. However, their simulations suggest this statistic is not very powerful, and they recommend against its use.
11. Of course, x_{1t} and x_{2t} can be $CI(2, 2)$ so that the linear combination $x_{1t} - \beta_2 x_{2t}$ is stationary. In this circumstance, $\gamma_1 = 0$ since z_t cannot be cointegrated with $x_{1t} - \beta_2 x_{2t}$. However, this is a rather unusual circumstance, so that the literature focuses on the case in which x_{1t} and x_{2t} are $CI(2, 1)$.
12. Wholesale prices and period average exchange rates were used in the study. Each series was converted into an index number such that each series was equal to unity at the beginning of its respective period (either 1960 or 1973). In the fixed exchange rate period, all values of $\{e_t\}$ were set equal to unity.
13. A second set of regressions of the form $p_t = \beta_0 + \beta_1 f_t + \mu_t$ was also estimated. The results using this alternative normalization are very similar to those reported here.
14. The numbers are slightly different from those reported by Johansen and Juselius (1990) due to rounding.
15. As discussed in Appendix 6.1, the Johansen procedure consists of the matrix of vectors of the squared canonical correlations between the residuals of x_t and Δx_{t-1} regressed on lagged values of Δx_t. The cointegrating vectors are the rows of the normalized eigenvectors.
16. You can normalize with respect to any variable with only one caveat. If the true value of $\beta_i = 0$ (so that x_{it} does not enter the cointegrating relationship), normalizing with respect to x_{it} will wildly inflate the other coefficients and their significance levels. In essence, you will have divided all other magnitudes by a coefficient that is estimated to be near zero.
17. In the general case of an n-variable system, let rank(π) = r and define the $n \times r$ matrices α and β such that $\alpha\beta' = \pi$. Also, let $\alpha_\perp$ and $\beta_\perp$ denote the orthogonal complements of α and β. Johansen (1995) obtains the value of s from $\alpha_\perp\Gamma\beta_\perp$ once the value of r has been chosen.
18. Johansen shows that this two-step procedure has the following properties: (i) if the rank of π is r and there are no $I(2)$ components, the procedure picks out the true value of r with a high probability, (ii) a value of r that is too low is selected with a limiting probability of zero, and (iii) if there are $I(2)$ components, the procedure will accept no $I(2)$ components with a small probability. Jorgensen, Kongsted, and Rahbek (1996) showed how to simultaneously select the values of r and s.
19. Hence, the characteristic roots are those of $\alpha_\perp{}'\Gamma\beta_\perp$.
20. Engle, Hendry, and Richard (1983) provided a comprehensive analysis of various types of exogeneity. In general, a variable x_{it} is weakly exogenous for the parameter set P if the marginal distribution of x_{it} contains no useful information for conducting inference on P. Hence, x_{it} can be exogenous in one econometric model but not another.
21. As summarized in Ericsson and MacKinnon (2002), there are other variants of the ECM test. For example, if $\alpha_1 = 0$, it follows that $\beta_1 = \alpha_1$ and $\beta_2 = \alpha_1\beta$ should both equal zero. As such, it is also possible to test whether the restriction $\beta_1 = \beta_2 = 0$ is binding on (6.67). However, this type of test becomes more difficult when z_t is actually a vector of weakly exogenous variables.

APPENDIX 6.1: CHARACTERISTIC ROOTS, STABILITY, AND RANK

Characteristic Roots Defined

Let A be an $(n \cdot n)$ square matrix with elements a_{ij} and let x be an $(n \cdot 1)$ vector. The scalar λ is called a characteristic root of A if

$$Ax = \lambda x \tag{A6.1}$$

Let I be an $(n \cdot n)$ identity matrix so that we can rewrite (A6.1) as

$$(A - \lambda I)x = 0 \tag{A6.2}$$

Since x is a vector containing values not identically equal to zero, (A6.2) requires that the rows of $(A - \lambda I)$ be linearly dependent. Equivalently, (A6.2) requires that the determinant $|A - \lambda I| = 0$. Thus, we can find the characteristic root(s) of (A6.1) by finding the values of λ that satisfy

$$|A - \lambda I| = 0 \tag{A6.3}$$

Example 1
Let A be the matrix

$$A = \begin{bmatrix} 0.5 & -0.2 \\ -0.2 & 0.5 \end{bmatrix}$$

so that

$$|A - \lambda I| = \begin{vmatrix} 0.5 - \lambda & -0.2 \\ -0.2 & 0.5 - \lambda \end{vmatrix}$$

Solving for the value of λ such that $|A - \lambda I| = 0$ yields the quadratic equation

$$\lambda^2 - \lambda + 0.21 = 0$$

The two values of λ which solve the equation are $\lambda = 0.7$ and $\lambda = 0.3$. Hence, 0.7 and 0.3 are the two characteristic roots.

Example 2
Now change A such that each element in column 2 is twice the corresponding value in column 1. Specifically,

$$A = \begin{bmatrix} 0.5 & 1 \\ -0.2 & -0.4 \end{bmatrix}$$

Now,

$$|A - \lambda I| = \begin{bmatrix} 0.5 - \lambda & 1 \\ -0.2 & -0.4 - \lambda \end{bmatrix}$$

Again, there are two values of λ which solve $|A - \lambda I| = 0$. Solving the quadratic equation $\lambda^2 - 0.1\lambda = 0$ yields the two characteristic roots $\lambda_1 = 0$ and $\lambda_2 = 0.1$.

Characteristic Equations

Equation (A6.3) is called the characteristic equation of the square matrix A. Notice that the characteristic equation will be an nth-order polynomial in λ. The reason is that the determinant $|A - \lambda I| = 0$ contains the nth degree term λ^n resulting from the expression

$$(a_{11} - \lambda)(a_{22} - \lambda)(a_{33} - \lambda) \ldots (a_{nn} - \lambda)$$

As such, the characteristic equation will be an nth-order polynomial of the form

$$\lambda^n + b_1\lambda^{n-1} + b_2\lambda^{n-2} + b_3\lambda^{n-3} + \ldots + b_{n-1}\lambda + b_n = 0 \tag{A6.4}$$

From (A6.4) it immediately follows that an $(n \cdot n)$ square matrix will necessarily have n characteristic roots. As we saw in Chapter 1, some of the roots may be repeating and some may be complex. In practice, it is not necessary to actually calculate the values of the roots solving (A6.4). The necessary and sufficient conditions for all characteristic roots to lie within the unit circle are given in Chapter 1 and in the *Supplementary Manual.*

Notice that the term b_n is of particular relevance because $b_n = (-1)^n|A|$. After all, b_n is the only expression resulting from $|A - \lambda I|$ that is not multiplied by λ. In terms of (A6.4), the expressions λ^n and b_n will have the same sign if n is even and opposite signs if n is odd. In Example 1, the characteristic equation is $\lambda^2 - \lambda + 0.21 = 0$ so that $b_2 = 0.21$. Since $|A| = 0.21$, it follows that $b_2 = (-1)^2(0.21)$. Similarly, in Example 2, the characteristic equation is $\lambda^2 - 0.1\lambda = 0$, so that $b_2 = 0$. Since it is also the case that $|A| = 0$, it also follows that $b_2 = (-1)^2|A|$. In Example 3 below, we consider the case in which $n = 3$.

Example 3
Let A be such that

$$|A - \lambda I| = \begin{bmatrix} 0.5 - \lambda & 0.2 & 0.2 \\ 0.2 & 0.5 - \lambda & 0.2 \\ 0.2 & 0.2 & 0.5 - \lambda \end{bmatrix}$$

The characteristic equation is

$$\lambda^3 - 1.5\lambda^2 + 0.63\lambda - 0.081 = 0$$

and the characteristic roots are

$$\lambda_1 = 0.9, \lambda_2 = 0.3, \text{ and } \lambda_3 = 0.3$$

The determinant of A is 0.081 so that $b_3 = -0.081 = (-1)^3|A|$.

Determinants and Characteristic Roots

The determinant of an $(n \cdot n)$ matrix is equal to the product of its characteristic roots, that is,

$$|A| = \prod_{i=1}^{n} \lambda_i \qquad \text{(A6.5)}$$

where $\lambda_1, \lambda_2, \ldots, \lambda_n$ are the n characteristic roots of the $(n \cdot n)$ matrix A.

The proof of this important proposition is straightforward since the values λ_1, $\lambda_2, \ldots, \lambda_n$ solve (A6.4). Yet, from the algebra of polynomials, the product of the factors of (A6.4) is equal to $(-1)^n b_n$:

$$\prod_{i=1}^{n} \lambda_i = (-1)^n b_n$$

From the discussion above, we also know that $(-1)^n b_n = |A|$. Hence (A6.5) must hold in that the product $(\lambda_1)(\lambda_2) \dots (\lambda_n) = (-1)^n b_n = |A|$.

Examples 1 to 3, Continued

In Examples 1 and 2, the characteristic equation is quadratic of the form $\lambda^2 + b_1\lambda + b_2 = 0$. To find the roots of this quadratic equation, we seek the factors λ_1 and λ_2 such that

$$(\lambda - \lambda_1)(\lambda - \lambda_2) = 0$$

or

$$\lambda^2 - (\lambda\lambda_1 + \lambda\lambda_2) + \lambda_1\lambda_2 = 0$$

or

$$\lambda^2 - (\lambda_1 + \lambda_2)\lambda + \lambda_1\lambda_2 = 0$$

Clearly, the value of $\lambda_1\lambda_2$ must equal b_2. To check the formulas in Example 1, recall that the characteristic equation is $\lambda^2 - \lambda + 0.21 = 0$. In this problem, the value of b_2 is 0.21, the product of the characteristic roots is $\lambda_1\lambda_2 = (0.7)(0.3) = 0.21$, and the determinant of A is $(0.5)^2 - (0.2)^2 = 0.21$. In Example 2, the characteristic equation is $\lambda^2 - 0.1\lambda = 0$ so that $b_2 = 0$. The product of the characteristic roots is $\lambda_1\lambda_2 = (0.0)(0.1) = 0.0$, and the determinant of A is $(0.5)(0.4) - (0.2) = 0$.

In Example 3, the characteristic equation is cubic: $\lambda^3 - 1.5\lambda^2 + 0.63\lambda - 0.081 = 0$. The value of b_3 is -0.081, the product of the characteristic roots is $(0.9)(0.3)(0.3) = 0.081$, and the determinant of A is 0.081.

Characteristic Roots and Rank

The rank of a square $(n \cdot n)$ matrix A is the number of linearly independent rows (columns) in the matrix. The notation rank$(A) = r$ means that the rank of A is equal to r. The matrix A is said to be of full rank if rank$(A) = n$.

From the discussion above, it follows that *the rank of A is equal to the number of its nonzero characteristic roots*. Certainly, if all rows of A are linearly independent, the determinant of A is not equal to zero. From (A6.5) it follows that none of the characteristic roots can equal zero if $|A| \neq 0$. At the other extreme, if rank$(A) = 0$, each element of A must equal zero. When rank$(A) = 0$, the characteristic equation degenerates into $\lambda^n = 0$ with the solutions $\lambda_1 = \lambda_2 = \dots = \lambda_n = 0$. Consider the intermediate cases wherein $0 < \text{rank}(A) = r < n$. Since interchanging the various rows of a matrix does not alter the absolute value of its determinant, we can always rewrite $|A - \lambda I| = 0$ such that the first r rows comprise the r linearly independent rows of A. The determinant of these first r rows will contain r characteristic roots. The other $(n - r)$ roots will be zeroes.

In Example 2, rank$(A) = 1$ since each element in row 1 equals -2.5 times the corresponding element in row 2. For this case, $|A| = 0$ and exactly one characteristic root is equal to zero. In the other two examples, A is of full rank and all characteristic roots differ from zero.

Example 4

Now consider a $(3 \cdot 3)$ matrix A such that $\text{rank}(A) = 1$. Let

$$|A - \lambda I| = \begin{bmatrix} 0.5 - \lambda & 0.2 & 0.2 \\ 1 & 0.4 - \lambda & 0.4 \\ -0.25 & -0.1 & -0.1 - \lambda \end{bmatrix}$$

The rank of A is unity since row 2 is twice row 1, and row 3 is -0.5 times row 1. The determinant of A equals zero and the characteristic equation is given by

$$\lambda^3 - 0.8\lambda^2 = 0$$

The three characteristic roots are $\lambda_1 = 0.8$, $\lambda_2 = 0$, and $\lambda_3 = 0$.

Stability of a First-order VAR

Let x_t be the $(n \cdot 1)$ vector $(x_{1t}, x_{2t}, \ldots, x_{nt})'$ and consider the first-order VAR

$$x_t = A_0 + A_1 x_{t-1} + \varepsilon_t \tag{A6.6}$$

where: $A_0 = (n \cdot 1)$ vector with elements a_{i0}
$A_1 = (n \cdot n)$ square matrix with elements a_{ij}
$\varepsilon_t = (n \cdot 1)$ vector of white-noise disturbances $(\varepsilon_{1t}, \varepsilon_{2t}, \ldots, \varepsilon_{nt})'$

To check the stability of the system, we need only examine the homogeneous equation

$$x_t = A_1 x_{t-1} \tag{A6.7}$$

We can use the method of undetermined coefficients and for each x_{it} posit a solution of the form

$$x_{it} = c_i \lambda^t \tag{A6.8}$$

where c_i is an arbitrary constant.

If (A6.8) is to be a solution, it must satisfy each of the n equations represented by (A6.7). Substituting $x_{it} = c_i\lambda^t$ and $x_{it-1} = c_i\lambda^{t-1}$ for each of the x_{it} in (A6.7), we get

$$\begin{aligned} c_1\lambda^t &= a_{11}c_1\lambda^{t-1} + a_{12}c_2\lambda^{t-1} + \ldots + a_{1n}c_n\lambda^{t-1} \\ c_2\lambda^t &= a_{21}c_1\lambda^{t-1} + a_{22}c_2\lambda^{t-1} + \ldots + a_{2n}c_n\lambda^{t-1} \\ c_3\lambda^t &= a_{31}c_1\lambda^{t-1} + a_{32}c_2\lambda^{t-1} + \ldots + a_{3n}c_n\lambda^{t-1} \\ &\ldots \\ c_n\lambda^t &= a_{n1}c_1\lambda^{t-1} + a_{n2}c_2\lambda^{t-1} + \ldots + a_{nn}c_n\lambda^{t-1} \end{aligned}$$

Now, divide each equation by λ^{t-1} and collect terms to form

$$\begin{aligned} c_1(a_{11} - \lambda) + c_2a_{12} + c_3a_{13} \ldots + c_na_{1n} &= 0 \\ c_1a_{21} + c_2(a_{22} - \lambda) + c_3a_{23} \ldots + c_na_{2n} &= 0 \\ \ldots \quad \ldots \quad \ldots \quad \ldots \quad &\ldots \\ c_1a_{n1} + c_2a_{n2} + c_3a_{n3} \ldots + c_n(a_{nn} - \lambda) &= 0 \end{aligned}$$

so that the following system of equations must be satisfied:

$$\begin{bmatrix} (a_{11}-\lambda) & a_{12} & a_{13} & \dots & a_{1n} \\ a_{21} & (a_{22}-\lambda) & a_{23} & \dots & a_{2n} \\ \dots & \dots & \dots & \dots & \dots \\ a_{n1} & a_{n2} & a_{n3} & \dots & (a_{nn}-\lambda) \end{bmatrix} \begin{bmatrix} c_1 \\ c_2 \\ \dots \\ c_n \end{bmatrix} = \begin{bmatrix} 0 \\ 0 \\ 0 \\ 0 \end{bmatrix}$$

For a nontrivial solution to the system of equations, the following determinant must equal zero:

$$\begin{vmatrix} (a_{11}-\lambda) & a_{12} & a_{13} & \dots & a_{1n} \\ a_{21} & (a_{22}-\lambda) & a_{23} & \dots & a_{2n} \\ \dots & \dots & \dots & \dots & \dots \\ a_{n1} & a_{n2} & a_{n3} & \dots & (a_{nn}-\lambda) \end{vmatrix} = 0$$

The determinant will be an nth-order polynomial that is satisfied by n values of λ. Denote these n characteristic roots by $\lambda_1, \lambda_2, \dots \lambda_n$. Since each is a solution to the homogeneous equation, we know that the following linear combination of the homogeneous solutions is also a homogeneous solution:

$$x_{it} = d_1\lambda_1^t + d_2\lambda_2^t + \dots + d_n\lambda_n^t$$

Note that each $\{x_{it}\}$ sequence will have the same roots. The necessary and sufficient condition for stability is that all characteristic roots lie within the unit circle.

Cointegration and Rank

The relationship between the rank of a matrix and its characteristic roots is critical in the Johansen procedure. Using the notation from Section 7, let

$$x_t = A_1 x_{t-1} + \varepsilon_t$$

so that

$$\begin{aligned} \Delta x_t &= (A_1 - I)x_{t-1} + \varepsilon_t \\ &= \pi x_{t-1} + \varepsilon_t \end{aligned}$$

If the rank of π is unity, all rows of π can be written as a scalar multiple of the first. Thus, each of the $\{\Delta x_{it}\}$ sequences can be written as

$$\Delta x_{it} = s_i(\pi_{11}x_{1t-1} + \pi_{12}x_{2t-1} + \dots + \pi_{1n}x_{nt-1}) + \varepsilon_{it}$$

where $s_1 = 1$ and $s_i = \pi_{ij}/\pi_{1j}$.

Hence, the linear combination $\pi_{11}x_{1t-1} + \pi_{12}x_{2t-1} + \dots + \pi_{1n}x_{nt-1} = (\Delta x_{it} - \varepsilon_{it})/s_i$ is stationary since Δx_{it} and ε_{it} are both stationary.

The rank of π equals the number of cointegrating vectors. If rank$(\pi) = r$, there are r linearly independent combinations of the $\{x_{it}\}$ sequences that are stationary. If rank$(\pi) = n$, all variables are stationary.

The rank of π is equal to the number of its characteristic roots that differ from zero. Order the roots such that $\lambda_1 > \lambda_2 > \dots > \lambda_n$. The Johansen methodology allows you to determine the number of roots that are statistically different from zero. The relationship between A_1 and π is such that if all characteristic roots of A_1 are in the unit circle, π is of full rank.

Calculating the Characteristic Roots in Johansen's Method

Although commercially available software packages can obtain the characteristic roots of π, you might be interested in programming the method yourself (or at least understanding the method). First select the most appropriate lag length p in the VAR

$$x_t = A_1 x_{t-1} + \ldots + A_p x_{t-p} + \varepsilon_t$$

STEP 1: Estimate the VAR in first differences, that is, estimate

$$\Delta x_t = B_1 \Delta x_{t-1} + \ldots + B_{p-1} \Delta x_{t-p+1} + e_{1t}$$

STEP 2: Regress x_{t-1} on the lagged changes; that is, estimate a VAR of the form

$$x_{t-1} = C_1 \Delta x_{t-1} + \ldots + C_{p-1} \Delta x_{t-p+1} + e_{2t}$$

STEP 3: Compute the squares of the canonical correlations between e_{1t} and e_{2t}. In an n-equation VAR, the n canonical correlations are the n values of λ_i. The λ_i are obtained as the solutions to

$$|\lambda_i S_{22} - S_{12} S_{11}^{-1} S_{12}'| = 0$$

where $S_{ii} = T^{-1} \sum_{t=1}^{T} e_{it}(e_{it})'$, $S_{12} = T^{-1} \sum_{t=1}^{T} e_{2t}(e_{1t})'$

and e_{1t} and e_{2t} are the column vectors of residuals obtained in Steps 1 and 2.

STEP 4: The maximum-likelihood estimates of the cointegrating vectors are the n columns that are nontrivial solutions for

$$\lambda_i S_{22} \pi_i = S_{12} S_{11}^{-1} S_{12}' \pi_i$$

APPENDIX 6.2: INFERENCE ON A COINTEGRATING VECTOR

The Johansen procedure allows you to test restrictions on one or more cointegrating vectors. However, it is very tempting to use the t-statistics on a coefficient of a cointegrating vector estimated by OLS in the Engle–Granger methodology. Nevertheless, you must avoid this temptation since the coefficients *do not* have asymptotic t-distributions except in one special circumstance. The problem is that the coefficients are super-consistent but the standard errors are not. Nevertheless, it is typical for a published study to report the coefficients of the cointegrating vector and the associated t-statistics or standard errors. For example, the cointegrating relationship between y_t, z_t, and w_t in Section 5 was reported as (with t-statistics in parentheses)

$$\begin{aligned} y_t = {} & -0.0484 - 0.9273 z_t + 0.97687 w_t + e_{yt} \\ & (-0.575) \quad (-38.095) \quad (53.462) \end{aligned}$$

However, $\{e_{yt}\}$ may be serially correlated and z_t and w_t may not be exogenous variables. As in a traditional regression with stationary variables, you need to correct for serial correlation and the problem of endogenous regressors. To illustrate the *fully modified least squares* procedure developed by Phillips and Hansen (1990), consider the simple two-variable example

$$y_t = \beta_0 + \beta_1 z_t + e_{1t}$$
$$\Delta z_t = e_{2t}$$

The first equation is the cointegrating relationship and the second indicates that $\{z_t\}$ is the stochastic trend. The notation e_{1t} and e_{2t} is designed to illustrate the point that the residuals from both equations are stationary. However, they may be serially correlated and may be correlated with each other. As such, the second equation is actually quite general since Δz_t can be correlated with its own lags and with values of y_t.

Clearly, the relationship between the two errors is crucial. We begin with the simple case wherein

$$\begin{bmatrix} e_{1t} \\ e_{2t} \end{bmatrix} = Ni.i.d.\left(\begin{bmatrix} 0 \\ 0 \end{bmatrix}, \begin{bmatrix} \sigma_1^2 & 0 \\ 0 & \sigma_2^2 \end{bmatrix}\right).$$

Case 1: In this circumstance, the errors are serially uncorrelated and the cross-correlations are zero. Hence, the OLS regression of y_t on z_t and a constant is such that the explanatory variable (i.e., z_t) is independent of the error term e_{1t}. As indicated in the text, the OLS estimates of β_0 and β_1 can be tested using the normal distribution. Hence, t-tests and F-tests are appropriate. If the disturbances are not normally distributed, the asymptotic results are such that t-tests and F-tests are appropriate.

Case 2: In general, e_{1t} and e_{2t} will be correlated with each other so that $Ee_{1t}e_{2t} \neq 0$. In order to conduct inference on the parameters of the cointegrating vector, it is necessary to correct for the endogeneity of z_t. You do this by including leads and lags of $\{\Delta z_t\}$ in the cointegrating relationship. Hence, you estimate the equation

$$y_t = \beta_0 + \beta_1 z_t + \ldots + \gamma_{-1}\Delta z_{t+1} + \gamma_0 \Delta z_t + \gamma_1 \Delta z_{t-1} + \ldots + e_{1t}$$

In essence, you are controlling for innovations in z_t since the equation is equivalent to

$$y_t = \beta_0 + \beta_1 z_t + \ldots + \gamma_{-1} e_{2t+1} + \gamma_0 e_{2t} + \gamma_1 e_{2t-1} + \ldots + e_{1t}$$

Let var(e_{1t}) be denoted by σ_e^2. If $\{e_{1t}\}$ is serially uncorrelated, you can form a t-statistic to determine whether the estimated value of β_1 (i.e, $\hat{\beta}_1$) equals the hypothesized value β_1 using the t-statistic

$$t = (\hat{\beta}_1 - \beta_1)/\sigma_e$$

Case 3: In the most general case, $Ee_{1t}e_{2t} \neq 0$ and the residuals from the cointegrating vector (i.e., the estimated values of e_{1t}) are likely to be serially correlated. Hence, you also need to modify the t-statistic so that you use the appropriate estimate of the variance of e_{1t}. If the $\{e_{1t}\}$ series is serially correlated, you adjust the t-statistic using the following procedure:

STEP 1: Estimate the equation for y_t and obtain the estimated $\{e_{1t}\}$ series. Denote the t-statistic for the null hypothesis $\hat{\beta}_1 = \beta_1$ as t_0.

STEP 2: Estimate the $\{e_{1t}\}$ series as an AR(p) process to correct for autocorrelation. In particular, use the residuals from Step 1 to estimate the equation

$$e_{1t} = \alpha_1 e_{1t-1} + \ldots + \alpha_p e_{1t-p} + \varepsilon_t$$

Let σ^2 denote the estimated variance of ε_t so that σ is the standard deviation. Construct the value λ as

$$\lambda = \sigma/(1 - \alpha_1 - \ldots - \alpha_p)$$

STEP 3: Multiply t_0 by σ_e/λ. The resulting value is the appropriate t-statistic for the null hypothesis $\hat{\beta}_1 = \beta_1$. Compare the corrected t-statistic to that in a t-table. As you can see, the corrected t-statistic uses a more appropriate estimator for var(e_{1t}).

Little is altered if we allow z_t to be a vector of variables. However, a word of caution is in order. There are many possible sources of error in the three-step methodology outlined above. You could use too few or too many lags in Step 1. A similar problem arises in Step 2 because p is unknown. The Johansen procedure circumvents many of these problems in that all variables are treated as jointly endogenous and the VAR residuals are not serially correlated. Hence, you can conduct inference on the cointegrating vector(s) directly.

The procedure is not always as difficult as it sounds. As a practical matter, many researchers correct for serial correlation by adding lagged changes of Δy_t to the estimated equation. If the augmented equation eliminates the serial correlation, Steps 2 and 3 are unnecessary. The estimated equation has the form

$$y_t = \beta_0 + \beta_1 z_t + A_1(L)\Delta y_{t-1} + \ldots + \gamma_{-1}\Delta z_{t+1} + \gamma_0 \Delta z_t + \gamma_1 \Delta z_{t-1} + \ldots + \varepsilon_t$$

where $A_1(L)$ is a polynomial in the lag operator L and $\{\varepsilon_t\}$ is serially uncorrelated.

CHAPTER 7

NONLINEAR TIME-SERIES MODELS

Economic theory suggests that a number of important time-series variables should exhibit nonlinear behavior. The observation that wages display downward rigidity is a key feature of many macroeconomic models. Moreover, it has been established that downturns in the business cycle are sharper than recoveries in that key macroeconomic variables, such as output and employment, fall more sharply than they rise. Since the standard ARMA model relies on linear difference equations, new dynamic specifications are necessary to capture nonlinear behavior. In fact, research in this new area of time-series econometrics seems to be growing exponentially (itself a nonlinear process). This chapter has three aims:

1. Compare the ARMA model to various types of nonlinear models. Once the assumption of linearity is abandoned, it is possible to estimate any number of potential nonlinear processes. Several nonlinear forms have proved themselves to be especially useful. The text will focus on those nonlinear models that can be estimated by OLS methods, nonlinear least squares, or maximum-likelihood techniques.
2. Develop a number of tests that can detect the presence of nonlinear adjustment. The time-series literature contains a number of tests that are useful in determining whether or not a series is nonlinear. It will be shown that detecting nonlinearity is far simpler than trying to establish the precise nature of the nonlinearity.
3. Illustrate the process of estimating a nonlinear model. To a large extent, estimating a nonlinear process entails learning by doing. Toward this end, a number of series will be estimated using nonlinear techniques. Moreover, several studies that have appeared in the time-series literature will be reviewed. The issue of unit roots in a nonlinear setting will also be examined.

1. LINEAR VERSUS NONLINEAR ADJUSTMENT

On a long automobile trip to a new location, you might take along a road atlas. Since the earth is not flat, the maps contained in the atlas are a linear approximation of the actual path of your journey. However, for most trips, such a linear approximation is extremely useful. Try to envision the nuisance of a nonlinear road atlas. For other types of trips, the linearity assumption is clearly inappropriate. It would be disastrous for NASA to use a flat map of the earth to plan the trajectory of a rocket launch. Similarly, the assumption that economic processes are linear can provide useful approximations to the actual time paths of economic variables. Nevertheless, policy makers could make a serious error if they ignore the empirical evidence that unemployment increases more sharply than it decreases.

One example of a nonlinear model that has been used in the literature is the threshold autoregressive (TAR) model. To explain how it might be useful, let r_{Lt} and r_{St} be the long-term and short-term interest rates on two similar financial instruments. Suppose that the spread, defined as $s_t = r_{Lt} - r_{St}$, adjusts to the long-run value $\bar{s}$. A simple AR(1) representation of the dynamic adjustment mechanism might be

$$s_t = a_0 + a_1 s_{t-1} + \varepsilon_t \qquad \text{where } 0 < a_1 < 1$$

For our purposes, it is convenient to define $\bar{s}$ as the long-run value $a_0/(1 - a_1)$ and write the adjustment process as

$$s_t = \bar{s} + a_1(s_{t-1} - \bar{s}) + \varepsilon_t$$

If $s_t = \bar{s}$, the system is said to be in long-run equilibrium. In other circumstances, a_1 percent of the current period's deviation from the long-run value tends to persist into the next period. In fact, there is evidence that interest rate spreads display a nonlinear adjustment pattern. Periods in which the spread is low relative to its long-run value (so that $s_{t-1} - \bar{s} < 0$) are far more persistent than periods in which $s_{t-1} - \bar{s} > 0$. It is possible to model these differing degrees of persistence using

$$s_t = \begin{cases} \bar{s} + a_1(s_{t-1} - \bar{s}) + \varepsilon_{1t} & \text{when } s_{t-1} > \bar{s} \\ \bar{s} + a_2(s_{t-1} - \bar{s}) + \varepsilon_{2t} & \text{when } s_{t-1} \le \bar{s} \end{cases} \tag{7.1}$$

where ε_{1t} and ε_{2t} are white-noise processes.

In (7.1), when s_{t-1} is above the threshold value $\bar{s}$, the spread follows the AR(1) process $s_t = \bar{s} + a_1(s_{t-1} - \bar{s}) + \varepsilon_{1t}$, and when s_{t-1} is below the threshold, the spread follows the AR(1) process $s_t = \bar{s} + a_2(s_{t-1} - \bar{s}) + \varepsilon_{2t}$. As long as $|a_2| > |a_1|$, periods when $s_{t-1} < \bar{s}$ will tend to be more persistent than other periods.

To better illustrate the difference between linear and nonlinear adjustment, consider the homogeneous part of the first-order AR(1) model:

$$y_t = a_1 y_{t-1}$$

Given that $-1 < a_1 < 1$, we know that the long-run mean is such that $Ey_t = 0$. The nature of the adjustment process is such that a_1 percent of any current deviation from the long-run equilibrium persists into the next period. For example, if $a_1 = 0.5$ and the initial condition is such that $y_{t-1} = 1.0$, it immediately follows that $E_{t-1}y_t = 0.5$ and $E_{t-1}y_{t+1} = 0.25$. Let $\{y^*_{t-1}\}$ denote the specific sequence {1, 0.50, 0.25, 0.125, ...} generated by assuming the initial condition $y_{t-1} = 1$. The linearity of the adjustment process can be demonstrated by considering alternative values for y_{t-1}. If the initial condition is such that $y_{t-1} = 2$, the subsequent values of the new sequence are exactly twice those of the previous case. In fact, multiplying the initial value of y_{t-1} by any scalar λ results in the sequence $\{\lambda y^*_{t-1}\}$. The **phase diagram** shown in Panel (a)

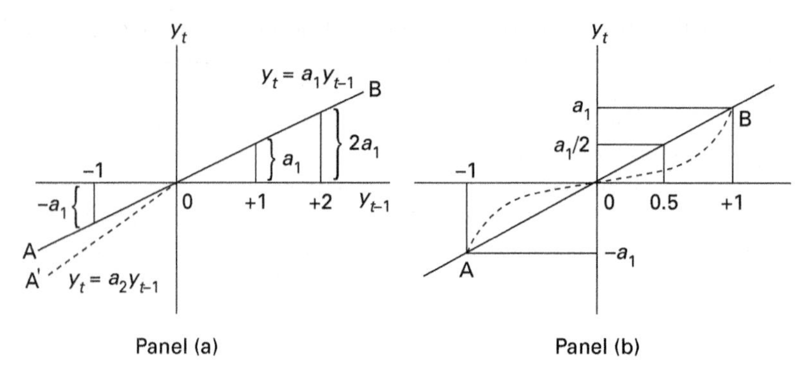

FIGURE 7.1 Two Nonlinear Adjustment Paths

of Figure 7.1 represents the linear nature of this adjustment process. The solid straight line labeled A0B is constructed to have a slope equal to a_1. Hence, for any value of y_{t-1}, you can obtain the next value in the sequence by using line A0B to project y_{t-1} onto the y_t axis. Since the slope is constant, any scalar multiple of y_{t-1} will result in the proportional value of y_t. As shown in the figure, if $y_{t-1} = 1$, the expected value of $y_t = a_1$ and if $y_{t-1} = 2$, the expected value of $y_t = 2a_1$.

Also note that adjustment is symmetric around zero. If $y_{t-1} = -1, E_{t-1}y_t = -0.5$ and $E_{t-1}y_{t+1} = -0.25$, and so on. Hence, for the linear model, multiplying the initial condition y_{t-1} by -1 results in the sequence $\{-y^*_{t-1}\}$.

Now suppose that the phase diagram is such that adjustment occurs along the kinked line passing through A'0B. Thus $y_t = a_1 y_{t-1}$ when $y_{t-1} > 0$ and $y_t = a_2 y_{t-1}$ when $y_{t-1} \leq 0$. Again, if $y_{t-1} = 1$, the next value in the sequence equals a_1. However, if $y_{t-1} = -1$, the next value in the sequence equals $-a_2$. Since $a_2 > a_1$, it should be clear that the $\{y_t\}$ sequence will approach zero more slowly when beginning from a negative value of y_{t-1} than a positive value. Hence, the adjustment process is not linear since the choice λy_{t-1} does not necessarily result in the sequence $\{\lambda y^*_{t-1}\}$. This is precisely the type of adjustment represented by equation (7.1); if $a_2 > a_1$ and $\bar{s} = 0$, the $\{s_t\}$ sequence has precisely the phase diagram shown by A'0B in Panel (a) of Figure 7.1.

A different type of nonlinear model is needed to represent the process of gravitational attraction. From elementary physics, we know that the speed of an object in space will increase as it falls toward the earth. We can represent the earth as being located at point 0 and suppose that the object in space is attracted to point 0. If y_t denotes the distance of the object from 0 at time t, gravitational attraction can be represented by the curve passing through A0B in Panel (b) of Figure 7.1. As shown in the figure, if we let $y_{t-1} = 1$, the value of y_t will be a_1. Instead, if y_{t-1} is 0.5, the value of y_t must be less than $a_1/2$. Since λy_{t-1} does not result in the sequence $\{\lambda y^*_{t-1}\}$, the adjustment process is not linear. The straight line $y_t = a_1 y_{t-1}$ passing through A0B does not capture this feature of the adjustment process.

Take a moment to imagine other types of nonlinear processes. For example, transport costs might deter arbitrage of a slight discrepancy between cotton prices in Alabama and Mississippi. In contrast, large price discrepancies might be eliminated almost immediately. You should be able to think of several other examples. The point is that once we decide to leave the realm of linearity, there are many potential types of nonlinearity. It can be especially important to determine the most appropriate form of the nonlinearity. After all, adopting an incorrect nonlinear specification may be more problematic than simply ignoring the nonlinearity altogether. Since selecting the proper nonlinear model can be difficult, it is not surprising that this remains an important area of current research. However, some special forms of nonlinearity have proven to be particularly useful in applied time-series research. We begin by presenting an overview of some simple nonlinear models.

2. SIMPLE EXTENSIONS OF THE ARMA MODEL

The simplest form of the nonlinear autoregressive (NLAR) model is

$$y_t = f(y_{t-1}) + \varepsilon_t$$

This is a first-order nonlinear autoregressive model, denoted by NLAR(1), in that the longest lag length is one. It is possible to reparameterize the model in a more interesting way:

$$y_t = a_1(y_{t-1}) \cdot y_{t-1} + \varepsilon_t \tag{7.2}$$

where $a_1(y_{t-1}) \cdot y_{t-1} \equiv f(y_{t-1})$.

Equation (7.2) looks exactly like an AR(1) model except for the fact that the autoregressive coefficient a_1 is allowed to be a function of the value of y_{t-1}. If we do not know the functional form of $f(\,)$, the usual dichotomy between nonlinearity in variables and time-varying parameters is not really clear-cut. It can be very difficult for a statistical test to detect the difference between a model in which some of the regressors are not raised to the power one and a model in which the parameters are varying over time. More generally, the pth order nonlinear autoregressive model is

$$y_t = f(y_{t-1}, y_{t-2}, \ldots, y_{t-p}) + \varepsilon_t \tag{7.3}$$

and is denoted by NLAR(p).

The difficulty in estimating (7.3) is that the functional form of $f(\,)$ is unknown. One way to proceed is to use a Taylor series approximation of the unknown functional form. For the NLAR(2) model $y_t = f(y_{t-1}, y_{t-2}) + \varepsilon_t$, the Taylor series approximation using terms no higher than order three is

$$\begin{aligned} y_t = a_0 &+ a_1 y_{t-1} + a_2 y_{t-2} + a_{12} y_{t-1} y_{t-2} + a_{11} y_{t-1}^2 + a_{22} y_{t-2}^2 \\ &+ a_{112} y_{t-1}^2 y_{t-2} + a_{122} y_{t-1} y_{t-2}^2 + a_{111} y_{t-1}^3 + a_{222} y_{t-2}^3 + \varepsilon_t \end{aligned}$$

For the more general NLAR(p) we need a more compact notation. A simple way of writing such a Taylor series approximation is

$$y_t = a_0 + \sum_{i=1}^{p} a_i y_{t-i} + \sum_{i=1}^{p} \sum_{j=1}^{p} \sum_{k=1}^{r} \sum_{l=1}^{s} a_{ijkl} y_{t-i}^k y_{t-j}^l + \varepsilon_t \tag{7.4}$$

where p is the order of the process and r and s are integers that are greater than or equal to 1. In order to avoid a very large number of parameters, the sum of r and s is usually restricted to be less than or equal to 4.

The GAR Model

Equation (7.4) is called the generalized autoregressive (GAR) model. GAR models extend the standard AR model by including various powers of the lagged values of y_{t-i} and cross-products of the powers of y_{t-i} and y_{t-j}. As a Taylor series approximation, the GAR model is capable of mimicking a wide variety of functional forms—all that is required is that the function be differentiable. Moreover, the model is easy to estimate; simply form the variables y_{t-i}^{j} and their cross-products, and estimate the model using OLS. A test for nonlinearity can be carried out directly since the linear model is nested with the GAR model. If it is not possible to reject the null hypothesis that all values of $a_{ijkl} = 0$, it can be concluded that the process is linear. On the downside, the resulting model is likely to be overparameterized. This is especially true if the number of lags in the model is more than two. You can use traditional t-tests and F-tests to pare down the number of parameters estimated. However, this can be tricky since the regressors are likely to be highly correlated. For example, the term y_{t-1}^{2} will clearly be correlated with y_{t-1}^{4}. As such, the usual practice is to pare down the equation using the AIC or SBC.

The Bilinear Model

Just as a parsimonious ARMA model can well approximate a high-order AR(p) process, it is possible to use moving-average terms in a nonlinear model. Consider the simple bilinear (BL) model

$$y_t = \alpha_0 + \alpha_1 y_{t-1} + \beta_1 \varepsilon_{t-1} + c_1 \varepsilon_{t-1} y_{t-1} + \varepsilon_t$$

The intent is to use moving-average terms and the interactions of autoregressive and MA terms to approximate a high-order GAR model. As such, bilinear models are a natural extension of ARMA models in that they add the cross-products of y_{t-i} and ε_{t-j} to account for nonlinearity. The general form of the bilinear model BL (p, q, r, s) is

$$y_t = \alpha_0 + \sum_{i=1}^{p} \alpha_i y_{t-i} + \varepsilon_t + \sum_{i=1}^{q} \beta_i \varepsilon_{t-i} + \sum_{i=1}^{r} \sum_{j=1}^{s} c_{ij} y_{t-i} \varepsilon_{t-j} \tag{7.5}$$

Notice that the linear ARMA(p, q) model is nested within (7.5); if all values of $c_{ij} = 0$, (7.5) is identical to an ARMA(p, q) model. As with the GAR model, the bilinear model can be viewed as having stochastic parameter variation. To understand the point, consider the BL model

$$y_t = \alpha_0 + \alpha_1 y_{t-1} + c_1 y_{t-1} \varepsilon_{t-1} + \varepsilon_t$$

so that

$$y_t = \alpha_0 + (\alpha_1 + c_1 \varepsilon_{t-1}) y_{t-1} + \varepsilon_t \tag{7.6}$$

Equation (7.6) looks like an autoregressive model except for the fact that the autoregressive coefficient is $\alpha_1 + c_1 \varepsilon_{t-1}$. In a sense, the autoregressive coefficient is a random variable with a mean equal to α_1. If c_1 is positive, the autoregressive coefficient will increase with ε_{t-1}. In this way, positive ε_{t-1} shocks will be more persistent than negative shocks.

Now for a little quiz. You cannot use OLS to estimate (7.5) or (7.6) since you cannot directly form the variables $y_{t-i}\varepsilon_{t-j}$. The question is: If you have the single time series $\{y_t\}$, how can you estimate the series as a bilinear process? The standard procedure is to use maximum-likelihood estimation. Many of the standard econometric software packages allow you to perform the estimation using a straightforward generalization of the method developed in Appendix 2.1 of Chapter 2 for the estimation on an MA process.

An Example

Rothman (1998) compared the in-sample fit and out-of-sample forecasting performance of a number of nonlinear models of the U.S. unemployment rate. Toward this end, he detrended the log of the unemployment rate and estimated the following three models over the 1948Q1 to 1979Q4 period:

AR $\quad u_t = \underset{(22.46)}{1.563u_{t-1}} - \underset{(-10.06)}{0.670u_{t-2}} + \varepsilon_t$

GAR $\quad u_t = \underset{(23.60)}{1.500u_{t-1}} - \underset{(-6.72)}{0.553u_{t-2}} - \underset{(-2.33)}{0.745u_{t-2}^3} + \varepsilon_t \qquad$ variance ratio = 0.965

BL $\quad u_t = \underset{(24.11)}{1.910u_{t-1}} - \underset{(-10.55)}{0.690u_{t-2}} - \underset{(-2.08)}{0.585u_{t-1}\varepsilon_{t-3}} + \varepsilon_t \qquad$ variance ratio = 0.936

where: u_t = the detrended log of the unemployment rate
variance ratio = the ratio of the residual variance of the estimated model to the residual variance of the AR model

The AIC was used to select the most appropriate values of p and q from the general class of ARMA(p, q) models. The AR(2) specification yielded the best fit from the class of linear ARMA models. A general specification search within the class of GAR models was undertaken and the AIC was used to select the one with the best fit. Simply, for the given lag length of two, all models in the form of (7.4) were estimated and the one with the lowest AIC was retained. Notice that only the cubic term on the second lag of the unemployment rate was deemed to be important. Since the GAR model incorporates the AR(2) models as a special case, it is not surprising that it has a smaller residual variance. As Rothman indicates, it is instructive to write the estimated GAR model as

$$u_t = 1.500u_{t-1} - [0.553 + 0.745u_{t-2}^2]u_{t-2} + \varepsilon_t$$

In this form, the GAR model can be viewed as an AR(2) process such that the coefficient on the second lag is $-[0.553 + 0.745u_{t-2}^2]$. As such, large deviations from trend unemployment (so that u_{t-2}^2 is large) are associated with lower autoregressive persistence than small deviations from trend. As such, the speed of adjustment is faster when unemployment is far from its trend value than when it is close to the trend. Hence, the speed of adjustment is opposite to that of gravitational attraction. As an exercise, you should sketch the phase diagram for this adjustment process and compare your answer to Panel (b) of Figure 7.1.

Of the three models, the BL model has the smallest residual variance. The general BL model in the form of (7.5) was estimated for various values of r and s. Again the AIC was used to select the best-fitting model from this class. Notice that the estimated bilinear model uses the cross-product $u_{t-1}\varepsilon_{t-3}$ even though the linear model contains only two lags (i.e., r and s were allowed to exceed the order of the linear portion of the equation). Rothman indicates that u_{t-1} and ε_{t-3} are positively correlated. Since the coefficient on $u_{t-1}\varepsilon_{t-3}$ is negative, large shocks to the unemployment rate imply a faster speed of adjustment than small shocks. As u_{t-1} and ε_{t-3} tend to move together, the larger $u_{t-1}\varepsilon_{t-3}$, the smaller the degree of persistence.

3. PRETESTING FOR NONLINEARITY

Before introducing other types of nonlinear models, it is important to develop several standard tests for the presence of nonlinearity. Pretesting for nonlinearity can help protect you from overfitting the data. This section will present a number of procedures that have been developed to determine if the data seem to be nonlinear and to help to determine the form of the nonlinearity. Be forewarned that no set of tests can actually pin down the proper form of nonlinearity. Rather, the tests can only suggest the form of the nonlinearity.

The ACF and the McLeod–Li Test

In estimating an ARMA model, the autocorrelation function can help you select the proper values of p and q, and the ACF of the residuals is an important diagnostic tool. Unfortunately, the ACF as used in linear models may be misleading for nonlinear models. The reason is that the autocorrelation coefficients measure the degree of *linear* association between y_t and y_{t-i}. As such, the ACF may fail to detect important nonlinear relationships present in the data. Consider the following example:

$$y_t = \varepsilon_{t-1}^2 + \varepsilon_t \tag{7.7}$$

where $\{\varepsilon_t\}$ is a normally distributed white-noise process.

Since y_{t-1} is a function of ε_{t-1}, the value of y_t is dependent on the value of y_{t-1}. Nevertheless, with a little bit of algebra, it is possible to show that all of the autocorrelations are equal to zero. To derive this result, call $\text{var}(\varepsilon_t) = \text{var}(\varepsilon_{t-i}) = \sigma^2$. If you take the conditional expectation of (7.7), it follows that $Ey_t = Ey_{t-1} = \sigma^2$. Thus, the autocorrelations are

$$\begin{aligned}
\rho_i &= E(y_t - \sigma^2)(y_{t-i} - \sigma^2) \\
&= E(\varepsilon_{t-1}^2 + \varepsilon_t - \sigma^2)(\varepsilon_{t-1-i}^2 + \varepsilon_{t-i} - \sigma^2) \\
&= E(\varepsilon_{t-1}^2\varepsilon_{t-1-i}^2 + \varepsilon_{t-1}^2\varepsilon_{t-i} - \varepsilon_{t-1}^2\sigma^2 + \varepsilon_t\varepsilon_{t-1-i}^2 + \varepsilon_t\varepsilon_{t-i} - \varepsilon_t\sigma^2 \\
&\quad - \sigma^2\varepsilon_{t-1-i}^2 - \sigma^2\varepsilon_{t-i} + \sigma^2\sigma^2)
\end{aligned}$$

Note that $E(\varepsilon_{t-i}^2\varepsilon_{t-j}^2) = \sigma^2\sigma^2$, $E(\varepsilon_t\varepsilon_{t-i}^2) = 0$, and $E(\varepsilon_t\sigma^2) = 0$. As such,

$$\rho_i = \sigma^2\sigma^2 + E\varepsilon_{t-1}^2\varepsilon_{t-i} - \sigma^2\sigma^2 - \sigma^2\sigma^2 + \sigma^2\sigma^2 = E\varepsilon_{t-1}^2\varepsilon_{t-i}$$

Clearly, all values of $E\varepsilon_{t-1}^2\varepsilon_{t-i} = 0$ if $i \neq 1$. Moreover, if ε_t is normally distributed, the third moment $E\varepsilon_t^3 = E\varepsilon_{t-1}^3 = 0$. Hence, all values of ρ_i $(i \neq 0)$ are equal to zero. Now suppose that you observe the sample ACF for $\{y_t\}$ but are unaware that the data were generated by (7.7). Based on the observation that the sample autocorrelations are small, you might mistakenly conclude that the series is white noise. You would not be the first person to fall into the trap of confusing a lack of correlation with statistical independence. Although the autocorrelations are zero, the value of y_t is clearly dependent on the value of y_{t-1}.

Since we are interested in nonlinear relationships in the data, a useful diagnostic tool is to examine the ACF of the squared or cubed values of a series. For example, the ACF of y_t^2 (or the squares of the residuals from an estimated equation) can reveal a nonlinear pattern. To illustrate the point, Granger and Teräsverta (1993) showed that the ACF from **chaos** may be indicative of white noise but that the ACF of squared values of the sequence may be large. A nonexplosive sequence is chaotic if it is generated from a deterministic difference equation such that it does not converge to a constant or to a repetitive cycle. Consider the following chaotic process:

$$y_t = 4y_{t-1}(1 - y_{t-1}) \quad \text{for } 0 < y_1 < 1 \tag{7.8}$$

In (7.8), y_t is related to the level and the squared value of y_{t-1}. However, the autocorrelations of $\{y_t\}$ will all be small but the ACF of $\{y_t^2\}$ will be large. To follow along using your software package, set $y_1 = 0.7$ and generate the next 99 values of $\{y_t\}$ using (7.8). Even though the sequence is perfectly predictable, you should find that the first six autocorrelations are

ρ_1	ρ_2	ρ_3	ρ_4	ρ_5	ρ_6
−0.074	−0.072	0.008	0.032	−0.016	−0.030

All of the correlations are less than one standard deviation from zero. However, the correlation coefficient between y_t^2 and y_{t-1}^2 is −0.281, and the autocorrelation coefficient between y_t^3 and y_{t-1}^3 is −0.386. With 100 observations, these two correlations are highly significant. The point of the example is to show that any neglected nonlinearity in your data can be checked using the ACF of the squared (or cubed) values of the series. To be a bit more formal, the McLeod–Li (1983) test sought to determine if there are significant autocorrelations in the squared residuals from a linear equation. To perform the test, estimate your series using the best-fitting linear model and call the residuals $\hat{e}_t$. As in a formal test for ARCH errors (see Section 2 of Chapter 3), form the autocorrelations of the squared residuals. Let ρ_i denote the sample correlation coefficient between squared residuals $\hat{e}_t^2$ and $\hat{e}_{t-i}^2$ and use the Ljung–Box statistic to determine whether the squared residuals exhibit serial correlation. Hence, form

$$Q = T(T + 2)\sum_{i=1}^{n} \rho_i/(T - i)$$

The value Q has an asymptotic χ^2 distribution with n degrees of freedom if the $\{\hat{e}_t^2\}$ sequence is uncorrelated. Rejecting the null hypothesis is equivalent to accepting that the model is nonlinear. Alternatively, you can estimate the regression

$$\hat{e}_t^2 = \alpha_0 + \alpha_1\hat{e}_{t-1}^2 + \ldots + \alpha_n\hat{e}_{t-n}^2 + v_t$$

If there are no nonlinearities, α_1 through α_n should be zero. With a sample of T residuals, if there are no nonlinearities, the test statistic TR^2 will converge to a χ^2 distribution with n degrees of freedom. In small samples you can use an F-test for the null hypothesis $\alpha_1 = \alpha_2 = \ldots = \alpha_n = 0$. If you are astute, you will remember that this test was used to detect ARCH-type errors. It turns out that the McLeod–Li (1983) test is the exact Lagrange multiplier (LM) test for ARCH errors. However, the test has substantial power to detect various forms of nonlinearity. Notice that the actual form of the nonlinearity is not specified by the test. Rejecting the null hypothesis of linearity does not tell you the nature of the nonlinearity present in the data.

The RESET

The Regression Error Specification Test (RESET) also posits the null hypothesis of linearity against a general alternative hypothesis of nonlinearity. If the residuals from a linear model are independent, they should not be correlated with the regressors used in the estimating equation or with the fitted values. Hence, a regression of the residuals on these values should not be statistically significant. To perform the RESET:

STEP 1: Estimate the best-fitting linear model. Let $\{e_t\}$ be the residuals from the model and denote the fitted values by $\hat{y}_t$.

STEP 2: Select a value of H (usually 3 or 4) and estimate the regression equation

$$e_t = \delta z_t + \sum_{h=2}^{H} \alpha_h \hat{y}_t^h \quad \text{for } H \geq 2.$$

where z_t is the vector that contains the variables included in the model estimated in Step 1. For example, if you estimate an ARMA(p, q) model, z_t will include a constant, y_{t-1} through y_{t-p}, and ε_{t-1} through ε_{t-q}. Note that the test can also be applied to a regression model. As such, z_t may also include exogenous explanatory variables.

This regression should have little explanatory power if the model is truly linear. As such, the sample value of F should be small. Hence, you can reject linearity if the sample value of the F-statistic for the null hypothesis $\alpha_2 = \ldots = \alpha_H = 0$ exceeds the critical value from a standard F-table. The RESET is easy to implement, does not require the estimation of a large number of parameters, and has reasonable power to detect some types of nonlinearities.

Other Portmanteau Tests

Portmanteau tests (translated as "a little brown suitcase") are residual-based tests that do not have a specific alternative hypothesis. The Ljung–Box Q-statistics are a good example of this type of catch-all test. Similarly, the popular Brock, Dechert,

Scheinkman, and LaBarron (1996) test, called the BDS test, is a portmanteau test for independence. In essence, the test examines the distance between different pairs of residuals. Let d represent a given distance and let ε_t and ε_{t-1} be two realizations of the $\{\varepsilon_t\}$ sequence. If all values of $\{\varepsilon_t\}$ are independent, then the probability that the distance between any pair of residuals $(\varepsilon_i, \varepsilon_j)$ is less than d should be the same for all i and j. Although very popular, the BDS test is able to detect serial correlation, parameter instability, neglected nonlinearity, structural breaks, and other misspecification problems. Hence, rejecting the null hypothesis of independence does little to help identify the nature of the problem. Also be aware that the BDS test does not have especially good small-sample performance unless you bootstrap the critical values.

The point is that the McLeod–Li Test, the RESET, and other portmanteau tests all have a very general alternative hypothesis. As such, the tests are helpful in determining whether a nonlinear model is appropriate but not in determining the nature of the nonlinearity. As noted by Clements and Hendry (1998, pp. 168–69), "parameter change appears in many guises and can cause significant forecast error when models are used in practice." They also establish that it can be difficult to distinguish model misspecification from the problem of nonconstant parameters. As such, it is worthwhile to examine Lagrange multiplier tests for nonlinearity since they have a specific null hypothesis and a specific alternative hypothesis.

Lagrange Multiplier Tests

Lagrange multiplier (LM) tests can be used to test for a specific type of nonlinearity. Thus, an LM test can help you to select the proper functional form to use in your nonlinear estimation. To keep the analysis simple, we will assume that $\text{var}(\varepsilon_t) = \sigma^2$ is constant. Let $f(\ )$ be the nonlinear functional form and let α denote the parameters of $f(\ )$. In these circumstances, the LM test can be conducted as follows:[1]

STEP 1: Estimate the linear portion of the model to get the residuals $\{e_t\}$.

STEP 2: Obtain all of the partial derivatives $\partial f(\ )/\partial \alpha$ evaluated under the null hypothesis of linearity. Typically, these partial derivatives will be nonlinear functions of the regressors used in Step 1. Estimate the **auxiliary regression** by regressing e_t on these partial derivatives.

STEP 3: The value of TR^2 has a χ^2 distribution with degrees of freedom equal to the number of regressors used in Step 2. If the calculated value of TR^2 exceeds the critical value from a χ^2 table, reject the null hypothesis of linearity and accept the alternative. With a small sample, it is standard to use an F-test.

One benefit of the method is that you need not estimate the nonlinear model itself. More importantly, the use of a number of LM tests can help you select the form of the nonlinearity. It could be the case, for example, that an LM test rejects the GAR model but accepts the BL model. Unfortunately, this is not the typical case. Instead, the two LM tests are likely to accept the GAR model and the BL model. Nevertheless, comparing the *prob*-values of the two can be helpful. Consider the two examples below.

TWO EXAMPLES

Example 1 Suppose you want to determine whether $\{y_t\}$ has the specific GAR form

$$y_t = \alpha_0 + \alpha_1 y_{t-1} + \alpha_2 y_{t-2} + \alpha_3 y_{t-1} y_{t-2} + \varepsilon_t \tag{7.9}$$

Of course, it would be straightforward to estimate (7.9) directly and obtain the t-statistic for the null hypothesis $\alpha_3 = 0$. However, the point of this section is to illustrate the appropriate use of the LM test. Toward this end, estimate the sequence as an AR(2) process and obtain the residuals $\{e_t\}$. Now, you need to find the partial derivatives of the nonlinear functional form. It should be clear that

$$\partial y_t/\partial\alpha_0 = 1;\ \partial y_t/\partial\alpha_1 = y_{t-1};\ \partial y_t/\partial\alpha_2 = y_{t-2};\text{ and }\partial y_t/\partial\alpha_3 = y_{t-1}y_{t-2}$$

Hence, Step 2 indicates that you regress e_t on a constant (i.e., a vector of 1's), y_{t-1}, y_{t-2}, and $y_{t-1}y_{t-2}$. Thus, the auxiliary regression is

$$e_t = a_0 + a_1 y_{t-1} + a_2 y_{t-2} + a_3 y_{t-1} y_{t-2} + \nu_t \tag{7.10}$$

Obtain the sample value of TR^2. If this value exceeds the critical value of χ^2 with four degrees of freedom, reject the null hypothesis of linearity and accept the alternative of the GAR model. Alternatively, you can perform an F-test for the joint hypothesis $a_0 = a_1 = a_2 = a_3 = 0$.

Example 2 A similar procedure can be used to determine whether $\{y_t\}$ has the BL form

$$y_t = \alpha_0 + \alpha_1 y_{t-1} + \alpha_2 y_{t-2} + \alpha_3 \varepsilon_{t-1} y_{t-2} + \varepsilon_t$$

Again, estimate the sequence as an AR(2) process and obtain the residuals $\{e_t\}$. The desired partial derivatives are

$$\partial y_t/\partial\alpha_0 = 1;\ \partial y_t/\partial\alpha_1 = y_{t-1};\ \partial y_t/\partial\alpha_2 = y_{t-2};\text{ and }\partial y_t/\partial\alpha_3 = \varepsilon_{t-1}y_{t-2}$$

so that auxiliary regression is

$$e_t = a_0 + a_1 y_{t-1} + a_2 y_{t-2} + a_3 \varepsilon_{t-1} y_{t-2} + \nu_t \tag{7.11}$$

Since the actual values of $\{\varepsilon_{t-1}\}$ are unobserved, use the estimated residuals to form $\varepsilon_{t-1}y_{t-2}$ in (7.11). If the sample value of TR^2 exceeds the critical value of χ^2 with four degrees of freedom, reject the null hypothesis of linearity and accept the alternative of the BL model. Alternatively, you can use an F-test for the null hypothesis $a_0 = a_1 = a_2 = a_3 = 0$.

Notice that (7.10) and (7.11) are very similar. Since ε_{t-1} will be highly correlated with y_{t-1}, the values of TR^2 from the two equations will be quite similar. Hence, the results of both tests should be quite similar; if (7.10) indicates that a GAR model is appropriate, (7.11) should indicate that a BL model is appropriate. Nevertheless, the two tests can be useful. If both accept the null hypothesis of linearity, you can be reasonably confident that the AR(2) model is adequate. If both reject the null hypothesis, you can be somewhat confident that a nonlinear model is appropriate. However, unless the *prob*-values of the two tests are quite different, the tests will not provide much guidance as to which nonlinear form is the most appropriate.

4. THRESHOLD AUTOREGRESSIVE MODELS

A **regime-switching model** allows the behavior of $\{y_t\}$ to depend on the state of the system. In a recession, the unemployment rate is likely to rise sharply and then slowly decline to its long-run value. However, the unemployment rate does not fall sharply in an economic expansion. As such, the dynamic adjustment equation for the unemployment rate depends on whether the economy is in an expansionary state (or regime) or in a recession. When the economy changes from an expansionary regime to a contractionary regime, the dynamic adjustment of the unemployment rate is likely to change. In other circumstances, regime switches might be due to the magnitude of the variable of interest, the result of an election that changes the behavior of policy makers, or may be completely unobservable. As you might expect, a number of regime-switching models have been developed to analyze these types of regime changes.

Before proceeding, you need to know that most regime-switching models can be quite difficult to estimate. Although many software packages allow you to estimate a linear model by appropriately clicking on a menu, this is not true for nonlinear models. In general, you need to use a statistical package that has its own programming language if you want to estimate a regime-switching model. Threshold autoregressive (TAR) models of the type developed by Tong (1983, 1990) can be estimated using OLS. Another type of threshold model allows for gradual regime change. Such smooth-transition autoregressive (STAR) models can be estimated using nonlinear least squares or maximum-likelihood methods. Other nonlinear models, such as the artificial neural network and Markov switching models, require methods that are more sophisticated. As such, discussion in the text emphasizes the threshold models. The *Programming Manual* that accompanies this text has a number of examples.

The Basic Threshold Model

Panel (a) of Figure 7.1 illustrates a simple TAR process. Recall that the degree of autoregressive persistence is a_1 when $y_{t-1} > 0$ and a_2 when $y_{t-1} \leq 0$. As in equation (7.1), if we include a disturbance term, the behavior of the $\{y_t\}$ sequence can be represented by

$$y_t = \begin{cases} a_1 y_{t-1} + \varepsilon_{1t} \text{ if } y_{t-1} > 0 \\ a_2 y_{t-1} + \varepsilon_{2t} \text{ if } y_{t-1} \leq 0 \end{cases} \tag{7.12}$$

You can think of the equation $y_{t-1} = 0$ as being a threshold. On one side of the threshold, the $\{y_t\}$ sequence is governed by one autoregressive process and on the other side of the threshold, there is a different autoregressive process. Although $\{y_t\}$ is linear in each regime, the possibility of regime switching means that the entire $\{y_t\}$ sequence is nonlinear. Shocks to $\{\varepsilon_{1t}\}$ or $\{\varepsilon_{2t}\}$ are responsible for regime switching. If, for example, $y_{t-1} > 0$, the subsequent values of the sequence will tend to decay toward zero at the rate a_1. However, a negative realization of ε_{1t} can cause y_t to fall by such an extent that it lies below the threshold. In the negative regime, the behavior of the process is governed by $y_t = a_2 y_{t-1} + \varepsilon_{2t}$. As you can infer, the larger the variance of $\{\varepsilon_{1t}\}$, the more likely is a switch from the positive regime to the negative regime.

Another common variant of the TAR model is to assume that the variances of the two error terms are equal [i.e., $\text{var}(\varepsilon_{1t}) = \text{var}(\varepsilon_{2t})$]. In this circumstance, (7.12) can be written as

$$y_t = a_1 I_t y_{t-1} + a_2(1 - I_t)y_{t-1} + \varepsilon_t \tag{7.13}$$

where $I_t = 1$ if $y_{t-1} > 0$ and $I_t = 0$ if $y_{t-1} \leq 0$.

In equation (7.13), I_t is an indicator function, or dummy variable, that takes on the value of 1 if y_{t-1} is above the threshold, and a value of 0 otherwise. When $y_{t-1} > 0$, $I_t = 1$ and $(1 - I_t) = 0$, so that (7.13) is equivalent to $a_1 y_{t-1} + \varepsilon_t$. When $y_{t-1} \leq 0$, $I_t = 0$ and $(1 - I_t) = 1$, so that (7.13) is equivalent to $a_2 y_{t-1} + \varepsilon_t$. Figure 7.2 provides a visual comparison of the AR, GAR, BL, and TAR models. A series of 200 random numbers was drawn from a standardized normal distribution so as to simulate the $\{\varepsilon_t\}$ sequence. The initial value y_1 was set equal to ε_1 and the next 199 values of $\{y_t\}$ were created according to the formula

$$y_t = 0.7y_{t-1} + \varepsilon_t$$

Panel (a) of Figure 7.2 shows the time path of this simulated AR(1) process. Notice that the series fluctuates around a mean of zero. Although it may not be possible to

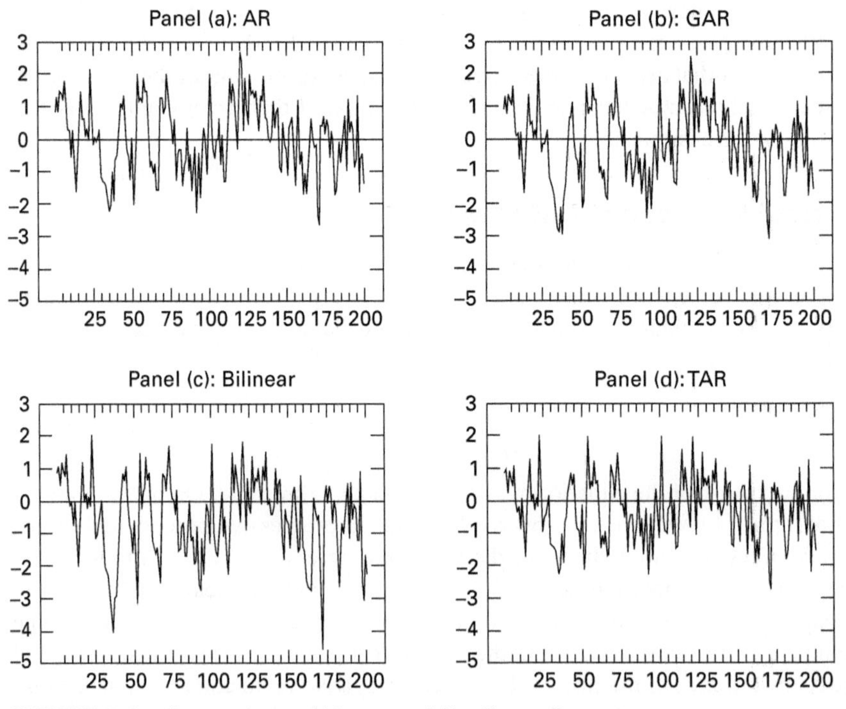

FIGURE 7.2 Comparison of Linear and Nonlinear Processes

discern with visual inspection alone, the degree of autoregressive decay is always the same; on average, 70 percent of the current value of y_t persists into the next period. Next, the same random numbers were used to generate the GAR process

$$y_t = 0.7y_{t-1} - 0.06y_{t-1}^2 + \varepsilon_t$$

or

$$y_t = [0.7 - 0.06y_{t-1}]y_{t-1} + \varepsilon_t$$

The nature of this particular GAR process is that it behaves as an AR(1) process with a random coefficient. The greater the value of y_{t-1}, the smaller the autoregressive coefficient. For values of $y_{t-1} = -2$, 0, and 2, the degrees of autoregressive persistence are 0.82, 0.7, and 0.58, respectively. This pattern can be seen in Panel (b) of Figure 7.2 because negative values of the simulated GAR process are far more persistent than positive values. Compare Panels (a) and (b) and note the values of the two series surrounding period 35 and period 85. You can clearly see that the GAR series returns to zero more slowly than the AR series.

The identical random numbers were used to construct the BL sequence shown in Panel (c). After initializing $y_1 = \varepsilon_1$, the remaining values of the sequence were generated from

$$y_t = 0.7y_{t-1} - 0.3y_{t-1}\varepsilon_{t-1} + \varepsilon_t$$

or

$$y_t = [0.7 - 0.3\varepsilon_{t-1}]y_{t-1} + \varepsilon_t$$

In the BL model, the degree of persistence depends on the value of ε_{t-1}; the larger is ε_{t-1}, the smaller the degree of persistence. In fact, for those periods in which $\varepsilon_{t-1} < -1.0$, the sequence behaves like an explosive process (since the value of $0.7 - 0.3\varepsilon_{t-1}$ exceeds unity). In Panel (c), you can see the extreme movements in the BL process if you examine the time intervals that surround period 55 and period 165. Nevertheless, the successive values of ε_t are more likely to exceed -1, so the BL process does not continue its decline.

Panel (d) of Figure 7.2 illustrates the time path of the TAR process

$$y_t = 0.3I_t y_{t-1} + 0.7(1 - I_t)y_{t-1} + \varepsilon_t$$

where $I_t = 1$ if $y_{t-1} > 0$ and $I_t = 0$ otherwise.

When $y_{t-1} \leq 0$, this TAR process behaves exactly like the AR(1) process shown in Panel (a). Thus, the lower portions of Panels (a) and (d) are nearly identical. However, for the TAR process, only 30 percent of the current value of y_t tends to persist into the subsequent period when $y_{t-1} > 0$. Hence, in contrast to the AR(1) process of Panel (a), the TAR process displays a substantial degree of mean reversion whenever $y_{t-1} > 0$.

Estimation

Estimation of a threshold model in the form of (7.13) can be performed by simple OLS. First construct the dummy variable I_t such that $I_t = 1$ if $y_{t-1} > 0$ and $I_t = 0$ if

$y_{t-1} \leq 0$. Then construct two variables, say y_{t-1}^{+} and y_{t-1}^{-}, such that $y_{t-1}^{+} = I_t y_{t-1}$ and $y_{t-1}^{-} = (1 - I_t)y_{t-1}$. Finally, use OLS to estimate the regression equation $y_t = a_1 y_{t-1}^{+} + a_2 y_{t-1}^{-} + \varepsilon_t$. It is straightforward to generalize the method such that there is a higher-order autoregressive process in each regime. For example, a more general version of (7.13) is

$$y_t = I_t\left[\alpha_{10} + \sum_{i=1}^{p} \alpha_{1i} y_{t-i}\right] + (1 - I_t)\left[\alpha_{20} + \sum_{i=1}^{r} \alpha_{2i} y_{t-i}\right] + \varepsilon_t \tag{7.14}$$

where $I_t = 1$ if $y_{t-1} > \tau$ and $I_t = 0$ if $y_{t-1} \leq \tau$.

In (7.14), there are two separate regimes defined by the value of y_{t-1}. When $y_{t-1} > \tau$, $I_t = 1$ and $(1 - I_t) = 0$, so that (7.14) is equivalent to $\alpha_{10} + \alpha_{11} y_{t-1} + \ldots + \alpha_{1p} y_{t-p} + \varepsilon_t$. When $y_{t-1} \leq \tau$, $I_t = 0$ and $(1 - I_t) = 1$, so that (7.14) is equivalent to $\alpha_{20} + \alpha_{21} y_{t-1} + \ldots + \alpha_{2r} y_{t-r} + \varepsilon_t$. Unlike the TAR models depicted in Figures 7.1 and 7.2, the value of the threshold τ is allowed to differ from zero. Moreover, the particular phase diagram shown in Panel (a) of Figure 7.1 was a special type of TAR model in that it is continuous. The specification in equation (7.14) allows the two segments of the phase diagram to be discontinuous at the threshold. If τ is known, the estimation of a TAR model is straightforward. Create the dummy variable I_t according to whether y_{t-1} is above or below the threshold τ and form the variables $I_t y_{t-i}$ and $(1 - I_t)y_{t-i}$. You can then estimate the equation using OLS. To use a specific example, suppose that the first seven observations of a time series are:

t	1	2	3	4	5	6	7
y_t	0.5	0.3	−0.2	0.0	−0.5	0.4	0.6
y_{t-1}	NA	0.5	0.3	−0.2	0.0	−0.5	0.4

If the threshold $\tau = 0$, you should be able to verify that the time path of the indicator function I_t and the values of $I_t y_{t-1}$, $I_t y_{t-2}$, $(1 - I_t)y_{t-1}$, and $(1 - I_t)y_{t-2}$ are those shown in the table.

t	1	2	3	4	5	6	7
y_t	0.5	0.3	−0.2	0.0	−0.5	0.4	0.6
y_{t-1}	NA	0.5	0.3	−0.2	0.0	−0.5	0.4
y_{t-2}	NA	NA	0.5	0.3	−0.2	0.0	−0.5
I_t	NA	1	1	0.0	0.0	0.0	1
$I_t y_{t-1}$	NA	0.5	0.3	0.0	0.0	0.0	1
$(1- I_t)y_{t-1}$	NA	0.0	0.0	−0.2	0.0	−0.5	0
$I_t y_{t-2}$	NA	NA	0.5	0.0	0.0	0.0	−0.5
$(1 - I_t)y_{t-2}$	NA	NA	0.0	0.3	−0.3	0.0	0.0

To estimate a model with two lags in each regime, you estimate the six values of α_{ij} from the regression equation

$$y_t = \alpha_{10}I_t + \alpha_{11}I_t y_{t-1} + \alpha_{12}I_t y_{t-2} + \alpha_{20}(1 - I_t) + \alpha_{21}(1 - I_t)y_{t-1} + \alpha_{22}(1 - I_t)y_{t-2} + \varepsilon_t$$

Hence, when $y_{t-1} > 0$, $I_t = 1$ and $(1 - I_t) = 0$, so that

$$y_t = \alpha_{10} + \alpha_{11}y_{t-1} + \alpha_{12}y_{t-2} + \varepsilon_t$$

Similarly, when $y_{t-1} \leq 0$, $I_t = 0$ and $(1 - I_t) = 1$, so that

$$y_t = \alpha_{20} + \alpha_{21}y_{t-1} + \alpha_{22}y_{t-2} + \varepsilon_t$$

The estimation is only a bit more complicated if you want to allow the variances of the error terms to differ across regimes. A more general version of equation (7.14) is the two-regime TAR model

$$y_t = \begin{cases} \alpha_{10} + \alpha_{11}y_{t-1} + \ldots + \alpha_{1p}y_{t-p} + \varepsilon_{1t} \text{ if } y_{t-1} > \tau \\ \alpha_{20} + \alpha_{21}y_{t-1} + \ldots + \alpha_{2r}y_{t-r} + \varepsilon_{2t} \text{ if } y_{t-1} \leq \tau \end{cases} \tag{7.15}$$

If τ is known, you can separate the observations according to whether y_{t-1} is above or below the threshold. Each segment of (7.15) can then be estimated using OLS. The lag lengths p and r can be determined as in an AR model. Hence, you can determine the lag lengths using t-tests on the individual coefficients, F-tests on groups of coefficients, or the AIC and/or SBC.

For example, for $\tau = 0$, sort the observations into two groups according to whether y_{t-1} is greater than or less than zero. Since values when $y_{t-1} = 0$ are included with those when $y_{t-1} < 0$, the two regimes using the seven sample observations listed above would look like this:

Positive		**Negative**	
y_t	y_{t-1}	y_t	y_{t-1}
0.3	0.5	0.0	−0.2
−0.2	0.3	−0.5	0.0
0.6	0.4	0.4	−0.5

The two separate AR(1) processes can be estimated for each regime. For each (y_t, y_{t-1}) pair, the first regression would use (0.3, 0.5), (−0.2, 0.3), and (0.6, 0.4) and the second regression would use (0.0, −0.2), (−0.5, 0.0), and (0.4, −0.5). It is only a bit more complicated to estimate an AR(2) model for each regime. For an AR(2), the first regression would use the (y_t, y_{t-1}, y_{t-2}) values (−0.2, 0.3, 0.5) and (0.6, 0.4, −0.5) and the second regression would use (0.0, −0.2, 0.3), (−0.5, 0.0, −0.2), and (0.4, −0.5, 0.0).

Regardless of whether you restrict the residual variances to be equal, OLS gives consistent estimates of the intercept and slope coefficients conditional on the threshold being correct.

Unknown Threshold

In most instances, the value of the threshold is unknown and must be estimated along with the other parameters of the TAR model. Fortunately, Chan (1993) showed how to obtain a super-consistent estimate of the threshold τ. To best explain the logic of the procedure, consider the TAR series shown in Figure 7.3. If the threshold is to be meaningful, the series must actually cross the threshold. It would be nonsense to use a threshold of four to estimate the TAR model since the series never crosses that threshold. Thus, τ must lie between the maximum and minimum values of the series. In practice, the highest and lowest 15 percent of the values are excluded from the search to ensure an adequate number of observations on each side of the threshold. Your estimates will be very imprecise if, for example, one regime has only twenty observations. If you have a very large number of observations, you may want to exclude only the highest and lowest 10 percent of the observations as potential thresholds.

In the example at hand, τ should lie within the band containing the middle 70 percent of the observations. Each data point within the band has the potential to be the threshold. Thus, try a value of $\tau = y_1$ (i.e., the first observation in the band) and estimate an equation in the form of (7.14) or (7.15). As you can see in the figure, y_2 lies outside the band. Hence, there is no need to estimate a regression using $\tau = y_2$. Next, estimate TAR models using $\tau = y_3$ and $\tau = y_4$ since these two values lie within the band. Continue in this fashion for each observation within the band. With 200 observations, there should be about 141 estimates of the TAR model. The regression containing the smallest residual sum of squares contains the consistent estimate of the threshold.

Now you can see why you need to use a software package that contains a programming language. Instead of estimating the 141 equations one at a time, as illustrated in the *Programming Manual*, you could embed the estimations within a Do-End loop or a For-Next loop.

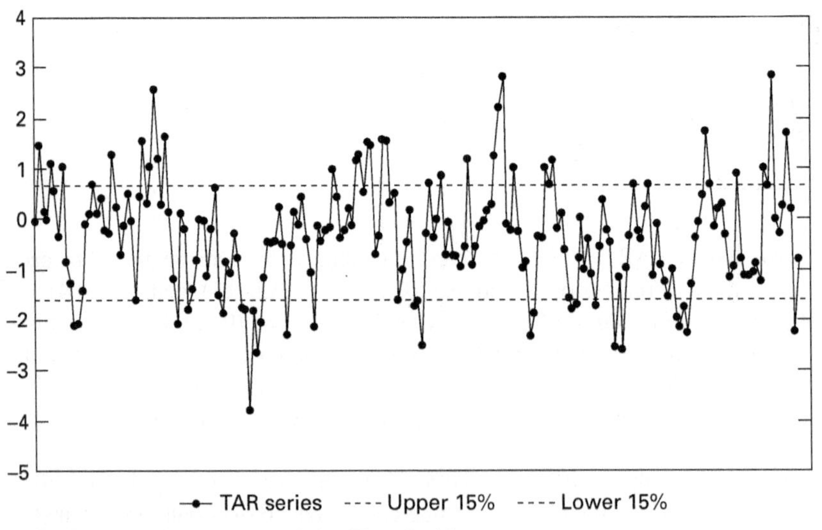

FIGURE 7.3 Estimation of the Threshold

Rothman's (1998) TAR estimate of the U.S. unemployment rate tells an interesting story. His two-regime model in the form of (7.15) is

$$\begin{aligned} u_t &= \underset{(3.46)}{0.0529} + \underset{(16.03)}{1.349}u_{t-1} - \underset{(-9.37)}{0.665}u_{t-2} + \varepsilon_{1t} && \text{if } u_{t-1} \geq 0.062 \\ u_t &= \underset{(14.27)}{1.646}u_{t-1} - \underset{(-6.37)}{0.733}u_{t-2} + \varepsilon_{2t} && \text{if } u_{t-1} < 0.062 \end{aligned}$$

There is a high-unemployment and a low-unemployment regime separated by the value $u_{t-1} = 0.062$. Rothman notes that unemployment is more persistent in the high-unemployment regime than the low-unemployment regime in that shocks that increase unemployment do not decay to zero. The variance ratio for the TAR model is 0.942. As measured by the residual sum of squares, the TAR model fits the data better than the AR(2) and GAR models, but not as well as the BL model. Notice that the estimated AR(2), GAR, BL, and TAR models contain 2, 3, 3, and 6 parameters (remember that τ is an estimated parameter in the TAR model). As such, a different pattern emerges if AIC is used to select the most appropriate model. The BL model has the lowest value of the AIC followed by the AR, GAR, and TAR models. Based on the AIC, most applied econometricians would discard the TAR model in favor of the BL model.

5. EXTENSIONS OF THE TAR MODEL

Note that there is something very different about the TAR model versus the GAR and BL models. The latter two are designed to be useful when the functional form of the nonlinear process is unknown. Nevertheless, some researchers view specifications based on a Taylor series approximation as being somewhat ad hoc. In contrast, the TAR model posits a type of adjustment mechanism that corresponds to the state of the economic system. This has led to a growing popularity of TAR models and a number of interesting extensions.

Selecting the Delay Parameter

In the TAR models considered thus far, the regime is determined by the value of y_{t-1}. However, it might be that the timing of the adjustment process is such that it takes more than one period for the regime switch to occur. In such circumstances, we could allow the regime switch to occur according to the value of y_{t-d} where $d = 1, 2, 3, \ldots$. Thus, the system would be in regime 1 if $y_{t-d} > \tau$ and in regime 2 if $y_{t-d} \leq \tau$. There are several procedures available to select the value of the **delay parameter** d. The standard procedure is to estimate a TAR model for each potential value of d. The one with the smallest value of the residual sum of squares yields a consistent estimate of the delay parameter. Alternatively, you can choose the delay parameter that leads to the smallest value of the AIC or the SBC. This second approach is most useful when the optimal values for p and r (i.e., the lag lengths in the various regimes) depend on the choice of d.

Multiple Regimes

In some instances, it may be reasonable to assume that there are more than two regimes. For example, if we assume that the shocks do not differ across regimes, we can write (7.1)—the TAR model of the interest rate spread—in the form

$$s_t = \begin{cases} \bar{s} + a_1(s_{t-1} - \bar{s}) + \varepsilon_t & \text{when } s_{t-1} > \bar{s} \\ \bar{s} + a_2(s_{t-1} - \bar{s}) + \varepsilon_t & \text{when } s_{t-1} \leq \bar{s} \end{cases}$$

Now suppose that there is a transaction cost c that prevents complete adjustment of the spread to $\bar{s}$. If the gap between s_{t-1} and $\bar{s}$ is less than the cost of undertaking the transaction, it would not be profitable to switch funds between the securities. As such, there may be a neutral band within which the spread may fluctuate. Within this band, there are no economic incentives to act in a way that equates the spread with $\bar{s}$. Outside of the band, however, there may be strong incentives for individuals to act in a way that drives the spread toward $\bar{s}$. A simple way to model this behavior is with the **band-TAR** model:

$$\begin{aligned} s_t &= \bar{s} + a_1(s_{t-1} - \bar{s}) + \varepsilon_t && \text{when } s_{t-1} > \bar{s} + c \\ s_t &= s_{t-1} + \varepsilon_t && \text{when } \bar{s} - c < s_{t-1} \leq \bar{s} + c \\ s_t &= \bar{s} + a_2(s_{t-1} - \bar{s}) + \varepsilon_t && \text{when } s_{t-1} \leq \bar{s} - c \end{aligned}$$

For this specification, there tends to be no tendency for mean reversion unless s_{t-1} lies outside the neutral band formed by adding and subtracting the transaction cost from the long-run value of the spread. Hence, inside the band, the behavior of the spread is a random walk. Balke and Fomby (1997) used this type of band threshold process to estimate a model of the term structure of interest rates. If c is unknown, it can be estimated using the type of grid search discussed above.

In a more general multiple-regime model, each regime can be represented by a distinct autoregressive process. As discussed in the next section, graphical techniques can be used to detect the presence of multiple thresholds.

More on Estimating the Threshold

The discussion in Section 4 gave an overview of Chan's (1993) method of finding the consistent estimate of the threshold. However, there are some graphical techniques that can be helpful in fine-tuning the estimate. The general point is that we can think of the sum of squared residuals from any TAR model as being a function of the particular threshold value used in the estimation, i.e., $ssr = ssr(\tau)$. The closer we come to the true threshold value τ, the smaller should be the sum of squared residuals. Hence, ssr should be minimized at the true value of the threshold. Moreover, the sum of squared residuals will have several distinct local minima if there are several thresholds. This suggests the following method to detect the thresholds:

STEP 1: Sort the threshold variable (i.e., sort y_{t-d}) from the lowest to the highest value. Let y^i denote the ith value of the sorted series. Hence, in a sample with T observations, y^1 is the smallest value of y_{t-d} and y^T is largest value.

STEP 2: Estimate a TAR model in the form of (7.14) or (7.15) using the successive values of $\{y^i\}$ as thresholds. Save the sum of squared residuals associated with each model. Since you want to maintain 15 percent of the observations on each side of the threshold, use only the middle 70 percent of the values of y^i. For example, if you have 200 observations, estimate 141 TAR models beginning with $\tau = y^{30}$ and ending with $\tau = y^{170}$. When you are done, you will have 141 values of the sum of squared residuals.

STEP 3: Create a graph of the successive values of the sum of squared residuals. If $ssr(30)$ is the sum of squared residuals using $\tau = y^{30}$ and $ssr(170)$ is the sum of squared residuals using $\tau = y^{170}$, plot the values of $ssr(30)$ through $ssr(170)$.

In the absence of threshold behavior, there should be no clear relationship between the sum of squared residuals and the potential thresholds. However, if there is a single threshold, there should be a single trough in the graph you create in Step 3. For example, if there is a distinct trough at $ssr(132)$, the consistent estimate of the threshold is y^{132}. After all, $\tau = y^{132}$ results in the TAR model with the best fit. If there are two troughs, there are two potential thresholds. To explain in a bit more detail, consider the model that was used to generate the 200 values shown in Figure 7.3:

$$y_t = 0.3I_t y_{t-1} + 0.7(1 - I_t)y_{t-1} + \varepsilon_t$$

where $I_t = 1$ if $y_{t-1} > 0$ and $I_t = 0$ otherwise.

Figure 7.4 reproduces the same numerical values sorted from low to high. As shown in both figures, the first value lying within the 70 percent band (i.e., y^{30}) is -1.623. Hence, the first estimate of the TAR model uses $\tau = -1.623$. The second estimation uses the next sorted observation as the threshold. This second value happens to be -1.601; thus, the second estimation uses $\tau = -1.601$. Continuing in this fashion brings us closer to the true threshold value of zero. As such, the fit of the TAR model should continue to improve as we move from threshold values of -1.623 toward $\tau = 0$. However, once we cross the true threshold and use values

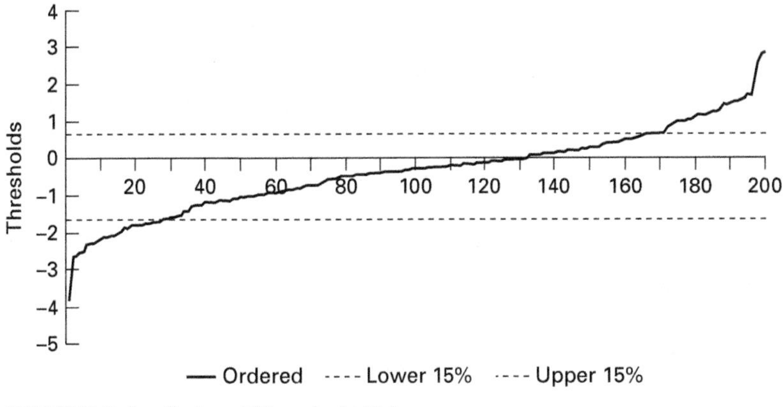

FIGURE 7.4 Ordered Threshold Values

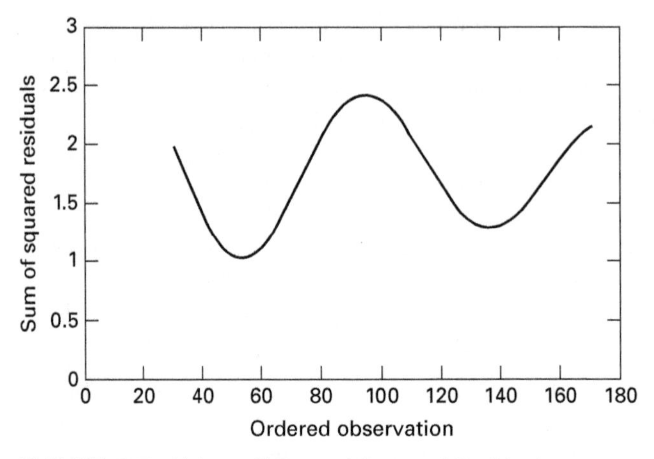

FIGURE 7.5 Values of Sum of Squared Residuals

of τ that are greater than zero, the sum of squared residuals should begin to increase. As such, the plot of the residual sum of squares should reach a minimum at $\tau = 0$. If you examine Figure 7.4, you can see that $\tau = 0$ corresponds to y^{132}. You can reproduce these results using the data in the file SIM_TAR.XLS; respectively, the second and third columns of the file contain the 200 values of the simulated series along with their sorted values.

What would happen if the true model contained two thresholds? In particular, suppose one threshold is -1 and the other is zero. As you can see from Figure 7.4, the value of $y^{55} = -1$ and $y^{132} = 0$. Now consider the idealized plot of the residual sum of squares shown in Figure 7.5. As we estimate TAR models beginning with y^{30} and proceed toward y^{55}, the sum of squared residuals declines. Also depicted is the fact that the sum of squared residuals should begin to increase as we use threshold values in excess of -1. This increase continues until you near the second threshold. In the example at hand, the sum of squared residuals begins another decline as we approach the second threshold of 0. The second trough at ordered observation 132 indicates the second threshold $\tau = 0$. In order to estimate a two-threshold model, many researchers would simply use the trough values as shown in Figure 7.5. In practice, there is a degree of subjectivity since the troughs might not be as distinct as those shown in the figure. What might appear to be a trough to one researcher might appear to be a small decline to another.

Threshold Regression Models

It has also become popular to use a threshold in the context of a traditional regression model. Consider the following specification:

$$y_t = a_0 + (a_1 + b_1 I_t)x_t + \varepsilon_t$$

where $I_t = 1$ if $y_{t-d} > \tau$ and $I_t = 0$ otherwise.

Here, a_1 measures the effect of x_t on y_t when $y_{t-d} \leq \tau$. However, when $y_{t-d} > \tau$, the effect of x_t on y_t is $a_1 + b_1$. Hence, if a_1 and b_1 are positive, changes in x_t have a greater effect on y_t when $y_{t-d} > \tau$ than when $y_{t-d} \leq \tau$. You can estimate the value of the threshold using Chan's method described above. For example, Shen and Hakes (1995) estimated a nonlinear reaction function for the central bank of Taiwan. The idea is that the central bank will respond differently to changes in economic variables in a high-inflationary environment than in a low-inflationary environment. Similarly, Galbraith (1996) showed that for Canada and the United States, the effect of money on output depends on whether credit conditions are already tight or loose.

There is no requirement that the threshold variable be given by y_{t-d}. For example, the threshold variable can be x_{t-d} where the delay parameter is any nonnegative integer. The threshold variable can even be a variable that does not appear directly in the regression equation. Two examples of threshold regression models are provided in Section 6.

Pretesting for a TAR Model

A Lagrange multiplier test cannot be used for a threshold model since it is not differentiable. For example, suppose you have TAR model in the form

$$y_t = \alpha_1 I_t y_{t-1} + \alpha_2(1 - I_t)y_{t-1} + \varepsilon_t \tag{7.16}$$

where $I_t = 1$ of $y_{t-1} > \tau$ and $I_t = 0$ if $y_{t-1} \leq \tau$.

The derivative $\partial y_t/\partial\alpha_1$ is discontinuous at τ in that $\partial y_t/\partial\alpha_1 = y_{t-1}$ when $y_{t-1} > \tau$ and $\partial y_t/\partial\alpha_1 = 0$ when $y_{t-1} \leq \tau$. Nevertheless, the appropriate test for threshold behavior is straightforward if the threshold value is known. Under the null hypothesis of linearity, (7.16) is the AR(1) process

$$y_t = \alpha y_{t-1} + \varepsilon_{2t}$$

As such, it is possible to estimate (7.16) and use a standard F-test to determine whether $\alpha_1 = \alpha_2$. However, if the threshold is unknown, another method must be used since you have searched over all possible values of τ to estimate the values of α_1 and α_2. You need to account for the fact that the estimated value of τ is such that it makes the two regimes as different as possible. Hence, the sample value of F will be as large as possible.

A number of papers, including Davies (1987) and Andrews and Ploberger (1994), consider the distributions of testing procedures that use the maximized value of a test statistic. Tests using such distributions are called **supremum tests.** Hansen (1997) showed how to obtain the appropriate critical values using a bootstrapping procedure. Search over all possible values of τ to find the best-fitting TAR model. Let SSR_u denote the unrestricted sum of squared residuals from the estimated threshold model. Similarly, let SSR_r denote the sum of squared residuals obtained from restricting the model to be linear. If you have T usable observations, a traditional F-statistic could be constructed as

$$F = \frac{(SSR_r - SSR_u)/n}{(SSR_u/(T - 2n))}$$

where n is the number of parameters estimated in the linear version of the model. In the example at hand, $n = 1$.

However, this sample value of F cannot be compared to the critical value found in a table for F. Instead, to use Hansen's (1997) bootstrapping method, you need to draw T normally distributed random numbers with a mean of zero and a variance of unity; let e_t denote this set of random numbers. You treat e_t as the dependent variable. Regress e_t on the actual values of y_{t-1} to obtain an estimate of SSR_r called SSR_r^*. Similarly, for each potential value of τ, regress e_t on $I_t y_{t-1}$ and $(1 - I_t)y_{t-1}$ [i.e., estimate a regression in the form $e_t = \alpha I_t y_{t-1} + \beta(1 - I_t)y_{t-1}$] and use the regression providing the best fit. Call the sum of squared residuals from this regression SSR_u^*. Use these two sums of squares to form

$$F^* = \frac{(SSR_r^* - SSR_u^*)/n}{(SSR_u^*/(T - 2n))}$$

Repeat this process several thousand times to obtain the distribution of F^*. If the value of F from your sample exceeds the 95th percentile for F^*, you can reject the null hypothesis of linearity at the 5 percent significance level.

The method generalizes to testing threshold regressions and the higher-order processes given by (7.14). Create SSR_u by estimating (7.14) and SSR_r by estimating the linear model that constrains all values of $\alpha_{1i} = \alpha_{2i}$. Create SSR_r^* by regressing e_t on all of the regressors in the linear model, and SSR_u^* by regressing e_t on all of the regressors in (7.14). After several thousand replications, you should have a good approximation to the distribution of F^*. A number of software packages can readily perform such a test. A detailed example of the testing procedure is provided immediately below.

TAR Models and Endogenous Breaks

If you have been paying careful attention, you might have recognized that the threshold model is equivalent to a model with a structural break. The only difference is that in a model with structural breaks, *time* is the threshold variable. In Chapter 2 (see Figure 2.10 and the file Y_BREAKS.XLS), we analyzed the simulated series $y_t = 1 + 0.5y_{t-1} + \varepsilon_t$ for $1 \leq t \leq 100$ and $y_t = 2.5 + 0.65y_{t-1} + \varepsilon_t$ for $101 \leq t \leq 150$. When we treated the break date as known, we were able to form the dummy variable D_t and the variable $D_t y_{t-1}$, and estimate:

$$\underset{}{y_t} = \underset{(7.22)}{1.6015} + \underset{(2.76)}{0.2545y_{t-1}} - \underset{(-0.39)}{0.2244D_t} + \underset{(4.47)}{0.5433D_t y_{t-1}}$$

where $D_t = 1$ if $t < 101$ and $D_t = 0$ otherwise. Since the coefficient on $D_t y_{t-1}$ was highly significant, we were able to verify the presence of a break in the series. Of course, this model of a breaking series is equivalent to the threshold form

$$y_t = (\underset{(2.60)}{1.3771} + \underset{(10.10)}{0.7977y_{t-1}})I_t + (1 - I_t)(\underset{(7.22)}{1.6015} + \underset{(2.72)}{0.2545y_{t-1}})$$

where $I_t = 1$ if $t < 101$ and $I_t = 0$ otherwise.

If we pretend that the break date is unknown, we can illustrate the use of a supremum test. It turns out that $t = 100$ yields the model with the smallest sum of squared residuals. Using this value as the threshold, the sum of squared residuals is 138.63. If you estimate the model under the null hypothesis of linearity, you should find

$$y_t = \underset{(2.64)}{0.4442} + \underset{(22.76)}{0.8822}y_{t-1}$$

The sum of squared residuals is 195.18. Since there are 149 usable observations, and two extra coefficients in the threshold model, the sample value of the F-statistic is given by

$$F = \frac{(195.18 - 138.63)/2}{138.63/(149 - 4)} = 29.57$$

Next, draw a sequence of 150 normally distributed random numbers with a standard deviation of unity to represent the e_t series. For each t in the interval $22 < t < 128$, create the indicator function I_t and estimate a threshold regression in the form

$$e_t = (\alpha_{10} + \alpha_{11}y_{t-1}) + I_t(\alpha_{20} + \alpha_{21}y_{t-1}) + \varepsilon_t$$

Use this regression to construct the sample F-statistic (i.e., construct F^*) for the null hypothesis $\alpha_{20} = \alpha_{21} = 0$. Repeat this process several thousand times to obtain the distribution of F^*. Compare your distribution to the value of $F = 29.75$. If you perform this process using the data in the file Y_BREAKS.XLS, approximately 95 percent of the constructed F^* values should be below 3.15. As such, the null hypothesis of linearity is clearly rejected.

There is a more general point to be made from this example. Carrasco (2002) showed that the usual tests for structural breaks (i.e., those using dummy variables) have little power if the data are actually generated by a threshold process. Her observation is that the multiplicity of regime changes in a TAR model cannot be adequately captured by the dummy variables. However, a test for a threshold process using y_{t-d} as the threshold variable has power to detect both threshold behavior *and* structural change. Even if there is a single structural break at time period t, using y_{t-d} as the threshold variable will mimic this type of behavior. After all, if the series suddenly increases at t, values of y_{t-d} will tend to be low before date t and high after date t. As such, she recommends using the threshold model as a general test for parameter instability.

6. THREE THRESHOLD MODELS

Perhaps the best way to understand the nature of threshold models is to consider a few specific examples. This section illustrates the estimation of a threshold autoregressive model and two threshold regression models.

The Unemployment Rate

In addition to Rothman (1998), many papers have indicated that the U.S. unemployment rate displays nonlinear behavior. You can follow along the estimation process using the data set UNRATE.XLS. Figure 7.6 shows the monthly values of the rate over

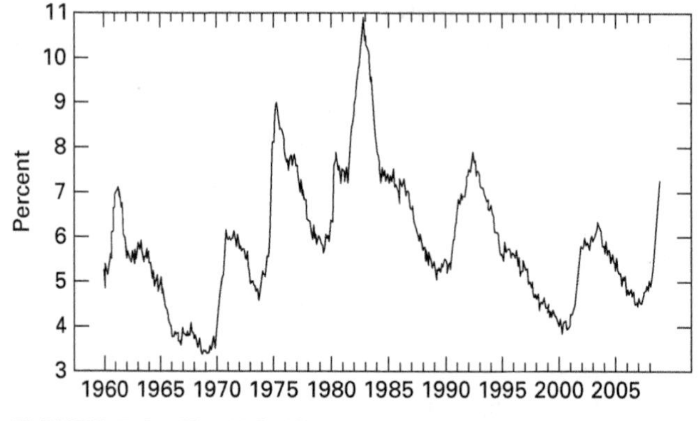

FIGURE 7.6 The U.S. Unemployment Rate

the period January 1960 through December 2008. In November 1982 the rate rose to as high as 10.8 percent although there were also sharp increases in 1970, 1973, 1991, 2001, and 2008. The mean of the 588 values is 5.846 percent and the standard deviation is 1.43 percentage points. After some experimentation, you can convince yourself that it is reasonable to difference the series and estimate:

$$\begin{aligned}\Delta u_t = {} & \underset{(0.07)}{0.001} - \underset{(-0.56)}{0.023}\Delta u_{t-1} + \underset{(4.77)}{0.193}\Delta u_{t-2} + \underset{(5.17)}{0.209}\Delta u_{t-3} + \underset{(3.95)}{0.162}\Delta u_{t-4} \\ & - \underset{(-2.96)}{0.116}\Delta u_{t-12}\end{aligned} \tag{7.17}$$

where $SSR = 15.889$, AIC = 1602.25, and SBC = 1628.38.

The residual autocorrelations are all quite small. For example, the first 10 autocorrelations are

ρ_1	ρ_2	ρ_3	ρ_4	ρ_5	ρ_6	ρ_7	ρ_8	ρ_9	ρ_{10}
−0.01	−0.02	−0.01	0.00	0.06	0.01	−0.03	0.07	0.03	−0.02

The RESET is not supportive of nonlinearity. Let e_t denote the regression residuals from (7.17). If we regress the residuals on the regressors and the powers of the fitted values, we obtain

$$e_t = \underset{(0.46)}{0.004} - \underset{(0.91)}{1.172}\Delta\hat{u}_t^2 - \underset{(0.37)}{3.648}\Delta\hat{u}_t^3 + \underset{(0.67)}{17.791}\Delta\hat{u}_t^4 + \sum_i \alpha_i \Delta u_{t-i} \quad i = 1, 2, 3, 4, 12$$

The F-statistic for the joint restriction that the coefficients on $\Delta\hat{u}_t^2$, $\Delta\hat{u}_t^3$, and $\Delta\hat{u}_t^4$ jointly equal zero is 0.304. With three degrees of freedom in the numerator and 566 in the denominator, the *prob*-value is 0.822. Hence, the RESET does not detect the presence of nonlinear behavior. Notice that the RESET has a very general alternative hypothesis; as such, it does not have power against all types of nonlinearity.

However, other diagnostic checks indicate a potential problem with the linear specification. The McLeod–Li (1983) test is such that

$$e_t^2 = \underset{(8.33)}{0.021} + \underset{(3.63)}{0.152}e_{t-1}^2 + \underset{(1.84)}{0.077}e_{t-2}^2$$

The sample value of F for the restriction that the coefficients on e_{t-1}^2 and e_{t-2}^2 are jointly equal to zero is 9.67; this value is highly significant. It is also interesting that other variants of the test also suggest nonlinearity. Consider the regression

$$e_t = \underset{(-1.64)}{-0.013} + \underset{(3.40)}{0.462}e_{t-1}^2 \tag{7.18}$$

Equation (7.18) suggests that a large error (either positive or negative) in the previous period is associated with a *positive* error in the current period. In a linear model, the adjustment is symmetric so that the residuals should not be correlated with the lagged squared residuals.

If you set $d = 1$, and estimate a model in the form of (7.14), you should find that the threshold value yielding the smallest residual sum of squares is such that $\tau = 0.073$. Figure 7.7 shows the value of the sum of squared residuals for each threshold value considered. You can see the single sharp trough in the scatter plot of $ssr(\tau)$ for $\tau = 0.073$. Although there is a second trough in the scatter plot near $\tau = 0.18$, it is rather small. As such, it makes sense to ignore the possibility of multiple thresholds. Also note that other delay parameters do not fare as well as $d = 1$. For example, the sum of squared residuals with $d = 1$, 2, and 3 are 15.0257, 15.491, and 15.55, respectively. (The estimated values of τ for $d = 2$ and 3 are 0.053 and 0.200, respectively.) Hence, we can be confident that a delay parameter of unity is appropriate.

If you set $d = 1$ and $\tau = 0.073$, you should find

$$\begin{aligned}\Delta u_t = {} & I_t(\underset{(-2.34)}{-0.055} + \underset{(2.86)}{0.294}\Delta u_{t-1} + \underset{(5.34)}{0.322}\Delta u_{t-2} + \underset{(1.72)}{0.116}\Delta u_{t-3} + \underset{(2.14)}{0.145}\Delta u_{t-4} \\ & - \underset{(-1.64)}{0.120}\Delta u_{t-12}) + (1 - I_t)(\underset{(-1.73)}{-0.018} - \underset{(-3.12)}{0.225}\Delta u_{t-1} + \underset{(0.96)}{0.052}\Delta u_{t-2} \\ & + \underset{(4.35)}{0.217}\Delta u_{t-3} + \underset{(3.37)}{0.168}\Delta u_{t-4} - \underset{(-2.73)}{0.124}\Delta u_{t-12})\end{aligned}$$

$$SSR = 15.031, \text{AIC} = 1582.32, \text{SBC} = 1634.57$$

where $I_t = 1$ when $\Delta u_{t-1} > 0.073$, and $I_t = 0$ when $\Delta u_{t-1} \leq 0.073$.

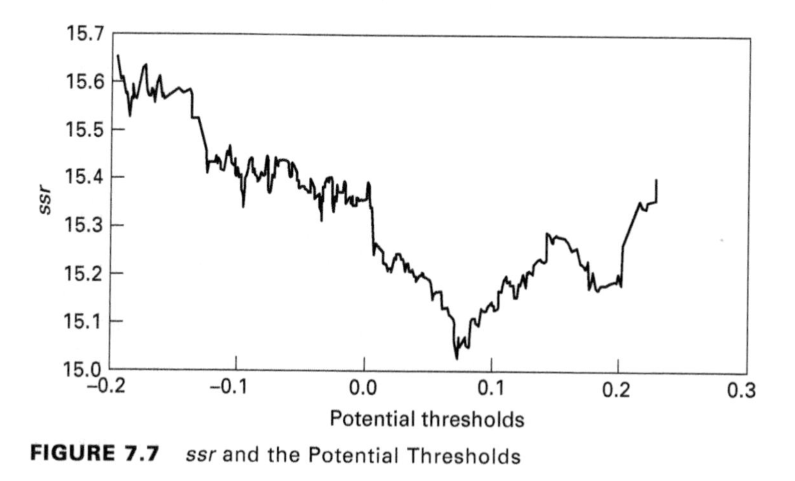

FIGURE 7.7 *ssr* and the Potential Thresholds

Note that the AIC selects the threshold model while the SBC selects the linear model in (7.17). However, the threshold model contains a number of parameters that are small relative to their standard errors. Clearly, it makes sense to test for the presence of threshold behavior. Toward this end, we can construct the sample F-statistic for the null hypothesis of linearity as

$$F = [(15.889 - 15.031)/6]/[15.031/(575 - 12)] = 5.36$$

Since it was necessary to estimate the threshold value, it is not appropriate to compare 5.36 to a standard table of F. However, if you use Hansen's (1997) bootstrapping method, you should find that 5.36 is significant at the 0.001 level so that we can conclude there is threshold behavior.

Inference on the coefficients in a threshold model is not straightforward since it was necessary to search for τ. The t-statistics yield only an approximation of the actual significance levels of the coefficients. The problem is that the coefficients on the various Δu_{t-i} are multiplied by I_t or $(1 - I_t)$ and that these values are dependent on the estimated value of τ. Nevertheless, both model selection criteria indicate that you can pare down the model by eliminating $I_t \Delta u_{t-4}$, and $(1 - I_t)\Delta u_{t-2}$. Also note that the coefficients on $I_t \Delta u_{t-12}$ and $(1 - I_t)\Delta u_{t-12}$ are almost identical. Thus, it makes sense to simply include Δu_{t-12} in the model. Paring down the model in this fashion results in AIC = 1581.99 and SBC =1621.18. As an exercise, try to verify these results. You might also find it interesting to estimate the series using a second threshold value near 0.18.

The point estimates are such that there is far more persistence when $\Delta u_{t-1} > \tau$ than when $\Delta u_{t-1} \leq \tau$. This result strongly suggests that increases in unemployment are far more persistent than decreases in unemployment.

Asymmetric Monetary Policy

Much of the literature concerning the behavior of the Federal Reserve is based on the type of feedback rule introduced by Taylor (1993). The so-called Taylor rule has the form

$$i_t = \gamma_0 + \pi_t + \alpha_1(\pi_t - \pi^*) + \beta y_t + \gamma_1 i_{t-1} + \varepsilon_t$$

or, setting $\alpha_0 = \gamma_0 - \alpha\pi^*$ and $\alpha = 1 + \alpha_1$, we can form

$$i_t = \alpha_0 + \alpha\pi_t + \beta y_t + \gamma_1 i_{t-1} + \varepsilon_t$$

where i_t is the nominal federal funds rate, π_t is the inflation rate over the last four quarters, π^* is the target inflation rate, y_t is output gap measured as percentage deviation of real GDP from its trend, and $\alpha_1, \beta, \gamma_0, \gamma_1$, and γ_2 are positive parameters.

The intuition behind the rule is that the Federal Reserve wants to keep inflation at the target level and to stabilize real GDP around its trend. Since high interest rates discourage spending, the Taylor rule posits that the Federal Reserve will increase i_t when inflation is above its target level and when the output gap is positive. The lagged value of the interest rate creates some inertia in the system and represents the desire of the Federal Reserve to smooth interest rate changes over time.

The file labeled TAYLOR.XLS contains the variables necessary to estimate the Taylor rules reported below. Specifically, the interest rate (i_t) is the quarterly average of the monthly values of the federal funds rate. The four-quarter inflation rate (π_t) is constructed as

$$\pi_t = 100^*(\ln p_t - \ln p_{t-4})$$

where p_t is the chain-weighted GDP deflator.

In order to account for the fact that real GDP is often subject to substantial revisions, it is standard to use the real-time values of GDP available at the Philadelphia Federal Reserve Bank's Web site (http://research.stlouisfed.org/fred2). The notion is that the Federal Reserve makes decisions using then-current values of GDP. Revised values are only available after a substantial delay. The output gap is obtained by detrending the real output data with a Hodrick–Prescott (HP) filter as described in Chapter 4. Specifically, beginning with t = 1963Q2, the HP filter is applied to the real-time output series running from 1947Q1 through t. The filtered series represents the trend values of real GDP. Call y_t^f the last observation of the filtered series. We construct the output gap for time period t (y_t) as the percentage difference between real-time output at t and the value of y_t^f. We then increase t by one period and repeat the process. The aim is not to ascertain the way that real output evolves over the long run. Instead, the goal is to obtain a reasonable measure of the pressure felt by the Federal Reserve to use monetary policy to affect the level of output.

In applied work it is typical to estimate the Taylor rule over a number of sample periods reflecting the fact that a change in the Federal Reserve's operating procedures occurred in 1979Q4, the Volker disinflation ended by 1983Q1, Alan Greenspan became Fed chairman in August 1987, and Ben Bernanke became chairman in February 2006. Consider the estimated model for the 1979Q4−2007Q3 sample period:

$$\underset{}{i_t} = \underset{(-1.47)}{-0.269} + \underset{(6.05)}{0.464\pi_t} + \underset{(5.16)}{0.345y_t} + \underset{(21.83)}{0.810i_{t-1}} \quad \text{AIC} = 500.75 \text{ and SBC} = 511.63$$

The estimated model appears to be reasonable in that the coefficients on inflation and the output gap are both positive and significant at conventional levels. The coefficient on the lagged interest rate (i.e., γ_1 = 0.810) suggests a substantial amount of interest rate smoothing. In the long run, i_t responds more than proportionally to changes in π_t [since 0.464/(1 − 0.810) = 2.44] so that the real interest rate rises (falls) when inflation increases (decreases).

A number of authors have questioned the linear form of the Taylor rule and have argued that the Federal Reserve's reactions to π_t and y_t are best modeled as a nonlinear process. For example, it is likely that the Federal Reserve prefers inflation to be below the target rather than above the target. Moreover, it is probable that the Federal Reserve prefers a positive output gap rather than a negative one.

The point is that interest rate changes should be more dramatic when inflation is high and/or output is low. As such, it seems natural to estimate the Taylor rule as a threshold regression using either the inflation rate or the output gap as the threshold variable. Since we do not know the delay factor, we can estimate four threshold regressions with π_{t-1}, π_{t-2}, y_{t-1}, and y_{t-2} as the threshold variables. For each regression, the consistent estimate of τ is obtained using a grid search over all potential thresholds using a trimming value of 15 percent. The estimated threshold value, sum of squared residuals (*SSR*), AIC, and SBC for each of the four regressions are:

	τ	*SSR*	*AIC*	*BIC*
π_{t-1}	3.527	50.80	455.93	477.67
π_{t-2}	3.527	50.42	455.08	476.83
y_{t-1}	−1.183	63.97	481.75	503.49
y_{t-2}	−1.565	53.41	461.53	483.28

Notice that all of the threshold regressions have a better fit than the linear model. Moreover, if you bootstrap the sample F-statistics, you will find that all are highly significant. Since π_{t-2} provides the best fit, we should use it as the threshold variable. As such, the estimated Taylor rule is

$$\underset{}{i_t} = \underset{(3.02)}{1.383} + \underset{(10.56)}{1.055}\pi_t + \underset{(6.25)}{0.472}y_t + \underset{(5.75)}{0.374}i_{t-1} \text{ when } \pi_{t-2} \geq 3.527$$

and

$$i_t = \underset{(-1.39)}{-0.440} + \underset{(1.88)}{0.227}\pi_t + \underset{(3.85)}{0.305}y_t + \underset{(24.98)}{0.967}i_{t-1} \text{ when } \pi_{t-2} < 3.527$$

Notice that the coefficients on π_t and y_t are much greater in the high-inflation regime than in the low-inflation regime. Moreover, the interest rate–smoothing coefficient is far greater when inflation is low than when inflation is high. In essence, in the high-inflation regime, the Federal Reserve is far more policy-active than in the low-inflation regime. Also notice that the linear variant of the rule seems to "average" the responses of the Federal Reserve across the high- and low-inflation regimes.

Capital Stock Adjustment with Multiple Thresholds

Boetel, Hoffman, and Liu (2007) estimated an interesting model that contains three regimes. The problem addressed in the paper is that pork producers do not always adjust their capital input in the face of changing market conditions. However, there are times when even a very small change in market conditions induces a large adjustment in the capital stock. Their model asserts that there is a "normal" range for the price of hogs and that price changes within this range will induce a sluggish investment response. For our purposes, the key variables in the model are

$$K_t - K_{t-1} = \underset{(3.30)}{4569} + \underset{(5.59)}{6360}I_{1t} + \underset{(5.20)}{6352}I_{2t} + \underset{(1.84)}{452}p_{Ht-1} - \underset{(-3.66)}{2684}p_{Ft-1} + \ldots + \varepsilon_t$$

where K_t is the size of the breeding stock, p_{Ht-1} is a measure of the output price of hogs, and p_{Ft-1} is a measure of the price of feed. The indicator functions are such that $I_{1t} = 1$ if $p_{Ht-1} > \tau_{high} = 1.1185$ and $I_{2t} = -1$ if $p_{Ht-1} < \tau_{low} = 1.1105$. The use of lagged values for the dependent variables is designed to reflect a one-period delay between the time of the investment decision and its realization.

It should not be surprising that the net acquisition of the breeding stock $(K_t - K_{t-1})$ is positively related to the price of hogs and negatively related to the price of feed. An appealing feature of the model is that the indicator functions multiply the intercepts but not the variable p_{Ft-1}. Boetel, Hoffman, and Liu (2007) noted that allowing all variables to have asymmetric effects on $K_t - K_{t-1}$ would entail estimating a large number of parameters with a consequent loss of degrees of freedom.

Notice that the three regimes are distinguished by the value of p_{Ht-1} relative to two threshold values. When p_{Ht-1} is between τ_{high} and τ_{low}, I_{1t} and $I_{2t} = 0$ so that the intercept is 4569. Instead, when $p_{Ht-1} > \tau_{high}$, $I_{1t} = 1$ and $I_{2t} = 0$ so that the intercept is 10929. When $p_{Ht-1} < \tau_{low}$, $I_{1t} = 0$ and $I_{2t} = -1$ so that the intercept is 8. Thus, there is a high, a sluggish, and a disinvestment regime whose presence is dependent on the value of p_{Ht-1}. As such, it would be a mistake to conclude that the slope coefficient 452 measures the full effect of a price change on net investment. When the value of p_{Ht-1} crosses one of the thresholds, the change in investment is enhanced since the intercept changes along with the price. Also note that price changes within the interval τ_{high} to τ_{low} will have little effect on investment.

Boetel, Hoffman, and Liu (2007) used a different method than the one described above to estimate the two threshold values appearing in their model. First, they performed a grid search to find the single threshold value that provides the smallest value of the sum of squared residuals. Let τ_1 denote this threshold value. Next, maintaining the value of τ_1, they estimated a second threshold—say, τ_2—so as to further minimize the residual sum of squares. Although Hansen (1999) showed that this second threshold estimate is efficient, the first is not since it was estimated in the absence of the second threshold. Finally, they fixed the value of τ_2 and reestimated the threshold value of τ_1 so as to provide the smallest value of the sum of squared residuals. An alternative would have been to use the graphical method discussed in Section 5.

7. SMOOTH-TRANSITION MODELS

For some processes, it may not seem reasonable to assume that the threshold is sharp. Instead, the speed of adjustment may be the type of nonlinear process shown in Panel (b) of Figure 7.1. Smooth-transition autoregressive (STAR) models allow the autoregressive parameters to change slowly. Consider the special NLAR model given by

$$y_t = \alpha_0 + \alpha_1 y_{t-1} + \beta_1 y_{t-1} f(y_{t-1}) + \varepsilon_t$$

If $f(\ )$ is a smooth continuous function, the autoregressive coefficient $(\alpha_1 + \beta_1)$ will change smoothly along with the value of y_{t-1}. There are two particularly useful forms of the STAR model that allow for a varying degree of autoregressive decay. The logistic version of the STAR model (called the LSTAR model) generalizes the standard autoregressive model such that the autoregressive coefficient is a logistic function:

$$y_t = \alpha_0 + \alpha_1 y_{t-1} + \ldots + \alpha_p y_{t-p} + \theta[\beta_0 + \beta_1 y_{t-1} + \ldots + \beta_p y_{t-p}] + \varepsilon_t \tag{7.19}$$

where

$$\theta = [1 + \exp(-\gamma(y_{t-1} - c))]^{-1} \tag{7.20}$$

Note that γ is called the *smoothness* parameter. In the limit, as $\gamma \to 0$ or ∞, the LSTAR model becomes an *AR*(p) model since the value of θ is constant. For intermediate values of γ, the degree of autoregressive decay depends on the value of y_{t-1}. As $y_{t-1} \to -\infty$, $\theta \to 0$ so that the behavior of y_t is given by $\alpha_0 + \alpha_1 y_{t-1} + \ldots + \alpha_p y_{t-p} + \varepsilon_t$. Similarly, as $y_{t-1} \to +\infty$, $\theta \to 1$ so that the behavior of y_t is given by $(\alpha_0 + \beta_0) + (\alpha_1 + \beta_1)y_{t-1} + \ldots + \varepsilon_t$. Thus, the intercept and the autoregressive coefficients smoothly change between these two extremes as the value of y_{t-1} changes.

The exponential form of the model (ESTAR) uses (7.19), but replaces (7.20) with

$$\theta = 1 - \exp[-\gamma(y_{t-1} - c)^2] \qquad \gamma > 0$$

Notice that θ contains a squared term so that the coefficients for the ESTAR model are symmetric around $y_{t-1} = c$. As y_{t-1} approaches c, θ approaches 0 so that the behavior of y_t is given by $\alpha_0 + \alpha_1 y_{t-1} + \ldots + \alpha_p y_{t-p} + \varepsilon_t$. As y_{t-1} moves further from c, θ approaches 1 so that the behavior of y_t is given by $(\alpha_0 + \beta_0) + (\alpha_1 + \beta_1)y_{t-1} + \ldots + \varepsilon_t$. The ESTAR model has proven to be useful for periods surrounding the turning points of a series (i.e., periods in which y_{t-1}^2 will be extreme) in that such periods have different degrees of autoregressive decay than others. Since the ESTAR model is symmetric around $y_{t-1} = c$, it can approximate the type of gravitational attraction shown in Figure 7.1. Also note that as γ approaches zero or infinity, the model becomes an AR(p) model since θ is constant. Otherwise, the model displays nonlinear behavior.

You can see the difference between the LSTAR and ESTAR models by examining Figure 7.8. The top panel constructs $\theta = [1 + \exp(-\gamma(y_{t-1} - c))]^{-1}$ for $c = 0$ and values of $\gamma = 1$ and 2. As y_{t-1} ranges from -5 to $+5$, the value of θ ranges from 0 to 1. Note that as the S-shape of the transition is sharper, the greater is γ. For large values of γ, the adjustment is so sharp that LSTAR model acts as a TAR process. The bottom panel also uses $c = 0$ and values of $\gamma = 1$ and 2, but constructs the transition function using the ESTAR formula $\theta = 1 - \exp[-\gamma(y_{t-1} - c)^2]$. You can see that the U-shape becomes sharper as γ increases.

Michael, Nobay, and Peel (1997) made the point that transaction costs are an important feature of international trade. Such costs may include the purchase of foreign exchange or forward cover, the payment of tariffs and import licensing fees, and transportation costs. As in the band-TAR model, small deviations from PPP will not be corrected through the process of commodity arbitrage. Larger discrepancies are expected to be mean-reverting such that speed of adjustment is an increasing function of the size of the discrepancy. The idea is that very large discrepancies are quickly eliminated but mid-size discrepancies are eliminated more slowly.

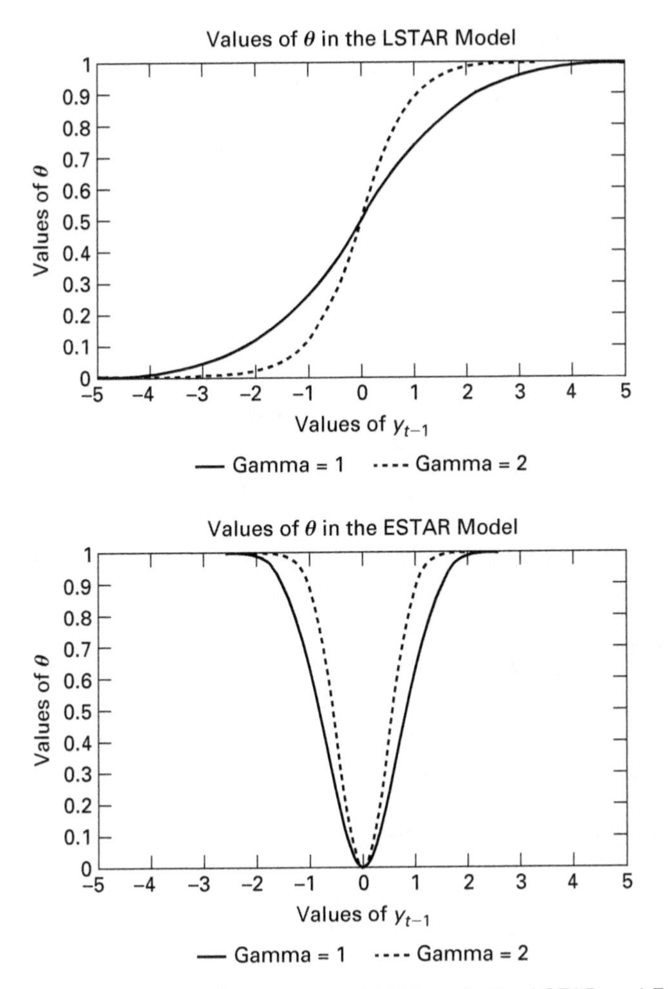

FIGURE 7.8 A Comparison of θ Values in the LSTAR and ESTAR Models

This type of behavior can be captured by an ESTAR process. The particular form of the ESTAR model they consider is

$$\Delta y_t = \alpha_0 + a_1 y_{t-1} + \sum_{i=1}^{p-1} \alpha_i \Delta y_{t-i}$$
$$+ [1 - \exp(-\gamma(y_{t-d} - c)^2]\left(\beta_0 + b_1 y_{t-1} + \sum_{i=1}^{p-1} \beta_i \Delta y_{t-i}\right) + \varepsilon_t$$

where y_t is the real exchange rate.

When $y_{t-d} = c$, the adjustment process is given by

$$\Delta y_t = \alpha_0 + a_1 y_{t-1} + \sum_{i=1}^{p-1} \alpha_i \Delta y_{t-i} + \varepsilon_t$$

and as $y_{t-d} \to \pm\infty$, the adjustment process is given by

$$\Delta y_t = (\alpha_0 + \beta_0) + (a_1 + b_1)y_{t-1} + \sum_{i=1}^{p-1} (\alpha_i + \beta_i)\Delta y_{t-i} + \varepsilon_t$$

The nature of transaction costs implies that a_1 may be very small (or zero). After all, when $y_{t-d} \approx c$, there is little incentive to arbitrage the market. However, since large deviations are mean-reverting, b_1 should be negative. Their estimate of the monthly United States–United Kingdom real exchange rate over the 1921:1−1925:5 period is (with t-statistics in parentheses):

$$\begin{aligned}\Delta y_t = \underset{(3.37)}{0.40}\Delta y_{t-1} &+ [1 - \exp(-\underset{(2.44)}{532.4}(y_{t-1} - \underset{(7.21)}{0.038})^2)] \\ &(-y_{t-1} + \underset{(3.90)}{0.59}\Delta y_{t-2} + \underset{(2.89)}{0.57}\Delta y_{t-4} - \underset{(5.17)}{0.017})\end{aligned}$$

The point estimates imply that when the real rate is near 0.038, there is no tendency for mean reversion since $a_1 = 0$. However, when $(y_{t-1} - 0.038)^2$ is very large, the speed-of-adjustment coefficient is quite rapid. Hence, the adjustment of the real exchange rate is consistent with the presence of transaction costs.

Pretests for STAR Models

It is not possible to directly perform an LM test for the presence of ESTAR or LSTAR behavior. Consider the LSTAR model

$$y_t = \alpha_0 + \alpha_1 y_{t-1} + (\beta_0 + \beta_1 y_{t-1})[1 + \exp(-\gamma(y_{t-1} - c))]^{-1} + \varepsilon_t$$

For this model, the null hypothesis that the model is linear is identical to setting $\gamma = 0$. You should be able to see the problem with using the LM test. If $\gamma = 0$, the magnitudes of β_0, β_1, and c are completely irrelevant! Clearly, if $\gamma = 0$, it makes no difference whether β_1 is -0.5 or $+1$. The point is that the values of β_0, β_1, and c are unidentified under the null hypothesis that the model is linear.[2] As such, it is not possible to perform an LM test. It is worth a few minutes of your time to try the following exercise. Find the partial derivatives of the LSTAR model and evaluate each under the null hypothesis $\gamma = 0$. Indicate the functional form of the resulting auxiliary regression. (*Hint:* $\partial y_t/\partial c$ evaluated at $\gamma = 0$ is zero.)

Since the LM test fails for LSTAR (and ESTAR) adjustment, other means are necessary to detect the presence of a smooth-transition model. Teräsverta (1994) developed a framework that can often detect the presence of nonlinear behavior. Moreover, the method can be used to determine whether a series is best modeled as an LSTAR or an ESTAR process. The test is based on a Taylor series expansion of the general STAR model. For the LSTAR model, we can write θ as

$$\theta = [1 + \exp(-\gamma(y_{t-d} - c))]^{-1} \equiv [1 + \exp(-h_{t-d})]^{-1}$$

so that $h_{t-d} = \gamma(y_{t-d} - c)$.

Now, the trick is to take a third-order Taylor series approximation of θ with respect to h_{t-d} evaluated $h_{t-d} = 0$. Of course, this is identical to evaluating the expansion at $\gamma = 0$. Although taking the partial derivatives is a bit tedious, after a bit of manipulations it is possible to obtain[3]

	Equals	Evaluated at $h_{t-d} = 0$
$\partial\theta/\partial h_{t-d}$	$\exp(-h_{t-d})/[1 + \exp(-h_{t-d})]^2$	1/4
$\partial^2\theta/\partial h_{t-d}^2$	$-\exp(-h_{t-d})[1 - \exp(-h_{t-d})]/[1 + \exp(-h_{t-d})]^3$	0
$\partial^3\theta/\partial h_{t-d}^3$	$\exp(-h_{t-d})[1 + \exp(-2h_{t-d}) - 4\exp(-h_{t-d})]/[1 + \exp(-h_{t-d})]^4$	$-1/8$

Thus, since the second derivative is zero, the desired expansion has the form $\theta = h_{t-d}/4 - h_{t-d}^3/48 = \gamma(y_{t-d} - c)/4 - \gamma^3(y_{t-d} - c)^3/48$. Hence, we can write the approximation of the LSTAR model in the form

$$y_t = \alpha_0 + \alpha_1 y_{t-1} + \ldots + \alpha_p y_{t-p} + (\beta_0 + \beta_1 y_{t-1} + \ldots + \beta_p y_{t-p})$$
$$(\pi_1 h_{t-d} + \pi_3 h_{t-d}^3) + \varepsilon_t$$

Because h_{t-d} depends only on the value of y_{t-d}, we can write the model in the more compact form

$$y_t = c_0 z_t + c_1 z_t y_{t-d} + c_2 z_t \ \ldots \ + c_3 z_t \ \ldots \ + \varepsilon_t$$

where $z_t = (\alpha_0, y_{t-1}, y_{t-2}, \ldots, y_{t-p})$ and the c_i are coefficients.

Thus, you form the products of the regressors and the powers of y_{t-d} (i.e., y_{t-d}^0, y_{t-d}^1, y_{t-d}^2, and y_{t-d}^3). Then, you can test for the presence of LSTAR behavior by estimating an auxiliary regression:

$$\begin{aligned} e_t = a_0 &+ a_1 y_{t-1} + \ldots + a_p y_{t-p} + a_{11} y_{t-1} y_{t-d} + \ldots \\ &+ a_{1p} y_{t-p} y_{t-d} + a_{21} y_{t-1} y_{t-d}^2 + \ldots + a_{2p} y_{t-1} y_{t-d}^2 \\ &+ a_{31} y_{t-1} y_{t-d}^3 + \ldots + a_{3p} y_{t-p} y_{t-d}^3 + \varepsilon_t \end{aligned} \quad (7.21)$$

The test for linearity is identical to testing the joint restriction that all nonlinear terms are zero (i.e., $a_{11} = \ldots = a_{1p} = a_{21} = \ldots = a_{2p} = a_{31} = \ldots = a_{3p} = 0$). You can perform the test using a standard F-test with $3p$ degrees of freedom in the numerator. If you are not sure of the delay factor, the recommendation is to run the test using all plausible values of d. The value of d that results in the smallest *prob*-value (i.e, the value of d providing the best fit) is the best estimate of d.

With all of the background work completed, it is straightforward to rework the details for an ESTAR model. Let θ be

$$\theta = 1 - \exp(-h_{t-d}^2)$$

so that $h_{t-d} \equiv \gamma^{1/2}(y_{t-d} - c)$. Now, the partial derivatives are given by

	Equals	Evaluated at $h_{t-d} = 0$
$\partial\theta/\partial h_{t-d}$	$2h_{t-d}\exp(-h_{t-d}^2)$	0
$\partial^2\theta/\partial h_{t-d}^2$	$2\exp(-h_{t-d}^2) - 4h_{t-d}^2\exp(-h_{t-d}^2)$	2
$\partial^3\theta/\partial h_{t-d}^3$	$-12h_{t-d}\exp(-h_{t-d}^2) + 8h_{t-d}^3\exp(-h_{t-d}^2)$	0

Unlike the LSTAR model, the expansion for the ESTAR model has the quadratic form $\theta = \pi_2 h_{t-d}^2$. Thus, we can write the expansion of the ESTAR model without h_{t-d} and h_{t-d}^3. Hence, the Taylor series approximation has the form

$$y_t = \alpha_0 + \alpha_1 y_{t-1} + \ldots + \alpha_p y_{t-p} + (\beta_0 + \beta_1 y_{t-1} + \ldots + \beta_p y_{t-p})(\pi_2 h_{t-d}^2) + \varepsilon_t$$
$$= z_t + z_t y_{t-d} + z_t y_{t-d}^2 + \varepsilon_t$$

The key insight in Teräsverta (1994) is that the auxiliary equation for the ESTAR model is nested within that for an LSTAR model. If the ESTAR is appropriate, it should be possible to exclude all of the terms in the cubic expression $z_t y_{t-d}^3$ from (7.21). Hence, the testing procedure follows these steps:

STEP 1: Estimate the linear portion of the AR(p) model to determine the order p and to obtain the residuals $\{e_t\}$.

STEP 2: Estimate the auxiliary equation (7.21). Test the significance of the entire regression by comparing TR^2 to the critical value of χ^2. If the calculated value of TR^2 exceeds the critical value from a χ^2 table, reject the null hypothesis of linearity and accept the alternative hypothesis of a smooth-transition model. (Alternatively, you can perform an F-test.)

STEP 3: If you accept the alternative hypothesis (i.e., if the model is nonlinear), test the restriction $a_{31} = a_{32} = \ldots = a_{3n} = 0$ using an F-test. If you reject the hypothesis $a_{31} = a_{32} = \ldots = a_{3n} = 0$, the model has the LSTAR form. If you accept the restriction, conclude that the model has the ESTAR form.

8. OTHER REGIME-SWITCHING MODELS

The artificial neural network and the Markov switching model represent other types of regime-switching models that appear in the literature. Although they cannot be readily estimated by OLS, it is worthwhile to review their properties.

The Artificial Neural Network

The artificial neural network (ANN) can be useful for nonlinear processes that have an unknown functional form. The simple form of the ANN model is

$$y_t = a_0 + a_1 y_{t-1} + \sum_{i=1}^{n} \alpha_i f_i(y_{t-1}) + \varepsilon_t \tag{7.22}$$

where the function $f_i(y_{t-1})$ is a cumulative distribution or a logistic function such as that in (7.20). For the case of the logistic function, we can write

$$y_t = a_0 + a_1 y_{t-1} + \sum_{i=1}^{n} \alpha_i [1 + \exp(-\gamma_i (y_{t-1} - c_i))]^{-1} + \varepsilon_t$$

Although the ANN is very similar to the LSTAR model, there are some important differences. First, the ANN allows only the intercept to be time-varying; the autoregressive coefficient a_1 is constant. As such, the level of the series is changing over time. Second, the ANN uses n different logistic functions (called **nodes**). Kuan and White (1994) proved that, for sufficiently large n, this type of model can approximate any first-order nonlinear model arbitrarily closely. As such, the ANN is particularly useful for estimating nonlinear relationships that have an unknown functional form.

Although the model can fit the data extraordinarily well, there is an obvious difficulty in that the model does not have a clear economic interpretation. Since the ANN can be extended to high-order autoregressive processes, it can have an extremely large number of parameters. As such, there is a danger of overfitting the data. If you let n become too large (i.e., if you use too many nodes), you will wind up fitting the noise component of the data. The fact that $R^2 \rightarrow 1$ as n grows increasingly large should not be especially comforting if the goal is to forecast subsequent values of the series. Many researchers would select the value of n using the parsimonious SBC.

Notice that the parameters are not globally identified for $n > 1$. Numerical optimization routines have difficulty finding the parameter values that minimize the sum of squared residuals since many local minima often exist. To circumvent the problem, a number of different routines are used to estimate the parameter values. Although the details are not necessary for our purposes, it is instructive to consider the "recursive learning" method discussed in White (1989). Suppose you use the first t observations of your data set to obtain the nonlinear least squares estimates of the parameters. Let $\hat{\theta}_t$ denote the vector of estimated parameters using these t observations and let $\hat{y}_{t+1}$ denote the predicted value y_{t+1} The value of $\hat{\theta}_t$ acts as an initial condition in the difference equation

$$\hat{\theta}_{t+1} = \hat{\theta}_t + \eta_t(y_{t+1} - \hat{y}_{t+1})$$

where η_t is generally taken to be a multiple of the vector of partial derivatives of (7.22) with respect to the parameters evaluated at the point estimates of θ_t. The successive values of $\hat{\theta}_{t+1}$ are obtained until all the parameter estimates converge.

We can follow White (1989) and explore the ability of the ANN to mimic **chaos.** Recall that a sequence $\{y_t\}$ is said to be chaotic if it is generated from a deterministic difference equation such that it does not explode or converge to a constant or to a repetitive cycle. Thus, the sequence may appear to be random even though it is completely deterministic. In particular, let $y_1 = y_2 = 0.5$ and suppose that the next 98 values of the $\{y_t\}$ sequence are generated according to

$$y_t = 1 - 1.4y_{t-1}^2 + 0.3y_{t-2}$$

The actual and fitted values of the series are shown in Figure 7.9. Although just two nodes were used to estimate the series, the fit of the ANN is quite reasonable. The example illustrates the point that the ANN is capable of capturing a highly nonlinear process when the functional form is completely unknown.

The Markov Switching Model

The basic threshold model allows the regime switch to depend on the magnitude of an observable variable. If y_{t-d} exceeds some threshold value, the system is in regime one; otherwise, the system is in regime two. Although regime switching is more gradual in

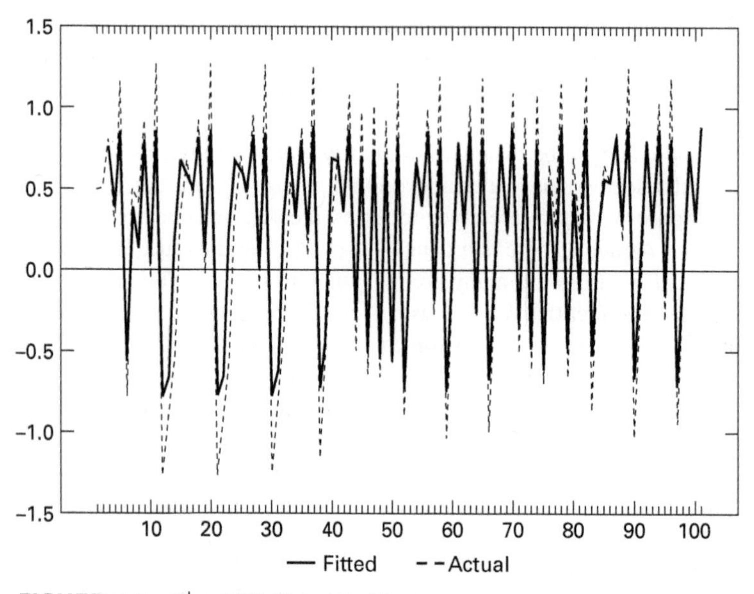

FIGURE 7.9 The ANN Fitted to Chaos

the STAR and ANN models, the adjustment process depends on the current state of the system. In contrast, the Markov switching model developed by Hamilton (1989) posits that regime switches are exogenous. To take a simple example, suppose there are two regimes (or states of the world) and that the autoregressive process for y_t is regime-dependent. In particular, let

$$y_t = a_{10} + a_1 y_{t-1} + \varepsilon_{1t} \quad \text{if the system is in regime 1}$$
$$y_t = a_{20} + a_2 y_{t-1} + \varepsilon_{2t} \quad \text{if the system is in regime 2}$$

At this point, the model looks very much like a TAR model of (7.15) in that the autoregressive coefficient is a_1 in regime 1 and a_2 in regime 2. However, in contrast to the TAR model, there are fixed probabilities of a regime change. If p_{11} denotes the probability that the system remains in regime one, $(1 - p_{11})$ denotes the probability that the system switches from regime 1 to regime 2. Similarly, if p_{22} denotes the probability that the system remains in regime 2, $(1 - p_{22})$ is the probability that the system switches from regime 2 to regime 1. Thus, the switching process is actually a first-order Markov process. No attempt is made to explain the reason that regime changes occur and no attempt is made to explain the timing of such changes. There are several important features of the Markov switching model:

1. Since the transition probabilities (i.e., p_{11} and p_{22}) are unknown, they need to be estimated along with the coefficients of the two autoregressive processes. As in the TAR model, if one of the regimes rarely occurs, the coefficients for that regime will be poorly estimated.
2. The overall degree of persistence depends on the autoregressive parameters *and* the transition probabilities. For example, if $a_1 > a_2$ and p_{11} is large, the

process will tend to remain in the regime with substantial autoregressive persistence. Moreover, if p_{22} is small, the system will have a tendency to switch into regime 1 from regime 2.

3. The probabilities p_{11}, $(1 - p_{11})$, p_{22}, and $(1 - p_{22})$ are all *conditional* probabilities. For example, if the system is in regime 2, $(1 - p_{22})$ is the conditional probability that the system switches into regime 1. It is also of interest to calculate the *unconditional* probability that the system is in regime 1 (p_1) and in regime 2 (p_2). In Exercise 3 at the end of this chapter, you are asked to show that

$$p_1 = (1 - p_{22})/(2 - p_{11} - p_{22})$$
$$p_2 = (1 - p_{11})/(2 - p_{11} - p_{22})$$

Thus, if $p_{11} = 0.75$ and $p_{22} = 0.5$, $p_1 = 2/3$ and $p_2 = 1/3$.

4. A number of papers, including Clements and Krolzig (1998), try to use various statistical means to distinguish between a Markov switching model and a STAR model. It is very difficult to do so, especially if the Markov switching model is modified to allow the transition probabilities to depend on the variables in the model.

Usually, Markov switching models are applied to estimate the level of a series. However, Edwards and Susmel (2000) used a regime-switching model to examine the interest rate volatility in emerging markets. It is argued that the standard GARCH model is not applicable to emerging markets because of the occurrence of large shocks. Although a GARCH model estimated using a t-distribution could account for fat-tailed returns, such models will typically predict too much volatility persistence. As illustrated in Chapter 3, the sum of the coefficients in a GARCH model is often close to unity. As an alternative, consider a three-state model containing a low-volatility regime, a moderate-volatility regime, and a high-volatility regime. If the probability of switching out of a high-volatility state is large, high volatility does not need to be extremely persistent.

Edwards and Susmel use weekly interest rate data for Argentina, Brazil, Chile, Hong Kong, and Mexico. They begin by estimating an AR(1) equation for the model of the mean and a GARCH(1, 1) model for the variance. Consider the estimated set of equations for Brazil (with standard errors in parentheses) over the April 18, 1994, through April 16, 1999, period:

$$\underset{}{\Delta r_t} = \underset{(0.04)}{-0.0133} - \underset{(0.10)}{0.217\Delta r_{t-1}} + \varepsilon_t$$

$$h_t = \underset{(0.03)}{0.058} + \underset{(0.25)}{1.321\varepsilon_{t-1}^2} + \underset{(0.05)}{0.395h_{t-1}}$$

where r_t is the Brazilian short-term interest rate and h_t is the conditional variance. The model of the mean is in first differences since r_t is a unit root process.

Although the coefficients are significant at the 5 percent level, there is a disturbing feature of the model. Notice that the sum of the coefficients in the equation for h_t exceeds unity. As such, the model predicts that volatility is explosive. As an alternative,

Edward and Susmel consider the volatility-switching ARCH (SWARCH) model. The basic form of the model is

$$h_t/\gamma_s = \alpha_0 + \sum_{i=1}^{q} \alpha_i(\varepsilon_{t-i}^2/\gamma_s)$$

where $s = 1, 2$, or 3 refers to the current state (i.e., low, moderate, or high).

Note that one of the values of γ_s must be normalized to equal unity. Moreover, if $\gamma_1 = 1$, the other values of γ_s measure the ratio of the conditional variance in state s relative to that in state 1. The estimated SWARCH model for Brazil is

$$\underset{}{\Delta r_t} = \underset{(0.03)}{-0.087} + \underset{(0.05)}{0.016\Delta r_{t-1}} + \varepsilon_t$$

$$h_t/\gamma_s = \underset{(0.03)}{0.131} + \underset{(0.10)}{0.068\varepsilon_{t-1}^2/\gamma_s}$$

and

$$\gamma_1 = 1, \gamma_2 = 4.851, \text{and } \gamma_3 = 128.51$$

It is striking that the high-volatility state is more than 128 times more volatile than the low-volatility state. Nevertheless, the probability of a switch from the high-volatility state to the other states was found to be high. Hence, the high-volatility state was found to be short-lived.

9. ESTIMATES OF STAR MODELS

This section illustrates a number of techniques used in the estimation of regime-switching models. The goal is to demonstrate a number of practical issues that arise in applied work.

An LSTAR Model

To illustrate the process of estimating an LSTAR model, 250 realizations of the following sequence were generated:

$$y_t = 1 + 0.9y_{t-1} + (-3 - 1.7y_{t-1})/(1 + \exp(-10(y_{t-1} - 5))) + \varepsilon_t \quad (7.23)$$

If you compare (7.23) to (7.20) you will see that the smoothness parameter $\gamma = 10$ and that $\theta = 1/[1 + \exp(-10(y_{t-1} - 5)]$. As $y_{t-1} \to -\infty$, the behavior of y_t is governed by the autoregressive process $1 + 0.9y_{t-1} + \varepsilon_t$; and as $y_{t-1} \to +\infty$, the behavior of y_t is governed by $-2 - 0.8y_{t-1} + \varepsilon_t$. Note that in the neighborhood of $y_{t-1} = 0$, the value of θ is approximately equal to zero. The 250 realizations are shown in Figure 7.10 and are contained in the file LSTAR.XLS. The simulated sequence has a sample mean of 0.62 and a standard deviation of 3.43. The first six autocorrelations are:

ρ_1	ρ_2	ρ_3	ρ_4	ρ_5	ρ_6
0.552	0.270	0.067	−0.039	−0.136	−0.161

The first few autocorrelations seem to exhibit geometric decay and those for lags 5 and 6 have *prob*-values near 5 percent ($2 \cdot 250^{1/2} = 0.1265$). If we did not know the

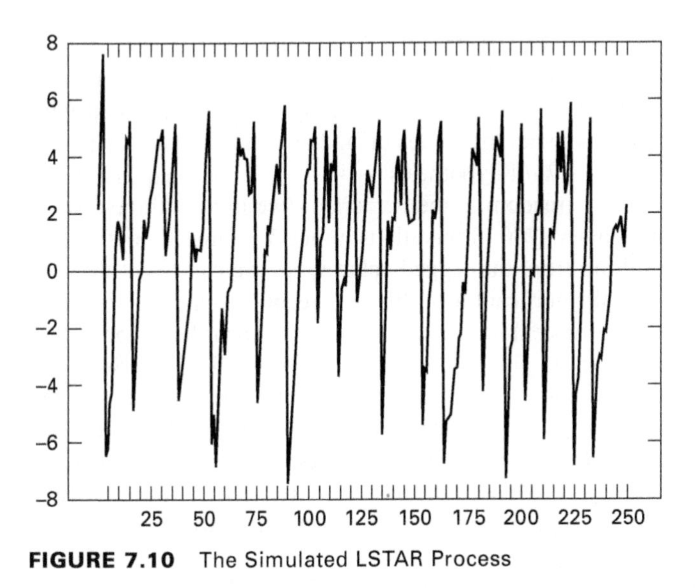

FIGURE 7.10 The Simulated LSTAR Process

actual data-generating process, we might be tempted to estimate the series as a linear AR(1) process. In fact, the estimated linear model looks to be quite plausible; consider

$$y_t = \underset{(1.50)}{0.278} + \underset{(10.42)}{0.552}y_{t-1} + e_t \tag{7.24}$$

and AIC = 1901.18885 SBC = 1908.22375

The residual autocorrelations are such that there is no *linear* relationship in the residuals. The first 12 autocorrelations of the residuals are:

ρ_1	ρ_2	ρ_3	ρ_4	ρ_5	ρ_6	ρ_7	ρ_8	ρ_9	ρ_{10}	ρ_{11}	ρ_{12}
0.03	0.01	−0.06	−0.02	−0.1	−0.04	−0.11	−0.09	0.07	−0.00	−0.05	−0.06

The Ljung–Box *Q*-statistics are such that the *prob*-values for the first four, eight, and 12 lags are 0.900, 0.347, and 0.471, respectively. Since the autocorrelations of the residuals are not significant at conventional levels, you might be tempted to conclude that the true data-generating process was an AR(1). However, a battery of nonlinear diagnostic testing reveals a very different picture. Note first that the autocorrelations of the squared residuals also suggest that the linear model is adequate. The autocorrelations of the squared residuals are:

ACF of the squared residuals

ρ_1	ρ_2	ρ_3	ρ_4	ρ_5	ρ_6	ρ_7	ρ_8	ρ_9	ρ_{10}	ρ_{11}	ρ_{12}
0.03	−0.04	−0.07	−0.10	−0.09	−0.10	−0.08	−0.07	0.14	0.00	−0.02	−0.05

In contrast, the RESET test indicates a nonlinear relationship. Call e_t and $\hat{y}_t$ the residuals and the fitted values from the linear model, respectively. Given that the best-fitting model is an AR(1), we can use the residuals from (7.24) to obtain

$$e_t = \underset{(4.24)}{0.932} + \underset{(9.04)}{0.710}y_{t-1} + \underset{(0.64)}{0.058}\hat{y}_t^2 - \underset{(-9.39)}{0.157}\hat{y}_t^3 - \underset{(-4.84)}{0.034}\hat{y}_t^4$$

Notice that most of the individual coefficients appear to be statistically significant. However, you should not rely on the individual t-statistics because the regressors are highly correlated; for example, large values of $\hat{y}_t^2$ will be associated with large values of $\hat{y}_t^4$. The issue is whether the values of $\hat{y}_t^i$ have any explanatory power as a group. The F-statistic for the null hypothesis $\alpha_1 = \alpha_2 = \alpha_3 = 0$ equals 95.60. Since there are three degrees of freedom in the numerator (we impose three restrictions) and 244 in the denominator (250 observations minus 5 estimated coefficients and 1 lost observation resulting from the lagged value y_{t-1}), we can reject the null hypothesis at any conventional significant level (the 1 percent critical value is 3.86). Hence, we conclude that the series exhibits some form of nonlinear behavior.

It is quite a bit more difficult to pin down the form of the nonlinearity. Since the data are simulated, there is no possibility of using economic theory to suggest the most probable form of nonlinearity. Hence, one way to proceed is to estimate a number of nonlinear models and select the one that fits the best. However, the danger of this procedure is that you are likely to overfit the data. A more prudent way to proceed is to perform a number of Lagrange multiplier tests to determine which models are likely to be the most plausible.

One test that can be useful to select the functional form is Teräsverta's test for LSTAR versus ESTAR behavior. Pretend that we do not know the value of the delay parameter d. It seems natural to begin with $d = 1$. From the Taylor series expansion for a first-order LSTAR model, we need to regress the residuals from the linear model on the regressors (i.e, a constant and y_{t-1}) and on y_{t-1}, y_{t-1}^2, and y_{t-1}^3 multiplied by the regressors. The estimated auxiliary regression is

$$\underset{(4.35)}{e_t = 0.933} + \underset{(9.21)}{0.706y_{t-1}} - \underset{(-0.987)}{0.027y_{t-1}^2} - \underset{(-11.52)}{0.039y_{t-1}^3} - \underset{(-4.84)}{0.003y_{t-1}^4}$$

The F-statistic for the entire regression is 71.70; with four numerator and 244 denominator degrees of freedom, the regression is highly significant. Moreover, the F-statistic for the presence of the nonlinear terms y_{t-1}^2, y_{t-1}^3, and y_{t-1}^4 is 95.60; with three numerator and 244 denominator degrees of freedom, we can conclude that there is threshold behavior. Next, we can determine if LSTAR or ESTAR behavior is the most appropriate. Given the t-statistic on the coefficient for y_{t-1}^4, we cannot exclude this expression from the auxiliary equation. Hence, we can rule out ESTAR behavior in favor of LSTAR behavior. It is possible that the delay parameter is 2, even though y_{t-2} does not directly appear in the model. To determine whether the y_{t-1} or y_{t-2} is the most appropriate threshold variable, you can estimate the following auxiliary equation using $d = 2$:

$$e_t = 0.738 + 0.047y_{t-1} - 0.158y_{t-1}y_{t-2} - 0.005y_{t-1}y_{t-2}^2 + 0.003y_{t-1}y_{t-2}^3$$

The F-value for this regression was only 5.73. Since the $d = 1$ yields a substantially better fit than the $d = 2$, we can conclude that the y_{t-1} is the most appropriate threshold variable. Thus, it seems reasonable to estimate a nonlinear model of the form

$$y_t = \alpha_0 + \alpha_1 y_{t-1} + (\beta_0 + \beta_1 y_{t-1})/(1 + \exp(-\gamma(y_{t-1} - c))) + \varepsilon_t$$

Since the coefficients are multiplicative, OLS cannot be used to obtain the least squares estimates of the coefficients. Instead, it is standard to use nonlinear least squares (NLLS) to obtain the estimates of the parameter values. Consider the estimated model

$$\begin{gathered} y_t = \underset{(14.43)}{0.941} + \underset{(45.15)}{0.923}y_{t-1} + (\underset{(-2.07)}{-5.86} - \underset{(-2.45)}{1.18}y_{t-1})(1 + \exp(-\underset{(6.77)}{11.206}(y_{t-1} - \underset{(312.33)}{5.01}))) + \varepsilon_t \\ \text{AIC} = 1365.22 \quad \text{SBC} = 1386.33 \end{gathered} \tag{7.25}$$

The point estimates that all the parameters except β_0 are every close to their true values. Moreover, the AIC and the SBC both select the LSTAR model over the linear model. Note that you need to be wary of the t-statistics for several reasons. First, the nonlinear least squares estimates do not make the assumptions that the estimated coefficients have normal distributions. Second, the estimates are all performed using numerical methods. Third, the equation is not linear in the coefficient so that their magnitudes are interdependent. Given these caveats, the estimated model does capture the essential features of (7.23).

In many circumstances, the numerical methods used to estimate the parameters of STAR models have difficulty in simultaneously finding γ and c. It is crucial to provide the numerical routine with very good initial guesses. If there are problems, a popular modification of Haggan and Ozaki's (1981) method is to estimate γ using a grid search. Fix γ at its smallest possible value and estimate all of the remaining parameters using NLLS. Slightly increase the value of γ and reestimate the model. Continue this process until the plausible values of γ are exhausted. Use the value of γ yielding the best fit. Note that if γ is large, the transition is sharp in the neighborhood of $y_{t-d} = c$ so that the LSTAR model acts like a TAR model—in fact, if γ is large and convergence to a solution is a problem, it could be easier to estimate a TAR model instead of the LSTAR model. Teräsverta (1994) noted that rescaling the expressions in θ can aid in finding a numerical solution. With an LSTAR model, he found it useful to standardize by dividing $\exp[-\gamma(y_{t-d} - c)]$ by the standard deviation of the $\{y_t\}$ series. With an ESTAR model, he standardized by dividing $\exp[-\gamma(y_{t-d} - c)^2]$ by the variance of the $\{y_t\}$ series. In this way, the threshold value c is measured in standardized units so that a reasonable value for the initial guess (e.g., $c = 1$ standard deviation) can be readily made. An example is shown in Question 5 at the end of this chapter.

The Real Exchange Rate as an ESTAR Process

As indicated earlier, Michael, Nobay, and Peel (1997) argued that transaction costs should make real exchange rates behave as ESTAR processes. For our purposes, the series of interest is now the annual observations of the U.K.–U.S. real rate over the 1791 to 1992 period. The first issue is to determine whether or not the rates are stationary; after all, if the rates are unit root processes, the theory of PPP fails. As such, they use augmented Dickey–Fuller tests to determine whether the series contains a unit root. The use of annual data results in very short lags. If we ignore the intercept, the estimated equation for the U.K.–U.S. rate is

$$\Delta y_t = \underset{(-3.62)}{-0.12}y_{t-1} + \underset{(1.75)}{0.12}\Delta y_{t-1} + \varepsilon_t \tag{7.26}$$

In absolute value, the t-statistic of -3.62 exceeds the critical value reported in the Dickey–Fuller table; as such, it is possible to reject the null hypothesis of a unit root in the real exchange rate. The point estimate of -0.12 implies a fairly slow speed of adjustment; approximately 88 percent of the current period's discrepancy from PPP is expected to persist into the next year. Nevertheless, this linear model forces the speed of adjustment to be constant. (Some of the issues concerned with unit roots and nonlinearity are discussed in detail in Section 11.)

Given that the series is stationary, the next issue is to determine whether Teräsverta's four-step methodology indicates the presence of ESTAR adjustment. Given the lag lengths, the most plausible value of the delay parameter d is unity. Nevertheless, the authors follow the standard procedure and select the value of d that results in the best fit of the auxiliary equation. As suspected, the value $d = 1$ fits the data better than the alternatives $d = 2$ or $d = 3$. The auxiliary regression has the form of (7.21). The *prob*-value of the F-statistic for the null hypothesis that all values of $a_{ij} = 0$ in the U.K.–U.S. auxiliary equation is 0.076. Hence, there is weak evidence of nonlinear behavior in the U.K.–U.S. rate.

Given the presence of threshold adjustment, the next issue is to test for LSTAR versus ESTAR adjustment. The F-test for the null hypothesis that all values of $a_{3i} = 0$ has a *prob*-value 0.522; as such, it is possible to reject the null hypothesis of LSTAR adjustment. Notice that the auxiliary equation for nonlinear adjustment has coefficients for both the LSTAR and ESTAR models. If there is ESTAR adjustment, a number of the coefficients are unnecessary. Hence, the authors constrain all values of $a_{3i} = 0$, and test whether the remaining coefficients are zero. The F-statistic for this test has a *prob*-value of 0.028. Hence, this test with enhanced power suggests ESTAR versus linear adjustment.

10. GENERALIZED IMPULSE RESPONSES AND FORECASTING

This section presents two different estimated threshold models. Each was selected to emphasize a different aspect of the general methodology. First, Potter's (1995) TAR model of U.S. GNP is presented. The interesting feature of Potter's study is the calculation of impulse responses from a TAR model. Second, Enders and Sandler's (2005) forecast function for the number of casualties caused by transnational terrorists is examined.

Nonlinear Estimates of GNP Growth

Potter (1995) argues that a nonlinear model of U.S. GNP growth performs much better than a linear one. To begin, Potter estimates the following AR(5) model of the logarithmic change in the quarterly values of real U.S. gross national product (GNP) growth over the 1947Q1 to 1990Q4 period:

$$\begin{aligned} y_t = {} & \underset{(4.42)}{0.540} + \underset{(4.23)}{0.330}y_{t-1} + \underset{(2.35)}{0.193}y_{t-2} - \underset{(-1.27)}{0.105}y_{t-3} \\ & - \underset{(-1.12)}{0.092}y_{t-4} - \underset{(-0.308)}{0.024}y_{t-5} + \varepsilon_t \qquad \text{AIC}' = 8.00 \end{aligned}$$

where $y_t = 100^*[\log(\text{GNP}_t) - \log(\text{GNP}_{t-1})]$

Potter also estimates a two-regime TAR model allowing the variances to differ across regimes. He states that pretesting yields a delay factor of 2 (i.e., $d = 2$) and a threshold of zero. After purging the threshold regression of insignificant coefficients, Potter reports the following TAR model:

$$\underset{}{y_t} = \underset{(3.21)}{0.517} + \underset{(3.74)}{0.299y_{t-1}} + \underset{(1.77)}{0.189y_{t-2}} - \underset{(-16.57)}{1.143y_{t-5}} + \varepsilon_{1t} \qquad y_{t-2} > 0$$

$$y_t = \underset{(-1.91)}{-0.808} + \underset{(2.79)}{0.516y_{t-1}} - \underset{(-2.68)}{0.946y_{t-2}} - \underset{(-1.63)}{0.352y_{t-5}} + \varepsilon_{2t} \qquad y_{t-2} \leq 0$$

The presence of the AR(5) terms is unusual because the data are seasonally adjusted and there is no particular reason to suppose that the fifth lag (but not lags 3 and 4) affects the contemporaneous value of GNP. However, Potter reports that the AR(3) and AR(4) coefficients are not statistically different from zero at the 5 percent significance level. There are 37 observations in the contractionary regime ($y_{t-2} \leq 0$) and 133 observations in the expansionary regime ($y_{t-2} > 0$). The estimated variance of ε_{1t} equals 0.763 and the estimated variance of ε_{2t} equals 1.50. Thus, the magnitudes of shocks while in the contractionary regime tend to be quite large. The large negative coefficient in the AR(2) term in the contractionary regime has an interesting economic implication. When $y_{t-2} < 0$, there tends to be a sharp reversal in the contraction of output since the product of -0.946 and y_{t-2} is positive.

RECURSIVE FORECASTS The AIC was constructed by combining the residual sums of squares from the two segments of the TAR model. This value of the AIC' $(= -4.89)$ clearly selects the TAR model over the linear model. In order to compare the out-of-sample forecasts, the following procedure was used. Beginning with the sample period 1947*Q*1 through 1960*Q*1, linear and TAR models were estimated. For each model, the one-step-ahead forecast was obtained. Then, the sample period was updated by one quarter and new linear and TAR models were estimated. These updated models were used to obtain one-step-ahead forecasts. Repeating this procedure through the end of the sample yielded two sets of one-step-ahead forecasts. The correlation of the forecasts with the actual values of output growth was 0.23 for the linear model and 0.35 for the TAR model. As such, the forecasting performance of the TAR model exceeds that of the linear model.

Impulse Responses

In a linear model, the impulse responses are not history-dependent and the magnitude of the shock does not alter the time profile of the responses. For example, in the linear AR(1) model $y_t = \rho y_{t-1} + \varepsilon_t$, the impulse responses are given by

$$y_t = \sum_{i=0}^{\infty} \rho^i \varepsilon_{t-i}$$

Hence, the effect of a one-unit shock on y_t is 1, the effect of the shock on y_{t+1} is predicted to be ρ (i.e., $\partial y_{t+1}/\partial \varepsilon_t = \partial y_t/\partial \varepsilon_{t-1} = \rho$), the effect of the shock on y_{t+2} is predicted to be ρ^2, and so forth. Moreover, the effects of a two-unit shock are simply twice those for the one-unit shock and the effects of a negative shock are simply the negative of those for positive shocks. However, the interpretation of impulse response functions for a nonlinear model is not straightforward. The reason is that the impulse responses are history-dependent. The effect of an ε_t shock on the time path of the system depends on the magnitudes of the current and subsequent shocks. Clearly, the sign of the shocks can matter. To take a simple example, in a TAR model with $\tau = 0$, the impulse responses for a one-unit positive shock will have a different time path than a one-unit negative shock. Moreover, the size of the shocks matters; if you are in the contractionary regime, a small positive shock can imply a different time profile than a very large shock, since the small shock is less likely to induce a regime change. Thus, to calculate impulse responses, it is necessary to specify the history of the system and the magnitude of the shock. Moreover, the effects of a shock to ε_t on y_{t+10} will depend on the magnitudes of the shocks that take place in periods $(t + 1)$ through $(t + 9)$. There are several ways to attack the problem. Potter considers shocks of four different magnitudes: -2 percent, -1 percent, 1 percent, and 2 percent. Moreover, he considers several different histories. Consider:

- In the three quarters of 1983Q3, 1983Q1, and 1984Q1, real GNP growth at an annual rate was a remarkable 7.1 percent, 8.2 percent, and 8.2 percent, respectively. As such, even a -2 percent shock would cause GDP growth to remain in the positive regime. Hence, the responses are very similar to those that are obtained from a linear model. Since there is no regime switching, the 1 percent and 2 percent shocks are multiples of each other. The four impulse responses for this particular history are shown in Panel (a) of Figure 7.11.
- The situation was very different in 1970Q2 in that the economy experienced a mild downturn. For 1969Q4, 1970Q1, and 1970Q2, GNP growth measured at an annual rate was -1.9 percent, -0.46 percent, and 0.91 percent, respectively. For periods with negative growth, positive shocks can push GNP growth across the threshold of zero. Panel (b) of Figure 7.11 shows the asymmetric responses. Given that the contractionary regime has an AR(2) coefficient that is nearly -1.0, negative shocks are less persistent than positive shocks. As such, you can see the rather quick turnaround in GNP growth predicted to begin in 1970Q3. Also notice that the effects of the -1 percent and -2 percent shocks are not proportional to each other.

Notice that these impulse response functions trace out the effects of different-sized ε_t shocks (t = 1984Q1 and 1970Q2), assuming subsequent shocks are all zero. Using the methods discussed below, it is possible to generalize the impulse response functions to allow for the effects of any ensuing shocks. To preview the method, consider Potter's -1 percent shock used as a value for ε_{1972Q2}. It is possible to draw eight random values to simulate a possible sequence for ε_{1972Q3} through ε_{1974Q3}. This set of shocks can then be used to obtain the impulse responses through 1974Q3. If this procedure is repeated several thousand times, it is possible to obtain the mean values of the impulse responses along with the various percentiles. Hence, it is possible to obtain a 5 percent or 10 percent confidence band around the calculated mean values of the impulses.

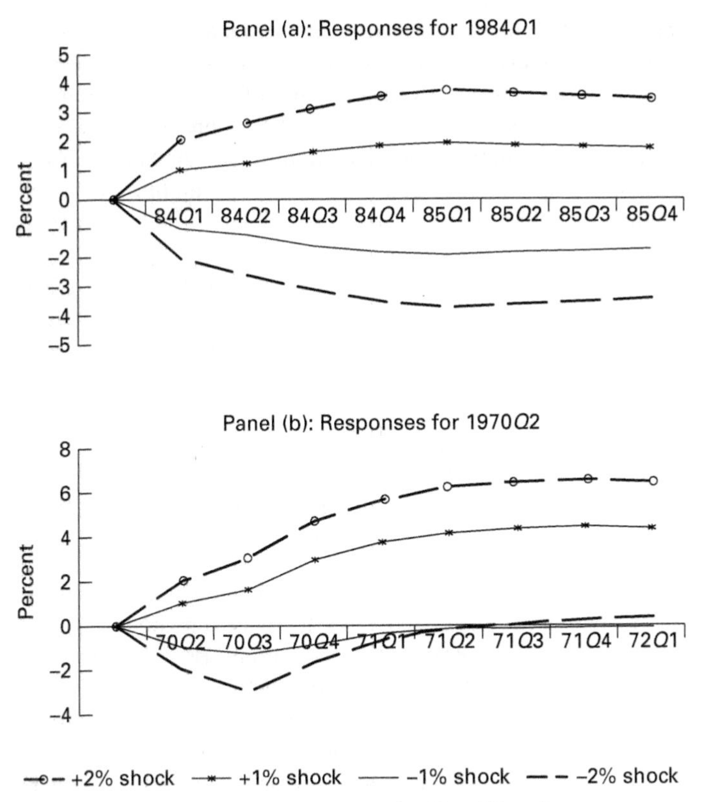

FIGURE 7.11 Impulse Responses for Two Histories

Terrorist Incidents with Casualties

A realistic way to capture the nature of terrorist campaigns is to use a two-regime TAR model. In relatively tranquil regimes, terrorists can replenish and stockpile resources, recruit new members, raise funds, and plan for future attacks. Terrorism can remain low until an event occurs that switches the system into the high-terrorism regime. Because each terrorist attack utilizes scarce resources, high-terrorism states are not anticipated to exhibit a high degree of persistence when a shock raises the level of terrorism. On the other hand, periods with little terrorism can be highly persistent to shocks. In order to measure the differing persistence across the two states, Enders and Sandler (2005) acquired quarterly data on the number of incidents containing one or more casualties over the 1968Q1 to 2000Q4 period. We first estimated the number of incidents with casualties (*cas*) as the linear AR(3) autoregressive process:

$$cas_t = \underset{(2.83)}{5.91} + \underset{(2.98)}{0.261}cas_{t-1} + \underset{(3.59)}{0.310}cas_{t-2} + \underset{(2.40)}{0.209}cas_{t-3} + \varepsilon_t \qquad \text{AIC} = 1205.72$$

where cas_t represents the number of incidents with casualties.

The model appears adequate in that it satisfies the standard diagnostic tests. All t-statistics are significant at conventional levels and the point estimates of the autoregressive coefficients imply stationarity. The results of a Dickey–Fuller test allow us to reject the null hypothesis of a unit root at the 5 percent significance level. Moreover, the Ljung–Box Q-statistics indicate that the residuals are serially uncorrelated. For example, the Q-statistics using the first four, eight, and twelve lags of the residual autocorrelations have *prob*-values of 0.98, 0.52, and 0.72, respectively.

Correlation coefficients are measures of linear association and may not detect nonlinearities in the data. We begin a search for the most appropriate TAR representation by estimating a model in the form of (7.14) using a value of $p = 3$. By applying Chan's method to search over all potential thresholds, we obtained:

$$\begin{aligned} cas_t = {} & \underset{(2.89)}{(17.05} + \underset{(1.59)}{0.173cas_{t-1}} + \underset{(1.72)}{0.236cas_{t-2}} + \underset{(0.458)}{0.053cas_{t-3})}\, I_t \\ & + \underset{(0.08)}{(2.57} + \underset{(2.28)}{0.361cas_{t-1}} + \underset{(0.91)}{0.207cas_{t-2}} + \underset{(2.71)}{0.362cas_{t-3})}(1 - I_t) + \varepsilon_t \end{aligned} \tag{7.27}$$

where the estimates of the threshold and the delay are $\tau = 25$ and $d = 2$.

Equation (7.27) is clearly overparameterized, since there are a number of coefficients with t-statistics less than 1.96 in absolute value. A problem is that tabulated t-statistics are actually an approximation of the actual distribution since we searched for the best-fitting threshold function.[4] In terms of (7.14), the distribution of coefficient a_{ij} depends on the accuracy of the estimated threshold. To purge the model of the "unimportant" coefficients, we use the following strategy. If the t-statistics indicate that one or more of the coefficients has a *prob*-value below 5 percent, we reestimate the model excluding the "least significant" term. Thus, (7.27) is reestimated without $I_t cas_{t-3}$. If the AIC of the resulting model is smaller than that of the original model, the term is purged. The exceptions to this procedure are the two intercept terms, because the absence of an intercept implies a zero mean. This procedure yields the final TAR estimate:

$$\begin{aligned} cas_t = {} & \underset{(3.19)}{(17.87} + \underset{(1.83)}{0.189cas_{t-1}} + \underset{(1.83)}{0.237cas_{t-2})}\, I_t \\ & + \underset{(1.48)}{(3.92} + \underset{(2.97)}{0.423cas_{t-1}} + \underset{(3.12)}{0.398cas_{t-3})}(1 - I_t) + \varepsilon_t \end{aligned} \tag{7.28}$$

where AIC $= 1205.07$, $\tau = 25$, and $d = 2$.

Diagnostic checking indicates that the model is appropriate. For example, the first twelve autocorrelations of the residuals are less than 0.14 in absolute value and the *prob*-values for the Ljung–Box $Q(4)$, $Q(8)$, and $Q(12)$ statistics are 0.98, 0.57, and 0.81, respectively. Even though the TAR model contains seven parameters (i.e., six coefficients plus τ), the AIC selects it over the linear model.

The threshold model yields very different implications about the behavior of the cas_t series than the linear model. The latter indicates that the cas_t sequence converges to the long-run mean of approximately 26.9 incidents per quarter [5.91 ÷ (1.0 − 0.261 − 0.301 − 0.209)]. Since the linear specification makes no distinction between high- and low-terrorism states, the degree of autoregressive decay is always constant. Regardless of whether the number of incidents is above or below the mean, the degree of persistence is quite large; the largest characteristic root of the linear model is 0.88.

A useful way to conceptualize the time path of the TAR model is to refer to the **skeleton.** The skeleton is the model excluding the error term; hence, it captures the properties of the underlying nonlinear difference equation. For example, the skeleton of (7.28) is

$$cas_t = [17.87 + 0.189cas_{t-1} + 0.237cas_{t-2}]I_t + [3.92 + 0.423cas_{t-1} + 0.398cas_{t-3}](1 - I_t)$$

In the high-terrorism regime (i.e., when the number of incidents is 25 or more), $I_t = 1$, so that $cas_t = 17.87 + 0.189cas_{t-1} + 0.237cas_{t-2}$. Thus, the number of incidents gravitates toward the attractor 31.1 [= 17.87 ÷ (1 − 0.189 − 0.237)]. As measured by the largest characteristic root, the speed of adjustment is 0.59; when terrorism is high, approximately 60 percent of each incident is expected to persist into the next period.

In the low-terrorism state, $I_t = 0$, so that $cas_t = 3.92 + 0.423cas_{t-1} + 0.398cas_{t-3}$. The attractor is 21.9 and the largest characteristic root is 0.88. When the number of incidents is below the threshold value of 25, there is little tendency to return to a long-run mean value. Perhaps a better way to understand the nature of the system is to consider the forecast function. As analyzed in Koop, Pesaran, and Potter (1996), the forecasts and impulse responses from a nonlinear model are state-dependent. In terms of (7.28), a positive shock when $y_{t-2} > 25$ will be less persistent than the same shock when y_{t-2} is far below the threshold. Since we are interested in comparing short-run and long-run forecasts in the two states (rather than a generalized impulse response function), we use a modified version of Koop, Pesaran, and Potter's methodology.

For a model with three lags, we select a particular history for y_t, y_{t-1}, and y_{t-2}. For example, in the last three quarters of 1985—a high-terrorism regime—the numbers of casualty incidents were 33, 50, and 40, respectively. Hence, to forecast the subsequent number of incidents from the perspective of 1985Q4, we let $y_{t-2} = y_{1985Q2} = 33$, $y_{t-1} = y_{1985Q3} = 50$, and $y_t = y_{1985Q4} = 40$. We then select 25 randomly drawn realizations of the residuals of (7.28). Since the residuals may not have a normal distribution, the residuals are selected using standard bootstrapping procedures. In particular, the residuals are drawn with replacement using a uniform distribution. Call these residuals $\varepsilon^*_{t+1}, \varepsilon^*_{t+2}, \ldots, \varepsilon^*_{t+25}$. We then generate y^*_{t+1} through y^*_{t+25} by substituting these bootstrapped residuals into (7.26) and setting I_t appropriately for high- or low-terrorism states. In essence, y^*_{t+1} is one possible realization of the cas_t series for 1986Q1, y^*_{t+2} is one possible realization of the cas_t series for 1986Q2, and so on. For this particular history, we repeat the process 1,000 times. Under very weak conditions, the Law of Large Numbers guarantees that the sample average of the 1,000 values of

y^*_{t+1} converges to the conditional mean of y_{t+1} denoted by $E_t y_{t+1}$. Similarly, the sample means of the various $y^*_{t+i}(k)$, where $y^*_{t+i}(k)$ is the result for draw k, converge to the true conditional i-step-ahead forecasts; that is:

$$\lim_{n\to\infty}\left[\sum_{k=1}^{N} y^*_{t+i}(k)/N\right] = E_t y_{t+i}$$

The essential point is that the sample averages of y^*_{t+1} through y^*_{t+25} yield the one-step- through 25-step-ahead conditional forecasts of the cas_t series from the perspective of 1985*Q*4. Intuitively, because the number of casualty incidents exceeds the threshold, the value of cas_t should quickly decline from 40 toward the attractor of 31.1. Nevertheless, the long-run forecast need not equal the attractor, which can be seen by examining the conditional forecasts (indicated by the solid line) shown in Panel (a) of Figure 7.12. Although the expected number of cas_t incidents does decline

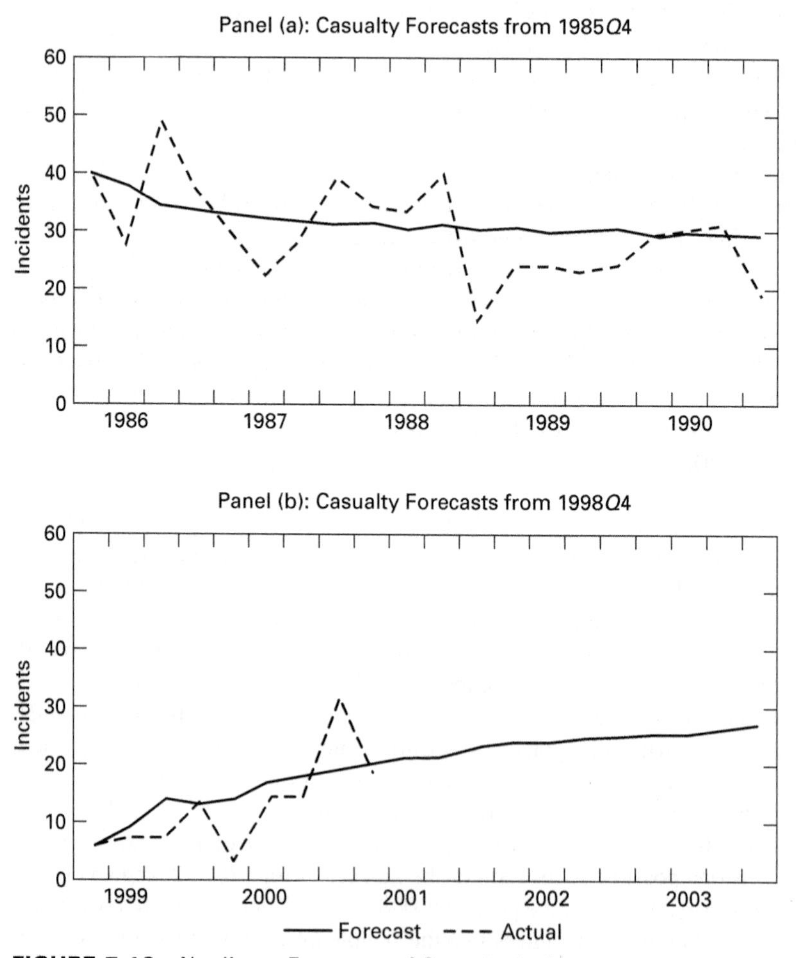

FIGURE 7.12 Nonlinear Forecasts of Casualty Incidents

toward 31.1, there are two reasons why the long-term forecasts continue to decline. Since incidents below the threshold are (on average) more persistent than those above, the system's mean will be below the attractor. Moreover, the forecasts allow for the possibility of a regime switch into the low-terrorism state. As shown in Panel (a) of the figure, the long-run expected value is about 28.5 casualty incidents per quarter. When the number of incidents is high, there is a rapid decline to the threshold, as terrorist networks cannot maintain high-level, resource-using offensives. A comparison of the forecasts with the actual number of casualty incidents (the dashed line in the figure) is instructive. The close fit is remarkable given that the forecasts are not the successive one-step-ahead forecasts. Instead, the figure traces out the one-step- through 25-step-ahead forecasts from the perspective of 1985Q4.

In contrast, the number of terrorist incidents in the last three quarters of 1998 was quite low; $y_{1998Q2} = 5, y_{1998Q3} = 15$, and $y_{1998Q4} = 6$. As shown in Panel (b) of Figure 7.12, reversion back toward the attractor of 21.9 is quite slow in the low-terrorism state. In fact, conditional on the history of 1998Q4, the forecasts remain below 21.9 until the third quarter of 2001. The forecasts seem to track the actual number of incidents occurring through the end of our data set reasonably well and ultimately converge to those for Panel (a). In either case, these long-run forecasts are relatively close to the attractor for the high-terrorism state (31.1).

11. UNIT ROOTS AND NONLINEARITY

Suppose you were convinced that the interest rate spread displays the type of nonlinear adjustment given by (7.1). Before estimating the TAR model directly, you might want to determine whether the series does revert to a long-run equilibrium value (called an **attractor**). However, the established tests for the presence of an attractor assume a linear adjustment process. For example, the Dickey–Fuller (1979) test for a unit root uses a linear adjustment process of the form

$$y_t = a_1 y_{t-1} + \varepsilon_t \quad [\text{or } \Delta y_t = \rho y_{t-1} + \varepsilon_t] \tag{7.29}$$

If the null hypothesis $a_1 = 1$ can be rejected in favor of the alternative $-1 < a_1 < 1$, it can be concluded that the $\{y_t\}$ sequence decays to the attractor $y^* = 0$. However, if the $\{y_t\}$ sequence is generated from a nonlinear model, the Dickey–Fuller test might fail to detect an attractor since it is misspecified. Although (7.29) can be augmented with deterministic regressors and lagged changes of $\{y_t\}$, the crucial point to note is that the dynamic adjustment process is assumed to be linear. The issue is important since Pippenger and Goering (1993) and Balke and Fomby (1997) showed that tests for unit roots have low power in the presence of asymmetric adjustment. After all, (7.29) does not appropriately capture the dynamic adjustment process of a nonlinear model.

Notice that the discussion above is directly applicable to the findings of Michael, Nobay, and Peel (1997). Recall that their aim was to determine whether real exchange rates should be modeled as an ESTAR processes. Nevertheless, the dynamic equation used to determine whether the U.K.–U.S. real exchange rate was stationary [i.e., equation (7.26)] assumes a linear adjustment process. As it turned out, they were able to reject the null hypothesis of a unit root. However, in other circumstances, a linear test may not be able to detect the presence of an attractor for a nonlinear process.

To circumvent this problem, there is a large and growing body of literature designed to test for the presence of an attractor in the presence of nonlinear adjustment. For example, in Enders and Granger (1998), we generalized the Dickey–Fuller methodology to consider the null hypothesis of a unit root against the alternative hypothesis of a threshold autoregressive (TAR) model. The simple version TAR model is

$$\Delta y_t = I_t \rho_1 (y_{t-1} - \tau) + (1 - I_t)\rho_2 (y_{t-1} - \tau) + \varepsilon_t \tag{7.30}$$

$$I_t = \begin{cases} 1 \text{ if } y_{t-1} \geq \tau \\ 0 \text{ if } y_{t-1} < \tau \end{cases} \tag{7.31}$$

As shown by the phase diagram illustrated in Figure 7.13, when $y_{t-1} = \tau$, $\Delta y_t = 0$. However, Δy_t equals $\rho_1 (y_{t-1} - \tau)$ if the lagged value of the series is above τ, and equals $\rho_2 (y_{t-1} - \tau)$ if the lagged value of the series is below τ. The attractor is τ since Δy_t has an expected value of zero when $y_{t-1} = \tau$. Hence, if $y_{t-1} = a$, Δy_t equals the distance ab.

If we use the specification given by (7.30) and (7.31), it is possible to test for an attractor even though the adjustment process is nonlinear. Notice that if $\rho_1 = \rho_2 = 0$, the process is a random walk. A sufficient condition for the $\{y_t\}$ sequence to be stationary is $-2 < (\rho_1, \rho_2) < 0$.[5] Also notice that the Dickey–Fuller test emerges as the special case in which $\rho_1 = \rho_2$. If it is possible to reject the null hypothesis $\rho_1 = \rho_2 = 0$, it can be concluded that there is an attractor. However, as in the Dickey–Fuller test, it is not possible to use a classical F-statistic to test the null hypothesis $\rho_1 = \rho_2 = 0$. Instead, the F-statistics for the null hypothesis $\rho_1 = \rho_2 = 0$ are reported in Table G at the end of the text.

If the null hypothesis of nonstationarity is rejected, it is possible to test for symmetric versus asymmetric adjustment. In particular, if the null is rejected (so that the sequence has an attractor), then you can perform the test for symmetric adjustment (i.e., $\rho_1 = \rho_2$) using a standard F-distribution. If the threshold is unknown (but estimated consistently using Chan's method), the conjecture is that you can also use a

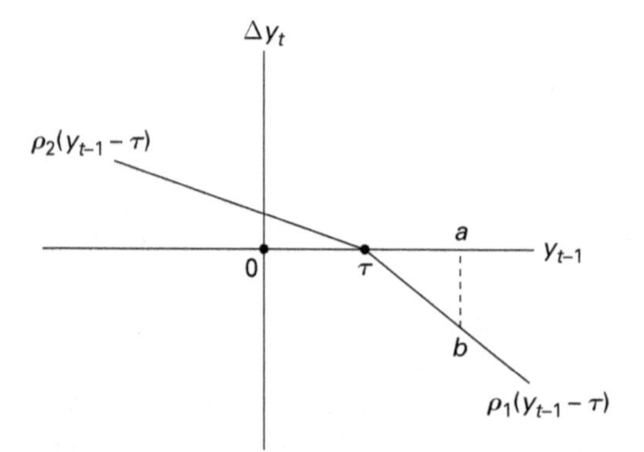

FIGURE 7.13 Phase Diagram for the TAR Model

standard F-test. However, Hansen shows that small sample properties of the OLS estimates of the individual ρ_1 and ρ_2 values have inflated standard errors and the convergence properties of the OLS estimates can be poor. To avoid this problem, you can use Hansen's (1997) bootstrapping method that was described at the end of Section 5.

An alternative to the basic TAR model is the momentum threshold autoregressive (M-TAR) model. Since the exact nature of the nonlinearity may be unknown, it is possible to allow the adjustment to depend on the change in y_{t-1} (i.e., Δy_{t-1}) instead of the level of y_{t-1}. In this case, the model becomes (7.30) along with the indicator function

$$I_t = \begin{cases} 1 \ \textit{if} \ \ \Delta y_{t-1} > 0 \\ 0 \ \textit{if} \ \ \Delta y_{t-1} \le 0 \end{cases} \tag{7.32}$$

This variant of the basic model, used by Enders and Granger (1998) and Caner and Hansen (1998), allows a variable to display differing amounts of autoregressive decay depending on whether it is increasing or decreasing. This specification is especially relevant when the adjustment is such that the series exhibits more momentum in one direction than the other; the resulting model is called a momentum-threshold autoregressive (M-TAR) model. The F-statistic for the null hypothesis $\rho_1 = \rho_2 = 0$ using the M-TAR specification is called Φ_M. As there is generally no presumption as to whether to use the TAR or the M-TAR model, the recommendation is to select the adjustment mechanism (7.31) or (7.32) by a model-selection criterion such as the AIC or SBC.

To perform the test, follow these steps:

STEP 1: If you know the value of τ (for example, $\tau = 0$), estimate (7.30). Otherwise, use Chan's method; for each potential threshold τ, set the indicator function using (7.31). Estimate (7.30) for each potential threshold value and select the value of τ from the regression containing the smallest value for the sum of squared residuals.

STEP 2: If you are unsure as to the nature of the adjustment process, repeat Step 1 using the M-TAR model. For each potential threshold τ, set the indicator function using (7.32). Select the value of τ resulting in the best fit. Use the AIC or SBC to select the TAR or M-TAR specification.

STEP 3: Use the model selected from Step 1 or Step 2 to calculate the F-statistic for the null hypothesis $\rho_1 = \rho_2 = 0$. For the TAR model, compare this sample statistic with the appropriate critical value in Table G. The critical values depend on sample size (T) and whether you augment the model with lagged changes. Use Panel (a) if you estimate τ for a TAR model and Panel (b) if you estimate τ for an M-TAR model. If you know the threshold and estimate an M-TAR model, use Panel (c). In the case of a TAR model with a known threshold, the test seems to have low power relative to the Dickey–Fuller test. As such, the critical values for this case are not reported.

STEP 4: If the alternative hypothesis is accepted (i.e., if there is an attractor), it is possible to test for symmetric versus asymmetric adjustment since the asymptotic joint distribution of ρ_1 and ρ_2 converges to a multivariate normal. As such, the restriction that the adjustment is symmetric (i.e., the null hypothesis $\rho_1 = \rho_2$) can be tested using Hansen's (1997) bootstrapping method or using a standard F-test as an approximation.

STEP 5: Diagnostic checking of the residuals should be undertaken to ascertain whether the estimated $\{\varepsilon_t\}$ series could reasonably be characterized by a white-noise process. If the residuals are correlated, return to Step 1 and re-estimate the model in the form

$$\Delta y_t = I_t \rho_1 (y_{t-1} - \tau) + (1 - I_t)\rho_2 (y_{t-1} - \tau) + \sum_{i=1}^{p} \alpha_i \Delta y_{t-i} + \varepsilon_t \quad (7.33)$$

In working with this specification, it is possible to use diagnostic checks of the residuals and/or the various model selection criteria to determine the lag length. In principle, it is possible to allow the lagged changes to appear asymmetrically. However, this particular extension has not been thoroughly explored in the literature.

An Example

Enders and Granger use quarterly values of the 10-year government securities (r_{Lt}) and the federal funds rate (r_{St}) over the period 1958Q1 through 1994Q1. You can find the data used in the study in the file labeled GRANGER.XLS. The issue is to determine how to model the relationship between the two interest rates. First form the interest rate spread as $s_t = r_{Lt} - r_{St}$. After a bit of experimentation, the most appropriate equation for the Dickey–Fuller test is

$$\underset{(1.52)}{\Delta s_t = 0.120} \underset{(-3.56)}{- 0.156 s_{t-1}} + \underset{(1.94)}{0.162 \Delta s_{t-1}} + \varepsilon_t$$

AIC = 669.79 SBC = 678.68

The coefficient on s_{t-1} has a t-statistic of -3.56; hence, the null hypothesis of a unit root can be soundly rejected. Since the point of this section is to illustrate the test for threshold adjustment, we can pretend that the results of the Dickey–Fuller test are ambiguous. Nevertheless, diagnostic checking reveals that the equation is inadequate. For example, the RESET with $H = 3$ and $H = 4$ has *prob*-values of 0.0016 and 0.00009, respectively. Hence, there is substantial evidence of neglected nonlinearity.

Next, estimate a TAR model in the form of (7.30) and (7.31). The value of τ yielding the best fit is -0.27, so that the resulting TAR model of the spread is

$$\Delta s_t = \underset{(-1.59)}{-0.066 I_t (s_{t-1} + 0.27)} - \underset{(-3.67)}{0.286(1 - I_t)(s_{t-1} + 0.27)} + \underset{(2.07)}{0.172 \Delta s_{t-1}} + \varepsilon_t$$

AIC = 669.12 SBC = 680.97

For Step 2, we can estimate an M-TAR model by replacing (7.31) with (7.32). The value of τ yielding the best fit is 1.64, so that the resulting M-TAR model of the spread is

$$\Delta s_t = \underset{(-4.75)}{-0.299 I_t (s_{t-1} + 1.64)} - \underset{(-0.145)}{0.007(1 - I_t)(s_{t-1} - 1.64)} + \underset{(1.183)}{0.016 \Delta s_{t-1}} + \varepsilon_t$$

AIC = 662.55 SBC = 674.40

Notice that the AIC and the SBC both select the M-TAR model even though it has two coefficients that are statistically insignificant. You might want to experiment and estimate the model without these two extraneous coefficients. The F-statistic for the null hypothesis that $\rho_1 = \rho_2 = 0$ is 11.44. If we compare this to the critical values for Φ_M, we can reject the null hypothesis of no attractor. As such, we can test whether the adjustment is symmetric or asymmetric. The F-statistic for the null hypothesis $\rho_1 = \rho_2$ is 12.24 with a *prob*-value of 0.0006. Hence, we can conclude that the M-TAR best captures the adjustment process of the interest rate spread. The point estimates suggest that the equilibrium value of the spread is 1.64. When the spread is increasing (i.e., when $\Delta s_{t-1} > 0$), the speed of adjustment is fairly rapid. However, when the spread is decreasing (so that the long-term rate is falling relative to the referral funds rate), the adjustment of -0.007 is almost nonexistent. This is in contrast to the linear model; the linear model suggests that the speed of adjustment is -0.158 regardless of whether the spread is increasing or decreasing. Moreover, the linear specification suggests that the long-run equilibrium value of the spread is zero since the intercept has a t-statistic of 1.52.

NONLINEAR ERROR-CORRECTION If you experiment with the data set, you will find that both r_{Lt} and r_{St} act as $I(1)$ processes. Since there is a linear combination of these two $I(1)$ variables that is stationary, the Granger representation theorem indicates that there is an error-correction model. However, there is nothing requiring that the dynamic adjustment mechanism must be linear. Instead, it seems plausible that the error-correction model has the M-TAR form

$$\begin{aligned}\Delta r_{Lt} = {} & \underset{(-0.766)}{-0.03} I_t(s_{t-1} - 1.64) \underset{(-2.11)}{- 0.07}(1 - I_t)(s_{t-1} - 1.64) \\ & + A_{11}(L)\Delta r_{Lt-1} + A_{12}(L)\Delta r_{St-1} + \varepsilon_{1t} \\ & \quad F_{11} = 0.087 \qquad F_{12} = 0.521\end{aligned}$$

$$\begin{aligned}\Delta r_{St} = {} & \underset{(2.67)}{0.21} I_t(s_{t-1} - 1.64) \underset{(-0.67)}{- 0.04}(1 - I_t)(s_{t-1} - 1.64) \\ & + A_{21}(L)\Delta r_{Lt-1} + A_{22}(L)\Delta r_{St-1} + \varepsilon_{2t} \\ & \quad F_{21} = 0.001 \qquad F_{22} = 0.844\end{aligned}$$

where t-statistics are in parentheses, two lags of each variable are used in each equation, F_{ij} is the *prob*-value that all coefficients in the polynomial $A_{ij}(L) = 0$, and I_t is the M-TAR indicator given by (7.32).

The t-statistics suggest an interesting adjustment process toward the long-run equilibrium. Increases in the spread tend to be accompanied by changes in the federal funds rate while decreases are accompanied by changes in the 10-year rate. When the spread is increasing (i.e., if $\Delta s_{t-1} > 0$), we would expect the fed funds rate to increase by 21 percent of the discrepancy between s_{t-1} and the long-run value of 1.64. When the spread is decreasing, the long-term rate declines by 7 percent of the discrepancy.

The linear error-correction model tells a very different story. If we use the type of linear error-correction model used in Chapter 6, we obtain

$$\begin{aligned}\Delta r_{Lt} &= -0.114\hat{e}_{t-1} + A_{11}(L)\Delta r_{Lt-1} + A_{12}(L)\Delta r_{St-1} + \varepsilon_{1t}\\ &\quad (-3.30) \qquad F_{11} = 0.062 \qquad F_{12} = 0.288\end{aligned}$$

$$\begin{aligned}\Delta rs_t &= -0.002\hat{e}_{t-1} + A_{21}(L)\Delta r_{Lt-1} + A_{22}(L)\Delta r_{St-1} + \varepsilon_{2t}\\ &\quad (-0.04) \qquad F_{21} = 0.000 \qquad F_{22} = 0.333\end{aligned}$$

where $\hat{e}_{t-1}$ is the residual from the regression of r_{Lt} on a constant and r_{st}. Hence, $\hat{e}_{t-1}$ is the estimated deviation from the long-run relationship obtained by using the Engle–Granger technique.

In contrast to the threshold model, the linear model implies that only the 10-year interest rate responds to the discrepancy from the long-run equilibrium.

12. SUMMARY AND CONCLUSIONS

There are many important economic variables that exhibit nonlinear behavior. The difficulty is to properly capture the form of the nonlinearity. Once you abandon the linear framework, you must address the specification problem. As surveyed in this chapter, there are many nonlinear models and there is no clear way to decide which nonlinear specification is the best. The issue is important since the use of an incorrect nonlinear specification may be worse than ignoring the nonlinearity. Moreover, a linear model can always be viewed as a local approximation of a nonlinear process. There are some standard recommendations for estimating a nonlinear process. The most important is to use a specific-to-general modeling strategy. In particular:

1. Always start by plotting your data. Visual inspection of the data can help you detect the nature of the nonlinearity. Moreover, an outlier or a structural break will appear as a departure from the standard linear model with constant coefficients. You can save yourself substantial modeling time if you inspect the data for these features.
2. Fit the series of interest using the best linear model possible. For example, you might fit $\{y_t\}$ as an ARMA process using the Box–Jenkins methodology. The coefficients should be well estimated and the residuals should show no evidence of any serial correlation.
3. There are a number of tests designed to detect nonlinear behavior. The McLeod–Li, RESET, and various Lagrange multiplier tests can be used to detect nonlinear behavior. A Lagrange multiplier test has a specific nonlinear model as its alternative hypothesis. You can also test for coefficient stability using the methods discussed in Chapter 2. Nevertheless, even a battery of such tests is not able to reveal the precise nature of the nonlinearity.
4. If nonlinearity is detected, you have to decide on the appropriate form of the nonlinear specification. There is no substitute for an underlying theoretical model of the adjustment process. For example, if your model suggests that prices increase more readily than they decrease, some form of threshold model is likely to be the most appropriate.

5. The estimated nonlinear model(s) should fit the data better than the linear specification, and all coefficients should be statistically significant. In most instances, you will search over a number of plausible specifications. As such, the individual t-statistics and F-statistics are likely to be misleading. After all, you are examining the t-statistic on the best-fitting specification. If you examine 10 different specifications, on average, you should find one that is significant at the 10 percent level. Because overfitting is a distinct possibility, many researchers would use the parsimonious SBC as a measure of fit. Moreover, traditional t-tests and F-tests may not be appropriate in some nonlinear settings. Hansen (1997) considered the issue of inference in TAR models.
6. The generalized impulse response function can help you detect whether the nonlinear model is plausible. A useful diagnostic check is to use a Granger–Newbold or Diebold–Mariano test (see Chapter 2) to check the out-of-sample forecasting performance of the various models.

The nonlinear models discussed in this chapter were used to estimate a series $\{y_t\}$. However, it is possible to apply nonlinear models to the equation for the conditional variance. For example, the TARCH model discussed in Chapter 3 is an example of a nonlinearity applied to the equation for the conditional variance. Hamilton and Susmel (1994) showed how to apply the Markov switching model to the conditional variance of a time series. Higgens and Bera (1992) developed a nonlinear ARCH (NARCH) model that posits a "constant elasticity of substitution" functional form for the model of the conditional variance.

In addition, a large body of literature is growing concerning the presence of unit roots and cointegration in the presence of nonlinearities. For example, Granger, Inoue, and Morin (1997) developed some of the issues in terms of a nonlinear error-correction model. Enders and Siklos (2001) extended the TAR unit root test discussed in Section 11 to allow for a cointegrated system. Tsay (1998) developed a test that can be used to detect threshold cointegration. The appropriate use of the test is illustrated using spot and futures prices. Caner and Hansen (2001) developed a maximum-likelihood method to test for a threshold unit root, and Hansen and Seo (2002) extended the analysis to a cointegrated system. Kapetanios, Shin, and Snell (2003) developed a simple way to test for a unit root against the alternative of an ESTAR model. Another way to think about nonlinear models is in the frequency domain. Granger and Joyeux (1980) provided an introduction to the notion that a series may be integrated of some order other than an integer. Such nonlinear processes may be mean-reverting yet can behave similarly to unit root processes.

QUESTIONS AND EXERCISES

1. Let p_A and p_M denote the price of cotton in Alabama and Mississippi, respectively. The price gap, or discrepancy, is $p_A - p_M$. For each part, present a nonlinear model that captures the dynamic adjustment mechanism given in the brief narrative.
   a. A large price gap (in absolute value) tends to be eliminated quickly as compared to a small gap.
   b. The price gap is closed more quickly if it is positive than if it is negative.
   c. It costs ten cents to transport a bale of cotton between Alabama and Mississippi. Hence, a price discrepancy of less than 10 cents will not be eliminated by arbitrage.

However, 50 percent of any price gap exceeding 10 cents will be eliminated within a period.

d. The value of p_A, but not the value of p_M, responds to a price gap.

2. Draw the phase diagram for each of the following processes:

a. The GAR model: $y_t = 1.5y_{t-1} - 0.5y_{t-1}^3 + \varepsilon_t$

b. The TAR model: $y_t = 1 + 0.5y_{t-1} + \varepsilon_t$ if $y_{t-1} > 2$ and $y_t = 0.5 + 0.75y_{t-1} + \varepsilon_t$ if $y_{t-1} \leq 2$

c. The TAR model: $y_t = 1 + 0.5y_{t-1} + \varepsilon_t$ if $y_{t-1} > 0$ and $y_t = -1 + 0.5y_{t-1} + \varepsilon_t$ if $y_{t-1} \leq 0$

Notice that this model is discontinuous at the threshold. Show that $y_{t-1} = +2$ and $y_{t-1} = -2$ are both stable equilibrium values for the skeleton.

d. The TAR model: $y_t = -1 + 0.5y_{t-1} + \varepsilon_t$ if $y_{t-1} > 0$ and $y_t = +1 + 0.5y_{t-1} + \varepsilon_t$ if $y_{t-1} \leq 0$

Show that there is no stable equilibrium for the skeleton.

e. The LSTAR model: $y_t = 0.75y_{t-1} + 0.25y_{t-1}/[1 + \exp(-y_{t-1})] + \varepsilon_t$

f. The ESTAR model: $y_t = 0.75y_{t-1} + 0.25y_{t-1}[1 - \exp(-y_{t-1}^2)] + \varepsilon_t$

3. In the Markov switching model, let p_1 denote the *unconditional* probability that the system is in regime one, and let p_2 denote the *unconditional* probability that the system is in regime two. As in the text, let p_{ii} denote the probability that the system remains in regime i. Prove the assertion

$$p_1 = (1 - p_{22})/(2 - p_{11} - p_{22})$$
$$p_2 = (1 - p_{11})/(2 - p_{11} - p_{22})$$

4. The file labeled LSTAR.XLS contains the 250 realizations of the series used in Section 9.

a. Verify that (7.25) represents the best-fitting linear model for this process.

b. Perform the RESET using $H = 3$. How does this compare to the result using $H = 4$?

c. If your software package can perform the BDS test, determine whether the residuals from (7.25) pass the BDS test for white noise.

d. Perform the LM tests for LSTAR adjustment and for ESTAR adjustment.

e. If you estimate the process as a GAR process, you should find

$$\underset{}{y_t} = \underset{(8.97)}{2.03} + \underset{(6.97)}{0.389y_{t-1}} + \underset{(3.48)}{0.201y_{t-2}} - \underset{(-10.57)}{0.147y_{t-1}^2} + \varepsilon_t$$

All of the t-statistics imply that the coefficients are well estimated. Show that all of the residual autocorrelations are less than 0.1 in absolute value.

How would you determine whether the GAR model or the LSTAR model is preferable?

5. The file INDPROD.XLS contains the monthly values of U.S. industrial production (ip_t) over the period January 1972 through October 2008. Form the growth rate of the series as $y_t = \Delta\ln(ip_t)$:

a. Show that the AR(3) model is a plausible representation of the series.

b. Verify that the RESET test does not detect any nonlinearity. You should find that for $H = 4$, the sample F-statistic is only 0.534.

c. Use the residuals from the AR(3) model to perform the test for LSTAR behavior. Specifically, estimate an equation in the form of (7.21) and test the restriction that the coefficients on the variables $(y_{t-1})(y_{t-1})^3$, $(y_{t-2})(y_{t-1})^3$, and $(y_{t-3})(y_{t-1})^3$ are jointly equal to zero. You should find that the sample value of F is 10.47. With three numerator degrees of freedom and 423 denominator degrees of freedom, do not reject the null hypothesis of LSTAR behavior.

d. It is quite difficult to estimate the series as an LSTAR process. If you fix c at -0.0025, and estimate the remaining parameters using nonlinear least squares, you should obtain results similar to

$$\begin{aligned} y_t = {} & 0.0009 + 0.209y_{t-1} + 0.184y_{t-2} + 0.144y_{t-3} + \\ & (0.032 + 1.92y_{t-1} - 0.958y_{t-2} - 0.143y_{t-3}) / \\ & (1 + 1000^* \exp(-409.39(y_{t-1} + 0.0025))) + \varepsilon_t \end{aligned}$$

Notice that the term $\exp(-409.39(y_{t-1} + 0.0025))$ was multiplied by 1,000 in order to achieve convergence. An alternative would have been to follow Teräsverta's (1994) recommendation and multiply by $144.05 \approx \text{var}(y_t)^{-1/2}$. Experimentation indicated that several software packages converged more readily using the somewhat arbitrary value of 1,000 than using the value 144.05.

e. Experiment with other values of c and the delay parameter.

f. Construct the F-statistic for threshold behavior using y_{t-1} as the threshold variable. You should find that $F = 2.16$. If your software package can use Hansen's (1997) bootstrapping method, verify that the null hypothesis of no-threshold behavior is not rejected (i.e., conclude that the series is not a TAR process).

6. The file GRANGER.XLS contains the interest rate series used to estimate the TAR and M-TAR models in Section 11.

a. Estimate the TAR and M-TAR models reported in Section 11.

b. Estimate the M-TAR model without the two insignificant coefficients.

c. Calculate the AIC and the SBC for the TAR model and the M-TAR model without the insignificant coefficients. In your calculations, be sure to adjust the two model-selection criteria to allow for the fact that you estimated the threshold.

d. Calculate the multivariate AIC for the linear error-correction model. How does this value compare to the multivariate AIC for the nonlinear error-correction model?

7. Consider the linear process $y_t = 0.75y_{t-1} + \varepsilon_t$. Given $y_t = 1$, find E_ty_{t+1}, E_ty_{t+2}, and E_ty_{t+3}.

a. Now consider the GAR process $y_t = 0.75y_{t-1} - 0.25y_{t-1}^2 + \varepsilon_t$. Given $y_t = 1$, find E_ty_{t+1}. Can you find E_ty_{t+2} and E_ty_{t+3}? [*Hint*: $(E_ty_{t+1})^2 \neq E_t(y_{t+1}^2)$]

b. Use your answer to part a to explain why it is difficult to perform multi-step-ahead forecasting with a nonlinear model.

8. Chapters 1.4 and 1.5 of the *Programming Manual* that accompanies this text contain a discussion of nonlinear least squares and maximum-likelihood estimation. If you have not already done so, download the manual from the Wiley Web site. Also download the data set MONEYDEM.XLS and the programs.

a. Program 1.4 uses nonlinear least squares to estimate the logarithmic change in M3 ($dlm3_t$) as an LSTAR model. Use the program to reproduce the results shown in the *Programming Manual*.

b. Program 1.4 estimates the same LSTAR model using maximum-likelihood methods. Compare the results of the two estimations. Why are there slight discrepancies between the two estimation methods?

c. Use your software package to estimate the logarithmic change in M3 as a bilinear process. You should find that the best-fitting model is

$$dlm3_t = 0.977dlm3_{t-1} + \varepsilon_t - 0.273\varepsilon_{t-1} + 9.12dlm3_{t-1}\varepsilon_{t-1}$$

9. Section 4.3 of the *Programming Manual* contains a discussion of the appropriate way to program a TAR model. If you have not already done so, download the manual from the Wiley Web site. As shown in Program 4.3, use the data in MONEYDEM.XLS to construct the logarithmic change in real GDP ($dlrgdp_t$).

a. Estimate $dlrgdp_t$ as an AR(2) process.

b. Estimate the series as a TAR process such that $d = 1$ and $\tau = 0$. Your answer should be consistent with that reported in the manual.

c. Use Chan's method to find the consistent estimate of the threshold. You should find that $\tau = 0.01724$. Create the variables I_t, $I_t y_{t-1}$, $I_t y_{t-2}$, $(1 - I_t)$, $(1 - I_t)y_{t-1}$, and $(1 - I_t)y_{t-2}$. You should find

$$\begin{aligned} y_t = {} & (0.022 - 0.444y_{t-1} - 0.037y_{t-2})I_t \\ & + (1 - I_t)(0.005 + 0.0238y_{t-1} + 0.0142y_{t-3}) + \varepsilon_t \end{aligned}$$

Eliminate all insignificant coefficients and use the AIC to compare you answer to the model found in part a.

d. As shown in the *Programming Manual,* there might be a second threshold at 0.0025. Use the two thresholds to estimate a three-regime TAR model.

10. Section 5.4 of Chapter 4 in the *Programming Manual* shows you how to generate the critical values used in Enders and Granger (1998).

a. Reproduce the results for a sample size of 100, but use only 2,000 Monte Carlo replications. How do your answers compare to those reported in the *Programming Manual*?

b. How would you obtain critical values for other sample sizes?

11. The file labeled SIM_TAR.XLS contains the 200 observations used to construct Figure 7.3.

a. Show that it is reasonable to estimate the series as $y_t = -0.162 + 0.529y_{t-1} + \varepsilon_t$.

b. Verify that the RESET does not indicate any nonlinearities. In particular, show that the RESET (using the second, third, and fourth powers of the fitted values) yields an $F = 1.421$.

c. Plot the residual sum of squares for each potential threshold value. What is the most likely value of the threshold(s)?

d. Estimate the model $y_t = (0.057 + 0.260y_{t-1})I_t + (-0.464 + 0.402y_{t-1})(1 - I_t)$ where $I_t = 1$ if $y_{t-1} > -0.4012$ and zero otherwise.

e. Show that the performance of the model is improved if the intercepts are eliminated.

ENDNOTES

1. In these circumstances, the log likelihood for observation t is $-(1/2)\ln(2\pi) - (1/2)\ln\sigma^2 - (1/2\sigma^2)\varepsilon_t^2$. Hence, the partial derivatives of the log likelihood with respect to α do not depend on the variance.

2. If a parameter is not identified under the null hypothesis, its partial derivative has no effect on the log likelihood function evaluated under the null hypothesis. Hence, in this circumstance, any type of likelihood test fails.
3. Recall that the third-order Taylor series expansion of $y = f(x)$ around the point $x = x_0$ is $y = f(x_0) + f'(x_0)(x - x_0) + (1/2)f''(x_0)(x - x_0)^2 + (1/6)f'''(x_0)(x - x_0)^3$ where ' denotes differentiation.
4. Hansen (1999) and Enders, Falk, and Siklos (2007) consider the issue of inference on the coefficients (and the threshold τ) in a TAR model. Although the confidence intervals obtained from a conventional t-distribution are only approximations of the actual distributions, often the distributions obtained by bootstrapping do not perform better.
5. Petrucelli and Woolford (1984) showed that a weaker set of sufficient conditions for the stationarity of $\{y_t\}$ is $\rho_1 < 0$, $\rho_2 < 0$, and $(1 + \rho_1)(1 + \rho_2) < 1$.

STATISTICAL TABLES

Table A **Empirical Cumulative Distribution of τ**

	Significance Level			
Sample Size T	**0.01**	**0.025**	**0.05**	**0.10**
The τ Statistic: No Constant or Time Trend ($a_0 = a_2 = 0$)				
25	–2.66	–2.26	–1.95	–1.60
50	–2.62	–2.25	–1.95	–1.61
100	–2.60	–2.24	–1.95	–1.61
250	–2.58	–2.23	–1.95	–1.62
300	–2.58	–2.23	–1.95	–1.62
∞	–2.58	–2.23	–1.95	–1.62
The τ_μ Statistic: Constant but No Time Trend ($a_2 = 0$)				
25	–3.75	–3.33	–3.00	–2.62
50	–3.58	–3.22	–2.93	–2.60
100	–3.51	–3.17	–2.89	–2.58
250	–3.46	–3.14	–2.88	–2.57
500	–3.44	–3.13	–2.87	–2.57
∞	–3.43	–3.12	–2.86	–2.57
The τ_τ Statistic: Constant + Time Trend				
25	–4.38	–3.95	–3.60	–3.24
50	–4.15	–3.80	–3.50	–3.18
100	–4.04	–3.73	–3.45	–3.15
250	–3.99	–3.69	–3.43	–3.13
500	–3.98	–3.68	–3.42	–3.13
∞	–3.96	–3.66	–3.41	–3.12

Source: The table is reproduced from Fuller (1976). It is used by permission of John Wiley & Sons, Inc.

Table B **Empirical Distribution of ϕ**

Sample Size T	Significance Level 0.10	0.05	0.025	0.01
		ϕ_1		
25	4.12	5.18	6.30	7.88
50	3.94	4.86	5.80	7.06
100	3.86	4.71	5.57	6.70
250	3.81	4.63	5.45	6.52
500	3.79	4.61	5.41	6.47
∞	3.78	4.59	5.38	6.43
		ϕ_2		
25	4.67	5.68	6.75	8.21
50	4.31	5.13	5.94	7.02
100	4.16	4.88	5.59	6.50
250	4.07	4.75	5.40	6.22
500	4.05	4.71	5.35	6.15
∞	4.03	4.68	5.31	6.09
		ϕ_3		
25	5.91	7.24	8.65	10.61
50	5.61	6.73	7.81	9.31
100	5.47	6.49	7.44	8.73
250	5.39	6.34	7.25	8.43
500	5.36	6.30	7.20	8.34
∞	5.34	6.25	7.16	8.27

Source: Dickey, David, and Wayne A. Fuller (1981). Used by permission of the Econometric Society.

Table C **Critical Values for the Engle–Granger Cointegration Test**

T	1%	5%	10%	1%	5%	10%
	Two Variables			**Three Variables**		
50	–4.123	–3.461	–3.130	–4.592	–3.915	–3.578
100	–4.008	–3.398	–3.087	–4.441	–3.828	–3.514
200	–3.954	–3.368	–3.067	–4.368	–3.785	–3.483
500	–3.921	–3.350	–3.054	–4.326	–3.760	–3.464
	Four Variables			**Five Variables**		
50	–5.017	–4.324	–3.979	–5.416	–4.700	–4.348
100	–4.827	–4.210	–3.895	–5.184	–4.557	–4.240
200	–4.737	–4.154	–3.853	–5.070	–4.487	–4.186
500	–4.684	–4.122	–3.828	–5.003	–4.446	–4.154

The critical values are for cointegrating relations (with a constant in the cointegrating vector) estimated using the Engle–Granger methodology.

Source: Critical values are interpolated using the response surface in MacKinnon (1991).

Table D **Residual-Based Cointegration Test with $I(1)$ and $I(2)$ Variables**

		Intercept Only				**Linear Trend**			
m_1	T	$m_2 = 1$ *prob*-value		$m_2 = 2$ *prob*-value		$m_2 = 1$ *prob*-value		$m_2 = 2$ *prob*-value	
		0.01	0.05	0.01	0.05	0.01	0.05	0.01	0.05
0	50	–4.18	–3.51	–4.70	–4.02	–4.66	–4.01	–5.14	–4.45
	100	–4.09	–3.42	–4.51	–3.86	–4.55	–3.90	–4.93	–4.31
	250	–4.02	–3.38	–4.35	–3.80	–4.41	–3.83	–4.81	–4.20
1	50	–4.65	–3.93	–5.15	–4.40	–5.11	–4.42	–5.62	–4.89
	100	–4.51	–3.89	–4.85	–4.26	–4.85	–4.26	–5.23	–4.62
	250	–4.39	–3.80	–4.71	–4.18	–4.73	–4.19	–5.11	–4.50
2	50	–4.93	–4.30	–5.54	–4.77	–5.47	–4.74	–5.98	–5.17
	100	–4.81	–4.25	–5.29	–4.59	–5.21	–4.58	–5.59	–4.93
	250	–4.77	–4.16	–5.06	–4.49	–5.07	–4.51	–5.35	–4.80
3	50	–5.38	–4.71	–5.76	–5.08	–5.89	–5.13	–6.23	–5.48
	100	–5.20	–4.56	–5.58	–4.92	–5.52	–4.91	–5.97	–5.25
	250	–5.05	–4.48	–5.44	–4.83	–5.38	–4.78	–5.69	–5.07
4	50	–5.81	–5.09	–6.24	–5.48	–6.35	–5.47	–6.64	–5.82
	100	–5.58	–4.93	–5.88	–5.20	–5.86	–5.20	–6.09	–5.50
	250	–5.39	–4.28	–5.64	–5.07	–5.66	–5.08	–5.95	–5.34

Note: m_1 is the number of $I(1)$ variables and m_2 is the number of $I(2)$ variables on the right-hand side of the multicointegrating relationship.

Source: The critical values for the intercept-only case are from Haldrup (1994) and critical values for the linear trend are from Engsted, Gonzalo, and Haldrup (1997).

Table E **Empirical Distributions of the λ_{max} and λ_{trace} Statistics**

Significance Level	10%	5%	2.5%	1%	10%	5%	2.5%	1%
	λ_{max} and λ_{trace} Statistics Without Any Deterministic Regressors							
$n-r$	λ_{max}				λ_{trace}			
1	2.86	3.84	4.93	6.51	2.86	3.84	4.93	6.51
2	9.52	11.44	13.27	15.69	10.47	12.53	14.43	16.31
3	15.59	17.89	20.02	22.99	21.63	24.31	26.64	29.75
4	21.56	23.80	26.14	28.82	36.58	39.89	42.30	45.58
5	27.62	30.04	32.51	35.17	54.44	59.46	62.91	66.52
	λ_{max} and λ_{trace} Statistics with Drift							
$n-r$	λ_{max}				λ_{trace}			
1	2.69	3.76	4.95	6.65	2.69	3.76	4.95	6.65
2	12.07	14.07	16.05	18.63	13.33	15.41	17.52	20.04
3	18.60	20.97	23.09	25.52	26.79	29.68	32.56	35.65
4	24.73	27.07	28.98	32.24	43.95	47.21	50.35	54.46
5	30.90	33.46	35.71	38.77	64.84	68.52	71.80	76.07
	λ_{max} and λ_{trace} Statistics with a Constant in the Cointegrating Vector							
	λ_{max}				λ_{trace}			
1	7.52	9.24	10.80	12.97	7.52	9.24	10.80	12.95
2	13.75	15.67	17.63	20.20	17.85	19.96	22.05	24.60
3	19.77	22.00	24.07	26.81	32.00	34.91	37.61	41.07
4	25.56	28.14	30.32	33.24	49.65	53.12	56.06	60.16
5	31.66	34.40	36.90	39.79	71.86	76.07	80.06	84.45

Source: Osterwald–Lenum (1992). The tables reported here are reproduced with permission from Blackwell Publishers.

Table F Critical Values for $\beta_1 = 0$ in the Error-Correction Model

k		$T^a = 50$	$T^a = 100$	$T^a = 200$	$T^a = 500$
		No Intercept or Trend ($d = 0$)			
2	1%	−3.309	−3.259	−3.235	−3.220
	5%	−2.625	−2.609	−2.602	−2.597
	10%	−2.273	−2.268	−2.266	−2.265
3	1%	−3.746	−3.683	−3.652	−3.633
	5%	−3.047	−3.026	−3.016	−3.009
	10%	−2.685	−2.680	−2.677	−2.675
4	1%	−4.088	−4.015	−3.979	−3.957
	5%	−3.370	−3.348	−3.337	−3.331
	10%	−3.000	−2.997	−2.995	−2.994
		Intercept but No Trend ($d = 1$)			
2	1%	−3.954	−3.874	−3.834	−3.811
	5%	−3.279	−3.247	−3.231	−3.221
	10%	−2.939	−2.924	−2.916	−2.911
3	1%	−4.268	−4.181	−4.138	−4.112
	5%	−3.571	−3.538	−3.522	−3.512
	10%	−3.216	−3.205	−3.199	−3.195
4	1%	−4.537	−4.446	−4.401	−4.374
	5%	−3.819	−3.789	−3.774	−3.765
	10%	−3.453	−3.447	−3.444	−3.442
		Intercept and Trend ($d = 2$)			
2	1%	−4.451	−4.350	−4.299	−4.269
	5%	−3.778	−3.733	−3.710	−3.696
	10%	−3.440	−3.416	−3.405	−3.398
3	1%	−4.712	−4.605	−4.552	−4.519
	5%	−4.014	−3.971	−3.949	−3.935
	10%	−3.662	−3.643	−3.634	−3.629
4	1%	−4.940	−4.831	−4.776	−4.743
	5%	−4.221	−4.182	−4.162	−4.150
	10%	−3.857	−3.846	−3.840	−3.837

Note: T^a is the adjusted sample size equal to $T - (2k - 1) - d$ where T is the usable sample size, d is the number of deterministic regressors, and k is the number of $I(1)$ variables in the model. The critical values are calculated using equation (26) in Ericsson and MacKinnon (2002).

Table G **Critical Values for Threshold Unit Roots**

Panel (a): Consistent Estimate of the Threshold Using the TAR Model

T	No Lagged Changes				One Lagged Change				Four Lagged Changes			
	90%	95%	97.5%	99%	90%	95%	97.5%	99%	90%	95%	97.5%	99%
50	5.15	6.19	7.25	8.64	5.55	6.62	7.66	9.10	5.49	6.55	7.59	9.00
100	5.08	6.06	6.93	8.19	5.39	6.34	7.30	8.54	5.38	6.32	7.29	8.56
250	5.11	6.03	6.88	8.04	5.26	6.12	6.99	8.14	5.36	6.29	7.15	8.35

Panel (b): Consistent Estimates of the Threshold Using the M-TAR Model

T	No Lagged Changes				One Lagged Change				Four Lagged Changes			
	90%	95%	97.5%	99%	90%	95%	97.5%	99%	90%	95%	97.5%	99%
50	5.02	6.05	7.09	8.59	4.98	6.07	7.15	8.56	4.93	5.96	7.01	8.48
100	4.81	5.77	6.73	7.99	4.77	5.71	6.56	7.90	4.74	5.70	6.67	7.97
250	4.70	5.64	6.51	7.64	4.64	5.54	6.40	7.56	4.64	5.54	6.39	7.61

Panel (c): Known Threshold Value in the M-TAR Model

T	No Lagged Changes				One Lagged Change				Four Lagged Changes			
	90%	95%	97.5%	99%	90%	95%	97.5%	99%	90%	95%	97.5%	99%
50	4.21	5.19	6.15	7.55	4.12	5.11	6.05	7.25	3.82	4.73	5.65	6.84
100	4.11	5.04	5.96	7.10	4.08	4.97	5.87	7.06	3.81	4.72	5.63	6.83
250	4.08	4.97	5.83	6.91	4.05	4.93	5.78	6.83	3.69	4.71	5.63	6.78

REFERENCES

Amsler, Christine, and Junsoo Lee. "An LM Test for a Unit Root in the Presence of Structural Change." *Econometric Theory* 11 (June 1995), 359–68.

Anderson, L., and J. Jordan. "Monetary and Fiscal Actions: A Test of Their Relative Importance in Economic Stabilization." *Federal Reserve Bank of St. Louis Review* (Nov. 1968), 11–24.

Andrews, Donald Werner Ploberger. "Optimal Tests When a Nuisance Parameter Is Present Only under the Alternative." *Econometrica* (1994), 1383–1414.

Ashley, Rick. "Statistically Significant Postsample Forecasting Improvements: How Much Out-of-Sample Data Is Likely Necessary?" *International Journal of Forecasting* 19 (2003), 229–39.

Baba, Yoshihisa, David F. Hendry, and Ross M. Starr. "The Demand for M1 in the U.S.A., 1960–1988." *Review of Economic Studies* 59 (Jan. 1992), 25–61.

Bai, J., and Pierre Perron. "Estimating and Testing Linear Models with Multiple Structural Changes." *Econometrica* 66 (1998), 47–78.

———. "Computation and Analysis of Multiple Structural Change Models." *Journal of Applied Econometrics* 18 (2003), 1–22.

Balke, Nathan S., and Thomas B. Fomby. "Threshold Cointegration." *International Economic Review* 38 (1997), 627–43.

Baltagi, Badi. "Nonstationary Panels, Panel Cointegration, and Dynamic Panels." *Advances in Econometrics* 15. Amsterdam: Elsevier Science (2000).

Bell, W., and S. Hilmer. "Issues Involved with the Seasonal Adjustment of Economic Time Series." *Journal of Business and Economic Statistics* 2 (1984), 291–320.

Ben-David, Dan, and David Papell. "The Great Wars, the Great Crash, and Steady State Growth: Some New Evidence About an Old Stylized Fact." *Journal of Monetary Economics* 36 (1995), 453–75.

Bernanke, Ben. "Alternative Explanations of Money-Income Correlation." *Carnegie-Rochester Conference Series on Public Policy* 25 (1986), 49–100.

Beveridge, Stephen, and Charles Nelson. "A New Approach to Decomposition of Economic Time Series into Permanent and Transitory Components with Particular Attention to Measurement of the Business Cycle." *Journal of Monetary Economics* 7 (March 1981), 151–74.

Blanchard, Oliver, and Danny Quah. "The Dynamic Effects of Aggregate Demand and Supply Disturbances." *American Economic Review* 79 (Sept. 1989), 655–673.

Bollerslev, Tim. "Generalized Autoregressive Conditional Heteroscedasticity." *Journal of Econometrics* 31 (1986), 307–27.

Box, George, and D. Cox. "An Analysis of Transformations." *Journal of the Royal Statistical Society,* Series B.26 (1964), 211–52.

Box, George, and Gwilym Jenkins. *Time Series Analysis, Forecasting, and Control.* San Francisco, Calif.: Holden Day, 1976.

Box, George, and D. Pierce. "Distribution of Autocorrelations in Autoregressive Moving Average Time Series Models." *Journal of the American Statistical Association* 65 (1970), 1509–26.

Brock, William, Davis Dechert, Jose Scheinkman, and Blake LeBaron. "A Test for Independence Based upon the Correlation Dimension." *Econometric Reviews* 15 (1997), 197–235.

Bureau of the Census. *X-11 Information for the User, U.S. Department of Commerce.* Washington, D.C.: U.S. Government Printing Office, 1969.

Cagan, Phillip. "The Monetary Dynamics of Hyperinflation." Milton Friedman, ed., *Studies in the Quantity Theory of Money.* Chicago, Ill.: University of Chicago Press, 1956, pp. 25–120.

Campbell, John Y., and Pierre Perron. "Pitfalls and Opportunities: What Macroeconomists Should Know About Unit Roots." Technical Working Paper 100, NBER Working Paper Series, April 1991.

Caner, Mehmet, and Bruce Hansen. "Threshold Autoregression with a Unit Root." *Econometrica* 69 (2001), 1555–96.

Carrasco, Marine. "Misspecified Structural Change, Threshold, and Markov-switching Models." *Journal of Econometrics* 109 (2002), 239–73.

Chan, K. S. "Consistency and Limiting Distribution of the Least Squares Estimator of a Threshold Autoregressive Model." *The Annals of Statistics* 21 (1993), 520–33.

Clarida, Richard H., and Jordi Gali, "Sources of Real Exchange Rate Fluctuations: How Important Are Nominal Shocks?" *Carnegie-Rochester Conference Series on Public Policy* 41 (1994), 1–56.

Clark, Todd, and Michael McCracken. "Tests of Equal Forecast Accuracy and Encompassing for Nested Models." *Journal of Econometrics* 105 (2001), 85–110.

Clark, Todd, and Kenneth West. "Using Out-of-Sample Mean Square Prediction Errors to Test the Martingale Difference Hypothesis." *Journal of Econometrics* 135 (2006), 155–86.

———. "Approximately Normal Tests for Equal Predictive Accuracy in Nested Models." *Journal of Econometrics* 138 (2007), 291–311.

Clements, Michael P., and David Hendry. *Forecasting Economic Time Series.* Cambridge, Mass.: MIT Press, 1998.

Clements, Michael P., and Hans-Martin Krolzig. "A Comparison of the Forecast Performance of Markov-switching and Threshold Autoregressive Models of US GNP." *The Econometrics Journal* 1 (1998), 47–75.

Cover, James, Walter Enders, and C. James Hueng. "Using the Aggregate Demand–Aggregate Supply Model to Identify Structural Demand-Side and Supply-Side Shocks: Results Using a Bivariate VAR." *Journal of Money, Credit and Banking* (April 2006), 777–90.

Cutler, Harvey, Stephen Davis, and Martin Smith. "The Demand for Nominal and Real Money Balances in a Large Macroeconomic System." *Southern Economic Journal* 63 (April 1997), 947–61.

Davidson, J., D. Hendry, F. Srba, and S. Yeo. "Econometric Modeling of the Aggregate Time-Series Relationship Between Consumers' Expenditure and Income in the United Kingdom." *The Economic Journal* 88 (1978), 661–92.

Dickey, David, W. Bell, and R. Miller. "Unit Roots in Time Series Models: Tests and Implications." *American Statistician* 40 (1986), 12–26.

Dickey, David, and Wayne A. Fuller. "Distribution of the Estimates for Autoregressive Time Series with a Unit Root." *Journal of the American Statistical Association* 74 (June 1979), 427–31.

———. "Likelihood Ratio Statistics for Autoregressive Time Series with a Unit Root." *Econometrica* 49 (July 1981), 1057–72.

Dickey, David, and S. Pantula. "Determining the Order of Differencing in Autoregressive Processes." *Journal of Business and Economic Statistics* 15 (1987), 455–61.

Diebold, Francis, and Roberto Mariano. "Comparing Predictive Accuracy." *Journal of Business and Economic Statistics* 13 (1995), 253–63.

Dimitrios, Thomakos, and John Guerard. "Naïve, ARIMA, Nonparametric, Transfer Function and VAR Models: A Comparison of Forecasting Performance." *International Journal of Forecasting* 20 (2004), 53–67.

Doan, Thomas. *RATS User's Manual.* Evanston, Ill.: Estima, 2009.

Doldado, Juan, Tim Jenkinson, and Simon Sosvilla-Rivero. "Cointegration and Unit Roots." *Journal of Economic Surveys* 4 (1990), 249–73.

Dornbusch, Rudiger. "Expectations and Exchange Rate Dynamics." *Journal of Political Economy* 84 (1976), 1161–76.

Edwards, Sebastian, and Raul Susmel. "Interest Rate Volatility and Contagion in Emerging Markets: Evidence from the 1990s." National Bureau of Economic Research Working Paper 7813, 2000.

Efron, Bradley. "Bootstrap Methods: Another Look at the Jackknife." *Annals of Statistics* 7 (1979), 1–26.

Efron, Bradley, and Robert Tibshirani. *An Introduction to the Bootstrap.* New York: Chapman and Hall, 1993.

Elliott, Graham, Thomas Rothenberg, and James Stock. "Efficient Tests for an Autoregressive Unit Root." *Econometrica* 64 (1996), 813–36.

Elliott, Graham, and Allan Timmermann. "Economic Forecasting." *Journal of Economic Literature* 46 (2008), 3–56.

Enders, Walter. "ARIMA and Cointegration Tests of Purchasing Power Parity." *Review of Economics and Statistics* 70 (Aug. 1988), 504–8.

Enders, Walter, Barry Falk, and Pierre Siklos. "A Threshold Model of Real U.S. GDP and the Problem of Constructing Confidence Intervals in TAR Models." *Studies in Nonlinear Dynamics and Econometrics* 11 (Sept. 2007), Article 4.

Enders, Walter, and C.W. J. Granger. "Unit-root Tests and Asymmetric Adjustment with an Example Using the Term Structure of Interest Rates." *Journal of Business and Economic Statistics* 16 (1998), 304–11.

Enders, Walter, and Stan Hurn. "Identifying Aggregate Demand and Supply Shocks in a Small Open Economy." *Oxford Economic Papers* 59 (July 2007), 411–29.

Enders, Walter, and Bong-Soo Lee. "Accounting for Real and Nominal Exchange Rate Movements in the Post-Bretton Woods Period." *Journal of International Money and Finance* 16 (1997), 233–54.

Enders, Walter, and Todd Sandler. "Causality Between Transnational Terrorism and Tourism: The Case of Spain." *Terrorism* 14 (Jan. 1991), 49–58.

———. "Transnational Terrorism 1968–2000: Thresholds, Persistence and Forecasts." *Southern Economic Journal* 71 (2005), 467–83.

Enders, Walter, Todd Sandler, and Jon Cauley. "Assessing the Impact of Terrorist-Thwarting Policies: An Intervention Time Series Approach." *Defense Economics* 2 (Dec. 1990), 1–18.

Enders, Walter, Todd Sandler, and Gerald F. Parise. "An Econometric Analysis of the Impact of Terrorism on Tourism." *Kyklos* 45 (1992), 531–54.

Enders, Walter, and Pierre Siklos. "Cointegration and Threshold Adjustment." *Journal of Business and Economic Statistics* 19 (2001), 166–76.

Engle, Robert F. "Dynamic Conditional Correlation—A Simple Class of Multivariate GARCH Models." *Journal of Business and Economic Statistics* 20 (2002), 339–50.

———. "Autoregressive Conditional Heteroscedasticity with Estimates of the Variance of United Kingdom Inflation." *Econometrica* 50 (July 1982), 987–1007.

Engle, Robert F., and Tim Bollerslev. "Modelling the Persistence of Conditional Variances." *Econometric Reviews* 5 (1986), 1–50.

Engle, Robert F., and Clive W. J. Granger. "Cointegration and Error-Correction: Representation, Estimation, and Testing." *Econometrica* 55 (March 1987), 251–76.

Engle, Robert F., David Hendry, and Jean-Francois Richard. "Exogeneity." *Econometrica* 51 (March 1983), 277–304.

Engle, Robert F., and David Kraft. "Multiperiod Forecast Error Variances of Inflation Based on the ARCH Model." A. Zellner, ed., *Applied Time Series Analysis of Economic Data*. Washington, D.C.: Bureau of the Census, 1983, pp. 293–302.

Engle, Robert F., David Lilien, and Russell Robins. "Estimating Time Varying Risk Premia in the Term Structure: The ARCH-M Model." *Econometrica* 55 (March 1987), 391–407.

Engsted, Tom, Jesus Gonzalo, and Neils Haldrup. "Testing for Multicointegration." *Economic Letters* 56 (1997), 259–66.

Ericsson, Neil, and James G. MacKinnon. "Distributions of Error Correction Tests for Cointegration." *Econometrics Journal* 5 (2002), 285–318.

Evans, G., and N. Savin. "Testing for Unit Roots: 1." *Econometrica* 49 (1981), 753–79.

Farmer, Roger E. *The Macroeconomics of Self-Fulfilling Prophecies*. Cambridge, Mass.: MIT Press, 1993.

Friedman, Benjamin, and Kenneth Kuttner. "Money, Income, Prices, and Interest Rates." *American Economic Review* 82 (June 1992), 472–92.

Galbraith, John W. "Credit Rationing and Threshold Effects in the Relation Between Money and Output." *Journal of Applied Econometrics* 11 (1996), 419–29.

Granger, Clive, T. Inoue, and N. Morin. "Nonlinear Stochastic Trends." *Journal of Econometrics* 81 (1997), 65–92.

Granger, Clive, and R. Joyeux. "An Introduction to Long Memory Time Series Models and Fractional Differencing." *Journal of Time Series Analysis* 1 (1980), 15–29.

Granger, Clive, and Tae-Hwy Lee. "Multicointegration." *Advances in Econometrics* 8 (1990), 77–84.

Granger, Clive, and Paul Newbold. "Spurious Regressions in Econometrics." *Journal of Econometrics* 2 (1974), 111–20.

———. "Forecasting Transformed Series." *Journal of the Royal Statistical Society* B 38 (1976), 189–203.

Granger, Clive, and Timo Teräsvirta. *Modelling Nonlinear Economic Relationships*. Oxford: Oxford University Press, 1993.

Gregory, Allen, and Bruce Hansen. *Residual Based Tests for Cointegration in Models with Regime Shifts*. Kingston: Queens University, 1992.

Haggan, V., and T. Ozaki. "Modeling Nonlinear Random Vibrations Using an Amplified-Dependent Autoregressive Time Series Model." *Biometrica* 86 (1981), 189–96.

Haldrup, Neils. "The Asymptotics of Single-Equation Cointegration Regressions with $I(1)$ and $I(2)$ Variables." *The Journal of Econometrics* 63 (1994), 153–81.

Hamilton, James. "A New Approach to the Economic Analysis of Nonstationary Time Series and the Business Cycle." *Econometrica* 57 (1989), 357–84.

Hamilton, James, and Raul Susmel. "Autoregressive Conditional Heteroskedasticity and Changes in Regime." *Journal of Econometrics* 64 (1994), 307–33.

Hansen, Bruce. "Inference in TAR Models." *Studies in Nonlinear Dynamics and Econometrics* 2 (1997), 1–14.

———. "Inference When a Nuisance Parameter Is Not Identified under the Null Hypothesis." *Econometrica* 64 (1996), 413–30.

———. "Threshold Effects in Non-Dynamic Panels: Estimation, Testing, and Inference." *Journal of Econometrics* 93 (1999), 345–68.

———. "The Grid Bootstrap and the Autoregressive Model." *Review of Economics and Statistics* 81, 594–607.

Hansen, Bruce, and Byeongseon, Seo. "Testing for Two-Regime Threshold Cointegration in Vector Error-Correction Models." *Journal of Econometrics* 110 (2002), 293–318.

Harvey, Andrew. *Forecasting, Structural Time Series Models, and the Kalman Filter*. Cambridge: Cambridge Univ. Press, 1989.

Hendry, David, A. Neale, and N. Ericsson. *PC-NAIVE: An Interactive Program for Monte Carlo Experimentation in Econometrics*. Institute for Economics and Statistics: Oxford University, 1990.

Higgens, M., and Anil Bera. "A Class of Nonlinear ARCH Models." *International Economic Review* 33 (1992), 137–58.

Hillebrand, Eric. "Neglecting Parameter Changes in GARCH Models." *Journal of Econometrics* 129 (2005), 121–38.

Hodrick, Robert, and Edward Prescott. "Postwar Business Cycles." *Journal of Money, Credit and Banking* 29 (1997), 1–16.

Holt, Matthew, and Satheesh Aradhyula. "Price Risk in Supply Equations: An Application of GARCH Time-Series Models to the U.S. Broiler Market." *Southern Economic Journal* 57 (July 1990), 230–42.

Horvath, Michael, and Mark Watson. *Testing for Cointegration When Some of the Cointegrating Vectors Are Known*. Chicago, Ill.: Federal Reserve Bank of Chicago, 1993.

Hylleberg, S., R. Engle, C. Granger, and B. Yoo. "Seasonal Integration and Cointegration." *Journal of Econometrics* 44 (1990), 215–38.

Johansen, Soren. "Statistical Analysis of Cointegration Vectors." *Journal of Economic Dynamics and Control* 12 (June–Sept. 1988), 231–54.

———. "Estimation and Hypothesis Testing of Cointegrating Vectors in Gaussian Vector Autoregressive Models." *Econometrica* 59 (Nov. 1991), 1551–80.

———. "The Role of the Constant and Linear Terms in Cointegration Analysis of Non-Stationary Variables." *Econometric Reviews* 13 (1994), 205–30.

Johansen, Soren, and Katerina Juselius. "Maximum Likelihood Estimation and Inference on Cointegration with Application to the Demand for Money." *Oxford Bulletin of Economics and Statistics* 52 (1990), 169–209.

———. "Testing Structural Hypotheses in a Multivariate Cointegration Analysis of PPP and the UIP for UK." *Journal of Econometrics* 53 (1992), 211–44.

Kapetanios, George, Yongcheol Shin, and Andy Snell. "Testing for a Unit Root in the Nonlinear STAR Framework." *Journal of Econometrics* 112 (2003), 359–79.

King, Robert, Charles Plosser, James Stock, and Mark Watson. "Stochastic Trends and Economic Fluctuations." *American Economic Review* 81 (Sept. 1991), 819–40.

Koop, Gary, M. Hashem Pesaran, and Simon Potter. "Impulse Response Analysis in Nonlinear Multivariate Models." *Journal of Econometrics* 74 (1996), 119–47.

Kuan, Chung-Ming, and Halbert White. "Artificial Neural Networks: An Econometric Perspective." *Econometric Reviews* 13 (1994), 1–91.

Kwiatkowski, D., Peter Phillips, Peter Schmidt, and Y. Shin. "Testing the Null Hypothesis of Stationarity Against the Alternative of a Unit Root: How Sure Are We That Economic Time Series Have a Unit Root?" *Journal of Econometrics* 54 (1992), 159–78.

Lee, Junsoo, and Mark Strazicich. "Minimum Lagrange Multiplier Unit Root Test with Two Structural Breaks." *Review of Economics and Statistics* 85 (Nov. 2003), 1082–89.

Litterman, Robert. *A Bayesian Procedure for Forecasting with Vector Autoregressions.* Minneapolis: Federal Reserve Bank of Minneapolis, 1981.

Liu, Yamei, and Walter Enders. "Out-of-Sample Forecasts and Nonlinear Model Selection with an Example of the Term-Structure of Interest Rates." *Southern Economic Journal* 69 (2003), 520–40.

Liu, Yu, W. Enders, and R. Prodan. "Forecasting Series Containing Offsetting Breaks: Old School and New School Methods of Forecasting Transnational Terrorism." *Defense and Peace Economics*, forthcoming, 2009.

Ljung, G., and George Box. "On a Measure of Lack of Fit in Time Series Models." *Biometrica* 65 (1978), 297–303.

Lutkepohl, H., and Hans-Eggert Reimers. "Impulse Response Analysis of Cointegrated Systems." *Journal of Economic Dynamics and Control* 16 (Jan. 1992), 53–78.

Ma, Jun, Charles Nelson, and Richard Startz. "Spurious Inference in the GARCH(1, 1) Model When It Is Weakly Identified." *Studies in Nonlinear Dynamics and Econometrics* 11 (2007), Article 1.

McLeod, A., and W. Li. "Diagnostic Checking ARMA Time Series Models Using Squared Residual Correlations." *Journal of Time Series Analysis* 4 (1983), 269–73.

Meese, R., and Ken Singleton. "On Unit Roots and Empirical Modeling of Exchange Rates." *Journal of Finance* 37 (1982), 1029–35.

Michael, Panos, A. Robert Nobay, and David Peel. "Transactions Costs and Nonlinear Adjustment in Real Exchange Rates: An Empirical Investigation." *Journal of Political Economy* 105 (1997), 862–79.

Morley, James, Charles Nelson, and Eric Zivot. "Why Are the Beveridge-Nelson and Unobserved Components Decompositions of GDP So Different?" *Review of Economics and Statistics* 85 (2003), 235–43.

Muth, John. "Optimal Properties of Exponentially Weighted Forecasts." *Journal of the American Statistical Association* 55 (1960), 299–306.

Nelson, Charles, and Charles Plosser. "Trends and Random Walks in Macroeconomic Time Series: Some Evidence and Implications." *Journal of Monetary Economics* 10 (1982), 130–62.

Osterwald-Lenum, Michael. "A Note with Quantiles of the Asymptotic Distribution of the Maximum Likelihood Cointegration Rank Test Statistics." *Oxford Bulletin of Economics and Statistics* 54 (1992), 461–71.

Perron, Pierre. "The Great Crash, the Oil Price Shock, and the Unit Root Hypothesis." *Econometrica* 57 (Nov. 1989), 1361–1401.

———. "Further Evidence on Breaking Trend Functions in Macroeconomic Variables." *Journal of Econometrics* 80 (1997), 355–85.

Perron, Pierre, and Timothy Vogelsang. "Nonstationary and Level Shifts with an Application to Purchasing Power Parity." *Journal of Business and Economic Statistics* 10 (1992), 301–20.

Petrucelli, J., and S. Woolford. "A Threshold AR(1) Model." *Journal of Applied Probability* 21 (1984), 270–86.

Phillips, Peter. "Understanding Spurious Regressions in Econometrics." *Journal of Econometrics* 33 (1986), 311–40.

Phillips, Peter, and Bruce Hansen. "Statistical Inference in Instrumental Variables Regression with *I*(1) Processes." *Review of Economic Studies* 57 (1990), 99–125.

Phillips, Peter, and Pierre Perron. "Testing for a Unit Root in Time Series Regression." *Biometrica* 75 (June 1988), 335–46.

Pippenger, Michael K., and Gregory E. Goering. "A Note on the Empirical Power of Unit Root Tests under Threshold Processes." *Oxford Bulletin of Economics and Statistics* 55 (1993), 473–81.

Potter, Simon. "A Nonlinear Approach to US GNP." *Journal of Applied Econometrics* 10 (1995), 109–25.

Quah, Danny. "The Relative Importance of Permanent and Transitory Components: Identification and Some Theoretical Bounds." *Econometrica* 60 (Jan. 1992), 107–18.

Romer, C. "Changes in Business Cycles: Evidence and Explanations." *Journal of Economic Perspectives* 13 (1999), 23–44.

Rothman, Philip. "Forecasting Asymmetric Unemployment Rates." *Review of Economics and Statistics* 80 (1998), 164–68.

Said, S., and David Dickey. "Testing for Unit Roots in Autoregressive-Moving Average Models with Unknown Order." *Biometrica* 71 (1984), 599–607.

Samuelson, Paul. "Interactions Between the Multiplier Analysis and Principle of Acceleration." *Review of Economics and Statistics* 21 (May 1939), 75–78.

———. "Conditions That the Roots of a Polynomial Be Less than Unity in Absolute Value." *Annals of Mathematics Statistics* 12 (1941), 360–64.

Schmidt, Peter, and Peter Phillips. "LM Tests for a Unit Root in the Presence of Deterministic Trends." *Oxford Bulletin of Economics and Statistics* 54 (1992), 257–87.

Shen, Chung Hua, and David Hakes. "Monetary Policy as a Decision-Making Hierarchy: The Case of Taiwan." *Journal of Macroeconomics* 17 (1995), 357–68.

Sims, Christopher. "Macroeconomics and Reality." *Econometrica* 48 (Jan. 1980), 1–49.

———. "Are Forecasting Models Usable for Policy Analysis?" *Federal Reserve Bank of Minneapolis Quarterly Review* (Winter 1986), 3–16.

———. "Bayesian Skepticism on Unit Root Econometrics." *Journal of Economic Dynamics and Control* 12 (1988), 463–74.

Sims, Christopher, James Stock, and Mark W. Watson. "Inference in Linear Time Series Models with Some Unit Roots." *Econometrica* 58 (1990), 113–44.

Souki, Kaouthar, and Walter Enders. "Assessing the Importance of Global Shocks Versus Country Specific Shocks." *Journal of International Money and Finance* 27 (2008), 1420–29.

Stock, James. "Asymptotic Properties of Least-Squares Estimators of Cointegrating Vectors." *Econometrica* 55 (1987), 1035–56.

Stock, James, and Mark Watson. "Testing for Common Trends." *Journal of the American Statistical Association* 83 (Dec. 1988), 1097–1107.

———. "Has the Business Cycle Changed and Why?" *NBER* working paper No. 9127, 2002.

Suits, Daniel, and Gorden Sparks. "Consumption Regressions with Quarterly Data." T. S. Duesenberry, G. Fromm, L. R. Klein, and E. Kuh, eds., *The Brookings Quarterly Econometric Model of the United States.* Chicago, Ill.: Rand McNally, 1965, pp. 203–23.

Taylor, Mark. "Estimating Structural Macroeconomic Shocks Through Long-Run Recursive Restrictions on Vector Autoregressive Models: The Problem of Identification." *International Journal of Finance and Economics* 9 (2004), 229–44.

Teräsvirta, Timo. "Specification, Estimation and Evaluation of Smooth Transition Autoregressive Models." *Journal of the American Statistical Association* 89 (1994), 208–18.

Tong, H. *Threshold Models in Nonlinear Time Series Analysis.* New York: Springer Verlag, 1983.

Tsay, Ruey. "Testing and Modeling Multivariate Threshold Models." *Journal of the American Statistical Association* 93 (1998), 1188–1202.

Vogelsang, Timothy, and Pierre Perron. "Additional Tests for a Unit Root Allowing for a Break in the Trend Function at an Unknown Time." *International Economic Review* 39 (1998), 1073–1100.

Watson, Mark. "Univariate Detrending Methods with Stochastic Trends." *Journal of Monetary Economics* 18 (1986), 49–75.

West, Mike, and Jeff Harrison. *Bayesian Forecasting and Dynamic Models.* New York: Springer Verlag, 1989.

White, Halbert. "Some Asymptotic Results for Learning in Single Hidden-Layer Feedforward Network Models." *Journal of the American Statistical Association* 84 (1989), 1003–13.

Zellner, Arnold. "Bayesian Analysis in Econometrics." *Journal of Econometrics* 37 (1988), 27–50.

Zivot, Eric, and Donald Andrews. "Further Evidence on the Great Crash, the Oil-Price Shock, and the Unit-Root Hypothesis." *Journal of Business and Economic Statistics* 20 (2002), 25–44.

INDEX